# ATTENTION and PERFORMACE XIII

*Proceedings of the Thirteenth International Symposium on Attention and Performance*
*Arc-et-Senans, France, June 27–July 2, 1988*

# ATTENTION and PERFORMANCE XIII

## Motor Representation and Control

*Edited by*
**M. Jeannerod**
*Université Claude Bernard and INSERM U 94, Lyon, France*

Routledge
Taylor & Francis Group
LONDON AND NEW YORK

First Published 1990 by
Lawrence Erlbaum Associates, Inc.
365 Broadway
Hillsdale, New Jersey 07642

Published 2009 by Routledge
2 Park Square, Milton Park, Abingdon, Oxfordshire OX14 4RN
711 Third Avenue, New York, NY 10017, USA

First issued in paperback 2016

*Routledge is an imprint of the Taylor & Francis Group, an informa business*



Library of Congress Catalog Card No. 89-646104

ISBN 13: 978-1-138-98765-4 (pbk)
ISBN 13: 978-0-8058-0565-9 (hbk)

ISSN 1047-0387

**Publisher's Note**
The publisher has gone to great lengths to ensure the quality of this reprint but points out that some imperfections in the original may be apparent.

# Contents

# Preface

The thirteenth Symposium of the Association for Attention and Performance was held at the Salines Royales in Arc-et-Senans (France), on June 27 to July 2, 1988. The architecture of the Salines Royales, built at the end of the 18th century by Claude Nicolas Ledoux, was intended (somewhat utopically) to model the social order of the time. The size of the buildings reflected the hierarchical status of the people who lived within. Their relative spatial location reflected their function for specific activities, such as work, pleasure, and reproduction. No doubt, this symbolic atmosphere influenced the quality and the level of discussions during the meeting. Such abstract concepts as *hierarchy, category,* or *representation,* often used for modelling psychological functions, suddenly became directly and visually accessible.

The general theme of the meeting was the "Organization of Action." Several basic aspects of this topic were considered, including: anatomical and physiological constraints on motor preparation and execution; influence of control (proprioceptive, cutaneous, visual, oculomotor) signals; contribution of kinematics to the understanding of the underlying mechanisms; and role of cognitive constraints such as attention or learning in goal selection. Theoretical aspects played a large part, specially for clarifying the notions of motor program, representation, or schema. Perspectives from developmental and pathological studies were present in several contributions.

This book has been organized into sections in order to provide a comprehensive view of the main issues raised during the meeting. Following the Association Lecture by Saul Sternberg (Part I), four tutorials on organization principles of motor representations have been grouped in Part II. The other sections deal with initiation of movement and specification of movement parameters (Part III), structure of learned motor patterns (Part IV), sensorimotor transformation and the representation of action coordinates (Part V), sensory-based adjustments (Part VI), and motor learning and development (Part VII).

The help of the Scientific Committee of the Association in preparing the scientific program and of the referees in improving the quality of the contributions is gratefully acknowledged. Financial support was provided by grants from the U. S. Office of Naval Research (ONR) and the Air Force Office of Scientific Research (AFOSR) to Sylvan Kornblum, and by a grant from IBRO to Marc Jeannerod. Françoise Girardet, Hugues Jeannerod, and Olivier Martin collaborated on the many technical aspects before and during the meeting. Michele Busmey provided secretarial assistance during the editing process.

**M. Jeannerod**

# List of Contributors and Participants

**J.H. Abbs,** Speech and Motor Control Laboratories, Waisman Center, 1500 Highland Avenue, University of Wisconsin, Madison, WI 53705-2280, USA

**R. A. Abrams,** Department of Psychology, Washington University, St. Louis, MO 63130, USA

**C. Abry,** Institut de Phonétique, Institut de la Communication Parlée, CNRS UA 368-Grenoble III, B. P. 25X, 38040 Grenoble cedex, France

**S. Aglioti,** Istituto di Fisiologia, Università di Verona, 37134 Verona, Italy

**M. A. Arbib,** Computer Science Department, University of Southern California, University Park, Los Angeles, CA 90089-0782, USA

**H. J. Barnes,** Department of Psychology, University of Massachusetts, Amherst, MA 01003, USA

**A. Basso,** Università di Milano, Istituto di Clinica Neurologica, via Francesco Sforza, 35, I-21122 Milano, Italy

**C. Botti,** Centro Interdipartimentale di Calcolo Automatico e Informatica Applicata, Università di Modena, Modena, Italy

**H. Bouma,** Institute for Perception Research, P.O.B. 513, 5600 Eindhoven, The Netherlands

**M. Brouchon,** Laboratoire de Neuropsychologie Humaine, 4, rue de la Vieille Charité, 13002 Marseille, France

**G. Butterworth,** University of Stirling, Department of Psychology, Stirling FK9 4LA Scotland, United Kingdom

**L. Carlton,** Freer Hall, University of Illinois, Urbana, IL 61801, USA

**A. Cohen,** Department of Psychology, University of Oregon, Eugene, OR 97403, USA

**M. Coltheart,** Department of Psychology, School of Behavioral Sciences, MacQuacle University, Sydney, New South Wales, Australia

**M. L. Cook,** Department of Psychology, Australian National University, Australia

**B. M. Daniel,** University of Edinburg, Department of Psychology, 7, George Square, Edinburgh EH8 9J2 Scotland, United Kingdom

**C. Darlot,** Laboratoire de Physiologie Neurosensorielle, 15, rue de l'Ecole de Médecine, 75006 Paris, France

**B. Davis,** The University of Texas at Austin, Department of Linguistics, 510 Calhon Hall, Austin, TX 78712, USA

**J. D. DelColle,** Center for Complex Systems, Department of Psychology, Florida Atlantic University, Boca Raton, FL 33431, USA

**J. Droulez,** Laboratoire de Physiologie Neurosensorielle, 15, rue de l'Ecole de Médecine, 75006 Paris, France

**P. Faglioni,** Clinica Neurologica, Università di Modena, Modena, Italy

**M. Favilla,** Istituto di Fisiologia Umana, Università di Ferrara, U. Fossato di Mortara, 44100 Ferrara, Italy

**M. Gentil,** 90, avenue du Maine, 75014 Paris, France

**A. P. Georgopoulos,** The Philip Bard Laboratory of Neurophysiology, Department of Neuroscience, The Johns Hopkins University, 725 North Wolfe Street, Baltimore, MD 21205, USA

**C. Ghez,** Center for Neurobiology and Behavior, Columbia University College of Physicians and Surgeons, 722 West 168th Street, New York, NY 10032, USA

**J. C. Gilhodes,** Laboratoire de Neurobiologie Humaine, UA CNRS 372, University de Provence, 13397 Marseille cedex 13, France

**R. Gottsdanker,** University of California, Santa Barbara, CA, USA

**L. Grover,** Eastleigh College of Technology, Eastleigh, England

**W. Hening,** Department of Neurology (127), Lyons VAMC, Lyons, NJ 07939, USA

**H. Heuer,** Gachbereich Psychologie der Philipps-Universität, Gutenberg Strasse 18, D-3550 Marburg, F.R.G.

**C. Von Hofsten,** University of Umea, Department of Psychology, S-90187 Umea, Sweden

**H. Honda,** Department of Behavioral Sciences, Faculty of Humanities, Niigata University, Niigata 950-21, Japan

**R. Ivry,** University of California, Santa Barbara, CA, USA

**R. Jagacinski,** Office State University, Department of Psychology, 404 B West 17th Avenue, Columbus, OH 43210, USA

**M. Jeannerod,** Unité 94, INSERM, 16, avenue du Doyen Lépine, 69500 Bron, France

**R. S. Johansson,** Department of Physiology, University of Umea, S-90187 Umea, Sweden

**J. Jonides,** Human Performance Center, The University of Michigan, Ann Arbor, MI 48109, USA

**M. I. Jordan,** Department of Brain and Cognitive Sciences, Massachusetts Institute of Technology, Cambridge, MA 02139

**M. J. Jorgensen,** Department of Psychology, University of Massachusetts, Amherst, MA 01003, USA

**S. W. Keele,** Department of Psychology, University of Oregon, Eugene, OR 97403, USA

**J. E. Keith-Smith,** University of Michigan, Department of Psychology, Human Performance Center, 330 Packard Road, Ann Arbor, MI 48104-2994, USA

**J. A. S. Kelso,** Florida Atlantic University, Center for Complex Systems, P.O. Box 3091, Boca Raton, FL 33431-0991, USA

**R. L. Knoll,** AT&T Bell Laboratories, Room MH 30G-004, 600 Mountain Avenue, Murray Hill, NJ 07974, USA

**S. Kornblum,** Mental Health Research Institute, University of Michigan, 205 Washtenaw Place, Ann Arbor, MI 48109-0010, USA

**J. R. Lackner,** Ashton Graybiel Spatial Orientation Laboratory, Brandeis University, Waltham, MA 02554, USA

**D. N. Lee,** University of Edinburgh, Department of Psychology, 7, George Square, Edinburgh EH8 9JZ Scotland, United Kingdom

**C. L. MacKenzie,** Department of Kinesiology, University of Waterloo, Waterloo, Ontario, Canada N2L 3G1

**P. F. MacNeilage,** The University of Texas at Austin, Department of Linguistics, 510 Calhon Hall, Austin, TX 78712, USA

**J. L. MacClelland,** Department of Psychology, Carnegie Mellon University, Pittsburgh, PA 15213, USA

**F. Marchak,** Department of Psychology, University of Massachusetts, Amherst, MA 01003, USA

**R. Marteniuk,** Department of Kinesiology, University of Waterloo, Waterloo, Ontario, Canada N2L 3G1

**D. E. Meyer,** University of Michigan, Department of Psychology, Human Performance Center, 330 Packard Road, Ann Arbor, MI 48104-2994, USA

**S. Monsell,** Department of Experimental Psychology, University of Cambridge, Downing Street, Cambridge CB2 3EB, United Kingdom

**P. Morasso,** Dipartimento di Informatica, Sistemistica e Telematica, Dist-Università di Genova, Via Opera Pia 11A, 1-16145 Genova, Italy

**P. Mounoud,** F.P.S.E., Université de Genève, 24, rue Général Dufour, CH-1211 Genève, Switzerland

**Y. Paulignan,** Unité 94, INSERM, 16, avenue du Doyen Lépine, 69500 Bron, France

**D. Pélisson,** Unité 94, INSERM, 16, avenue du Doyen Lépine, 69500 Bron, France

**G. Pellizzer,** F.P.S.E., Universitér de Genève, 24, rue Général Dufour, CH-1211 Genève, Switzerland

**M. T. Perenin,** Unité 94, INSERM, 16, avenue du Doyen Lépine, 69500 Bron, France

**C. Prablanc,** Unité 94, INSERM, 16, avenue du Doyen Lépine, 69500 Bron, France

**G. Rizzolatti,** Istituto di Fisiologia, Università di Parma, I-43100 Parma, Italy

**J. P. Roll,** Laboratoire de Neurobiologie Humaine, Université de Provence,

Centre St-Jérôme, Avenue Escadrille Normandie-Niémen, 13397 Marseille cedex 13, France

**R. Roll,** Laboratoire de Neurobiologie Humaine, UA CNRS 372, Université de Provence, 13397 Marseille Cédex 13, France

**D. A. Rosenbaum,** University of Massachusetts, Department of Psychology, Tobin Hall, Amherst, MA 01003, USA

**C. Saetti,** Clinica Neurologica, Università di Modena, Modena, Italy

**J. N. Sanes,** National Institutes of Health, Department of Health and Human Services, Human Motor Control Section, Bldg. 10, Room 5N–226, Bethesda, MD 20892, USA

**R. A. Schmidt,** University of California, Department of Psychology, Los Angeles, CA 90024, USA

**G. Schoner,** Center for Complex Systems, Florida Atlantic University, Boca Raton, FL 33431, USA

**J. Seal,** LNF 1, CNRS, 31, chemin Joseph Aiguier BP 71, 13402 Marseille Cedex 9, France

**A. Semjen,** LNF 1, CNRS, 31, chemin Joseph Aiguier BP 71, 13402 Marseille cedex 9, France

**J. D. Slotta,** Department of Psychology, University of Massachusetts, Amherst, MA 01003, USA

**J. F. Soechting,** Department of Physiology, 5-257 Millard Hall, Minneapolis, MN 55455, USA

**S. Sternberg,** University of Pennsylvania, Department of Psychology, 3815 Walnut Street, Philadelphia, PA 19104–6196, USA

**S. Swinnen,** Katholieke Universiteit Leuven, Instituut voor Lichamelyke Opleiding, Tervuuse vest 101, B-3030 Heverlee, Belgium

**C. A. Terzuolo,** Neurophysiology Laboratory, University of Minnesota Medical School, 334 Millard Hall, Minneapolis, MN 55455, USA

**B. Tuller,** Florida Atlantic Jniversity, Center for Complex Systems, P.O. Box 3091, Boca Raton, FL 33431–0991, USA

**J. Turnbull,** University of Edinburgh, Department of Psychology, 7, George Square, Edinburgh EH8 9J2 Scotland, United Kingdom

**D. L. Turok,** Bell Communications Research, Red Bank, NJ, USA

**D. L. Van Eerd,** Department of Kinesiology, University of Waterloo, Waterloo, Ontario N2L 3G1 Canada

**J. Vaughan,** Psychology Department, Hamilton College, Clinton, NY 13323, USA

**W. S. Vaughan,** Office of Naval Research, 800 North Queency Street, Arlington, VA 22219–5000, USA

**J. L. Velay,** Laboratoire de Neurobiologie Humaine, UA CNRS 372, University de Provence, 13397 Marseille Cedex 13, France

**P. Viviani,** Faculté de Psychologie et Sciences de l'Education, Université de Genève, 24, rue Général Dufour, CH–1211 Genève, Switzerland

**G. Westling**, Department of Physiology, University of Umea, S-901 87 Umea, Sweden

**M. Wiesendanger**, Institut de Physiologie, Université de Fribourg, 5, rue du Musée, CH-1700 Fribourg, Switzerland

**C. J. Winstein**, Speech and Motor Control Laboratories, Waisman Center, 1500 Highland Avenue, University of Wisconsin, Madison, WI 53705-2280, USA

**C. E. Wright**, Department of Psychology, Columbia University, 406 Shermerhorn Hall, New York, NY 10027, USA

**D. E. Young**, Motor Behavior Laboratory, Department of Physical Education, California State University, Long Beach, CA 90840, USA.

1. H. Honda
2. R. S. Johansson
3. C. Von Hofsten
4. S. Monsell
5. B. Davis
6. R. A. Schmidt
7. S. Sternberg
8. P. F. MacNeilage
9. C. Ghez
10. D. N. Lee
11. C. Abry
12. O. Martin
13. P. Mounoud
14. R. Knoll
15. C. E. Wright
16. A. P. Georgopoulos
17. J. Seal
18. M. A. Arbib
19. G. Butterworth
20. M. Wiesendanger
21. M. Jeannerod
22. D. E. Meyer
23. P. Viviani
24. C. A. Terzuolo
25. J. N. Sanes
26. D. Turock
27. J. Vaughan
28. J. F. Soechting
29. D. A. Rosenbaum
30. Y. Paulignan
31. J. A. S. Kelso
32. S. Kornblum
33. M. I. Jordan
34. P. Morasso
35. R. Marteniuk
36. S. W. Keele
37. J. R. Lackner
38. H. Bouma
39. G. Pellizzer
40. R. Jagacinski
41. (unidentified)
42. C. Prablanc
43. B. Tuller

# I ASSOCIATION LECTURE

# 1 Hierarchical Control in the Execution of Action Sequences: Tests of Two Invariance Properties

Saul Sternberg
*University of Pennsylvania*

Ronald L. Knoll
*AT&T Bell Laboratories*

David L. Turock
*Bell Communications Research*

ABSTRACT

What might it mean for execution of an action sequence to be controlled hierarchically? We argue that if production of a sequence consists of the execution of nested constituent subsequences, then it should be characterized by two invariance properties—properties that limit the effects of one part of the sequence on another. Because one such constituent structure merely partitions the stream of action into *action units*, these properties have wide applicability. According to *low-level invariance*, the process that executes a constituent should not be influenced by changes in any higher level constituent. According to *high-level invariance*, changes in a constituent should have at most limited and local effects on higher level constituents. We report on tests of these two properties in the rapid production of brief utterances and short strings of keystrokes, in which we examine the effects of sequence length, serial position, and unit size on measures of timing. The tests support the existence of hierarchical constituents at the level of the stroke in typing and the stress group in speech, but provide only limited evidence for deeper hierarchical structure.

## INTRODUCTION

In this chapter we investigate one sense in which execution of action sequences might be hierarchical, by virtue of their being composed of separately controlled constituent subsequences. We argue that to claim merely that a continuous stream of action can be decomposed into a string of concatenated *units* is to assert a simple form of such hierarchical structure; properties that distinguish the constituents in such a structure should thus characterize purported action units. We focus on two invariance properties that flow from this sense of hierarchical control—properties that reflect the idea that different control levels function autonomously and thus impose limits on the effects of sequential context—and report attempts to determine the extent to which these properties characterize timing in the rapid production of speech and keystroke sequences. The spirit of our inquiry is to minimize the number of ancillary assumptions and thus avoid strong models. Our aim is to illustrate some alternative approaches to testing for hierarchical structure, applying them in most cases to data collected for other purposes.

In our examination of evidence we consider instances where short action sequences prescribed well in advance are correctly produced under time pressure, and where time measurements thus indicate performance constraints. We justify this choice of procedure by our desire to separate the execution of planned sequences from the process by which they are planned. Performance measurements indicate that under these conditions a plan or "program" for the entire sequence exists before it is initiated (Rosenbaum, Kenny, & Derr, 1983; Sternberg, Monsell, Knoll, & Wright, 1978). We deliberately do not investigate cases where choice of sequence is free rather than prescribed (e.g., Fentress, 1983), or where errors are of central interest (e.g., Shattuck-Hufnagel, 1983): The choice of action element (when choice is free) or errors in such choice (when the sequence is prescribed) could occur during the planning process as well as the execution process, whereas temporal effects in the execution of prescribed sequences seem more likely to be associated with the latter. Because our concern is with how the control mechanism selects successive actions, we consider primarily sequences whose successive actions are distinct rather than repeating.

## PROPERTIES OF SEQUENCES UNDER HIERARCHICAL CONTROL

### Concepts of Hierarchy

One example of the numerous ways[1] in which the term *hierarchy* has been used is to denote a simple ordering on some dimension, often described as a set of *levels*. More interesting are tree-like branching structures, con-

[1]See, for example, Dawkins (1976); Estes (1972); Gallistel (1980); Johnson (1970); Lehrdahl and Jackendoff (1983); Narmour (1983); Nelson (1973); and Wilson (1969).

sisting of a set of elements (*nodes*) at different levels, partially ordered by a relation (*branches*), usually antisymmetric and transitive (Wall, 1972). Examples of relations are *inclusion* and *control*. If the relation is inclusion, we have a classification hierarchy in which each class (at one level) consists of a set of subclasses (at the next), and a subclass can belong to only one class. If the relation is control, each element (at one level) controls a set of elements (at the next), and an element can be controlled by at most one higher level element.

In applying this idea to action sequences we are interested in strings of rapid actions that can be said to contain one or more separately controlled substring *constituents* (or *units*), larger than a "single" action (that is, susceptible to further analysis), but smaller than the whole string (Gallistel, 1980, pp. 288–290). The validity of such an assertion hinges theoretically on the specification of criteria that define a substring as a constituent, and empirically on the demonstration that such criteria are satisfied. This sense of hierarchy should be distinguished from the idea of different levels of specificity, such that more detailed aspects of the same action are controlled autonomously at lower levels, closer to its execution (Szentagothai & Arbib, 1975), and where control branches can converge and cross (Gallistel, 1980).

Perhaps the most familiar domain where hierarchically organized sequences have been used to characterize human behavior is language. The constituent structure of sentences is represented in linguistics by the recursively branching phrase marker (Wall, 1972), which we introduce to add precision to the idea of hierarchical structure. The phrase marker can be expressed as either an *ordered tree*, (Fig. 1.1A), or a *labeled-bracket structure* (Fig. 1.1B). An ordered tree consists of a set of *nodes* (from a *root* at the highest level to *leaves* at the lowest, or first) connected by diverging and non-crossing *branches*. Any pair of nodes is related either by *dominance* (for instance, in Fig. 1.1 *VP* dominates the right-hand *N*, and is thus at a higher "level") or by *precedence* (for instance, the upper *NP* precedes *V*) but not by both. Each node corresponds to a substring *constituent* of the sentence. The nesting of constituents is more evident in the labeled-bracket structure, in which dominance becomes inclusion.

A hierarchically organized string may thus be defined as a set of constituent (sub)strings, partially ordered by inclusion, such that any string either fully contains, or is fully contained in, or is disjoint from any other string, and such that disjoint constituents are temporally ordered. A string of "action unit" substrings defines the most shallow structure that can be described as hierarchical; in a deeper hierarchy the substrings would be further partitioned into smaller disjoint substrings.

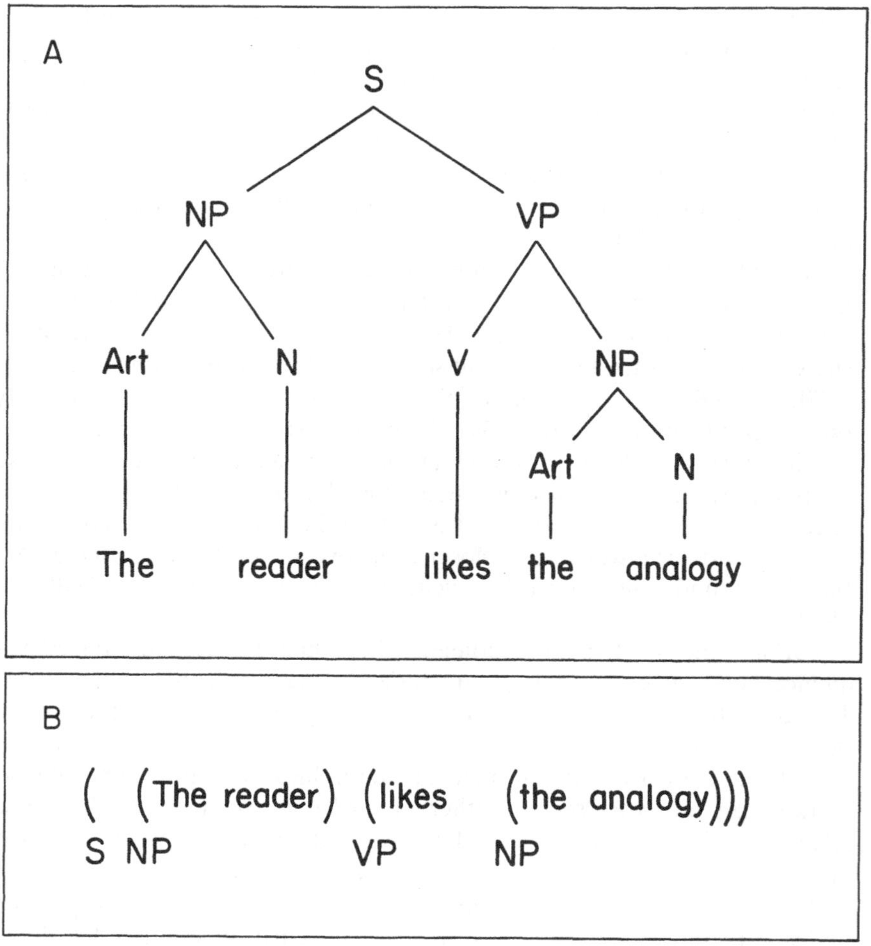


FIG. 1.1. Two forms of the phrase marker representation of a sentence. A: An ordered tree composed of nodes and branches. B: A partially equivalent structure of nested labeled brackets.

## Two Invariance Properties of Hierarchically Controlled Sequences

*Augmenting of hierarchical structure with hierarchical control.* The phrase marker is a hierarchical structure with no commitment to any particular embodiment in real time or to any specification of the flow of control. To convert such a structure into one that can actually control the execution of the sequence it describes, we must augment it with an interpretive process (Greeno & Simon, 1974) that uses or operates on the structure to create an action sequence. In a plausible process of this kind, the nodes, which correspond to nested constituents, also correspond to loci of control. Let us use *program* to refer to the structure that produces the entire sequence when the process is applied to it. Then a constituent subsequence is produced by invoking a *subprogram* that produces the subsequence and is associated with the constituent's node; such actuating of subprograms proceeds recursively until first-level nodes are reached.

In the present chapter we focus on the implications of independence between control levels and the resulting constraints on context effects. Such implications are especially clear if the process that operates on the structure does so by *transfer of control* from program (at one level) to subprogram (at a lower level) and back.[2] Production of a constituent results from execution of a (sub)program that, we suggest, should be insensitive to properties of the superordinate program that invoked it. That is, once initiated, the subprogram is executed autonomously until completion, without reference to information stored at a higher level; control is returned to that level upon completion. Similarly, the superordinate program is insensitive to details of the subsequence; it "knows" at most whether or not the subprogram has completed its job.

*Low-level and high-level invariance as properties of hierarchical control.* Two consequences flow from these considerations. The first is a property we term *low-level invariance*: The process that executes a constituent substring should be invariant relative to changes in disjoint or higher level parts of the control structure. (When these two loci of change are distinguishable we have two kinds of low-level invariance: For example, if we altered only the content of a disjoint constituent we could test for invariance relative to a disjoint part of the control structure; if we altered the number of constituents we could test for invariance relative to a higher level part of the control structure.) Given low-level invariance, performance characteristics of all elements within a constituent should thus depend only on the content of that constituent; they should not be influenced by changes in the constituents of which it is a substring. An analogy in language arises in context-free phrase-structure grammars, in which the rewriting rule according to which a node is elaborated downward depends only on

[2]Tree-traversal models of sequence production (Collard & Povel, 1982; Greeno & Simon, 1974; Rosenbaum, Kenny, & Derr, 1983; Volpert, 1982) exemplify processes with transfer of control, processes that are natural in single-processor systems and supported by some of our analyses. An alternative (command/signal transmission) that is possible in multiprocessor systems and favored by one of our analyses is considered later.

the label of that node, and not on any other nodes. Low-level invariance, or an approximation thereof, is inherent in the idea of a *unit,* an entity that must preserve some or all of its properties across different contexts.

The second consequence is *high-level invariance:* relative to changes in a constituent substring, the higher level process that controls its execution along with the execution of disjoint other constituents, and hence those other constituents themselves, should be invariant. Changes in the content of a constituent should thus have only local effects (limited to the constituent itself) on the higher level constituents of which it is a substring. An analogy in language is the existence of syntactic dependence, such as number agreement between subject and verb, that is unaltered by expanding an intervening constituent and thus increasing their separation.

It might be argued that these invariance criteria are too stringent; the hypothesis of hierarchical structure should perhaps not be rejected in every instance in which one or the other criterion is found wanting. We might instead ask to what degree of approximation these properties obtain, and for which aspects of performance, in cases where we have other reasons to argue for hierarchical structure. Thus, Simon (1962) suggests that hierarchical systems might be "nearly decomposable," rather than displaying full invariance. On the other hand, it seems reasonable to start with the expectation of full invariance, and discover how far it will carry us.

*The testing of invariance.* To test either form of invariance one assesses the effects of manipulating an experimental *factor* on some *measure.* To provide an interesting test, the factor should influence *some* aspect of the performance under study. For this reason we refer to a factor used in this way as an *influencer.* Although the invariance properties apply to any aspect of performance, our tests to date are limited to duration measures. To emphasize this limitation we sometimes use the term *duration invariance.*

*Command/signal transmission as an alternative to transfer of control.* Strict interpretation of "transfer of control" requires that there be no temporal overlap of operations associated with different levels (Broadbent, 1977). Durations of operations at different levels must then contribute additively to sequence duration, and time measurements become especially revealing. Given such a process, however, we must reconcile the apparent continuity of action not only with the discreteness of the action unit, discussed later, but also with the possibility of time-consuming operations at high levels, which (on this account) should lead to epochs during which action is frozen. It may thus be necessary to at least augment a process of transfer of control such that action controlled by a low-level subprogram can outlast the execution of that subprogram, filling the gap during which the next subprogram is being selected and invoked, but without delaying its execution (Sternberg, Knoll, Monsell, & Wright, 1989, Section 4.3).

These considerations, together with aspects of coordination of the hands in typewriting described later, also suggest that we consider a fundamentally different class of control processes that might operate on a hierarchical structure, in which transfer of control is replaced by *command/signal transmission:* A program invokes a subprogram by transmitting a command downward rather than relinquishing control; completion (or forthcoming completion) may be indicated by an upward signal, but return of control is not needed. Processes at different levels, or even at the same level, may thus overlap in time and proceed concurrently. Our invariance properties should still obtain, but their consequences may be obscured, depending on the extent of temporal overlap and other details; the interpretation of time measurements may be more complicated. If higher level operations must await a signal of completion from below, such a process could mimic transfer of control.

### Fundamental Discreteness of Continuous Action

Action is in general continuous; even when it is punctuated by a series of distinctive time markers, such as keystroke actuation points in typing or articulatory closures in speech, the system is unlikely to be static during intervals associated with these events, nor do we wish to limit our analyses to those cases where it might be. Given such continuity, it is well to bring out a common but often implicit assumption: The apparently continuous action stream is fundamentally discrete. Without this assumption we could not readily consider the planning of action sequences, and we would not be attempting to identify action units. The idea is commonplace in thinking about speech production, which is often thought to have phonemes as low-level units, with their concatenation smoothed by coarticulation (Fowler, 1985; Perkell, 1980). In relation to manual movement trajectories, especially in handwriting, Morasso and others have shown how an inherently discrete and discontinuous mechanism can produce smooth movements by a particular form of coarticulation: the partial time overlap of consecutive units, achieved by superposition of their kinematic characterizations.[3] If such smoothing were produced by a process that operated on the output of a hierarchical control mechanism, then the resulting perturbations at constituent boundaries might obscure instances of expected invariance; If

[3]See Morasso (1983) and Morasso, Mussa Ivaldi, and Ruggiero (1983). In an experimental test using planar arm-movements, such superposition failed to describe the resultant trajectory (Massey, Schwartz, & Georgopoulos, 1986).

such effects were widespread it would be difficult to proceed further. We encounter cases in both speech and typing where we observe such contamination, but fortunately find it to be limited.

## Alternative Criteria for Action Units

The decomposition of a continuous stream of action into units is an assertion of a particular hierarchical structure. Any criteria that might be diagnostic for action units are thus likely to be helpful in relation to constituents at any level in a hierarchical structure, and vice versa; this is especially evident when one recognizes that units at one level of description of a sequence are also likely to be both compounds at a lower level and subunits at a higher level.

Unit criteria that have been discussed or implied in previous work should be considered in relation to those here under investigation.[4] Ultimately we may discover that several of these properties tend to co-occur. Some are as follows:

1. Stereotypy of unit execution. Execution of a unit should be invariant in the presence of contextual variation (as in low-level invariance) or from occasion to occasion. For example, time intervals between its parts or ratios of such intervals should be relatively constant. (See Barlow, 1977; Gentner, 1987; Terzuolo & Viviani, 1980.) A related criterion arises from Vorberg and Hambuch's (1978, 1984) definition of hierarchical timing discussed later: Time intervals between features within a unit should display smaller variances than intervals (of approximately the same mean duration) between features in different units.
2. Correlation of component durations. Durations of components of the same unit, but not of different units,should be correlated over repeated productions (Benecke, Rothwell, Day, Dick, & Marsden, 1986).
3. Temporal separation. Intervals separating successive component actions should be longer across than within unit boundaries (Gee & Grosjean, 1983; Machlis, 1977; Reitman & Reuter, 1980; Rosenbaum, et al., 1983).

---

[4]Given an explicit theory of the production of action sequences, definition of a unit or constituent may emerge naturally. For example, in the Rumelhart and Norman (1982) model of transcription typing, the parser transforms a long string of input characters into a sequence of substring constituents. And in the subprogram selection model developed for the production of short action sequences in speech and typing by Sternberg, Monsell, et al. (1978; see also Sternberg, Wright, Knoll, & Monsell, 1980; Sternberg, Knoll, Monsell, & Wright, 1983; and Sternberg, Knoll, et al., 1989) a unit at the level of the action sequence is defined as that segment of action controlled by a single subprogram, or associated with a single rendition of the subprogram-selection process.

4. Equivalence of effects of units on other actions. Any effect on the remainder of a sequence of adding a unit should be independent of the size or content of that unit (Sternberg, Knoll, et al., 1989). Note that this is equivalent to high-level invariance, according to which one constituent can be replaced by another, with no effect on the remainder of the control structure.[5]

5. Equivalence of effects of other actions on units. An effect *on* a unit of lengthening a sequence should be independent of the content of that unit (Sternberg, Knoll, et al., 1989). Note that although this requirement appears at first glance to be equivalent to low-level invariance, it is weaker: Low-level invariance requires *no* effect of changing the embedding structure, rather than merely an invariant effect.

6. Preferential influence. Segments within the same unit should influence each other more than segments of different units. Fowler (1981, 1985) uses this criterion to support the idea of a stress-group unit in speech production. In a perceptual analog, Palmer (1977) proposes that elements within units of a hierarchically organized percept have relatively more influence on each other. Simon (1962) describes hierarchical systems as "nearly decomposable," displaying stronger interactions within than between units.

7. Internal coherence. Parts of a unit should be difficult to separate. Segments within the same unit should thus never (or seldom) participate independently in errors of omission or misordering (Drummond, 1981; Fentress, 1983; Fromkin, 1981; Gallistel, 1980), and it should be more difficult to produce a prescribed rearrangement of a prepared sequence that breaks up units than one that leaves them intact (Gordon & Meyer, 1987).

8. Process discreteness. A unit is a segment of performance that results from a single rendition of some process. This criterion is, of course, more model-dependent than the others because in general, a process is hypothetical and requires a model to be related to observables (Sternberg, Knoll, et al., 1989, Section 2.5).

9. Bell-shaped velocity profile. In arm and hand movements, a unit can be distinguished by a bell-shaped velocity profile, and is thus a segment of the trajectory delimited by velocity minima (Abend, Bizzi, & Morasso, 1982).

---

[5]The equivalence can be seen as follows: Suppose equal effects on some univariate property $P_a$ of sequence $a$, of adding (concatenating or incorporating) constituents $u_x$ and $u_y$. That is, $P_a(a+u_x) - P_a(a) = P_a(a+u_y) - P_a(a)$. It follows that $P_a(a+u_x) = P_a(a+u_y)$, so that replacing $u_x$ by $u_y$ leaves $P_a$ invariant. Conversely, it can be seen that invariance under replacement implies equal effects.

## Existing Evidence for Hierarchical Control in Sequence Production

The idea that fluently performed action sequences tend to be structured hierarchically, and that skill learning includes the creation of higher-level action units, dates back at least to the turn of the century (Book, 1908). However, there is a disparity between the enthusiasm for the hypothesis of hierarchical organization in the skilled execution of action sequences (e.g., Gallistel, 1980, Ch. 12; Johnson, 1970; Keele, 1987; MacKay, 1982; Miller, Galanter, & Pribram, 1960) and the persuasiveness of much of the cited evidence. For example, contrary to frequent citation, Bryan and Harter's (1899) claims were for hierarchical structure in receiving, rather than sending Morse code. Other evidence does not exclude a perceptual locus (Leonard & Newman, 1964). The often-cited conclusions of Book (1908) about multiple-stroke units in typewriting depend exclusively on introspective evidence. Furthermore, they have been only partially confirmed by the effects on transcription speed at different skill levels of the length of familiar sequences incorporated in the copy (Fendrick, 1937; West, 1969, Ch. 3).[6] More recently, some evidence consistent with hierarchical structure of the representation that controls the execution of action sequences has begun to emerge: Gordon and Meyer (1987) have applied the internal coherence criterion listed in the previous section to speech production, and Rosenbaum et al. (1983), whose work we discuss later, have applied the temporal separation criterion to finger tapping (but see also Klein, 1983, and Rosenbaum, 1983).

Because their problem is the timing of "identical" actions rather than the choice of distinct ones, Vorberg and Hambuch's (1978, 1984) attempts to determine conditions (if any) for hierarchical control in the timing of repeated rhythmic patterns fall outside the domain of central interest here; they are nonetheless important in any review. Suppose a series of taps is produced under instructions to make them approximately isochronic but to group them in triple meter. Denote the times of occurrence of each of the successive taps as

$$t_{0,0},\ t_{0,1},\ t_{0,2},\ t_{1,0},\ t_{1,1},\ t_{1,2},\ \ldots,\ t_{n,0},\ t_{n,1},\ t_{n,2},\ \ldots$$

According to a serial-timing model, each tap is timed from the preceding one. In a typical hierarchical-timing model, the timing of one or more

[6]Inferences in these studies about the size of the action unit depend on two questionable assumptions: First, skill level is assumed to have no influence on the perceptual or translation processes in transcription; second, given no change in unit size, skill level is assumed to have a multiplicative effect on transcription speed for different kinds of copy.

intervals [e.g., $(t_{n,0}, t_{n,1})$ and $(t_{n,1}, t_{n,2})$] can be nested within the timing of a longer interval between nonsuccessive events [e.g., $(t_{n,0}, t_{n+1,0})$] such that long-interval and short-interval "timers" can start in synchrony and operate in parallel. If so, the durations of one or more (short) intervals [$(t_{n,2}, t_{n+1,0})$ in our example] are determined indirectly, as the difference between other, directly timed intervals.

As shown by Vorberg and Hambuch's elegant analyses, numerous properties of the resultant time intervals distinguish these two models; perhaps the most obvious is *phase-dependent variability*: For serial timing, the variance, $s_i^2 = Var_n(T_{n,i})$ of the intervals $\{T_{n,i}\} = \{t_{n+1,i} - t_{n,i}\}$ between corresponding taps is independent of $i$, because each such interval is generated by a sum of intervals from the same three timers. In contrast, for hierarchical timing this variance depends systematically on $i$, because for $i = 0$ it is generated by one timer whereas for $i = 1$ or $i = 2$ it is generated by a linear combination of intervals from three timers. Under the instructions described above, subjects show clear evidence of grouping in the periodicity of the mean intervals $\{t_{\cdot,i+1} - t_{\cdot,i}\}$ between successive taps. This might tempt one to conclude in favor of a hierarchical structure, but the analysis of patterns of variance and covariance instead indicates serial timing. Under instructions to produce taps in an uneven rhythm, subjects' phase-dependent variance provides evidence against serial timing, but their covariance data argue against hierarchical timing. Like evidence for hierarchical structure in the selection of actions, the evidence for hierarchical timing has thus been elusive.[7]

Patterns of error have sometimes been cited as evidence for hierarchical structure and for the existence of particular units, especially in speech production (see Dell, 1984; Fromkin, 1971, 1981; Shattuck-Hufnagel, 1983). In evaluating such evidence, however, the distinction should be made between the idea that *some* representation during the evolution of a movement pattern contains intermediate-size subsequence units, and the idea that this is true of the one finally used in executing the pattern. Given separability of planning and execution processes, it is important to distinguish between *units of planning* and *units of execution* (Monsell & Sternberg, 1981; Rosenbaum, et al., 1983). Errors may arise during planning or execution. Our hope is that *temporal* aspects of the execution of a sequence prescribed in advance are more likely than errors to reflect the representation used during its execution. It is partly for this reason that despite the generality of the two invariance properties under consideration,

---

[7]Jagacinski, Marshburn, Klapp, and Jones (1988) have recently produced an analysis that favors a modified version of Vorberg and Hambuch's hierarchical model, in which concurrently operating timers produce positively correlated (rather than independent) intervals.

our tests thus far are restricted to the *duration* invariance of elements and sequences.

## TESTS OF LOW-LEVEL INVARIANCE

### Existence of an Action Unit in Rapid Speech Production

The analysis we initially present sets the stage for some of the others and also leads to an important elaboration of the idea of low-level invariance. Subjects were asked to recite word sequences of different composition and length, specified well in advance of a "go" signal, so as to complete the utterance as soon as possible (Sternberg, Knoll, et al., 1989). The utterances were either simple counting sequences of lengths one through five ("one, two" or "one, two, three, four," for example; *digit condition*), or the same strings with *and* interpolated ("one and two and three," for example; *and-digit condition*) or the same strings with *minus* interpolated ("one minus two minus three," for example; *minus-digit condition*). Within each condition one can regard the strings of increasing length as being generated by starting with *one* and adding elements of a particular "size," the elements being *digit*, *and digit*, and *minus digit*, for the three conditions, respectively, where *digit* represents a digit name and receives a primary stress.

Rather than increasing linearly with the number of elements, $n$, the durations of utterances produced with this procedure (*sequence durations*) are usually well fitted by a quadratic function:

$$\hat{D}_n = \hat{\alpha} + \hat{\beta}n + \hat{\gamma}n(n - 1). \tag{1}$$

The nonlinearity reflects the fact that longer sequences are produced at lower rates. We adjust for end effects (such as prepausal lengthening) by forming $\hat{D}_n - \hat{\alpha}$, divide by number of elements, $n$, and add $\hat{\gamma}$, to arrive at the *mean element duration*:[8]

$$\hat{d}_{\cdot n} = \frac{\hat{D}_n - \hat{\alpha}}{n} + \hat{\gamma} = \hat{\beta} + \hat{\gamma}n. \tag{2}$$

That is, the mean time between corresponding points of successive elements increases linearly with the number of elements in the utterance (*utterance length*), by an amount given by the parameter $\gamma$. Quadratic functions also

[8]Duration of an element here includes the durations of all events that occur from its beginning to the beginning of its successor or (for the last element) to the end of the utterance. The addition of $\hat{\gamma}$ is unimportant for our present purposes; it is explained in Sternberg, Knoll, et al. (1989).

account well for the durations of sequences of $n$ distinct keystrokes in typing (Sternberg, Monsell, et al., 1978) where a similar analysis also leads to a linear function of $n$ for mean element duration.[9]

Next we consider the relation between the arbitrary *element*, (such as the word or syllable) in terms of which we design experiments and analyze the data, and the action *unit*, a theoretical entity. We do so here because the result enables us to develop an argument for the existence and identity of an action unit in these rapid utterances: Decomposition of a sequence into units defines a minimal hierarchical structure, and thus permits a test of low-level invariance. Suppose there exists a unit of speech production, an entity that is defined by the scope of an execution process and that is invariant across utterances of different lengths and compositions. We define the *unit period* as the time from the beginning of one unit to the beginning of the next; note that the unit period includes both the time to execute the unit (a low-level process) and the time to invoke its successor (a higher level process). Now consider the effects of increasing the number of equal-size elements in an utterance. Suppose the following equivalence requirement for action units: Regardless of the elements comprising the utterance or of the size of the unit (e.g., the number of phonemes it contains) *mean unit period is either invariant or is increased by the same amount per added unit*. (If unit execution time is itself altered, we have a failure of low-level invariance. If any increase in unit period can be identified with the higher level process, however, then low-level invariance applies to unit execution time itself, and the requirement of a hierarchical control structure is satisfied. The invariance relative to unit size expressed by the equivalence property is then an expression of high-level invariance.) We wish to determine whether there exists such a unit and, if so, to identify it.

Let the entities we designate as the elements of a particular utterance each contain $k$ such units ($k > 0$). Let $\delta$ be the mean increase in unit period per added unit. Then the mean increase in unit period per added *element* (i.e., per added $k$ units) is $k\delta$, and the mean increase in *element* duration (i.e., in the summed periods of $k$ units) per added element, or the slope of the element-duration function, is $k(k\delta) = k^2\delta$. (For example, suppose the unit is the syllable and elements are disyllabic words, so that $k = 2$. Let 10 msec be the mean increase in the time from the beginning of one syllable to the next, per added syllable. Because $k^2 = 4$ the mean increase

[9]For typewriting (see Fig. 1.11) we use a slightly different parametrization of the quadratic function [$D_n = \alpha + \beta(n - 1) + \gamma n(n - 1)$] and definition of mean element duration [$d._n = (D_n - \alpha) / (n - 1) = \beta + \gamma n$]. These differences are dictated by an interpretation of the duration of an utterance as running approximately from the *beginning* of the first element to the *end* of the last, whereas in those typing experiments where only strokes (taken to be the ends of elements) are recorded, we consider that sequence duration runs approximately from the *end* of the first element to the end of the last.

TABLE 1.1.
Expected and observed slopes ($\gamma$) of element-duration functions for three element sizes. Elements are shown in bold type. The slope in the control condition, shown in parentheses, is used to estimate $\delta$.

| | *Units/Element and Expected Slope* | | | | | | |
|---|---|---|---|---|---|---|---|
| *Condition* | *Syllable Unit* | | *Word Unit* | | *Stress-Group Unit* | | |
| *(msec/element)* | *k* | $k^2\delta$ | *k* | $k^2\delta$ | *k* | $k^2\delta$ | *Observed Slope* |
| 1 **2** . . . | 1 | $\delta = 8$ | 1 | $\delta = 8$ | 1 | $\delta = 8$ | (8.0) |
| 1 **and 2** . . . | 2 | $4\delta = 32$ | 2 | $4\delta = 32$ | 1 | $\delta = 8$ | 6.2 |
| 1 **minus 2** . . . | 3 | $9\delta = 72$ | 2 | $4\delta = 32$ | 1 | $\delta = 8$ | 8.7 |

in the time from one disyllable element to the next is 40 msec per added disyllable.) For utterances composed of different elements, as in the three conditions of the experiment described above, alternative hypotheses about the nature of the action unit lead to alternative sets of $k$-values (the numbers of units per element), and thus to alternative expectations about relations among slopes of the element-duration functions. By comparing these expectations to the observed slopes we thus develop evidence that bears on the existence of the action unit and its identity.

In Table 1.1 we show the implications for slope ratios of three simple assumptions about the action unit, given that $n$ represents utterance length in elements. If the unit is the *syllable*, then for the *digit*, *and-digit*, and *minus-digit* conditions the numbers of units per element (values of $k$) are 1, 2, and 3, respectively. If the unit is the *word*, these values are 1, 2, and 2. Finally, if the unit is the *stress group* (composed of a stressed syllable and any associated unstressed syllables), then the values are 1, 1, and 1. If we analyze the data in terms of number of elements, then the predicted ratios of slopes of the element-duration functions generated by our three definitions of the unit are 1:4:9 (syllable unit), 1:4:4 (word unit) and 1:1:1 (stress-group unit). The observed values (Table 1.1; functions themselves in Fig. 1.2) clearly favor the stress-group unit.[10] That these values are consistent with expectations based on any one of these simple assumptions supports the existence of an action unit, as defined above. A stress-group unit is also consistent with the finding of parallel element duration functions for utterances composed of unrelated monosyllabic and disyllabic nouns (Sternberg, Monsell, et al., 1978) and for utterances composed of monosyllabic words and phonologically matched nonwords (Sternberg, Wright, et al., 1980). Approximate equality of slopes of the element-duration functions (Equation 2) when $n$ denotes the number of units implies approximate

[10]Thus the number $n$ of units happens to equal the number of those elements in terms of which (in this case) we elected to manipulate the length of the utterance. For further discussion of the stress group as a unit see Fowler (1981, 1985) and Sternberg, Knoll, et al. (1989).

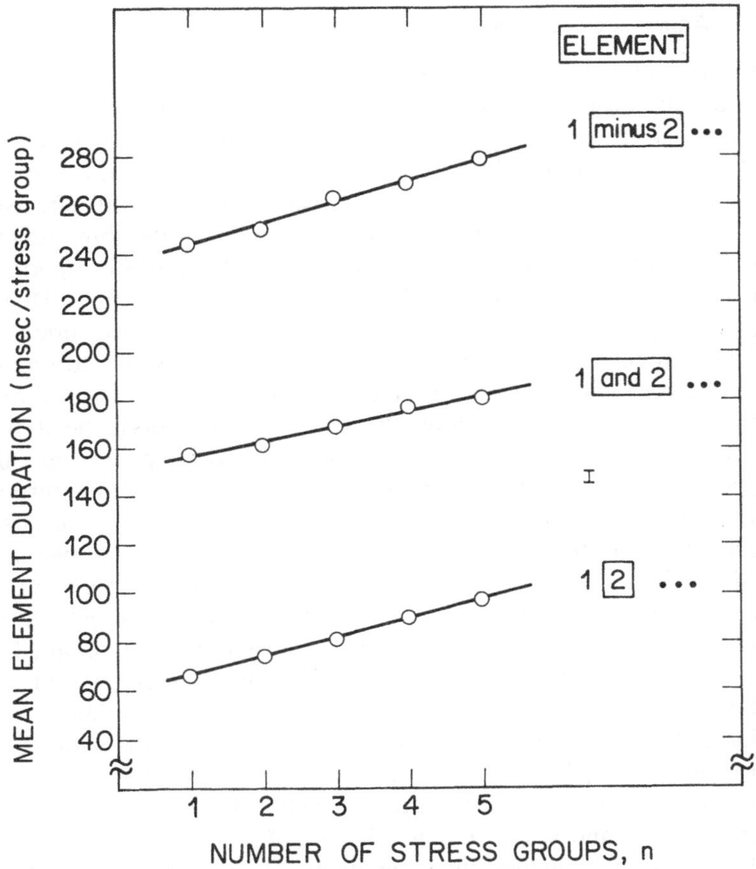


FIG. 1.2. Mean element-duration functions based on stress-group elements for control utterances (*digit* condition) and for utterances with interpolated unstressed words (*and* or *minus*), with estimate of $\pm SE$ and fitted linear functions: $58 + 8.0n$ (*digit* condition); $151 + 6.2n$ (*and-digit* condition); $235 + 8.7n$ (*minus-digit* condition). Results are averaged over four subjects and three days; about 140 (70) observations per point for control (other) conditions.

equality of the nonlinear part of the sequence-duration function (Equation 1). Sequence-duration functions for utterances containing units of different size can then differ by at most a linear function. We see later how this constraint follows from our two invariance properties.

For our present purposes the main point is that we have a set of data that is accounted for very nicely by a particular definition of the action unit. Having identified a unit, we now ask whether that unit functions as a constituent in a hierarchical structure. We first consider whether it obeys low-level invariance; later we discuss a test of high-level invariance, as units of different size are exchanged.

## Further Analysis by Segmentation of Single Speech Units

For the test of low-level invariance, the property of the unit that we consider is its mean duration; the influencer is the number $n$ of units in the utterance. It is clear that if we identify unit duration with unit period, as plotted in Fig. 1.2, we have a violation of low-level invariance. The three approximately parallel functions indicate two effects on unit period, effects that combine additively. First, the vertical displacement of the three functions shows, unsurprisingly, that the *size* of the unit—the number of phonemes or syllables it contains—has a substantial effect on its period. That this effect is independent of $n$ suggests the possibility that unit execution time, a component of the unit period, might be invariant relative to sequence length. Second, the fact that instead of being flat, the element-duration functions rise systematically, shows that the *number* of units in the utterance also has an effect on unit period. Here we have what appears initially to be a clear violation of low-level invariance: a characteristic of a purported hierarchical constituent depends on its context. However, the analysis of how the effect of utterance length on unit period is distributed within that period provides a means by which low-level invariance can be preserved.

In two experiments conducted under conditions similar to those already described (Sternberg, Wright, et al., 1980; Sternberg, Knoll, et al., 1989), utterances were composed of words selected such that segmentation based on the acoustic signal would be feasible: Words were initially stressed disyllables (e.g., *copper, token*) in which each syllable starts with an articulatory closure. Each word in the second experiment (1989) was decomposed into 6 segments, corresponding roughly, for each syllable, to the closure, release, and vowel: *C*1, *R*1, *V*1, *C*2, *R*2, *V*2. The upper panel of Fig. 1.3 displays the element-duration function from that experiment, comparable to the functions shown in Fig. 1.2, here with a slope of 11 msec/word. The lower panel shows the segment-duration functions, which indicate how the effect of the number of units in the utterance is distributed within the stress-group unit.[11] Rather than pervading the unit, the effect of utterance length is almost entirely restricted to *V*2, the vowel of the second syllable; durations of the 5 other segments are virtually invariant. It is possible that the second vowel is prolonged until higher level control structures have invoked the subprogram for the subsequent unit. Low-level invariance is required of the time to execute a subprogram, not of the time to invoke it. We thus have what appears to be a qualified form of low-

[11]The segment-duration function for segment $i$ (here, $1 \leq i \leq 6$) is obtained by starting, for each $n$, with the summed durations of the $n$ $i$th segments in utterances of length $n$, and using the resultant set $\{D_{ni}\}$ to determine a set of $n$ mean segment durations $\{d_{\cdot ni}\}$ for the $i$th segment, just as utterance durations $\{D_n\}$ are used to determine mean element durations $\{d_{\cdot n}\}$. The observed and fitted element-duration functions are then sums of the observed and fitted segment-duration functions, respectively.

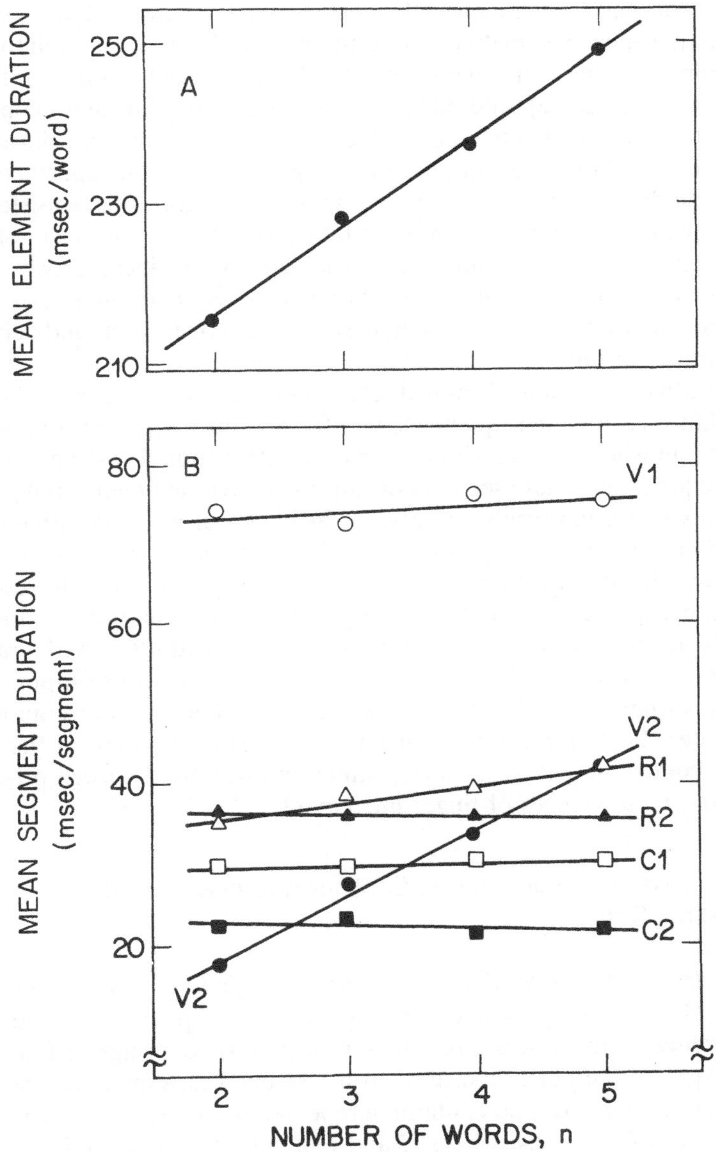


FIG. 1.3. Mean element duration (with fitted function 194 + 11*n msec*) and mean segment durations of two-syllable spoken words. Results are averaged over three subjects and two days; about 100 (50) observations per point for lists of $n > 2$ ($n = 2$) words.

level invariance which, on reflection, is what we might have expected: Given that speech is relatively continuous, that the time required to invoke a subprogram may vary with sequence length (because length influences the control structure), and that smoothing occurs to provide for continuity, not all speech segments would be expected to demonstrate duration invariance. That the effect of utterance length is seen in one epoch per stress group (indicating one invoking process per stress group) is consistent with the conclusion about unit size based on properties of the mean element-duration function. The unit period thus appears to include two additive components, one attributable to the unit and dependent on unit size alone, and the other attributable to higher level control operations and dependent on number of units.

A difficulty remains, however. Although we find approximate duration invariance relative to sequence length for 5 of the 6 segments of the stress-group unit when the mean is taken over serial position, low-level invariance also requires that these segment durations be invariant over serial position; and this is not uniformly the case: instead, serial position influences the duration of *V*1, the first syllable's vowel, which tends to increase with serial position (Sternberg, Wright, et al., 1980). One possible way to rescue the invariance property would be to argue that prosodic effects—including effects of serial position on timing—are introduced *after* the hierarchical control structure has done its work, so as to modulate its output.[12] Low-level invariance would then characterize a hypothetical representation that antedates the action sequence itself. With this elaboration of the hierarchical control model we turn our attention away from speech production and toward sequences of finger movements.

## Tests in Keystroke Sequences Containing Repeated Pairs of Distinct Strokes

*Evidence of hierarchical control in strings of eight finger taps.* Rosenbaum, et al. (1983) have provided what appears to be persuasive evidence of hierarchical control in the production of patterned strings of finger taps. They studied the performance of strings of eight taps made by the fingers on alternating hands and containing repeated pairs. An example is $R_1$, $L_1$, $R_1$, $L_1$, $R_2$, $L_2$, $R_2$, $L_2$, where $R$ and $L$ denote left hand and right hand, and subscripts 1 and 2 denote index finger and middle finger, respectively. Subjects were asked on each trial to produce one of a set of such strings repeatedly without interruption and as fast as possible, for six renditions.

---

[12]The idea that the motor system temporally modulates or "interprets" the output of a higher level control system by introducing delays that might have some structure has been suggested for isochronic taps by Wing (1977).

Evidence for hierarchical control was derived mainly from this pattern of mean intervals between successive taps: The mean intertap intervals (*ITI*s) defined by the four pairs of taps 1-2, 3-4, 5-6, and 7-8 (*first set*) were similar and shortest; they were longer and similar for the two intervals defined by taps 2-3 and 6-7 (*second set*); and they were longest for the two intervals defined by pairs 8-1 and 4-5 (*third set*). Furthermore, the increases in mean *ITI* from first set to second and from second set to third were approximately equal, consistent with a recursive process.

The control structure suggested by Rosenbaum et al. is an ordered binary tree (Fig. 1.4, $n = 8$) a special case (restricted to binary) of the phrase marker discussed earlier. Thus, each of the two nodes below the first bifurcation in the tree representation of the eight-tap sequence specifies a four-tap constituent, each of the four nodes below the next bifurcations specifies a two-tap constituent, and each of the eight final, bottom-level nodes specifies a single tap. The tree structure is augmented, for control purposes, by a ("preorder") tree-traversal process (Aho, Hopcroft, & Ullman, 1983) that starts at the top (root) node and proceeds to the lowest level nodes from left to right, in turn. To repeat the sequence the process passes from rightmost node to root, and then proceeds as before. The intervals between successive taps are assumed to increase linearly with the number of nodes that must be traversed; the resulting model fits the mean *ITI* data reasonably well. (In an isomorphic conceptualization of such a tree and its control process, each node corresponds to a subprogram of the program associated with its parent node, and tree traversal is replaced by a "subprogram-retrieval process" with properties as in Sternberg, Monsell, et al., 1978.)

Tree traversal is one control process associated with a hierarchical structure that operates by transfer of control and requires low-level and high-level invariance of the sequences it produces. Thus (low-level invariance) once control is passed to a lower level node the traversal process for that node is independent of the structures connected to it through higher level nodes (from which control was derived earlier and to which it is transferred later); similarly (high-level invariance) the traversal processes in those other higher and disjoint parts of the structure are independent of what the process does in relation to the lower level node. Although strings of taps known to have the suggestive pattern of mean *ITI*s described above form a very limited set, it is worth testing the invariance properties in such strings, partly because hierarchical control is strongly indicated by the *ITI* data, and partly because the model advocated by Rosenbaum et al. (1983) requires these properties.

*Modifications and elaborations in a keystroke experiment.* We therefore proceeded to test low-level invariance in strings with repeated pairs, by modifying and elaborating the experiment of Rosenbaum et al. (1983).

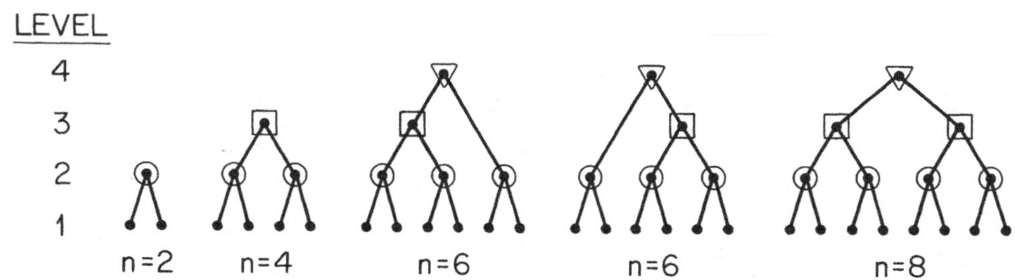


FIG. 1.4. Binary trees for strings of 2, 4, 6, and 8 letters. Nodes at second, third, and fourth control levels are designated by circles, squares, and triangles, respectively.

TABLE 1.2.
Repeating-subsequence letter strings of four lengths. The letters *j, f, k,* and *d* are typed with the right index finger, left index finger, right middle finger, and left middle finger, respectively.

| | *Repeating-Subsequence Strings* | |
|---|---|---|
| $n = 2$ | jf (. . .) | kd (. . .) |
| $n = 4$ | jfkd (. . .) | kdjf (. . .) |
| $n = 6$ | jfkdkd (. . .) | kdjfjf (. . .) |
| $n = 8$ | jfjfkdkd (. . .) | kdkdjfjf (. . .) |

The principal differences were as follows: First, we studied typed rather than tapped sequences. One reason was to more closely approximate the procedure we had used in a series of studies of short keystroke bursts deliberately designed to contain no repeats (Sternberg, Monsell, et al., 1978). Second, in addition to strings containing obvious substrings of length four (*pair-quad strings*), we also used strings containing no such substrings; an example of such a *pair-only string*, using the notation above, is $R_1$, $L_1$, $R_1$, $L_1$, $R_1$, $L_1$, $R_1$, $L_2$. Third, in addition to a *recycling* condition in which each string was performed six times with no break, we also used a *one-cycle* condition with the pair-quad strings, in which subjects had to produce only a single rendition of the string on each trial. (Although this variation had an effect, it had no important bearing on the hypotheses under test.) Finally, we used sequence length and serial position as influencers in tests of low-level invariance.[13]

Examples of the strings we used are shown in Table 1.2. Those of length $n = 8$ required finger patterns similar to those used by Rosenbaum et al.; the shorter ones ($n = 2, 4, 6$) were substrings selected to have no period smaller than $n$.[14]

---

[13]Although expected from several considerations, it had not been known that sequence length influences mean performance rate in such strings. For sequence length to be determinate under recycling conditions, a string of length $n$ must have no period smaller than $n$. Thus, whereas *jfkdkd* or *jfjfjd* are acceptable length-6 strings, *kdkdkd* is not: 6 renditions of the latter string is equivalent to 18 renditions of a length-2 string.

[14]In the first part (A) of the experiment (4 practiced subjects) we examined only recycling performance of the 8 strings in Table 1.2. Subjects practiced a string ad lib before initiating the trial, and were signaled after 6 renditions, with no penalty for overrunning. Data were taken in 4 brief sessions of 48 trials per subject. In the second part (B) of the experiment (3 practiced subjects) we examined the same strings in both recycling and one-cycle conditions, and also studied pair-only strings of 4 lengths in the recycling condition, with similar procedures. One-cycle data were taken in sessions both before and after the recycling data for pair-quad strings. Data from pair-only strings (*jf, jfjd, jfjfjd, jfjfjfjd* and *kd, . . ., kdkdkdkf*) were taken in later sessions.

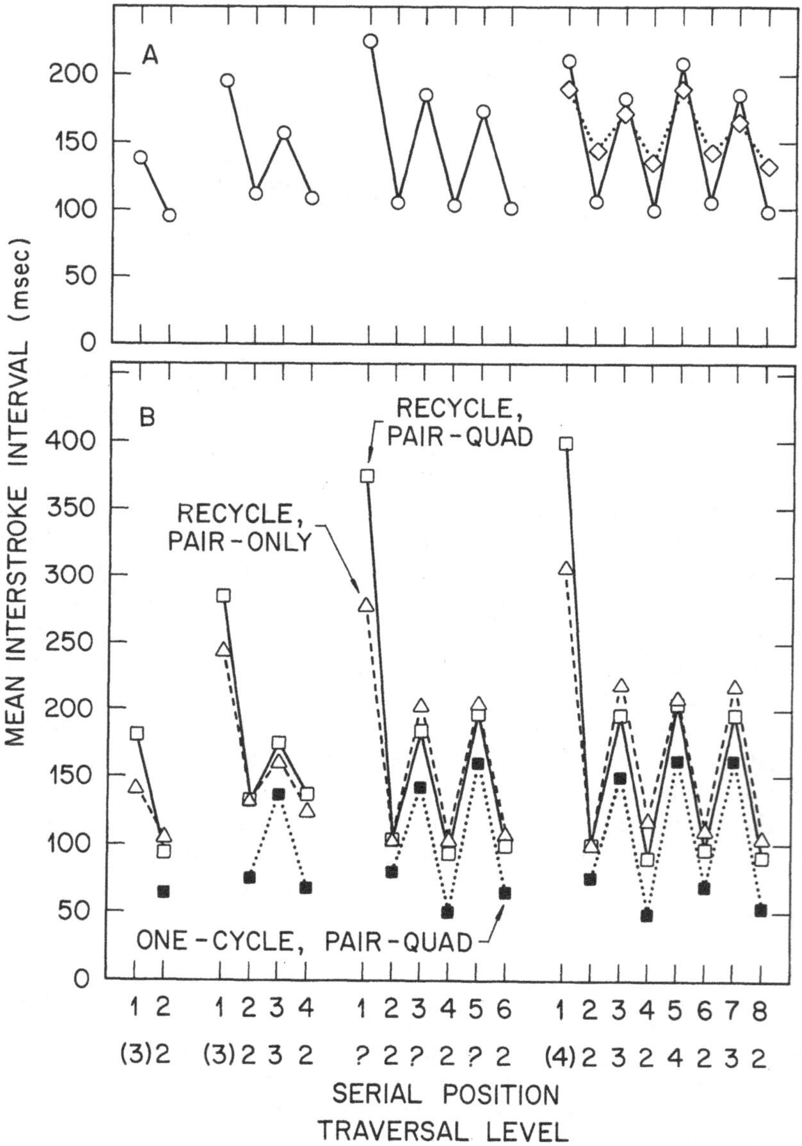


FIG. 1.5. Mean Interstroke Interval (*ISI*) as a function of serial position for strings of 2, 4, 6, and 8 letters. The serial position of an *ISI* denotes the stroke with which it terminates. Numbers in the bottom row indicate *traversal level*: the level of the highest node traversed in generating each *ISI*, according to the binary tree-traversal model. Traversal levels in parentheses depend on an auxiliary assumption in the model. The time interval from the "ready" signal to the initial stroke has been excluded.
A: Mean *ISI* from the four subjects in Part A of the experiment, with 48

*Results of the modified experiment.* Before considering tests of low-level invariance we comment on some other aspects of the data. Mean *ISI* as a function of serial position in strings of four lengths is shown in Figs. 1.5A and 1.5B. Also shown, for comparison, are mean *ITI*s in the study of Rosenbaum et al. (1983), for pair-quad strings with $n = 8$. Qualitatively our findings for $n = 8$ strings agree with those of Rosenbaum et al.: First-set *ISI*s (see definition above; second-level traversal) are the shortest, and third-set *ISI*s (fourth-level traversal) are the longest; magnitudes of the differences are greater and the effect of assumed traversal level less linear in the typing than the tapping data.

One difficulty presented by these data for the model of Rosenbaum et al. is the systematically large value of the first *ISI* under recycling conditions. Although not required by the spirit of the model, it attributes the time between the end of one rendition and the beginning of the next to traversal of the tree from the rightmost to the leftmost node. For $n = 2$, for instance, this traversal covers the same distance as the one presumably required between the first and second stroke, so the marked serial-position effect is unexpected. Similarly, the large difference (especially for subjects in Part B) for strings of length $n = 4$ between the first and third *ISI*, which also should involve equal traversal distances, is unexpected. Structural ambiguity for $n = 6$ precludes this comparison; the corresponding effect for $n = 8$ (difference between first and fifth *ISI*) is negligible in Part A but large in Part B. Because there appears to be a systematic inflation of the first *ISI*, perhaps attributable to a mechanism that counts renditions, we exclude it from further tests.

*Two tests of low-level invariance.* Given this experiment, there are two different tests of low-level invariance that we might perform, of differing generality. In the *binary-hierarchy test* (the less general), we assume that if the control structure is hierarchical then it must be binary, as proposed by Rosenbaum et al. (1983). The resulting structures for strings of four lengths are shown in Fig. 1.4. The figure shows that longer strings are distinguished by having higher level nodes; indeed, any slowing of performance with increasing length—any increase in mean *ISI*—is due entirely to the increasing depth of the tree, that is, to the augmentation of

---

trials per sequence length per subject; about 600 observations per point. Also shown, for $n = 8$ are the tapping data from Rosenbaum et al. (1983), from five subjects; about 6000 observations per point.

B: Mean *ISI* as a function of serial position from the three subjects in Part B of the experiment, under three conditions: Recycling pair-quad strings (as in Part A; about 850 observations per point); One-cycle pair-quad strings (about 500 observations per point); Recycling pair-only strings (about 850 observations per point).

the set of traversals required to execute the sequence by traversals of increasing length (and duration) as well as additional traversals of the same lengths. As we move from $n = 2$ to $n = 4$, for example, we add, within the string, an additional traversal of a node at the second level (associated with the second pair), as well as a traversal of a third-level node (between the two pairs). The structure is ambiguous for $n = 6$; whereas we use *ISI*s from hypothesized second-level nodes in $n = 6$ strings to test second-level invariance, we therefore limit our tests of third-level invariance to *ISI*s generated by hypothesized traversals of third-level nodes in strings of lengths $n = 4$ and $n = 8$.

In a *general-hierarchy test* we do not constrain the control structure to be binary, or the control process to be tree-traversal. As we vary an influencer such as sequence length in a test of low-level invariance, however, we cannot permit the hierarchical structure to be completely free. Recall that the property in question is the invariance across changes in the control structure of the process that executes a constituent substring. Testing this property requires that the *same* constituent exist under different levels of the influencer. Given a binary-tree control structure, this property follows *a fortiori*: the lowest level constituents of the hierarchy are *pairs* of elements for strings of all (even) lengths. However, if we impose no constraints on the structure, then the lowest level constituents might change as the influencer is varied. Thus, the execution of strings of several lengths might indeed by hierarchical, but the length difference might correspond to differences in the size of the lowest level constituents, or to differences in the size of constituents at all levels, rather than to differences in the depth (extent of bracket nesting) of the structure. In that case if we assumed fixed low-level constituents we might conclude against low-level invariance for the wrong reasons. Especially for strings constructed from repeated pairs as in the present experiment, but even for strings with no repeats, such flexibility in the organization of low-level constituents is implausible, whereas increasing the size of the structure by increasing its depth is in the spirit of the recursiveness of hierarchical control.

In connection with the present experiment we restrict ourselves to the binary-hierarchy test. Insofar as this test succeeds, the constraint is unimportant; insofar as this test fails, one from the more general class might succeed.

*Application of the binary-hierarchy test.* For *ISI*s after the first in pair-quad strings we note that although Part B reveals large and significant differences between one-cycle and recycling conditions ($p < .05$ for $n = 2, 6, 8$ strings, despite the small number of subjects), the effects of serial position at each length ($n = 4, 6, 8$) are statistically indistinguishable (interactions with one-cycle versus recycling not significant); we therefore examine these effects for the two conditions combined. Considering all positions after the first, Parts A and B both show large and significant effects of serial position, inviting us to test low-level invariance relative to this influencer.

Combined with the binary-hierarchy model, low-level invariance relative to serial position requires that for each string length, mean *ISI*s for second traversal-level (first-set) stroke pairs are equal, regardless of serial position. Instead there is systematic perturbation in all pair-quad strings, in which the *ISI* tends to be reduced by immediate repetition of a pair, which reaches significance in Part B ($p = .05$ for $n = 6, 8$). This effect is small relative to the large range of variation of *ISI* with serial position, so should perhaps not be given great weight; it reminds us of the difficulties for low-level invariance presented by prosodic (serial position) effects in speech. Low-level invariance relative to serial position also requires equality of the two third-level *ISI*s in $n = 8$ strings (indexed by 3 and 7 in Fig. 1.5), a requirement that is nicely satisfied.

We test low-level invariance with respect to sequence length by examining *ISI*s associated with second and third traversal levels. Before considering the effect of sequence length on these *ISI*s, we note that the overall mean *ISI* increases from $n = 2$ to $n = 4$, and then rises only slightly.[15] Sequence length does not therefore qualify as an influencer from $n = 4$ to $n = 8$; low-level invariance observed over that range would not be very persuasive. Means over the *ISI* values associated with purported second-, third-, and fourth-level traversals are shown in Fig. 1.6 as a function of sequence length, separated by condition and part of experiment. We see a systematic effect of sequence length at the second level, in the form of a peak at $n = 4$; the effect is significant in both Parts A ($p = .04$) and B ($p = .03$). At the third level we also appear to have a failure of invariance, with the mean *ISI* increasing with sequence length.[16]

We thus have three pieces of evidence against low-level invariance in the performance of patterned keystroke strings: Effects of serial position on first-set *ISI*s, and of sequence length on first-set and second-set *ISI*s. Even if we use these findings to reject the binary model, however, we

[15]Mean *ISI*s in msec over $n - 1$ serial positions (excluding *ISI*s, in recycling conditions, that terminate with the first stroke) for pair-quad strings of lengths $n = 2, 4, 6$, and 8, respectively, are (Part A, recycling) 95, 126, 134, 141; (Part B, recycling) 94, 148, 136, 139; and (Part B, one cycle) 64, 94, 100, 104. For pair-only strings (Part B, recycling) they are 106, 140, 146, 156.

[16]The increase is shown by 6 of the 7 subjects ($p = .06$ by a one-tailed binomial test), but there is wide variation in the size of the effect, which thus needs further substantiation.

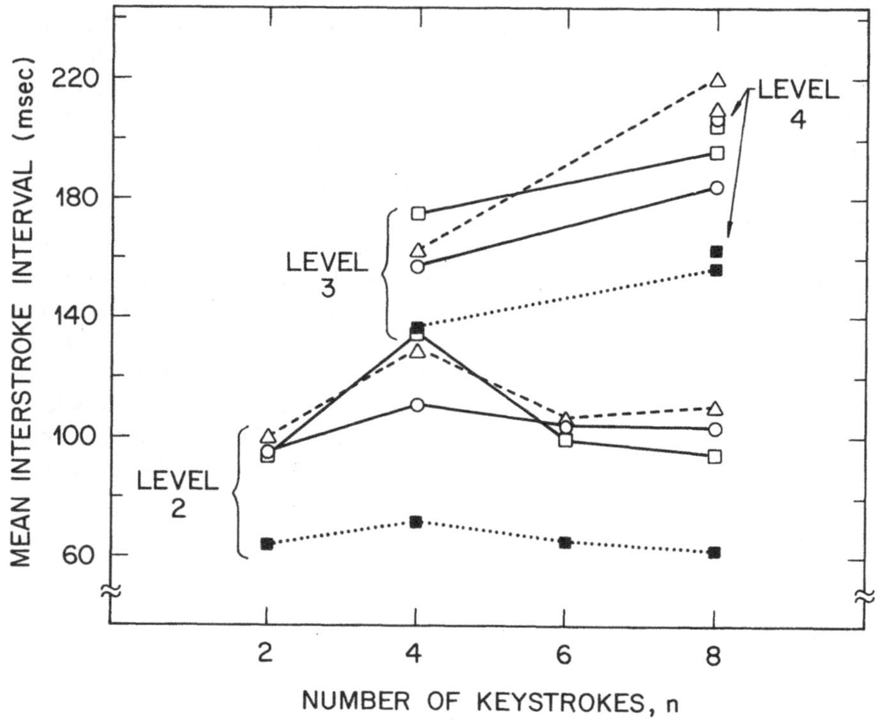


FIG. 1.6. Mean *ISI* as a function of sequence length and traversal level (see Fig. 1.5). The four conditions are designated as follows. From Part A: pair-quad, recycling, open circles. From Part B: pair-quad, recycling, open squares; pair-quad, one-cycle, filled squares; pair-only, recycling, open triangles.

might be able to avoid rejecting the more general hierarchical model. For example, the failure at the third level, relative to sequence length, could be explained if the constituents at that level for $n = 8$ contained one pair and three pairs rather than two and two, and if the number of pairs controlled by a third-level node had a disproportionate effect on the associated *ISI*.

## Test of Constituent Invariance in Strings of Distinct Strokes

*Assumptions for restricted and general tests; expected linkage proportions.* In the case of patterned strings such as those examined above there are plausible mappings of the string onto a hierarchical control structure. For a string of distinct strokes (with no repeats), however, there is little basis for specifying the mapping between string and structure. For such strings we would thus like to test the hypothesis that some stroke pairs are linked together in constituent substrings (at the second level), but without specifying which strokes are contained in these constituents. Our test in this case is therefore applied to the distribution of *ISI*s pooled over all pairs within the string; we ask about the proportion of the adjacent stroke pairs in the string that are linked together in low-level substrings.

Again, we consider both a restricted and a more general test. In the restricted test we assume, with Rosenbaum et al. (see Footnote 20) that the hierarchy is binary—even for strings with no repeats. For even-length strings the lowest level substrings are thus composed exclusively of pairs; for odd-length strings they contain pairs plus a single one-element constituent (in an unknown position). For this restricted class of hierarchical structures we can thus specify the number of pairs of adjacent strokes that must be linked together in low-level (two-stroke) substrings: In a string containing either $2k$ strokes ($2k-1$ possible pairings; $k \geq 1$) or $2k+1$ strokes ($2k$ possible pairings), a binary hierarchy will contain $k$ actual linked pairs at its lowest level. The linkage proportions are thus $k/(2k-1) > 0.5$ and $k/2k = 0.5$, respectively, or at least 0.5 regardless of string length.

As mentioned earlier, any general test of low-level invariance must jointly test the assumption that identities of low-level constituents are fixed over levels of the influencer. The specific form of the assumption that we require here is weak, however: We assume that at least one of the lowest level substrings of size $s \geq 2$ (containing $s-1$ linked pairs) is of the same size in long and short strings. Again, we justify this assumption by the recursiveness naturally associated with the idea of hierarchy: Changes in the size of a structure are thus expressed by changes in its depth rather than in the sizes of its low-level constituents. Given the assumption, the linkage proportion for a string of length $n$ ($n-1$ possible pairings) is no less than $(s-1)/(n-1)$. Because $s \geq 2$, the linkage proportion is thus bounded below by $1/(n-1)$.

In summary, the linkage proportion for a binary hierarchy can be no less than 0.5; for a general hierarchy and for strings of no more than five strokes, it can be no less than 0.25.

Our tests require one additional assumption: The interstroke times of pairs linked at the lowest (second) level tend to be shorter than other interstroke times. This is obviously true for a tree-traversal model and is plausible for any hierarchy model.

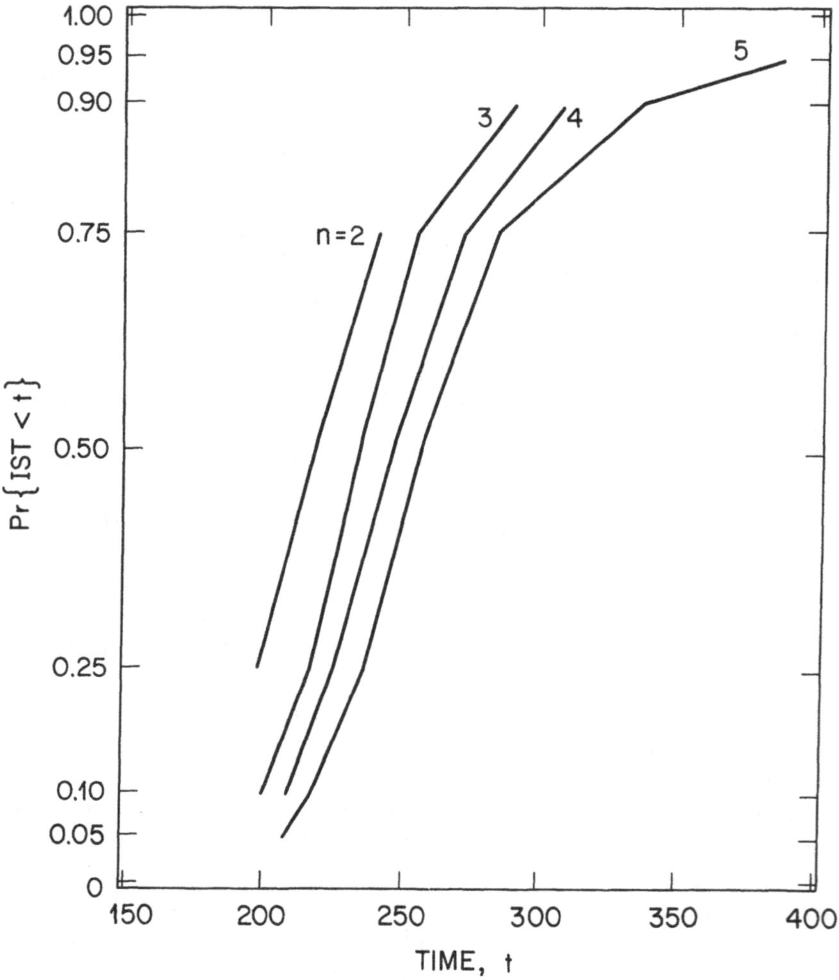


FIG. 1.7. Mean interstroke-time distributions for sequences of length $2 \leq n \leq 5$. See Footnote 17 for details of estimation.

*Observed linkage proportion: Upper bound.* The test is based on the following idea: If strings of two lengths (for example, $n = 2$ and $n = 5$ with *ISI* distributions $H_2$ and $H_5$ respectively) contain invariant substrings, then the shortest *ISI*s in $H_5$ should be as short as the shortest *ISI*s in $H_2$. The two cumulative distributions should therefore diverge or fan out from the same low-probability point. What we observe, however, are distributions better described as being displaced on the time axis, or translated, even at low probability values, which suggests that the shortest *ISI*s in $H_5$ are longer than the shortest in $H_2$. Fig. 1.7 shows mean *ISI* distributions for sequences of four different lengths ($n = 2, 3, 4, 5$) in an experiment in which subjects typed strings of distinct consonants.[17] In can be seen that these distributions are better described as being displaced than as diverging from a common point.

Without additional assumptions we wish to quantify the extent to which an invariant substring (and hence a hierarchical structure) is precluded by such findings. One way to do so is to express the pooled distribution of *ISI*s for a string of length $n$ as a mixture of two distributions. One component of the mixture is the distribution of *ISI*s associated with invariant substrings; this distribution, which we call $G_I(t)$, must be independent of $n$; its proportion in the mixture (its mixing probability $p_n$) is the same as the linkage proportion. The other component of the mixture is the distribution of *ISI*s for which invariance is permitted to fail and which can therefore vary with $n$; we call it $G_n(t)$; its mixing probability is $1 - p_n$. Our strategy is to determine what the relations among observed distributions can tell us about $p_n$. The observed distribution of *ISI*s pooled over serial positions corresponds to the mixture distribution, $H_n(t)$, which may be written in terms of its components as follows:

$$H_n(t) = p_n G_I(t) + (1 - p_n) G_n(t). \tag{3}$$

The assumption that linked pairs tend to have shorter *ISI*s than unlinked pairs is expressed by a dominance relation: $G_I(t)$ dominates $G_n(t)$. That is, for any $t$, $G_I(t) \geq G_n(t)$.

Given the invariance of $G_I(t)$ with respect to sequence length $n$, and the dominance assumption, it is easy to show that for any two sequence

---

[17]The data were drawn from a subset of the trials in a large experiment conducted for different purposes. They are the interstroke times of 6 bigrams, each of which appeared in all possible $(n - 1)$ serial positions in strings of length $n = 2, 3, 4$, and 5. Each string was typed by the fingers of one hand; the two hands were used equally often. We first pooled the data for a bigram across serial positions. Next we pooled the pairs of distributions for the two fastest bigrams (*jm, jh*), the two medium-speed bigrams (*fg, hm*), and the two slowest bigrams (*fb, rb*). This procedure provided sample sizes of $6(n-1)$ for each bigram pair and each subject. We then used linear interpolation to estimate quantiles for a specified set of probability values ($p = 0.25, 0.50, 0.75$ for $n = 2$, these plus $p = 0.10, 0.90$ for $n = 3, 4$, and these plus $p = 0.05, 0.95$ for $n = 5$). Means of the resulting quantiles were then obtained over bigram pairs and subjects to provide the plotted values.

lengths, $r$ and $s$ (e.g., $r = 2$ and $s = 5$), and for all values of $t$,

$$p_s \leq \frac{H_s(t)}{H_r(t)}, \tag{4}$$

and hence that

$$p_s \leq \min_t \left[ \frac{H_s(t)}{H_r(t)} \right]. \tag{5}$$

With only these minimal and plausible assumptions the relation between pooled *ISI* distributions for strings of different lengths can therefore provide us with an upper bound on the mixing probability, and thus on the linkage proportion.[18] It follows that $p_5 \leq \min[H_5(t)/H_2(t)]$, where the minimum is taken over all values of $t$. For example, if there is any (small) *ISI*, $t_0$, such that $H_2$ contains some *ISI*s that small $[H_2(t_0) > 0]$ but $H_5$ contains none that small $[H_5(t_0) = 0]$, then the $H_5/H_2$ ratio is zero for $t = t_0$, which implies that $p_5$ must be zero. Of course, this discussion concerns population distributions, and we must work with samples, which creates an estimation problem; we chose the estimation method so as to be conservative, in the sense of increasing the generosity of the upper bound.[19]

Upper bounds on $p_5$ estimated for each of eight subjects are represented by the filled circles in Fig. 1.8; The mean such bound (0.21) and $\pm SE$ are also shown. (Open circles are derived from a condition to be described later for which other data suggest a high linkage proportion and a hierarchical structure.) We checked the method by applying it to simulated data from mixtures of several known distributions with specified linkage proportions, and found that the upper bound was substantially greater than the actual linkage proportion in all cases, suggesting that it is a generous bound. For example, an estimated upper bound of 0.2 corresponded to an actual linkage proportion of 0.04 or less.

Even without taking the generosity of the bound into account, these findings clearly permit us to reject the restricted (binary) hierarchy model, which, as we have said, requires a linkage proportion for $n = 5$ of 0.5.

---

[18]To prove the inequality, first set $n = s$ and note that $H_s(t) \geq p_s G_I(t)$ for all $t$. (The mixture dominates one of its weighted components.) Now set $n = r$ and note that $G_I(t) \geq H_r(t)$ for all $t$. [Because the invariant distribution dominates the non-invariant, it also dominates any mixture of itself and the non-invariant, and therefore dominates $H_r(t)$.] Now, combine the two inequalities so as to eliminate $G_I(t)$: $H_s(t) \geq p_s H_r(t)$, for all $t$. The desired inequality follows.

[19]A discussion of estimation methods for essentially the same problem is provided by Sternberg and Knoll (1986).

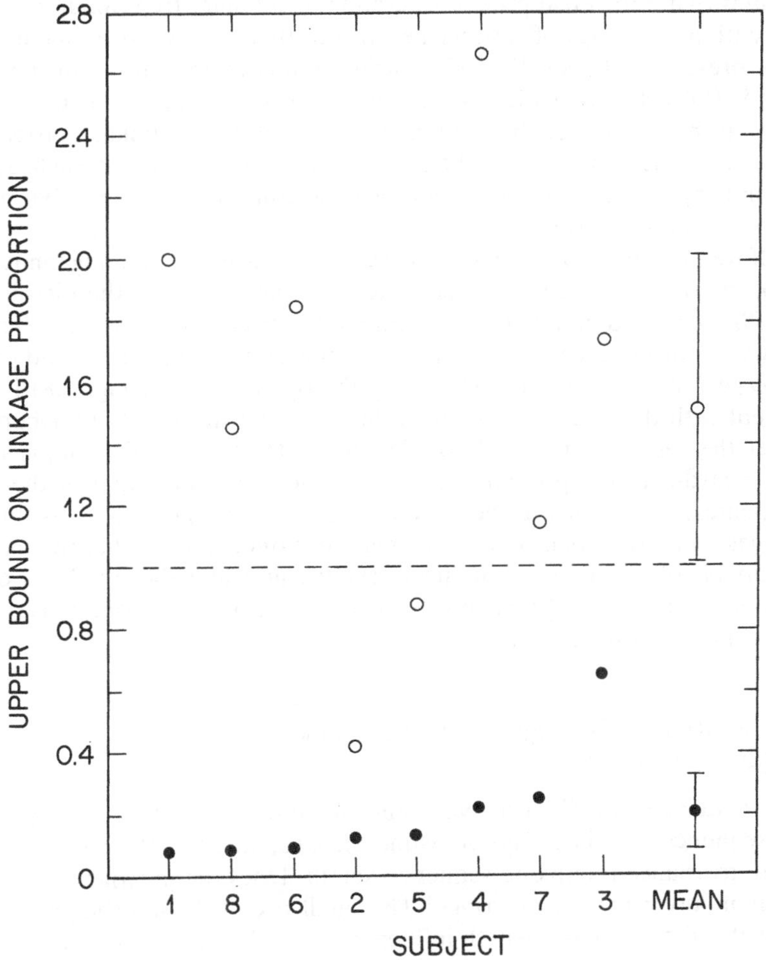


FIG. 1.8. Estimates for eight subjects of the upper bounds for the linkage proportion in keystroke strings of length $n = 5$. Filled circles represent values for strings of distinct strokes. Open circles represent estimates for the same subjects of upper bouncs on the linkage proportion for the two strokes of doublets embedded in strings of length $n = 5$. Also shown in each case is the mean and $\pm SE$.

Claims that a traversal process operating on a binary tree can describe the execution of strings of distinct keystrokes thus appear to be incorrect.[20] The present analysis is limited to strings typed by the fingers of the same hand; *ISI* distributions for alternating-hand strings suggest a similar conclusion, however. For the more general model, whose linkage proportion must be at least 0.25, the obtained values of the bound as such do not permit rejection. However, given the generosity of the bound, that model also becomes questionable.

If we turn from strings of distinct strokes to strings containing immediate repeats, or *doublets*, (strings such as *sccrz*) and examine the distribution of *ISI*s for the doublet (the time interval between the first and second *c* in *sccrz*, for instance) in the same way, the estimated lower bound on the linkage proportion (values shown by the open circles in Fig. 1.8) is consistent with the doublet forming an invariant substring.[21] We further consider the case of embedded doublets in a later section. Two implications are relevant to the present argument: First, the finding that of the eight estimates, six are above the maximum possible proportion (1.0), further suggests that the upper bound is a generous one; second, that the method is capable of revealing hierarchical structure increases the credibility of the evidence it provides against low-level invariance of multistroke constituents in strings of distinct strokes.

## Further Analysis by Segmentation of Single-Stroke Finger Trajectories

The evidence described above argues against the existence of invariant subsequences within strings of distinct strokes, based on the failure of low-level invariance relative to sequence length. Does the invariance property apply in any way to such strings? The small size of the linkage proportion indicates that sequence length influences the *ISI* between essentially every stroke and its successor. Can we find any decomposition of the stream of

---

[20]See Pew and Rosenbaum (1988); Rosenbaum (1985); and Rosenbaum (1987) for such claims, which also cover the production of utterances of unrelated words and of pseudowords. Other features of performance that such a model has difficulty explaining include the observed linearity of the effects of utterance length on latency and mean element duration (Sternberg, Monsell, et al., 1978; Sternberg, Wright, et al., 1980; Sternberg, Knoll, et al., 1989), for which a binary-hierarchy model with constant mean traversal time per node produces decelerating effects, and the small magnitude of the effects on element duration of serial position relative to length.

[21]This observation is consistent with the lower tail of the *ISI* distribution for the doublet being approximately invariant relative to sequence length; we see in a later section that the mean *ISI* is invariant also.

actions between one stroke and the next that reveals invariant segments? The movements associated with each stroke in such a string are complex, and thus susceptible to further analysis. Such analysis requires more information than can be obtained from just the sequence of time points at which keys are actuated, however. We report a study in which we attempted to decompose the trajectories of individual strokes, as we varied the length of the sequence containing them.[22]

The design of one experiment in which we examined strings of consonants typed by the fingers of alternating hands, together with examples of the strings, is shown in Table 1.3. In these examples, the trajectory examined is the one followed by the index finger of the left hand between the strokes corresponding to *f* and *v*; between these two strokes the right index finger strikes the *m* key. Subjects performed such sequences in isolation ($n = 3$), and at all possible serial positions in strings of length $n = 4$ and $n = 5$. The trajectory starts with a *hold* segment, during which the finger depresses a key (*f* in our example). There follows a *lift* segment during which the finger rises from the key and moves in the plane of the keyboard towards its next target. (It is typically during the lift of one finger that the hold segment of the other fingers' trajectory begins.) Finally there is an *attack* segment, during which the finger continues to move in the plane while it moves down to strike and depress the next key (*v* in our example). We defined these three segments by means of (low) *velocity* criteria. We used (zero) *acceleration* criteria to further segment the lift and attack into *early* and *late* components, giving us a total of five segments.

Fig. 1.9A shows mean duration of the full trajectory, from the start of the hold through the end of the attack. Because the hands alternate, the trajectory covers two interstroke times. The stroke rate for strings of length $n = 3$, for example, is therefore about 9 strokes/sec. We see the usual effect of sequence length on production rate. Fig. 1.9B shows durations of the five segments. Four of the five segments show the sought-for low-

---

[22]To measure such trajectories we used an apparatus that permits continuous measurement of the tips of two fingers in three dimensions. For each of the fingers to be measured it incorporates a separate lightweight linkage attached to the fingernail and connected to three low-torque potentiometers. The moment-to-moment settings of the potentiometers are uniquely determined by the position of the fingertip in three dimensions. The signal from each potentiometer was filtered, sampled at 200Hz, digitized, and stored for later analysis. We found that attachment of the linkages to two index fingers slowed the mean typing rate by only 4% and induced no change in error rate; the measurement procedure thus appears not to destroy the behavior under investigation. The analysis we present here (Sternberg, Knoll, et al., 1983) was restricted to the vertical component of the trajectories of the two index fingers.

TABLE 1.3.
Examples of letter sequences used for analysis of index-finger trajectories in typing. The examples shown were selected to permit assessment of effects of sequence length on durations of trajectory segments of the left index finger.

| *Sequence* | *Other Finger(s)* | | *Measured Index Fingers* | | | *Other Finger(s)* | |
|---|---|---|---|---|---|---|---|
| *Length* | *Left* | *Right* | *Left* | *Right* | *Left* | *Right* | *Left* |
| $n = 3$ | | | f | m | v | | |
| $n = 4$ | | | f | m | v | k | |
| | | p | f | m | v | | |
| $n = 5$ | | | f | m | v | l | s |
| | | k | f | m | v | p | |
| | d | p | f | m | v | | |

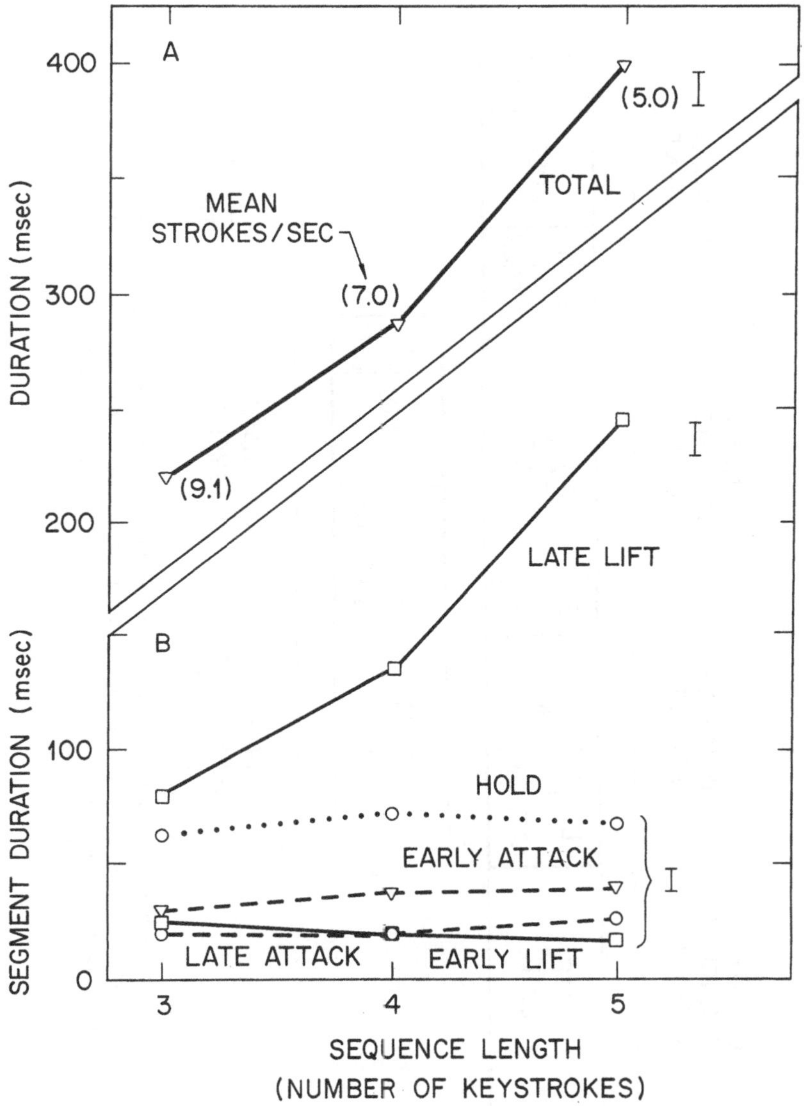


FIG. 1.9. Trajectory analysis for sequences of keystrokes produced by alternating hands. A: Mean total duration of the time interval between two strokes typed by the same index finger, and estimates of $\pm SE$. [Three subjects; about $18(n - 2)$ observations per point, where $n$ = sequence length.] Because these two strokes are separated by another stroke typed by the other hand, this duration is the sum of two *ISI*s. B: Mean duration of five trajectory segments whose sum is the total duration shown in Panel A, and estimates of $\pm SE$.

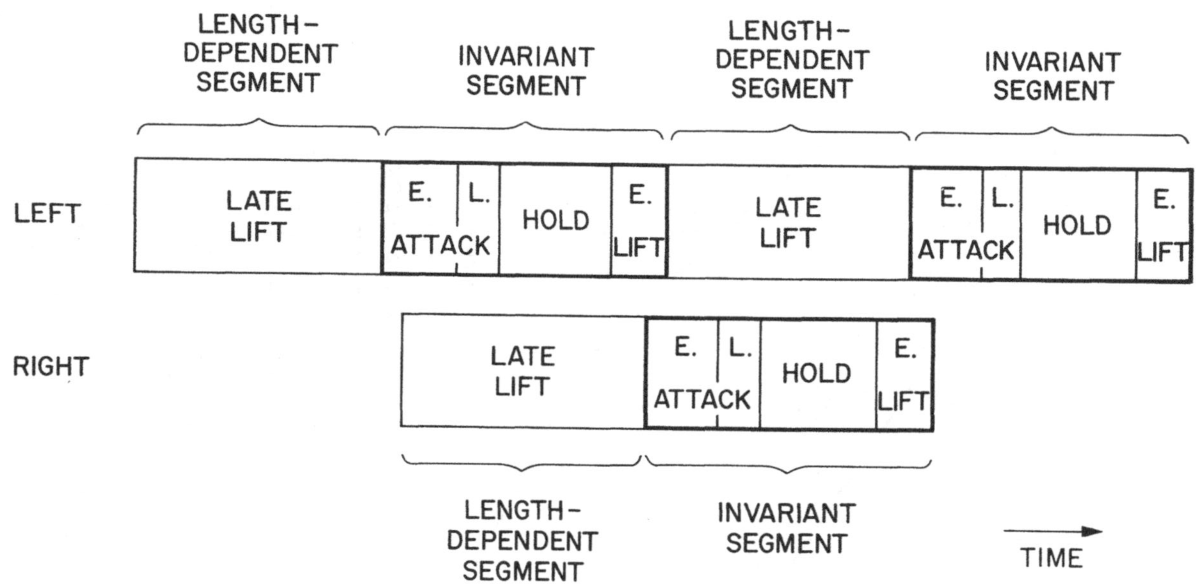


FIG. 1.10. Temporal overlap of finger-trajectory segments in a typed sequence produced by alternating hands. While one hand produces segments whose durations are independent of sequence length (*early attack*, *late attack*, *hold*, and *early lift*), the other hand produces its length-dependent segment (*late lift*).

level invariance at the level of individual strokes; only the late lift shows any substantial effect of sequence length.[23]

This promising analysis of segment durations thus reveals a qualified form of low-level invariance at the level of individual strokes, similar to what we have already described for speech; together with the near-zero value of the linkage proportion, it also indicates that a separate action unit is associated with each stroke. It is tempting to argue that as in the case of speech, we need only elaborate a process with transfer of control by adding a mechanism that permits actions controlled by a subprogram to outlast the execution of that subprogram, so as to provide behavioral continuity. The situation in this case is more complicated, however, as indicated in Fig. 1.10. While the index finger of one hand is in its late lift phase (whose duration depends on $n$, presumably because it occurs while higher level processes select and invoke the next subprogram for that hand), the other hand executes other segments (whose durations are independent of $n$). With each stroke, then, as the hands alternate, a high-level control process for one hand (sensitive to $n$) proceeds concurrently with execution of a low-level subprogram for the other (not sensitive to $n$), producing alternation of autonomously executed subprograms. One possibility is that we may have two separate but concurrent (and coordinated) hierarchical control processes, one per hand, each operating by transfer of control. Other data, however, show that the timing of each hand is influenced by the total number of strokes in a sequence and not just by the number of strokes assigned to that hand, which suggests an integrated control structure for two-hand sequences.[24] If we wish to preserve the idea that this structure is hierarchical, we are forced to conclude that rather than using transfer of control, the associated process functions by command/signal transmission, which, as discussed earlier, permits operations at different levels to proceed concurrently.

## Invariance Test of Embedded Doublets

For our final test of low-level invariance we turn to the experiment mentioned earlier on the performance of keystroke strings containing a doublet. As shown by the examples in Table 1.4 we used strings of lengths $n = 2$,

[23]In unpublished work we have replicated this finding, and also extended it to strings of letters typed by the fingers of one hand. We have also found that for another influencer—serial position—the same segments are again duration-invariant. That is, the effect of serial position on segment duration is also restricted to the late lift. In this last respect these findings in typing differ from those obtained in speech production, discussed earlier.

[24]These data arise in the production of letter strings that consist of two sucessivley executed substrings, one typed by the fingers of each hand.

TABLE 1.4.
Examples of letter strings containing the embedded doublet *cc,* and control strings.

| *Sequence Length* | *Doublet Strings Serial Position of Doublet* | | | | *Corresponding Control Strings* | | | | |
|---|---|---|---|---|---|---|---|---|---|
| | *1* | *2* | *3* | *4* | *1* | *2* | *3* | *4* | *5* |
| *n* = 2 | cc | | | | cf | xc | | | |
| *n* = 3 | ccf | xcc | | | czf | rcz | zsc | | |
| *n* = 4 | cczf | rccz | zscc | | csrz | zcfx | tscz | szfc | |
| *n* = 5 | ccsrz | zccfx | tsccz | szfcc | cxzfs | tcvsz | zfcvx | fxzcs | szxtc |

3, 4, and 5 typed by the fingers of one hand; for each length, each of several doublets (such as *cc*) appeared in each possible serial position; distinct-letter strings (examples in Table 1.4) were among the controls. We reported earlier on our use of the *ISI* distributions for doublets in strings of lengths $n = 2$ and $n = 5$ to test our estimation method, which suggested that the linkage probability for doublets was substantially higher than for nondoublet *ISI*s. Other analyses of the production of strings that include a doublet indicate that the two strokes of the doublet are contained in the same action unit, when the definition of *unit* is provided by a model of rapid sequence production (Sternberg, Knoll, et al., 1983). It follows that the doublet should display low-level invariance. In Fig 1.11 we provide a test, based on the mean *ISI* rather than its distribution: The element duration function for control strings (with a slope of 29 ± 6 msec/letter) is compared to the mean *ISI* for doublets (with a small and nonsignificant slope of 2 ± 3 msec/letter), which we thus find to be essentially invariant relative to string length, the first persuasive instance of low-level invariance above the level of the individual stroke in typing or tapping.

Insofar as the doublet is a low-level constituent, we expect its duration also to be invariant relative to serial position. Consistent with this expectation we find the effects of serial position on doublet duration to be small and nonsignificant at all lengths ($n = 3, 4, 5$).[25]

If the doublet is an action unit, then adding an immediate repeat of one of the strokes in a string of *n* distinct consonants increases the number of strokes to $n+1$ without increasing the number of *units*. This observation leads to the tests of high-level invariance that we consider later.

### Summary of Evidence For and Against Low-Level Invariance

In the preceding sections we have reported on our search for duration-invariance of subsequences of elements and of element segments, relative to sequence length and serial position, where the elements are keystrokes and stress-groups.

In speech production, we find low-level invariance relative to sequence length at the level of the stress group. To make this claim, however, we must modify the invariance property to permit occasional epochs for which invariance fails, a deviation that we explain by the smoothing of action between the end of one unit and the beginning of the next. That serial

[25]Significance levels for effects of serial position on nondoublet bigrams ($n = 3, 4, 5$) are (*ns*, .001, .003) in distinct-letter strings, and (.006, .05, *ns*) in doublet strings. Significance levels for the *difference* between effects of serial position on doublet and non-doublet bigrams are (*ns*, .01, .001) and (.02, *ns*, *ns*) for distinct-letter and doublet strings, respectively.

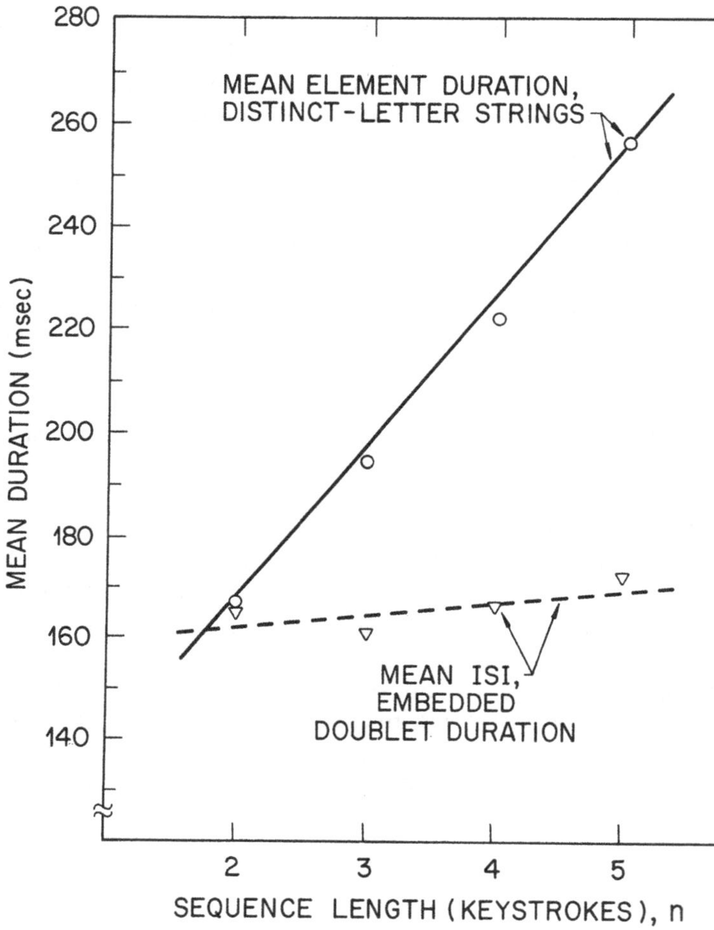


FIG. 1.11. Mean element duration for typed sequences of distinct letters [fitted function 110 + 29*n* msec; eight subjects; about 96($n-1$) observations per point; slope estimate *SE* = 6 msec] and mean doublet duration as a function of the length of the sequence in which it is embedded [fitted function 157 + 2*n* msec; about 96($n$ − 1) observations per point; slope estimate *SE* = 3 msec]. See Footnote 9.

position has an effect on the very segments that are invariant relative to sequence length is a serious challenge, however; this violation could perhaps be dealt with by invoking a process of temporal modulation, possibly related to the imposition of prosodic organization, that operates between a hierarchical control structure and the performance itself. Unless we can specify constraints on such a process, however, the invariance property loses much of its power. In speech production we have not yet searched for low-level invariance above the level of the stress-group.

In the typing of strings (such as *jfjfkdkd*) that contain repeated stroke-pairs, we find evidence against invariance at the level of the pair relative to both sequence length and serial position, and an indication of failure above that level as well. It would be especially interesting if further work strengthens these findings, given that performance of such sequences has been taken as a paradigmatic case of hierarchical control, based on the pattern of *ISI* means for strings of length $n = 8$. For strings (such as *zccfx*) that contain a repeated stroke (doublet), we find low-level invariance of the doublet itself. Finally, at the level of the single stroke in strings of distinct consonants we find duration invariance of segments of the stroke trajectory, relative to both sequence length and serial position; this supports the invariance property, again modified as mentioned above.

Insofar as low-level duration invariance is regarded as a critical property of action sequences produced under hierarchical control, we thus have evidence for shallow hierarchies: At this writing, only the doublet experiment provides clear evidence for constituents that incorporate more than one stroke (in typing or tapping) or one stress-group (in speech).

## TESTS OF HIGH-LEVEL INVARIANCE

### Sequence Duration Under Singlet-Doublet Exchange

Both tests of high-level invariance that we describe make use of the doublet experiment, and both assume constraints on the extent of temporal overlap of high-level and low-level operations equivalent to that provided by a process with transfer of control. With respect to the *ISI* of the doublet itself, we provided evidence in earlier sections that neither the lower tail of its distribution nor its mean depends on the number of strokes in the string in which it is embedded, and also that its mean is invariant with respect to serial position. These findings support the hypothesis that the doublet is executed by a subprogram that functions autonomously relative to the control structure that invokes it, as we would expect if that structure were hierarchical.

Combined with high-level invariance, the hypothesis leads to the following complementary property: If we replace a single stroke (*singlet*) in a string of length $n$ by a doublet, creating a string of length $n + 1$, we should have exchanged one autonomous subprogram in the control structure (associated with the singlet) by another one (associated with the doublet), without changing other aspects of the structure. In the simplest case, the exchange would leave all the *ISI*s unaltered, merely inserting the invariant doublet *ISI* among them. Because the execution of the subprogram may start before the timepoint of its (first) stroke, and may end after the timepoint of its (last) stroke, the exchange might also affect the adjacent *ISI*s. Any such effects, however, should be independent of the embedding structure, and hence of the number of additional strokes $(n-1)$ and of the serial position of the replacement.

One implication is that the exchange should alter mean string duration by a constant, independent of $n$: If $D_n$ is the mean duration of a string of $n$ distinct strokes, and $D_n^*$ is the mean duration of a string of $n$ strokes containing an embedded doublet, then we expect that

$$D_n^* = D_{n-1} + k, \qquad (n \geq 2). \tag{6}$$

That is, replacement of singlet by doublet should leave invariant, up to a possible additive constant, the *sum* of the mean *ISI*s for stroke pairs that are common to the two strings (as well as adding the constant mean *ISI* of the doublet).

There are conditions under which validation of Equation 6 would be of little interest, however: Suppose that augmenting a string by one stroke incremented its mean duration by the same constant, regardless of $n$ and of whether the string contained a doublet. Then $D_n^*$ and $D_n$ would be linear with equal slopes, and Equation 6 would follow, as would Equation 7:

$$D_n^* = D_n + c, \qquad (n \geq 2), \tag{7}$$

where $c$ is a different additive constant. Equation (7), however, would be expected to obtain if the doublet functioned simply as a stroke pair with an unusual *ISI*, rather than having the same status in a hierarchical control structure as a singlet. It is by virtue of the nonlinearity of $D_n$ that it is possible to determine that only one of these relationships obtains.

A second implication of doublet-singlet equivalence is that the exchange should leave the *individual* mean *ISI*s invariant, up to possible additive constants, for those stroke pairs adjacent to the replacement. In this section we report a test of the first implication; in the next section, the second.

Results of a test of the first implication are shown in Fig. 1.12. The leftmost function is $D_n$ $(2 \leq n \leq 5)$, the sequence-duration function for non-doublet strings. Translated one unit to the right, it becomes the $D_{n-1}$ function, $(3 \leq n \leq 6)$. The leftmost broken curve is $D_n^*$ $(2 \leq n \leq 5)$, the sequence-duration function for doublet strings. This function is parallel to the $D_{n-1}$ function (i.e., vertically displaced, as described by Equation 6) and not to the $D_n$ function (Equation 7), in accordance with the implication

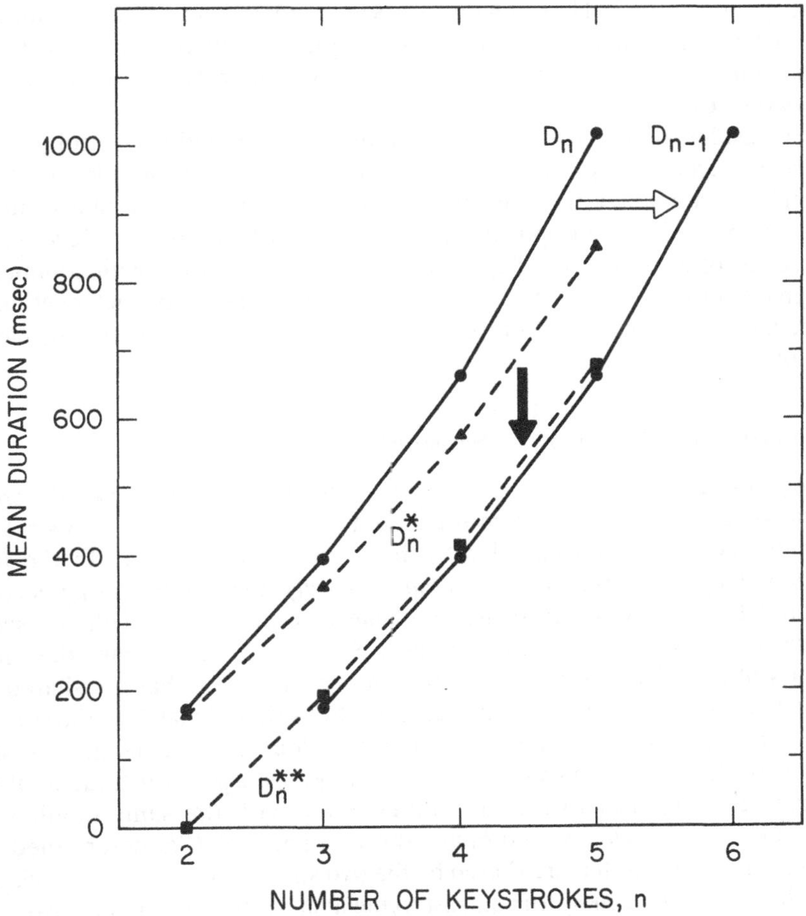


FIG. 1.12. Sequence durations $D_n$ and $D_{n-1}$ for distinct-letter strings and $D_n^*$ for doublet strings, each containing $2 \leq n \leq 5$ strokes. Also shown is $D_n^{**} = D_n^* - ISI_n^*$, the sum of the non-doublet *ISI*s in doublet strings. $D_{n-1}$ is the same as $D_n$, but translated one unit to the right. See text for explanation.

under test. To obtain the lower of the two broken curves, we have subtracted the mean doublet duration $ISI_n^*$ (shown in Fig. 1.11) for strings of length $n$ to give the mean sum of *ISI*s for non-doublet stroke pairs in doublet strings, $D_n^{**} = D_n^* - ISI_n^*$. Agreement of $D_n^{**}$ with $D_{n-1}$ is not perfect; apparently there is an effect of the singlet-doublet exchange on contextual *ISI*s. However, this effect is both small (about 20 msec) and independent of $n$.[26] Thus, not only does its context not influence the doublet (low-level invariance), but also, except for a small constant effect, exchanging singlet for doublet does not appear to alter its context (high-level invariance).

Consider the implications of this finding for the kind of control process that underlies the production of one-handed letter strings. Replacing a singlet by a doublet increases the time required for subprogram execution, providing additional opportunity for overlap of high- and low-level processes insofar as such overlap is possible, as in a process with command/signal transmission. As we see essentially no indication of such overlap, a process with transfer of control (in which no overlap is permitted) easily explains these results.

### Serial-Position Effect Under Singlet-Doublet Exchange

For this second test of high-level invariance we performed two different analyses. We exemplify them for a doublet string $S_n^*$ with $n = 5$ strokes, whose doublet is associated with the third *ISI*: $S_5^* = zxttc$. In the first analysis (comparison by *subprogram position*), the doublet string is compared to a non-doublet string with one less stroke, but with the same number of hypothetical subprograms: $S_4 = zxtc$; we compare the three pairs of *ISIs* determined by the bigrams *zx, xt,* and *tc* that are shared by the two strings. *ISIs* in corresponding positions should match, if the control structure satisfies high-level invariance as doublet is exchanged for singlet. In the second analysis (comparison by *stroke position*), the doublet string is compared to a non-doublet string with the same number of strokes: $S_5 = zxtfc$; we compare the two pairs of *ISIs* determined by bigrams *zx* and *xt* that are shared by the two strings, and the pair associated with *tc* in $S_5^*$ and *fc* in $S_5$. In contrast to the analysis by subprogram position, invariance provides no reason to expect *ISIs* in corresponding stroke positions to match, even if we adjust for the difference between overall mean interstroke times.

To perform the analyses we combined data over doublet strings with

---

[26]Analysis of variance shows a significant ($p = .01$) interaction of sequence length $n$ with $D_n^*$ versus $D_n$, ($2 \leq n \leq 5$), but none ($p > .25$) for $D_n^{**}$ versus $D_{n-1}$; the latter analysis reveals a significant ($p < .05$) main effect, however. Another approach to the data dramatizes the usefulness of the nonlinerarity of $D_n$: We find $(D_4^* - D_2^*) - (D_4 - D_2) = -78.6 \pm 13.0$ msec ($p < .001$), where the 7 *df SE* is based on variability over the eight subjects, whereas $(D_5^{**} - D_3^{**}) - (D_4 - D_2) = 490.7 - 488.4 = 2.3 \pm 6.9$ msec, supporting the remarkable parallelism of means displayed in the figure.

the doublet in different serial positions.[27] Results are shown in Fig. 1.13.

For the stroke-position analysis (upper panel) we had strings of three lengths to consider. In each case ($n = 3, 4, 5$), the shapes of the serial-position functions being compared were reliably different ($p \approx .001$). When we move from the stroke-position to subprogram-position analysis shown in the lower panel, we find a reliable ($p < .001$) improvement in agreement, in the direction of what is implied by high-level invariance. For $n = 5$, however, the shapes for doublet and distinct-letter strings are still significantly different ($p \approx .007$), indicating a failure of high-level invariance in the distribution of duration over serial positions.

The exchange of doublet for singlet thus appears to leave the summed durations of other constituents invariant, in accordance with high-level invariance, but alters the way in which the sum is distributed.

## JOINT TESTS OF HIGH-LEVEL AND LOW-LEVEL INVARIANCE

Given that the action unit in the rapid articulation of short utterances is the stress group, an experiment in speech production suggests itself that is analogous to the doublet experiment examined above. One could compare utterances of various lengths composed exclusively of monosyllabic units with the same utterances after replacing one of the units by a multisyllabic unit. To our knowledge, no such experiment under rapid-speech conditions has been performed. However, there are studies of both typing and speech production in which each string is homogeneous with respect to unit size (in syllables/unit in speech, for example) and in which pairs of strings can be regarded as being related by exchange of *all* units for others of a different size. (For example, consider *bay-rum-limb* and *baby-rumble-limit.*) For such experiments, the combination of high-level and low-level invariance properties, given a process with transfer of control, has a surprisingly strong implication, which can thus be used to test these properties in combination.

---

[27]The combination rule differed for the two analyses. For the analysis by subprogram position, we assigned positions to *ISI*s by ignoring the doublet *ISI*; for the doublet string $S_5^{\bullet}$ = *zxttc* the nondoublet *ISI*s would then be numbered 1, 2, and 3. We then averaged correspondingly numbered *ISI*s within lengths. For the analysis by stroke position, we assigned positions to *ISI*s that took into account the presence of the doublet *ISI*; for $S_5^{\bullet}$ = *zxttc* the nondoublet *ISI*s would then be numbered 1, 2, and 4. We excluded the doublet *ISI* itself from the analysis, and again averaged correspondingly numbered *ISI*s.

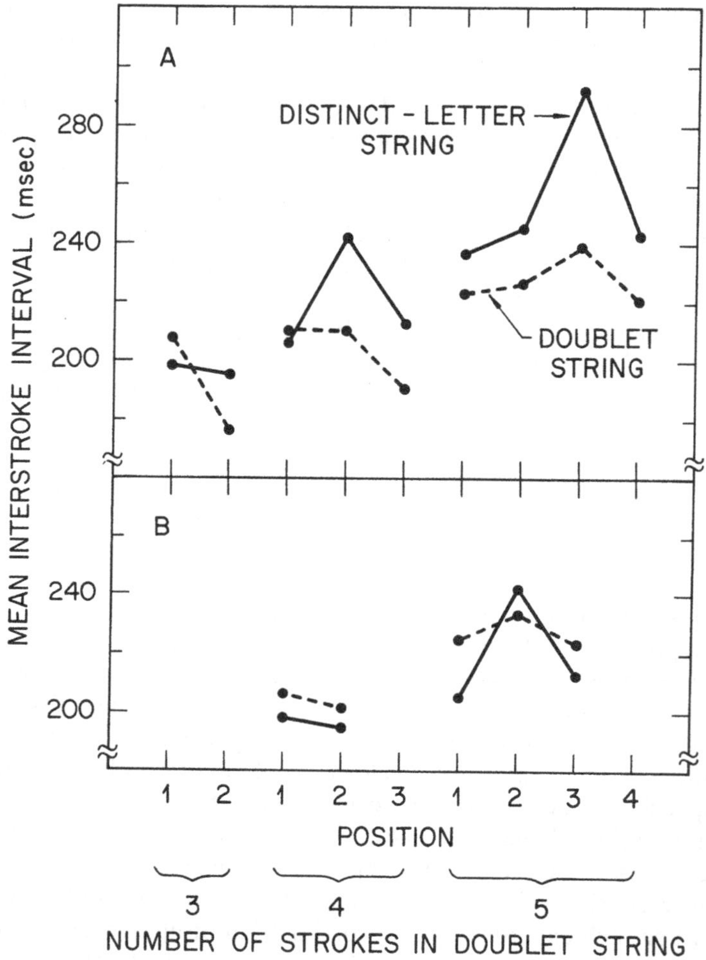


FIG. 1.13. Serial-position effects in distinct-letter strings and doublet strings of three lengths. A: Mean *ISI*s matched by stroke position. B: Mean *ISI*s matched by subprogram position.

We develop the argument with respect to sequence duration. Given high-level invariance, we can replace one unit ($u_x$) of an $n$-unit string by another ($u_y$), leaving the remainder of the control structure invariant. Because the execution times of $u_x$ and $u_y$ differ, in general, such an exchange would alter sequence duration. Low-level invariance implies that the execution times, and hence their difference, are independent of the higher level control structure, and hence are independent of sequence length, $n$. We can therefore describe the difference as a constant rather than a function of $n$: Let $u_y$ be the unit with the longer execution time, and let $\delta_{xy}$ represent the mean execution-time difference. Given a process with transfer of control, $\delta_{xy}$ would also describe the mean sequence-duration difference due to the exchange. High-level invariance implies that we can make additional exchanges, until all $n$ $u_x$-units have been replaced by $u_y$-units, again leaving the remainder of the control structure invariant, and thus generating a total sequence-duration increment $n\delta_{xy}$. The difference between the durations of sequences of $n$ $u_x$-units and $n$ $u_y$-units can thus be at most linear in $n$:

$$D_n(u_y) - D_n(u_x) = n\delta_{xy}. \tag{8}$$

It follows that *any nonlinear term in the sequence-duration functions* $D_n(u_x)$ *and* $D_n(u_y)$ *must be the same, regardless of the sizes of units* $u_x$ *and* $u_y$. If the duration functions are quadratic, as they appear to be, their quadratic coefficients must therefore be invariant under unit exchange. Equivalently, the mean element-duration functions, $d._n(u_x)$ and $d._n(u_y)$, must have equal slopes.

### Effects of Utterance Length Under Speech-Unit Exchange

Results of one experiment, with unit size 1 versus 2 versus 3 syllables, and typical results, were described in an earlier section. There we reported that a plausible unit definition (the stress group) led to mean element-duration functions with equal slopes (Fig. 1.1, Table 1.1). Similar results have been reported for an experiment comparing utterances of monosyllabic and disyllabic nouns (Sternberg, Monsell, et al., 1978). Because such evidence favors an implication of the two invariance properties jointly, together with a process of transfer of control, it lends support to the properties and process.[28]

### Effects of Number of Strokes Under Keystroke Exchange

For skilled transcription typing it is well known that *ISI*s are shorter between strokes typed by different hands than by the same hand (Gentner, 1983). For a letter string transcribed by one hand the duration is thus greater than for a string transcribed by alternating hands; this is also true for strings prescribed and prepared in advance. If we assume that the action

---

[28]It also follows from the present argument that the shapes of serial-position functions for utterances with a specified number of units should be independent of unit size. This implication has not yet been tested in speech production.

unit for both hand patterns is the single stroke, then the difference in interstroke transition perhaps corresponds to a difference in unit size, and the relation between the two element-duration functions is of special interest. The functions in one comparison were essentially parallel, which again supports the invariance properties; findings on this point are mixed, however.[29]

### Summary of Evidence For and Against High-Level Invariance

Fewer data are available that permit testing high-level than low-level invariance, but here, also, results are mixed. In both typing and speech production, where we have compared sequences in which constituents of different size are exchanged, we find striking support for high-level invariance (together with transfer of control) in measures of sequence duration. But where we have examined the distribution of sequence duration over serial positions (in typing) we find that it is altered by the exchange. Again, as in our suggestion of how low-level invariance might be salvaged given disallowed effects of serial position, advocates of hierarchical control may be forced to argue for a process of temporal modulation that intervenes between the output of a hierarchical control structure and the actions it controls.

## CONCLUSIONS

High-level and low-level invariance seem to us to be inherent in the idea of hierarchical control in the execution of action sequences. Furthermore, we have found instances in both speech and typing where these properties apply, especially if we allow for the smoothing of action at constituent

[29]Sternberg, Monsell, et al. (1978) reported element-duration slopes of 15.2 and 14.1 msec/stroke for alternating- and one-hand strings, respectively. Although we have replicated this finding, we have also observed violations, with a tendency for a larger slope in the production of alternating-hand strings; more work is needed. Note also that even in the study mentioned, the magnitude of the effect of serial position is larger for alternating-hand strings, instead of being invariant as required. A second difficulty arises from the analysis of finger trajectories as described, which suggests that the change from one-hand to alternating-hand strings is more complicated than merely a change of unit size, because it requires concurrent processes at different levels of control. Indeed, this analysis induces skepticism about the finding of parallel element-duration functions, and suggests attempting alternative manipulations of unit size in typing.

boundaries. Within the limited domain of duration invariance there is also a pattern of violations, however, associated with the distribution of duration over serial position—violations that suggest that significant control operations must intervene between the output of a hierarchical structure and the actions it controls.

In our limited search we have found little evidence for anything other than shallow hierarchies, even though our procedures were designed to maximize both the opportunity and the incentive for advance planning. Until further work is done, including the use of other measures and the study of other factors, it seems fair to conclude that the largest constituents smaller than the whole sequence in rapid utterances or keystroke strings are seldom larger than the elements we have described as action units—individual strokes in typing and stress groups in speech.

In much of our analysis we have assumed that the process that operates on the hierarchical structure does so by transfer of control, and some of our findings in both speech and one-hand typing strongly support this assumption. Our measurements also suggest, however, that a different control process must be operating when coordination of the hands is required, one that permits concurrent operations at different levels, such as control/signal transmission.

Among the questions our findings suggest are the following: Are there conditions of sequence production not yet considered that would reveal deeper hierarchies? What other interesting properties might a hierarchical control structure confer on the sequences it produces? Are there other conceptions of hierarchical control in the execution of action sequences to which the invariance properties do not apply, but that are nonetheless testable?

## ACKNOWLEDGMENTS

We are grateful to the International Association for the Study of Attention and Performance and to Marc Jeannerod and Sylvan Kornblum for encouraging us to prepare this report. We also thank A. S. Coriell and Charles E. Wright for technical consulting, Stephen Monsell, Ruth M. Colwill, and Anthony S. Kroch for useful discussions, and especially Felice Bedford, C. R. Gallistel, Lila Gleitman, Stephen Monsell, and Seth Roberts for helpful comments on an earlier draft. Stephen Monsell and Charles E. Wright collaborated in related research that had a strong impact on this work, and from which some of the data used in these analyses were drawn. Saul Sternberg's participation was supported in part by the Linguistics and Artificial-Intelligence Research Department of AT&T Bell Laboratories.

## REFERENCES

Abend, W., Bizzi, E., & Morasso, P. (1982). Human arm trajectory formation. *Brain, 105*, 331–348.

Aho, A. V., Hopcroft, J. E., & Ullman, J. D. (1983). *Data structures and algorithms*. Reading, MA: Addison Wesley.

Barlow, G. W. (1977). Modal action patterns. In T. A. Sebeok (Ed.), *How animals communicate* (pp. 98–134). Bloomington: Indiana University Press.

Benecke, R., Rothwell, J. C., Day, B.L., Dick, J. P. R., & Marsden, C. D. (1986). Motor strategies involved in the performance of sequential movements. *Experimental Brain Research, 63*, 585–595.

Book, W. F. (1908). *The psychology of skill with special reference to its acquisition in typewriting*. Missoula, MT: University of Montana.

Broadbent, D. (1977). Levels, hierarchies, and the locus of control. *Quarterly Journal of Experimental Psychology, 29*, 181–201.

Bryan, W. L., & Harter, N. (1899). Studies on the telegraphic language: The acquisition of a hierarchy of habits. *Psychological Review, 6*, 345–375.

Collard, R., & Povel, D.-J. (1982). Theory of serial pattern production: Tree traversals. *Psychological Review, 89*, 693–707.

Dawkins, R. (1976). Hierarchical organization: A candidate principle for ethology. In P. P. G. Bateson & R. A. Hinde (Eds.), *Growing points in ethology* (pp. 7–51). Cambridge: Cambridge University Press.

Dell, G. S. (1984). The representation of serial order in speech: Evidence from the repeated phoneme effect in speech errors. *Journal of Experimental Psychology: Learning, Memory, and Cognition, 10*, 222–233.

Drummond, H. (1981). The nature of and description of behavior patterns. In P. P. G. Bateson & P. H. Klopfer (Eds.), *Perspectives in ethology: Vol. 4*. New York: Plenum.

Estes, W. K. (1972). An associative basis for coding and organization in memory. In A. W. Melton & E. Martin (Eds.), *Coding processes in human memory* (pp. 161–190). Washington, DC: Winston.

Fendrick, P. (1937). Hierarchical skills in typewriting. *Journal of Educational Psychology, 28*, 609–620.

Fentress, J. C. (1983). Ethological models of hierarchy and patterning of species-specific behavior. In E. Satinoff & P. Teitelbaum (Eds.) *Handbook of behavioral neurobiology* (*Vol. 6*) (pp. 185–234). New York: Plenum.

Fowler, C. A. (1981). A relationship between coarticulation and compensatory shortening. *Phonetica, 38*, 35–50.

Fowler, C. A. (1985). Current perspectives on language and speech production: A critical review. In R. Daniloff (Ed.), *Speech science: Recent advances* (pp. 193–278). San Diego: College-Hill.

Fromkin, V. A. (Ed.). (1971). The non-anomalous nature of anomalous utterances. *Language, 47*, 27–52.

Fromkin, V. A. (Ed.). (1981). *Errors of linguistic performance: Slips of the tongue, ear, pen, and hands*. New York: Academic.

Gallistel, C. R. (1980). *The organization of action: A new synthesis*. Hillsdale, NJ: Lawrence Erlbaum Associates.

Gentner, D. R. (1983). Keystroke timing in transcription typing. In W. E. Cooper (Ed.), *Cognitive aspects of skilled typing* (pp. 95–120). New York: Springer-Verlag.

Gentner, D. R. (1987). Timing of skilled motor performance: Tests of the proportional duration model. *Psychological Review, 94*, 255–276.

Gee, J. P., & Grosjean, F. (1983). Performance structures: A psycholinguistic and linguistic appraisal. *Cognitive Psychology, 15,* 411–458.

Gordon, P. C., & Meyer, D. E. (1987). Control of serial order in rapidly spoken syllable sequences. *Journal of Memory and Language, 26,* 300–321.

Greeno, J. G., & Simon, H. A. (1974). Processes for sequence production. *Psychological Review, 81,* 187–198.

Jagacinski, R. J., Marshburn, E., Klapp, S. T., & Jones, M. R. (1988). Tests of parallel vs. integrated structure in polyrhythmic tapping. *Journal of Motor Behavior, 20,* 416–442.

Johnson, N. F. (1970). The role of chunking and organization in the process of recall. In G. H. Bower (Ed.) *The psychology of learning and motivation* (*Vol. 4*) (pp. 172–247). New York: Academic.

Keele, S. W. (1987). Sequencing and timing in skilled perception and action: An overview. In A. Allport, D. G. MacKay, W. Prinz, & E. Scheerer (Eds.), *Language perception and production: Relationships between listening, speaking, reading, and writing* (pp. 463–487). London: Academic.

Klein, R. (1983). Nonhierarchical control of rapid movement sequences: A comment on Rosenbaum, Kenny, and Derr. *Journal of Experimental Psychology: Human Perception and Performance, 9,* 834–836.

Lehrdahl, F., & Jackendoff, R. (1983). An overview of hierarchical structure in music. *Music Perception, 1,* 229–252.

Leonard, J. A., & Newman, R. C. (1964). Formation of higher habits. *Nature, 203,* 550–551.

Machlis, L. (1977). An analysis of the temporal pattern of pecking in chicks. *Behaviour, 63,* 1–70.

MacKay, D. G. (1982). The problems of flexibility, fluency, and speed-accuracy trade-off in skilled behavior. *Psychological Review, 89,* 483–506.

Massey, J. T., Schwartz, A. B., & Georgopoulos, A. P. (1986). On information processing and performing a movement sequence. *Experimental Brain Research* (*Supplement*) *15,* 242–251.

Miller, G. A., Galanter, E., & Pribram, K. H. (1960). *Plans and the structure of behavior.* New York: Holt, Rinehardt, & Winston.

Monsell, S., & Sternberg, S. (1981). *Speech programming: A critical review, a new experimental approach, and a model of the timing of rapid utterances* (*Part 1*). Bell Laboratories Technical Memorandum, June. Murray Hill, NJ: AT&T Bell Laboratories.

Morasso, P. (1983). Three dimensional arm trajectories. *Biological Cybernetics, 48,* 187–194.

Morasso, P., Mussa Ivaldi, F., & Ruggiero, C. (1983). How a discontinuous mechanism can produce continuous patterns in trajectory formation and handwriting. *Acta Psychologica, 54,* 83–98.

Narmour, E. (1983). Some major theoretical problems concerning the concept of hierarchy in the analysis of tonal music. *Music Perception, 1,* 129–199.

Nelson, K. (1973). Does the holistic study of behavior have a future? In P. G. Bateson & P. H. Klopfer, *Perspectives in Ethology* (pp. 281–328). New York: Plenum.

Palmer, S. E. (1977). Hierarchical structure in perceptual representation. *Cognitive Psychology, 9,* 441–474.

Pew, R. W., & Rosenbaum, D. A. (1988). Human movement control: Computation, representation, and implementation. In R. C. Atkinson, R. J. Herrnstein, G. Lindzey, & R. D. Luce (Eds.) *Stevens' handbook of experimental psychology, 2nd edition: Vol. 2. Learning and cognition* (pp. 473–509). New York: Wiley.

Perkell, J. S. (1980). Phonetic features and the physiology of speech production. In B. Butterworth (Ed.), *Language production: Vol. 1, Speech and talk* (pp. 135–173). London: Academic.

Reitman, J. S., & Reuter, H. H. (1980). Organization revealed by recall orders and confirmed by pauses. *Cognitive Psychology, 12,* 554–581.

Rosenbaum, D. A. (1983). Hierarchical versus nonhierarchical models of movement sequence control: A reply to Klein. *Journal of Experimental Psychology: Human Perception and Performance, 9,* 837–839.

Rosenbaum, D. A. (1985). Motor programming: A review and scheduling theory. In H. Heuer, U. Kleinbeck, & K.-H. Schmidt (Eds.), *Motor behavior: Programming, control, and acquisition* (pp. 1–33). Berlin: Springer-Verlag.

Rosenbaum, D. A. (1987). Hierarchical organization of motor programs. In S. P. Wise (Ed.), *Higher brain functions: Recent explorations of the brain's emergent properties* (pp. 45–66). New York: Wiley.

Rosenbaum, D. A., Kenny, S. B., & Derr, M. A. (1983). Hierarchical control of rapid movement sequences. *Journal of Experimental Psychology: Human Perception and Performance, 9,* 86–102.

Rumelhart, D. E., & Norman, D. A. (1982). Simulating a skilled typist: A study of skilled cognitive-motor performance. *Cognitive Science, 6,* 1–36.

Shattuck-Hufnagel, S. (1983). Sublexical units and suprasegmental structure in speech production planning. In P. F. MacNeilage (Ed.), *The production of speech* (pp. 109–136). New York: Springer-Verlag.

Simon, H. A. (1962). The architecture of complexity. *Proceedings of the American Philosophical Society, 106,* 467–482.

Sternberg, S., & Knoll, R. L. (1986). Determination of the unit in rapid action sequences. AT&T Bell Laboratories Technical Memorandum. Murray Hill, NJ: AT&T Bell Laboratories.

Sternberg, S., Knoll, R. L., Monsell, S., & Wright, C. E. (1983). Control of rapid action sequences in speech and typing. Bell Laboratories Technical Memorandum. Murray Hill, NJ: AT&T Bell Laboratories.

Sternberg, S., Knoll, R. L., Monsell, S., & Wright, C. E. (1989). Motor programs and hierarchical organization in the control of rapid speech. *Phonetica, 45,* 175–197.

Sternberg, S., Monsell, S., Knoll, R. L., & Wright, C. E. (1978). The latency and duration of rapid movement sequences: Comparisons of speech and typewriting. In G. E. Stelmach (Ed.), *Information processing in motor control and learning* (pp. 117–152). New York: Academic.

Sternberg, S., Wright, C. E., Knoll, R. L., & Monsell, S. (1980). Motor programs in rapid speech: Additional evidence. In R. A. Cole (Ed.), *The perception and production of fluent speech* (pp. 507–534). Hillsdale, NJ: Lawrence Erlbaum Associates.

Szentagothai, J., & Arbib, M. A. (1975). *Conceptual models of neural organization.* Cambridge, MA: MIT Press.

Terzuolo, C. A., & Viviani, P. (1980). Determinants and characteristics of motor patterns used for typing. *Neuroscience, 5,* 1085–1103.

Volpert, W. (1982). The model of the hierarchical-sequential organization of action. In W. Hacker, W. Volpert, & M. von Cranach (Eds.), *Cognitive and motivational aspects of action* (pp. 35–51). Amsterdam: North Holland.

Vorberg, D., & Hambuch, R. (1978). On the temporal control of rhythmic performance. In J. Requin (Ed.), *Attention and Performance VII.* Hillsdale, NJ: Lawrence Erlbaum Associates.

Vorberg, D., & Hambuch, R. (1984). Timing of two-handed rhythmic performance. In J. Gibbon & L. Allan (Eds.), *Timing and time perception. Annals of the New York Academy of Sciences, 423,* 390–406.

Wall, R. (1972). *Introduction to mathematical linguistics.* Englewood Cliffs, NJ: Prentice-Hall.

West, L. J. (1969). *Acquisition of typewriting skills*. Belmont, CA: Pitman.
Wilson, D. (1969). Forms of hierarchy: A selected bibliography. In L. L. Whyte, A. G. Wilson, & D. Wilson (Eds.), *Hierarchical structures* (pp. 287–314). New York: Elsevier.
Wing, A. M. (1977). Effects of type of movement on the temporal precision of response sequences. *British Journal of Mathematical and Statistical Psychology, 30,* 60–72.

# II TUTORIALS ON THE BASIC ORGANIZATION PRINCIPLES OF MOTOR REPRESENTATIONS

# 2 The Motor Cortical Areas and the Problem of Hierarchies

M. Wiesendanger
*Institut de Physiologie*
*Université de Fribourg*
*Fribourg, Switzerland*

ABSTRACT

The historical development of some ideas concerning the hierarchical organization of the brain is traced back to Spencer and Jackson. Concepts such as *humunculus* representation, *upper motor neuron*, and *higher motor deficits*, all still widely used in neurological thinking, emerged from the original ideas of Jackson. The process of movement initiation and execution is seen as a cascadelike activation, beginning in the "highest centers" with propagation to "lower centers." This view, in its extreme, has been challenged in recent years as it became apparent that many structures at all levels of the neuraxis may operate in parallel, with many interconnections and interactions as well as with many opportunities for back-propagation. In a second part of this chapter the evidence for parallel processing in motor cortical areas is reviewed and compared with more advanced insights in the domain of the visual system.

> *There are undoubtedly animals that perform more forceful, more precise, even more delicate movements that those of the human being; human motor acts are, however, uncomparably superior to those of animals with respect to their richness and diversity.*
> —von Monakow, 1911

Because the problem of hierarchy is a recurrent theme of this book, it might be appropriate to deal generally with the meaning of this term. Quite obviously, it can mean different things in the neurosciences. Relatively loosely defined in neurology, it acquired new, more stringent meaning in computer science and in its application to neuronal network models. In the first part of this chapter, I attempt to review the historical origin of the hierarchy concept as it emerged toward the end of the 19th century and how the concept was applied to clinical neurology and to anatomical neuronal structures. The latter field concerns mainly the issue of serial versus parallel processing. I also briefly review newer ideas that arose from single-unit studies in the cerebral cortex leading to the concept of nerve cell populations in "higher" brain structures encoding emergent complex features. In its extreme, this led to the now discredited notion of *pontifical* neurons or *grandmother* cells (Hubel, 1988). In a second part of this chapter, I summarize some anatomical and physiological features of sensory and motor cortical areas of primates that, I believe, reveal both hierarchical and parallel processing.

## THE DEVELOPMENT OF IDEAS ON NEURONAL HIERARCHIES

### Jackson's Doctrines and His Criteria for Hierarchical Classification (Jackson 1932)

The concept of hierarchy as an organizational principle of the brain was introduced by Jackson in the aftermath of Spencer's and Darwin's theories on evolution and dissolution in biology (Richards, 1987). The ideas were attractive for clinical neurology because in the process of "dissolution" or "taking to pieces" by "destroying lesions" when evolutionary new structures are damaged by disease, one can observe that the behavior is placed under the rule of lower structures. In Jackson's words: "To undergo dissolution is to be reduced to a lower level of evolution" (p. 46). In its more naive form, it resulted in the still widely used concept of the *humunculus* made up of *upper motor neurons* that—like a keyboard—govern the lower motor neurons. If we want to avoid a simplistic and absurd cascadelike or "pyramidal" interpretation of hierarchy, we need clearly defined criteria for classifying hierarchical levels. Jackson's criteria are not sharply defined and are sometimes expressed in a vocabulary that needs interpretation. Essentially, *dissolution* in Jackson's sense meant:

1. Loss of the least organized or retention of the more organized.
2. Loss of the most complex or retention of the less complex.
3. Loss of the most voluntary or retention of the more automatic.

It may be necessary to briefly consider the meaning of these terms. The *automaticity criterion* is perhaps the best known, but in my opinion also the most problematic one. What Jackson may have had in mind was the automaticity of breathing, stepping, chewing. Modern neurophysiological work indeed proved that these performances are organized, in their basic form, at a low level of the neuraxis. But what about the automatization of complex motor acts when we practice them and when they become habits? We don't know whether Jackson thought that with practice the control of such habits is delegated from higher to lower structures. I don't think that there is enough evidence to suggest that this is so; some evidence in fact pleads against this notion. For example, cells of the motor cortex continue to discharge in relation to very simple elbow movements in overtrained monkeys (e.g., Conrad, Wiesendanger, Matsunami, & Brooks 1977).

At first sight, the criterion of *organization* is also surprising: Why should high-level controls be the least organized? Presumably, what Jackson had in mind concerned the degrees of freedom. Fixed-action patterns would be the "best organized" ones, resulting in "machine-like behaviour" to use Sherrington's expression. The least organized movements would thus be those showing a high flexibility, or those that are the least organized.

And then there is the criterion of *complexity*. In Jackson's terminology, evolution from the lowest to the highest level means " . . . passage from the least complex to the most complex." Complexity is primarily a matter of the number of neural elements and connections. At a low level, anatomically speaking at the spinal cord segment, the neural network concerned with hand movements has much fewer elements than the cortical network dedicated to this task. Complexity is less well defined in functional terms. It has probably also to do with the degrees of freedom with which, for instance, hand movements can be performed. A complex movement disorder of the hand would be one in which the hand can be moved for one act, say greeting, but not for another, say writing. Complexity thus includes the context or the plan of the purposeful behavioral act in which the hand is used. Jackson was referring to this principle when talking of "re-representation" of movements in the cerebral cortex.

Interesting is also Jackson's proposition that "higher centres" are "unifying centres." Presumably, this refers to Jackson's view that "the cortex is organized in terms of synergies." A concatenation of activity in many muscles is needed for a purposeful act that might involve eye, head, trunk, and hand movements. The detailed implementations, what Jackson calls the "local affairs" would be delegated to the lower control levels that are " . . . thereby raised to much higher power."

Jackson's ideas, centered on the concept of dissolution, had a strong and worldwide impact on clinical neurology. For example, the concept of dissolution (*Abbau der Funktionen*) became also central in the work of von Monakow (1914). This happened at a time when many new discoveries on structural relationships in the brain were made, thanks to the availability of new histological techniques. These allowed for the first time a more precise correlation of the clinical deficits with the site of brain lesions. This coincidence much contributed to the hierarchic conceptualization in the neurological sciences (for a stimulating critique cf. Kelso & Tuller, 1981).

## Attempts to Identify the Anatomical Structures in a Serial Schema of Motor Control Processes

Lesions within the precentral motor strip, that is in the primary motor cortex (MI), result in the well-known paretic deficits. This is in contrast to the complex apraxic disorders observed following lesions outside the Rolandic cortex. Recognizing the oversimplification, one may say that the former deficit is one of movement execution, the latter one of movement programming (for example, inability to perform properly sequenced movements without loss of muscular strength or of the ability to perform individual components of a sequence). Jackson conjectured that the highest level of movement organization occupies cortical areas in front of the Rolandic cortex, whereas movement execution would be the task of primary motor cortex, being thus at a middle level of the hierarchy. Subsequently Campbell (1905), who constructed the first cytoarchitectonic map of the human cortex, was of the opinion that the highest level occupies what he termed the *Intermediate Precentral Cortex*, which roughly corresponds to the ensemble of premotor areas. To quote him "I am of the opinion that this particular stretch of cortex is specially designed for the execution of complex movements in which consciousness or volition takes an active part" (p. 217). To Liepmann (1920) goes the credit for having systematized the various forms of apraxias with the underlying cortical substrates.

At about the same time, von Monakow (1914), in his treatise on brain localization in which he extensively deals with motor disorders including apraxias and aphasias, also addressed the problem of cortical localization of motor cortical centers. Here is my free translation of his summary:

> One can envisage that a purposeful behavioural act is produced in successive, partly overlapping steps:
>
> 1. If we aim toward a given target, we first construct a mental image or plan, on how we will reach this target. This hypothetical stage cannot yet be correlated with any anatomical structure-it has probably no sharply defined localization in the cortex.
>
> 2. The next step is a translation of this general plan into instructions for elemental movements that are chained together. At this stage the prefocal areas are called into action via associational impulses coming from the various cortical areas. We can now identify these areas, but the situation is complex.

The activity is not confined to one particular island of cortex, rather it includes a number of disjunctive areas in various constellations. These areas extend far beyond the Rolandic cortex.

3. Now follows the focal activity within the Rolandic motor cortex. It is always a variable combination of activated foci which depends on the body part to be moved, but also on the purpose of the movement. It is only during this phase that we have well-localized, that is focal activity.

4. The last step then concerns corticofugal transmission from the motor cortex to the subcortical and spinal motor centers and the final translation of this impulse traffic into muscular activity. (pp. 620–621).

Thus, it might be concluded from von Monakow that, at the beginning of this century, the idea of a sequential processing for the initiation of voluntary, purposeful movements was established. Interestingly, one can already envisage a differentiation within the highest levels, with cortical association areas (frontal and parietal) being superior to premotor areas (the Intermediate Precentral Cortex of Campbell or the "prefocal areas" of von Monakow). The former were held responsible for the general movement plan ("ideation of movements," "*Bewegungsentwurf*"), the latter for more detailed information ("the kinetic formula") that would then be transmitted to the executive zone of the motor cortex.

This anatomical schema of movement initiation has been revived in recent years as refined anatomical tracing methods became available allowing a more precise identification of the pathways connecting cortical association areas with the motor cortex. The cortico-subcortico-cortical

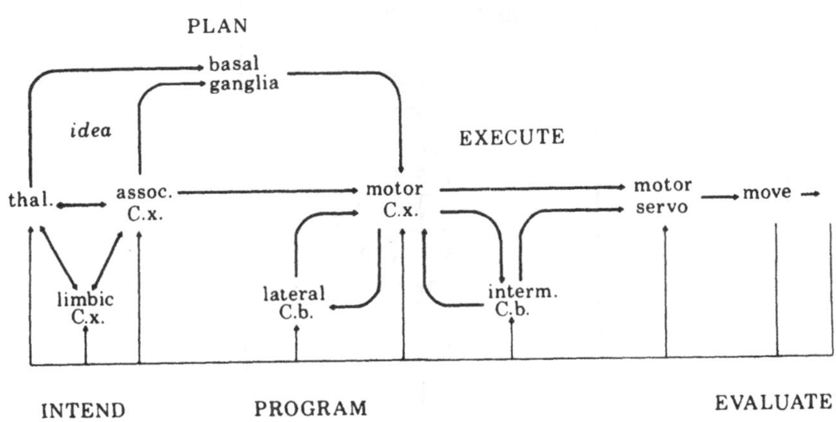


FIG. 2.1. Schematic representation of a serial activation of brain structures in the course of movement initiation as proposed by Paillard (1982). This is a revised and completed schema based on Allen and Tsukahara (1974). Note the association of anatomical structures to the phases of INTENTION, PLANNING, PROGRAMMING, EXECUTION, EVALUATION.

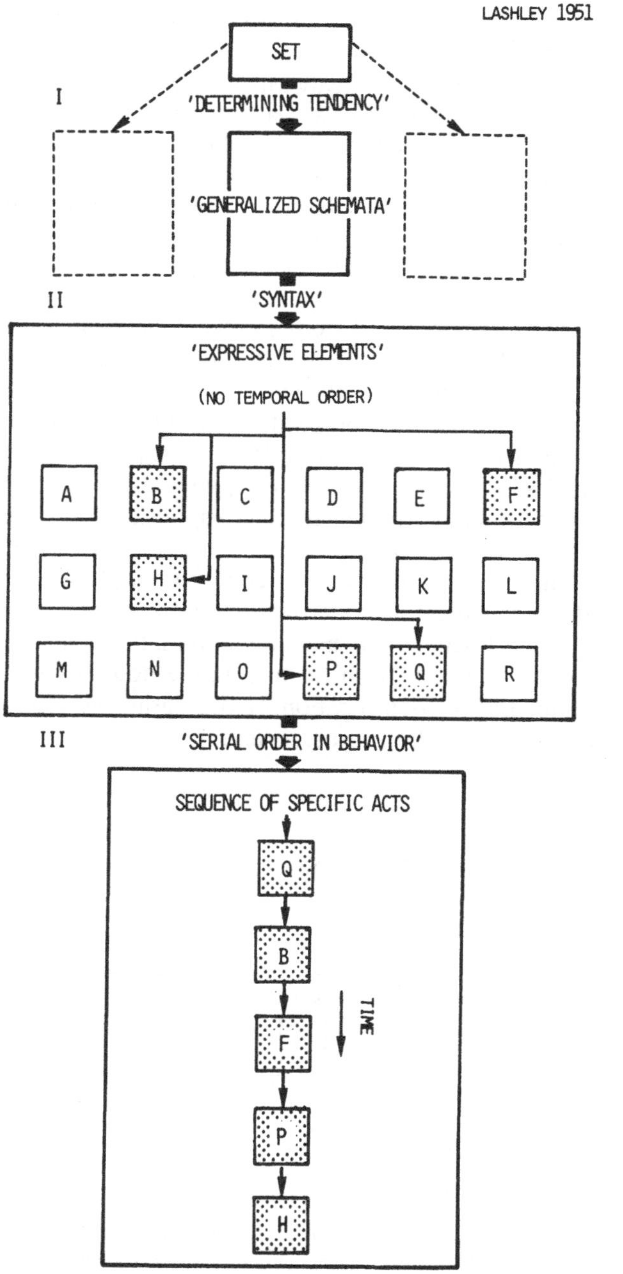


FIG. 2.2. Schematic representation of Lashley's ideas on SERIAL ORDER IN BEHAVIOR (1951).

circuits have been summarized in the well-known diagrams of Kemp and Powell (1971) and of Allen and Tsukahara (1974). These anatomical diagrams are widely used as reference for classifying brain areas in the framework of a hierarchical and serial organization of movement initiation (e.g., Roll, 1983). An excellent recent treatment of this general proposition of movement initiation can be found in the review of Paillard (1982) from which I borrowed the updated diagram of Allen and Tsukahara (Figure 2.1).

A serial ordering in the process of movement initiation was also conceived by Lashley (1951), but in a more abstract and nonanatomical sense. The key idea is that the central system has in store the basic building bricks of movements (expressive elements). Temporal ordering is achieved by a syntax of the act that Lashley describes as a "habitual order mode of relating the expressive elements (in time)." The appropriate 'syntax' is called into action by the determining tendency or set (Figure 2.2). In Lashley's words: "This is the essential problem of serial order: the existence of generalized schemata of action which determine the sequence of specific acts . . . ." Is there concrete evidence for the existence of generalized schemata? The question cannot be answered in physiological terms, but it is well known that the cerebellum has an important function in the correct timing of movements, perhaps also the supplementary motor cortex, as recently suggested by Deecke et al. (1985).

## HIERARCHIC LEVELS OF CORTICAL AREAS

### Multiplicity of Sensory Cortical Areas

Recent, detailed investigations of cortical sensory areas revealed a complexity of subareas for each modality that are richly interconnected. For example, "Twenty areas are candidates for being largely or exclusively visual in function; 11 of these are demonstrably visual and have been identified with a reasonable degree of confidence" (Van Essen, 1985, p. 268). Some of these areas have a full representation of the visual field (Hubel, 1987). Most important in the current context is the observation that typically these visual areas have the tendency to be reciprocally interconnected. Interestingly, within an interconnecting pair of fiber systems, the terminations are often asymmetric. This is shown schematically in Fig. 2.3, taken from the work of Maunsell and van Essen (1983). It was proposed that the asymmetric distributions of cells of origin and of terminals reveal forward and feedback projections. Layer IV in the cortex is typically the recipient of specific ascending fiber systems. Therefore, the corticocortical fibers terminating in layer IV are considered to be also *ascending*, that is,

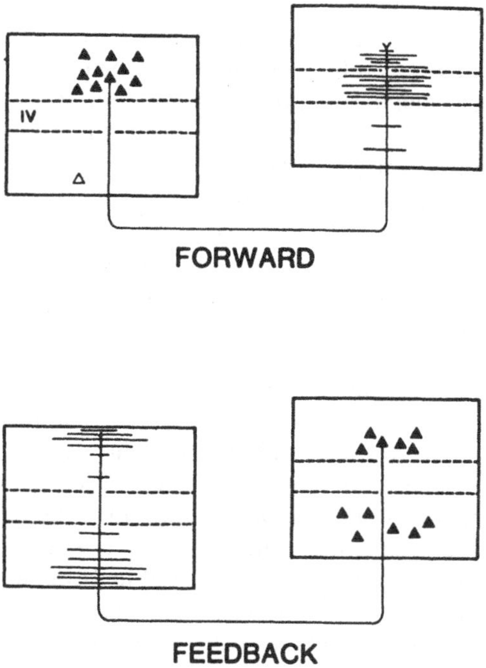


FIG. 2.3 Differential laminar distributions of neuronal cell bodies and terminals in feedforward and feedback pathways of visual cortical areas (Maunsell and Van Essen 1983). According to this scheme and the anatomical results of Primrose and Strick (1985) in motor cortical areas, the primary motor cortex would have a feedforward projection to secondary (premotor) areas which, in turn, project back to the primary motor cortex.

feedforward—the target area becoming thus hierarchically superior to the donor area. This hierarchic scheme, based on anatomical data, is still hypothetical. Some physiological properties are, however, in line with the hierarchical hypothesis in that receptive fields become larger and more complex in higher areas (Van Essen, 1985).

A hierarchic level may thus be defined by the ensemble of inputs and outputs that are common to a neuron population. In this respect, the visual areas are not homogenous; rather it turns out that there are specifications in the cortical layers and in their intrinsic connectivity. The thalamic sources (from the magnocellular, parvocellular, and interlaminar portions of the lateral geniculate nucleus) project to different strata in the visual area V1 and have separate, cascaded projections within the cortex, thus determining the intra-areal hierarchic level. Again, this anatomical segregation appears to match a segregation of neurons with particular functional attributes. At least in a few situations, the orderly processing from thalamic nuclei to a cascade of intra-areal laminae appears to reflect the anatomical basis of functional transformations (leading to complex receptive field organizations, to orientation selectivity, to disparity selectivity, and so forth). Even further transformations occur in visual areas outside the striate cortex with neuronal populations "recognizing" familiar objects and faces (Bruce, Desimone, & Gross, 1981). The details of the circuits that lead to such transformations are, however, unknown or hypothetical, and most likely depend on cooperative actions of distributed populations of neurons.

This recent, combined anatomical and microelectrophysiological approach to reveal the step-by-step processing of visual information from the ascending inputs to the different layers of the visual cortex and the further area-to-area processing is still at its beginning, but nevertheless far ahead from the situation in motor cortical areas.

## Sequential Processing in Other Cortical Areas

In the somatosensory system, it was found that signals transmitted to area SI are further transmitted to SII and thence to insular and retroinsular areas (Friedman, Murray, O'Neill, & Mishkin, 1986). In accord with this, responses to cutaneous inputs were abolished in SII when SI was lesioned in the adult monkey (Pons, Garraghty, and Mishkin, 1988). In a brief note, Primrose and Strick (1985) reported on the pattern of connections in motor cortical areas. They also found the reciprocal interconnections in motor areas to be asymmetrical with the primary motor cortex MI feeding forward to the premotor cortex, that in turn feeds back to MI. Given that the interpretation of the corticocortical connectivity, as revealed in visual areas, is correct and also applicable to motor areas, it would confirm that motor cortical areas, too, are hierarchically organized. Interestingly, the observed patterns of connections would be in line with the notion that the secondary somato- sensory and -motor areas are hierarchically superior to primary cortical areas. One has to realize that the hierarchization of cortical areas on the basis of corticocortical connections is an *ascending* one. In speculations about the hierarchic superior features of premotor areas, it was on the contrary assumed that the hierarchy is descending: that is, that the hierarchically superior premotor areas "instruct" the motor cortex, being "downstream" of the former areas. I think that we touch here on a difficult, but crucial, problem of defining hierarchic levels in sensory systems and in motor systems.

The problem of corticocortical relationships in the various motor areas clearly requires much more detailed investigations. Whatever we learn in terms of hierarchic levels between motor cortical subareas, it is important to keep in mind the established fact that the connections are not unidirectional and that the simple unidirectional hierarchic view in the higher motor control centers does not apply.

Before returning to the motor cortical areas, it might be worthwhile to ask whether there are perhaps other criteria that may help to distinguish hierarchic levels of neuronal systems.

## Emergence of Higher Properties by Convergence and Divergence

Anatomically speaking, axonal branching and synaptic convergence on the somatodendritic membranes is one of the most striking properties of all neuronal systems, sensory as well as motor. Intuitively, it seems obvious that such divergent and convergent connections are at the root of emergent features in the central nervous system. Emergence is, by definition, more than summation: Accordingly, it is hardly possible to explain emergent, complex neuronal features by analyzing features of "lower" neurons converging to the complex neuron. The best known complex and emergent encoding features of single neurons are those of the visual cortical areas alluded to already. They include features that contribute to stereopsis, color, form, movement, and even face perception. In recent years, much progress has been achieved in unravelling the specific functions of visual areas, subserved by multiple afferent streams of visual information. Although there is clear evidence for a considerable division of labor, there are also many indications that visual processing in the various areas is not independent. Thus, there are areal interconnections providing opportunities for concurrent processing by convergence and divergence. The situation in visual areas has recently been reviewed by DeJoe and van Essen (1988) who came to the following conclusion that might well be relevant also for the multiple motor cortical areas: ". . .One should not anticipate. . . a simple one-to-one relationship between a particular stream and a particular computational strategy. Rather, it seems likely that some strategies will be handled in parallel by more than one stream or will involve significant crossover between streams." (p. 225)

Emergent properties of sensory neurons by convergence find their counterpart in emergent features of projection neurons in motor centers by virtue of their divergent collateralization (Evarts 1984). A focal excitation in the motor cortex inevitably is widely distributed via collateral transmission to spinal as well as to supraspinal targets (Wiesendanger, 1986a).

## Is There Evidence for Parallel Processing and for Division of Labor in Motor Cortical Areas?

The multiplicity of visual areas is paralleled by a multiplicity of motor areas in the frontal cortex, although classically only a few areas were identified: the primary motor cortex (MI), the premotor cortex (PMC), and the supplementary motor area (SMA). The two latter areas were also termed secondary motor areas (see Wiesendanger, 1981), as opposed to the primary motor cortex considered to constitute the major control area of the motor apparatus. More recent hodological investigations, however, led to a more detailed parcellation of motor areas, a situation that might turn out to be more akin to that of visual areas (Wiesendanger & Wise, in press). The motor areas, like the visual areas, are heavily interconnected, with MI not only receiving fibers from secondary areas, but also projecting to these areas as mentioned above. A somatotopic organization, clearly demonstrable with microstimulation in MI, is also maintained in the SMA (Mitz & Wise, 1987), although the topological efferent microzones appear to be more intermingled in the SMA (Hummelsheim, Wiesendanger M., Bianchetti, Wiesendanger R., and Macpherson, 1986; Macpherson, Marangoz, Miles, & Wiesendanger, 1982). The PMC, at least in its posterior portion near the arcuate sulcus, contains also corticospinal neurones, and microstimulation effects may be obtained, usually with higher thresholds than in MI. There is also evidence for somatotopy in the corticocortical connections between primary and secondary motor areas (Muakkassa & Strick, 1979), although these relations are presumably not organized in a one-to-one fashion, allowing for convergence and divergence.

Single-unit recordings in somatotopically appropriate locations of MI in trained monkeys revealed cellular activity that was usually tightly related to the motor task. This work has been well reviewed (e.g., Evarts, Shinoda, & Wise, 1984; Fetz, 1981; Hepp-Reymond, 1988), and it may suffice here to recall that the technique of spike-triggered averaging (Fetz & Cheney, 1987; Lemon, Mantel, & Muir, 1986) clearly demonstrates the direct transmission to sets of motoneurons. Moreover, it was shown that the activity of MI neurons may encode specific parameters of the performed movement, such as movement direction (Georgopoulos, in this volume). It is, however, equally important to note that activity changes in MI have often been reported to occur also in the preparatory period "while the monkey waits" (Evarts & Tanji, 1974; Lecas, Requin, Anger, & Vitton, 1986).

The emphasis of current reports on secondary motor areas concerns observations of relatively complex relationships of cellular activity with movements. Particularly revealing examples are those from the SMA reported by Tanji and coworkers (Tanji & Kurata, 1983) that all indicate that many neurons encode aspects of movement preparation rather than of the movement per se, such as cue signals or preparatory set. Most recently, Tanji, Okano, and Sato (1988) reported on SMA cells that had specific relationships with contralateral, ipsilateral, or bilateral finger movements. Such neurons discharged, for example, exclusively in association with bilateral finger movements, but not when the same movements were performed either with the right or with the left hand. This type of neuron, therefore, appears to encode higher order commands, for instance, in a coordinated bimanual task. Pointing in the same direction are recent results reported by Mann, Thau, and Schiller (1988) who discovered SMA neurons covarying either with goal-directed saccades or with goal-directed arm movements.

Gentilucci, Fogassi, Luppino, Matelli, Camarda, and Rizzolatti (1988) and Rizzolatti, Camarda, Fogassi, Gentilucci, Luppino, and Matelli (1988) recorded from PMC neurons from monkeys performing natural behavioral acts and also found relations between the neuronal activity and the complex movement sequence. Such neurons were classified descriptively as *bringing-to-the-mouth-neurons*, *grasping-with-the-hand-neurons*, *precision-grip-neurons*, and so on, the interpretation being that ". . .all together these neurons form a vocabulary where proximal and distal movements necessary for reaching, grasping, holding and bringing food to the mouth are represented."

In our own investigations on SMA neurons from monkeys performing arm movements in a choice–reaction paradigm (Hyland, Chen, Maier, Palmeri, & Wiesendanger, in press; Wiesendanger M., Hummelsheim, Bianchetti, Chen, Hyland, Maier, & Wiesendanger R. 1987), we could confirm the presence of neurons with activity changes during the waiting period (interval between *warning* signal and *go* signal). This activity may have been associated with various aspects of preparation, including postural preparation as suggested by concomitant electromyographic changes in some of the proximal muscles. However, we were struck by the surprisingly large population of SMA neurons that had properties indistinguishable from those of MI neurons. Out of 431 task-related SMA neurons, 67% were classified as *short-lead neurons* because they displayed a burst of activity preceding movement onset, often with shifts toward longer response latencies to the *go* signal as the reaction time increased, thus demonstrating a clear relation with the executed movement. These neurons may still transmit instructions to MI. However, we obtained no evidence for an earlier involvement of the SMA cells in the task when compared with a similar population of MI neurons recorded in the same animal under the same conditions, and the histograms of lead times of MI and SMA neurons were completely overlapping.

Taken together with the well-established facts that the SMA has direct connections with the spinal cord and that microstimulation effects were obtained from this area, we come to the conclusion that the parallel model of the SMA, as proposed long ago by Woolsey, Settlage, Meyer, Spencer, Hamuy, and Travis (1952) still holds. Of course, this does in no way exclude the hierarchical model because it emerged particularly from neurological studies on the SMA of the human brain (see Wiesendanger, 1986b), and from single-unit recordings in performing monkeys, as briefly discussed.

Assuming that at least in part there is parallel processing in the various subfields of motor cortical areas, the question naturally arises whether there is a division of labor among these fields. The search for area-specific attributes is lagging far behind that in visual cortical areas. The PMC has a rich supply from teleceptive (Godschalk, Lemon, Nijs, & Kuypers, 1981; Weinrich, Wise, & Mauritz, 1984) as well as somatosensory afferent sources (Rizzolatti, Scandolara, Matelli, & Gentilucci, 1981). It was therefore suggested that the PMC might be "specialized" for sensory, and especially visually guided movements (see Wise, 1984). On the other hand, clinical observations of patients with SMA lesions suggested that the SMA might be specialized for self-initiated movements (cf. review by Goldberg, 1985). However, recent single-unit studies, in which the involvement of PMC and SMA neurons was compared, hardly supported this notion (Okano & Tanji, 1987). Our own studies revealed a preponderance of proprioceptive inputs to single neurons of the SMA, whereas the PMC neurons were more often responding to light cutaneous stimuli (Hummelsheim, Bianchetti, Wiesendanger M., & Wiesendanger R., 1988). The reported difficulty of monkeys with SMA lesions to perform a self-paced, goal-directed movement in the dark (Passingham, 1987), together with the demonstration of proprioceptive inputs to the SMA, may be taken to suggest that the SMA is important for movements guided by proprioceptive signals. It is also noteworthy that the SMA is an important target of basal ganglia outflow via thalamus (Schell & Strick, 1984; Wiesendanger & Wiesendanger, 1985), whereas the PMC appears to be more dominated by cerebellar outflow (Schell & Strick, 1984).

The functional role of two further motor regions in the cingulate sulcus that are occupied by corticospinal neurons, is virtually unknown. As suggested by Hutchins, Martino, and Strick (1988), their proximity to the cingulate cortex and thus to the limbic system points to a role in transmitting signals from the limbic system to the spinal cord via their corticospinal projections. Evidently, present knowledge about areal specializations is still scarce, but so far not incompatible with the concept of division of labor.

## CONCLUDING REMARKS

The problem was posed whether hierarchic organization, a cornerstone in neurological thinking, holds for those brain centers involved in movement control. Not long ago, the question was hardly asked, and the model of hierarchic organization became well established in the 100 years of modern neurology. I have attempted to briefly sketch the story about hierarchies in the brain that had its root in theories of evolution and in the concept of *dissolution*. I believe that if hierarchies are seen mainly as levels of increasing complexity and of increasing abstraction of the sensory representation and of the motor commands, then the principle of hierarchical organization might still be a fruitful concept. As it became apparent that populations of single cortical neurons may be specialized to extract relatively complex features of the visually perceived world, time has come to search for similar complex and abstract features of *motor commands* (a term used for lack of a better one). A few insights gained in recent years were briefly discussed here that point in this direction, such as the population coding of movement direction, independently of how and with which effectors the goal is reached. Such neuronal codes would clearly represent higher level commands with higher information content, as compared to the lower level motoneuronal commands.

On the other hand, it was precisely the single-unit approach to investigating the functional attributes of cortical cells, together with the explosion of new anatomical knowledge about corticocortical connectivity, that led to the somewhat different view of *parallel distributed processing* that was clearly expounded by Mountcastle (e.g., Edelman & Mountcastle, 1978) on the basis of his extensive work on the "modular construction" in the somatosensory and associational cortical areas. It is rather obvious that perception or movement plans are not encoded in single sets of neurons, but in widely distributed, multiple sets, each encoding different aspects.

Progress in unravelling the language of cortical cells in motor areas has been much slower than in the visual areas. But I wish to emphasize that more and more evidence is accumulating that indicates that a similar modular organization exists in these motor fields as in sensory fields, that a number of subfields address directly the spinal cord, and that there are massive reciprocal cortical interconnection among the fields. There is thus ground to believe that within the cortical motor areas much of the processing occurs parallel as well as serially.

Theoretical approaches in studies of neural networks has become very popular. Provided that they rely on reasonably realistic anatomy and physiology, they may provide some impetus and help for understanding at least the basic principles of how the cerebral cortex is organizing and controlling movement (see, for example, Kawato, Furakawa, & Suzuki, 1987).

If the tenet today is that multiple-cortical areas have developed to allow for division of labor, it might be appropriate to remember that it was Spencer (cf. Richards, 1987) who proposed, more than 100 years ago, that division of labor is a leading organizational principle in brain evolution.

## ACKNOWLEDGMENTS

Research performed in the laboratory of the author was supported by the Swiss National Science Foundation, Grant 3.509-0.86, the Thomas Stanley Johnson Foundation, and the Swiss Multiple Sclerosis Society. I thank Mrs. S. Rossier for secretarial help.

## REFERENCES

Allen, G. I., & Tsukahara, N. (1974). Cerebrocerebellar communications systems. *Physiological Review, 54*, 957–1006.

Bruce, C. J., Desimone, R., & Gross, C. G. (1981). Visual properties of neurons in a polysensory area in superior temporal sulcus of the macaque. *Journal of Neurophysiology, 46*, 369–384.

Campbell, A. W. (1905). *Histological studies on the localization of cerebral function* (p. 360). Cambridge, England: University Press.

Conrad, B., Wiesendanger, M., Matsunami, K., & Brooks, V. B. (1977). Precentral unit activity related to control of arm movements. *Experimental Brain Research, 29*, 85–95.

Deecke, L., Kornhuber, H. H., Lang, W., Lang, M., & Schreiber, H. (1985). Timing function of the frontal cortex in sequential motor and learning tasks. *Human Neurobiology, 4*, 143–154.

DeJoe, E. A., & Van Essen, D. C. (1988). Concurrent processing streams in monkey visual cortex. *Trends in Neurosciences, 11*, 219–226.

Edelman, G. M., & Mountcastle, V. B. (1978). *The mindful brain.* Cortical organization and the group-selective theory of higher brain function. Cambridge, MA: MIT Press.

Evarts, E. (1984). Hierarchies and emergent features in motor control. In G. M. Edelman, W. E. Gall, & W. M. Cowan, (Eds.), *Dynamic aspects of neocortical function* (pp. 557–579). New York: Wiley.

Evarts, E. V., Shinoda, Y., & Wise, S. P. (1984). *Neurophysiological approaches to higher brain functions* (pp. 1–198). New York: Wiley.

Evarts, E. V., & Tanji, J. (1974). Gating of motor cortex reflexes by prior instructions. *Brain Research, 71*, 479–494.

Fetz, E. E. (1981). Neuronal activity associated with conditioned limb movements. In G. M. Edelman, W. E. Gall, & W. M. Cowan (Eds.), *Motor coordination, Handbook behavioral neurobiology* (Vol. 5, pp. 493–526). New York: Plenum.

Fetz, E. E., & Cheney, P. D. (1987). Functional relations between primate motor cortex cells and muscles: Fixed and flexible. In Ciba Foundation Symposium 132 *Motor areas of the cerebral cortex* (pp. 98–117). New York: Wiley.

Friedman, D. P., Murray, E. A., O'Neill, J. B., & Mishkin, M. (1986). Cortical connections of the somatosensory fields of the lateral sulcus of macaques: Evidence for a corticolimbic pathway for touch. *Journal of Comparative Neurology, 252*, 323–347.

Gentilucci, M., Fogassi, L., Luppino, G., Matelli, M., Camarda, R., & Rizzolatti, G. (1988). Functional organization of inferior area 6 in the macaque monkey. I. Somatotopy and the control of proximal movements. *Experimental Brain Research, 71*, 475–490.

Godschalk, M., Lemon, R. N., Nijs, H. G. T., & Kuypers, H. G. J. M. (1981). Behavior of neurons in monkey peri-arcuate and precentral cortex before and during visually guided arm and hand movements. *Experimental Brain Research, 44*, 113–116.

Goldberg, G. (1985). Supplementary motor area structure and function: Review and hypotheses. *Behavioral Brain Sciences, 8*, 567–616.

Hepp-Reymond, M.-C. (1988). Functional organization of motor cortex and its participation in voluntary movements. In H. D. Steklis & J. Erwin (Eds.), *Comparative primate biology* (Vol. 4, pp. 501–624). New York: Alan R. Liss.

Hubel, D. H. (1988). *Eye, brain and vision.* New York: Science American Library.

Hummelsheim, H., Bianchetti, M., Wiesendanger, M., & Wiesendanger, R. (1988). Sensory inputs to the agranular motor fields: A comparison between precentral, supplementary-motor and premotor areas. *Experimental Brain Research, 69*, 289–298.

Hummelsheim, H., Wiesendanger, M., Bianchetti, M., Wiesendanger, R., & Macpherson, J. (1986). Further investigations of the efferent linkage of the supplementary motor area (SMA) with the spinal cord in the monkey. *Experimental Brain Research, 65*, 75–82.

Hutchins, K. D., Martino, A. M., & Strick, P. L. (1988). Corticospinal projections from the medial wall of the hemisphere. *Experimental Brain Research, 71*, 667–672.

Hyland, B., Chen, D. F., Maier, V., Palmeri, A., & Wiesendanger, M. (in press). What is the role of the supplementary motor area in movement initiation? *Progress in Brain Research*.

Jackson, J. H. (1932). In J. Taylor (Ed.), *Selected writings of John Hughlings Jackson* (Vol. II). London: Hodder & Stoughton.

Kawato, M., Furakawa, K., & Suzuki, R. (1987). A hierarchical neural-network model for control and learning of voluntary movement. *Experimental Brain Research, 57*, 169–185.

Kelso, J. A. S., & Tuller, B. (1981). Toward a theory of apractic syndromes. *Brain and Language, 12*, 224–245.

Kemp, J., & Powell, T. P. S. (1971). The connexions of the striatum and globus pallidus: synthesis and speculation. *Philosophical Transactions of the Royal Society*. London, Series B, *262*, 441–457.

Lashley, K. S. (1951). The problem of serial order in behavior. In L. Jeffress (Ed.), *Cerebral mechanisms in behavior* (pp. 112–136). New York: Interscience.

Lecas, J. C., Requin, J., Anger, C., & Vitton, N. (1986). Changes in neuronal activity of the monkey precentral cortex during preparation for movement. *Journal of Neurophysiology, 56,* 1680–1702.

Lemon, R. N., Mantel, G. W. H., & Muir, R. B. (1986). Corticospinal facilitation of hand muscles during voluntary movement in the conscious monkey. *Journal of Physiology (London), 381*, 497–527.

Liepmann, H. (1920). Apraxie. In T. Brughsh & A. Entenburg (Eds.), *Real-Encyclopädie der gesamten Heilkunde*. Ergebnisse der gesamten Medizin, vol. 1 (pp. 116–143). Berlin: Urban & Schwarzenberg.

Macpherson, J. M., Marangoz, C., Miles, T. S., & Wiesendanger, M. (1982). Microstimulation of the supplementary motor area (SMA) in the awake monkey. *Experimental Brain Research, 45,* 410–416.

Mann, S. E., Thau, R., & Schiller, P. H. (1988). Conditional task-related responses in monkey dorsomedial frontal cortex. *Experimental Brain Research, 69,* 460–468.

Maunsell, J. H. R., & Van Essen, D. C. (1983). The connections of the middle temporal visual area (MT) and their relationship to a cortical hierarchy in the Macaque monkey. *Journal of Neuroscience, 3,* 2563–2586.

Mitz, A. R., & Wise, S. P. (1987). The somatotopic organization of the supplementary motor area: Intracortical microstimulation mapping. *Journal of Neurophysiology*, 1010–1021.

Muakkassa, K. F., & Strick, P. L. (1979). Frontal lobe inputs to primate motor cortex: Evidence for four somatotopically organized 'premotor' areas. *Brain Research, 177,* 176–182.

Okano, K., & Tanji, J. (1987). Neuronal activities in the primate motor fields of the agranular frontal cortex preceding visually triggered and self-paced movement. *Experimental Brain Research, 66*, 155–166.

Paillard, J. (1982). Apraxia and the neurophysiology of motor control. *Philosophical Transactions of the Royal Society (London), 298*, 111–134.

Passingham, R. E. (1987). Two cortical systems for directing movement. In *Motor areas of the cerebral cortex*. Ciba Foundation Symposium, *132*, 151–161.

Pons, T. P., Garraghty, P. E., & Mishkin, M. (1988). Lesion-induced plasticity in the second somatosensory cortex of adult macaques. *Proceedings of the National Academy of Science, 85,* 5279–5281.

Primrose, D. C., & Strick, P. L. (1985). The organization of interconnections between the premotor areas in the primate frontal lobe and the arm area of primary motor cortex. *Society of Neuroscience Abstract, 11,* 1274.

Richards, R. J. (1987). *Darwin and the emergence of evolutionary theories of mind and behaviour*. Chicago, London: University Chicago Press.

Rizzolatti, G., Camarda, R., Fogassi, L., Gentilucci, M., Luppino, G., & Matelli, M. (1988). Functional organization of inferior area 6 in the macaque monkey. II Area F5 and the control of distal movements. *Experimental Brain Research, 71,* 491–507.

Rizzolatti, G., Scandolara, C., Matelli, M., & Gentilucci, M. (1981). Afferent properties of periarcuate neurons in macque monkeys. 1. Somatosensory responses. *Behavioural Brain Research, 2,* 125–146.

Rolls, E. T. (1983). The initiation of movements. In J. Massion, J. Paillard, W. Schultz, & M. Wiesendanger (Eds.), *Neural coding of motor performance* (pp. 77–113). *Experimental Brain Research Supplementum.*

Schell, G. R., & Strick, P. L. (1984). The origin of thalamic inputs to the arcuate premotor and supplementary motor area. *Journal of Neuroscience, 4,* 539–560.

Tanji, J., & Kurata, K. (1983). Functional organization of the supplementary motor area. In J. E. Desmedt (Ed.), *Motor control mechanisms in health and disease* (pp. 421–431). New York: Raven.

Tanji, J., Okano, K. & Sato, K. C. (1988). Neuronal activity in cortical motor areas related to ipsilateral, contralateral and bilateral digit movements of the monkey. *Journal of Neurophysiology, 60,* 325–343.

Van Essen, D. C. (1985). Functional organization of primate visual cortex (chapter 7). In A. Peters & E. G. Jones (Eds.), '*Cerebral cortex*', (vol. 3, pp. 259–329). New York, London: Plenum.

von Monakow, C. (1911). In Bericht IV Kongress für Experimentelle Psychologie Insbruck. *Aufbau und Lokalisation der Bewegungen beim Menschen* (pp. 1–28). Leipzig: J. A. Barth.

von Monakow, C. (1914). *Die Lokalisation im Grosshirn und der Abbau der Funktion durch kortikalė Herde.* Wiesbaden: Bergmann.

Weinrich, M., Wise, S. P., & Mauritz, K. H. (1984). A neurophysiological study of the premotor cortex in the Rhesus monkey. *Brain, 107,* 385–414.

Wiesendanger, M. (1981). Organization of secondary motor areas of cerebral cortex. In V. B. Brooks (Ed.), *Handbook of Physiology* (vol. II, part I, pp. 1121–1147). Bethesda, MD: American Physiological Society.

Wiesendanger, M. (1986a). Redistributive function of the motor cortex. *Trends in Neuroscience, 9,* 120–125.

Wiesendanger, M. (1986b). Recent development in studies of the supplementary motor area of primates. *Reviews of Physiology, Biochemistry and Pharmacology, 103,* 1–59.

Wiesendanger, R., & Wiesendanger, M. (1985). The thalamic connections with medial area 6 (supplementary motor cortex) in the monkey (Macaca fascicularis). *Experimental Brain Research, 59,* 91–104.

Wiesendanger, M., Hummelsheim, H., Bianchetti, M., Chen, D. F., Hyland, B., Maier, V., & Wiesendanger, R. (1987). Input and output organization of the supplementary motor area. In *Motor areas of the cerebral cortex* (pp. 40–62). Ciba Foundation Symposium 132. New York: Wiley.

Wiesendanger, M., & Wise, S. P. (in press). Current issues concerning the functional organization of motor cortical areas in nonhuman primates. *Advances in Neurology, Frontal lobe seizures and epilepsy.*

Wise, S. P. (1984). The non-primary motor cortex and its role in the cerebral control of movement. In G. Edelman, W. Cowan, & E. Gall (Eds.), *Dynamic aspects of cortical function.* New York: Wiley.

Woolsey, C. N., Settlage, P. H., Meyer, D. R., Spencer, W., Hamuy, T. P., & Travis, A. M. (1952). Patterns of localisation in precentral and 'supplementary' motor areas and their relation to the concept of a premotor area. *Proceedings of the Association for Research in Nervous Mental Disease, 39,* 238–264.

# 3 Motor Programs: Concepts and Issues

Steven W. Keele
Asher Cohen
Richard Ivry
*University of Oregon*

ABSTRACT

Lashley introduced a number of arguments that sequences of human motor activity are guided by plans and not by peripheral or central associations between one event and the next. Although there is some evidence in favor of an associationist position, a large body of evidence has been supportive of Lashley's seminal ideas. Moreover, a variety of paradigms including reaction time, pauses between successive motor acts, and transfer of learning have emphasized two concepts that characterize these plans or motor programs. First, the plans that guide activity are hierarchical in nature. Second, many sequence representations can be regarded as modular in that they are separable from the muscular system of initial learning and can be interfaced with various other modules to produce novel actions. Hierarchic and modular components are thus viewed as important components for any model of sequencing. One particular computational model developed by Jordan is described and analyzed in terms of how well it meets this criterion.

In a seminal paper published in 1951, Lashley laid out a series of issues regarding the representation of serial order. He noted the centrality of the serial order problem to understanding human intelligence. "Temporally integrated actions do occur even among insects," he stated, "but they do not reach any degree of complexity until the appearance of the cerebral cortex. They are especially characteristic of human behavior, and they contribute as much as does any single factor to the superiority of man's intelligence" (p. 114). A little reflection makes clear why Lashley was of this opinion. Humans are extraordinary in learning new sequences. They learn to speak, to write, and to type. They play the violin, knit sweaters, and dance. In all of these cases, people rearrange a small number of elementary movements to produce new skills. For example, in language, basic elements—be they letters, phonemes, or words—continually are produced in new arrangements. The ubiquity of complex sequencing leads one to speculate that whatever brain mechanisms evolved for allowing the general sequencing of action also provided the foundation for specialized systems of language.

The present paper is concerned with the same level of analysis as was Lashley, the representations of the orders of actions rather than their elementary movements. Such a dichotomy between sequence and elementary movement is overly simple. Even elementary movements, such as reaching a particular finger toward a piano key, involves sequential action of a variety of muscles over time. Indeed, in some models, such as the one of Jordan (1986a, b; this volume) to be described later, the sequence of successive movements is not divorced from the control of single movements. Nevertheless, the distinction is worth making, because it is not clear that the same principles of sequencing apply always to both levels. At the level of action, where people can learn new and rather arbitrary sequences, it might be more appropriate to talk of plans and motor programs. At the level of movement, other concepts such as mass-spring systems (Bizzi, Polit, & Morasso, 1976; Feldman, 1986) may help explain the sequence or trajectory of movement over time.

Lashley was largely responsible for introducing the concept of a plan or motor program that guides the sequence of action. He contrasted that idea with one of an associative chain, especially a peripheral chain. The basic idea of a peripheral chain was that feedback from one movement stimulated the next in chainlike fashion. He pointed out several problems with such a concept. Earlier, Lashley (1917) had studied a patient with a gunshot wound to the spinal cord in which kinesthetic appreciation of leg movement was lost. The patient nonetheless could move his leg at various speeds and extents, even without vision, suggesting that kinesthesis was not responsible for the sequential activation of the involved muscles. Lashley further noted that sometimes successive movements occurred too close together in time, separated by less than $\frac{1}{10}$ second, for feedback from one movement to stimulate the next. Indeed, a plan is often revealed by the fact that one movement in a series sometimes erroneously occurs prior to a movement that was supposed to be the stimulant. Finally, he articulated a basic problem with a simple association chain explanation. The same elementary movements occur in different orders. For example, one may speak the word *tire* or the word *rite*. In the one case, *r* follows *i*. In the other, *t* follows *i*. Such sequencing, Lashley argued, is inconsistent with serial chaining because two different events are associated with the same stimulus. Correct production instead requires a plan to guide phoneme order.

Since Lashley's paper, his basic points have been substantiated in numerous experiments. Sequences of movements can be generated in the absence of peripheral feedback. Movements from late in a series sometimes make their presence known early in the series or even before movement begins. In the first part of this paper, we briefly mention some representative examples of this work. The main body further explores the nature of program representation. Three general topics are discussed. First, programs appear to be hierarchical. Second, some programs may be modular in that they can be interfaced to different effector systems. The third topic concerns computational theories of how sequencing is actually implemented.

## Movement Without Feedback

Investigations by Taub and Berman (1968; see, also, Taub, Perella, & Barro, 1973) demonstrated that monkeys deprived of kinesthetic feedback could grasp, walk, run, climb, and jump, even when blindfolded or not looking at their deafferented limbs. Such a demonstration is contrary to a peripheral chaining theory in which feedback from one movement stimulates the next, but it has not provided the best animal model for learned sequential skills of humans. Locomotor skills very likely have an innate basis, and the basic locomotor patterns appear generated at the spinal level (see Grillner, 1981, for a review). Even human bipedal locomotion may have a strong innate basis, as the basic stepping pattern can be demonstrated in infants of a few weeks of age (Forrsberg & Wallberg, 1980; Thelen & Fisher, 1982).

More useful animal models have been provided by the analysis of bird songs and of grooming patterns in rodents. Birds of many species learn the details of their songs. Once the song is acquired, it appears under program control. It persists even when the bird is deafened (Konishi, 1965). When one side of the vocal apparatus is denervated, some portions of the song drop out, depriving the bird of both auditory and kinesthetic feedback from the missing segments. The remaining portions still occur at their normal times, invalidating a peripheral chaining theory (Nottebohm, 1970). With respect to rodents, mice and rats exhibit interesting grooming and feeding patterns (Fentress & Stillwell, 1973; Berridge, Fentress, & Parr, 1987). Two phenomena are especially noteworthy. The patterns are not completely stereotyped. But they are not random either. Rules of "grammar" constrain which movements follow which. Second, some of the elementary movements in feeding and grooming are identical, but they occur in different orders in the two activities. Both of these features are characteristic of many learned human skills. Thus, it is of particular interest that the feeding and grooming patterns can be skillfully executed in the absence of peripheral feedback from the limbs or face (Berridge & Fentress, 1987; Fentress, 1973).

Recent work with neurological patients has also shown that sequential movements can occur in the absence of kinesthetic and visual feedback. Rothwell, Traub, Day, Obeso, Thomas, and Marsden (1982, see also Marsden, Rothwell, & Day, 1984, and Sanes, this volume) describe a patient who suffered a loss of senses of position, movement, and touch in the hands caused by peripheral nerve damage. Nonetheless, the patient was able to trace out figures like squares, circles, and figure 8s with the index finger and perform a series of thumb-to-finger touches. Of special interest is the observation that with repeated cycles of thumb-to-finger touches, the basic sequence was maintained, but the thumb missed accurate contact with the fingers. Such observations indicate that feedback is not used so much in producing the sequence, suggested by peripheral chaining theory, as in calibrating the positions of movements as they drift off target.

Similar results on patients have been described by Volpe, LeDoux, and Gazzaniga (1979). In our own lab, Ivry (1986) has shown that a patient with kinesthetic loss from a peripheral neuropathy can produce a regular series of finger taps with no more variability of timing than many normal subjects.

## Anticipatory Movement Effects

The observations of skilled movement from deafferented animals and humans support Lashley's proposition that the control of many sequential movements is not due to chaining of feedback to movement. Conceivably, though, the sequence representation could be a chain of internal representations of movement, one associated with the next. Other observations, however, pose problems for this idea as well. Consider some results from a study by Gentner, Grudin, and Conway (1980) of expert typists. Fig. 3.1

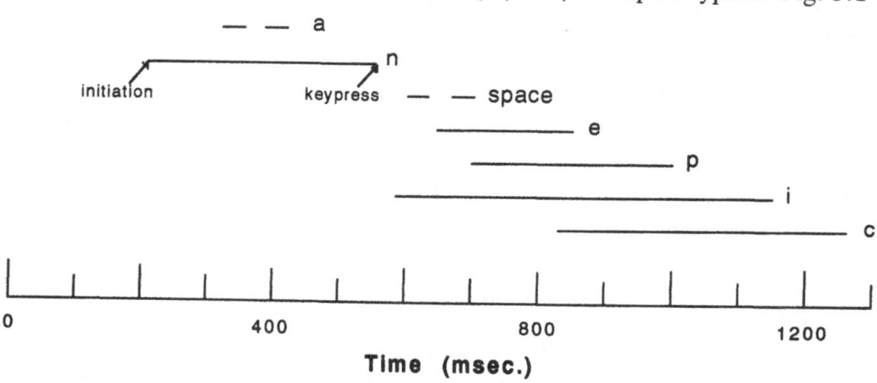


FIG. 3.1 Typing of the phrase *an epic*. The beginning of a solid line indicates the time at which a finger that will produce a particular key strike begins to move toward the key (the beginning of the a and the space bar movements are not indicated). The end of the line indicates the time of the key strike. Note that a later letter, the *i*, begins before letters that are completed earlier. Based on Gentner, Grudin, & Conway, 1980.

shows the onsets and offsets of different strokes in a particular typing of the phrase *an epic*. What is notable is that sometimes the stroke for a following letter starts before the beginning of the preceding letter even though the strokes are finished in the proper order. Three points are worth making about this observation. It illustrates a kind of parallel activity in that several fingers for different letters may simultaneously be active even though their ending points are serial. Although not apparent from only these data, it illustrates coarticulation in typing, a phenomenon well known in speech, in which nearby motor elements interact with each other. Some of the movements in the figure show anticipation, whereas others do not. In changed contexts, the movements that show anticipation are altered. Such dependence on context is the phenomenon of coarticulation. And third, with respect to Lashley's claim, the results rule out a particular form of internal chaining, namely a chain in which the only stimulus for a movement is the preceding movement. Instead the results indicate preknowledge of forthcoming movements, implicating a plan that guides them.

Another commonly used paradigm also indicates preknowledge of movements before they actually occur. Reaction time prior to movement onset is sensitive to the complexity of the movement series to follow, suggesting preplanning of the series. Klapp, Anderson, and Berrian (1973) presented a seminal experiment in developing this paradigm. Five-letter words, some one-syllable and some two-syllable, were exposed on a screen. In one condition, subjects named the words. Reaction times were 14 msec longer for two-syllable words than for one-syllable words. Conceivably, the result could reflect not preplanning of the articulatory movements but longer perceptual times for two-syllable words. To test this possibility, in a second experiment subjects categorized one- and two-syllable words as animals or objects by saying yes or no. Here, where the number of input syllables bears no relation to the output, reaction time differences between one- and two-syllable words disappeared. In a third experiment, pictures of objects were presented and subjects named them. In this case there are no syllables in the input, but syllables are produced in the output. The syllabic effect on reaction time reappeared and again was 14 msec. The effect thus appears tied to the response end rather than the perceptual end. Something about a two-syllable response produces a slightly longer preparation time. The major point is that an upcoming movement series is planned in anticipation of actual movement.

Thus, a variety of evidence supports Lashley's conjecture that well-learned movement series are guided by programs rather than unfolding in accordance with association chains. This is not to say that no movement sequences are guided by an association chain. Indeed, later we will suggest that when sequences are learned with impoverished attention, they may primarily consist of association chains. Moreover, by some models, such as that of Jordan (1986a, 1986b; this volume), behavior is guided by a compound mechanism that involves both plans and associations. Thus, the evidence so far is best taken to indicate that plans typically are critical in the guidance of an action, but association mechanisms may also operate.

## HIERARCHICAL CONCEPTIONS OF PROGRAMS

The idea that a plan guides a sequence of action has implicit in it the additional idea of hierarchic representation. In a hierarchy, a superordinate element is instrumental in controlling the course of lower order elements. To return to the example of *tire* versus *rite*, whether an *r* or a *t* follows the *i* is governed by whether the plan is to produce *tire* or *rite*. A superordinate thus controls the ordering of the subordinates. This idea too was expressed by Lashley (1951), as in his statement that ". . .whether *gh*, *ga*, or *gu* is written depends upon a set for a larger unit of action, the word" (p. 116). However, a variety of evidence suggests that the plan itself has more than one level, with higher levels controlling lower levels. In this section we illustrate this point with a wide variety of experiments.

Among the first of forceful arguments for hierarchical representation of motor sequences was an experiment by Restle and Burnside (1972). They presented subjects with repeating patterns of six lights that were arranged in a row. One such sequence was 1234666662323543. The lights were presented at a fixed pace of one light every .7 sec. After observing the lights for some time, subjects attempted to follow them by tapping keys. Because of the fixed pace, subjects sometimes made errors. The error rates are shown in Fig. 3.2. The error pattern suggests that the sequence is coded as a series of four subordinate patterns. The first pattern is the run 1234, the second is the repetition 66666, the third is the trill 2323, and the final pattern is the reverse run of 543. The errors tended to be over- or underextensions of the subpattern in question, as though people were uncertain of the termination point of one rule-governed chunk and the start of the next. In an ingenious set of experiments in which rats pressed bars under lights, Fountain (1987) showed that rats also code Restle-type sequences in groups.

When subjects perform tapping sequences at their own rates, the hierarchic organization may be manifest in the intertap intervals. A particularly compelling example is provided by Povel and Collard (1982; a very similar demonstration has been presented by Geoffroy & Norman, 1982). Povel and Collard showed subjects short sequences of numbers, such as 123321 in which the numbers refer to fingers to be tapped. The subjects tapped out the sequences as rapidly as possible, recycling from end to beginning.

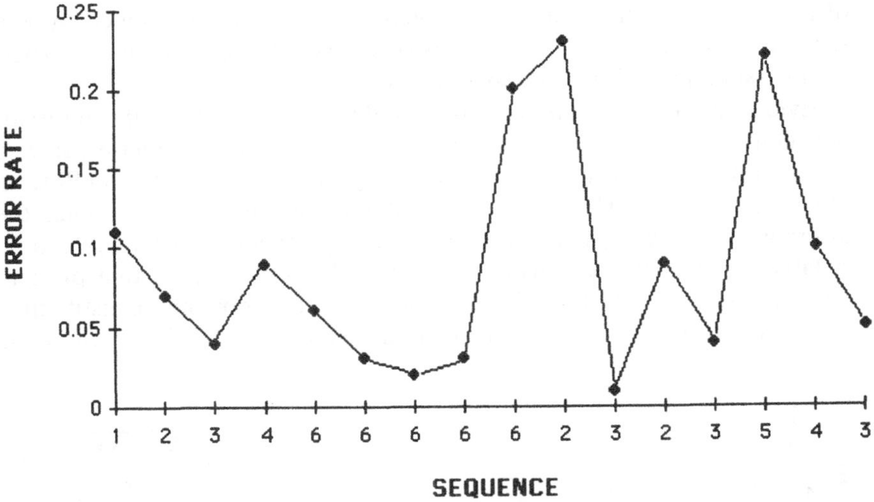


FIG. 3.2. Error rates for responses at varying parts of a sequence. Based on Restle & Burnside, 1972.

The mean rank of intertap intervals for this sequence is shown in the top, left graph of Fig. 3.3, where it is seen that the longest intervals usually occur between the two 3s and between the last 1 and the starting 1. Contrast that profile with the one for the sequence 332112, which is shown in the bottom, left graph of Fig. 3.3. There, the longest intervals tend to fall between the first 2 and the first 1 and between the last 2 and the first 3. What may have escaped notice is that the two sequences are the same except for starting positions. In each case, fingers follow one another in the same order. Nonetheless, the intertap profiles differ, suggesting that different plans guide the two sequences. It is as though each sequence is coded as two subpatterns. Below the subpattern level there is further division into more elementary components. The right panel of Fig. 3.3 shows the organizations for the two patterns inferred from the intertap intervals.

Povel and Collard explain their data by a tree traversal mechanism. They presume that the act of retrieval prior to each elementary response involves a passage between levels of control. To produce the first response requires traversal from the top node to the second level and then to the terminal level. Once within the first subgroup, retrieval requires passage only between the last two levels until that subgroup is exhausted. To retrieve the first element of the next subgroup requires passage back to the top level, retrieval of the second subgroup, and then retrieval of the first item in that subgroup. Presumably, total retrieval time reflects the number of passages between levels. This model has also been developed by Rosenbaum and colleagues in an extensive series of finger-sequencing experiments (summarized in Rosenbaum, 1987).

Evidence for hierarchic coding is reflected not only in the intervals between successive actions but also in the reaction time phase prior to actual movement. Recall the experiment by Klapp, Anderson, and Berrian (1973) described earlier. They showed that reaction time to begin the utterance of a two-syllable word was slightly longer than that to a one-syllable word. How can this be explained? It would appear that prior to an utterance, a word must be unpacked into its phonemic constituents. The unpacking passes through an intermediate syllabic level when the

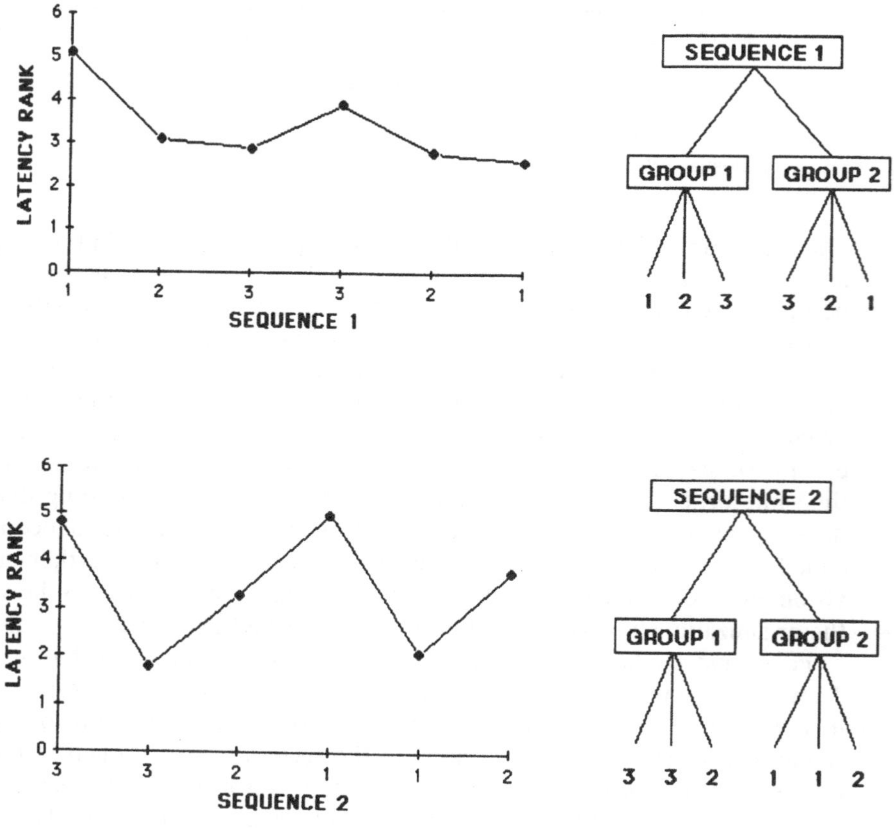


FIG. 3.3. Mean latency ranks for interlap intervals at different points in a sequence of taps. The top panel A is for the sequence 123321. . . The bottom of Panel A is for the sequence 332112. . . Panel B shows the hypothecial memory structures for the two sequences. Based on Povel & Collard, 1982.

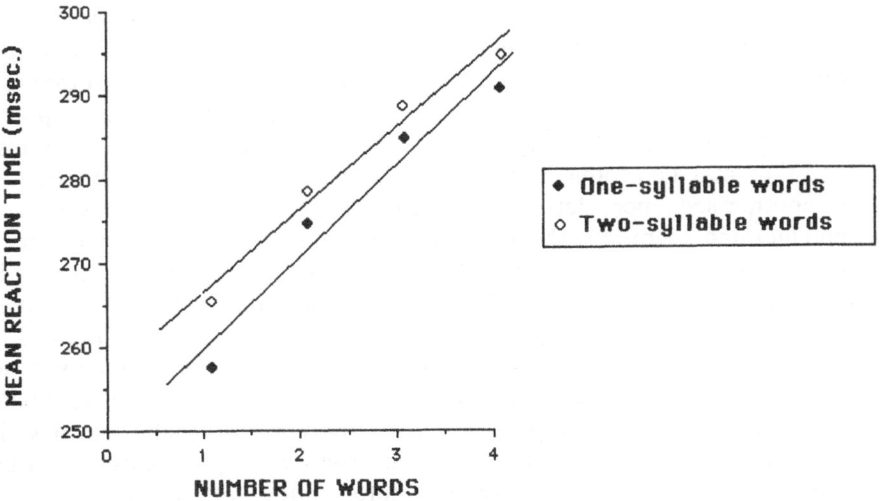


FIG. 3.4. Mean reaction time to begin the utterance of 1- or 2-syllable strings of 1, 2, 3, or 4 words. Based on Sternberg, et al. 1978.

presented work is multisyllabic. This, in fact, is the explanation given by Sternberg, Monsell, Knoll, and Wright (1978) in a similar experiment. They gave subjects short lists of 1, 2, 3, or 4 words to produce upon occurrence of a signal to start. Some lists contained one-syllable words and others contained two-syllable words. Fig. 3.4 shows that mean reaction time to begin the utterance increased with the number of words in the list, and the intercept of the function increased with the number of syllables per word.

Two similar explanations have been given for the Sternberg, et al. results. One, proposed by Sternberg, et al., supposes that the words are entered into a buffer, one slot per word, regardless of number of syllables. When the signal to respond occurs, the first word is retrieved from the buffer. The time required for retrieval depends on the number of words in the buffer. When the first word is retrieved, it is unpacked into its phonemic units. If the word has two syllables, the unpacking proceeds through an extra layer, which takes the extra 5 msec or so. A slightly different explanation is provided by Rosenbaum (1987). He supposed that the entire sequence of words is represented hierarchically, with the number of levels tending to increase with the number of words. Having two syllables in a word adds an additional level. Both of these explanations, that of Sternberg, et al. and that of Rosenbaum, predict an additive effect between sequence length and number of syllables, because in each case the extra syllable adds one more level to a representation hierarchy that must be traversed before the beginning of an utterance.

One aspect of the syllable effect seems especially worthy of notice. Some may claim that simple tapping paradigms, such as those studied by Restle and Burnside, Povel and Collard, and Rosenbaum, produce data that are not representative of highly learned sequences of activity. The tapping paradigms often involve sequences that are learned in a matter of seconds, and they may be produced only for a few cycles before a subject switches to another sequence. Perhaps such sequences are represented in symbolic and hierarchic form. On the other hand, sequences that have been practiced over a lifetime may be represented in some other way. Word naming is certainly a highly practiced skill. Yet the reaction time studies of Klapp et al. and Sternberg et al. suggest that words, too, are encoded hierarchically.

The fact that hierarchic representation is manifest both in reaction time and in pauses during the actual execution suggests that two passes are made through a hierarchical representation (Rosenbaum, 1987). One pass is preparatory in setting up the execution program. A possibility would be that this pass would load movements into a buffer where they would be read out in serial fashion. Such seems not to be the case, however, because the pauses in execution suggest that on line production also involves passage back and forth between levels.

A recent reaction time study by Gordon and Meyer (1987) provides additional insight into hierarchic representation. Subjects memorized two sequences of four nonsense syllables. They were told to prepare one of the sequences to speak, but sometimes the signal to respond called for production of the unprepared sequence. The crucial issue concerned the effect of the relationship between the prepared and unprepared sequences. Suppose the prepared sequence is Bee Bay Bah Boo. Further, suppose the subject must instead produce the unprepared sequence. These were of three types: Bay Bah Boo Bee; Bah Boo Bee Bay; and Bah Bay Boo Bee. Notice that the first of the unprepared sequences preserves much of the linear order of the primary sequence. The first element of the prepared sequence is simply shifted to the end. If sequences are represented as association chains, then savings in reaction time might occur if that unprepared sequence is called for. Now consider the second of the unprepared sequences. Here some of the item-to-item associations are also preserved, but the crucial point is that the first and last halves of the sequence are switched. If the prepared sequence is represented as a binary tree, then switching to the second of the unprepared sequences should produce reaction time savings because it maintains the subgroup structures. The third unprepared sequence is a control in which no associations are maintained and against which performance on the other two unprepared sequences can be compared.

Gordon and Meyer found a savings in reaction time when the unprepared sequence preserved low levels of a putative hierarchy. The savings were also manifest in error rates and duration of production.

The results of Gordon and Meyer make additional, important points about hierarchic representation. There are no logical relationships between the successive items as in the Restle and Burnside work involving runs, repetitions, and trills. One might expect that a short list of four unrelated items would be learned in strictly linear fashion. Gordon and Meyer's results indicate, however, that subjects represent even arbitrary sequences as a branching tree. This result is consistent both with Rosenbaum's (1987) interpretation of the results of Sternberg, et al. and additional data provided by Rosenbaum himself.

A second point extracted from the Gordon and Meyer study illustrates a major advantage of hierarchic representation. New sequences can be constructed by rearranging higher order groupings without the necessity for relearning the lower level groupings. This, of course, is a notable feature about language. When learning a new word, one does not need to relearn the letters. When learning a new sentence, one does not need to relearn the words.

This point regarding transferability of sections of a hierarchic sequence has also been made in a very different way by MacKay (1982; see also MacKay, 1981, and MacKay & Bowman, 1969). German–English bilinguals practiced saying sentences, half of the sentences in one language and half in the other. Production times systematically improved over 12 successive reproductions of a sentence spoken in one language. After 12 trials, subjects switched to the other language. Sometimes they continued to produce the same meaning, though in the other language, and in other cases the sentence changed in meaning. The former case exhibited *perfect* transfer from one language to the other. That is, all the production time benefits from the 12 practice repetitions switched over to the new language despite an almost entirely different series of speech sounds. In a variant of the experiment, subjects practiced repeating a random string of words. Once randomized, the string remained constant for the entire 12 productions. After initial practice, when each word was translated to the other language, there was absolutely no evidence of transfer.

MacKay's explanation of these two results—perfect transfer with the sentences and no transfer with the random word strings—was in terms of hierarchic representation. In MacKay's view, the order of words in a sentence is represented first as a whole sentence. At the next level the sentence is parsed into noun and verb phrases. Below that level, further subdivisions provide such things as articles and nouns. These levels are all conceptual and are in common to German and English. Once the conceptual word level is reached, the next level of the hierarchy involves decoding the concept into a particular word in a particular language. Given that the subjects were very well practiced at this translation because they were all fluent bilinguals, there was little to be gained in 12 practice repetitions compared to previous years of practice with each individual word. All the novelty, on which practice had its effect, lies in the order of the concepts. Because the order was in common to the two languages, perfect transfer occurred.

To understand why with random word strings no transfer occurred from one language to the other, one must consider that many words outside a sentence context are quite ambiguous. The word *right* for example, could mean the opposite of *left*, *correct*, *to turn upright*, *conservative*, and so on. The hierarchic representation of a sentence specifies the meaning of the constituents. Without a sentence, little basis for transfer exists from one language to the other because the sequential structure does not convey the meaning of items.

MacKay's study leads into the next topic of modularity. Essentially, what MacKay has shown is that a high-order description of a sequence is independent of the particular implementation. The same sequence representation can be interfaced with two different implementations, German in one case and English in the other.

## MODULARITY

The term *modularity* conveys two separate ideas. One idea is that separable systems provide specialized computations. For example, Arbib (1981; this volume) has suggested that the act of reaching for and grasping an object depends on separable computations of location, shape, and orientation. The other closely related idea of modularity is that the computation can be interfaced with other modules. Thus, one may suppose that the visual system that locates an object can be used not only to reach for the object but to verbally describe where in space the object is located. Likewise, it could be that the system responsible for computing shape can be used not only for forming the grasp but also for naming the object.

Although experimental (e.g., Jeannerod, 1981) and neurological (e.g., Perinin & Vighetto, 1982, 1988) evidence exists for the separable computations that Arbib has proposed, the possibility that the modules can be inserted into different tasks is largely untested. With respect to computation of time, however, we have produced evidence for a module. We have shown that the ability to produce regular time intervals is correlated across different effectors—fingers, arm, and foot. Moreover, the ability to regulate motor time is correlated with acuity in perceiving durations of auditory events (Keele, Pokorny, Corcos, & Ivry, 1985). In contrast, timing control correlates little or not at all with the ability to regulate force of movement, even when the same effectors are used (Keele, Ivry, and Pokorny, 1987). Such results are consistent with the notion of a system spe-

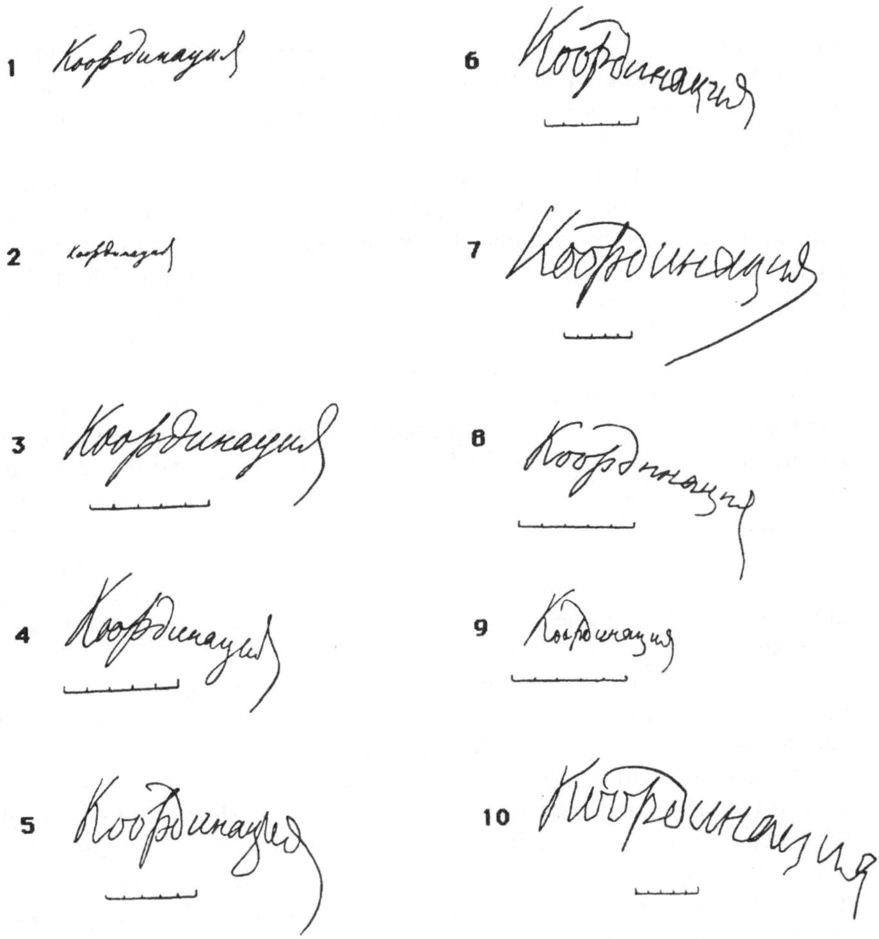

FIG. 3.5. Writing samples produced by a single subject in Bernstein's laboratory in 1940. Samples 1 and 2 were produced by normal right-hand writing. Sample 3 was produced with a pencil in the fingers of the right hand, but the hand was moved as a whole unit. In 4, the pencil was attached at the styloid process of the right radial bone, which is just above the wrist joint. In 5, the pencil was attached at the medial epicondyle, which is just above the elbow joint. In 6, the pencil was attached to the shoulder. In 7, the pencil was attached to the toe of the right shoe. In 8 the pencil was held in the teeth and bandaged to the head. In 9, the pencil was held in the fingers of the left hand, and in 10, the pencil was attached to the toe of the left shoe.

In each case the scale is marked in centimeters and the total length is 5 cm. From Bernstein, 1947.

cialized for timing, independent of force control, and capable of interfacing with a variety of effectors and tasks. The analysis of neurological patients has further suggested that the lateral cerebellum is the primary neural structure responsible for the timing computation (Ivry & Keele, 1989; Ivry, Keele, & Diener, 1988).

A similar question can be asked about sequencing. To what degree is the representation of sequences of action separable from and interfaceable with different effector systems? The basic issue is illustrated by the similarity of writing samples produced by hand, arm, mouth, and foot. Fig. 3.5 shows a historical example from Bernstein (1947). The similarity of style across effectors suggests that the sequence representation is independent of muscles that execute them.

Despite the heuristic value of the writing demonstrations, they are not conclusive regarding whether sequence knowledge is completely effector independent. Some effectors yield poor quality writing, and that could be caused either by lack of practice with those effectors on letters and words or by lack of practice with fine spatial movement in general. Moreover, even if writing were fluent but slightly different in form with different effectors, differences could be caused either by effector-specific sequence knowledge or by peripheral, mechanical effects. This issue is thoroughly discussed by Wright (this volume).

One approach to the problem of sequence representation makes use of neurological patients. Margolin (1984) has concluded from a survey of patients with handwriting disorders that sequence representation of letters is shared with various modes of production. Margolin's proposed architecture for processing systems involved in writing is shown in Fig. 3.6. A sequence of letters to be written is stored in an orthographic buffer. The same buffer is shared by oral spelling and typing. In this model, speech itself derives not from the orthographic buffer but from phonological buffers. Those phonological buffers also are modular in the sense that not only can they be interfaced with speech, but they can be interfaced with the orthographic buffer. Handwriting is not directly parasitic on a phonological buffer, because the orthographic buffer can also be activated by nonphonological routes, such as from semantics. A very similar model has been presented by Caramazza, Miceli, & Villa (1986).

Margolin arrived at the model by noting a variety of dissociations. For example, one patient, described by Bub and Kertez (1982), could write words, but not nonwords, suggesting that orthography comes via either lexical phonology or semantics. Yet another patient, described by Shallice (1981), could *repeat* nonwords, as well as words, but could not *write* nonwords. The patient could write words, but often made semantic substitutions, such as *clock* for *time*. Such results suggest that the orthographic

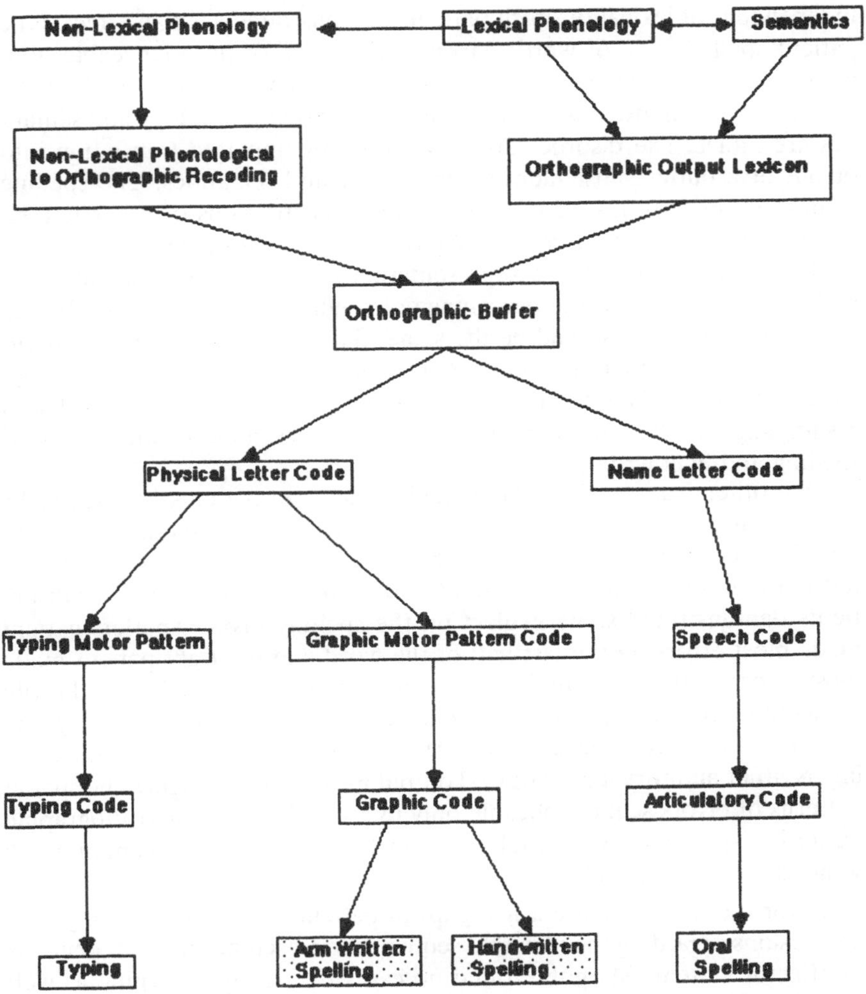


FIG. 3.6. Hypothesized modules involved in spelling by typing, writing, and oral production. Based on Margolin (1984) except for the suggestion that both arm writing and hand writing stem from the same graphic code.

buffer was derived from semantic inputs, but access to orthography from phonology was cut off.

Recently a remarkable pair of patients described by Hillis and Caramazza (1988) have provided striking confirmation of Margolin's idea that an orthographic buffer provides a common computation for different input and output systems. If there is an orthographic buffer, separable from phonology and semantics, damage to that buffer should produce errors independent of lexicality. Both patients demonstrated such an effect. One patient spelled 50% of words correctly and 48% of pronounceable nonwords; the other patient spelled 36% of words correctly and 35% of the pronounceable nonwords. Such results suggest that phonology and semantics are intact. The disorder appears caused by impaired translation into or out of a buffer subsequent to phonology and semantics. The spelling errors that occurred were often omissions, substitutions of other letters, and reversals (such as *chucrh* for *church* and *rpiest* for *priest*).

If the same orthographic representation is used for both handwriting and oral spelling, then the percentage of spelling errors generated by the two effector systems should be the same. Such was the case. One patient was 52% correct in handwriting and a similar 54% correct in oral spelling. Corresponding values for the other patient were 35% and 36%. These results suggest that the same serial order representation underlies oral spelling and handwriting.

One other feature of the Hillis and Caramazza patients is noteworthy for the hint it provides regarding how the orthographic sequence is computed. One patient had left hemisphere damage including damage to the left parietal lobe. The other patient had right parietal damage. Both patients demonstrated some neglect on the spelling tasks. The left patient made most errors on the last half of the word and the right patient made most errors on the first half. The same error pattern occurred for auditorily presented words, for spelling the names of pictures, and for delayed copying of words. Moreover, the same pattern occurred for written and oral spelling. Neither auditory input nor oral output have a left and right. The spatial patterns of errors seem explicable only by assuming that the orthographic buffer is a spatial array of graphemes in which serial order is computed by some kind of left to right scan.

Unfortunately there are a few gaps in patient studies for drawing firm conclusions regarding the idea that sequence representation is independent of effector system. Margolin found no cases that analyzed typing as well as written and oral spelling except for one collosal patient described by Geschwind and Kaplan (1962). Collosal damage is not clearly analytic to the existence of an orthographic buffer. Thus, although Margolin suggests in his model that typing draws on the same orthographic representation as does handwriting and oral spelling, the proposition has not been tested. Another lack in patient studies concerns modules below the level of the orthographic buffer. We have suggested from writing similarities across effectors (Fig. 3.5) that the same representations of sequence are used for fingers and hand as for arm, mouth, and foot. Although an orthographic code may be in common to different effectors, such a code would not itself produce similarities in style of the graphemes. To explain similarities in style would require an addition to Margolin's model in which writing with different effectors would all be driven by the same graphic motor pattern. Though we are unaware of any patient studies that are analytic to this issue, we have made such an addition to the model portrayed in Fig. 3.6.

Recently we have embarked on an analysis of modularity of sequencing using simple key-pressing tasks (Cohen, Ivry, & Keele, in press; Keele, Cohen, Ivry, & Jennings, in preparation). The tasks we used borrow heavily from paradigms developed by Nissen and Bullemer (1987). The primary question is whether a learned sequence of key presses is independent of the effector system that executes the sequence. X-marks appeared at one of three horizontal locations on a CRT screen. Subjects pressed keys corresponding to the positions of the visual signals. Upon a key press, the signal disappeared, and 200 msec later another appeared. In one condition, the signals appeared at random with the restriction of no immediate repetition of position. In the other condition, the signals appeared in a particular order of five elements before recycling. An example is 32121, in which the numbers refer to the three stimulus and response positions. When a sequence occurred, it recycled without break for 20 cycles (a total of 100 keypresses) before a rest. If over practice subjects learn the sequence, reaction times should become faster in the sequence condition than when the signals occur at random.

In an initial training phase, all subjects practiced a sequence for eight blocks of 100 trials per block. They were never informed that the signals appeared in a sequence and instead simply responded to the signals as they occurred. During this period, half of the subjects responded with three fingers on the three keys. The other subjects hit the keys with their index fingers by moving their arms back and forth over the keys. The difference between the two response modes is much like touch typing versus hunt-and-peck typing. Blocks 9 through 12 were transfer blocks. Half of the subjects maintained the same effector as before, but the other half changed from fingers to arm or vice versa. Half of each of those groups continued the same sequence as before whereas the other half received signals at random with the exception of no immediate repetitions of positions.

The reaction times for this task are shown in Fig. 3.7. The top panel shows reaction times averaged over the two groups that maintained the same effector—finger-to-finger and arm-to-arm—from the training to the transfer phase. During transfer, the group that maintained the sequence continued to perform at the prior rate, but reaction times of the group shifted to the random condition slowed considerably. The difference in reaction times between the two groups indicates the degrees of sequence learning.

The central question, of course, concerns performance of the groups that changed effectors from training to transfer. Their results are shown in the bottom panel of Fig. 3.7, again averaged over the two groups—that

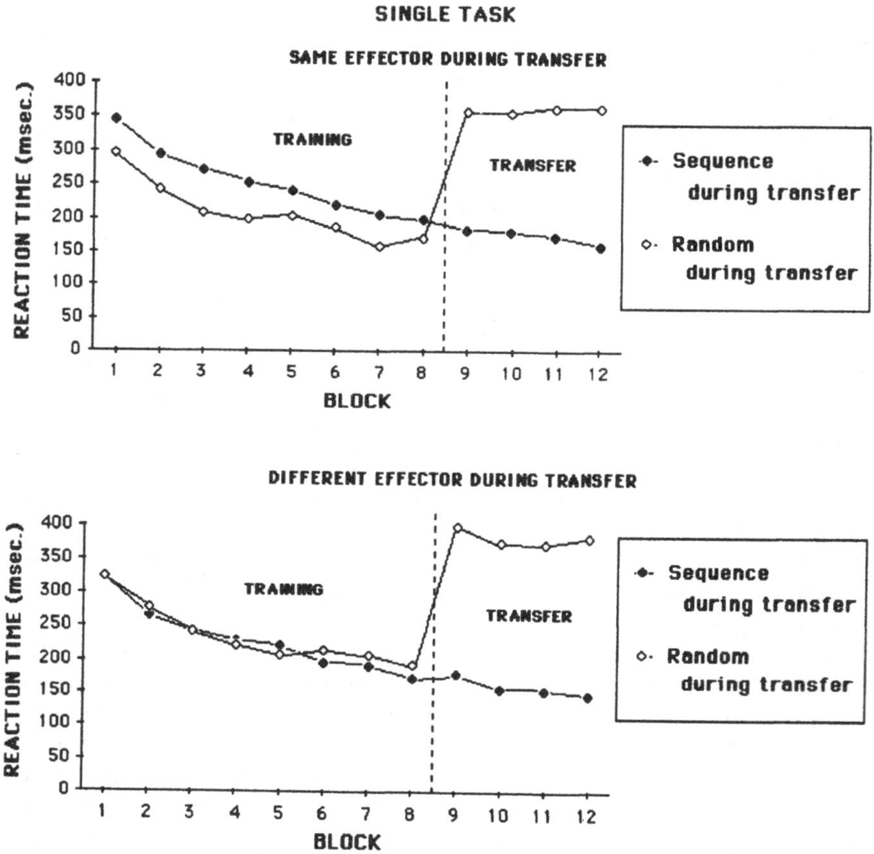


FIG. 3.7. Reaction time to visual signals appearing at one of three locations on a screen. In blocks 1–8, the signals appear in a fixed sequence. In blocks 9–12, one group retains the sequence, and for the other group, the signals appear at random. For data in Panel A., subjects use the same effector in all blocks. For data in Panel B, some subjects switch from arm to fingers and others from fingers to arm on blocks 9–12. All groups are run without a secondary task. Based on Keele, Cohen, Ivry, & Jennings, in preparation.

shifting from finger to arm and that from arm to finger. The results are virtually identical to those from the groups that maintained the same effector. The advantage of sequence maintenance over random targets is the same regardless of whether effectors change or not. These results suggest that knowledge of a sequence acquired in the course of performance with a particular effector system is completely independent of that effector system.

This experiment has been replicated with one change. Recall that 200 msec transpired between a response and the next visual signal. In the replication a tone occurred in each interval. At the end of a block of trials subjects reported how many of the tones were high in pitch. The purpose of this secondary task was to distract subjects from the sequence. In another experiment (Cohen, Ivry, & Keele, in press) we found that subjects, though unaware of the sequence under dual-task conditions, demonstrate learning when assessed by a performance criterion. In this replication the sequence was also learned in the presence of distraction, though with less learning than in the absence of distraction. Moreover, the sequence knowledge transferred from one effector to the other.

Both the single- and dual-task experiments suggest, therefore, that the system that implements the sequence does not itself encode the sequence. It would appear instead that the sequence representation is either a description of where in space successive signals should appear or where in space successive responses are to be directed. A similar conclusion arises from an earlier experiment by Pew (1974). He found that learning on a tracking task transferred in part to a mirror image of the track despite the fact that subjects were unaware that the track they had experienced was repeated on each trial. Mirror image required different muscles to execute the movements for the same portion of the track.

Given that the sequence representation is in spatial coordinates, one might speculate on its underlying neural base. A good candidate might be the parietal lobe, which is specialized for spatial encoding (e.g., Mishkin, Ungerleider, & Macko, 1983; Van Essen & Maunsell, 1983). We are not aware of any study of sequencing using procedures similar to ours that have assessed the role of parietal systems. However, the analysis of apraxia, a disorder of movement organization, is of some relevance.

Apraxic patients execute movements but have difficulty organizing them into appropriate patterns. They have difficulty producing gestures, such as a salute, or executing pantomimes, such as combing the hair. Apraxia can result from lesions in either frontal motor regions or parietal regions. Of special interest in this regard are studies by Heilman and colleagues (Gonzales & Heilman, 1985; Heilman, Rothi, & Valenstein, 1982). They have shown that apraxics with frontal lesions and apraxics with parietal lesions both have difficulty in executing proper movements. The parietal apraxics, but not frontal patients, have additional difficulty in recognizing gestures and pantomimes produced by others and in discriminating such movements from other similar ones of poorer quality.

Such results suggest that the parietal system encodes a spatial configuration. The description can be used either for purposes of recognition or for purposes of configuring the limbs to produce a gesture. It may be that frontal motor systems do not contain a gestural engram. Rather, the motor system simply places articulators in positions as dictated by the spatial module.

As we said, the possible role of parietal systems in sequencing has not been examined. Other neural systems, however, have been excluded as viable candidates. In a study after which our own were closely modeled, Nissen and Bullemer (1987) showed that Korsakoff patients were able to learn key-pressing sequences even though they expressed no awareness of them. Moreover, Nissen, Knopman, and Schacter (1987) found that administration of scopolamine to normal subjects impaired word-list learning and awareness of a sequence, but had no effect on the improvement of reaction time with practice to the sequence. Thus, the neural mechanisms for sequence representation appear independent of those for declarative memory.

In our studies (Cohen, Ivry, & Keele, in press), we have pursued an additional issue that relates to the concept of hierarchic representation. As mentioned, cyclic spatial sequences, such as 32121. . ., can be learned in the presence of a distraction task that not only diverts attention but also prevents awareness of the sequence. Not all simple sequences can be learned in the presence of distraction, however. We compared the learning of three kinds of sequence. One we called Linear, and an example is 15423. . . . Recall that a sequence recycles from beginning to end with no break for a total of 20 cycles. For the Linear sequences, signals appeared in 5 different positions, each position occurring once in a cycle. A second sequence type, called Repeat, is illustrated by 132312. . . . In a cycle of 6 items, each position occurred twice, but the orders were different in the two halves of a cycle. The third type of sequence we called Hybrid, because it involved a mixture of some elements that were repeated in a cycle and some that were unique. An example is 142312. . . .

Subjects rested their fingers on a keyboard and responded to whatever signal occurred by pressing the corresponding button. After each response, the next signal occurred 200 msec later. The first condition involved the dual-task situation. In each interval between response and stimulus a tone occurred. At the end of a block of trials, the subject was to report the number of high-pitched tones. This task served solely as a distraction from the primary reaction time task. Initially, all subjects performed two blocks of trials in which the key press signals occurred at random, with the exception of no immediate repetitions of the same signal. Next subjects performed 8 blocks of 20 cycles through the sequence. Different subjects were assigned to the three different sequence types. After 8 blocks of trials with the sequence, 2 blocks were given with the random order followed by 2 more with the original sequence.

The reaction times are shown in Fig. 3.8. Both the Linear and Hybrid

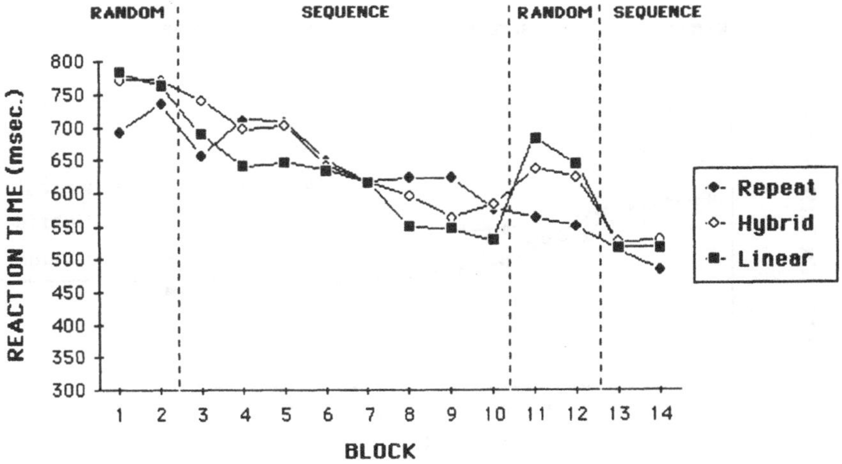


FIG. 3.8. Reaction times to three types of sequences under dual-task conditions. On blocks 1, 2, 11, and 12, the signals appear at random. On all other blocks the signals occur in sequence. From Cohen, Keele, & Ivry, in press.

groups show sequence learning as indicated by the increased reaction time when signals suddenly switched from sequence to random and then improved reaction time with the switch back to the sequence. The Repeat group, however, shows no evidence of sequence learning.

Before interpreting these results, consider the single-task condition which was identical to the dual-task condition except that the secondary task was not included. The results for the single-task condition are shown in Fig. 3.9. Here all groups, including the Repeat group, show evidence of sequence learning as indicated by the degradation in performance when the signals appear at random.

It appears that learning of some sequence types is preserved under attentional distraction. Indeed, we have shown (Cohen, Keele, & Ivry, in press) that whether the distraction is more or less severe makes no difference in the amount of learning. Full attention is needed, however, to learn the Repeat sequences. One interpretation of the results is that Repeat sequences are simply more difficult to learn than the other sequence types. There are two problems with such an interpretation. The first is that other results we report (Cohen, Keele, & Ivry, in press) suggest little difference in difficulty when subjects explicitly learn the sequence. There is especially little difference between the Repeat and Hybrid sequences. Differential difficulty appears under attentional distraction. Thus, the distraction task appears to block some process that is available with full attention. Second,

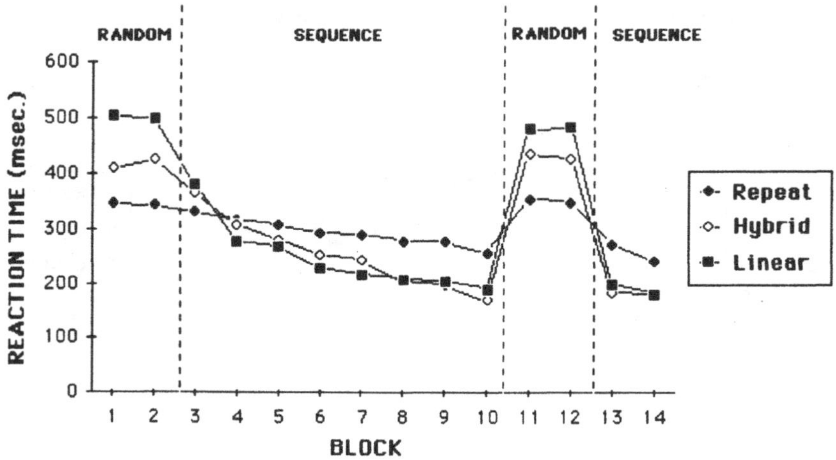


FIG. 3.9. Same as Figure 3.8 but without the secondary task.

the notion of difficulty is not an explanation. One must propose a mechanism responsible for the difficulty.

We hypothesize two kinds of learning mechanism to explain the results. One mechanism involves linear associations in which one element becomes associated with the next. Presumably that mechanism requires little or no attention to the relationship between successive elements for its operation. Such a mechanism has difficulty with sequences in which all the elements are repeated in different orders in different parts of the sequence. The reason is that each element is associated with two different elements, yielding ambiguity of what follows what. For those sequences we speculate that hierarchic coding is required. A parser subdivides the pattern into chunks and each chunk representation controls the order of elements within it. Moreover, the parser appears to require attention for its operation.

Our experiments thus lead to two major conclusions, both of which are extensions of the notion of hierarchic sequence representation. One conclusion is that sequences are represented in modular fashion. The sequence representation is independent of particular motor implementations. The second conclusion is that the representation may have two qualitatively different components. For sequences of the sort we have studied, one component may specify where in space successive movements are to be directed. We have speculated that parietal encoding may be involved here. The other component is a parsing system for sequences with many repeated elements. One might speculate that the parser involves a different neural system, perhaps frontal in origin, than does the primary sequence representation.

## ISSUES FOR A COMPUTATIONAL THEORY OF SEQUENCING

The preceding sections primarily have been concerned with describing the architecture of sequence representation. We have argued that a typical representation is hierarchical, and, for at least some sequences, the representation is modular. However, an architectural description does not specify exactly how a sequence is computed at a level of description appropriate for neural systems. Although we have made occasional speculations on what neural system might be involved in a particular computation, such as time or sequence, we have offered little suggestion on just how the computation might be effected. One exception is the writing and spelling analysis of two patients provided by Hillis and Caramazza (1988). That analysis suggested that the sequencing of letters is controlled by a left-to-right scan through a visual buffer.

The purpose of this section is to briefly describe a promising theory of serial order computation developed by Jordan (1986a, b; this volume) and to use that description as a vehicle for developing some issues about how computation and architecture may interact.

Two initial ideas on sequence computation have already been introduced and rejected. Recall the notion of peripheral chain associations, discussed by Lashley (1951), in which feedback resulting from the execution of one movement stimulates the next. That theory was rejected because movement series can occur without feedback, later movements sometimes begin before earlier movements, and reaction times indicate the planning of movements before the movement series begins. Lashley had also considered a notion of central chain associations in which successive events are internally associated one with the next. This idea is similar to what we have called a Linear associator. Such a theory also is not easily accommodated to results indicating the planning of an extensive series of movements prior to the beginning. Its greatest difficulty, however, is that it has no simple way to handle the problem of repeated elements that occur in different orders in different sequences.

Although a theory of central associations faces difficulties, there are reasons for not fully rejecting it. In our work just reviewed in the preceding section, we suggested that sequences learned under distraction might be heavily represented in associational form. Even when sequences are learned with full attention, there is evidence that an event at one position in a sequence primes events that follow it at distinctly different places in a sequence, resulting in occasional errors or increased reaction time (Keele & Summers, 1976). Thus, a complete model might contain an associational mechanism as a component.

One attempt to retain a basic Linear associator and to overcome the problem of repetition of elements, particularly with respect to speech, was made by Wickelgren (1969). He suggested that phonemes in different contexts, although similar, were not identical. In particular, a variant of a phoneme existed for every combination of the phoneme and the preceding and following elements. These elements could be silent intervals as well as other phonemes. If we return to the examples of *rite* and *tire*, the two instances of *i* are not identical. One is an *i* conditioned on the prior occurrence of *r* and the latter occurrence of *t* and the other *i* is conditioned on a prior *t* and following *r*. Wickelgren proposed that the two *i*'s are allophones. When a word, such as *tire* is to be pronounced, the word calls up in unordered form the constituent allophones. These allophones then associatively order themselves. For example, the occurrence of the *i* triggers *r* rather than *t* because the allophone for *i* would be associated with the allophone for *r* that is preceded by an *i*.

Coarticulation is a common feature of human sequential activity. One motor element is influenced by those elements that surround it. Wickelgren's model gives an explicit account of coarticulation by proposing different allophones for each context. But its particular form of coarticulation is the model's weakness. Wickelgren's model accounts for coarticulatory effects between immediately adjacent elements, but in fact coarticulatory effects span more elements than adjacent ones, and the span varies in different cases. Referring back to Fig. 3.1 for typing, it can be seen that sometimes items further separated than mere adjacency are simultaneously active. The degree to which this is the case depends on whether the items conflict with each other. Similar phenomena have been noted in speech. Such highly separated coarticulation is not consistent with Wickelgren's theory. It suggests, instead, that items to follow at far distances are being anticipated.

One motivation for Jordan's (1986a, b; this volume) model of serial order was to give a fuller account of coarticulation than other models have provided. The model, a version of which is portrayed in Fig. 3.10, uses a connectionist (Rumelhart & McClelland, 1986) architecture. The stimulating conditions for movement are provided by input units. The movement itself is reflected in the pattern of activation of output units. The inputs are of two types. One type, a plan, is a pattern across units that differs depending on what sequence is desired. During a sequence, the plan remains fixed. The other input units, called state units, vary during progression through the sequence. Whenever an output occurs, internal feedback from the output units changes the values of the state units. The current state additionally is influenced by the previous state. Suppose similar outputs occur in two different sequences. The states they produce are different because the states reflect their past history in the sequence. This mechanism allows similar outputs to be followed by different outputs depending on

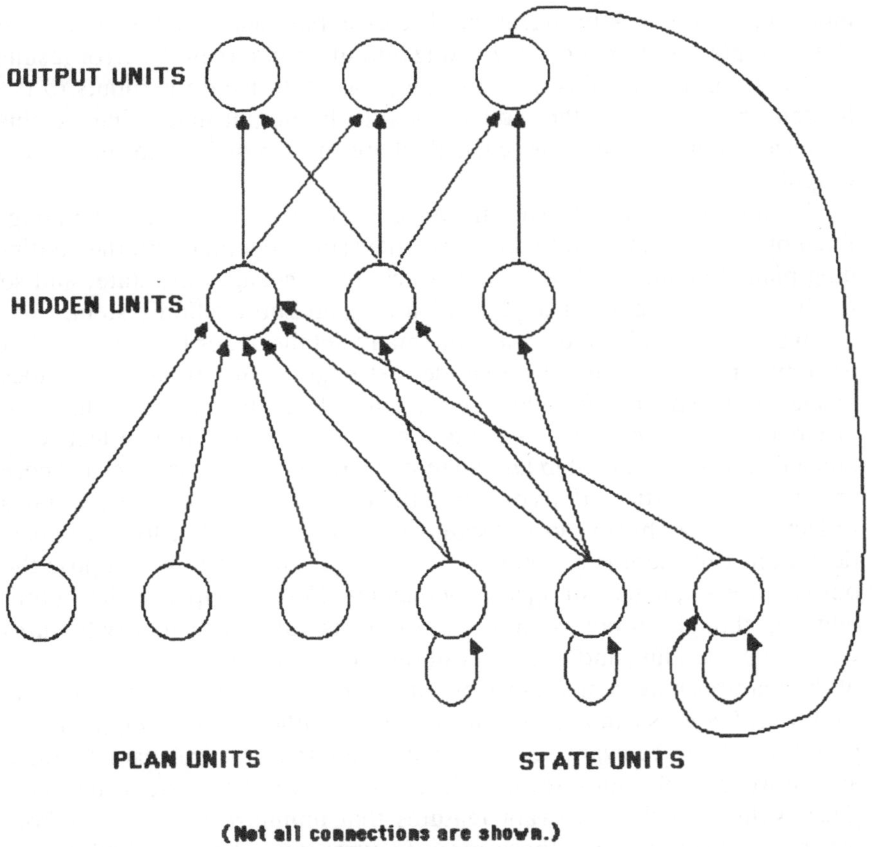


FIG. 3.10. A connectionist model of sequence learning. See text for explanation. Based on Jordan, 1986.

the context. A second factor that individuates similar outputs is the unique plans for different sequences. An output is a joint product of the current state and the plan in force at the time. From this view, a plan is very much like a superordinate node in a hierarchical representation in that it provides a controlling influence on the particular underlying events.

The joint influence of states and plans is implemented through an intermediate set of units, sometimes called hidden units. The representation they adopt is determined by learning. One way of thinking of hidden units is that they represent particular combinations of inputs, in this case plan and state combinations. This means that a particular output can be followed by very different configurations in the hidden layer. The configuration in the hidden layer actually determines the next output.

The way the model learns is by a formal rule used in connectionist modeling called back propagation. The essential idea is that if for a given state and plan combination, the wrong output occurs, then the error results in changes in the weights of the connections both from the inputs to the hidden units and from the hidden units to the output units. The weights stabilize when a situation is reached where the inputs lead to the correct outputs.

The Jordan system, once it starts a sequence, becomes self-perpetuating. Each output alters the state units. The new state together with the continuing plan stimulates a new output which again changes the state, and so on. Thus, the system is a sequence learning and production system.

This is not the place to review the many details of Jordan's model. Let us simply mention some of the model's strengths. First of all, it provides an elegant mechanism for allowing the same elements to occur in different orders in different contexts. Second, it produces coarticulatory effects very much like those observed in human motor sequences such as speech. There are two ways coarticulations occurs in the model. One is because the states reflect their past history in the sequence. This is a kind of forward coarticulation. But there is another, more important mechanism. Suppose the outputs are sequences of speech phonemes. Further suppose the output units are thought of not as unitary movements but as articulatory features such as lip rounding and tongue position. For a particular phoneme, some of the features may be necessary but others may be largely irrelevant. The irrelevant features can take a range of values without disrupting the interpretation of the relevant features. Suppose currently irrelevant features are, however, relevant elsewhere in the sequence. The system then produces values on the irrelevant features that minimize the changes from other occasions in which those same features are relevant. Similarly, to the degree that values of the relevant features can also occupy a range, values are chosen that minimize change from other occasions involving those features. The end result is extensive forward and backward coarticulation as in human movement.

Another aspect of the model's behavior that Jordan has stressed is its dynamic character that exists after the weights become stable from extensive practice. If a movement is perturbed, successive movements are automatically attracted back to the normal course. The reason is that a similar state produces an output close to the intended next output, which produces a new state even closer to the intended next state and so on.

A fourth strength of Jordan's model is that it may produce errors similar to some produced by animals and humans. As mentioned earlier, Fentress and colleagues have studied sequences of grooming and eating movements in rodents. Nearly identical movement elements can appear in both sequences, but they are followed by different events in the two cases. Jordan's model handles elements appearing in different sequences by the provision of different plans. Fentress (1983) has noted in infant mice what he calls *motor traps*. When a movement in feeding fortuitously resembles that of a movement in grooming, the young mouse sometimes shifts from feeding to grooming. Keele and Summers (1976) have noted a similar phenomenon in learned human sequences. An item that repeats within a sequence tends to prime an item that follows it in a different place in the sequence. Such priming of an incorrect element both slows reaction time and increases the likelihood of an error. Such behavior can be explained by Jordan's model by supposing that the plan can vary in strength. A weaker plan can leave sequence direction under the strong influence of state similarity. In essence, the model provides a mixture of plan and association.

How does Jordan's model fare with phenomena emphasized in this chapter, including those that argue for hierarchical coding? As Jordan (1986) himself has suggested, hierarchic levels can easily be implemented. Recall that a plan stays constant until the sequence it controls is finished, changing only at the end. A sequence of plans can be implemented in exactly the same way as a sequence of elements. Such a model would mimic the effects of complexity of the upcoming movement series on reaction time. No overt movement would begin until the plan sequencer produced the first subplan, and the first subplan readied the first movement. The time for such preparation would depend on the number of plan levels. Likewise, when a plan was exhausted, the system would have to produce the next plan before it could ready the next movement. Thus, transitions between "chunks" would take more time exactly as seen in data such as that of Povel and Collard (1982) and as specified in Rosenbaum's (1987) hierarchical editor model.

Given these strengths of Jordan's model of sequencing, let us now raise an issue. A major theme of this paper is modular organization of sequencing. Sequence representations are independent of the effector systems responsible for execution. Such seems to be the case for handwriting (Hillis & Caramazza, 1988; Margolin, 1984) and for the key-pressing sequences that we have studied. In Jordan's model, contrary to the notion of modularity, sequence learning is bound to a particular effector system: For speech, the output units are part of the articulatory system. For arm movements to targets, the output units are joint positions. Sequence learning involves building a network of connections between the hidden units and motor output units.

A first impulse may be to think of output units as occupying an earlier level in the system. For spelling, one could suppose that the output corresponds to an orthographic buffer that specifies the order of letters. That output could be transmitted to writing, to typing, or to oral spelling. The major problem with this solution, however, is that it strips Jordan's model of one of its prominent features, namely, an account of coarticulation. Recall that when motor features necessary in some parts of a sequence are irrelevant in other parts, those features carry over or are anticipated during the phases in which they are irrelevant. Such behavior minimizes abrupt changes in effector features. In a sense, practice tunes sequence performance to details of the articulators.

To further appreciate how Jordan's model incorporates the motor system into sequence representation, consider typing. The word *they* is typed by alternating fingers on the two hands: *t* is produced by the left index finger, *h* by the right index finger, *e* by the left middle finger, and *y* by the right index finger. With this arrangement, fingers on the two hands can move at the same time with little conflict. Jordan's system would come to make use of that capability as sequence learning progresses. Now, suppose the keyboard were rearranged, putting all four letters of *they* on the right hand. The constraints on movement would differ, and, as a result, a different sequential representation would emerge with practice.

It would appear, then, that the architecture of Jordan's model is contradictory to our conclusions regarding modularity. What resolutions are possible?

One possibility is that sequential representation at low levels of skill is effector independent. At that stage of skill there may be very little coarticulation except for that between adjacent movements as a result of mechanical interactions. As practice proceeds, a sequential representation may arise within the effector system that predominantly executes the sequence. Indeed, Grudin and LaRochelle (1982) have made such an argument for typing. They claim that with excessive practice, specialized motor routines emerge for particular digraph combinations. Their argument is twofold. First, more-frequent digraphs are typed faster than less-frequent ones. Second, particular letters conflict with one another when produced in succession, and typists may develop special movement routines to compensate. Neither argument, however, is convincing. The digraph frequency effects may emerge at the level of an orthographic buffer (Margolin, 1984) and may not be specific to typing. With respect to special routines for difficult letter combinations, Grudin and LaRochelle provide no evidence that the special routines are learned rather than due to peripheral motor interactions.

Another possible resolution to the conflict between Jordan's computational model and the notion of modularity is to place the sequencing mechanism more central and to explain coarticulation by more peripheral mechanisms. Consider a model of typing developed by Rumelhart and Norman (1982) and illustrated in Fig. 3.11. In this model, a word representation, say *very*, activates its constituent letters *v*, *e*, *r*, and *y*. One may suppose that these letters are activated in an effector-independent buffer such as the orthographic buffer of Margolin's model. Although all letters are activated simultaneously, they are active to varying degrees because each letter partially inhibits those that follow. At the start *v* is most active, *e* next most active, and so on. After *v* is typed, it becomes inhibited,

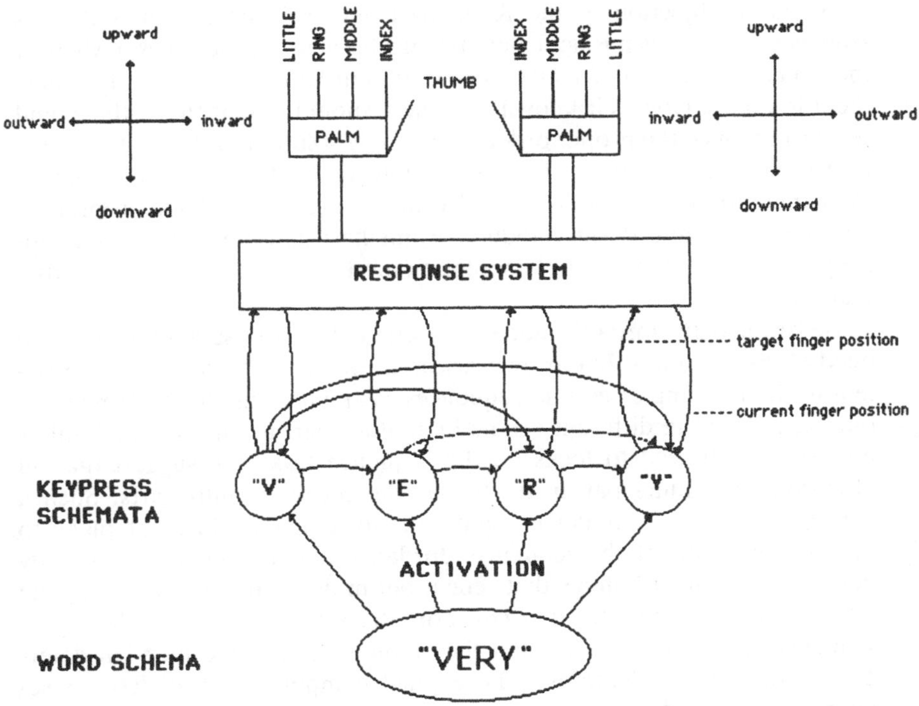


FIG. 3.11. A model of typing. See text for explanation. Based on Rumelhart & Norman, 1982.

releasing its inhibiting effect on the remaining letters, and *e* becomes the most active. Below the level of the orthographic buffer, the letters are interfaced with an effector system.

Each letter from the higher level is connected to a specific effector. The effector system itself knows nothing about the sequence. It simply becomes active to the degree to which its input is active. Activity in the letter *v* tends to pull the left hand toward the bottom row of the keyboard and shorten the index finger that is to strike the *v*. The letter *e* pulls the left hand toward the upper keyboard row and extends the left middle finger; *r* drives the left hand upward and extends the index finger; *y* drives activity of the right hand and the right index finger. The influences of *e* and *r* on the left hand tend to counter the influence of *v* on that hand, because the influences are in opposite direction. Likewise, the *r* influence on the index finger partially counters the index finger movement for the *v*. In short, extensive coarticulation occurs and that coarticulation spans several letters. Such coarticulation is *not* the result of the central sequencing mechanism, however, but a result of implementation at the motor level.

A major objection to the Rumelhart and Norman model is that its sequencing mechanism is not computational because the mechanism whereby the concept of a word activates its constituent letters and properly orders them is vague. How is it known that the *v* should inhibit the *e*, the *r*, and the *y* to produce the proper order? A possible solution to the problem may be to merge the sequence-learning system of Jordan with the sequence implementation of Rumelhart and Norman. Jordan's model appropriately describes a serial-ordering module. Rumelhart and Norman's model appropriately describes an implementation module that accounts for coarticulation.

Before leaving Jordan's model altogether, let us raise another issue in need of investigation. In our own research, we have shown a special problem in the learning of certain sequences. Sequences in which elements in one part repeat in different orders in another part are not learned unless attention is allowed to focus on the sequence task. We suggest that an attention-demanding parser subdivides the sequence into parts. Such a concept in Jordan's model is equivalent to assigning different plans to different portions of the sequence. Jordan's model, however, is already designed to deal with items that repeat but in different contexts. An issue we are currently investigating concerns the interaction of the length of sequence spanned by a plan and the amount of repeated items in determining the ease of learning. The explicit computational analysis of sequencing provided by Jordan provides a powerful model around which to direct research issues.

## SUMMARY AND CONCLUSIONS

We began this article by paying homage to the influential ideas of Lashley. It was perhaps Lashley more than any other person who noted the inadequacies of an associational theory of sequencing and developed as an alternative the idea that a sequence of actions is guided by a plan. Subsequent research has largely shown this to be correct: A plan is apparent by orderly activity in the absence of peripheral feedback and by anticipatory activity in which later portions of a movement series make their presence known before the occurrence of preceding portions.

Since the time of Lashley, it has become more clear that a plan is hierarchic, consisting of successively more-specific implementations. Hierarchic arrangement is documented by a large variety of phenomena, including reaction times prior to the start of movement, pauses during production, and transfer of performance.

A second theme to have emerged since Lashley's work, and one on which some of our own work has concentrated, is a notion of modularity. Some evidence suggests that the representation of a sequence is independent of the effectors that execute it. The same representation can be interfaced to various motor systems for execution. The modular decomposition of the total motor act holds promise for identifying neural systems responsible for particular computations.

The notion of modularity also implies that the succession of representations is not hierarchic in the strictest sense, but also has a heterarchic character. A plan at one level does not by itself dictate the action at a lower level. Rather, converging influences effect specifics of action. Thus, if a module represents sequence independent of effectors, other sources of information, such as size of the movement, instructions about which effectors to use, availability of various effectors, desired tempo, and the like also influence movement details.

A theory of plans and of modularity is essentially a partial theory of the architecture of interacting units. At a lower level of description a particular theory of computation is needed. A recent and promising computational theory has been described by Jordan. An interesting feature of Jordan's theory is that it contains both planning elements and associational elements. It provides an explanation of how a movement can both be guided by a plan, yet allow apparent associational effects such as "motor traps." Thus, it may be premature to dismiss central chaining theories in their entirety because the actual system of computation may involve both plans and central associations of a sort.

One aspect on which Jordan's computational model appears at odds with empirical work is with regard to modularity. In Jordan's model, sequence representation is closely entwined with the motor system that produces movement. Some possible resolutions to this conflict are raised, but it remains for future research to give a full reconciliation.

## ACKNOWLEDGMENT

We wish to thank Peggy Jennings, Clifford Keele, and Jim Tanaka for their discussion and help on parts of this project. The financial assistance of the Office of Naval Research (Contract No. N00014-87-K-0279) and the John Guggenheim Memorial Foundation is gratefully acknowledged. Richard Ivry is now at the University of California, Santa Barbara.

## REFERENCES

Arbib M. A. (this volume) The hierarchical control of skilled hand movements.

Arbib, M. A. (1981). Perceptual structures and distributed motor control. In V. B. Brooks (vol. Ed.) *Handbook of Physiology: The Nervous System: Motor control: Vol. II.* (pp. 1449–1480). Baltimore: American Physiological Society. Distributed by Williams & Wilkins.

Bernstein, N. A. (1947). [On the formation of movement]. Published in Russian by State Medical Literature Publishers (Medgiz).

Berridge, K. C., & Fentress, J. C. (1987). Deafferentation does not disrupt natural rules of action syntax. *Behavioural Brain Research, 23*, 69–76.

Berridge, K. C., Fentress, J. C., & Parr, H. (1987). Natural syntax rules control action sequences of rats. *Behavioural Brain Research, 23*, 59–68.

Bizzi, E., Polit, A., & Morasso, P. (1976). Mechanisms underlying achievement of final head position. *Journal of Neurophysiology, 39*, 435–444.

Bub, D., & Kertez, A. (1982). Evidence for lexicographic processing in a patient with preserved written over oral single word naming. *Brain, 105*, 697–717.

Caramazza, A., Miceli, G., & Villa, G. (1986). The role of the (output) phonological buffer in reading, writing, and repetition. *Cognitive Neuropsychology, 3*, 37–76.

Cohen, A., Ivry, R. I., & Keele, S. W. (in press). Attention and structure in sequence learning. *Journal of Experimental Psychology: Learning, Memory, and Cognition.*

Feldman, A. G. (1986). Once more on the equilibrium-point hypothesis (Lamda model) for motor control. *Journal of Motor Behavior, 18*, 17–54.

Fentress, J. C. (1973). Development of grooming in mice with amputated forelimbs. *Science, 179*, 704–705.

Fentress, J. C. (1983). Hierarchical motor control. In M. Studdert-Kennedy (Ed.) *The Psychobiology of Language* (pp. 40–61). Cambridge, MA: MIT Press.

Fentress, J. C., & Stillwell, F. P. (1973). Grammar of a movement sequence in inbred mice. *Nature, 244*, 52–53.

Forrsberg, H., & Wallberg, H. (1980). Infant locomotion: A preliminary movement and electromyographic study. In K. Berg & B. Eriksson (Eds.) *International series on sport sciences, Vol. 10: Children and exercise IX* (pp. 32–40). Baltimore: University Park Press.

Fountain, S. B. (1987). *Do multiple neurobehavioral systems underlie serial-pattern learning?* Talk presented at Psychonomics Society, Seattle, Washington.

Gentner, D. R., Grudin, J., & Conway, E. (1980). *Finger movements in transcription typing* (Tech. Rep. No. 8001). La Jolla, CA: University of California, San Diego, Center for Human Information Processing.

Geoffroy, A., & Norman, D. A. (1982). *Ease of tapping the fingers in a sequence depends on the mental encoding* (ONR Report 8203). La Jolla, CA: University of California, San Diego, Center for Human Information Processing.

Geschwind, N., & Kaplan, E. (1962). A human cerebral deconnection syndrome: A preliminary report. *Neurology, 12*, 675–685.

Gonzalez, L. J., & Heilman, K. M. (1985). Ideomotor apraxia: Gestural discrimination, comprehension and memory. In E. A. Roy (Ed.), *Neuropsychological studies of apraxia* (pp. 65–74). New York: North Holland.

Gordon, P. C., & Meyer, D. E. (1987). Control of serial order in rapidly spoken syllable sequences. *Journal of Memory and Language, 26*, 300–321.

Grillner, S. (1981). Control of locomotion in bipeds, tetrapods, and fish. In V. Brooks (Ed.), *Handbook of physiology: Section 1: The nervous system. Volume II: Motor control, Part 2* (pp. 1179–1236). Baltimore: American Physiological Society.

Grudin, J. T., & Larochelle, S. (1982). *Digraph frequency effects in skilled typing* (CHIP Rep. 110). La Jolla, CA: University of California, San Diego, Center for Human Information Processing.

Ivry, R. I. (1986). *Components of coordination: The cerebellum as an internal clock.* Unpublished doctoral dissertation, University of Oregon, Eugene.

Ivry, R. I., & Keele, S. W. (1989). Timing functions of the cerebellum. *Cognitive Neuroscience, 1*, 136–152.

Ivry, R. I., Keele, S. W., & Diener, H. (1988). Dissociation of the lateral and medial cerebellum in movement timing and movement execution. *Experimental Brain Research*, *73*, 167–180.

Heilman, K. M., Rothi, L. J., & Valenstein, E. (1982). Two forms of ideomotor apraxia. *Neurology*, *32*, 342–346.

Hillis, A. E., & Caramazza, A. (1988). *The graphemic buffer and attentional mechanisms* (Report No. 30). Reports of the Cognitive Neuropsychology Laboratory, Johns Hopkins University, Baltimore, Maryland.

Jeannerod, M. (1981). Intersegmental coordination during reaching at natural visual objects. In J. Long & A. Baddeley (Eds.) *Attention and Performance IX*. Hillsdale, NJ: Lawrence Erlbaum Associates.

Jordan, M. I. (1986a) *Serial order: A parallel distributed processing approach* (ICS Report 8604). Institute for Cognitive Science, University of California, San Diego.

Jordan, M. I. (1986b) Attractor dynamics and parallelism in a connectionist sequential machine. *Eighth Annual Conference of the Cognitive Science Society* (pp. 531–546). Hillsdale, NJ: Lawrence Erlbaum Associates.

Jordan, M. I. (this volume). Learning to articulate: Sequential networks and distal constraints.

Keele, S. W., Cohen, A., Ivry, R. I., & Jennings, P. (in preparation) [*On the modularity of sequence representation*].

Keele, S. W., Ivry, R. I., & Pokorny, R. A. (1987). Force control and its relation to timing. *Journal of Motor Behavior*, *19*, 96–114.

Keele, S. W., Pokorny, R. A., Corcos, D. M., & Ivry, R. I. (1985). Do perception and production share common timing mechanisms: A correlational analysis. *Acta Psychologica*, *60*, 173–191.

Keele, S. W., & Summers, J. G. (1976). Structure of motor programs. In G. E. Stelmach (Ed.), *Motor control: Issues and trends* (pp. 109–142). New York: Academic Press.

Klapp, S. T., Anderson, W. G., & Berrian, R. W. (1973). Implicit speech in reading, reconsidered. *Journal of Experimental Psychology*, *100*, 368–374.

Konishi, M. (1965). The role of auditory feedback in the control of vocalization in the white-crowned sparrow. *Zeitschrift fur Tierpsychologie*, *22*, 770–783.

Lashley, K. S. (1917). The accuracy of movement in the absence of excitation from the moving organ. *American Journal of Physiology*, *43*, 169–194.

Lashley, K. S. (1951). The problem of serial order in behavior. In L. A. Jeffress (Ed.), *Cerebral Mechanisms in Behavior* (pp. 112–136). New York: Wiley.

MacKay, D. G. (1981). The problem of mental practice. *Journal of Motor Behavior*, *13*, 274–285.

MacKay, D. G. (1982). The problem of flexibility and fluency in skilled behavior. *Psychological Review*, *89*, 483–506.

MacKay, D. G., & Bowman, R. W. Jr. (1969). On producing the meaning in sentences. *American Journal of Psychology*, *82*, 23–39.

Margolin, D. I. (1984). The neuropsychology of writing and spelling: Semantic, phonological, motor, and perceptual processes. *Quarterly Journal of Experimental Psychology*, *36A*, 459–489.

Marsden, C. D., Rothwell, J. C., & Day, B. L. (1984). The use of peripheral feedback in the control of movement. *Trends in Neuroscience*, *7*, 253–257.

Mishkin, M., Ungerleider, L. G., & Macko, K. A. (1983) Object vision and spatial vision: Two cortical pathways. *Trends in Neuroscience*, *6*, 414–417.

Nissen, M. J., & Bullemer, P. (1987). Attentional requirements of learning: Evidence from performance measures. *Cognitive Psychology*, *19*, 1–32.

Nissen, M. J., Knopman, D. S., & Schacter, D. L. (1987). Neurochemical dissociation of memory systems. *Neurology*, *37*, 789–794.

Nottebohm, F. (1970). Ontogeny of bird song. *Science*, *167*, 950–956.

Perinin, M.-T., & Vighetto, A. (1982). Optic ataxia: a specific disorder in visuomotor coordination. In A. Hein & M. Jeannerod (Eds.), *Spatially Oriented Behavior* (pp. 305–326). New York: Springer.

Perenin, M.-T., & Vighetto, A. (1988). Optic ataxia: A specific disruption in visuomotor mechanisms. *Brain*, *111*, 643–674.

Pew, R. W. (1974). Levels of analysis in motor control. *Brain Research*, *71*, 393–400.

Povel, D.-J., & Collard, R. (1982). Structural factors in patterned finger tapping. *Acta Psychologica*, *52*, 107–123.

Restle, F., & Burnside, B. L. (1972). Tracking of serial patterns. *Journal of Experimental Psychology*, *95*, 299–307.

Rosenbaum, D. A. (1987). Successive approximations to a model of human motor programming. In G. Bower (Ed.), *The psychology of learning and motivation* (pp. 153–182). New York: Academic Press.

Rothi, L. J. Gonzalez, Heilman, K. M., & Watson, R. T. (1985). Pantomine comprehension and ideomotor apraxia. *Journal of Neurology, Neurosurgery, and Psychiatry*, *48*, 207–210.

Rothwell, J. C., Traub, M. M., Day, B. L., Obeso, J. A., Thomas, P. K., & Marsden, C. D. (1982). Manual motor performance in a deafferented man. *Brain*, *105*, 515–542.

Rumelhart, D. E., & McClelland, J. L. (Eds.) (1986). *Parallel distributed processing: Explorations in the microstructure of cognition*. Cambridge, MA: MIT Press/Bradford Books.

Rumelhart, D. E., & Norman, D. A. (1982). Simulating a skilled typist: A study of skilled cognitive–motor performance. *Cognitive Science*, *6*, 1–36.

Sanes, J. N. (this volume) *Motor representation in deafferented humans*.

Shallice, T. (1981). Phonological agraphia and the lexical route in writing. *Brain*, *104*, 413–429.

Sternberg, S., Monsell, S., Knoll, R. L., & Wright, C. E. (1978). The latency and duration of rapid movement sequences: Comparisons of speech and typewriting. In G. E. Stelmach (Ed.), *Information processing in motor control and learning*. (pp. 117–152). New York: Academic Press.

Taub, E., & Berman, A. J. (1968). Movement and learning in the absence of sensory feedback. In S. J. Freedman (Ed.), *The neuropsychology of spatially oriented behavior*. Homewood, IL: Dorsey Press. (pp. 173–192).

Taub, E., Perella, P., & Barro, G. (1973). Behavioral development after forelimb deafferentation on day of birth in monkeys with and without blinding. *Science*, *181*, 959–960.

Thelen, E., & Fisher, D. (1982). Newborn stepping: An explanation for a "disappearing" reflex. *Developmental Psychology*, *18*, 760–775.

Van Essen, D. C., & Maunsell, J. H. R. (1983). Hierarchical organization and functional streams in the visual cortex. *Trends in Neuroscience*, *6*, 370–375.

Volpe, B. T., LeDoux, J. E., & Gazzaniga, M. S. (1979). Spatially oriented movements in the absence of proprioception. *Neurology*, *29*, 1309–1313.

Wickelgren, W. A. (1969). Context-sensitive coding, associative memory, and serial order in (speech) behavior. *Psychological Review*, *76*, 1–15.

Wright, C. E. (this volume) *Generalized motor programs: Reevaluating claims of effector independence*.

# 4 Programs, Schemas, and Neural Networks for Control of Hand Movements: Beyond the RS Framework

Michael A. Arbib
*University of Southern California*

ABSTRACT

The present chapter argues for the general utility of coordinated control programs as a setting for the study of motor control in terms of programs involving the dynamic interweaving of activation of perceptual and motor schemas rather than a sterotypical list of movements. The notions of "virtual finger" and "opposition space" are introduced for the analysis of hand movements, with special emphasis on a theory of preshaping the hand in reaching for an object. The RS language for robot schemas is then outlined to make clear the notions of schema instance, instantiation, deinstantiation, and schema assemblage, which last includes the notion of coordinated control program. The rest of the chapter then addresses the issue of how schemas and schema assemblages may be implemented in neural networks. In some cases, a schema assemblage may be seen as no more than the anatomy linking the networks that implement the constituent schemas. In the case of learned behaviors, however, a more plastic implementation is required, and the notion of motor set is examined for clues as to how schema assemblages might be dynamically implemented in monkey and human cerebral cortex. Finally, an explicit "winner take all" neural network is presented for the task of positioning the wrist to achieve a desired opposition of a pair of virtual fingers; while back-propagation is used to train two simple artificial neural networks in choosing the appropriate (pad or palm) opposition and in allocating the appropriate set of real fingers to a virtual finger in grasping any one of a variety of objects.

## INTRODUCTION

Because some kinesiologists still use the word "program" to denote a fixed, stereotyped, sequence—as in a theater program—it is worth stressing (and stressing the utility of) the computer scientist's notion of a program as comprising a "text" that, because it incorporates tests, may yield quite different courses of action in different circumstances. Certainly, neural networks may provide a varying time course of motor output on the basis of current sensory and internal variables. However, neural networks are parallel computers, and thus merit a notion of program distinct from that appropriate for a serial, one instruction at a time, computer. Instead, I offer the notion of a *coordinated control program* that has some of the properties of a conventional computer program (in that activation can be passed from one part of the program to another) but also some of the properties of a control system (in that many of the boxes can be active at the same time, passing data back and forth as in a feedback loop). Such a program is made up of *perceptual* and *motor schemas*.

Before turning to the specific issues of the control of hand movements, I wish to place the relationship of schemas and neural networks in an overall perspective. As in the analysis of any complex system, a computational account of cognition or action requires a methodology that allows an overall task to be decomposed into subtasks, with further stepwise refinement (and various processes of iteration up and down the levels of complexity) being required before subtasks are derived for which implementation is a straightforward process. Schema theory provides the beginnings of such a methodology for parallel and distributed computation. A top-down analysis starts with the choice of some overall behavior for study. Proceeding with a functional analysis, we explain behaviors in terms of the interaction of functional units called *schemas*. Then, as we seek to bridge from large unit to neuron, we may elaborate our schemas into networks of interacting subschemas until finally we may realize our constructs in terms of arrays, columns or modules implemented as neural networks.

Schema theory provides a knowledge representation protocol that is in some ways related to the frames (Minsky 1975) and scripts (Schank and Abelson 1977) of artificial intelligence, but is distinguished by its special attention to perceptual structures and distributed motor control (Arbib, 1975, 1981). A *perceptual schema* embodies the process that allows the organism to recognize a given domain of interaction. Various schema parameters represent properties such as size, location, and motion. Given a schema, we may need several schema instances, each suitably tuned, to subserve our perception of several instances of its domain. A *schema assemblage*—an assemblage of perceptual schema instances—provides an estimate of environmental state with a representation of goals and needs. New sensory input as well as internal processes update the schema assemblage. The internal state is also updated by knowledge of the state of execution of current plans made up of *motor schemas* that are akin to control systems but distinguished in that (as suggested above) they can be combined with perceptual schemas to form *coordinated control programs* that control the phasing in and out of patterns of coactivation, with mechanisms for the passing of control parameters from perceptual to motor schemas.

Although work on schemas has to date yielded no single formalism, we do see the evolution of a theory of schemas as programs for a system that has continuing perception of, and interaction with, its environment, with concurrent activity of many different schemas passing messages back and forth for the overall achievement of some goal. In terms of current computer science, there is a convergence of interest between schema theory and object-oriented concurrent programming (e.g., Yonezawa & Tokoro, 1987). In terms of artificial intelligence and cognitive psychology, issues in schema theory are related to studies in distributed artificial intelligence (Bond & Gasser, 1988) rooted in work on the HEARSAY speech understanding system (Erman et al., 1980) and distributed problem solving (Davis & Smith, 1983).

The view of schema theory espoused here was introduced into the study of hand movements with the following example. Not only does the hand move toward an object that it is to grasp, but it *preshapes* itself in a way that is clearly adapted to the shape of the target (Jeannerod & Biguer, 1982), although the preshape is somewhat larger than the object itself. The corresponding coordinated control program (Fig. 4.1, Arbib, 1981) shows a perceptual schema that can locate the object and then activate perceptual subschemas that recognize the object's size and orientation. The parameters describing where the object is, how big it is, and its orientation can then be passed to the motor schemas. One motor schema moves the arm to carry the hand in a ballistic movement that, when completed, triggers a slow adjustment movement to the final position of the hand. At the same time, two motor subschemas of the grasp schema are activated to achieve the preshape. However, once the preshape is achieved, the grasp schema engages in no further processing until the hand-reaching schema activates the "actual grasp" subschema when the hand is in the right place for tactile guidance of the hand as it grasps the object.

But how is the preshape determined? The answer advanced here is that the task of the perceptual schemas is not so much to recognize the object as a ball or pencil in any generic sense as it is to anatomize the object in terms of parameters such as size and orientation that are crucial to the task

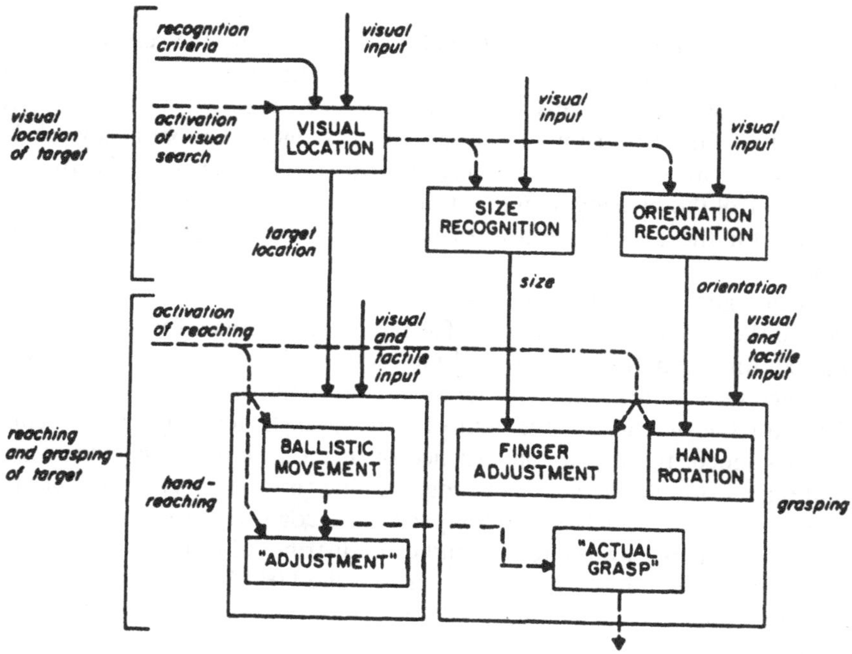


FIG. 4.1. Control of the movements for a prehensile task can be described by a coordinated control program of interacting perceptual and motor schemas. Visual location and parameterization of the target are performed by three perceptual schemas. Control of the arm movements is done by two motor schemas (on the bottom left) and control of the finger and wrist movements is done by three motor schemas (bottom right). Solid lines convey data between concurrently active schemas; dashed lines convey activation signals. (Arbib, 1981)

of grasping. They can then pass these parameters to the various motor schemas for moving and preshaping the hand.

In further attempting to understand the schemas controlling hand movements, Arbib, Iberall, and Lyons (1985) introduced the concept of the *virtual finger*. We analyzed the task of picking up a mug not directly in terms of what the five fingers do, but rather in terms of three "virtual fingers." The first (always the thumb) places itself on top of the handle. Virtual finger 2 goes through the handle, and can contain one, two, three, or even four fingers. Whatever fingers remain constitute virtual finger 3. We tested our ideas in a more conceptual way by developing a language called RS (Robot Schema language) in which we could specify coordinated control programs to control a hand to exhibit the behavior. The important point is that we do not have a separate program explicitly telling the five fingers what to do in different situations. A perceptual schema takes the parameter of handle size and on that basis partitions the five actual fingers into three virtual fingers. From then on, all the control is in terms of commands being passed, and feedback being received from, the three virtual fingers.

The concept of the virtual finger tells us how to replace analysis of hand movements directly based on the mechanical degrees of freedom of individual fingers by analysis of the functional roles of the forces being applied in carrying out some task. But, having agreed to analyze the hand in terms of these virtual fingers, how do we specify the movement of these units? Iberall, Bingham, and Arbib (1986) argue that *opposition space* provides the appropriate coordinate systems. We focused on three forms of opposition space for specifying the relative movement of a pair of virtual fingers (Fig. 4.2; the study of these oppositions is carried out extensively in Iberall 1987):

a. *Pad opposition* : Here, the pad of the thumb opposes the pad of the virtual finger—in a two-finger task we oppose just the index finger; in general several fingers may work together as "virtual finger 2" in pad opposition. The opposition space is the axis along which the finger pads move to provide that opposition.
b. *Palm opposition* : This is a power grip, as in gripping the shaft of a hammer. Here several fingers work together as a virtual finger to oppose the palm, which acts as a second virtual finger. The axis determines the movement of the virtual finger towards the palm.
c. *Side opposition* : This opposes the thumb to the side of another finger, or may be used to grasp an object between two adjacent fingers.

Of course, when an object is grasped, the virtual fingers moving along the opposition axis may come to rest their opposing surfaces on the object between them, rather than making direct contact with each other. This makes possible a theory of preshaping. One task of vision is to determine, from the retinal input, an opposition space embedded in the object that is to be the target for the positioning of the appropriate opposition space of the hand. As shown in Fig. 4.3, there is a safety margin extending the opposition space beyond the boundaries of the object (see Arbib, Iberall, & Lyons, 1985, 1987 for a discussion of the crucial role of the safety margin in the analysis of movement). Preshaping forms the hand so that the opposing surfaces of the virtual fingers are correspondingly separated. An approach vector determines the orientation and distance of the hand relative to the object. Figure 4.3 shows how to characterize this. A virtual

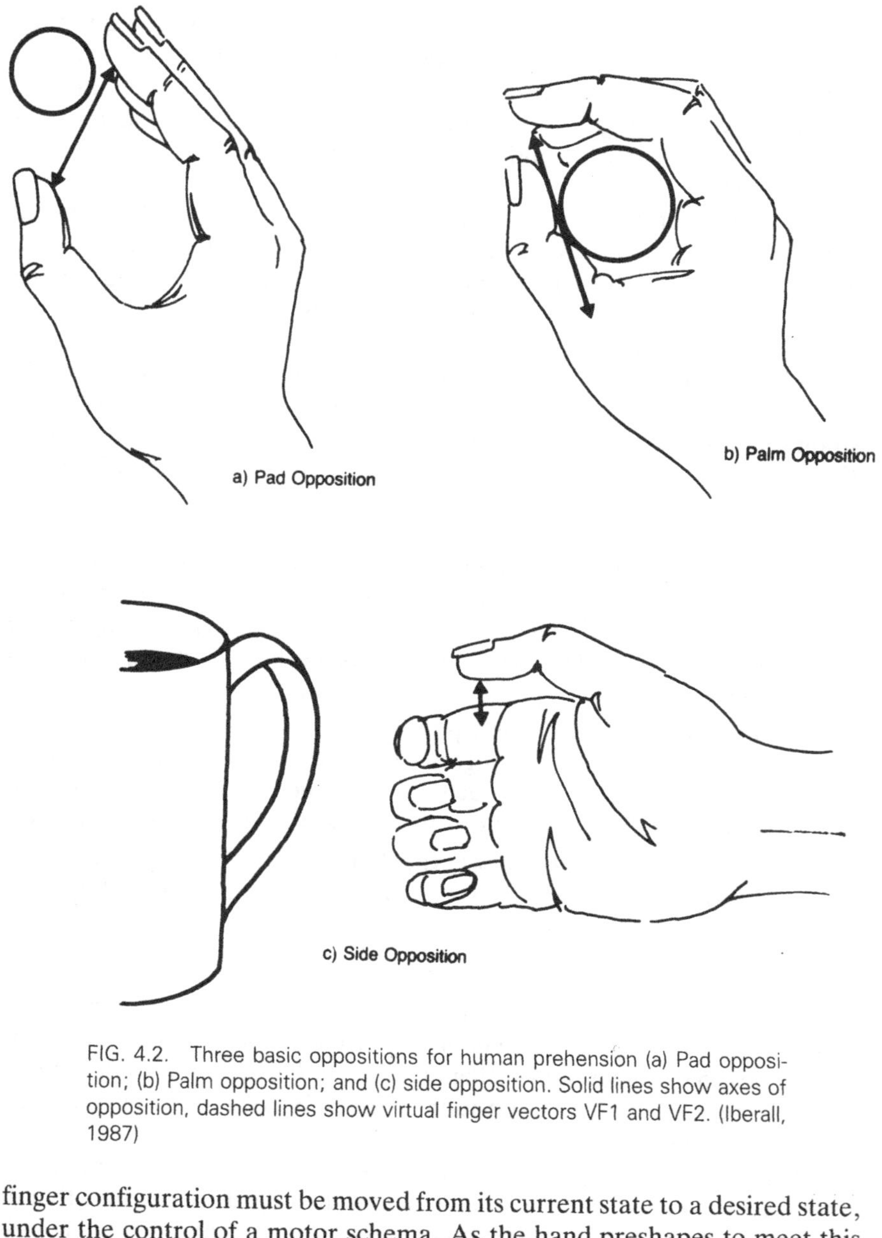


FIG. 4.2. Three basic oppositions for human prehension (a) Pad opposition; (b) Palm opposition; and (c) side opposition. Solid lines show axes of opposition, dashed lines show virtual finger vectors VF1 and VF2. (Iberall, 1987)

finger configuration must be moved from its current state to a desired state, under the control of a motor schema. As the hand preshapes to meet this specification (the Preshape Schema), the arm transports it (Move Arm Schema) and the wrist rotates it (the Orient Schema) to approximately the correct position. Note that *the target position of the wrist depends on the proposed preshape of the hand*—the arm controller must know the offset

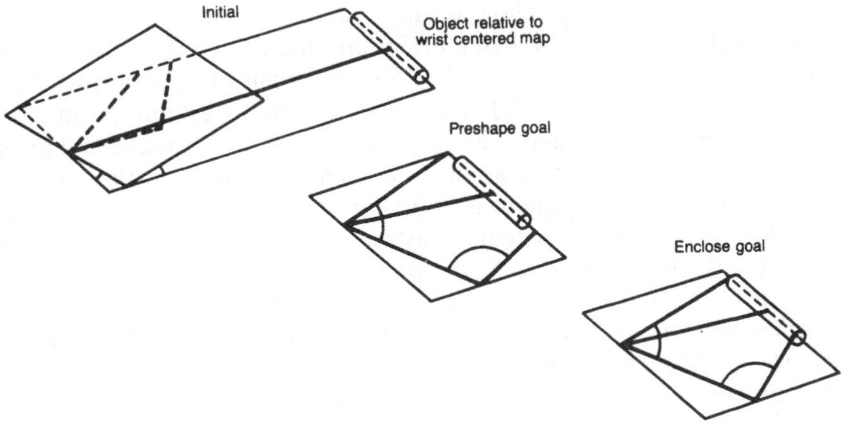


FIG. 4.3. The opposition axis embedded in the object sets goals for the preshape of the hand. After the opposition axis of hand and object are aligned, enclosing can take place. (Iberall, 1987)

of the wrist from the center of the opposition space as embedded in the object if it is to transport the hand successfully to its goal. Thereafter, the position of the hand is adjusted to align the two opposition spaces (that in the hand with that in the object) and the virtual fingers then close along the aligned axes to firmly grasp the object under tactile control (the Enclose Schema).

What gives this language of opposition spaces its expressive power is that the hand is not limited to one opposition at a time. For example, in holding a screwdriver to lever open a can, a simple palm opposition alone suffices, whereas in using it to turn a screw, the thumb and index finger form a side opposition to direct the shaft, and the other fingers grasp the handle in a palm opposition. However, expressive though this language may be, it is not exhaustive. It does not describe the hand movements involved in playing the piano or using sign language. Yet it does describe a *vast* range of manipulations and grasping, and opposition space provides a theory of hand movement that *explains* preshaping (Iberall, 1987), rather than merely describing it as before.

## SCHEMA ASSEMBLAGES

The present section outlines the **RS** language for robot schemas—not as an end in itself, but rather as a means to clarifying the notion of an assemblage of schema instances, which is now developed in a general form that includes the notion of coordinated control program. The next section then addresses the issue of how assemblages might be embedded in artificial neural networks or in the living brain, and the final section offers neural networks for specific schemas. (For further details of **RS** itself, see Lyons 1986, Lyons & Arbib, 1989). The aim is simply to extend the intuitions provided by Fig. 4.1. There we saw each schema as a box that could receive inputs and generate outputs—in a moment we formalize this by saying that the semantics of a schema is a port automaton. However, it is important to add that, whereas a schema constitutes the "long-term memory" of a perceptual and/or motor skill, the process of perception or action is controlled by an active copy of the schema, called a *schema instance*. For certain behaviors, there may be no distinction between schema and instance—a single neural network may embody the skill memory and provide the processor that implements it. However, in more complex behaviors, the different mobilizations of a given "skill unit" must be carefully distinguished. I thus formalize a schema as an automaton that is augmented with instantiation and deinstantiation operations. This is to capture the notion that, as action and perception progresses, certain schema instances need no longer be active (they are deinstantiated), while new ones are added as new objects are perceived and new plans of action are elaborated (schemas are instantiated as new schema instances). Two more points in preparation for what follows: A schema assemblage formed as a network of schema instances itself may be considered a schema for further processes of assemblage formation; and the network itself may be dynamic, growing and shrinking as various instantiations and deinstantiations occur.

A *port automaton* generalizes the concept of a discrete-time system (Fig. 4.4a). Such a system is characterized by sets $X$ of inputs, $Q$ of states, and $Y$ of outputs, together with a state-transition function $\delta;Q \times X \rightarrow Q$ and an output function $\beta;Q \rightarrow Y$. Each time the system is in state $q$ and receives input $x$, it changes state to $\delta(q,x)$ and emits output $\beta(y)$. A port automaton is generalized from this as follows (Fig. 4.4b): Instead of one input line, it may have several input ports. Instead of one output line, it may have several output ports. As we see later, the reading of inputs and writing of outputs are more subtle than the $\delta$ and $\beta$ of Fig. 4.4a. Moreover, these ports are to be thought of as linked to communication channels that may or may not contain a message en route from an output port to some input port (Fig. 4.4c)—I adopt here the convention that each channel can hold at most one message. For a port automaton, then, $\delta$ and $\beta$ are replaced by a set of transitions.

(a configuration of messages on some of its input ports, an old state)

$\Rightarrow$(a new state, a configuration of messages on some of its output ports)

The port automaton changes state when it can apply one of these quadruples, and because there are many ports, more than one may be applicable at a given time, that is, the automaton is nondeterministic: If the given

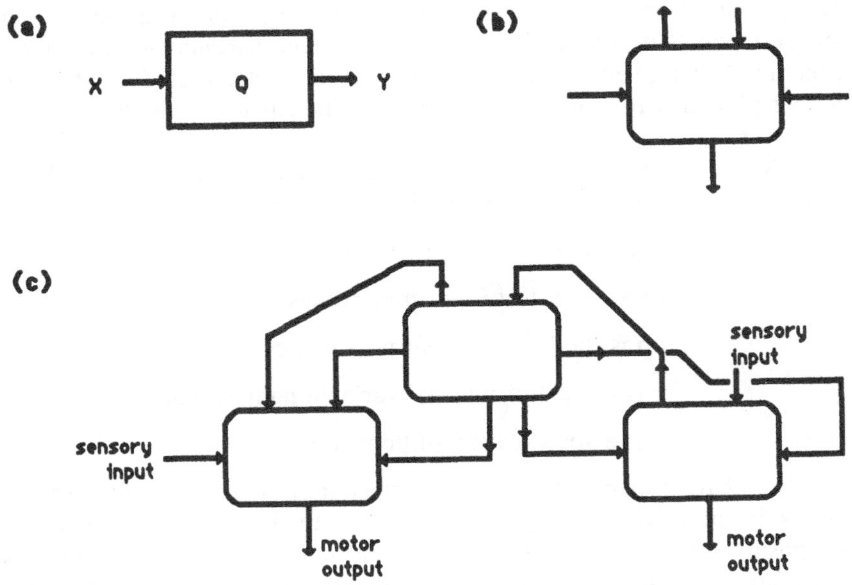


FIG. 4.4. (a) A discrete-time automation. (b) A port automation. (c) A port automation embedded in a network. Each channel from an output to an input port can hold at most one symbol at a time.

configuration of messages appears on its input lines, the automaton is in the specified "old" state, and the channels connected to the necessary output lines are free, then the automaton will read the messages from the specified input ports (thus "emptying" the attached channels, but leaving other input ports unaffected), change to the new state, and write the output messages to the specified output ports.

The semantics of a schema is a port automaton augmented with instantiation and deinstantiation operations: In addition to the above transitions, the augmented port automaton may execute transitions of the form

(a configuration of messages on some of its input ports, an old state)

⇒(a new state, W)

where W is a specific instance of either an instantiation operation or a deinstantiation operation.

More specifically, think of a *schema* as the formal specification of such an automaton, whereas a *schema instance* is an "active" copy of this schema, in a particular state, with each port connected to that of another schema instance or to the "external world." The *instantiation operation*, then, can only be executed by an instance of a schema. It forms a new schema instance by instantiating a schema with specified initial values for internal variables, and a connection map specifying connections of ports on this schema instance to ports on other schema instances. The *deinstantiation operation* specifies that a given schema instance is to be deinstantiated, so that it no longer plays any role in the transactions of the current network of schema instances.

In **RS**, a schema may be either a basic schema or a schema assemblage. The syntax of a *basic schema* definition is [N (ip) (op) (v) (b)] where

N is the identifying name for the schema

ip is the list of input port names

op is the list of output port names

v is the list of internal variable names; and

b is a specification of behavior.

The *behavior specification* makes explicit how an instance of the schema reads inputs, writes outputs, changes internal state, and executes (de)instantiations. In **RS**, it is a sequence of statements forming a Pascal-like program augmented by instantiation and deinstantiation operations and "Forall" statements. The Pascal-like program comprises assignments that either evaluate an expression based on internal variables and/or values read in on input ports to update an internal variable, or evaluates such an expression to write a value to an output port. It also contains conventional conditional statements to link the assignments via appropriate tests.

Instantiation and deinstantiation are novel. *Instantiation* creates a new instance of a schema and couples its ports to the extant schema instance in the specified way. It thus names the schema to be instantiated, and pairs certain input ports of the instance to output ports of the schema instance calling for the instantiation, and vice versa. *Deinstantiation* names a schema instance and simply removes the specified instance from the schema instance network. Another novel instruction is the *Forall* (which we do not discuss further here) that allows all instances of a given schema to simultaneously instantiate their own similar but private schema instances.

The above definition establishes a basic schema in terms of an explicitly programmed behavior specification. We also build up schemas recursivley. An *assemblage* is a schema in which the behavior is defined in terms of the behavior of a network of communicating schema instances. The syntax of an assemblage definition is [N (ip) (op)(v) (s)(ib)(p)(n)] where

N is the identifying name for the assemblage

ip is the list of input port names

op is the list of output port names

v is the list of internal variable names

s is a list of component schemas

ib lists the commands to form instances

p renames the ports of the component schema instances as ports of the assemblage,

n defines the port connection mappings between component schema instances.

The idea is this: Because several copies of a given schema in the list *s* may be used in defining the network that constitutes *N*, *ib* is required to spell out and name just which instances are needed. Some of the ports of the resulting network will serve as ports of the newly defined assemblage. *p* then provides the facility to give those ports names that can be used in specifying communication with the assemblage, and that differ from their original names (this is especially important when we must distinguish corresponding ports from different instances of a given schema). Finally, *n* makes explicit the cases where the output port of one schema instance must be connected to an input port of another in forming *N*.

Where our analysis of perceptual structures and distributed motor control (Arbib 1981) distinguishes a *schema assemblage* of perceptual schemas as the basic perceptual structure from a *coordinated control program* for distributed motor control, **RS** offers a single syntactic structure, the assemblage, that enables us to construct both types of compound schema. This tells us how to put schemas together in a way that does not depend on how the behavior specification is given, whether directly as in a basic schema, or indirectly, as when the schemas constituting an assemblage are themselves schema assemblages. It is thus simple to extend **RS** to a neural net specification by making it possible, in the *basic* schemas, to define the behavior directly in terms of a neural network rather than a Pascal-like program. Of course, this defines a research program rather than completes it, for it is an open question whether the neural analogs of assemblage formation in **RS** can be found. We offer an initial approach to this question in the next Section.

## MOTOR SET AND THE NEURALIZATION OF SCHEMA ASSEMBLAGES

We have now distinguished a *basic schema* (and seen that it may be instantiated in a neural network) from a *schema assemblage*. But how are we to relate these notions to—or constrain them by—data on the brain? Iberall and Arbib (1989) have postulated that the various functions of Fig. 4.1 may, after their refinement in terms of our subsequent theory of preshaping based on opposition space (Fig. 4.3), be distributed in the primate brain as shown in Fig 4.5:

1. Area 7 sends task features to Area 5, cues the Approach-Vector-Selector Perceptual Schema (hypothesized to be in FEF) and cues the Orient-Trunk and Move-Arm Schemas (hypothesized to be in the premotor area PM).

2. The Frontal Eye Fields FEF (Areas 8 and 9) are posited not only to monitor eye movements but also to collect visual information about objects from peristriate cortex and Area 7. They seem able to suppress reaching movements until the trunk is oriented correctly. We hypothesize that FEF serves as the approach vector selector, with Area 7 as requestor.

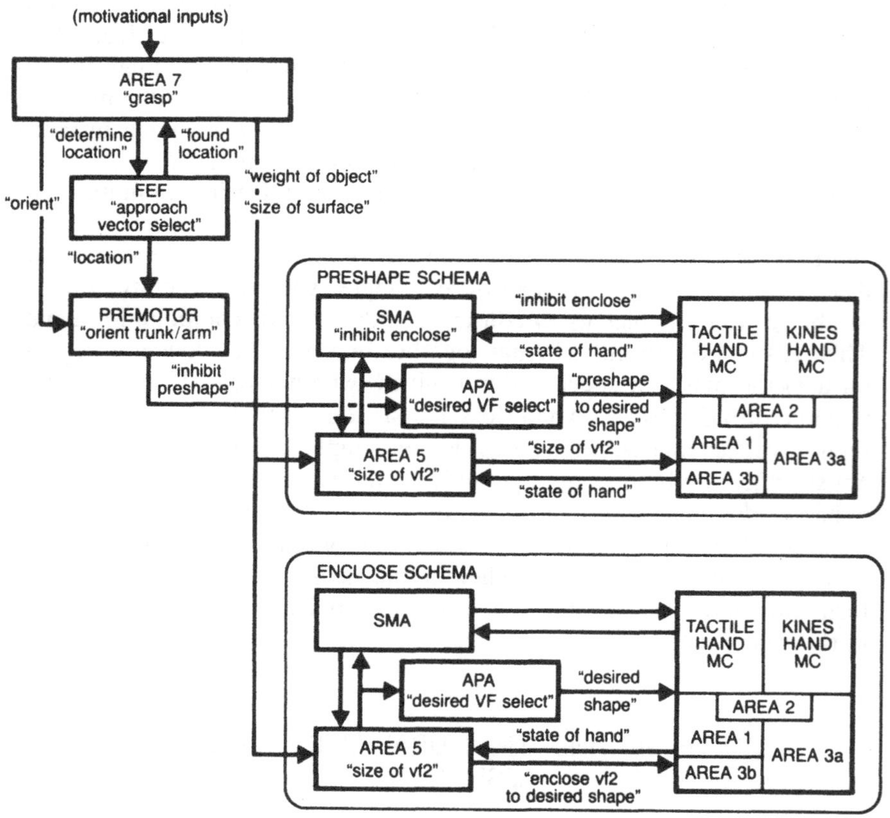


FIG. 4.5. Hypothetical mapping of the coordinated control program for grasping onto cerebral cortex (subcortical contributions are not included). We identify the roles various cortical areas might have in prehensile movements, but especially in pad opposition. (Iberall, 1987)

The output of FEF would be the desired wrist location to be sent to PM for orienting arm and trunk. To do this, FEF would maintain a map of object locations and suppress reaching until the body is oriented; at which time it would trigger the Orient-Trunk and Move-Arm schemas (PM).

3. Area 6 comprises PM and the supplementary motor area SMA. PM seems to be involved in the transport component of reaching using FEF to direct proximal areas of the arm. A large hand region in PM known as the APA, the arcuate premotor area, is involved in the grasping component. Lesion of SMA yields tactile following, an inability to let go of objects. We thus posit that SMA inhibits enclosing as part of the Preshape Schema; whereas SMA, APA, and Area 5 work as a unit for gating tactile and kinesthetic information through the transcortical somatosensory cortex/motor cortex loop.

4. We suggest that Area 5 with the APA can gate kinesthetic information from somatosensory cortex to motor cortex to flex the fingers during the closing of the hand. After contact is made, Area 5 and the APA can then gate the cutaneous information to motor cortex for tactile following of the object.

5. We postulate that APA, SMA, and Area 5 contribute to both the Preshape and Enclose Schemas. APA facilitates preshaping, using virtual finger size information from Area 5 to gate desired VF configurations to motor cortex. We thus posit that APA contains knowledge about opposition space, using desired wrist location to compute the desired preshape state based on VF information from Area 5. SMA ensures success of the preshape by inhibiting gating of finger flexion information to cortex until the wrist is correctly oriented and located. When the wrist is correctly located, inhibition is released and the Preshape Schema closes the fingers using VF information from Area 5.

Here, then, the connections in the schema assemblage are exactly those given by the anatomy—the schema assemblage is simply a description of the innate wiring of a number of subsystems in the brain. However, what I want to emphasize here is that in **RS** the different schemas may be precompiled, and that *separate code* is required to specify how the activity of these schemas is to be coordinated in a schema assemblage or coordinated control program—as in the assemblage specification [N (ip) (op)(v) (s)(ib)(p)(n)] of the previous section that calls on the prior specification of the schemas in the list *s*. The lesson for brain theory is that we must distinguish (without denying that they might overlap) the circuitry that implements the schemas from the circuitry that implements the "assemblage code" that specifies how these are coordinated. In some cases, as in Fig. 4.5, such "code" is really implicit—the simple linkage of the output of one brain region to the input of another provides fixed "port connections" (to return to the language of Fig 4.4c) that **RS** describes but that *in this case* do not require explicit circuitry to implement. However, when we learn a skill, we learn to deploy schemas that are already available—to provide an assemblage that may then be tuned by experience. The latter point shows where the instantiation in neural nets takes us beyond the reach of any programming language. The language tells us how to assemble a network for a task; but synaptic plasticity may then reshape the neural net to achieve properties on the basis of experience that were not explicit in the original high-level specification. We thus turn now to a body of data on "motor set" that suggest one way in which patterns in one brain region (implementing the assemblage code) may coordinate schemas implemented in the neural circuitry of other brain regions.

In describing learned behavior, the schema assemblage is to express the way in which a number of already available skills (expressed as perceptual or motor schemas) may be combined, on the basis of instruction and/or experience, in diverse ways to constitute a variety of novel skills. Here the *effective connectivity* between the neural networks implementing the schemas *is to be changed depending on which skill is being exercised*; moreover, given our stress on instantiation and deinstantiation, that pattern is to change dynamically during a single episode of skilled behavior. The issue, then, is how—going beyond the case of fixed connections—the connections between a set of neural circuits can be under the dynamic control of neural activity (probably elsewhere in the brain) encoding a schema assemblage. I thus turn to a *preliminary* account, based on Dakin and Arbib (1986), of one way in which this might be done. The starting point is the notion that the coordinated interweaving of motor schema activations might be enforced through the neural mechanism of *motor set*.

Evarts and Tanji (1976; see also Evarts, Shinoda, & Wise, 1984) monitored premovement neural activity in the primary motor cortex of monkeys. They found evidence that sensory information is selectively gated to pyramidal tract neurons (PTNs), depending on the premovement signals from *set cells* that relay the current intention of the animal to primary motor cortex. The set cells, located perhaps in area 6, a high-level motor center, are able to facilitate or block, depending on the desired movement, sensory input to primary motor cortex regions concerned with specific movements. An example of this neural gating mechanism is illustrated in Fig. 4.6 that uses single cells to represent cell populations in the monkey brain. Here SI is not "schema instance" but "somatosensory area one," the primary area of cortex for receipt of tactile and kinesthetic information; correspondingly, we follow the neuroscientist's usage of MI for primary motor cortex. When set cell A is highly active, it stimulates an inhibitory neuron that blocks the flow of tactile information (from somatosensory

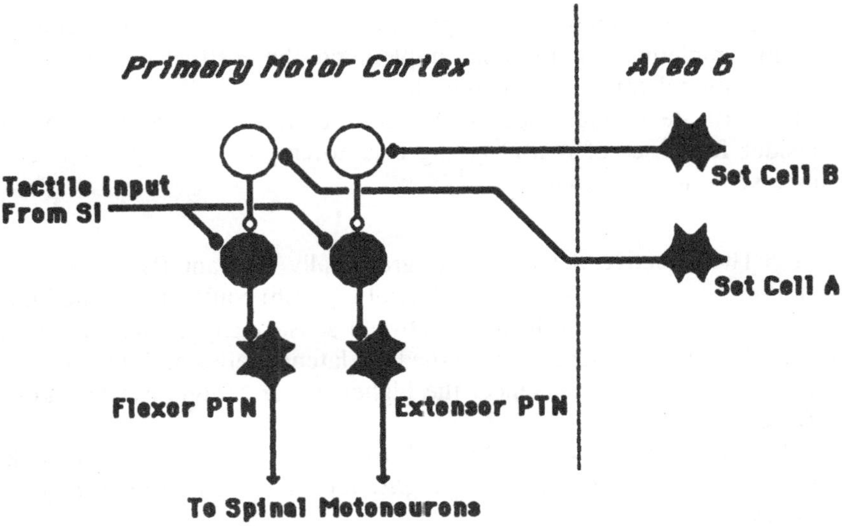


FIG. 4.6. Neural gating in motor cortex.

cortex SI) to the arm flexor PTN (or, more properly, the arm flexor *pool* of PTNs, pyramidal tract neurons, in MI). Because sensory input to the arm extensor PTN remains unchecked, a tactile stimulus then produces an arm extension. On the other hand, if set cell A is dormant and set cell B is active, the arm extensor PTN's access to the sensory input will be blocked, leaving the arm flexor PTN's input open, and a tactile signal triggers an arm flexion. We thus see how set cells can change the effective connectivity of the underlying circuitry for tactile control of arm movement.

In what follows, then, we view each basic motor schema as having its own neural representation, and ignore the problem of concurrent activity of multiple instances of a given schema. We explore the hypothesis that higher motor centers implement an assemblage/coordinated control program by controlling the activity of motor schemas through the regulation of sensory information to each motor schema's neural representation. (I explicitly focus on motor control here. The building up of a complex perceptual representation appears to involve quite different processes in which multiple instances of a given schema must be coordinated, as when we see a scene with several houses.) Through the same neural gating mechanism, the higher motor centers might also regulate excitatory or inhibitory flow among the schemas.

It is the pattern established in the neural circuitry of the "higher center" that encodes the schema assemblage; it is the interaction of these patterns with the circuitry encoding the schemas that enables the brain as a whole to execute the assemblage. A (possibly dynamic) "port connection" to circuitry implementing a schema implements the abstract use of pointers in conventional computer languages.

To illustrate various aspects of such "administrative control" of MI, consider four uses of neural gating in enforcing an orderly sequence of submovement executions:

1. The selective denial of topographically relevant PTN's access to different sources of sensory input (recall Fig. 4.6) could enable the higher motor centers to establish loops between sensory trigger stimuli and arbitrary functional synergies, as in the long latency reflex and the intention-related grasping reflex. Hence, the higher motor centers can "program" reflexes, as needed.

The crucial point here is that rich excitatory pathways exist to link a large variety of circuits; *much of higher level control is provided by "inhibitory sculpting."*

2. The mutual exclusion of any given pair of motor schema activations could be "implemented" by setting up, via neural gating, symmetric inhibition between the schemas' networks. Conversely, schemas can be made to "cooperate" by establishing lines of excitatory communication (or by blocking lines of inhibitory communication). If the circuits embodying the motor schemas' characteristic patterns overlap, this cooperation would lead to a blending of the sites of two patterns, to form a composite pattern and movement. In this case, the blending of the two prototypical movements would take place in MI. If, on the other hand, two schemas' patterns are topographically distinct, the blending of movements would take place at the periphery.

3. If asymmetric inhibitory communication is set up between two motor schemas' patterns of activity, the recipient schema will be inhibited whenever the other is activated. Thus, the higher motor centers could enforce the sequential ordering of any pair of submovements by priming their schemas for activity and by establishing an asymmetric channel of inhibitory communication between them.

4. The higher motor centers would be in a position to leave much of the decision making regarding movement selection to the intermediate motor structures, such as MI, effectively easing the planning burden of the former: By priming of a chosen set of motor schemas, either through direct neural excitation of PTNs or by the gating of sensory stimuli to PTNs, coupled with the establishment of inhibitory lines of communication among those motor schemas (to ensure that only one schema activates), the higher motor centers may initiate competition between a set of "favored" motor schemas, leaving the final decision to the intermediate motor level. This

distributes the processing of sensorimotor information through the motor hierarchy, allowing efficient use of resources through parallelism.

It should now be apparent that the interweaving of motor schema activations can be accomplished through various neural gating tactics involving the temporary establishment of excitation or inhibition among the patterns of neural activity corresponding to different motor schemas. We might envision the motor hierarchy as a directed graph structure, with coordinated control programs for complex skills parenting low-level coordinated control programs and motor schemas. Each coordinated control program has a characteristic pattern of neural activity that involves the imposition of motor set on the neural structures subordinate to the coordinated control program, effectively interweaving the activations of "child" coordinated control programs, and ultimately, the basic motor schemas. What remains to be discovered is how the various assemblages that exercise such "restructuring" are themselves acquired and updated through experience. We also need simulation studies to probe what patterns of activity in the competing networks do and do not trigger action, and how the "decision time" for an act increases with the complexity of competitive interactions.

An important distinction between the hierarchical organization of motor activity and the perceptual organization of complex object categories, of course, is the temporal–sequential nature of a skilled action. A skill's motor hierarchy unfolds in the course of its submovements, instead of "all at once" during the preparatory phase. Indeed, during the execution of movement sequences, the motor hierarchy is in a constant state of flux. Motor schemas in the hierarchy do not activate until necessary, and each motor schema remains active only as long as necessary. There is now evidence, in fact, that neural activity corresponding to the upper echelons of the motor hierarchy subsides whereas lower structures remain active and carry out the actual movements:

Tanji and Kurata (1985) trained monkeys to perform key presses in response to different stimuli. In the tactile mode, an instruction stimulus (the Tactile Instruction) warned the animal to perform a key press in response to a tactile trigger stimulus (Tactile Stimulus) and to ignore an audio signal. In the audio mode, a different instruction warned the animal to respond instead to an audio stimulus and to ignore a tactile signal. The experimenters recorded the activity of neurons in MI and area 6 during the trials. In area 6, 128 neurons responded to the instruction, and of these, 44 responded with greater magnitude to one instruction than the other (i.e., some neurons responded to the tactile instruction but not the auditory instruction, and vice versa). All area 6 neurons that responded to an instruction resumed normal activity when the stimulus that triggered the key press arrived. Thus *the activity of neuronal populations associated with the high-level establishment of MI's motor set dissipated at the start of the movement's execution*, their influence being no longer required. Neurons in primary motor cortex responded to the instruction without preference to one instruction or the other.

Figure 4.7 illustrates the schema hierarchy for a schema-theoretic interpretation of these results: At level 1, we have the key press schema. At level 2, we have the Tactile coordinated control program (shown as Tactile CCP) that links tactile stimulation to the key press schema, and the Auditory coordinated control program linking auditory stimulation to the key press schema. At level 3, we have the Trial coordinated control program that can be activated (in ways not shown here) in response to the organism's goals and/or the environmental "situation."

Whereas the definition (learning) of these assemblages flows up the hierarchy, the activation flows down. (In the present example, we assume that instantiation of a schema or assemblage is identical with the attainment of some threshold level of distributed activity in the neural circuitry which embodies it). Activation of the Trial coordinated control program opens channels of communication from the source of the anticipated tactile instruction to the Tactile coordinated control program, and from the source of anticipated auditory instruction to the Auditory coordinated control program. In Fig. 4.7, arcs A and B represent the axons of set cells whose activity establishes the sensory communication pathways to the Tactile and Auditory coordinated control programs, respectively, via neural gating. Note that at this point, the active motor hierarchy for the task we are

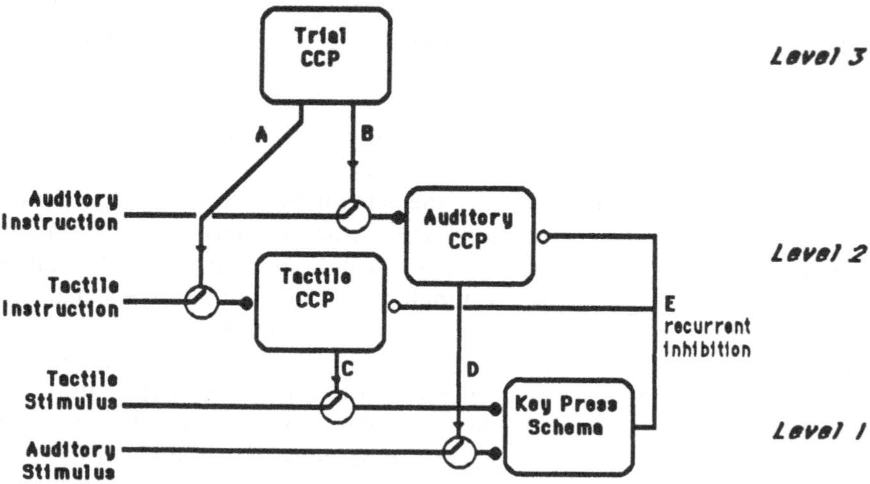


FIG. 4.7. Schema hierarchy for the Tanji-Kurata Experiment.

considering consists of only the Trial coordinated control program, the other components not yet having been activated. If a tactile instruction now occurs, the tactile input to the Tactile coordinated control program raises that coordinated control program's activation level above its activation threshold, allowing the Tactile coordinated control program to open (via arc C) a pathway of communication from the source of the tactile stimulus to the Key Press motor schema. The latter schema, upon receipt of the tactile stimulus, finally activates, producing the key press movement. An auditory stimulus, in this case, would not evoke any movement since a pathway has not been opened to the motor schema.

Alternatively, the event of an auditory instruction activates the Auditory coordinated control program, which establishes (via arc D) a pathway of communication from the auditory stimulus source to the KEY PRESS motor schema. Now, the event of an auditory stimulus, but not a tactile stimulus, will lead to the activation of the KEY PRESS motor schema, eliciting a key press response.

Continuing our schema-theoretic analysis of the Tanji–Kurata experimental data, the preferential responses of the area 6 neurons to one instruction or the other are interpreted here as having been due to the separate coordinated control programs (Tactile and Auditory) that activated in response to the respective stimuli. The nonpreferential nature of the primary motor cortex neurons to the different stimuli follows from the fact that whichever coordinated control program activated (Tactile or Auditory), the anticipated stimulus was channelled to the same motor schema.

Finally, we consider the evidence that coordinated control programs deactivate in the course of movement execution. Recall that 128 area 6 neurons resumed normal activity upon the arrival of the stimulus. Referring again to Fig. 4.7, this may be interpreted as follows: once the stimulus arrives and the motor schema activates, the motor schema suppresses the parent coordinated control program through inhibition (arc E), deactivating the coordinated control program, which has fulfilled its role. Whereas the descending arcs A, B, C and D are shaped by learning, it may be that the arcs of E represent an innate feedback path for broadcasting a "reset" to Area 6 whenever execution of the next step of a plan is initiated.

## ARTIFICIAL NEURAL NETWORKS FOR MOTOR CONTROL

Having briefly discussed a range of neurophysiological data to suggest how schema assemblages may be encoded in the brain, I devote this last section to a number of studies by Iberall and colleagues on the use of neural networks both to analyze data on human hand movements and to suggest strategies for robot hand control. Iberall (1987, see also Iberall & MacKenzie, 1988) conducted a pilot study in which the task of grasping a 4-cm dowel was analyzed. To compute the goal wrist location, Iberall used a pair of excitatorily coupled neural networks (House, 1984 used a similar structure for his Cue Interaction model for depth perception). Each network was a "winner take all" nonadaptive network consisting of a homogeneous two-dimensional excitatory neural field interacting with a one-dimensional inhibitory neuron pool (Amari & Arbib, 1977). Each node in the excitatory neuronal pool represented a location for the wrist in the space around the object. The nodes competed with each other in order to target the wrist to a good location for contacting a 4-cm dowel with two virtual fingers in pad opposition. The analyses for VF 1 (Virtual Finger 1, the thumb) and VF2 were separated: the left field of Figs. 4.8 and 4.9 represents the kinematic constraints of VF1, and the right those for VF2. The inhibitory processes constrain each field to converge to a single peak of activity; excitatory cross-coupling between corresponding points of the two fields allows gradual build up of an agreed upon solution between the nets. This solution signalled a "good" wrist location, that is, one where: (a) the Virtual Fingers could reach the ends of the dowel; (b) the pads were aligned with each other (as a rough approximation of how precision grasping occurs); and (c) the pads were aligned with the dowel. Some locations had more than one solution because the system is underconstrained. Thus, the input arrays are generated as a function of the present relation between the subject's shoulder and the opposition vector embedded in the target object. The graph of number of solutions for VF1 for each wrist position provides the network input shown on the left of Fig. 4.8; similarly for VF2 and the right-hand array. The generation of the input graphs of Fig. 4.8 is currently based on tabulated laboratory data; its automatic generation by a trained neural net is a matter for current research.

In each "winner take all" network, possible wrist location solutions compete, with distances and orientations outside the range of the currently targeted opposition space (the one that determines the current input arrays) dying out, and virtual finger configurations that correspond to an effective posture for the given opposition vector reinforcing each other. Starting from an initial zero excitation level, a peak forms indicating a solution for the wrist location that is within the constraints modeled (Fig. 4.9). In order to make the system converge, it was found necessary in tuning the cross-coupling to place stronger weights on the VF1 pool than on those of the VF2 pool. This supports the behavioral results of Wing and Fraser (1983) showing the thumb to be more constrained.

Although the above neural network is explicitly defined for its task, there is much interest in adaptive neural networks that can generalize a useful input–output transformation from a few sample points. Iberall (1988)

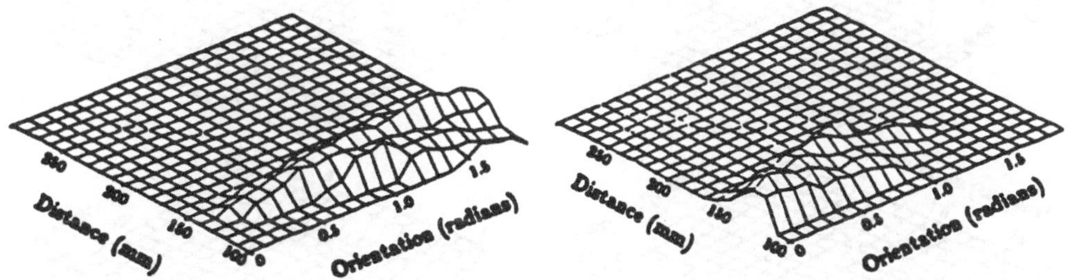


FIG. 4.8. The Approach Vector Selector involves the cooperation of two neural networks, each of which is a maximum selector for a 2-D excitatory array. This figure shows the input to the two excitatory arrays: the one on the left for VF1 (the thumb), that on the right for VF2. For a desired opposition, the corresponding graph shows the number of possible finger configurations for the given finger in the desired opposition when the wrist is in the indicated (distance, orientation) position.

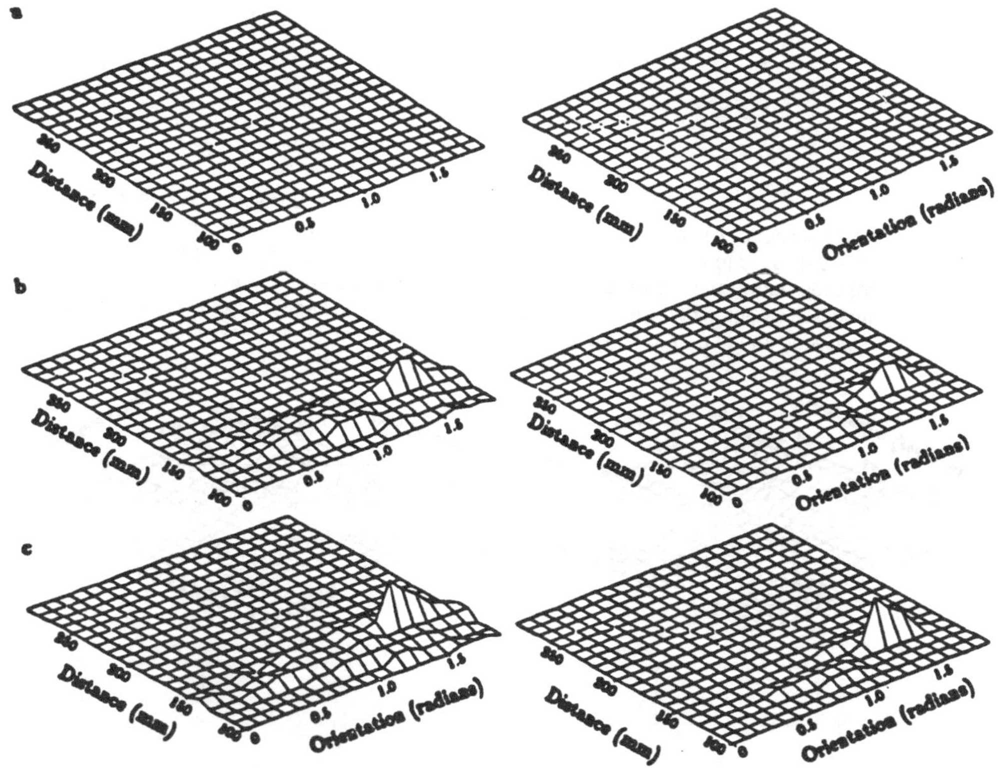


FIG. 4.9. This figure shows the time course of activity in the cross-coupled maximum selectors when driven by the input of Fig. 4.8. The activity for VF1 is shown at the left, that for VF2 at the right. (a) The initial quiescent states of the excitatory fields. (b) After 11 time steps, peaks are seen slowly developing. (c) At convergence, a peak is seen in both fields signalling the choice of wrist distance 140 mm and orientation 1.4 radians.

simulated a neural network for choosing an opposition space on the basis of input that encoded two object constraints (the length of the surface and magnitude of the opposition vector) and two task constraints (the precision needed in the task and the magnitude of anticipated forces). Two networks were used: the first to choose between pad and palm opposition, and the second to determine which real fingers should map into VF2. An iterative learning process of a few examples showed that it took hundreds of trials to assign the correct oppositions and mapping of virtual to real fingers given these inputs.

In order to solve this problem using adaptive neural networks, a set of data points were chosen, consistent with empirical evidence (Fig. 4.10). The values for input and output parameters are relative to the capability

| | INPUTS | | | | OUTPUTS | |
|---|---|---|---|---|---|---|
| Task Example | Surface Length | Object Width | Amount Forces | Task Prec. | Chosen Oppos. | VF2 Mapping |
| 1. Lift heavy beer mug | Gtr than four fingers | Large | Large | Little | Palm | imrl |
| 2. Place cylinder | Gtr than four fingers | Medium | Small | Lots | Pad | im |
| 3. Lift small disk | One finger | Small | Small | Little | Pad | i |
| 4. Lift med. disk | One finger | Medium | Small | Little | Pad | i |
| 5. Place med. disk | One finger | Medium | Small | Lots | Pad | i |
| 6. Throw med disk | One finger | Medium | Medium | Little | Pad | i |
| 7. Lift long steel cyl. | Gtr than four fingers | Medium | Large | Little | Palm | imrl |
| 8. Lift short steel cyl. | Three fingers | Medium | Large | Little | Palm | imr |
| 9. Place wide short steel cylinder | Two fingers | Large | Medium | Lots | Pad | im |
| 10. Lift large glass | Gtr than four fingers | Large | Medium | Med. | Pad | imr |

FIG. 4.10. The training set for OPPOSITION SELECTOR and VF-MAPPER specifies the appropriate opposition and set of fingers for inclusion in VF2 for each of 10 tasks on the basis of 4 measured characteristics. (i,m,r, and l denote the index, middle, ring and little fingers.)

of the hand (e.g., length is determined in terms of the number of real fingers that can fit along the surface length at the opposition vector, forces are relative to the magnitude of force that the hand can supply, and so on). The first data point is the example task of lifting a heavy beer mug by its body. In this case, the surface length is greater than four fingers of a hand, the width of the mug is generally large relative to the span of the hand, and the weight of the mug is relatively heavy. The lifting requires a large translation upward, involving little precision. The most reasonable posture, then, is a palm opposition involving all four fingers mapped into VF2.

The second task in Fig. 4.10 is the place cylinder task from Iberall, Bingham and Arbib (1986). In this task, subjects grasped a medium sized cylinder standing on its end. They used pad opposition with the index and middle fingers mapped into VF2, even though the cylinder was longer than the hand was wide. Because the object was light and not very wide, this was a sufficient posture for the task. The third and fourth tasks in the table have been studied by Marteniuk et al. (1987). In one experiment, subjects were asked to lift two different width disks. The task requirements (light object, little precision) dictated pad opposition, while the small height of the disk forced a one finger wide VF2. In the same study, Marteniuk et al. vary the task while maintaining the same object, asking subjects to either place or throw the disk. Although the first task used more precision than the second, the chosen opposition space remained the same.

The final four tasks in Fig. 4.10 were developed from commonly performed tasks. When forces are high in a task (either from the weight or from inertial forces), palm opposition is used with as many fingers as possible—Iberall et al. (1986) noted how the subject made a transition from a pad opposition used to lift an object to a palm opposition for using it in a task involving high inertial forces. However, when the object's width is a constraint (as in task 8), fewer real fingers are mapped into VF2. As well, tasks that involve short objects used in precise ways still involve pad opposition, even if they are somewhat wide and heavy. Finally, task 10 involves lifting a large object, somewhat heavy, using some precision. In this case, pad opposition is used, involving three of the four fingers (see Wing et al., 1986 for related data.)

These data provided a training sample for a network to perform supervised learning using back-propagation (Rumelhart, Hinton, & Williams, 1986). A multilayered network of simulated neurons is used in which input neurons connect to hidden layers of neurons that in turn are connected to one or more output neurons. Familiar algorithms (e.g., the Perceptron error-correction rule of Rosenblatt) tell how to adjust weights of connections to output neurons in such a way as to cumulatively reduce errors in output values—if such a weight setting exists. What the back-propagation algorithm adds (exploiting the use of a continuous, rather than binary, output function for every neuron) is a rule for apportioning the error of a neuron among those neurons from which it receives inputs. This allows the application of error-correction rules to successive layers of hidden units further and further back from the output units for which errors can be directly observed. As a result, each iteration of network activity involves two phases: Computation in the neural network first proceeds forward to yield overall output, then the output is evaluated and error correction propagates backward to update the weight settings.

In the simulation, two multilayered networks were used. The first one, OPPOSITION-SELECTOR (Fig. 4.11), determines if pad or palm opposition should be used, based on the requirements of the task. The second network, VF-MAPPER (Fig. 4.12), performs the virtual to real finger mapping for VF2 (the mapping for VF1 is fixed based on the selected

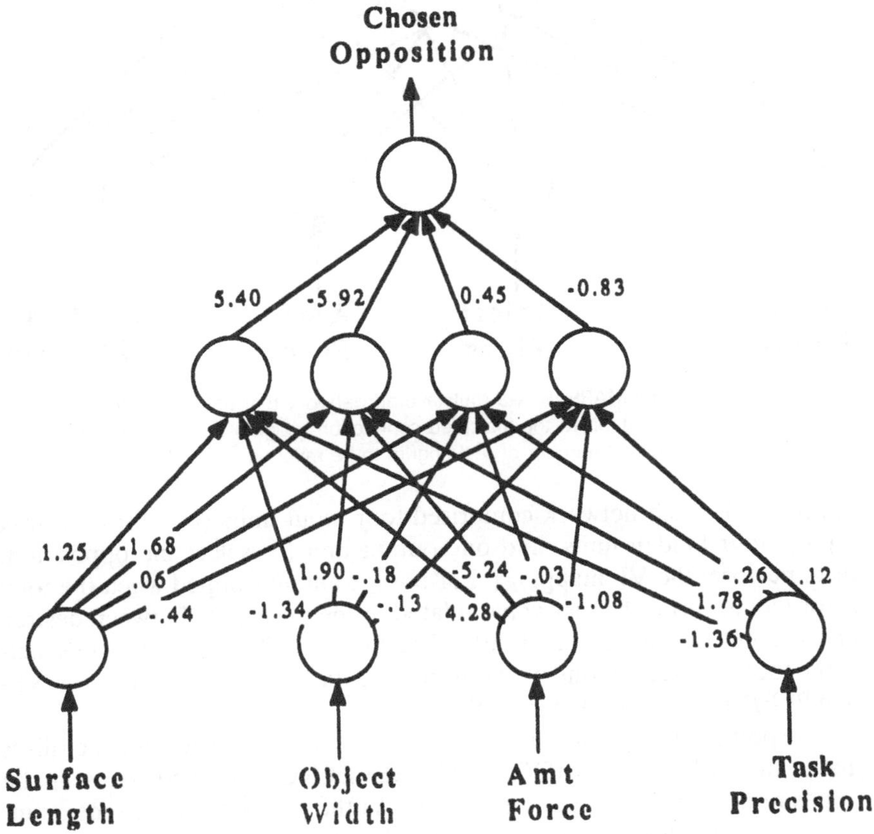


FIG. 4.11. OPPOSITION SELECTOR is a multilayered network that uses back-propagation to train itself to the indicated connection weights that enable it to select either pad or palm opposition based on the task requirements.

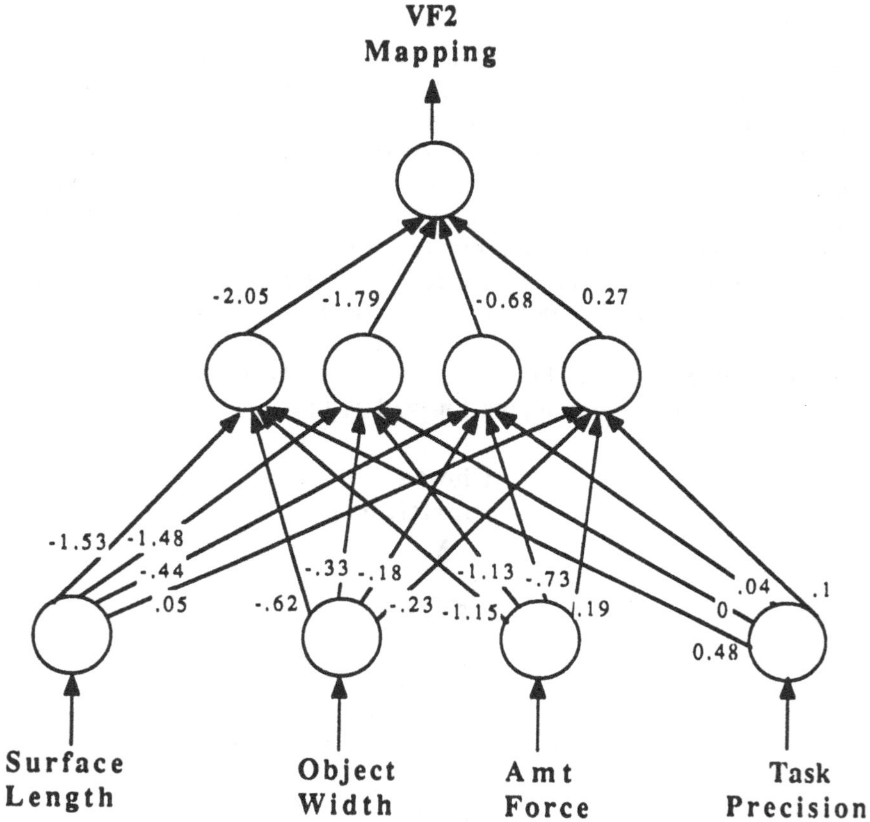


FIG. 4.12. VF-MAPPER is a multilayered network that uses back-propagation to train itself to the indicated connection weights that enable it to select the correct number of real fingers to be mapped into VF2.

opposition). Each network comprised four input units (one for each task input), four hidden units, and one output unit (to select the opposition, or to perform the VF mapping). For both, training stopped when the total error (some sort of history value related to the difference between desired and current output) reached 0.05. The OPPOSITION-SELECTOR network took 833 repetitions of the training data to reach criterion; the VF-MAPPER network took 531 trials.

The point of using a neural network approach is to see how it generalizes about the task space. OPPOSITION-SELECTOR learned to use palm opposition when the forces were large. As the precision requirements increased in the task, it went more toward pad opposition. Also, there was a tendency toward using palm opposition when the length of the object increased, particularly when the forces were increasing as well. These results mirror what is anecdotally known about human prehension. VF-MAP-

PER learned to use more fingers in VF2 as the task forces, object width, or object length increased. Preliminary results indicate that it did not correctly learn that the length of the object could present a true constraint on the number of real fingers mapped into VF2, particularly with a surface length of one finger.

Back-propagations is not a biological model, but does provide a useful way of training a neural network that permits us to study a whole range of questions including "What training sets permit ease of generalization?" and "What encodings by the hidden units seem to encapsulate essential features of the task repertoire?" With this I complete the present analysis of schemas and neural nets for the control of hand movements. I have endeavored to show how the concepts of opposition space and virtual finger have given new power to the definition of perceptual and motor schemas to be incorporated in coordinated control programs for human reaching and grasping; and have also suggested the (somewhat preliminary) state of the art in using neural network modeling to represent neurophysiological data and to provide a technology for teasing apart the basic ingredients of a learned behavior.

## ACKNOWLEDGMENTS

The research herein was supported in part by grants No. DMC-8719579 and IRI-8796249 from the National Science Foundation. My thanks to Gordon Dakin, Thea Iberall and Damian Lyons for all they contributed to the preparation of this paper.

## REFERENCES

Amari, S., & Arbib, M. A. (1977). Competition and cooperation in neural nets. In J. Metzler (Ed.), *Systems neuroscience* (pp. 119–165). New York: Academic Press.

Arbib, M. A. (1975). Artificial intelligence and brain theory: Unities and diversities. *Annals of Biomedical Engineering*, *3*, 238–274.

Arbib, M. A. (1981). Perceptual structures and distributed motor control. In V. B. Brooks (Ed.), *Handbook of physiology–the nervous system, II. Motor control*, (pp. 1449–1480). Bethesda, MD: American Physiological Society.

Arbib, M. A., Iberall, T., & Lyons, D. (1985). Coordinated control programs for movements of the hand. In A. W. Goodman & I. Darian-Smith (Eds.), *Hand function and the neocortex* (pp. 135–170). Berlin, Heidelberg, New York: Springer-Verlag.

Arbib, M. A., Iberall, T., & Lyons, D. (1987). Schemas that integrate vision and touch for hand control. In M. A. Arbib & A. R. Hanson (Eds.), *Vision, brain, and cooperative computation*, (pp. 489–510). Cambridge, MA: A Bradford Book/The MIT Press.

Bond, A. H., & Gasser, L. (Eds.). (1988). *Readings in distributed artificial intelligence*. Palo Alto: Morgan Kaufmann Publishers.

Dakin, G. A., & Arbib, M. A. (1986). The hierarchical control of skilled hand movements. (COINS Technical Report 86-35). Amherst: University of Massachusetts.

Davis, R., & Smith, R. G. (1983). Negotiation as a metaphor for distributed problem solving. *Artificial Intelligence, 20*, 63–109.

Erman, L. D., Hayes-Roth, F. A., Lesser, V. R., & Reddy, D. R. (1980). The Hearsay-II speech-understanding system: Integrating knowledge to resolve uncertainty. *Computing Surveys, 12*, 213–253.

Evarts, E. V., Shinoda, Y., & Wise, S. P. (1984). *Neurophysiological approaches to higher brain functions*. New York: Wiley.

Evarts, E. V., & Tanji, J. (1976). Reflex and intended responses in motor cortex pyramidal tract neurons of monkey. *Journal of Neurophysiology, 39*, 1069–1080.

House, D. (1984). Neural models of depth perception in frog and toad. Doctoral dissertation, Department of Computer and Information Science, University of Massachusetts, Amherst.

Iberall, T. (1987). A neural model of human prehension. Doctoral dissertation, Department of Computer and Information Science, University of Massachusetts, Amherst.

Iberall, T. (1988). A neural network for planning hand shapes in human prehension. *Proceedings of Automation and Controls Conference*. (pp. 2288–2293).

Iberall, T., & Arbib, M. A. (1989). Schemas for the control of hand movements: An essay on cortical localization. In M. Goodale, (Ed.), *Vision and action: the control of grasping*.

Iberall, T., Bingham, G., & Arbib, M. A. (1986). Opposition space as a structuring concept for the analysis of skilled hand movements. *Experimental Brain Research Series, 15*, 158–173.

Iberall, T., & MacKenzie, C. L. (1988). Behavioral and computational modelling using ballpark modelling for studying human prehension, *Society for Neuroscience Abstracts, 14*, 1234.

Jeannerod M., & Biguer, B. (1982). Visuomotor mechanisms in reaching within extra-personal space. In D. J. Ingle, R. J. W. Mansfield, & M. A. Goodale (Eds.), *Advances in the analysis of visual behavior* (pp. 387–409). Cambridge, MA: MIT Press.

Lyons, D. M. (1986). **RS**: A formal model of distributed computation for sensory-based robot control. Doctoral dissertation and COINS Technical Report 86-43, Department of Computer and Information Science, University of Massachusetts, Amherst.

Lyons, D. M., & Arbib, M. A. (1989). A formal model of distributed computation for schema-based robot control. *IEEE Transactions Robotics and Automation, 5*, 280–293.

Marteniuk, R. G., MacKenzie, C. L., Jeannerod, M., Athenes, S., & Dugas, C. (1987). Constraints on human arm movement trajectories. *Canadian Journal of Psychology, 41*, 365–378.

Minsky, M. L. (1975). A framework for representing knowledge. In P. H. Winston (Ed.), *The psychology of computer vision* (pp. 211–277). New York: McGraw-Hill.

Rumelhart, D. E., Hinton, G. E., & Williams, R. J. (1986). Learning internal representations by error propagation. In D. Rumelhart & J. McClelland (Eds.), *Parallel distributed processing: Explorations in the microstructure of cognition* (Vol. 1, pp. 318–362). Cambridge, MA: MIT Press/Bradford Books.

Schank, R., & Abelson, R. (1977). *Scripts, plans, goals and understanding: An inquiry into human knowledge structures*. Hillsdale, NJ: Lawrence Erlbaum Associates, Inc.

Tanji, J., & Kurata, K. (1985). Contrasting neuronal activity in supplementary and premotor cortex of monkeys. I. Responses to instructions determining motor responses to forthcoming signals of different modalities. *Journal of Neurophysiology, 53*, 129–141.

Tanji, J., Taniguchi, K., & Saga, T. (1980). Supplementary motor area: Neuronal response to motor instructions. *Journal of Neurophysiology, 43*, 60–69.

Wing, A. M., & Fraser, C. (1983). The contribution of the thumb to reaching movements. *Quarterly Journal of Experimental Psychology, 35A*, 297–309.

Wing, A. M., Turton, A., & Fraser, C. (1986). Grasp size and accuracy of approach in reaching. *Journal of Motor Behavior*, 18(3), 245–260.

Yonezawa, A., & Tokoro, M. (Eds.). (1987). *Object-oriented concurrent programming*. Cambridge, MA: MIT Press.

# 5 Action-Perception as a Pattern Formation Process

J. A. S. Kelso, J. D. DelColle, and G. Schöner
*Program in Complex Systems and Brain Sciences*
*Center for Complex Systems*
*Florida Atlantic University, Boca Raton*

## ABSTRACT

In this chapter we aim to identify collective variables (or order parameters) and their dynamics (stability, loss of stability, hysteresis . . .) for perception–action patterns. To do so we extend earlier results on phase transitions in human bimanual coordination to a perception–action task, synchronizing or syncopating finger flexion with an auditory metronome whose frequency is varied in steps (.25 Hz) over a wide range (1.0 Hz to 3.5 Hz). The chief idea is to study the metronome–hand relation as an example of a pattern forming system. *Relative phase* between hand and metronome proves to be an adequate collective variable that characterizes empirically observed patterns. A theoretical model using the concepts of synergetics and dynamic pattern theory reproduces all observed patterns, among which are frequency locking and phase locking, phase transitions, and loss of entrainment (relative coordination). The general implications of these results are discussed and new predictions proposed.

## INTRODUCTION

Much has been written about the intimacy of the relation between perceiving and acting—that they evolved in parallel, and thus should be treated as a unitary system, and so forth (see e.g. Arbib, 1987; Gibson, 1979; Kelso & Kay, 1987; Kugler, Kelso, & Turvey, 1982; Lee & Young, 1986; Michaels & Carello, 1981; Warren, 1988 for recent treatments). Our understanding of this relation nevertheless remains quite limited. Although general intuitions about the coherence of action and perception are useful, it is probably more important to find a specific case where these intuitions can be given a concrete and precise meaning. From that case, it may then be possible to generalize to broader contexts. The present research presents a detailed experimental and theoretical analysis of a specific, but nevertheless potentially generalizable situation—synchronizing one's motions with a regularly occurring environmental event.

Very many skills are of this type: from skipping, where one must synchronize one's action with the periodicity of the rope, to dribbling a basketball, to dancing, to stepping onto an escalator, and so on. In biology, numerous examples exist of external periodic stimuli entraining biological pacemakers—from firefly entrainment (Hanson, 1978) to regularly spaced synaptic inputs on "pacemaker" neurons (e.g., Perkel, Schulman, Bullock, Moore, & Segundo, 1964) to synchronizing central neurons in locust flight ganglia by peripheral sinusoidal driving of the wings (Wendler, 1974). We consider this problem as part of a broader research program dealing with dynamic pattern formation in complex, biological systems (Kelso, Schöner, Scholz, & Haken, 1987; Kelso & Schöner, 1987; Schöner & Kelso, 1988a,b,c). Our experimental results, nevertheless, may be understood in detail as corresponding to the process of entrainment of a nonlinear oscillator by an external harmonic source of excitation. Macroscopically, the patterns formed by the action–perception system correspond to empirically observed stable phasing (relative timing) relations and their breakdown. We believe these results to be quite relevant not only to a wider spectrum of skilled behavior (see General Discussion) but to a deeper understanding of information in the sense of information that is meaningful for biological functions and tasks (Schöner & Kelso, 1988c,e)—in short, semantic information (see also Shimizu, Yamaguchi, & Satoh, 1988).

What does it mean to understand action–perception as a pattern-formation process? From our perspective the key idea is to characterize action–perception patterns in terms of their dynamics. This means, fundamentally, identifying collective variables that correspond to these patterns and mapping them onto attractors of the collective variable dynamics. We can determine the dynamics of the collective variables by studying the stability and loss of stability of the patterns. Thus this dynamic pattern strategy (e.g., Kelso, et al., 1987; Kelso & Schöner, 1987; Schöner & Kelso, 1988a,b,c) can be implemented best near points of pattern change because these allow: (a) the clear differentiation of one pattern from another; and (b) the study of the relative stability of different patterns. More important, theoretically, is that such phase transitions enable us to formulate the (nontrivial) nonlinear collective dynamics underlying the patterns that may then be derived from a lower (individual component) level of observation. In fact, this strategy has been successfully applied (in the sense of an explicit theory–experiment relation that can be formulated in closed mathematical form) to behavioral patterns that arise spontaneously without any specific environmental information (e.g., Haken, Kelso, & Bunz, 1985; Kelso, 1984; Schöner, Haken, & Kelso, 1986). More recently, the theory has been demonstrated to include certain environmentally elicited and learned patterns of movement coordination (Schöner & Kelso, 1988a,b) as well as intended behavioral patterns (see Kelso, Scholz, & Schöner, 1988; Schöner & Kelso, 1988c,e). The operational aspect of the theory cannot be overemphasized: In each case, dynamics are formulated for empirically observed variables and predictions made that can be experimentally checked. As a consequence of this work, a further, quite explicit meaning of the concept of information has materialized: Information is considered meaningful and specific only to the extent that it contributes to the collective variable dynamics, attracting the system to the required (or learned, or memorized, or intended) pattern (Schöner & Kelso, 1988c). Further clarification of this notion is provided in the theoretical modeling section of this paper, where we model explicitly the various perception–action behavioral patterns observed in the present study.

The present experimental configuration is somewhat similar to temporal tracking paradigms that have been studied spasmodically by psychologists and physiologists for over a hundred years. The task is to synchronize a movement, such as that of a finger, with an external stimulus such as a metronome or a strobe (see, e.g., Bartlett & Bartlett, 1959; Dunlap, 1910; Woodrow, 1932). Many of these older experiments were purely empirically motivated. For instance, Dunlap (1910) sought to determine the relationship between the length of an interstimulus interval in a metronome-following task and the directional error of response. More theoretically based efforts were conducted by Bartlett and Bartlett (1959) and more recently, (though in the same spirit of identifying central and peripheral processes involved in timing patterns) by Wing and Kristofferson (1973) and Hary and Moore (1987).

Based on previous theoretical and experimental work regarding the stability and change of coordinated rhythmic movements without explicit environmental information present, our interest was in the temporal order formed by the action–perception system. Can one determine the collective dynamics of this system also by studying the stability and (putative) loss of stability of the patterns formed? Might loss of stability be seen as a failure or entrainment limit, of synchronizing one's motion to the frequency of a periodically occurring event? And so on.

To investigate these kinds of questions, subjects were situated in a finger recording apparatus (see Kelso & Holt, 1980) and instructed to synchronize flexion movements (specifically, peak flexion) either "on the beat" or "off the beat" of an auditory metronome. The latter condition required subjects to produce a flexion movement *between* (i.e., in the interstimulus interval) each metronome pulse whose frequency was varied either in an *ascending* order from 1.0 Hz to 3.5 Hz in steps of .25 Hz every 10 cycles or a corresponding *descending* order that began at 3.5 Hz.

The aim of the "off the beat" condition was to observe phase transitions—phenomenologically thought to be present[1]—in perception–action patterns, as a means of determining the underlying dynamics of these patterns. Thus, as mentioned briefly before, finding points at which the metronome-movement pattern changes not only allow one to distinguish one pattern from another, but the *relative stability* of the patterns can also be evaluated and quantified. Moreover, the mechanism thought to underlie changes in patterns is *loss of stability*, which can be measured given appropriate time scale relations (see Schöner, Haken, & Kelso, 1986, Kelso et al., 1987; Schöner & Kelso, 1988a,c). Other characteristic features of dynamic patterns, such as hysteresis, may be studied through comparisons of ascending and descending conditions. In the latter case, it also becomes apparent how the intentional production of behavior (Kelso, Scholz, & Schöner, 1988; Schöner & Kelso, 1988e) is constrained dynamically by the stability of the collective variable that characterizes the perception–action pattern.

## METHODS

### Subjects

Seven normal adults, five male and two female, volunteered for this experiment and were not paid for their participation. Five of the seven subjects were naive to the purpose of the experiment; two were not (SG and KS).

### Apparatus

Flexion–extension movements of the subjects preferred index finger were recorded using an apparatus similar to the one described by Kelso and Holt (1980). The apparatus consisted of two manipulanda (one for each hand, although only one was used for each subject's preferred hand) connected to a metal sleeve into which the subject's finger was inserted. A linear potentiometer connected to the axis of rotation allowed for the direct transduction of the continuous angular position of the finger about the metacarpophalangeal joint as the finger cycled in the transverse plane.

[1]The idea comes from a "test" that music teachers sometimes use. The student who maintains an "off the beat" pattern at the highest metronome frequency wins the game.

The stimulus, a periodic acoustic signal of adjustable frequency, was generated by a computer driven metronome constructed by the engineering staff at the Center For Complex Systems. The signal was output through an RS232 port and then conditioned to produce a train of 50 ms square wave pulses. This signal was in turn used to drive a loudspeaker producing a series of clicklike sounds. Both the square wave signal of the metronome and the finger position signal were recorded on FM tape and subsequently digitized at 250 Hz for later analysis.

### Conditions

Subjects were instructed to synchronize peak flexion of the index finger (the point at which the movement of the finger changes from flexion to extension) with the metronome in either of two modes of coordination: (a) Off the beat (O)—so that peak flexion corresponds to the point in time midway between each metronome pulse, or (b) on the beat (I)—so that peak flexion occurred at the same time as each metronome pulse. The tasks were to be performed under two experimental conditions: (a) An ascending metronome series (A)—in which the frequency of the metronome was increased from 1 Hz to 3.5 Hz in the steps of 0.25 Hz, or (b) a descending metronome series (D)—where the metronome frequency was decreased from 3.5 Hz to 1 Hz in steps of 0.25 Hz. In all cases the metronome produced 10 cycles for each frequency plateau (e.g., at 1.25 Hz, 1.50 Hz, 1.75 Hz, . . .). Each mode of coordination was paired with each metronome series forming four conditions as follows: ascending–off the beat (AO), ascending–on the beat (AI), descending–off the beat (DO), and descending–on the beat (DI). Each subject performed a block of ten trials for each condition, and the order of conditions was randomized for all subjects.

### Procedures

Subjects were seated comfortably in a chair in front of the apparatus. The height of the chair was adjusted to permit the forearms to rest horizontally on a padded support at the front of the apparatus. A soft adjustable plastic splint secured around the subject's forearm provided additional support and also served to restrain any unwanted movements of the wrist and digits. A soft thin pad was wrapped around the intermediate phalange of the index finger that was then inserted into the metal sleeve of the manipulandum up to the first interphalangeal joint. Correctly positioned, the sleeve prevented movements of the two distal finger joints. The position of the hand was then adjusted and fixed so that the center of rotation of the metacarpophalangeal joint and that of the manipulandum were colinear.

After positioning the subject in the apparatus, instructions were issued. The subject was given a description of the conditions and instructed to cycle his/her finger once for each beat of the metronome maintaining the required mode of coordination as well as possible. Subjects were further instructed that should they feel the pattern begin to change they should not try to resist. It was emphasized that the primary task was to follow the metronome, maintaining one cycle of the finger for each cycle of the metronome. A demonstration of peak flexion on the beat and off the beat was given to insure that subjects clearly understood the task. A single trial run, 11 frequency plateaus, 10 cycles per plateau lasted approximately 1 min. Rest intervals were provided to the subject after each trial block of 10 runs, or on request. The time for the experiment varied between 1.5 and 2 hours.

### Analysis

Following digitization of the data, special in-house wave analysis software was employed to obtain the mean frequency of metronome and finger movement, the mean relative phase between the finger and the metronome, and the standard deviation of these variables at each frequency plateau for each subject performing in each of the four experimental conditions.

These four variables were then treated as individual scores in later analyses. In order to eliminate transient effects resulting from the change in metronome frequency (i.e., while subjects adjusted to the change) events occurring within the first 2 cycles of the metronome pulse at each frequency plateau were excluded in the calculations. The frequency of finger movement was obtained by computing peak flexion to peak flexion–cycle durations. Relative phase calculations were obtained by computing the position in time of peak flexion of the finger with respect to the local peak-to-peak cycle of the metronome. Depending on the relative phase behavior at a given plateau, one of the following modes of analysis was used to calculate the mean and standard deviation: (a) the 180 degree mode calculates relative phase with respect to a 0 to $2\pi$ reference interval; (b) the zero degree mode calculates relative phase between $-\pi$ and $+\pi$; and (c) the absolute mode calculates relative phase in the interval $-\pi$ to $\pi$ without regard to sign. If the relative phase at a given plateau was centered around 180 degrees, then the 180 degree mode was used. If relative phase centered around zero degrees, then the zero mode calculation was used. In cases where the relative phase was changing, the absolute mode was used.

Correlations between required (metronome) and actual (finger movement) frequencies were calculated for each subject in all four experimental conditions. Analyses of variance (ANOVA) were conducted separately for the ascending and descending conditions (see Results). For each subject, a mode by plateau ANOVA was conducted for pre- and post-transition plateaus treating trial runs as cases. Because most subjects differed in the exact frequency at which the switch from one pattern to another occurred, the number of pre- and post-transition plateaus varied in this analysis. Also, because of within-subject variation in switching frequency across trials, a minimum of 5 trials was required at each frequency plateau in order for that plateau to be included as a variable in the ANOVA.

## RESULTS AND DISCUSSION

### Accuracy of tracking

First, we establish that subjects were capable of tracking the metronome across all the frequencies used (1.0 to 3.5 Hz) and in all conditions. Pearson's product–moment correlations between actually produced and required frequency were calculated for all subjects. The means for ascending (in phase and antiphase) conditions were $r = .98$ ($SD = .024$) and $r = .99$ ($SD = .013$). For the descending conditions, the respective correlation values were $r = .996$ ($SD = .004$) and $r = .98$ ($SD = .015$). The range of correlations across subjects and conditions was $r = .933$ to $r = 1.0$. The overall correlations, though strong, can be a bit misleading under more detailed analysis in that under certain circumstances the frequency of finger movement does not conform to the frequency of the metronome. Obviously, when a transient in relative phase occurs, the frequencies must differ. Also, a necessary condition for the absence of phase-locking, that is, when the relative phase "drifts" or "wanders," is frequency detuning. We consider these cases in more detail in the next section.

### Patterns of Behavior: Ascending Conditions

Before providing quantitative results, we present a classification of the patterns produced by subjects when "prepared" initially in the in phase and antiphase modes of coordination. For the in phase synchronization condition this was quite easy. All 7 subjects on the majority of runs (68/70 or 96.6%) were capable of maintaining a synchronous, in phase relationship with the metronome across (at least) part of the range of frequencies. An example of the metronome and hand movement time series and the relative phase between them is shown in Fig. 5.1 (top). There it can be seen that finger flexion coincides (phase-locks) with the metronome

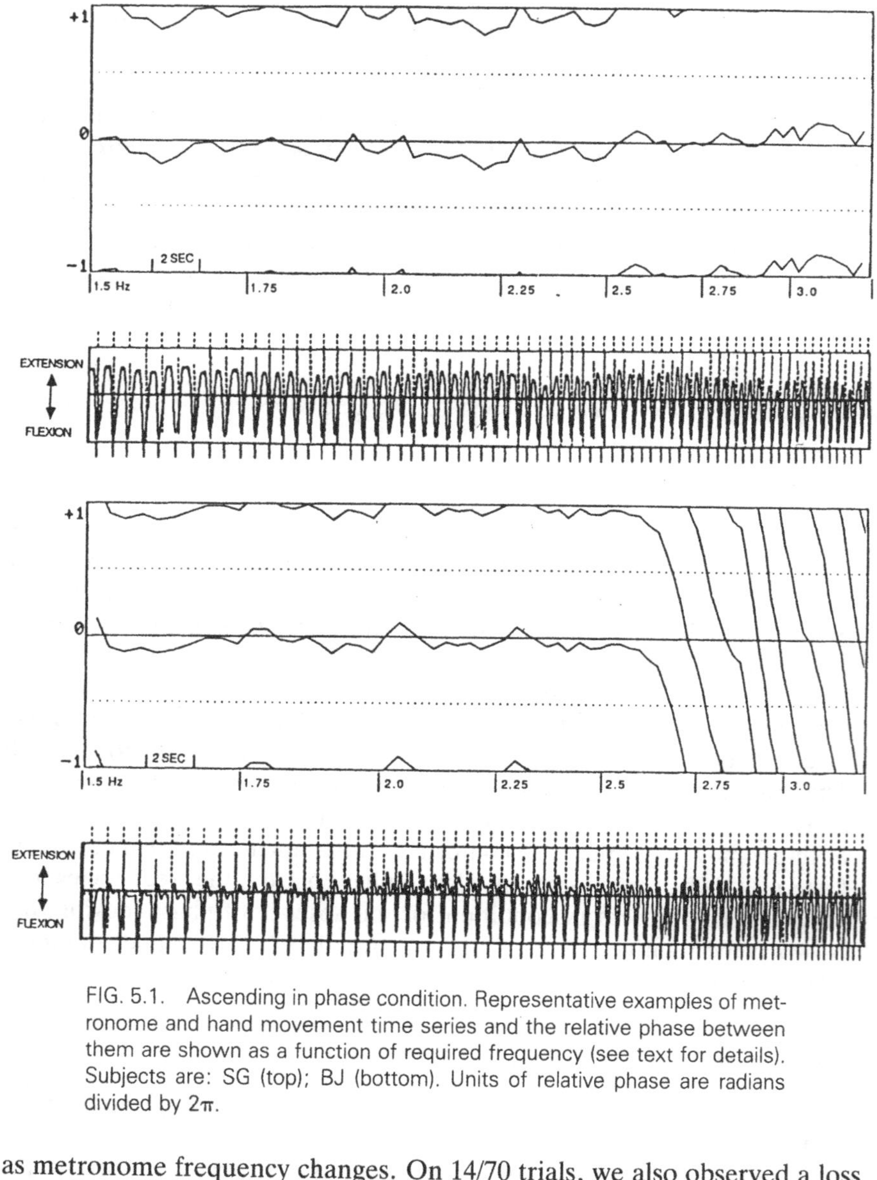


FIG. 5.1. Ascending in phase condition. Representative examples of metronome and hand movement time series and the relative phase between them are shown as a function of required frequency (see text for details). Subjects are: SG (top); BJ (bottom). Units of relative phase are radians divided by $2\pi$.

as metronome frequency changes. On 14/70 trials, we also observed a loss of phase locking at higher frequencies; that is, subjects (primarily one, 7/10 trials) were unable to synchronize in the in-phase mode at higher frequencies and phase-drift was observed. Only a very small frequency difference (or detuning) between the metronome and finger is necessary to induce phase drift. An example of wandering at higher frequencies is presented in Fig. 5.1 (bottom).

For the ascending antiphase condition, a variety of patterns were observed. Only one subject (LS) was able to preserve an antiphase relation with the metronome across all frequency plateaus. It turned out however, on postexperiment interview, that this subject discovered a special way to perform the task. Thus, instead of performing a *flexion* movement *between* metronome pulses as instructed, the subject produced an *extension* movement in synchrony with the metronome pulse. Using this strategy, the subject had little problem producing a flexion movement off the beat, that is, an antiphase relation with the metronome. Thus, in all ten trials, this subject stayed in the antiphase pattern with the metronome.

Sidestepping these results for the moment (see General Discussion), let us consider other action–perception patterns in the ascending and antiphase condition. On 37% of the trials, a spontaneous transition from an antiphase to in phase pattern was observed (see Fig. 5.2). Five of the six subjects (excluding LS) exhibited this pattern on three or more trial runs (Fig. 5.2A). On an additional 33% of the trials, subjects—after switching to the in phase pattern and stabilizing there—proceeded to display phase-wrapping behavior, in which the relative timing between metronome and finger was not stationary. Four of the six subjects showed this additional feature, one subject on all ten runs (see Fig. 5.2B).

On a proportion of trials (18%), we observed a direct pathway from the initial antiphase pattern to phase wrapping, mostly in one subject (6 out of 10 trials, see Fig. 5.2C). On a further 11% of trials, after phase wrapping, the pattern shifted to in phase and stabilized there for the remainder of the trial run (see Fig. 5.2D). In sum, on a total of 80% of the trials, subjects showed a transition from antiphase to in phase entrainment. Phase drift was observed *en route* to the in phase pattern but mostly following the in phase pattern.

### Patterns of Behavior: Descending Conditions

Once again, for the in-phase condition there was little problem in classifying patterns of relative phase between finger and metronome. On 65/70 trials (93%) subjects maintained an in phase relation at all frequencies. On 2/70 (3%) phase drift was observed before locking occurred. On a further 3 trials (2/10 for one subject and 1/10 for the other), the pattern went from in phase to antiphase and back again.

None of the subjects, save one, were able to maintain an antiphase relation with the metronome at higher frequencies. The "outlier" (LS), however, was able to do so because of adopting the strategy of producing extension on the beat, rather than flexion off the beat (see previous discussion). Even with this strategy, on 7/10 trials this subject displayed either phase drift before achieving the required phase relation, or adopted the

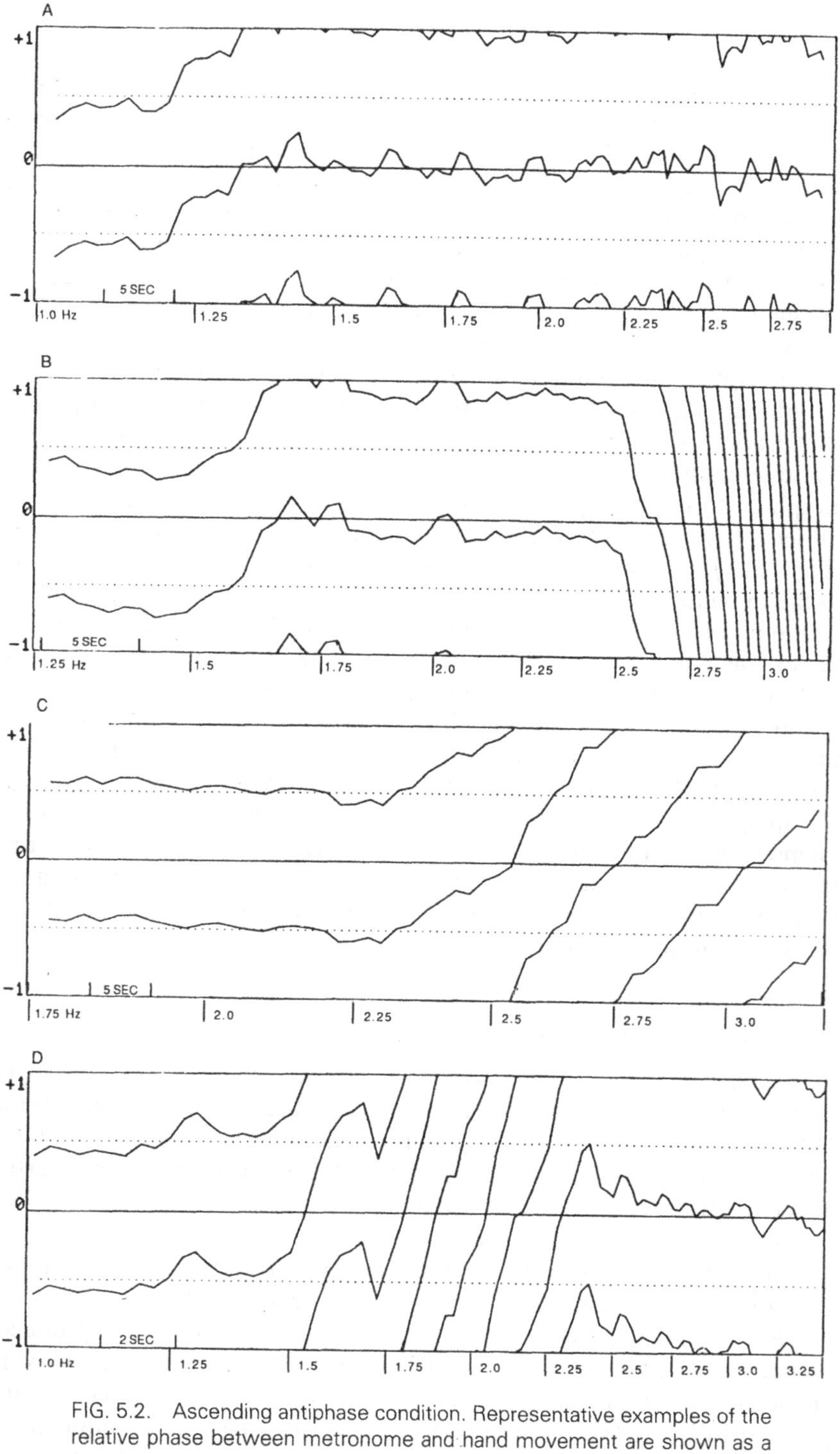


FIG. 5.2. Ascending antiphase condition. Representative examples of the relative phase between metronome and hand movement are shown as a function of required frequency for different subjects: (A) SG; (B) BJ; (C) JJ; (D) KS.

in-phase pattern first, that is, at higher frequencies, before shifting to antiphase.

All of the remaining subjects were able eventually to maintain an antiphase relation with the metronome. The pathways to this perception–action pattern varied, however. On 20% of trial runs, as frequency decreased, phase drift was followed by an in phase pattern followed by the antiphase pattern. In 8% of the cases, no phase drift was observed: The system changed from in phase to antiphase. A further 43% of trials showed a direct change from phase drifting to the antiphase pattern. Forty-eight percent of the latter runs were preceded by an in phase relation between movement and metronome. A few other patterns were observed, including ones in which several bouts of phase wandering occurred *en route* to the required pattern. Fig. 5.3 displays a number of examples of the descending patterns that are discussed more thoroughly in the Modelling section.

### Quantitative Analysis: Ascending Conditions

We examine the pretransition régime first for effects of mode or initial condition (in phase or antiphase), frequency plateau, and their interaction, on the measures of relative phase between hand and metronome, cycle duration, and their respective variabilities. Only significant effects are mentioned. As expected, the mode effect on mean relative phase was significant for all subjects. Individual F-values ranged from 166.35 in BJ to 5516.42 in JJ ($ps < .01$). In general, the longer the subject maintains the less-stable antiphase relation, the greater the level of relative phase fluctuations. Two of the subjects' data (TB and MC) tend to support this contention: A mode by plateau interaction was observed, $F(4,82) = 5.45$, $p < .01$ (TB) and $F(6,115) = 2.47$, $p < .03$ (MC) respectively. This interaction for TB (A) and MC (B) is shown in Fig. 5.4. For TB, the difference between modes is statistically significant on plateau 3 (1.5 Hz) and plateau 5, the pretransition plateau (2 Hz). For MC, the antiphase mode is more variable for plateaus 5 (2 Hz) and the pretransition plateau (2.5 Hz). None of the other subjects, however, showed a mode or mode by plateau interaction on relative phase variability. This is not surprising because subjects KS and BJ already revealed transition behavior on the third frequency plateau. In subject SG, nearly all transitions occurred on the second frequency plateau, making a detailed pretransition analysis impossible.

Subject LS who displayed no transitionlike behavior because of the strategy adopted, required a separate statistical treatment. Effects of mode, frequency plateau, and the interaction between these variables were found for mean relative phase, $F(1,194) = 2952.28$, $F(10,194) = 3.26$ and $F(10,194) = 3.44$, $p_s < .01$ respectively. The interaction was largely due to the strong main effects found for both frequency and mode (see Fig. 5.4, bottom).

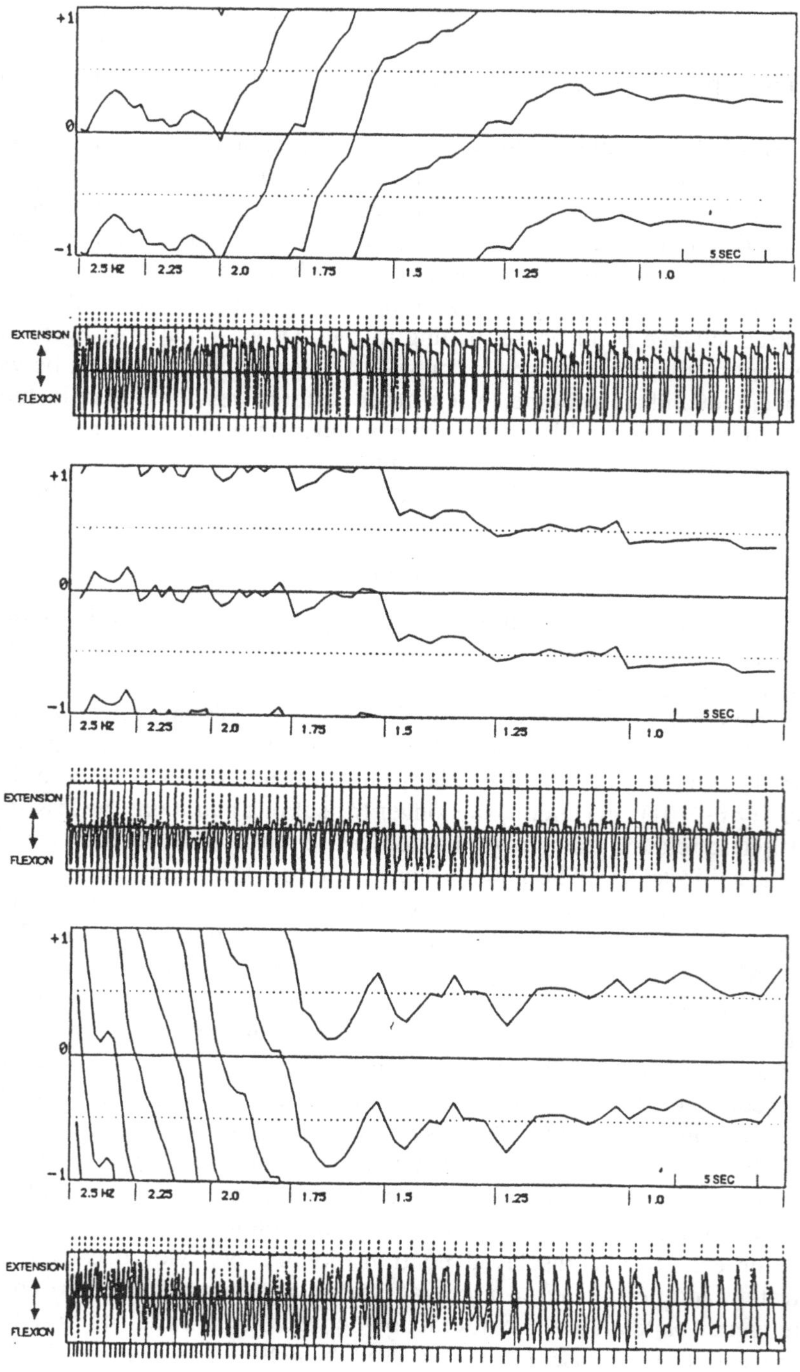


FIG. 5.3. Descending antiphase condition. Representative patterns as frequency decreases are shown for different subjects: SG (top), BJ (middle); TB (bottom) (see text for details).

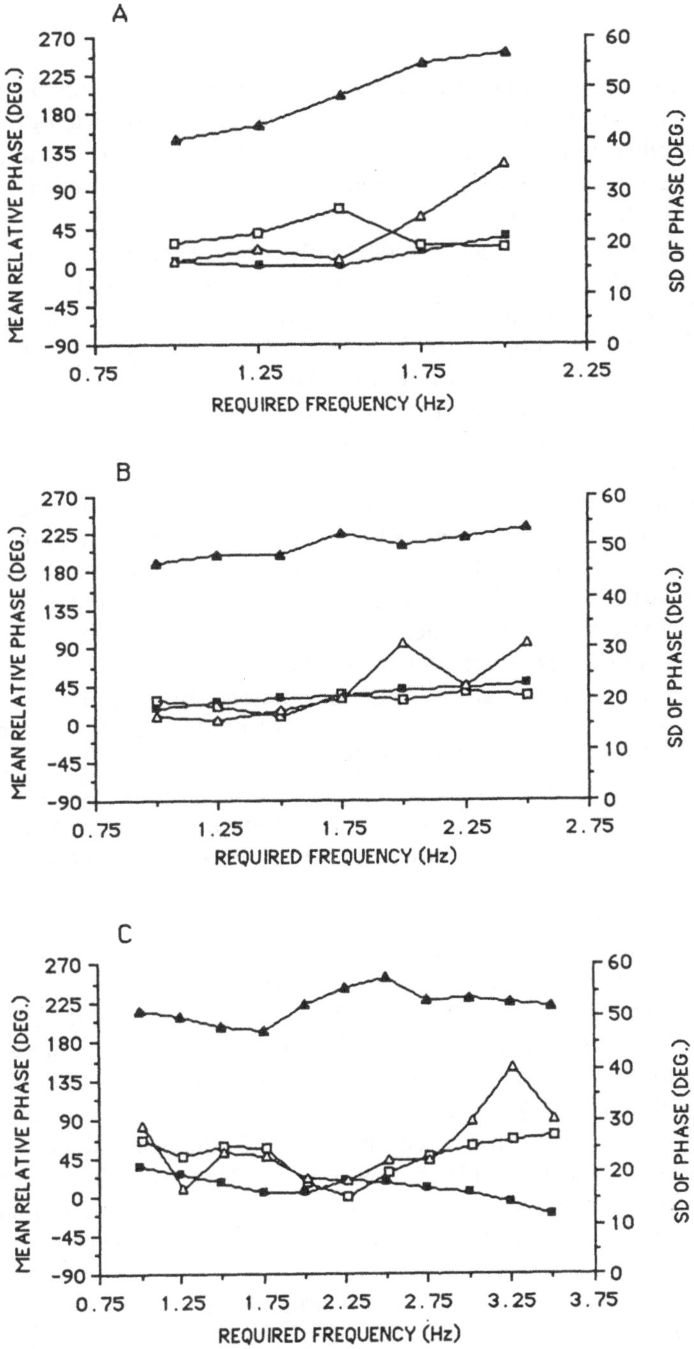


FIG. 5.4. Ascending antiphase (triangles) and corresponding in phase (squares) conditions: The mean relative phase (filled symbols) and the standard deviation of relative phase (open symbols) are shown as a function of the required frequency for the pretransition frequency plateaus. Subjects are: TB (a); MC (b); LS (c).

Phase variability increased with frequency, $F(10,194) = 6.05, p < .01$, but was not differential between modes. Typical effects on cycle duration and its variability with frequency plateau $Fs(10,194) = 12087.54$ and $31.94$, $p_s < .01$ were found, as well as a mode by plateau interaction, $F(10,194) = 2.30$ and $2.03$, $p < .05$.

For the 6 other subjects, modest influences of frequency on mean relative phase were observed as well as an interaction with mode. In 3 subjects (TB, BJ and JJ) the interaction was significant, $F(4, 82) = 8.30$, $F(1.32) = 10.05$, and $F(6,114) = 32.00$, $p_s < .01$ respectively. A main effect of frequency plateau was also noted for MC, $F(6,115) = 8.78$, $p < 01$. Although we expect the mean phase to be different between in-phase and anti-phase modes at all frequency plateaus, differential effects of frequency between the two modal conditions were not expected. *Frequency* as used here is meant to be nonspecific, in the sense of simply moving the metronome-hand system through different collective states or patterns. The reasons for systematic effects of frequency on relative phase need to be studied further, though we suspect they are largely artifactual and due, in part, to the classification scheme we employed: phase means designated to be equivalent across frequency may not have been equivalent after all.

It was no surprise that the frequency plateau influenced cycle duration in all subjects ($p_s < .01$). An interaction of mode and plateau was noted for two subjects (TB and MC) and a mode main effect on one (KS), $F(4, 82) = 3.31$, $p < .02$, $F(6, 115) = 2.26$, $p < .05$ and $F(1, 33) = 4.22$, $p < .05$ respectively. Overall, these effects are consistent with the phase results: Small differences in cycle duration between modes are indicative of frequency detuning between metronome (whose frequency is constant on a plateau) and hand, giving rise to phase variation. As cycle duration decreased with required frequency, its variability also decreased—a well known result (e.g., Deutsch, 1983), $F(1, 33) = 12.05$, $p < .01$ (KS); $F(6, 114) = 10.97$, $p < .01$ (JJ); $F(1, 35) = 5.52$ $p < .03$ (BJ); $F(6, 115) = 18.77$, $p < .01$ (MC); $F(4, 82) = 9.41$, $p < .01$ (TB). In some subjects, mode aslo had an effect on cycle duration variability, $F(1, 33) = 4.28$, $p < .05$ (KS); $F(1, 114) = 4.82$, $p < .03$ (JJ); and $F(1, 35) = 5.52$, $p < .03$ (BJ); and an interaction between mode and frequency plateau was found for another (MC), $F(6, 115) = 2.39$, $p < .03$. Variability in cycle duration tended to be greater in the antiphase condition than in the in phase condition, and was enhanced to a greater extent before the transition in subject MC.

To the extent subjects' behavior stabilized after changing pattern, we were able to perform a posttransition analysis analogous to the one described above. Small mode differences in mean phase and phase variability were noted in 2 subjects, both of whom exhibited a transition to the in phase pattern at a low frequency, $F(1, 55) = 10.95$ and 9.60, $p < .01$ (BJ) and $F(1, 114) = 47.17$ and 10.60 (SG) respectively. Still another subject (KS) exhibited a mode by frequency plateau interaction on both the mean and variability of relative timing, $F(5, 105) = 3.95$, $p < .01$ and 2.83, $p < .02$ respectively. Again, these results are linked to changes in cycle duration and its variability after the transition is over. The same subjects revealed a mode effect on cycle duration, $F(1, 55) = 10.67$, $p < .01$ (BJ) and $F(1, 114) = 12.86$, $p < .01$ (SG). In addition, the variability of cycle duration increased as frequency increased in posttransition regimes, $F(5, 105) = 5.18$ (KS), $F(6, 114) = 10.97$ (JJ), $F(3, 55) = 4.90$ (BJ), $F(6, 114) = 9.73$ (SG), $F(6, 115) = 18.77$ (MC) and $F(4, 82) = 9.41$, $p_s < .01$ respectively. The modes were a little different in cycle variability in one subject (JJ), $F(1, 114) = 4.82$, $p < .03$, after the transition even though they are supposed to be the same. One might interpret these differences as "history" dependent effects: Where you come from has an influence on the stability of the performed pattern.

### Quantitative Analysis: Descending conditions

As one might expect from the foregoing qualitative description of the descending patterns, it does not make much sense to try to perform the same kind of pre- versus posttransition analysis as we did for the ascending conditions. Nevertheless, it is possible to characterize the patterns quantitatively at lower frequencies, and there are a number of new aspects in the evolutionary path toward the timing patterns produced at lower frequencies.

It is important to recognize that the two cases—ascending and descending—are very different. In the former, it is possible to prepare the system in an antiphase pattern, for example, and watch it evolve as frequency increases. In the latter, as frequency decreases, the subject must "probe" for the antiphase pattern and once he/she finds it, the task is to maintain that pattern. There are thus many more transients and phase-wrapping phenomena in the antiphase descending condition than in any other condition. To perform this task, the subject must first track the metronome, that is, frequency lock, and then phase lock in an antiphase relation. Of course, as described earlier, some subjects do both in parallel, that is, they frequency- and phase-lock simultaneously, often after a bout of phase wandering or drifting. Fig. 5.5 shows an example of the phasing patterns between hand and metronome in the descending anti-phase condition, as they evolve over time (trials). This subject (JJ) had little or no problem synchronizing in phase "on the beat," in the descending condition, as can be seen in his very first trial run (see Fig. 5.5A)

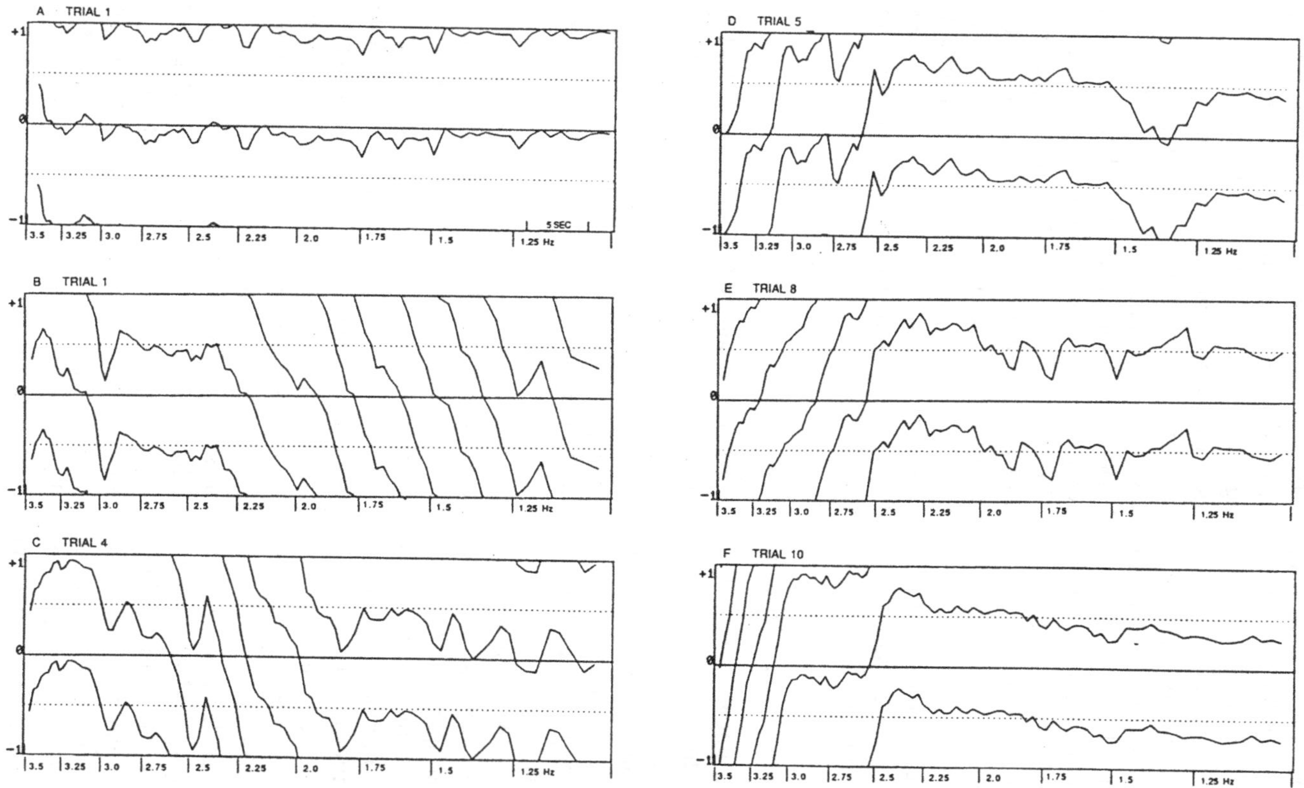


FIG. 5.5 Descending antiphase condition (Panels B through F). Time series of relative phase illustrate the patterns of behavior over subsequent trials of subject JJ. The decreasing frequency within each trial is indicated. Note the earlier onset of the required antiphase patterns for later trials. In Panel A a descending in phase trial is shown for comparison.

It is clear also from Fig. 5.5 that early trials are dominated by phase-drift, but that the extent of such drift diminishes over time (trials). Correspondingly, the amount of time spent on producing the required antiphase relation increases systematically. The degree of deviation or fluctuation around the antiphase state also decreases (see posttransition analysis below). The attraction of the in phase pattern is apparent throughout, but its influence also diminishes—even over the space of a single trial (compare e.g., trials 4 and 5 of Fig. 5.5). Very similar behavior is observed in other subjects, though perhaps performance improves more gradually than in subject JJ. The histogram shown in Figure 5.6A shows a steady increase across trials in the number of antiphase patterns produced, as well as their progressively earlier appearance in time (TB).

Still other subjects appear to be entirely frequency dependent, that is, it is only when a certain cycling frequency is reached that the antiphase pattern appears. An example is shown in Fig. 5.7 (KS). Phase wrapping is present (the slope of which reflects the frequency difference between hand and metronome) until, essentially, the penultimate frequency plateau is reached, when the antiphase pattern is observed. A very similar result is shown for two other subjects (BJ and SG) in histogram form in Fig. 5.6 (B and C). Note that regardless of the number of trials, these subjects appear to reach a certain frequency plateau before the antiphase relation is stably produced. Thus there is little or no "learning" or "adaptation"—in the sense of ability to perform the antiphase pattern at an earlier, higher frequency—as was observed for JJ and TB (Fig. 5.5 and 5.6A).

As one might expect, those subjects who changed to the antiphase pattern exhibited a strong mode effect on mean relative phase. For *F*-values ranged from 5819.54 (MC) to 314.07 (BJ), $p_s < .01$. In addition, phase variability is enhanced in the antiphase, relative to the in phase pattern in 3 of the subjects (including LS) as revealed by a mode by frequency plateau interaction, $F(7, 137) = 2.24, p < .03$ (MC), $F(3, 54) = 4.94, p < .01$ (JJ) and $F(9, 166) = 1.96, p < .05$ (LS). These interactions are plotted in Fig. 5.8 for all 3 subjects. For MC, the antiphase pattern is statistically more variable than the in phase pattern on the first three posttransition plateaus (Fig. 5.8A); for JJ, only the first posttransition plateau is statistically different (Fig. 5.8B); and for SL (Fig. 5.8C), although the variability of the antiphase pattern is greater in the first four posttransition plateaus, only on the first one is the variability statistically different between the two patterns.

Analogous to ascending conditions, as cycling frequency decreased over time in the posttransition régmie, its variability increased for all subjects, but one. Also, subject LS showed an interaction between plateau and mode, $F(9, 166) = 1.68, p < .01)$ but the difference between modes reached significance only at the slowest (1 Hz.) frequency.

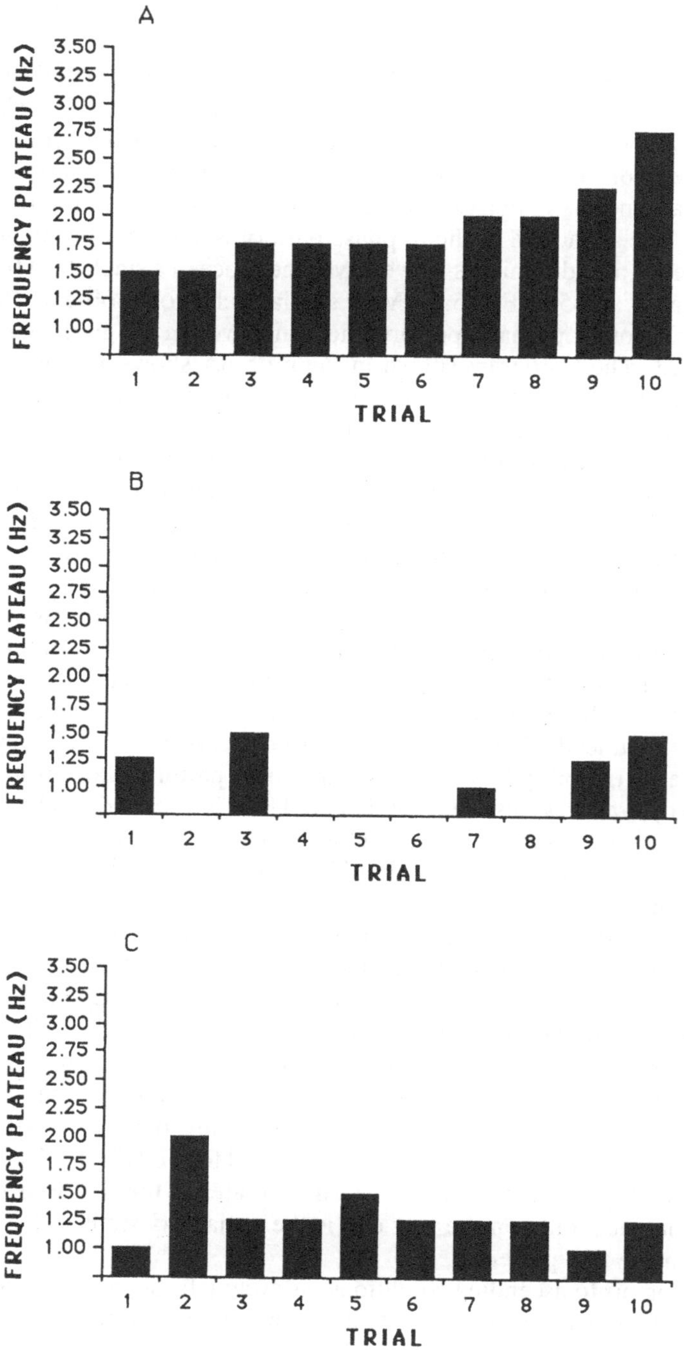


FIG. 5.6. Descending antiphase condition: The darkened bars designate those frequency plateaus at which subjects were able to maintain the antiphase pattern. Note that as a function of trial subject TB (panel A) reaches the antiphase pattern at progressively higher frequencies, whereas subjects SG (panel B) and BJ (panel C) show no such progression.

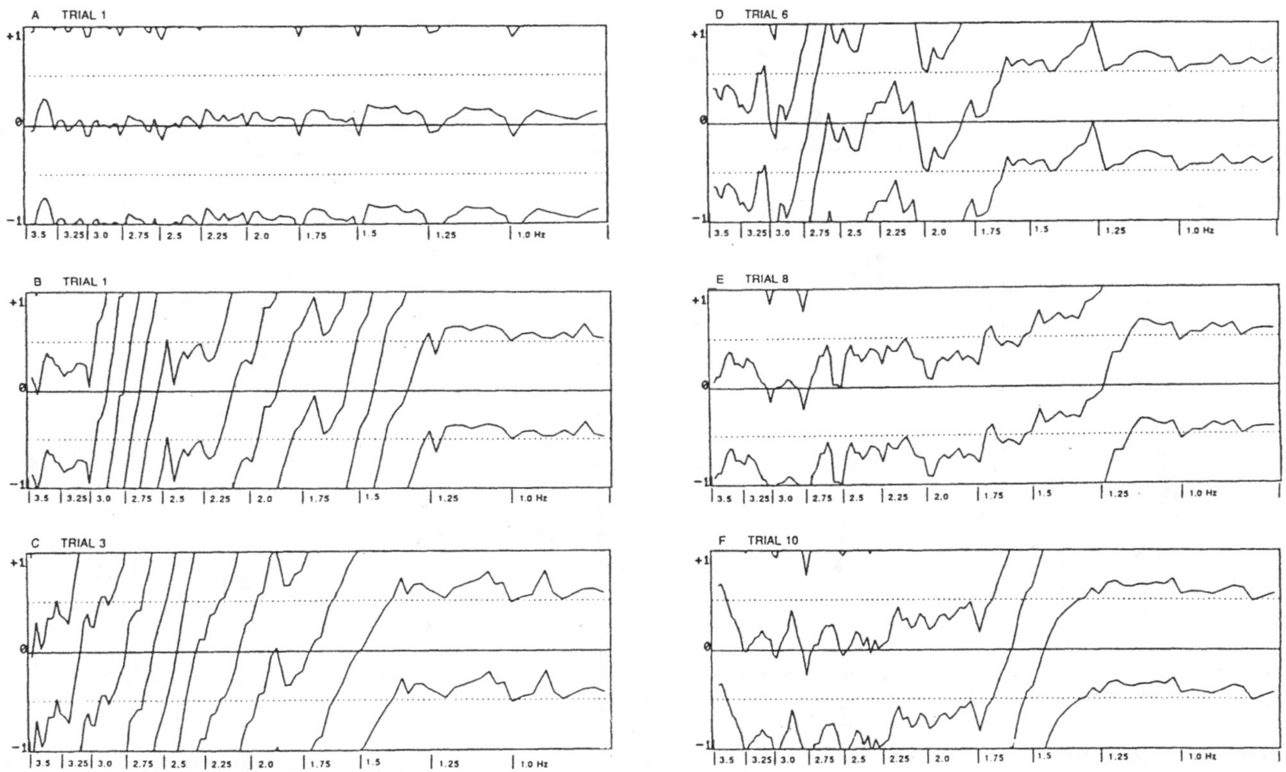


FIG. 5.7. Descending antiphase condition (Panels B through F). Time series of relative phase illustrate the patterns of behavior over subsequent trials of subject KS. The decreasing frequency within each trial is indicated. Note that the onset of the antiphase pattern occurs at a similar frequency across trial runs. In Panel A a descending in phase trial is shown for comparison.

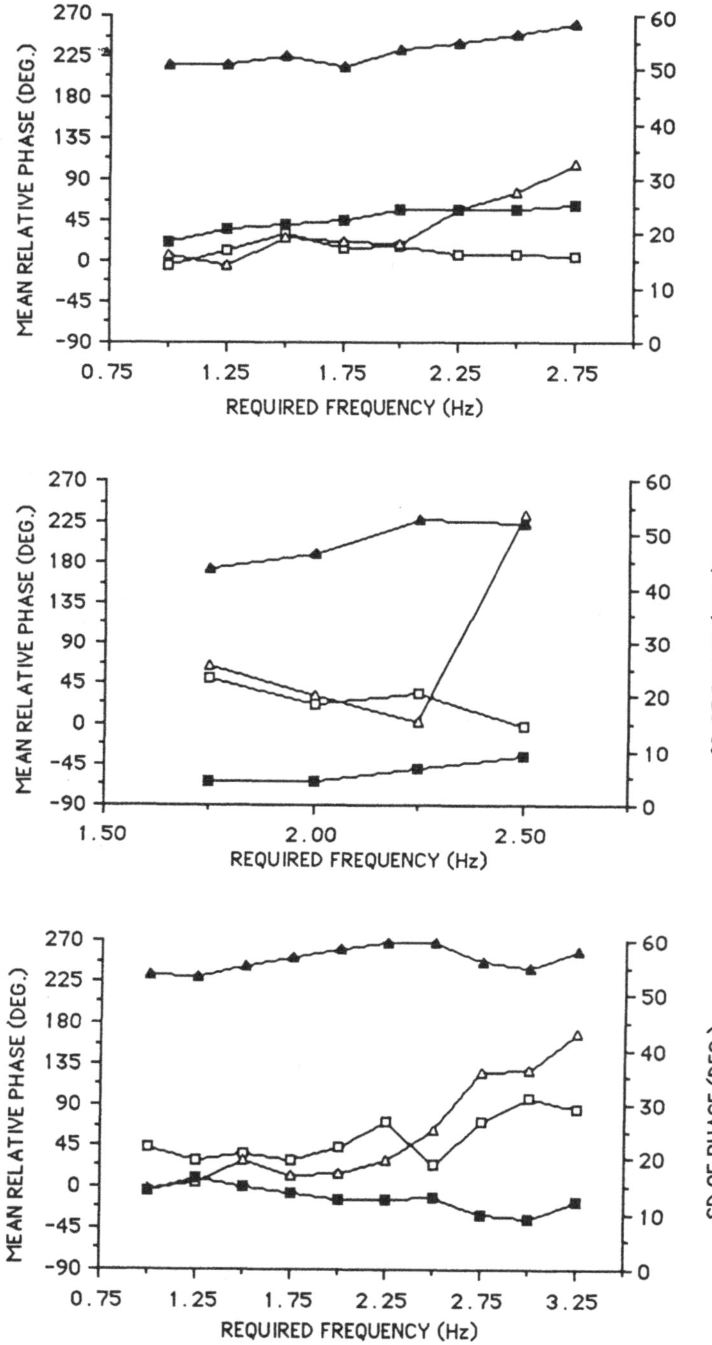


FIG. 5.8. Mean relative phase (filled symbols) and standard deviation of relative phase (open symbols) for posttransition frequency plateaus in the descending antiphase (triangles) and in phase (squares) conditions. Subjects are MC (top), JJ (middle) and LS (bottom).

### Hysteresis

It is intriguing to inquire, even though the tasks for ascending and descending conditions are very different, whether subjects switch patterns at the same frequency. The frequency plateau at which the subject switched permanently from one pattern to the other was noted and *t*-test comparisions conducted on each subject's data. The results were not very systematic: On the average 1 subject switched at a lower frequency plateau in the ascending condition than the descending condition, $t(12) = -3.52$, $p < .004$. The opposite occurred for another subject, $t(14) = 2.22$, $p < .04$. The remaining subjects showed no reliable difference between ascending and descending conditions.

## A THEORETICAL MODEL

To determine a theoretical model of the present action–perception system within the framework of the dynamic theory of behavioral patterns (c.f., Kelso & Schöner, 1987; Schöner & Kelso, 1988c) we identify in a first step an *order parameter* in the sense of synergetics (Haken, 1983) and its dynamics, thereby choosing a level of observation. As discussed earlier, the relative phase, $\phi$, between metronome and rhythmically moving limb is a good candidate. Because both metronome and moving limb are periodic in time, this relative phase is theoretically well defined.

An illuminating subtlety, not addressed so far, involves the exact definition of relative phase. Such definition determines, for example, which pattern is called in phase and which pattern antiphase. Is this choice purely conventional? The answer lies in the functional specificity of the order parameter dynamics. The task requires synchronization of a specific aspect of the movement—here, peak flexion—with the metronome signal. We define the relative phase measure such that temporal alignment of the metronome pulse with this specific movement event corresponds to an in phase pattern. Thus, in the present experiment, peak flexion is used as the reference event from which relative phase is calculated (see Methods). As a result, "on-the-beat" flexion corresponds roughly to in phase and "off-the-beat" flexion corresponds roughly to antiphase. We assume that the dynamics are specific to the task and independent of the synchronization strategy, whereas the *meaning* of relative phase depends on this strategy. If, for example, subjects were instructed to synchronize extension with the metronome rather than flexion, we would expect that the transition phenomenon would be inverted: Events that previously played the role of in phase now become antiphase and hence correspond to the less stable pattern. Whether this view is correct can obviously be tested experimentally.

Because oscillatory states for the collective variable, relative phase, appear not to be relevant, we assume that the dynamics of relative phase are of first order. To find a simple dynamical system that takes the angular character of relative phase into account, we expand the equation of motion into a Fourier series, keeping only the first few terms. It is important to note that we cannot assume symmetry of these dynamics under the operation $\phi \rightarrow -\phi$, because metronome and moving limb are, of course, functionally different. Limiting the expansion to second order we find:

$$\dot{\phi} = \Delta\omega - a \sin(\phi - \psi_1) - 2b \sin(2\phi - \psi_2) + \sqrt{Q}\xi_t \qquad (1)$$

where $\Delta\omega$, $a$, $b$, $\psi_1$ and $\psi_2$ are constants. For reasons discussed elsewhere (e.g., Schöner, Haken, & Kelso, 1986) we have added a Gaussian white noise process, $\xi_t$, as a stochatic force of strength $Q$.

A theoretical description of this perception–action system can also be obtained on the level of the component oscillator. On that level of description, the finger position, $x$, and its velocity, $\dot{x}$, play the role of order parameters. The observed rhythmic behavioral pattern is mapped onto a limit cycle of the $(x, \dot{x})$ dynamics (Haken, Kelso, & Bunz, 1985; Kay, Kelso, Saltzman, & Schöner, 1987). Using the concept of environmental information (Schöner & Kelso, 1988a,c) the influence of the metronome signal can be captured as a contribution to the oscillatory dynamics (see Schöner & Kelso, 1988b, for a related theoretical analysis). From such an underlying oscillator model we find that the parameter $\Delta\omega$ is due to the difference between the metronome frequency and the frequency that the moving limb generates spontaneously. As this frequency difference widens, synchronization is lost and running solutions are found. Futhermore, deriving the relative phase dynamics, Equation (1), from the oscillator model, we find $\psi_1 = \psi_2 = 0$. Therefore we neglect $\psi_1$ and $\psi_2$ in the following.

It is fairly easy to obtain an intuition about the solutions of Equation (1). In Fig. 5.9 we have plotted the right hand side of this equation of motion as a function of $\phi$ for various parameter values. The system has stationary states where $\dot{\phi}$ crosses the $\phi$-axis. These are stable states if the slope of $\dot{\phi}$ is negative, and are unstable states if the slope is positive at this point. The arrows in Fig. 5.9 indicate the direction of the flow. For $\Delta\omega << a$ and $\Delta\omega << b$ (Fig. 5.9, left) two stable states exist, one close to in phase and one close to antiphase. If, for example, $b$ is decreased, the amount of attraction to these stable states decrease until one (Fig. 5.9, center) and ultimately both (Fig. 5.9, right) cease to be stationary solutions. When stationary solutions are no longer found, "running" solutions ($\phi$ ever increasing or decreasing) or, if one applies the periodicity convention to relative phase, "wrapping" solutions occur. Note that for $\Delta\omega = 0$ the system is identical to the relative phase dynamics found earlier for interlimb coordination patterns (Haken, Kelso, & Bunz, 1985). In that case,

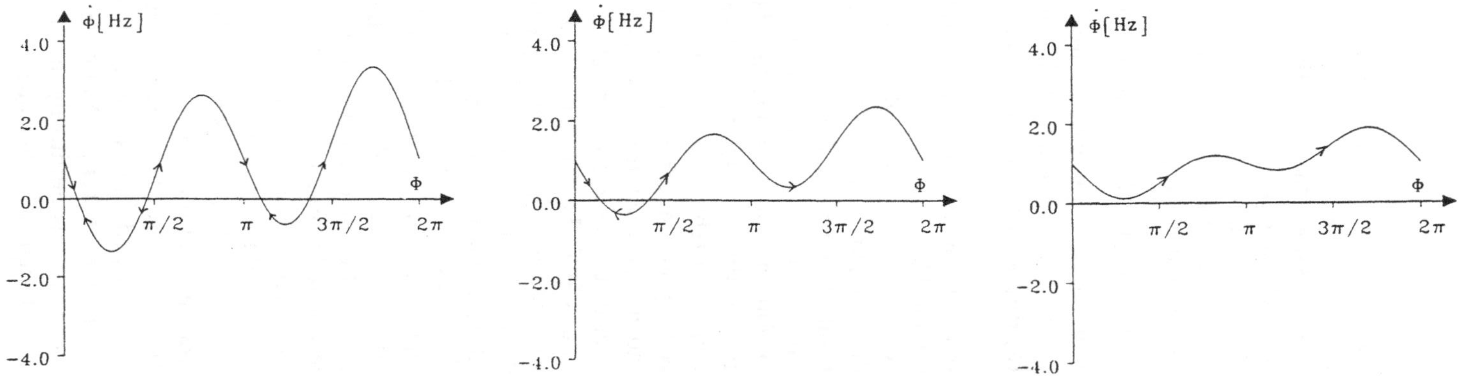


FIG. 5.9. The right hand side of the equation of motion Equation (1) plotted as a function of relative phase for $a$ = 0.5 Hz., $\Delta\omega$ = 1 Hz and $b$ = 1. Hz (left), $b$ = .5 Hz (center) and $b$ = .25 Hz (right). The arrows indicate the flow of $\phi$. Zeros of $\dot{\phi}$ are stationary solutions that are stable if the slope is negative and unstable if the slope is positive at the stationary state.

the system has either one attractor at $\phi = 0$ or two attractors at $\phi = 0$ and $\phi = \pm\pi$.

To obtain a survey over the possible states and their stability, we show in Fig. 5.10 a phase diagram of Equation (1), that is, those regions in parameter space in which the same number of states are stable as well as the boundary lines between different such regions. The stationary states, $\phi_0$, were determined by solving numerically

$$\dot{\phi} = \Delta\omega - a \sin(\phi_0) - 2b \sin(2\phi_0) = 0 \qquad (2)$$

Only attractors, that is, stationary states with positive relaxation rate, $\tau_{rel}^{-1} = a \cos(\phi_0) + 4b \cos(2\phi_0)$, are considered. In Fig. 5.10 the phase diagram is shown in the $(a, b)$-plane, while the parameter $\Delta\omega$ was kept fixed at $\Delta\omega = 1.$Hz.

When the parameters, $a$, $b$, or $\Delta\omega$, are changed, continuous changes of the attractor states are observed. However, if a boundary line of the phase diagram is crossed, phase transitions are observed that may involve abrupt state changes, loss of stability and hysteresis. Different paths in parameter space provide for different transition routes. Several such paths are shown in Fig. 5.10 corresponding to the transition routes that have been observed in our experimental system. On path number one the state close to antiphase first becomes unstable, so that a switch to in phase occurs. The in phase pattern likewise becomes unstable and wrapping solutions are observed. Path number two corresponds to cases in which an antiphase type solution is never stable and a transition from in phase to wrapping is observed. Along path number three the in phase solution persists, but the antiphase solution becomes unstable. Path number four leads directly to wrapping solutions from a régime in which both in phase and antiphase are stable. If we also consider path number four together with path number two in reverse direction, all experimentally observed transition scenarios in the ascending condition are covered (see Results). Similarly the four paths taken in the reverse direction describe all transition scenarios in the descending condition (except for the bouts of phase wandering).

The present theoretical formulation gives rise to several predictions that can be tested in further experimental work. The most important prediction concerns the loss of stability which is at the origin of the observed transitions. This loss can be observed through stability measures such as the relaxation time of relative phase, the time it takes to relax to the pattern after a small perturbation was applied for a short period of time, or the amount of fluctuations of relative phase, measured in terms of its variability. When a transition is approached and the system is prepared in the state that loses stability (cf. Schöner, Haken, & Kelso, 1986) these measures increase strongly. One can see this intuitively in Fig. 5.9: As a state loses stability the slope of $\dot{\phi}$ at that state becomes flatter, that is, the

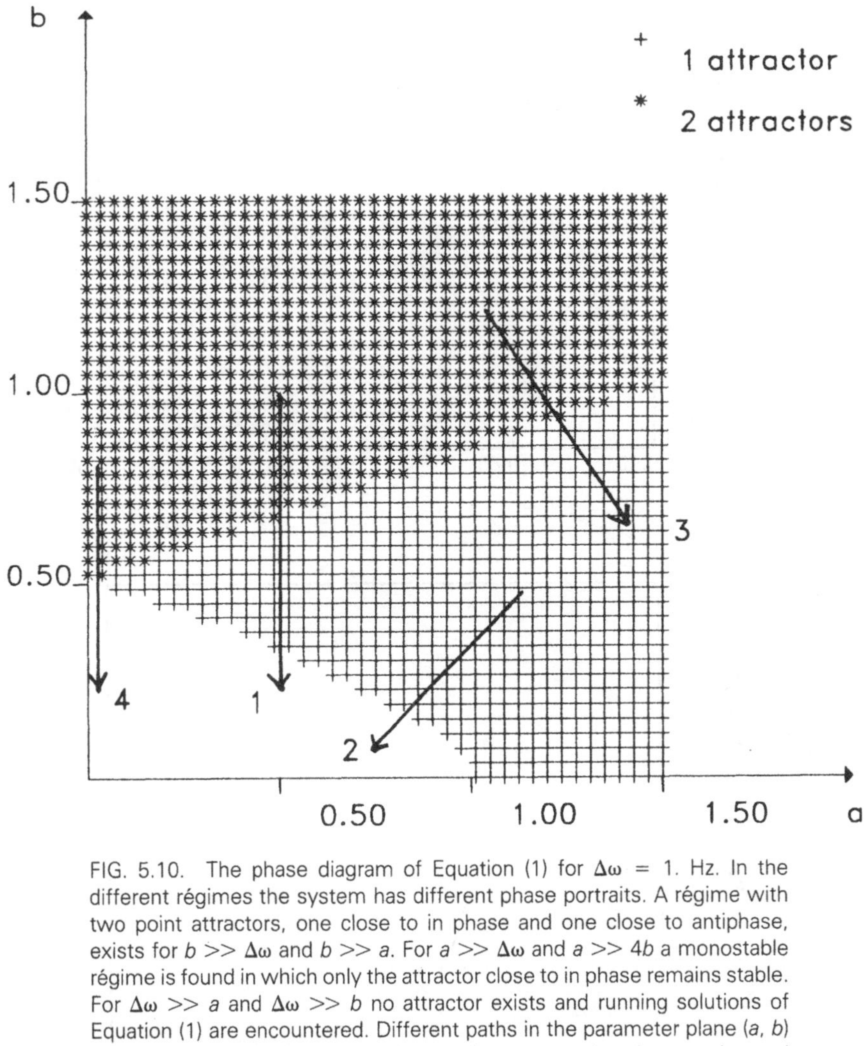


FIG. 5.10. The phase diagram of Equation (1) for $\Delta\omega = 1$. Hz. In the different régimes the system has different phase portraits. A régime with two point attractors, one close to in phase and one close to antiphase, exists for $b >> \Delta\omega$ and $b >> a$. For $a >> \Delta\omega$ and $a >> 4b$ a monostable régime is found in which only the attractor close to in phase remains stable. For $\Delta\omega >> a$ and $\Delta\omega >> b$ no attractor exists and running solutions of Equation (1) are encountered. Different paths in the parameter plane $(a, b)$ yield different transition routes that can be compared to the experimental results (see text). The dynamics depicted in Figure 5.9 correspond to different points on path number one.

system is "slowing down." Hints of a corresponding enhancement of fluctuations (cf. Kelso, Scholz & Schöner, 1986) are evident in the experimental data shown in Fig. 5.4. Fig. 5.11 shows the different solutions of Equation (1) as the parameter $b$ is decreased along path number one as shown in Fig. 5.10. The relaxation time of each solution is shown and exhibits loss of stability as the transition lines are crossed (Note that for the range of

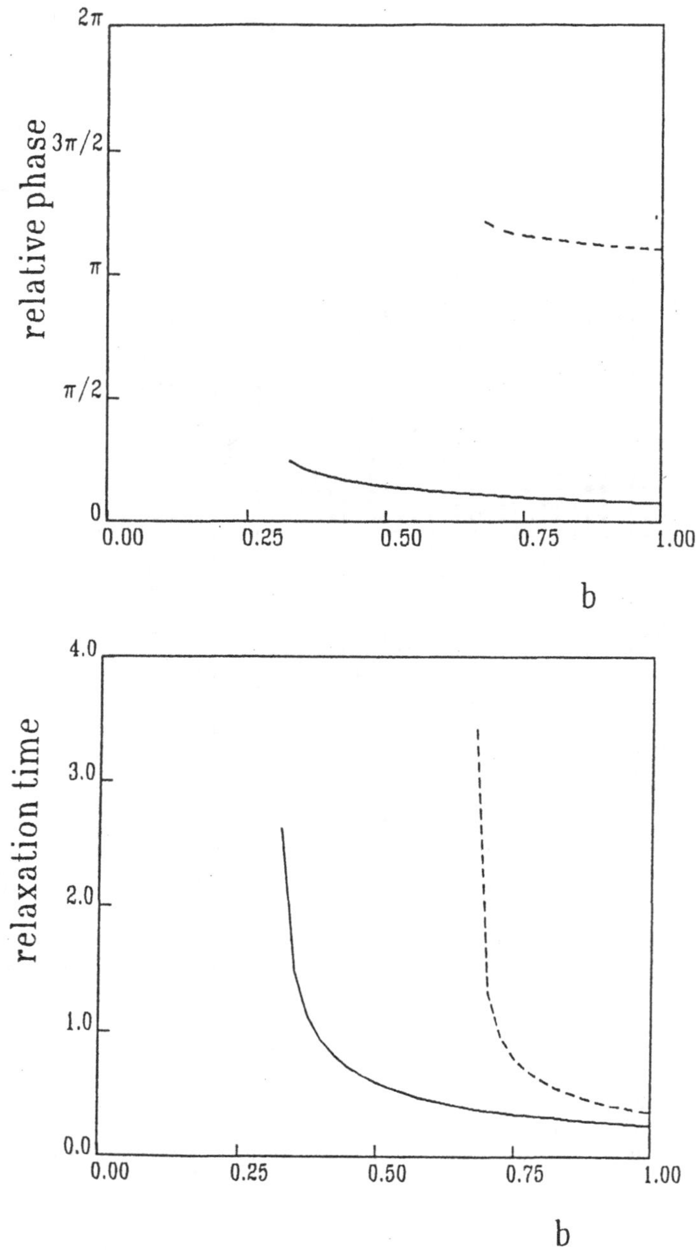


FIG. 5.11. (Top) The stationary solutions of Equation (1) and their relaxation time (bottom) as a function of the parameter $b$ (the parameters $a = 0.5$ Hz and $\Delta\omega = 1$ Hz are kept fixed). The transition route shown corresponds to the path number one in Figure 5.10. For large $b$ two attractors, one close to in phase ($\phi \approx 0$: dashed lines) and one close to antiphase ($\phi \approx \pi$: full line), exist (cf. left part of Figure 5.9). As $b$ is decreased first the attractor close to antiphase, then the attractor close to in phase become unstable. The loss of stability is indicated by an increase in the relaxation times of these attractors.

the parameter plane shown, the relaxation times of the different patterns are in a realistic order of magnitude, $\tau_{rel} \sim 1$ s.). The variability of relative phase shows the same increase as can be seen from the relation (Schöner, Haken, & Kelso, 1986)

$$SD = \sqrt{Q\tau_{rel}} \tag{3}$$

where *SD* is the standard deviation of relative phase around its mean state. For transitions among attractors, a further measure is the length of the transient of the actual switch—the so-called switching time—that can be calculated theoretically given the stability measures (Schöner, Haken, & Kelso, 1986) and measured experimentally (Scholz, Kelso, & Schöner, 1987).

An interesting aspect emerges when the so-called running solutions are examined. In the theory, these running solutions have a fine structure: They spend more time at those relative phase values at which the force $\phi$ is minimal. This is the phenomenon of *relative coordination* observed first by von Holst (1937/73) in studies of rhythmically moving limbs of several species. Fig. 5.12 shows a numerical simulation of Equation (1) in this parameter régime. The observed periodic modulation of relative phase is a signature of relative coordination and is due to the attraction to certain preferred phase relationships contained in the dynamics of Equation (1). The periodicity interval of these relative phase modulations can vary considerably due to noise, illustrating that this periodicity is not, in itself, a stable state.

## GENERAL DISCUSSION AND SUMMARY

Building on previous work that dealt with patterns of movement coordination, their stability and change, we studied here the question of how synchronization with a periodically structured environment occurs. We treated this problem as one of pattern formation, that is, as an attempt to understand the dynamic nature of the relationship between action and perception. A variety of different entrainment régimes between hand and metronome were observed and the pathways among patterns noted. A simple theoretical model—again developed from dynamic pattern theory—was able to reproduce the observed timing patterns. These are in turn derivable from the component level of description.

It is interesting to note that terms such as *in phase* or *antiphase* are *relative* in the system studied here. For example, as one subject (LS) so cleverly revealed, it is possible to form an *antiphase* pattern (defined experimentally as a flexion motion off-the-beat) as an *in phase* pattern, by producing an extension motion on-the-beat. The *abstract* dynamics of en-

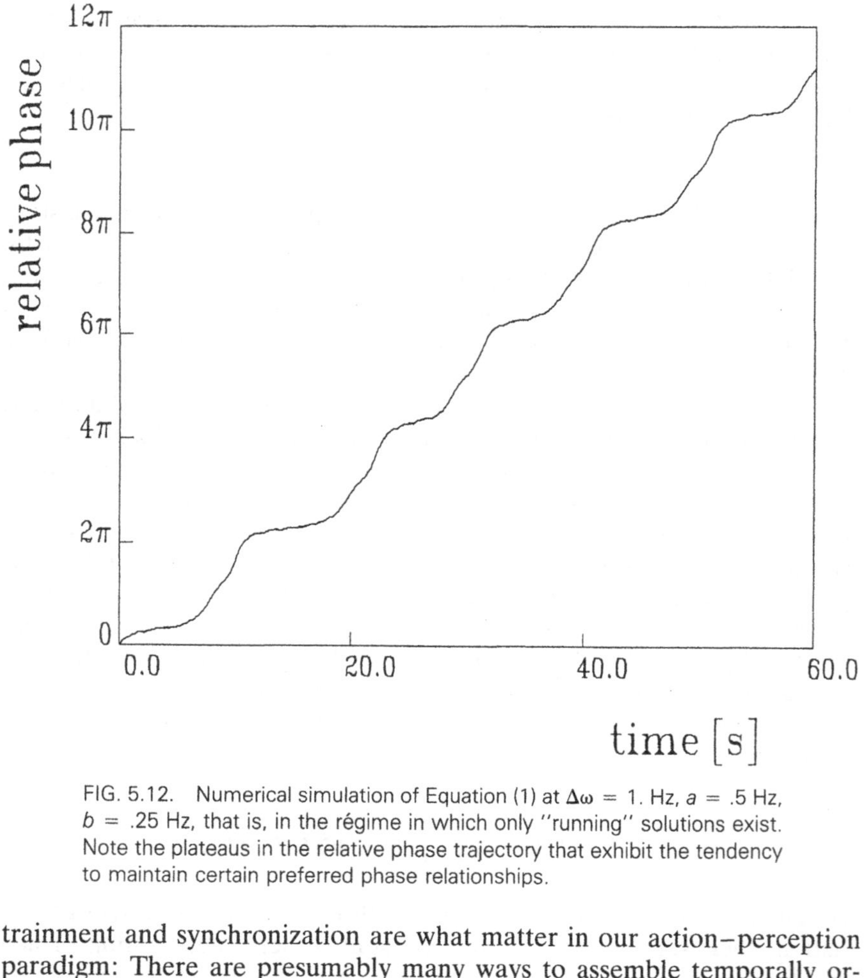


FIG. 5.12. Numerical simulation of Equation (1) at $\Delta\omega$ = 1. Hz, $a$ = .5 Hz, $b$ = .25 Hz, that is, in the régime in which only "running" solutions exist. Note the plateaus in the relative phase trajectory that exhibit the tendency to maintain certain preferred phase relationships.

trainment and synchronization are what matter in our action–perception paradigm: There are presumably many ways to assemble temporally ordered patterns. Nevertheless, we should emphasize that although the dynamics are abstract, they are always *measureable*, as an operational theory requires. Our modeling strategy of identifying collective variables for (in the present case) action–perception patterns, mapping them onto a minimum number of attractors and studying stability and its loss, offers a conceptual framework to understand how multiple, temporally ordered states may arise from different mechanisms. We treat the environment in a parametric fashion, that is, as that aspect of the physical environment that influences the behavioral patterns without itself being influenced by the system. For the physical environment to become meaningful to the biological system it is necessary that a dynamic relation exist between the system and its environment. We conceive this process as the formation of an action–perception pattern. Environmental information is viewed as meaningful and specific to the extent that is forms part of the behavioral dynamics. More precisely, environmental information perturbs the vector field of the intrinsic dynamics, such that it attracts these dynamics to the required behavioral pattern (cf. Schöner & Kelso, 1988a, b, c). In the present case, the limit cycle dynamics of the hand are "tunable," in the sense that they are modified by an external periodic signal. We see also that there are limits to this process—readily interpreted as loss of entrainment. When metronome frequency is scaled, the system shifts from one entrainment region to another. In fact, our simple model reveals a variety of pathways through which different states (stable and unstable) are achieved (see Theoretical Model and discussion therein). Often, loss of stability accompanies the shift from one entrainment régime to another as evident from enhanced fluctuations of the order parameter, relative phase. Such effects are predicted from dynamic pattern theory provided certain time scales relations exist (for a readable discussion, see Kelso et al., 1987, or Schöner & Kelso, 1988d).

What are the general implications of the present theoretical strategy for understanding perceptual–motor skills? Are we dealing with a very special case here, for example? We think not. Our psychological ancestors were deeply interested in the problem we have studied here, we think with good reason. As beautifully exemplified in Lee's work (e.g., Lee & Young, 1986, for review), synchronization and relative timing are crucial aspects of skilled behavior. Unfortunately, the parametric influences of environmental information are not always explored, as in Lee and Lishman's (1975) study of body sway, where only a single visual driving frequency was employed. The possibility of identifying multiple visual–postural dynamic patterns—analogous to those found here—and mapping them onto a theoretical model (with new predictions and different observables) remains open. Lee and Lishman (1975) for example, focus on the sway velocity of the *body* as their main dependent measure, rather than, as we have done here, the relative phasing between environment and body. Their Fig. 4, which shows the time series of the moving room and the moving body embedded in that environment, could clearly be subjected to the experimental and theoretical analysis conducted here in terms of collective variables spanning the action–perception system.

"Understanding" of the action–perception system depends, of course, on the language used. Inherent in our approach is an abstraction away from *what* material objects are producing temporal patterns, and toward defining the dynamical laws governing the patterns themselves. Concepts of collective variables characterizing entrained states, control parameters, multistability, loss of stability, and so forth have only recently been demonstrated to be useful in understanding temporally patterned phenomena at several scales of observation. A notable feature of the dynamic pattern approach is that it is possible to *derive* observed patterns and their dynamics from cooperative interactions at the component level. Our model also shows that the multiple observed patterns we have observed experimentally correspond to different régimes of the *same* underlying dynamics.

## ACKNOWLEDGMENTS

This research was supported by grant MH42900-01 from NIMH (Neurosciences Research Branch) and Contract No. N00014-88-J-1191 from the U.S. Office of Naval Research (Perception Sciences). We thank C. Madrid for help with the data analysis.

## REFERENCES

Arbib, M. A. (1987). A view of brain theory. In F. E. Yates (Ed.), *Self-organizing systems* (pp. 272–311). New York: Plenum Press.

Bartlett, N. R., & Bartlett, S. C. (1959). Synchronization of a motor response with an anticipated sensory event. *The Psychological Review*, *6*, 203–218.

Deutch, D. (1983). The generation of two isochronous sequences in parallel. *Perception & Psychophysics*, *34*, 331–337.

Dunlap, K. (1910). Reactions to rhythmic stimuli, with attempt to synchronize. *The Psychological Review*, *17*, 399–416.

Gibson, J. J. (1979). The ecological approach to visual perception. Boston: Houghton Mifflin.

Haken, H. (1983). *Synergetics: An introduction*. Springer-Verlag, Berlin, Heidleberg.

Haken, H., Kelso, J. A. S., & Bunz, H. (1985). A theoretical model of phase transitions in human hand movements. *Biological Cybernetics*, *51*, 347–356.

Hanson, F. (1978). Comparative studies of firefly pacemakers. *Federation Proceedings*, *37*, 2158–2164.

Hary, D., & Moore, G. P. (1987). Synchronizing human movement with an external clock source. *Biological Cybernetics*, *56*, 305–311.

Kay, B. A., Kelso, J. A. S., Saltzman, E. L., & Schöner, G. (1987). Space–time behavior of single and bimanual rhythmical movements. Data and limit cycle model. *Journal of Experimental Psychology: Human Perception and Performance*, *13*, 178–192.

Kelso, J. A. S. (1984). Phase transitions and critical behavior in human bimanual coordination. *American Journal of Physiology*, *246*, R1000–R1004.

Kelso, J. A. S., & Holt, K. (1980). Exploring a vibratory systems analysis of human movement production. *Journal of Neurophysiology*, *43*, 1183–1195.

Kelso, J. A. S., & Kay, B. A. (1987). Information and control: A macroscopic analysis of perception–action coupling. In H. Heuer & A. F. Sanders (Eds.), *Perspectives on perception and action* (pp. 3–32). Hillsdale, NJ: Lawrence Erlbaum Associates.

Kelso, J. A. S., Scholz, J. P., & Schöner, G. (1986). Non-equilibrium phase transitions in coordinated biological motion: Critical fluctuations. *Physics Letters A*, *118*, 279–284.

Kelso, J. A. S., Scholz, J. P., & Schöner, G. (1988). Dynamics governs switching among patterns of coordination in biological movement. *Physics Letters A*, *134*, 8–12.

Kelso, J. A. S., & Schöner, G. S. (1987). Toward a physical (synergetic) theory of biological coordination. *Springer Proceedings in Physics*, *19*, 224–237.

Kelso, J. A. S., Schöner, G., Scholz, J. P., & Haken, H. (1987). Phase-locked modes, phase transitions and component oscillators in biological motion. *Physica Scripta*, *35*, 79–87.

Kugler, P. N., Kelso, J. A. S., & Turvey, M. T. (1982). On the control and coordination of naturally developing systems. In J. A. S. Kelso & J. E. Clark (Eds.), *The development of movement control and coordination*. New York: Wiley.

Lee, D. N., & Lishman, J. R. (1975). Visual proprioceptive control of stance. *Journal of Human Movement Studies*, *1*, 87–95.

Lee, D. N., & Young, D. S. (1986). Gearing action to the environment. *Experimental Brain Research Series*, *15*, 217–230.

Michaels, C. F. & Carello, C. (1981). *Direct perception*. Englewood Cliffs, NJ: Prentice-Hall.

Perkel, D. H., Schulman, J. H., Bullock, T. H., Moore, G. P., & Segundo, J. P. (1964). Pacemaker neurons: Effects of regularly spaced synaptic input. *Science*, *145*, 61–63.

Scholz, J. P., Kelso, J. A. S., & Schöner, G. (1987). Non-equilibrium phase transitions in coordinated biological motion: Critical slowing down and switching time. *Physics Letters A*, *123*, 390–394.

Schöner, G., Haken, H., & Kelso, J. A. S. (1986). A stochastic theory of phase transitions in human hand movement. *Biological Cybernetics*, *53*, 247–257.

Schöner, G., & Kelso, J. A. S. (1988a). A synergetic theory of environmentally-specified and learned patterns of movement coordination. I. Relative phase dynamics. *Biological Cybernetics*, *58*, 71–80.

Schöner, G., & Kelso, J. A. S. (1988b). A synergetic theory of environmentally-specified and learned patterns of movement coordination. II. Component oscillator dynamics. *Biological Cybernetics*, *58*, 81–89.

Schöner, G., & Kelso, J. A. S. (1988c). Dynamic patterns of biological coordination: Theoretical strategy and new results. In. J. A. S. Kelso, A. J. Mandell, & M. F. Shlesinger (Eds.), *Dynamic patterns in complex systems* (pp. 77–102). Singapore: World Scientific.

Schöner, G., & Kelso, J. A. S. (1988d). Dynamic pattern generation in behavioral and neural systems. *Science*, *239*, 1513–1520.

Schöner, G., & Kelso, J. A. S. (1988e). A dynamic pattern theory of behavioral change. *Journal of Theoretical Biology*, *135*, 501–524.

Shimizu, H., Yamaguchi, Y., & Satoh, K. (1988). Holovision: A semantic information processor for visual perception. In. J. A. S. Kelso, A. J. Mandell, & M. F. Schlesinger (Eds.), *Dynamic patterns in complex systems* (pp. 42–76). Singapore: World Scientific.

von Holst, E. (1973). On the nature of order in the central nervous system. In R. Martin (Trans.), *The collected papers of Erich von Holst*. Coral Gables, Florida: University of Miami press (Original work published 1939).

Warren, W. H. (1988). Critical behavior in perception–action systems. In. J. A. S. Kelso, A. J. Mandell, & M. F. Shlesinger (Eds.), *Dynamic patterns in complex systems* (pp. 370–387). Singapore: World Scientific.

Wendler, G. (1974). The influence of proprioceptive feedback on locust flight coordination. *Journal of Comparative Physiology*, *88*, 173–200.

Wing, A. M., & Kristofferson, A. B. (1973). Response delays and the timing of discrete motor responses. *Perception & Psychophyics*, *14*, 5–12.

Woodrow, H. (1932). The effect of rate sequence upon the accuracy of synchronization. *Journal of Experimental Psychology*, *4*, 357–379.

# III MOVEMENT INITIATION AND MOTOR OUTPUT SPECIFICATION IN VOLUNTARY ACTION

# 6 Speed–Accuracy Tradeoffs in Aimed Movements: Toward a Theory of Rapid Voluntary Action

David E. Meyer
J. E. Keith Smith
Sylvan Kornblum
*University of Michigan*

Richard A. Abrams
*Washington University*

Charles E. Wright
*Columbia University*

## ABSTRACT

As the speed of rapid aimed movements increases, their spatial accuracy typically decreases. The mathematical form of the speed–accuracy tradeoff depends on the type of movement task being performed. Several alternative hypotheses have been proposed to account for this dependence, including ones that make assumptions about feedback-guided corrective submovements and about the stochastic variability of underlying neuromotor force pulses. A new hybrid class of stochastic optimized–submovement models provides a way to integrate these past accounts in a unified theoretical framework. From the vantage point of this framework, the present chapter reviews the evolution of speed–accuracy tradeoff research and shows how a fresh perspective regarding the properties of elementary movement mechanisms may be obtained.

## INTRODUCTION

A fundamental topic in the field of human motor performance concerns the characteristics of rapid aimed movements. Such movements are especially interesting because many physical skills involve sequences of actions during which a person must shift selected parts of the body both quickly and accurately from one place to another in space. Considerable research has therefore been devoted to studying how movement speed and accuracy are related to each other (Keele, 1968, 1981, 1986).

Significant progress has been made through this research. Investigators have found that precise quantitative tradeoffs occur in rapid aimed movements: As movement speed increases, the spatial accuracy of a movement systematically decreases; conversely, as the spatial accuracy of a movement increases, movement speed systematically decreases. The mathematical form of the speed–accuracy tradeoff depends on certain fundamental details of the movement task at hand (Fitts, 1954; Meyer, Abrams, Kornblum, Wright, & Smith, 1988; Meyer, Smith, & Wright, 1982; Schmidt, Zelaznik, Hawkins, Frank, & Quinn, 1979; Wright & Meyer, 1983; Zelaznik, Mone, McCabe, & Thaman, 1988). Given the nature of the dependence, there is now some hope that a unified theoretical framework for characterizing speed–accuracy tradeoff relations may soon emerge.

The present chapter provides a tutorial review of the research on movement speed and accuracy that has produced this progress. It contains three major parts. First, we outline key aspects of the methodology developed by psychologists, kinesiologists, and neurophysiologists to analyze speed–accuracy tradeoffs. Second, we survey some classic studies regarding the bases of these tradeoffs. Third, we summarize recent empirical and theoretical work conducted in our laboratory to clarify the psychophysical processes underlying various forms of tradeoff relations. In particular, a new hybrid class of stochastic optimized–submovement models (Meyer et al., 1988) is offered here as a way of explaining salient phenomena revealed by both our own experiments and those of other investigators. With these models, one may derive a comprehensive account of numerous details in rapid aimed movements.

Because of constraints on available space, the following review has a necessarily restricted scope. Our main concern is with rapid discrete aimed movements, not slow movements or continuous tracking. Although we recognize the importance of biomechanical and neurophysiological evidence regarding movement speed–accuracy tradeoffs, we primarily consider behavioral data to elucidate them. The range of data under consideration does not include all possible tradeoff relations. We discuss at length how movement speed and spatial accuracy are related to each other, but complementary relations between the speed and temporal accuracy of aimed movements are not addressed fully.[1] For further reviews of these topics,

[1]The relationship between the speed and temporal accuracy of aimed movements is not necessarily isomorphic with the relationship between movement speed and spatial accuracy. Newell (1980) and his colleagues (e.g., Hancock & Newell, 1985) have claimed that as the average velocity of movements increases, their durations may actually deviate less and less from a selected temporal goal. This claim still has to be assessed systematically, and not all investigators would agree completely with it (cf. Schmidt et al., 1979; Wright, 1983).

readers may consult a number of other sources (e.g., Georgopoulos, 1986; Hancock & Newell, 1985; Jagacinski, 1989; Jeannerod, 1988; Keele, 1981, 1986; Marteniuk & MacKenzie, 1980; Pew, 1974; Poulton, 1974; Sanes & Evarts, 1984).

## METHODOLOGICAL FRAMEWORK

To appreciate the substantive issues that surround movement speed-accuracy tradeoffs, some familiarity with the methodological framework used for pursuing them is essential. This framework embodies two important components: a family of experimental procedures designed to induce observed tradeoff relations, and a set of associated performance measures. The next subsections outline each of these in turn.

### Experimental Procedures

*Generic protocol.* Standard experimental procedures used to study speed-accuracy tradeoffs in discrete aimed movements involve a protocol with a series of individual test trials. On each trial, a marker is first displayed to specify the starting location for an aimed movement, and the subject must align a selected part of the body (e.g., hand) with the location that has been specified. Next, a target is displayed to indicate the desired stopping location for the movement. It may consist of either a point, line, or region placed some distance from the starting location. Then there is a response signal (e.g., auditory tone) that cues the subject to move quickly and stop accurately at the target.[2] The time between the response signal and the initiation of the subject's movement is the *movement latency*. The time between movement initiation and movement termination is the *movement duration*. Both of these times may be measured as a function of certain independent variables. An experimenter may also measure other aspects of the movement, such as its actual endpoint in space and its trajectory from start to stop. Finally, feedback may be provided to maintain motivation and improve performance.

*Movement characteristics.* If performance were perfect, then following the response signal on each trial, the subject would move in a continuous

[2]In certain special cases, the target may instead represent a location through which the movement must pass quickly at a specified time without stopping there (e.g., Schmidt et al., 1979). An example of this latter type is that of hitting a ball. Although such movements have considerable interest for a complete understanding of speed-accuracy tradeoffs, we do not dwell on them further here.

rapid fashion from the starting location and stop exactly at the target. There would be no hesitations or corrections along the way. However, this is not always what happens in actual practice. Even when the same movement should be made repeatedly, performance may vary from trial to trial, exhibiting both systematic and random fluctuations. On some trials, an observed movement may undershoot the target, whereas on other trials, an overshoot may occur. Also, the trajectories of the movements may contain salient discontinuities, manifesting the occurrence of component submovements (e.g., see Carlton, 1979, 1980; Crossman & Goodeve, 1963/1983; Jagacinski, Repperger, Moran, Ward, & Glass, 1980; Meyer et al., 1988). The properties of this between-trial variability depend on details of how the experimental procedure is implemented.

*Alternative movement tasks.* Subjects have been given several alternative tasks as part of experiments on movement speed–accuracy tradeoffs. These tasks are defined operationally by the spatial and temporal goals that the subjects must try to achieve with their movements. For example, suppose that a subject has the spatial goal of producing movements whose distances closely approximate a particular target value, *D*. Then an experimenter may combine this spatial goal with a variety of different temporal goals.

One procedure is to require that the subject produce movements whose durations match a specified target value *T*, while at the same time making their distances approximate the target value *D* as closely as possible. We call this a *time-matching movement task* with a spatial constraint (cf. Meyer et al., 1988, who have also called it a "temporally constrained" movement task). As discussed later, the time–matching movement task was the first to be used systematically for exploring speed–accuracy tradeoffs in discrete aimed movements (Woodworth, 1899).

In contrast, a second procedure requires subjects to minimize their average movement durations as much as possible while making their movement distances closely approximate a specified target value *D*. We call this a *time-minimization movement task* with a spatial constraint (cf. Meyer et al., 1988, who have also called it a "spatially constrained" movement task). The time-minimization movement task was introduced after the time-matching movement task (e.g., Searle & Taylor, 1948) and provides a potentially useful complement. We show subsequently that by comparing results from these two alternative types of task, considerable light can be shed on the processes underlying movement speed–accuracy tradeoffs.

## Performance Measures

For the time-matching and time-minimization movement tasks, several aspects of performance must be measured to assess induced speed–accuracy tradeoff relations. Obviously, measures of movement distance and duration are needed. Also, measures are needed of the degree to which the movements achieve their desired spatial and temporal goals. In each case, an experimenter may take several different approaches to meet these needs, and the chosen approach can influence the validity of resulting conclusions.

*Movement duration.* As indicated already, the duration of a movement is ordinarily defined to be the amount of time between movement initiation (i.e., the moment when a selected part of the body leaves its starting location) and movement termination (i.e., the moment when the stopping location is finally reached). This value may be measured either directly or indirectly. The direct approach involves separately recording both the initiation and termination of each individual movement as a function of time, for example, by extracting them from an explicit record of the full movement trajectory (Meyer et al., 1988). Alternatively, the indirect approach involves having subjects make a series of repetitive movements during a set time interval, and then counting how many movements are made from the moment when the interval starts until the moment when it ends. Given this count, an experimenter can estimate the mean movement duration by dividing the obtained number of movements into the overall length of the time interval (e.g., Fitts, 1954; Woodworth, 1899).

Although experimenters have sometimes taken the indirect approach to measuring movement durations, it seems less desirable than the direct approach. Hesitations having unknown and variable magnitudes may occur between successive repetitive movements, even though subjects are instructed to avoid them. Depending on how such hesitations vary as function of different factors, they can cause spurious biases in the estimated durations, leading to invalid conclusions.

*Movement distance.* Similarly, there are direct and indirect approaches to measuring movement distance. Using the direct approach, one would explicitly record the starting and stopping locations of each individual movement, treating the difference between them as the movement distance on a trial (e.g., Meyer et al., 1988). In contrast, using the indirect approach, one would simply approximate actual movement distances by assuming that they equal the nominal distance between the starting location and the target (e.g., Fitts, 1954).

Each of these approaches has advantages and disadvantages. The latter (indirect) approach is obviously easier to implement. However, if there are more undershoots than overshoots of the target, or vice versa, then this may introduce systematic biases in subsequent data analyses. Also, important effects associated with between-trial variability of actual movement endpoints may be obscured. Such problems may likewise be encountered to some extent when an experimenter tries to measure the distance of a movement directly. For example, background fluctuations in the motor system associated with tremor (Stein & Lee, 1981) and other springlike resonances (Bizzi, Dev, Morasso, & Polit, 1978; Cooke, 1980; Polit & Bizzi, 1979) can make the movement's starting and stopping locations somewhat ambiguous. Nevertheless, the direct approach still has the potential for providing more complete and precise information about movement accuracy than does the indirect approach.

*Movement accuracy.* Through direct measures of movement distance and duration, the spatial and temporal accuracy of aimed movements may be quantified in several ways. Suppose, in particular, that the target value for a subject's movements on some physical dimension (e.g., distance or duration) is $x^*$, and that while attempting to achieve this value with each of $n$ movements, the subject produces a set of observed values $x_i$ ($i = 1, 2, \ldots, n$). Then three popular quantitative measures of movement accuracy are available (Schutz & Roy, 1973).

The first of these is *constant error* (*CE*), the difference between the target value and the mean ($\bar{x}$) of the observed values for a set of movements (i.e., $CE = \bar{x} - x^*$, where $\bar{x} = \Sigma_{i=1}^{n} x_i/n$). *CE* indicates how close on average the movements come to the target value. With this measure, one can quantify systematic biases that the movements have relative to their spatial and temporal goals. If the biases are small, then *CE* should be essentially zero. On the other hand, if there is a greater tendency to overshoot than undershoot the target value, then *CE* would be positive, whereas *CE* would be negative when there is a greater tendency toward undershoots.

A second complementary measure is *variable error* (*VE*), the standard deviation of the observed values about their mean $\{$i.e., $VE = [\Sigma_{i=1}^{n} (x_i - \bar{x})^2/(n - 1)]^{1/2}\}$. With this measure, one can assess the amount of random noise in a distribution of repeated movements. The greater the noise, the greater *VE* will be. In principle, *VE* and *CE* are statistically independent, so experimenters should report both of them, because neither one alone provides a sufficient description of movement accuracy (Schutz & Roy, 1973).

Third, there is *mean absolute error* (*AE*). It involves subtracting the observed value for each movement from the corresponding target value and then averaging the absolute values of the resulting differences (i.e., $AE = \Sigma_{i=1}^{n} |x_i - x^*|/n$). With this measure, one can assess the average magnitude of a subject's errors regardless of whether they are positive or negative. As the errors increase in magnitude, so will *AE*. Many past investigators have used *AE* as their primary dependent variable (e.g., Woodworth, 1899), because it provides a convenient composite indicator of how well a subject has performed.

Unfortunately, *AE* is also potentially ambiguous and misleading (Schutz & Roy, 1973). It has a high correlation with the sum of $CE^2$ and $VE^2$. An experimenter who reports *AE* without looking separately at these other measures cannot tell to what extent performance has been affected by systematic biases or random noise. Such uncertainty may, in some cases, yield incorrect conclusions.

A similar caveat applies to other summary measures of movement accuracy. When the spatial goal for a set of movements is a target region, accuracy has occasionally been quantified in terms of the relative frequency of target hits (e.g., Fitts, 1954). Like *AE*, however, the target-hit frequency is actually a joint function of both *CE* and *VE* as well as higher-order moments of the movement-endpoint distribution, so it too may yield ambiguous results when reported alone. For this reason, a few investigators have recommended that a large battery of distributional statistics regarding movement endpoints, including not only *CE* and *VE* but also skewness, kurtosis, and so forth should be used to characterize a subject's accuracy of performance (e.g., Newell & Hancock, 1984).

## HISTORICAL SURVEY

With the preceding methodological points in mind, let us examine more closely some past research on speed–accuracy tradeoffs in discrete aimed movements. For our survey, a period of almost a century is covered, starting right before the year 1900 and progressing up through the late 1980s. The studies included here do not comprise an exhaustive sample of the literature, but they nevertheless illustrate some major trends that have occurred over time. We show how different theoretical accounts of relationships between movement speed and accuracy have emerged from results obtained through alternative movement tasks, and we show how inadequate performance measures have sometimes compromised the validity of these accounts.

### Woodworth's Legacy

The individual with whom we begin is R. S. Woodworth, one of the earliest pioneers in the behavioral study of aimed movements. He may be deemed the founding father of research on movement speed–accuracy tradeoffs. Although a few other investigators (e.g., Fullerton & Cattell, 1892) predated Woodworth, their work mainly concerned the effects of movement speed on the accuracy of sensory processes; they did not systematically study how the speed of movement affects the accuracy with which specified spatial and temporal goals (*viz.*, target distance and duration) are attained. Instead, the latter topic served as the primary focus of Woodworth's efforts, and he contributed the first comprehensive exploration of it. As Woodworth remarked, he sought to establish a psychophysics of human motor performance comparable in aim and scope to sensory psychophysics. Toward this end, his doctoral dissertation, which appeared in 1899, set the standard for many of the subsequent studies of speed–accuracy tradeoff relations.

*Research questions.* Several important questions guided Woodworth's (1899) work. The first of these concerned whether the spatial accuracy of aimed movements does indeed depend on movement speed. In Woodworth's day, the answer was not entirely clear yet. He was especially impressed by the fact that some rapid movements over long distances seemed to be consistently accurate despite their high speed. For example, in watching Italian construction workers swing a hammer repetitively, he observed that their swings were both fast and nonchalant, but they almost never missed the intended target. The striking quality of the workers' performance stimulated his interest in the sources of movement speed–accuracy tradeoffs and the circumstances under which such tradeoffs occur.

This immediately led Woodworth (1899) to raise a second question concerning what phase of an aimed movement might be affected by its speed. He suggested that rapid movements actually involve two successive phases, which he called *initial adjustment* and *current control*. According to him, the initial-adjustment phase transports a selected part of the body quickly toward a target location; the current-control phase subsequently corrects any errors made along the way, using sensory feedback to reach the target accurately. The issue, then, became one of whether movement speed affects the accuracy of initial adjustment or current control.

Consequently, part of Woodworth's (1899) work focused specifically on feedback processing in error corrections. For example, a third question that occurred to him dealt with what happens when visual feedback is precluded during rapid movements. He reasoned that by eliminating such feedback, one might learn more about the nature of the current-control phase. With this manipulation, his experiments yielded the first measurements of how quickly visual feedback can be processed during current control.

*Empirical approach.* To obtain relevant data on speed–accuracy tradeoffs in aimed movements, Woodworth (1899) introduced the time-matching movement task with a spatial constraint (*Methodological Framework*). He had subjects sit at a table holding a pencil in their dominant (right or left) hand. Through a slot in the table top, they repeatedly tried to draw lines of a specified length on a roll of moving paper placed beneath the slot. Their rate of movement was paced by a metronome set to produce beats at frequencies ranging from 20 to 200 per min. The subjects completed one movement back and forth for each beat of the metronome. Their movements were made under two different visual conditions. In one condition, the subjects had their eyes open, and visual feedback was available throughout each moment. In the other condition, their eyes were closed, and no visual feedback was available. Using a ruler, Woodworth measured the accuracy of the movements as a function of the visual condition and movement rate.

*Results and conclusions.* Some of Woodworth's (1899) results appear in Fig. 6.1, which shows mean absolute error versus movement rate (number of movements per minute) for each visual condition. Under the eyes-open condition (Fig. 6.1, solid curve), the AEs for slow movements were small, but gradually increased with movement rate. Under the eyes-closed condition (Fig. 6.1, dashed curve), the AEs were larger and remained relatively constant regardless of how quickly the subjects moved.

In evaluating these results, Woodworth (1899) assumed that when subjects had their eyes open, performance was mediated by both the initial-

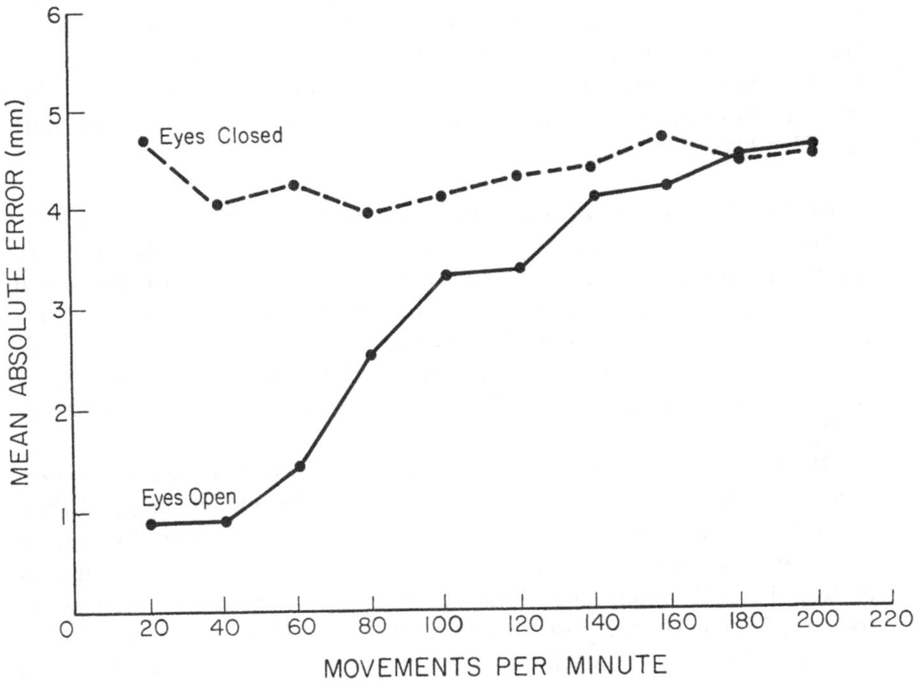


FIG. 6.1. Results from Woodworth's (1899) experiments. (The solid and dashed curves show mean absolute errors plotted versus movement rate, that is, the number of movements per minute, under conditions in which subjects had their eyes open or closed, respectively.)

adjustment and current-control phases of movement. This assumption was consistent with the pattern of errors under the eyes-open condition, where AEs were relatively small except for very fast movements. Moreover, Woodworth assumed that when subjects had their eyes closed, performance was mediated only by the initial-adjustment phase. This assumption was consistent with the pattern of errors under the eyes-closed condition, where AEs were generally larger than those under the eyes-open condition.

On the basis of his assumptions, Woodworth reached several tentative conclusions about the initial-adjustment and current-control phases of rapid aimed movements. He inferred that movement speed did not affect the accuracy of the initial adjustment per se, because under the eyes-closed condition, errors stayed about the same in magnitude as the rate of movement increased. The effect of movement speed on accuracy was attributed instead to the characteristics of current control. In particular, Woodworth inferred that increasing movement speed disrupts error corrections made through visual feedback, because under the eyes-open condition, errors increased in magnitude as the rate of movement increased.

Using the obtained pattern of AEs, Woodworth (1899) also estimated how much time subjects need to process visual feedback. His estimate came from the eyes-open condition, for which the minimum feedback-processing time was taken to be the duration of movements whose errors nearly equalled those in the eyes-closed condition. This yielded a value around 400 msec, about twice the size of an ordinary visual reaction time (cf. Woodworth, 1938).

*Evaluation.* Taken as a whole, Woodworth's (1899) work on movement speed and accuracy was a veritable *tour de force*. He asked some of the most basic questions about speed–accuracy tradeoffs in rapid aimed movements, collected extensive data to answer these questions, and proposed several interesting hypotheses to interpret the obtained results. His ideas were so insightful that they anticipated much of the research done by subsequent investigators on human motor performance (Keele, 1968, 1981).

Of course, this does not mean that Woodworth's (1899) work was perfect. There were several limitations in his experiments. Because of their instructive nature and ultimate theoretical consequences, these limitations should be kept firmly in mind, even though some reviewers (e.g., Hancock & Newell, 1985; Jeannerod, 1988) have tended to downplay them while focusing instead on the positive contributions that Woodworth made.

The first obvious limitation in Woodworth's (1899) experiments is that they only concerned performance in a time-matching movement task; they did not concern what happens when subjects try to minimize their movement durations while traveling a specified target distance. Perhaps under other circumstances, movement speed might have been found to affect accuracy during both the initial-adjustment and current-control phases.

A second limitation is that Woodworth (1899) did not measure individual movement durations directly. He estimated them indirectly from the number of movements made in time with a metronome. Subjects had considerable freedom to vary the details of the movements, and they may have occasionally gotten out of phase or hesitated along the way. This could have seriously contaminated the results obtained in some cases. Although a few investigators have replotted Woodworth's (1899) data in terms of average movement velocity (e.g., Hancock & Newell, 1985), such extrapolations are not entirely justified, given the vagaries of his experimental technique.

A third limitation is that Woodworth (1899) did not explicitly isolate the initial-adjustment phase of subjects' movements. His conclusions regarding it were based primarily on the endpoints of whole movements in the eyes-closed condition. As mentioned earlier, he assumed that this condition precluded current control by eliminating visual-feedback processing. However, performance there may still have been mediated at least partly through kinesthetic feedback (Crossman & Goodeve, 1963/1983; Keele, 1968). So, from the results obtained when subjects had their eyes closed, one cannot make strong inferences about the initial-adjustment phase per se.

A fourth limitation is that Woodworth (1899) analyzed movement mainly in terms of mean absolute errors. He seldom reported accuracy with respect to constant or variable errors. As a result, the speed–accuracy tradeoff curves derived from his results (*viz.*, Fig. 6.1) could reflect either systematic biases or random variability of subjects' movements. For example, there are two possible reasons why AEs in the eyes-closed condition stayed essentially the same regardless of movement speed. Perhaps the underlying CEs and VEs were both unaffected as movement speed increased, or perhaps the CEs decreased while the VEs increased, such that the joint changes in them roughly canceled each other. If the latter were the case, then Woodworth's (1899) claims about the initial-adjustment phase would be further suspect.

Given the way that the eyes-closed condition was arranged, it seems quite likely that the CEs and VEs obtained for it did change in unknown and uncontrolled ways as a function of movement rate. When subjects had their eyes closed, they received neither on-line visual feedback during the movements nor subsequent knowledge of results about the movement endpoints. Indeed, they could not even see the starting locations or targets of the movements with their eyes closed. Such uncertainties might have easily induced spurious negative correlations between CEs and VEs as movement speed and the memory traces of the targets varied over time. (For other criticisms of Woodworth's eyes-open and eyes-closed conditions, see Keele & Posner, 1968, and Carlton, 1981).

## Dormant Period

Although Woodworth's (1899) experiments had several limitations, a long time elapsed before other investigators followed them up extensively. Only sporadic scientific research on speed–accuracy tradeoffs in discrete aimed movements occurred from 1900 to 1945 (e.g., Garrett, 1922; Philip, 1936). During this dormant period, there were many studies of human motor performance, covering topics such as eye movements (Dodge & Cline, 1901), handwriting (Freeman, 1914), motor memory and movement reproduction (Hollingworth, 1909), sequential timing and rhythm (Langfield, 1915; Stetson, 1905), movement kinematics (Peters & Wenborne, 1936), muscle contraction (Fenn, 1938), and tremor (Travis, 1929). Yet, as these studies progressed, attention gradually shifted away from the issues and results that Woodworth introduced.

Several factors could explain why Woodworth's (1899) work was not followed immediately by more extensive studies of movement speed–accuracy tradeoffs. In light of how much he had discovered, it is possible that he seemed to have resolved most of the mystery surrounding these tradeoffs. As Fitts (1954) has noted, another possibility is that investigators were biased at the time by some emerging prejudices about the nature of movement speed. One case of such bias may be found, for example, in the work of Stetson, a contemporary of Woodworth. From research on rhythm and timing behavior, Stetson (1905) concluded that the duration of fast ballistic movements was independent of how far they traveled in space. He found, in particular, that baton movements made by musical conductors had the same duration regardless of whether their distances were short or long. A similar result was also reported for handwriting (Freeman, 1914). The belief that the duration of ballistic movements did not depend on movement distance then spread among numerous scientists in the motor-control community, later appearing as part of several influential review articles (e.g., Hartson, 1939; Stetson & McDill, 1923). Interest in studying the accuracy of the initial-adjustment phase in aimed movements may have been dampened, consequently, because the prospects for manipulating the duration of this phase to produce speed–accuracy tradeoffs appeared somewhat limited.

## Post-World War II Era

During World War II, however, problems associated with the manual operation of military vehicles and weapons stimulated considerable scientific interest in the relationship between movement speed and accuracy (Craik, 1943/1963a, 1944/1963b). Because of these problems, new research was initiated to learn more about the human perceptual-motor system. In this work, considerable effort was devoted to studying continuous tracking and to applying principles of mathematical control theory (Craik, 1947, 1948; Ellson, 1949). Along the way, experiments revealed that tracking behavior often consists of small discrete movements made at a rate of roughly four per sec. Investigators therefore began analyzing such movements more carefully, extending the study of speed–accuracy tradeoffs initiated originally by Woodworth (1899).[3]

*Vince.* An often-cited product of the latter developments was a study by Vince (1948). Using a procedure that required subjects to track discrete changes in the position of a target line, she replicated many of Woodworth's (1899) original findings, and introduced several new manipulations of visual feedback. Her results substantiated the importance of current control (i.e., feedback-based corrections) for movement accuracy. Still, Vince's study suffered from many of the same limitations as Woodworth's (1899) did. For example, she focused almost exclusively on performance in a time-matching movement task, and she only analyzed movement accuracy with respect to mean absolute errors, not constant or variable errors.

*Searle and Taylor.* Fortunately, about the same time, two other investigators, Searle and Taylor (1948), conducted an important study that overcame some of these limitations. Their study was one of the first to examine the relation between movement speed and spatial accuracy in a time-minimization movement task. Here subjects controlled a pointer by moving a knob back and forth horizontally. The pointer was supposed to be shifted as quickly as possible so as to position it near a target line whose distance varied from trial to trial. For these movements, Searle and Taylor (1948) carefully examined the initial-adjustment phase by recording and analyzing individual movement trajectories. In their analyses, constant and variable errors, not just mean absolute errors, were measured.

Some of the obtained results appear in Fig. 6.2, which shows VEs for the endpoints of the initial-adjustment phase as a function of average

---

[3]There was also some applied interest in movement speed–accuracy tradeoffs before World War II. This interest stemmed from industrial time-and-motion studies that were designed to create standardized systems for calculating the rate at which workers could perform various tasks with specified levels of accuracy (Bailey & Presgrave, 1958; Niebel, 1962). However, unlike the subsequent research outlined in our review, these studies did not yield a principled scientific foundation whereby the sources of speed–accuracy tradeoff relations might be better understood.

movement velocity. Searle and Taylor (1948) found that as movement velocity increased, so did the VEs. The monotonic form of the VE versus velocity function foreshadowed results reported by a number of subsequent investigators (e.g., Bainbridge & Sanders, 1972; Schmidt et al., 1979; Wright & Meyer, 1983; Zelaznik, Shapiro, & McColsky, 1981).

This outcome contradicted some of Woodworth's (1899) original conclusions and suggested that the accuracy of the initial-adjustment phase does depend on movement speed. The observed dependence could not be attributed simply to a disruption of error corrections during current control. Thus, it became clear that more exploration was needed regarding the mechanisms underlying speed–accuracy tradeoffs in aimed movements.

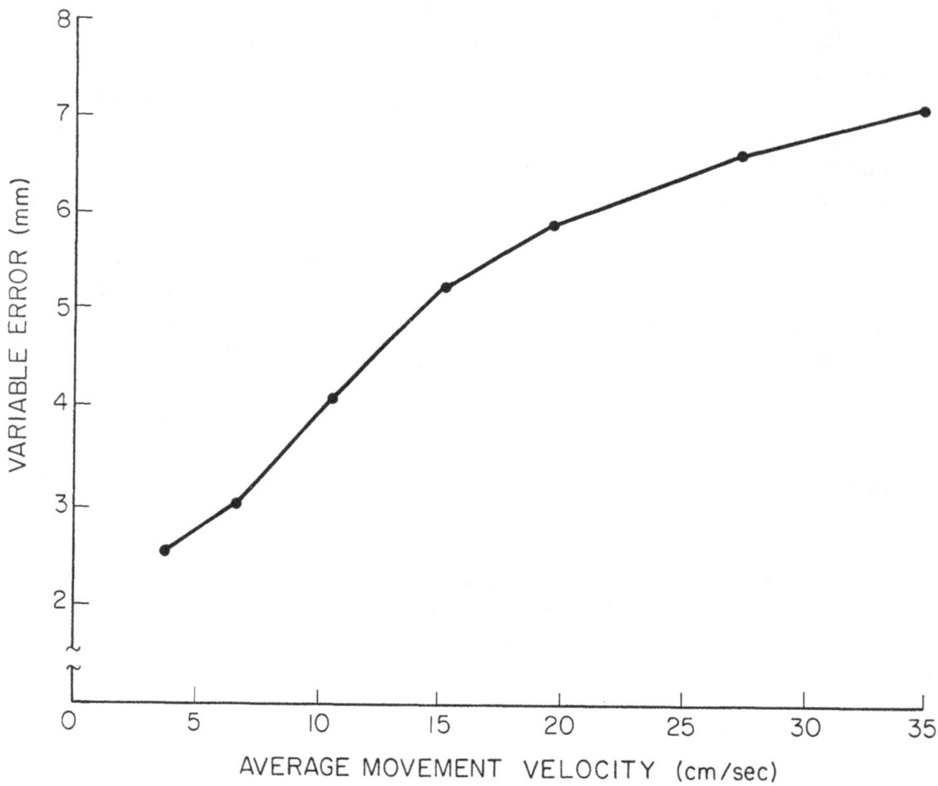


FIG. 6.2. Results from Searle and Taylor's (1948) study, which demonstrated an increasing monotonic relation between average movement velocity and variable errors (VE) in the endpoints of the initial-adjustment phase for rapid aimed movements.

*Brown and Slater-Hammel.* The next step along the way was taken soon after, through remarkable, though seldom cited, work by Brown and Slater-Hammel (1949). In their study, subjects again moved a pointer quickly and accurately toward a target line. The key new feature was that the constant and variable errors of subjects' movements were both required to be extremely small; the pointer always had to stop exactly on top of the target line. Movement durations were then measured as a function of target distance, with subjects being encouraged to reach the target as quickly as possible without missing it. Following this measurement, Brown and Slater-Hammel (1949) looked separately at both the initial-adjustment and current-control phases of the obtained movements by analyzing individual movement trajectories.

Some of the results can be seen in Fig. 6.3, which shows mean movement durations ($T$) from Brown and Slater-Hammel's (1949) study versus the

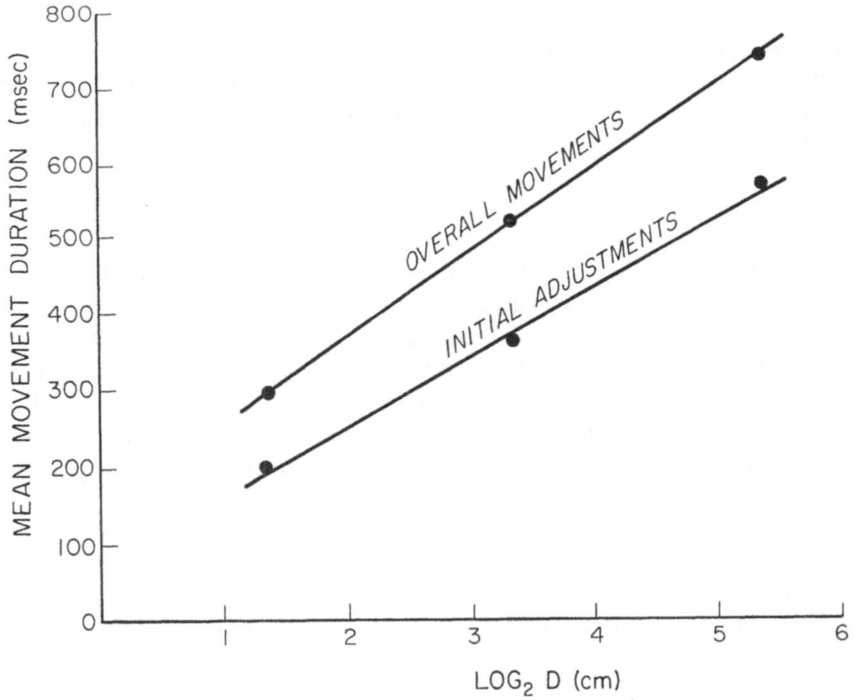


FIG. 6.3. Results from Brown and Slater-Hammel's (1949) study, which demonstrated a logarithmic relation between mean movement duration and target distance for initial adjustments and overall movements in a time-minimization movement task.

logarithm of the target distance ($\log_2 D$).[4] The data appear separately for initial adjustments and overall movements. There is an almost perfect linear relation between $T$ and $\log_2 D$ ($r$ = .999). In fact, most of the distance effect occurred during the initial-adjustment phase; the duration versus distance function for it essentially parallels the one for the overall movements. This suggests that subjects may have varied the duration of the initial adjustments in order to gain some significant benefit in terms of spatial accuracy. So, contrary to Woodworth's (1899) prior claims, one might further question whether the current-control phase provides the sole basis for movement speed–accuracy tradeoffs.

Another noteworthy fact about Brown and Slater-Hammel's (1949) study is that few citations of it appeared subsequently. There are at least two possible reasons for this turn of events. Although Brown and Slater-Hammel strictly controlled subjects' constant and variable errors, they did not measure movement durations as a function of movement precision per se; their experimental design included only one narrow target width (i.e., a line). They also offered no theoretical account for the underlying relation in their data between mean movement durations and the logarithm of target distance. Indeed, they were apparently unaware of this relation and its potential import.[5] As a result, further work by other investigators soon came to overshadow Brown and Slater-Hammel's (1949) significant empirical discoveries.

## Fitts' Law

The next major breakthrough regarding speed–accuracy tradeoffs in aimed movements came 5 years later in 1954 through work by Paul Fitts. His research constituted a direct complement and extension of Brown and Slater-Hammel's (1949) contributions. It provided a more thorough investigation of the functional relation between movement speed and accuracy in a time-minimization movement task, and it introduced a new theoretical framework for explaining how the initial-adjustment phase of aimed movements may mediate their ultimate accuracy.

---

[4]The choice of the base 2 for the logarithmic distance transformation in Fig. 6.3 is somewhat arbitrary. Other bases could be used instead without changing the linearity of the displayed function; they would only affect the function's slope. We have chosen 2 for the base simply because, in subsequent work (Fitts, 1954), it plays an important theoretical role.

[5]The present analysis of Brown and Slater-Hammel's (1949) results differs from the one that these investigators originally offered. In their initial report, average movement velocity (i.e., speed) was plotted as a power function of target distance, using log–log coordinates. This produced a good fit. Nevertheless, our own plot of mean movement duration versus the logarithm of target distance (Fig. 6.3) fits even better.

Fitts (1954) had subjects move a hand-held stylus rapidly back and forth between two target regions, tapping one of the targets at the end of each movement. The movements were supposed to be made as fast as possible without missing the targets. Movement durations were measured as a function of two independent variables; target distance ($D$), and target width ($W$). The manipulation of target width placed an explicit constraint on the constant and variable errors in subjects' movement endpoints. This allowed Fitts to determine more about how the intended precision of a movement affects resulting movement durations.[6]

*Logarithmic tradeoff relation.* Some of Fitts' results appear in Fig. 6.4, which shows mean movement duration ($T$) versus a logarithmic transformation of target distance divided by target width [*viz.*, $\log_2(2D/W)$]. Fitts (1954) called this transformation the *index of movement difficulty* (ID). He found that for stylus-tapping movements, there was a strong linear relation between the index of movement difficulty and mean movement duration, as expressed by the following equation:

$$T = A + B \log_2(2D/W), \qquad (1)$$

where $A$ and $B$ are constants. The same relation also emerged in studies that Fitts (1954) performed on other types of movement (e.g., pin transfer). His results generalized the speed–accuracy tradeoff obtained by Brown and Slater-Hammel (1949), whose work demonstrated a close linear relation between mean movement duration and the logarithm of target distance (cf. Fig. 6.3). Most important, Equation 1 implies that when an explicit constraint is placed on subjects' variable errors by specifying the target width, $W$, may influence $T$ systematically over and above the effect that $D$ has.[7]

*Information-theory hypothesis.* To interpret his results concerning movement speed and accuracy, Fitts (1954) adapted some concepts from information theory, which was popular at the time (Shannon, 1948; Shan-

[6]The original study by Fitts (1954) did have one limitation similar to those of Woodworth's (1899) study. Movement durations were not measured directly; they were measured indirectly be counting the number of repetitive taps made during a 15-sec interval, and then dividing this count into the length of the interval. However, subsequent experiments in which the durations of individual stylus-tapping movements were measured (e.g., Fitts & Peterson, 1964; Fitts & Radford, 1966) revealed that Fitts' (1954) results still held.

[7]Given the constraints imposed through $W$, theoretical accounts of Equation 1 should deal specifically with the nature of variable errors in rapid aimed movements. One cannot base such accounts solely on the nature of constant errors. Consequently, some recent models (e.g., Bullock & Grossberg, 1988) that attempt to explain the logarithmic tradeoff relation in terms of constant errors must be seriously questioned.

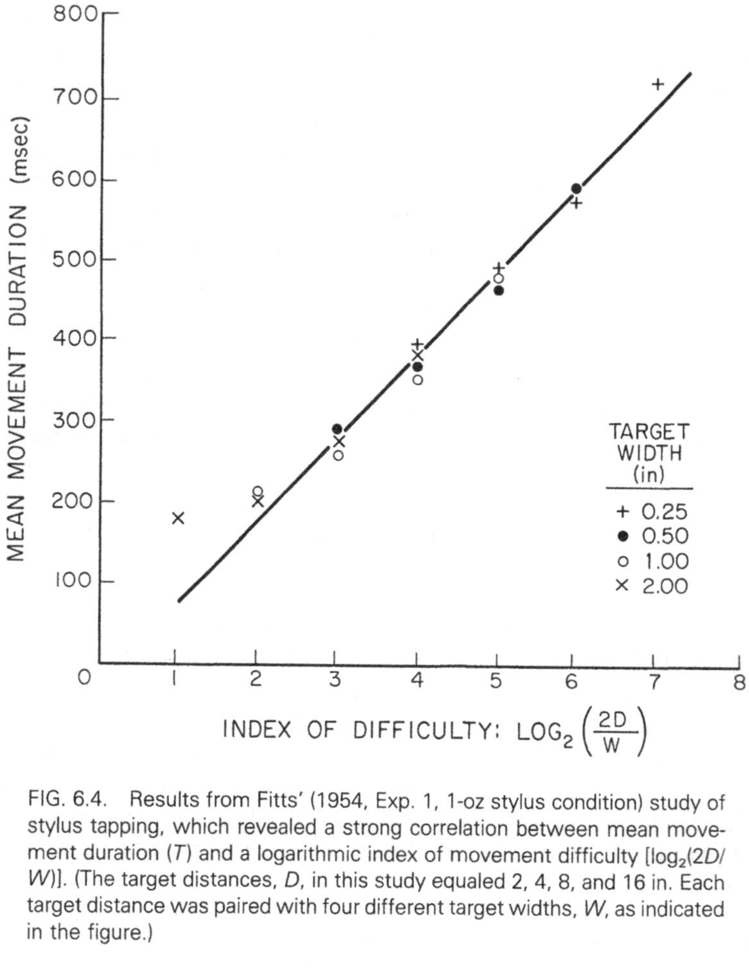


FIG. 6.4. Results from Fitts' (1954, Exp. 1, 1-oz stylus condition) study of stylus tapping, which revealed a strong correlation between mean movement duration (*T*) and a logarithmic index of movement difficulty [$\log_2(2D/W)$]. (The target distances, *D*, in this study equaled 2, 4, 8, and 16 in. Each target distance was paired with four different target widths, *W*, as indicated in the figure.)

non & Weaver, 1949; cf. Miller, 1953). He hypothesized that the human motor system functions like a stochastic communication channel through which information is transmitted in the presence of background noise. This hypothesis could conceivably explain the logarithmic tradeoff relation (Equation 1), assuming that the motor system has a fixed channel capacity, and that subjects move so as to transmit information at the maximum rate possible (Fitts, 1954; Fitts & Peterson, 1964).

For example, suppose one treats reciprocal movements like those in Fitts' (1954) stylus-tapping task as if they embody a waveform composed of a fundamental signal frequency perturbed by white Gaussian noise. Also, suppose one adjusts the noise power ($N$) to equal half the target width ($W/2$), the signal-plus-noise power ($S + N$) to equal the target distance

($D$), and the bandwidth ($B$) of the signal to equal half the reciprocal of the average movement duration ($1/2T$). Then it has been claimed that at the maximum rate of information transmission, the logarithmic tradeoff relation would emerge (Fitts & Peterson, 1964). This claim is based on a result of Shannon (1948, Theorem 17), who proved that $C = B \log_2[S + N)/N]$, where $C$ is channel capacity in bits per sec. From Shannon's theorem, Equation 1 can be derived by rearranging terms and replacing $B$, $N$, and $S + N$ with the preceding equivalents (i.e., $1/2T$, $W/2$, and $D$, respectively; however, see Kvålseth, 1979, for a critique of this derivation).

*Further studies.* Other investigators were therefore quick to test Fitts' (1954) ideas empirically. Further studies soon confirmed the logarithmic speed–accuracy tradeoff (Annett, Golby, & Kay, 1958; Crossman, 1957; Crossman & Goodeve, 1963/1983; Flowers, 1975; Jagacinski, Repperger, Moran, Ward, & Glass, 1980; Knight & Dagnall, 1967; Langolf, Chaffin, & Foulke, 1976; MacKenzie, Marteniuk, Dugas, Liske, & Eickmeier, 1987; Megaw, 1975). The reliability was so great that Equation 1 ultimately became known as *Fitts' Law* (Crossman & Goodeve, 1963/1983; Keele, 1968).

Fitts' Law has been found to hold for many different types of movement, manipulanda, environments, and subject populations (Anderson, 1987; Jagacinski, 1989; Keele, 1968, 1981). The logarithmic tradeoff relation occurs not only in stylus tapping but also in throwing movements (B. Kerr & Langolf, 1977), wrist movements (Crossman & Goodeve, 1963/1983; Jagacinski et al., 1980; Meyer et al., 1988), finger movements (Langolf et al., 1976), leg movements (Drury, 1975), and head movements (Jagacinski & Monk, 1985). Regardless of whether movements involve manipulating a hand-held stylus (Fitts, 1954), rotary handle (Crossman & Goodeve, 1963/1983), joystick (Hartzell, Dunbar, Beveridge, & Cortilla, 1982), computer mouse (Card, English, & Burr, 1978), keyboard (Card et al., 1978), or footpedal (Drury, 1975), the tradeoff relation still continues to be at least approximately logarithmic. Similarly, this relation is exhibited by young adults, children (Wallace, Newell, & Wade, 1978), elderly adults (Welford, Norris, & Shock, 1969), mental retardates (Wade, Newell, & Wallace, 1978), and drugged subjects (Kvålseth, 1977). Some experimenters have even demonstrated the validity of Fitts' Law in underwater environments (R. Kerr, 1973, 1978), aircraft flight (Hartzell et al., 1982), and situations where visual feedback is eliminated (Meyer et al., 1988; Prablanc, Echallier, Komilis, & Jeannerod, 1979; Wallace & Newell, 1983). Given these very robust results, modern systems designers have sought to incorporate Fitts' Law in standards of engineering practice (e.g., Card, Moran, & Newell, 1983; Karger & Hancock, 1982).

Still, some aspects of Fitts' (1954) work met with great doubt. Although his empirical results were easy to replicate, the theoretical framework that he proposed to account for them was not well accepted. Critics felt that the information-theory hypothesis was strained at best, and totally wrong at worst (e.g., Crossman & Goodeve, 1963/1983; Kvålseth, 1979). Consequently, this triggered a search for other ways of explaining the logarithmic speed–accuracy tradeoff.

## Deterministic Iterative-Corrections Model

Perhaps the best-known alternative to emerge as a result was developed by Crossman and Goodeve (1963/1983). In evaluating Fitts' (1954) ideas, they doubted that the initial-adjustment phase of rapid aim movements suffered from the noise assumed by the information-theory hypothesis. According to them (Crossman & Goodeve, 1963/1983, p. 283), there is an "empirical difficulty in establishing the existence of the postulated 'noise' . . . . Thus, the supposed 'noise' is apparently not present in the effector systems." So, returning to views that Woodworth (1899) had introduced earlier, Crossman and Goodeve (1963/1983) claimed instead that the current-control phase is the main source of the logarithmic speed–accuracy tradeoff.

Their account of Fitts' Law involves a *deterministic iterative-corrections model*. Under this model, movements intended to hit a target region quickly and accurately consist of several discrete submovements made in rapid succession. These submovements in combination satisfy three basic assumptions.

*Assumptions of the model.* The first assumption is that each submovement travels a constant proportion ($p$) of the distance between its starting location and the center of the target. For example, the first submovement would travel a distance $pD$, the second submovement would travel a distance $pD(1 - p)$, the third submovement would travel a distance $pD(1 - p)^2$, and so forth.

The second assumption is that each submovement takes the same constant amount of time ($t$) to be completed, regardless of how much distance the submovement travels. Consequently, the duration ($T$) of the overall submovement sequence should be proportional to the number of submovements in it; if there are $n$ such submovements, then $T$ would equal $nt$.

The third assumption is that the submovements are guided by sensory feedback. Either visual or kinesthetic feedback might be involved, depending on subjects' ability to see the target and the moving body part.

With whatever feedback is available, the submovement sequence supposedly continues until the target region has been reached.

In essence, the iterative-corrections model is completely deterministic. Unlike Fitts' (1954) information-theory hypothesis, it makes no claims about the presence of noise in the motor system during the initial-adjustment or current-control phase of an aimed movement. For a fixed target distance and width, the sequence of submovements would always be the same.[8]

*Predictions of the model.* The deterministic iterative-corrections model accounts for Fitts' Law in a straightforward manner. It implies that as the ratio of target distance to target width ($D/W$) increases, the number of submovements required to reach the target region also increases (Crossman & Goodeve, 1963/1983; Keele, 1968). To be specific, let $n$ denote the required number of submovements. Then, according to the model, $n$ approximately equals the index of movement difficulty, $\log_2(2D/W)$. Because each submovement supposedly takes the same constant amount of time ($t$) to be completed, the overall duration ($T$) of the submovement sequence should likewise increase logarithmically [i.e., $T = nt = t \log_2(D/W)$].

Furthermore, the model makes some other strong predictions. If it were valid, then one would expect to observe successive discrete corrections in the trajectories of rapid aimed movements toward a given target region. The duration of each submovement should be the same regardless of the target distance and width, and the proportion of the target distance traveled by the initial submovement should be independent of target width (Crossman & Goodeve, 1963/1983; Keele, 1968).

*Tests of the model.* To test some of these predictions, Crossman and Goodeve (1963/1983) performed experiments in which subjects moved a rotary pointer by making wrist rotations. Such movements were chosen for study because they have several desirable properties. In particular, wrist rotations involve a single axis of motion, making it relatively easy to record and analyze their trajectories, unlike those of complex multidimensional stylus-tapping movements (cf. Fitts, 1954). Also, wrist rotations have relatively low inertia and viscosity, so they may reveal details of

[8]In a later version of the iterative-corrections model, Keele (1968) informally considered the possibility that the distances traveled by the submovements might vary randomly from trial to trial, yielding occasional overshoots as well as undershoots of the target. However, the quantitative predictions that he derived from the model, like those of Crossman and Goodeve (1963/1983), all made the simplifying assumption that the submovements were completely deterministic.

underlying neuromotor control signals more clearly than do some other types of movement.

Crossman and Goodeve's (1963/1983) experiments on wrist rotations provided some support for the deterministic iterative-corrections model. When they plotted the durations of these movements as a function of the index of movement difficulty [$\log_2(2D/W)$], their results exhibited an approximate logarithmic tradeoff relation, consistent with Equation 1. Also, in associated plots of movement velocities, there were sequences of successive peaks whose amplitudes decreased progressively over time as subjects approached the target (e.g., Crossman & Goodeve, 1963/1983). The number of such peaks appeared to vary directly with the index of difficulty, as expected from the model's assumptions about the structure of submovement sequences.

Following the work of Crossman and Goodeve (1963/1983), other investigators were therefore encouraged to study the nature of corrective submovements in more detail. Keele (1968) extended the deterministic iterative-corrections model to incorporate results about the speed of visual-feedback processing. Based on one of his experiments (Keele & Posner, 1968), Keele (1968) estimated that visual feedback takes about 200 msec to be processed (cf. Carlton, 1981; Zelaznik, Hawkins, & Kisselburgh, 1983). Assuming that this time is approximately the same as the duration of a single submovement, he then showed that the model correctly predicts the slope of the logarithmic tradeoff relation (the constant $B$ in Equation 1) found by Fitts (1954) for stylus tapping. This demonstration was sufficiently impressive that the deterministic iterative-corrections model soon became accepted as the best available account of Fitts' Law.

*Failures of the model.* Since about 1970, however, it has become increasingly clear that the deterministic iterative-corrections model is seriously flawed. Not all of its key predictions have been upheld convincingly. These failures include the following:

1. On frequent occasions involving the time-minimization movement task, there are only one or two submovements per trial, even when the index of movement difficulty is relatively large. For example, Langolf et al. (1976) found numerous cases in which subjects reached a target region with a single submovement, but still exhibited a logarithmic tradeoff relation (Equation 1). The deterministic iterative-corrections model cannot deal with this outcome, because it attributes Fitts' Law solely to systematic variations in the number of submovements made as a function of $D/W$.
2. Submovements do not have constant durations. Contrary to the deterministic iterative-corrections model, the durations of individual submovements may vary greatly, depending on target distance and width.

Langolf et al. (1976) and Jagacinski et al. (1980) have found that initial submovements take significantly more time than subsequent corrective submovements do (cf. Annett et al., 1958). Also, initial-submovement durations appear to increase with target distance and decrease with target width (Jagacinski et al., 1980; Langolf et al., 1976). This suggests that the programming and execution of submovement sequences is considerably more flexible than Crossman and Goodeve (1963/1983) originally assumed.

3. Submovements do not travel a constant proportion of the distance between their starting location and the center of a target region. Jagacinski et al. (1976) have found considerable variability in the proportion of distance traveled by initial submovements, depending on the difficulty of the target. This violates one of the key assumptions made in the deterministic iterative-corrections model to account for Fitts' Law.

4. Subjects sometimes commit errors, missing a target region completely. Fitts (1954) and others (e.g., Fitts & Peterson, 1964; Meyer et al., 1988; Wallace & Newell, 1983) have found that error rates increase with the index of movement difficulty. The deterministic iterative-corrections model cannot explain this, because it assumes that a submovement sequence always continues until the target region is reached. On the other hand, without such continuity, the model would fail to explain the logarithmic speed–accuracy tradeoff.

5. Speed–accuracy tradeoffs in rapid aimed movements are not always logarithmic. They may have other forms instead, disobeying Fitts' Law and violating the main prediction of the deterministic iterative-corrections model (Beggs & Howarth, 1972; Ferrell, 1965; Hancock, Langolf, & Clark, 1973; Howarth, Beggs, & Bowden, 1971; Klapp, 1975; Kvålseth, 1980; Meyer et al., 1982; Schmidt et al., 1979; Sheridan & Ferrell, 1963; Wright & Meyer, 1983).

### Other Tradeoff Relations: Power Functions

Some potentially instructive exceptions to Fitts' Law involve power functions having the form

$$T = A + B(D/W)^p, \tag{2}$$

where $A$ and $B$ are non-negative constants, and $p$ is a fractional exponent ($0 < p \leq 1$). In evaluating results of past studies on speed–accuracy tradeoffs for the time-minimization movement task, Kvålseth (1980) discovered that mean movement durations can often be described better by Equation 2 than by a logarithmic tradeoff relation (Equation 1). Other investigators have noted a similar pattern as well (e.g., Ferrell, 1965; Gan & Hoffman, 1988; Hancock et al., 1973; Sheridan & Ferrell, 1963). The exponent of the best-fitting power function varies a bit from study to study, but usually has a value between .25 and .5. Thus, when $T$ is plotted versus Fitts' (1954) index of movement difficulty, $\log_2(2D/W)$, the data tend to exhibit a slight positive acceleration (i.e., have a U-shape) as $D/W$ increases.[9]

For example, consider Fig. 6.5, which shows some results from one of Crossman and Goodeve's (1963/1983) experiments on wrist rotations. Here we have plotted mean movement durations versus Fitts' logarithmic index of movement difficulty. The solid U-shape function represents a best fitting power function for these data [*viz.*, $T = 187\,(D/W)^{.31}$; $r^2 = .94$]. It comes closer to the mean movement durations than does Fitts' Law [*viz.*, $T = 64 + 83\log_2(2D/W)$; $r^2 = .91$].

A similar pattern can also be seen in Fig. 6.4, where a slight positive acceleration appears in the mean movement durations from Fitts' (1954) stylus-tapping experiment when they are plotted versus the logarithmic index of movement difficulty. The power function that comes closest there is $T = 13 + 165(D/W)^{.36}$. It accounts for about 99% of the variance in the mean movement durations ($r^2 = .988$), whereas Fitts' Law accounts for somewhat less variance ($r^2 = .966$).

The exact form of the tradeoff relation obtained in a given situation may depend on the type of movement task that a subject has to perform (Meyer et al., 1982, 1988; Wright & Meyer, 1983; Zelaznik et al., 1988). As mentioned already, there are different possible tasks, each having its own set of spatial and temporal goals. For the stylus-tapping task in which Fitts' Law was discovered, the subjects' goals were to hit a target region accurately and, given this constraint, to minimize the average movement duration. Under other conditions, however, a different tradeoff may emerge if subjects change their strategy of performance. One might, for example, expect this to happen in a time-matching movement task such as Woodworth (1899) used, where subjects are required to produce movements whose distances and durations both approximate specified target values.

## Linear Speed–Accuracy Tradeoff

Consistent with the latter expectation, further evidence that the form of the speed–accuracy tradeoff depends on the subjects' temporal and spatial

---

[9]In a plot of $T$ versus $D/W$ on an untransformed scale, the obtained curve would exhibit some negative acceleration (i.e., have an inverted U-shape), but the amount of curvature associated with it would be less for a power function whose exponent falls between zero and one than for a logarithmic function. Thus, if $T$ is actually a power function of $D/W$, and if $T$ is plotted versus the logarithm of $D/W$, then the logarithmic transformation would overcompensate for the original power function's negative acceleration, yielding an apparent positive acceleration instead.

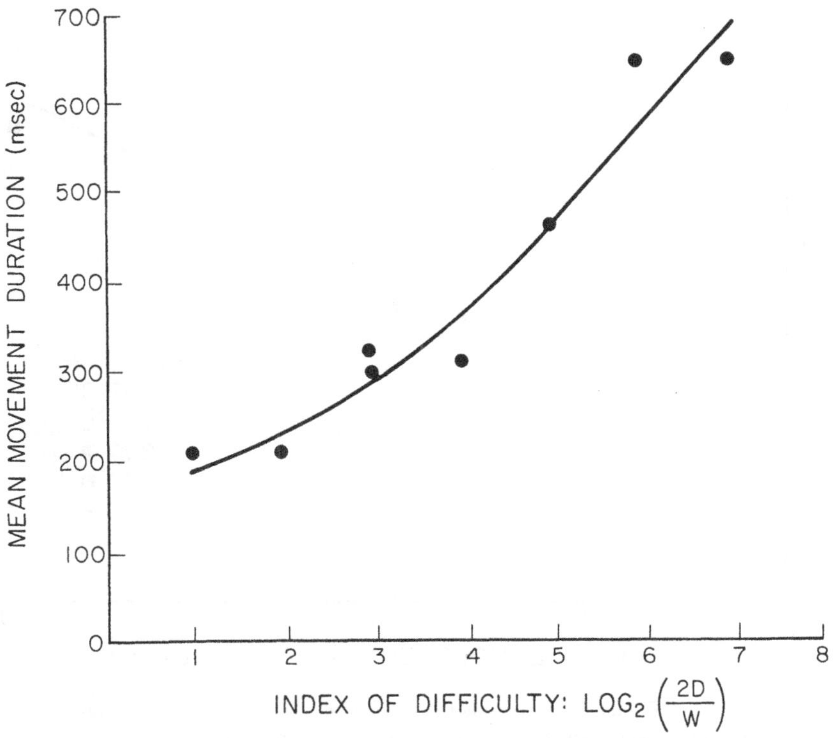


FIG. 6.5. Results from Crossman and Goodeve's (1963/1983) study of wrist rotations. (The solid curve represents a power function that relates mean movement durations to the target distance–width ratio better than Fitts' Law does. For further details, see text.)

goals has been reported by Schmidt et al. (1979). Like some other investigators, they became dissatisfied with the deterministic iterative-corrections model and the reciprocal stylus-tapping task in which Fitts' Law was originally obtained. This inspired them to take a different approach toward studying the relationship between movement speed and accuracy.

Pursuing Woodworth's (1899) lead, Schmidt et al.(1979) used a variant of the time-matching movement task. They had subjects make single aimed tapping movements whose distances and durations were both supposed to match specified target values. The distance and duration of each individual movement were measured directly, unlike in the original studies of Woodworth (1899), yielding better estimates of average movement velocity. Also, Schmidt et al. (1979) examined the constant and variable errors in subjects' performance more carefully.

Some results from one of their experiments (Schmidt et al., 1979, p. 427) appear in Fig. 6.6, which shows variable errors in movement distance

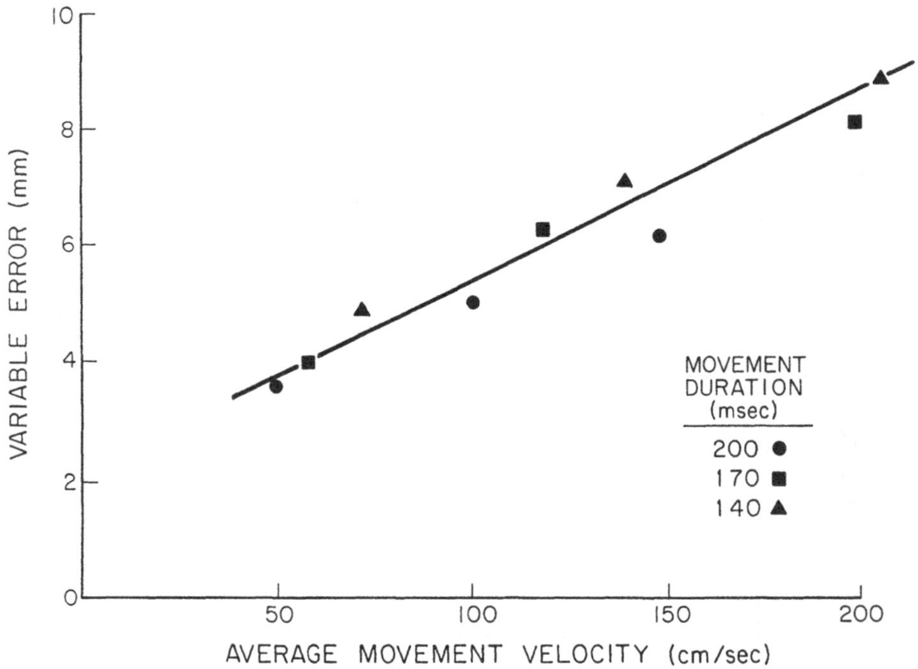


FIG. 6.6. Results from a study by Schmidt et al. (1979), showing a linear relation between variable error and average movement velocity in a time-matching movement task. (The closed triangles, squares, and circles correspond respectively to movements whose intended durations were 140, 170, and 200 ms. The solid line is the best-fitting linear function.)

versus average movement velocity. A straight line fits these data reasonably well, even though neither of the axes has been transformed logarithmically. We may characterize the obtained linear tradeoff in terms of the following equation:

$$S = A + B(D/T), \qquad (3)$$

where $S$ is the standard deviation (variable error) of the movement endpoints in space, $D$ is the mean movement distance, and $T$ is the mean movement duration.

The linear tradeoff relation embodied in Equation 3 differs markedly from Fitts' Law (Schmidt et al., 1979; Wright & Meyer, 1983). In essence, the endpoint standard deviation $S$ may be treated as corresponding to an "effective target width" ($W_e$). When this correspondence is substituted in Equation 3 and the variables there are rearranged to express $T$ as a function of $D/W_e$, a nonlogarithmic tradeoff emerges [*viz.*, $T = BD/(W_e - A)$].

For $A = 0$, the latter result is a power function of $D/W_e$ with an exponent of one.[10]

Extending the work of Schmidt et al. (1979), other investigators have found additional impressive evidence of a linear tradeoff relation between the standard deviation of movement endpoints and average movement velocity in the time-matching movement task (e.g., Zelaznik, Shapiro, & McColsky, 1981; Zelaznik et al., 1988). Equation 3 holds not only for stylus-tapping movements but also for wrist rotations (Wright 1983; Wright & Meyer, 1983) and saccadic eye movements (Abrams, Meyer, & Kornblum, 1989). When subjects must make their movement durations match a given temporal goal, this equation appears to have the same functional status as does the logarithmic speed–accuracy tradeoff (Fitts' Law) in the time-minimization movement task.

## Impulse-Variability Model

Clearly, the deterministic iterative-corrections model (Crossman & Goodeve, 1963/1983; Keele, 1968), which assumes constant movement endpoints, cannot account for the linear speed–accuracy tradeoff. On the basis of their results from the time-matching movement task and other related paradigms, Schmidt et al. (1979) therefore proposed a new theory called the *impulse-variability model*. It revives some of Fitts' (1954) earlier ideas about noise in the motor system, and adds more detail to them. These details are embodied in the following set of assumptions, which Fig. 6.7 graphically illustrates.

*Assumptions of the model.* The first assumption is that rapid movements in the time-matching movement task are generated by a pulse of force that has selected amplitude and time parameters. The amplitude parameter ($f$) supposedly controls the force pulse's height; the time parameter ($t$) controls its duration. Through these parameters, the force applied to a part of the body may be scaled in amplitude and duration to produce movements having desired average velocities and distances. Evidence of such scaling has been reported by a number of investigators (e.g., Abrams et al., 1989; Armstrong, 1970; Freund & Büdingen, 1978; Ghez, 1979; Ghez & Vicario, 1978; Gordon & Ghez, 1987).

The second assumption is that the amplitude and time parameters are

---

[10]Because of these differences between Fitts' Law (Equation 1) and the linear tradeoff relation (Equation 3), the former provides a much worse account than does the latter for the results of Schmidt et al. (1979). In particular, applying Fitts' Law to Fig. 6.6 yields $T = -.06 + .038\log_2(2D/W_e)$, $r^2 = .35$, whereas applying the linear tradeoff relation yields $W_e = .22 + .0031(D/T)$, $r^2 = .94$.

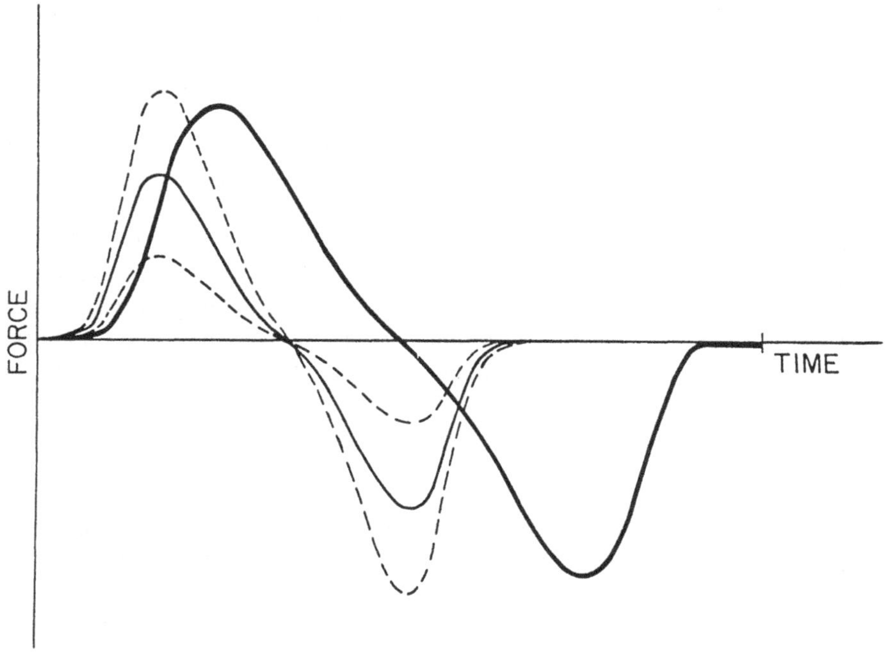


FIG. 6.7. Illustration of assumptions made under one version of the impulse-variability model for rapid movements in the time-matching movement task. (The thick solid curve represents a prototypical force–time function for an individual movement. The thin solid curve and dashed curves represent other functions that may emerge from trial to trial through noise associated with parameters for rescaling the prototype in the force and time dimensions, respectively. From "Models for the speed and accuracy of aimed movements" by D. E. Meyer, J. E. K. Smith, and C. E. Wright, *Psychological Review*, 1982, *89*, 449–482. Reprinted with permission of authors and publisher.)

stochastic variables. A subject supposedly sets them to have certain target values ($F$ and $T$, respectively), but they may fluctuate randomly about these values because of noise in the motor system. As a result, this produces variability in observed movement trajectories. Some evidence of the noise underlying such variability may be seen, for example, in records of electromyographic (EMG) activity during rapid movements, which typically exhibit agonist and antagonist bursts that fluctuate substantially from trial to trial (Angel, 1974; Bouisset, Lestienne, & Maton, 1977; Brown & Cooke, 1981; Ghez, 1979; Ghez & Gordon, 1987; Hallett, Shahani, & Young, 1975; Lestienne, 1979; Wadman, Denier van der Gon, Geuze, & Mol, 1979; Wallace, 1981).

The third assumption is that the noise in the amplitude and time parameters obeys Weber's Law. Accordingly, their standard deviations ($S_f$ and $S_t$) would increase proportionally with the parameters' desired magnitudes (i.e., $S_f = K_f F$; $S_t = K_t T$). Movements that must travel long distances at high velocities should, consequently, be more variable than short low-velocity movements. Evidence supporting Weber's Law here has been reported by Abrams et al. (1989), Schmidt et al. (1979), Wright (1983), and other investigators.

*Theoretical import.* Schmidt et al. (1979) claimed that, taken together, these assumptions of the impulse-variability model account adequately for the linear speed–accuracy tradeoff obtained in their experiments. However, soon after the model was first introduced, some controversy arose regarding whether the model is, in fact, consistent with results from the time-matching movement task. A few investigators have questioned the model's assumptions about the scaling of force pulses (Schmidt, Sherwood, Zelaznik, & Leikind, 1985; Zelaznik, Schmidt, & Gielen, 1986) and about the variability of the force pulse's amplitude and time parameters (Newell, 1980; Newell & Carlton, 1988; Sherwood & Schmidt, 1980). Questions have also arisen over the logic of Schmidt et al.'s (1979) formal theoretical derivations (Meyer et al., 1982; Tsiboulevsky, 1981). Nevertheless, Meyer et al. (1982) demonstrated mathematically that a refined version of the impulse-variability model does yield Equation 3. Some empirical results from saccadic eye movements strongly support their arguments (Abrams et al., 1989). In fact, the notion of impulse variability seems intuitively plausible and may have considerable merit.

## SUMMARY

In summary, past research on movement speed and accuracy has involved a variety of tasks from which different speed–accuracy tradeoffs have emerged, including both the logarithmic (Fitts' Law) and linear tradeoff relations. The obtained results have motivated a persistent struggle between two opposing theoretical camps. One camp has claimed that movement speed affects the accuracy of the current-control phase but not the initial-adjustment phase of movement. Its major founder, Woodworth (1899), hypothesized that current control based on visual feedback is disrupted as movement speed increases. In contrast, a second camp has claimed that neuromotor noise increases with movement speed, thereby affecting the accuracy of the initial-adjustment phase. Fitts (1954), the major founder of this camp, characterized the consequences of the noise in terms of information theory.[11] Despite the formal elegance of his information-theory hypothesis, later investigators rejected it and instead developed further models that attributed speed–accuracy tradeoffs to properties of the current-control phase (Crossman & Goodeve, 1963/1983; Keele, 1968). Yet, recent data have raised more doubts about whether the current-control phase, as embodied in Crossman and Goodeve's (1963/1983) deterministic iterative-corrections model, is the sole mediator of the relationship between movement speed and accuracy. To characterize this relationship more completely, some investigators have therefore revived and extended notions concerning motor noise, as in the impulse-variability model (Meyer et al., 1982; Schmidt et al., 1979).

What none of the preceding contributions does, however, is to provide a good account of Fitts' Law and of the differences in obtained speed–accuracy tradeoffs found for alternative movement tasks. The impulse-variability model has been applied so far only to results from the time-matching movement task, where there is a linear tradeoff relation. It is still unclear how this model, as originally formulated, would explain results from the time-minimization movement task where there is a logarithmic tradeoff relation. A major remaining question concerns whether the logarithmic tradeoff relation stems from the noisiness of the initial-adjustment phase or the structure of error corrections during current control.

The rest of this chapter presents some research that we have conducted to deal with these open issues in more detail. Our goal here is to show how different forms of speed–accuracy tradeoff obtained in alternative movement tasks may be explained with a shared set of principles for the initial-adjustment and current-control phases of rapid aimed movements. Some further efforts toward this goal also appear in Meyer et al. (1988).

## STOCHASTIC OPTIMIZED-SUBMOVEMENT MODELS

The theoretical framework for our research is based on a class of *stochastic optimized-submovement models* that we have developed (Meyer et al., 1988; Smith, 1988). These models synthesize and extend ideas introduced previously as part of the deterministic iterative-corrections model (Crossman & Goodeve, 1963/1983; Keele, 1968), impulse-variability model (Meyer et al., 1982; Schmidt et al., 1979), and normative motor-control theory

---

[11]One might argue that several other investigators (e.g., Brown & Slater-Hammel, 1949; Searle & Taylor, 1948) also contributed significantly to the formation of the second camp (see *Historical Survey*), but their contributions were less fundamental than those of Fitts (1954), because they did not explicitly articulate ideas about the role of motor noise as Fitts did.

(e.g., Hogan, 1984; Hollerbach, 1982; Kleinman, Baron, & Levinson, 1970; Nelson, 1983). Their principal domains of application include both the time-matching movement task in which the linear speed–accuracy tradeoff holds, and the time-minimization movement task in which the logarithmic speed–accuracy tradeoff (Fitts' Law) holds.

To illustrate some of the present theoretical framework's main features, we first consider one simple case called the *optimized dual-submovement model.* It applies specifically to the time-minimization movement task and involves either one or two submovements per trial. After assessing this model's goodness-of-fit, we turn to other, more complex, cases involving three or more submovements per trial.

## Assumptions of The Optimized Dual-Submovement Model

The optimized dual-submovement model makes several basic assumptions, as illustrated in Fig. 6.8.

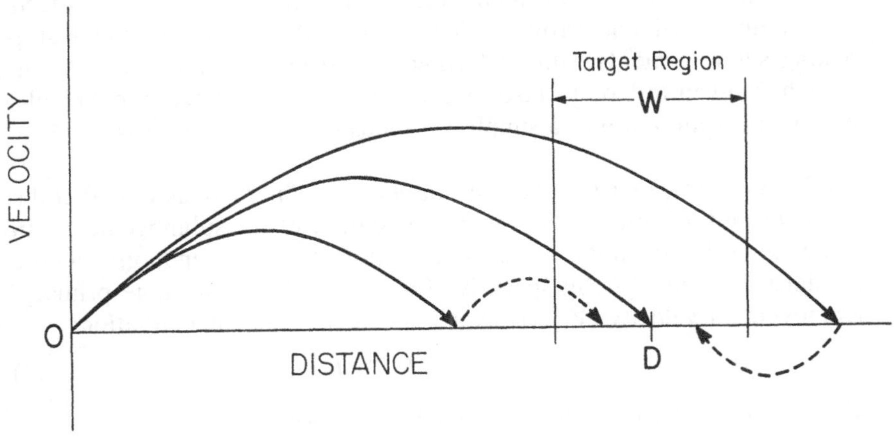


FIG. 6.8. Illustration of assumptions made under the optimized dual-submovement model. (The horizontal axis represents movement distance and the vertical axis represents movement velocity. The solid and dashed curves correspond respectively to hypothetical primary and secondary submovements for three trials on which there are movements between an initial home position [distance = 0] and a target region [bounded by vertical lines] whose center is *D* units from the home position and whose width is *W* units. From "Optimality in human motor performance: Ideal control of rapid aimed movements" by D. E. Meyer, R. A. Abrams, S. Kornblum, C. E. Wright, and J. E. K. Smith, *Psychological Review*, 1988, *95*, 340–370. Reprinted with permission of authors and publisher.)

*Primary submovements.* The first assumption is that on each trial of a time-minimization movement task, the subject begins with a primary submovement programmed to hit the center of the target region (Fig. 6.8, middle solid curve). If the primary submovement is successful, then action terminates without further ado.

*Motor noise and secondary submovements.* A second assumption is that noise in the motor system may affect the primary submovement, causing it either to overshoot or undershoot the target (Fig. 6.8, long and short solid curves). If a miss occurs, then a secondary corrective submovement based on sensory (e.g., visual) feedback is supposedly made (Fig. 6.8, dashed curves). The secondary submovement will usually hit the target, but even it may miss once in a while, because of further motor output variability caused by the noise.

For present purposes, no more corrections are assumed to occur after the secondary submovement. Several considerations motivate this restriction. Although a few investigators (e.g., Crossman & Goodeve, 1963/1983) have reported as many as five submovements per trial, this is not the norm. The findings of other investigators (e.g., Langolf et al., 1976) have revealed that the number of submovements may be rather limited. Also, subjects do commit occasional errors, which suggests that they sometimes stop making submovements without having hit the target. Later, we demonstrate how, even with only one or two submovements per trial, these results may be incorporated into a simple yet plausible account of Fitts' Law.

*Effects of noise on submovement endpoints.* Third, we assume that the effects of motor noise increase with the velocity of the submovements. In particular, primary submovements are assumed to have endpoints whose standard deviation ($S_1$) in space is proportional to the average primary-submovement velocity ($V_1$), as expressed by the following equation:

$$S_1 = KV_1 = K(D_1/T_1), \tag{4}$$

where $D_1$ is the mean distance traveled by the primary submovements, $T_1$ is their mean duration, and $K$ is a positive constant. This assumption is consistent with the results of Schmidt et al. (1979) and other investigators (e.g., Wright & Meyer, 1983; Zelaznik et al., 1981), who have shown that variable errors are a linear function of average movement velocity for rapid movements in the time-matching movement task (Equation 3).

Similarly, we assume that the standard deviation ($S_2$) of secondary-submovement endpoints in space increases proportionally with average secondary-submovement velocity ($V_2$), as expressed by the following equation:

$$S_2 = KV_2 = K(D_2/T_2), \tag{5}$$

where $D_2$ is the mean distance traveled by the secondary submovements, $T_2$ is their mean duration, and $K$ is the same constant as in Equation 4.[12]

*Minimization of movement durations.* Finally, another key assumption is that the average velocities of the primary and secondary submovements are programmed to minimize the average total movement duration ($T$). This assumption stems from the demands of the typical time-minimization movement task. Confronted with these demands, subjects presumably try to reach the target region as quickly as possible while attaining some set high proportion of final target hits. To achieve their aim, they must adopt an appropriate strategy for coping with the effects of motor noise. Such a strategy requires making an optimal compromise between the mean duration ($T_1$) of primary submovements and the mean duration ($T_2$) of secondary submovements, whose sum determines the average total movement duration (i.e., $T_1 + T_2 = T$).

In particular, the primary submovements should not take too much time. If they are very slow, then this would allow the noise in the motor system to be low, yielding greater spatial accuracy (Equation 4) without a need for secondary submovements. However, it would also tend to overinflate the average total movement duration because of an excessive increase in the mean primary-submovement duration ($T_1$).

On the other hand, the primary submovements should not take too little time either. If they are very fast, then this would generate lots of noise, causing them to miss the target frequently. As a result, many secondary corrective submovements would then have to be made, and the average total movement duration would again tend to be overinflated because of an excessive increase in the mean secondary-submovement duration ($T_2$).[13]

---

[12]More precisely, suppose that a primary submovement ends outside the target region at a point $\Delta$ units from the target center. Then secondary submovements made from this point are assumed to have a standard deviation $S_{2\Delta}$ such that $S_{2\Delta} = K\Delta/T_{2\Delta}$, where the mean distance ($\Delta$) and mean duration ($T_{2\Delta}$) of these submovements depend on their starting location. As the endpoints of the primary submovements get farther from the target, $\Delta$ and $T_{2\Delta}$ would also increase accordingly. $S_{2\Delta}$ is presumably controlled by adjusting the magnitude of $T_{2\Delta}$ to achieve a desired relative frequency of target hits (Meyer et al., 1988).

[13]Because some of the primary submovements may hit the target without further ado, the mean secondary-submovement duration would include contributions from some trials on which the duration of the secondary submovements is zero. The frequency of such trials depends on the accuracy of the primary submovements, which in turn depends on the mean primary-submovement duration as previously outlined. Thus, the calculation of the mean secondary-submovement duration must take into account a variety of factors, including the relative frequency of primary-submovement target hits, the distances of primary-submovement endpoints from the target when it is missed, and the desired relative frequency of accurate secondary submovements (Meyer et al., 1988).

So, under the optimized dual-submovement model, there is a putative ideal intermediate duration for the primary submovements, and associated with this ideal, there is also an ideal intermediate duration for the secondary submovements. The exact values of these ideals, which depend on target distance ($D$) and width ($W$), may be determined jointly through elementary differential calculus (Meyer et al., 1988). By adjusting the mean primary-submovement and secondary-submovement durations accordingly, a subject can simultaneously produce a desired overall relative frequency of target hits while minimizing the average total movement duration.

## Predictions of The Optimized Dual-Submovement Model

In essence, the present theoretical framework provides the basis for a normative psychophysics of motor control. It treats subjects as ideal actors who do the best they can, given the demands of their movement task and the limitations imposed by system noise. This perspective parallels one developed in sensory psychophysics, where signal-detection theory has defined optimal behavior for an ideal observer under noisy stimulus conditions (Green & Swets, 1966). Like signal-detection theory, our approach here leads to specific predictions about what the ideal is, and these predictions may then be compared with subjects' actual performance (Meyer et al., 1988). For example, we next consider four quantitative predictions made by the optimized dual-submovement model.

*Average total movement durations.* The first prediction concerns the average total movement durations ($T$) in the time-minimization movement task. According to the optimized dual-submovement model, $T$ is a quasi square-root function of the target distance–width ratio, $D/W$ (Meyer et al., 1988). In particular, Equation 6 should fit very well:

$$T = A + B\sqrt{D/W}, \tag{6}$$

where $A$ and $B$ are non-negative constants.

We therefore expect subjects to obey Fitts' Law approximately, but not exactly. This is consistent with results reported by a number of previous investigators. As mentioned earlier, Kvålseth (1980) has shown that power functions similar to Equation 6 actually come closer to average total movement durations than does a logarithmic function (also see Gan & Hoffman, 1988).[14]

---

[14]It should be noted, however, that Kvålseth (1980) had no detailed theory to motivate the use of a power function for characterizing movement speed–accuracy tradeoffs. His analyses only involved empirical curve fitting, and in some cases, he had more available parameters than Fitts' Law provides, so it is not too surprising that he obtained improved approximations to available data. In contrast, our approach has a principled theoretical basis; it does not succeed merely through the post hoc addition of extra parameters.

*Mean primary-submovement durations.* Second, the optimized dual-submovement model predicts that the mean duration of primary submovements ($T_1$) is a quasi square-root function of $D/W$ (Meyer et al., 1988). In particular, Equation 7 should fit very well:

$$T_1 = A_1 + B_1 \sqrt{D/W}, \qquad (7)$$

where $A_1$ and $B_1$ are non-negative constants ($A_1 < A$, $B_1 < B$). This follows directly from assuming that subjects program their primary submovements optimally. The optimality assumption is supported by results of investigators who have found that primary-submovement durations typically increase as target distance increases and width decreases (e.g., Brown & Slater-Hammel, 1949; Jagacinski et al., 1980; Langolf et al., 1976). In Brown & Slater-Hammel's (1949) study, for example, the durations of primary submovements came very close to having a square-root tradeoff relation with target distance (Fig. 6.3, dashed curve).[15]

*Proportion of secondary submovements.* A third prediction concerns the proportion of trials ($p_2$) on which there are secondary submovements. Under the optimized dual-submovement model, we expect $p_2$ to increase monotonically with $D/W$. When target distance increases or width decreases, the primary submovements should slow down somewhat (Equation 7), but the decrease in their velocity would not be sufficient to maintain a constant rate of target hits. Instead, given the nature of the optimization process, the primary submovements should miss the target more often, thereby requiring more corrections through secondary submovements. A precise mathematical formula describing the change predicted for $p_2$ appears in Meyer et al. (1988, Equation 7). It is consistent with results reported by Crossman and Goodeve (1963/1983) and other investigators (e.g., Jagacinski et al., 1980).

*Error rates.* Fourth, we can also make a prediction about error rates (i.e., the proportion of trials on which a subject finishes without having hit the target region). The optimized dual-submovement model assumes that a trial typically ends after a secondary submovement, even if the target has not been hit yet. Because of this and other related assumptions, the proportion of target misses ($p_e$) should increase monotonically with $D/W$. A precise mathematical formula for the predicted increase appears in Meyer

[15]A reanalysis of Brown and Slater-Hammel's (1949) primary-submovement and average total movement durations reveals that square-root functions of target distance fit them almost as well as logarithmic functions do. For primary submovements, $T_1 = 85 + 75\sqrt{D}$; $r^2 = .996$. For overall movements, $T = 183 + 91\sqrt{D}$;$r^2 = .966$. (Note: The time and distance units used in estimating the parameters of these equations are milliseconds and centimeters, respectively).

et al. (1988, Equation 9). It is consistent with error rates reported by Fitts (1954) and other investigators (e.g., Wallace & Newell, 1983).

### Tests of The Optimized Dual-Submovement Model

To test these predictions of the optimized dual-submovement model further, we have designed new experiments (Meyer et al., 1988) involving a wrist-rotation task similar to the one used by Crossman and Goodeve (1963/1983). Here the objective has been to analyze movement trajectories precisely for a variety of target distances and widths. From these analyses, the spatial and temporal properties of submovements may be assessed in detail as a function of movement durations, secondary-submovement frequencies, and error rates.

*Movement parsing.* Our approach uses a set of quantitative rules that parse the position, velocity, and acceleration records of each movement trajectory into component submovements. The parsing algorithm, which is described in Meyer et al. (1988), identifies the beginnings and ends of both primary and secondary submovements, as shown in Fig. 6.9. It also measures the spatial and temporal features of each submovement, including submovement distances and durations. The obtained results thus extend studies of movement kinematics conducted by a number of previous investigators (e.g., Brown & Slater-Hammel, 1949; Carlton, 1979, 1980; Crossman & Goodeve, 1963/1983; Jagacinski et al., 1980; B. Kerr, 1976; Langolf et al., 1976).

*Results.* Some results of our experiments appear in Fig. 6.10, which shows mean primary-submovement durations ($T_1$) and average total movement durations ($T$) versus the square root of the target distance-width ratio. The light and dark circles represent results from twelve different combinations of $D$ and $W$.[16] The solid and dashed lines represent the corresponding square-root functions that fit the data best. As predicted by the optimized dual-submovement model (Equations 6 and 7), the fit is reasonably good (for average total movement durations, $r^2$ = .92; for mean primary-submovement durations, $r^2$ = .90). In fact, we found a pattern similar to the one reported in Brown and Slater-Hammel (1949), whose data exhibited a montonic, negatively accelerated, relation between target

[16]The values of the target widths ($W_s$) used in Fig. 6.10 were estimated from subjects' actual performance, rather than simply being set equal to the values chosen by the experimenter (Meyer et al., 1988). We worked with subjective target widths instead of objective ones because subjects may respond to a target somewhat differently than its nominal physical dimensions would dictate (Sheridan, 1979; Welford, 1968).

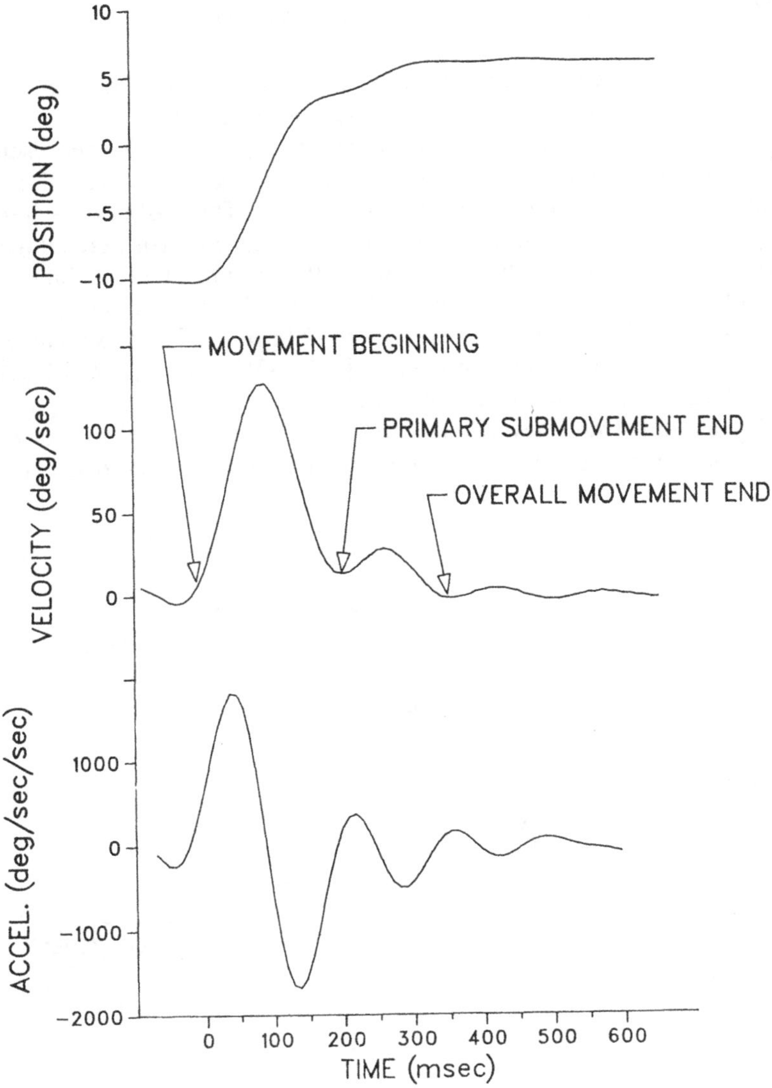


FIG. 6.9. Example of position, velocity, and acceleration records for a representative movement in an experiment on wrist rotations. (The target of the movement had an angular distance of 15.8° from the movement's starting location and an angular width of 6.3°. The arrows indicate the beginning of the movement [i.e., primary-submovement start], end of the primary submovement, and end of the secondary submovement [i.e., overall movement end] as identified by the movement-parsing algorithm cited in the text. From "Optimality in human motor performance: Ideal control of rapid aimed movements" by D. E. Meyer, R. A. Abrams, S. Kornblum, C. E. Wright, and J. E. K. Smith, *Psychological Review*, 1988, *95*, 340–370. Reprinted with permission of authors and publisher.)

distances and mean primary-submovement durations as well as average total movement durations (Fig. 6.3 and Footnote 15).

More results appear in Fig. 6.11. Here the dark circles represent observed versus predicted percentages of errors (target misses) for the twelve combinations of $D$ and $W$ in our experiment. The dashed line indicates where the circles should have fallen if the observed error percentages had exactly matched the predictions made by the optimized dual-submovement model. The fit is not perfect; some deviations did emerge. However, they were statistically insignificant [$\chi^2(12) < 12.0$; $p > .5$]. As expected, more errors occurred when $D/W$ was large [$F(11,33) = 4.09$; $p < .01$]. So the model again succeeded reasonably well.

We should stress, nevertheless, that some of the model's other predictions have been less successful. In particular, consider Fig. 6.12. Here the

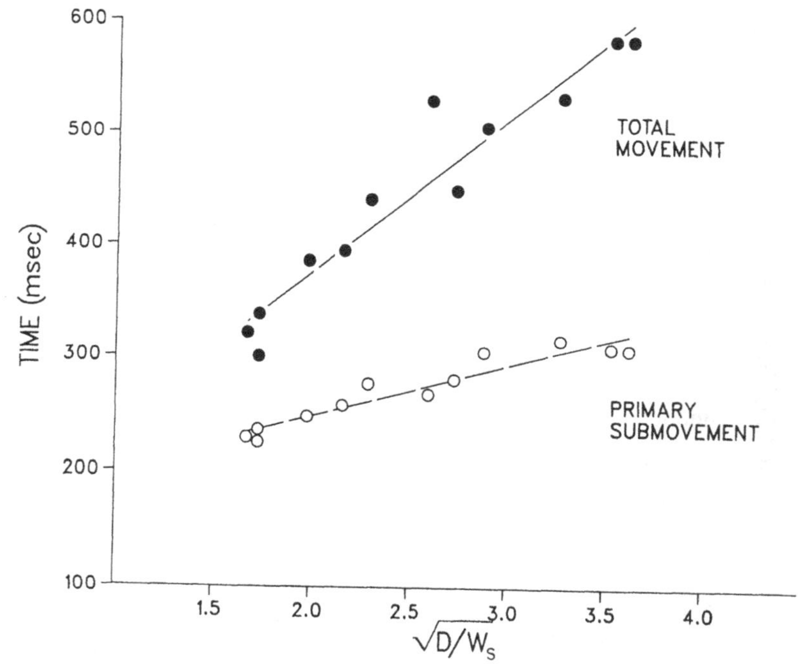


FIG. 6.10. Average total movement durations and mean primary-submovement durations versus the square root of target distance ($D$) divided by subjective target width ($W_s$) for a time-minimization movement task involving wrist rotations. (The solid and dashed lines represent the fit of predictions made by the optimized dual-submovement model. From "Optimality in human motor performance: Ideal control of rapid aimed movements" by D. E. Meyer, R. A. Abrams, S. Kornblum, C. E. Wright, and J. E. K. Smith, *Psychological Review*, 1988, *95*, 340–370. Reprinted with permission of authors and publisher.)

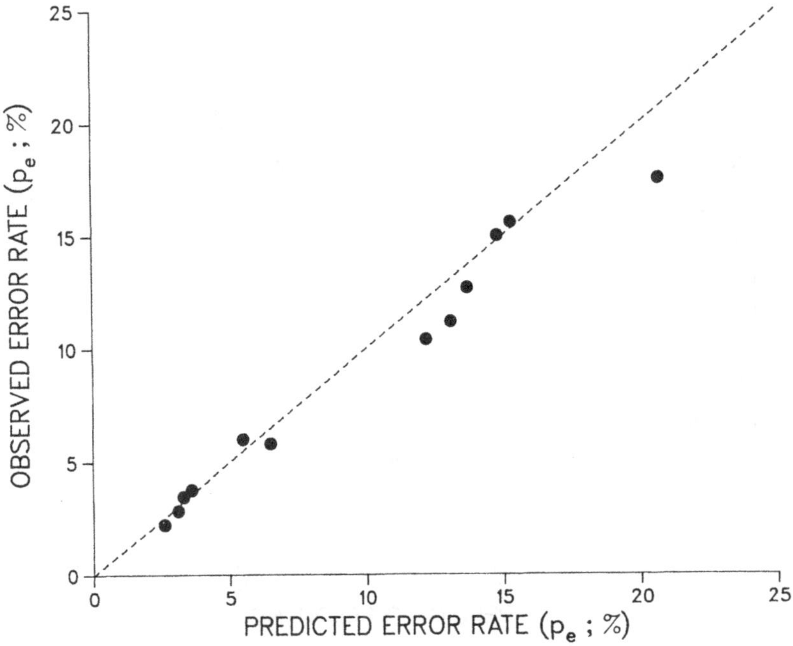


FIG. 6.11. Percentages of errors (target misses) from the experiment on wrist rotations in a time-minimization movement task. (The dark circles represent observed and predicted error rates for each of twelve target distance–width combinations. The dashed diagonal line indicates where the circles should have fallen if the predictions of the optimized dual-submovement model fit the observations perfectly. From "Optimality in human motor performance: Ideal control of rapid aimed movements" by D. E. Meyer, R. A. Abrams, S. Kornblum, C. E. Wright, and J. E. K. Smith, *Psychological Review*, 1988, *95*, 340–370.)

dark circles represent observed versus predicted percentages of trials on which secondary submovements occurred for each of the twelve target distance–width combinations. The dashed line indicates where the circles should have fallen if the observed secondary-submovement percentages had exactly matched the predicted ones.

Consistent with these predictions, more secondary submovements tended to occur as $D/W$ increased [$F(11,33) = 11.0$; $p < .001$]. The correlation between the observed and predicted secondary-submovement percentages was high ($r = .96$). Nevertheless, there were also some significant discrepancies [$x^2(12) = 86.0$; $p < .01$]. When the predicted secondary-submovement percentages were low, the observed percentages were even lower. In contrast, when the predicted percentages were high, the observed percentages were even higher.

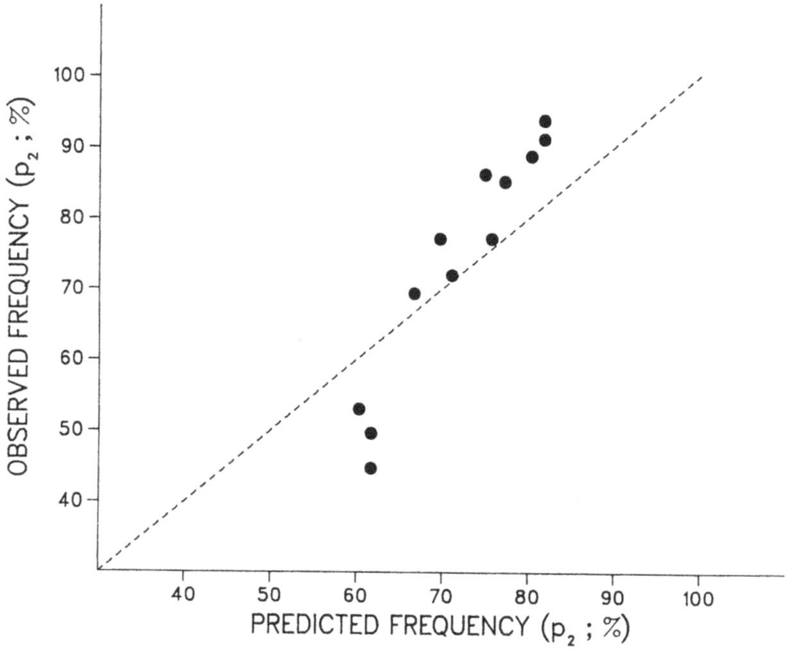


FIG. 6.12. Percentage of trials that yielded secondary submovements during the experiment on wrist rotations in a time-minimization movement task. (The dark circles represent observed and predicted secondary-submovement percentages for each of twelve target distance–width combinations. The dashed diagonal line indicates where the circles should have fallen if the predictions of the optimized dual-submovement model fit the observations perfectly. From "Optimality in human motor performance: Ideal control of rapid aimed movements" by D. E. Meyer, R. A. Abrams, S. Kornblum, C. E. Wright, and J. E. K. Smith, *Psychological Review*, 1988, *95*, 340–370.)

*Theoretical implications.* Our results concerning secondary-submovement percentages might be taken as evidence against the optimized dual-submovement model. Yet we are not ready to reject this model entirely. Its predictions about other dependent variables (e.g., movement durations and error rates) have been confirmed reasonably well. Such confirmation suggests that several of the model's key assumptions are at least approximately correct. In subsequent efforts, we have therefore focused on modifying and extending these assumptions to provide an even better theoretical account of rapid aimed movements in the time-minimization movement task.

## Extensions of The Optimized Dual-Submovement Model

One plausible extension of the optimized dual-submovement model concerns the number of submovements assumed to be made in reaching a target region. Although we previously presented an argument for assuming only one or two submovements (i.e., primary and secondary) per trial, perhaps our approach would work even better if it were not so restricted. This has led us to formulate a more general *optimized multiple-submovement model* in which each overall movement toward the target region may include one, two, three, or more submovements.

*Additional assumptions.* Under the optimized multiple-submovement model, subjects are assumed to make a total of $n$ or fewer submovements per trial. According to this assumption, each trial begins with an initial (primary) submovement toward the target. If by chance the initial submovement hits the target, then action supposedly terminates without further ado. Otherwise, subsequent corrective submovements (i.e., secondary, tertiary, and so on) would be executed, with each one having some intermediate probability of success. The submovement sequence may continue until either the target is hit or $n$ submovements have been completed.

We call $n$ the *maximum submovement number* and treat it as a parameter whose value is independent of target distance and width. If $n$ equals 3, for example, then regardless of $D/W$, up to 3 submovements may be executed per trial. Alternatively, $n$ could have some other preset value (e.g., $n$ = 4). For each value of $n$, there is a specific case of the optimized multiple-submovement model. Which case best fits a collection of data may, of course, vary from subject to subject and experiment to experiment, depending on particular details of the situation at hand. Thus, $n$ provides the model with a potentially powerful way to characterize a wide range of results.

The remaining assumptions of the optimized multiple-submovement model are similar to those introduced earlier (see *Optimized Dual-Submovement Model*). We still assume that noise exists in the motor system, and that the standard deviations of submovement endpoints increase with the average velocity of the submovements (e.g., Equations 4 and 5). Also, as before, it is assumed that subjects adjust the mean durations of the component submovements so as to minimize average total movement durations while maintaining some desired low error rate.

*Additional predictions.* On the basis of these assumptions, the optimized multiple-submovement model yields some additional striking predictions about average total movement durations ($T$) in the time-minimization

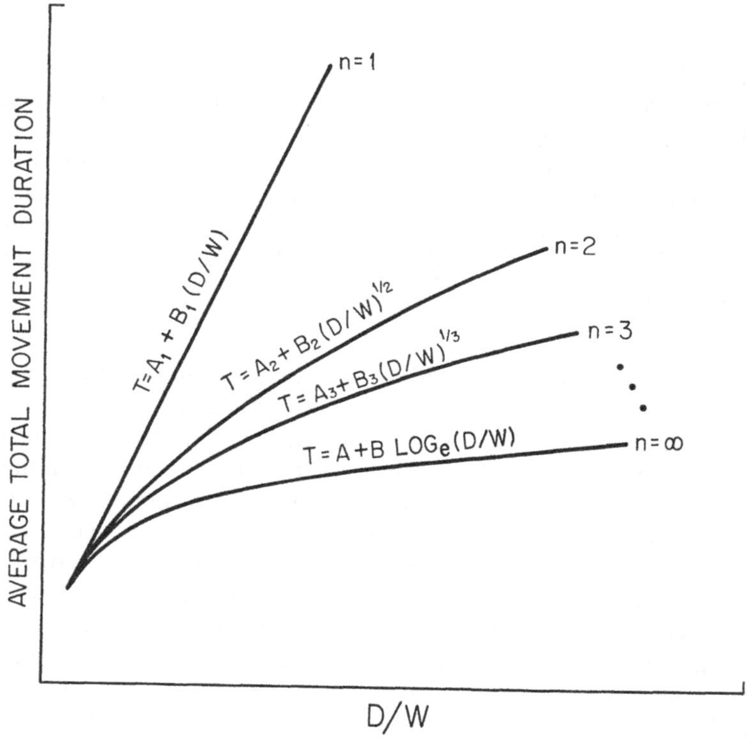


FIG. 6.13. Hypothetical average total movement durations (*T*) versus the target distance–width ratio, *D/W*. (The solid curves show predictions made by the optimized multiple-submovement model as a function of the maximum submovement number, *n*. In each case for which *n* has a finite value, the predicted relation is a quasi power function whose exponent equals $1/n$. As *n* grows larger, this relation approaches a logarithmic function, paralleling Fitts' Law.)

movement task. We have discovered through mathematical analyses (Smith, 1988) that for each value of the maximum submovement number, *T* should be a quasi power function of *D/W* with an exponent of $1/n$ (cf. Equation 2). This pattern is illustrated in Fig. 6.13. For example, if $n = 2$, as under the original optimized dual-submovement model (Fig. 6.8), then the predicted speed–accuracy tradeoff is a quasi square-root function (Equation 6). If instead, $n = 3$, then the predicted tradeoff relation is a quasi cube-root function. And so forth.

Some other details of the predicted tradeoff relations between *T* and *D/W* are interesting as well. For any intermediate value of *D/W* under the optimized multiple-submovement model, $T$ will generally be less when $n$ is large than when $n$ is small. Increasing the maximum number of submovements enhances the subject's ability to overcome the harmful effects of the assumed motor noise. The resulting benefits are especially salient when $D/W$ itself is large (Fig. 6.13).

Furthermore, as $n$ grows larger and larger, the predicted tradeoff relation approaches a lower bound corresponding to the natural logarithm of the target distance–width ratio [*viz.*, $T = A + B \log_e(D/W)$]. This result, which parallels Fitts' Law, emerges when no restriction is imposed on the maximum number of submovements per trial. So, in essence, the optimized multiple-submovement model encompasses one extreme case (*viz.*, $n = \infty$) that implies Fitts' Law exactly, whereas other intermediate cases (*viz.*, $n < \infty$) approximate it to varying degrees. From a theoretical standpoint, the logarithmic tradeoff relation constitutes the best of all possible solutions that a subject could achieve in terms of time minimization when aimed movements are made through a noisy motor system whose stochastic characteristics satisfy our previous assumptions.

*Additional analyses of movement durations.* Do subjects actually place a set limit on the number of submovements made in trying to reach a given target region? What particular value of the maximum submovement number ($n$) yields a predicted speed–accuracy tradeoff that comes closest to real data? The answers to these questions are not necessarily obvious in the light of our previous discourse. As we have seen already, both square root and logarithmic tradeoff relations may fit some data sets reasonably well. So, to obtain further needed insights, we have reanalyzed additional results from Fitts' (1954) original experiments on stylus tapping, which included movements made with a 1-oz stylus and a 1-lb stylus. Given the average total duration ($T$) of these movements (Fitts, 1954, Table 6.1), our analyses determined the goodness-of-fit obtained with the optimized multiple-submovement model, treating $n$ as a free parameter.

The results appear in Fig. 6.14 (solid and dashed curves). Here the horizontal axis represents the different possible values of $n$. For each of these, the vertical axis represents the percentage of variance that the optimized multiple-submovement model explains in the average total movement durations reported by Fitts (1954).

We found that the optimized multiple-submovement model with a maximum of three submovements per trial yielded the best fit. It accounts for about 99% of the variance in Fitts' (1954) data from both the 1-oz (Fig. 6.14, solid curve) and 1-lb (Fig. 6.14, dashed curve) stylus conditions. Other cases involving larger or smaller values of $n$ also fit reasonably well, but less so than the three-submovement case. In particular, when we let $n$ be

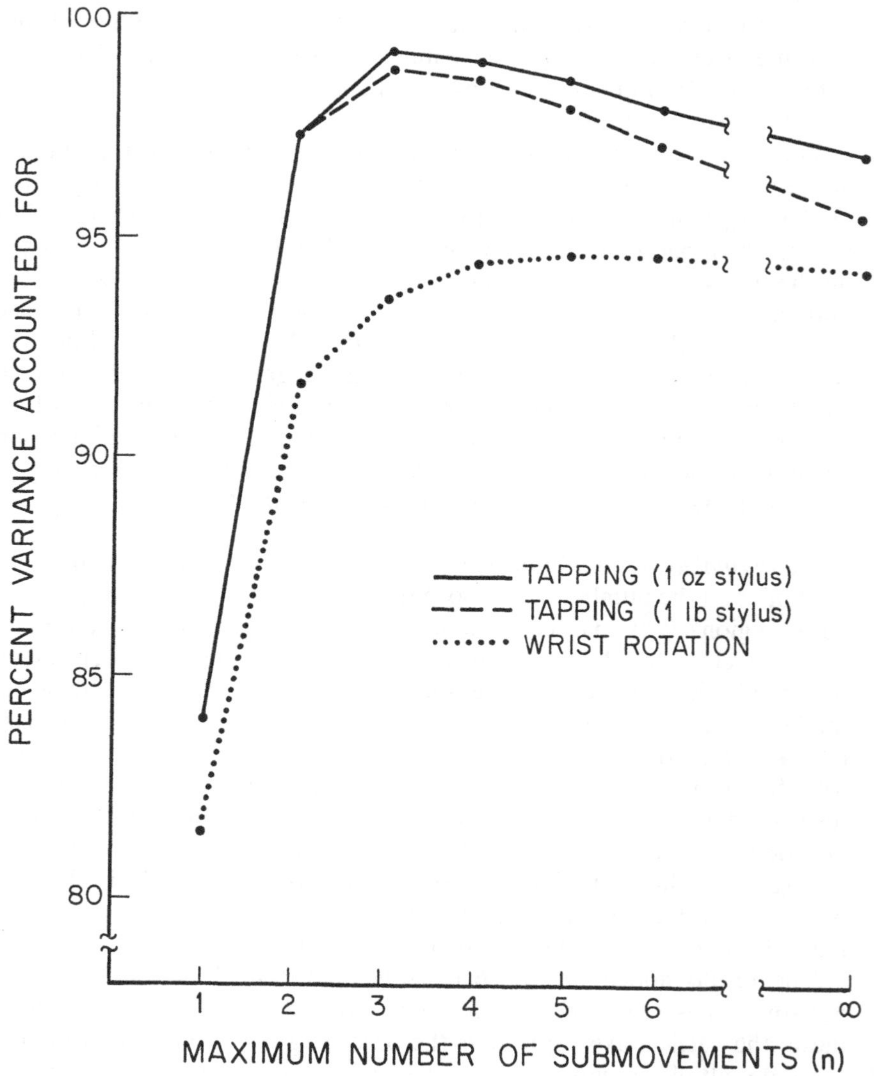


FIG. 6.14. Percentages of variance in average total movement durations that the optimized multiple-submovement model accounts for as a function of *n*, the maximum submovement number. [The solid and dashed curves apply respectively to results from a 1-oz stylus condition and a 1-lb stylus condition in Fitts' (1954, Exp. 1) study of stylus tapping. The dotted curve applies to results from Meyer et al.'s (1988, Exp. 1) study of wrist rotations.]

infinite, corresponding to a logarithmic tradeoff relation (Fig. 6.13), the resulting fit was relatively mediocre ($r^2 = .96$).[17]

Unfortunately, Fitts (1954) did not record his subjects' full movement trajectories, so we cannot tell whether they really always contained three or fewer submovements per trial. The present support for the three-submovement case of the optimized multiple-submovement model is only indirect. However, we may obtain further evidence about its general validity by looking more carefully at data from the experiments that we have performed on wrist rotations (Meyer et al., 1988).

The relevant data are summarized by the dotted curve in Fig. 6.14. For each possible value of the maximum submovement number, $n$, it shows the percentage of variance that the optimized multiple-submovement model explains in our subjects' average total movement durations. Unlike for Fitts' (1954) data on stylus tapping, here the most successful cases of the model are those with $n \geq 4$, not $n = 3$.

Several factors could explain this apparent difference between the values of the maximum submovement number that yielded the best accounts for our results (Fig. 6.14, dotted curve) and those of Fitts (Fig. 6.14, solid and dashed curves), respectively. One possibility involves neurophysiological and biomechanical considerations. Perhaps extra fine-grained corrective submovements are easier to make with wrist rotations than with stylus tapping, because of differences between the physical mechanisms underlying these two types of movement.

The value of $n$ that yields the best fit for the optimized multiple-submovement model may also depend on the level of accuracy (target hits) required by an experimenter. If the targets are very narrow, and if an extremely high hit rate is required, then a subject might be induced to make the maximum submovement number relatively large. For example, this could explain why a logarithmic tradeoff relation (Fig. 6.3) fit the results of Brown and Slater-Hammel (1949) so well. As mentioned before, they used very narrow targets (*viz.*, lines) and instructed subjects always to stop their movements right on target. According to our conjecture, such a procedure would encourage a large $n$ value, thereby biasing subjects toward a true logarithmic tradeoff relation (Fig. 6.13), which is what Brown and Slater-Hammel actually found.

*Additional analyses of submovement percentages.* Some other evidence likewise supports the present theoretical interpretation. In addition to reanalyzing the durations of our subjects' wrist rotations, we have examined

[17]As mentioned previously (Fig. 6.5), the results of Crossman and Goodeve (1963/1983) are also fit best by a power function whose exponent equals about 1/3, consistent with the optimized multiple-submovement model that has $n = 3$.

their movement trajectories again to identify and count submovements at a more microscopic level, using a refined version of the movement-parsing algorithm described earlier (cf. Fig. 6.9). This revealed that on each trial, subjects almost always produced either one, two, three, or four submovements before their activity terminated, consistent with the maximum submovement number of the optimized multiple-submovement model that best fit the average total movement durations (Fig. 6.14).

For example, consider Fig. 6.15. Here the open bars show observed percentages of trials involving one, two, three, and four submovements for a representative target region whose distance–width ratio ($D/W$) was 9.85. The dashed line shows corresponding percentages predicted by the optimized multiple-submovement model with the maximum number of submovements ($n$) equal to four. The model's predictions do not fit the data perfectly [$\chi^2(2) = 55.5$; $p < .001$]. Nevertheless, there is good qualitative agreement. Both the predicted and observed percentages exhibit an inverted U-shape pattern, in which a majority of the trials yielded intermediate numbers of submovements.

What caused the less than perfect quantitative fit in Fig. 6.15? One possibility is that our movement parser may have been somewhat inaccurate; perhaps it tended to misclassify single-submovement trials as dual-submovement ones, but tended to misclassify quadruple-submovement trials as involving only three submovements. This could have happened because passive damped oscillations at the end of strong initial submovements are themselves large and may appear much like real voluntary activity, whereas the final member of a submovement quartet is relatively weak and may appear much like residual tremor.

Another possibility is that subjects did not optimize their submovements completely, given a maximum submovement number ($n$) of 4. If, compared with the ideal, they made the velocities of the initial (primary) submovements too large and the velocities of the tertiary submovements too small, then this could produce discrepancies like those shown in Fig. 6.15. Still, we would infer that subjects came fairly close to ideal performance, because the optimized multiple-submovement model fit the average total movement durations remarkably well (Fig. 6.14) and the submovement percentages at least moderately well (Fig. 6.15).

## CONCLUSION

In conclusion, the class of stochastic optimized-submovement models provides a natural extension of previous research on speed–accuracy tradeoffs in rapid aimed movements. Through the theoretical framework embodied by these models, findings from a variety of movement tasks may be ra-

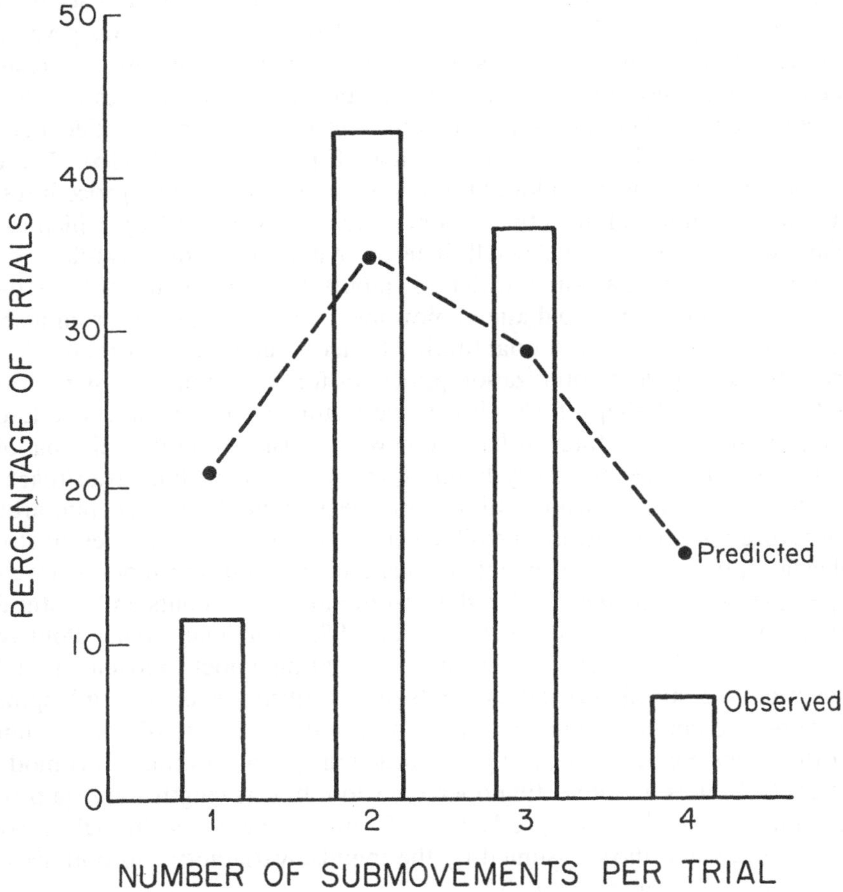


FIG. 6.15. Percentages of trials involving either 1, 2, 3, or 4 submovements in the study of wrist rotations conducted by Meyer et al. (1988, Experiment 1). [The results come from a condition in which $D/W = 9.82$. The solid vertical bars represent observed percentages, and the dashed curve represents predicted percentages for the optimized multiple-submovement model with a maximum submovement number ($n$) equal to 4.]

tionalized and integrated in a coherent fashion. The linear tradeoff between variable errors and average movement velocity in the time-matching movement task (Equation 3) manifests the underlying random processes assumed as part of this framework. Given the characteristics of these processes and the motor noise associated with them, the logarithmic speed–accuracy tradeoff (Equation 1, Fitts' Law) turns out to be an ideal solution that subjects may adopt for performing successfully in the time-minimization movement task.

As a result, the evolution of research on movement speed and accuracy has a striking parallel with the evolution of research on sensory psychophysics. When early psychophysicists began studying the nature of thresholds, stimulus discrimination, and recognition in sensory processes, they discovered a number of basic phenomena, including Weber's law, Fechner's law, and eventually Stevens' power law (Engen, 1971a, 1971b). These phenomena have been explored through various experimental procedures, culminating in signal-detection theory (Green & Swets, 1966), which explains many of the obtained results in terms of normative statistical-decision rules used as part of a system with noisy inputs. Similarly, studies of speed–accuracy tradeoffs in rapid aimed movements have progressed from a set of fundamental tasks and quantitative tradeoff relations to a theoretical framework involving optimization principles for regulating a noisy motor-output system. Perhaps if Woodworth were still alive today, he would find the present state of affairs at least somewhat satisfying, in that his goal of achieving a mature motor psychophysics may now be within easier reach.

Of course, more research still remains to be done. One important area for future research concerns possible connections between the present stochastic optimized-submovement models and mass–spring models developed previously to characterize slow positioning movements and postural maintenance (Bizzi et al., 1978; Cooke, 1980; Fel'dman, 1966; Polit & Bizzi, 1979; Sakitt, 1980). Under the mass–spring models, movement and posture are controlled by adjustments of the stiffnesses or resting lengths of opposing muscle groups, so as to attain a desired state of equilibrium in their tensions, corresponding to a targeted spatial location. This mode of control raises an interesting open question: how is length–tension programming related to the production of rapid movements through noisy force pulses like those assumed by the impulse-variability and optimized multiple-submovement models?

Perhaps the answer lies in yet-to-be-identified principles of coordination that govern a movement's initial pulselike stage and its final clamping stage, which together propel a selected part of the body toward a new target location and then stabilize it there upon completion of the movement (Bahill & Stark, 1979; Ghez, 1979). As Bahill and Stark (1979) have noted from data on saccadic eye movements, these two stages may not always mesh perfectly with each other, but the neuromotor output signals for each of them do tend to be positively correlated. Consequently, one might expect that similar features of stochastic operation apply in each stage. This similarity, together with considerations of movement optimization, could provide a theoretical bridge for linking models based on force–pulse programming and models based on length–tension programming, thereby resulting ultimately in a unified treatment of fast inaccurate and slow accurate movements.

## ACKNOWLEDGMENTS

This research was supported by grant R01 MH37145 from the U.S. National Institute of Mental Health to the University of Michigan. The authors thank R. Jagacinski, M. Jeannerod, S. Keele, and R. Schmidt for helpful comments. The technical assistance of Carol Huff is also greatly appreciated. Correspondence regarding this chapter should be addressed to: David E. Meyer, Human Performance Center, Dept. of Psychology, University of Michigan, 330 Packard Rd., Ann Arbor, MI, 48104, U.S.A.

## REFERENCES

Abrams, R. A., Meyer, D. E., & Kornblum, S. (1989). Speed and accuracy of saccadic eye movements: Characteristics of impulse variability in the oculomotor system. *Journal of Experimental Psychology: Human Perception and Performance*, *15*, 529–543.

Anderson, N. S. (1987, August). *The pervasiveness of Fitts' Law*. Presidential address to the Division of Applied Experimental and Engineering Psychology at the convention of the American Psychological Association, New York.

Angel, R. W. (1974). Electromyography during voluntary movements: The two-burst pattern. *Electroencephalography and Clinical Neurophysiology*, *36*, 493–498.

Annett, J., Golby, C. W., & Kay, H. (1958). The measurement of elements in an assembly task: The information output of the human motor system. *Quarterly Journal of Experimental Psychology*, *10*, 1–11.

Armstrong, T. R. (1970). *Training for the production of memorized movement patterns* (Tech. Rep. 26). Ann Arbor: University of Michigan, Human Performance Center.

Bahill, A. T., & Stark, L. (1979). The trajectories of saccadic eye movements. *Scientific American*, *241*, 108–117.

Bailey, G. B., & Presgrave, R. (1958). *Basic motion time study*. New York: McGraw-Hill.

Bainbridge, L., & Sanders, M. (1972). The generality of Fitts' Law. *Journal of Experimental Psychology*, *96*, 130–133.

Beggs, W. D. A., & Howarth, C. I. (1972). The accuracy of aiming at a target: Some further evidence for a theory of intermittent control. *Acta Psychologica*, *36*, 171–177.

Bizzi, E., Dev, P., Morasso, P., & Polit, A. (1978). Effect of load disturbances during centrally initiated movements. *Journal of Neurophysiology*, *41*, 542–556.

Bouisset, S., Lestienne, F., & Maton, B. (1977). The stability of synergy in agonists during the execution of a simple voluntary movement. *Electroencephalography and Clinical Neurophysiology*, *42*, 543–551.

Brown, J. S., & Slater-Hammel, A. T. (1949). Discrete movements in the horizontal plane as a function of their length and direction. *Journal of Experimental Psychology*, *39*, 84–95.

Brown, S. H., & Cooke, J. D. (1981). Amplitude- and instruction-dependent modulation of movement-related electromyogram activity in humans. *Journal of Physiology*, *316*, 97–107.

Bullock, D., & Grossberg, S. (1988). Neural dynamics of planned arm movements: Emergent invariants and speed-accuracy properties during trajectory formation. *Psychological Review*, *95*, 49–90.

Card, S. K., English, W. K., & Burr, B. J. (1978). Evaluation of mouse, rate-controlled joystick, step keys, and text keys for text selection. *Ergonomics*, *21*, 601–613.

Card, S. K., Moran, T. P, & Newell, A. (1983). *The psychology of human–computer interaction*. Hillsdale, NJ: Lawrence Erlbaum Associates.

Carlton, L. G. (1979). Control processes in the production of discrete aiming responses. *Journal of Human Movement Studies*, *5*, 115–124.

Carlton, L. G. (1980). Movement control characteristics of aiming responses. *Ergonomics*, *23*, 1019–1032.

Carlton, L. G. (1981). Processing visual feedback information for movement control. *Journal of Experimental Psychology: Human Perception and Performance*, *7*, 1019–1030.

Cooke, J. D. (1980). The organization of simple, skilled movements. In G. E. Stelmach and J. Requin (Eds.), *Tutorials in motor behavior* (pp. 199–212). Amsterdam: Elsevier-North Holland.

Craik, K. J. W. (1947). Theory of the human operator in control systems. Part I. The operator as an engineering system. *British Journal of Psychology*, *38*, 56–61.

Craik, K. J. W. (1948). Theory of the human operator in control systems. Part II. Man as an element in control systems. *British Journal of Psychology*, *38*, 142–148.

Craik, K. J. W. (1963a). Psychological and physiological aspects of control mechanisms with special reference to tank gunnery. Part I. *Ergonomics*, *6*, 1–33. Original work reported in 1943.

Craik, K. J. W. (1963b). Psychological and physiological aspects of control mechanisms with special reference to tank gunnery. Part II. *Ergonomics*, *6*, 419–440. Original work reported in 1944.

Crossman, E. R. F. W. (1957). The speed and accuracy of simple handmovements. In E. R. F. W. Crossman and W. D. Seymour (Eds.), *The nature and acquisition of industrial skills*. Report to M.R.C. and D.S.I.R. Joint Committee on Individual Efficiency in Industry.

Crossman, E. R. F. W., & Goodeve, P. J. (1983). Feedback control of hand-movement and Fitts' law. *Quarterly Journal of Experimental Psychology*, *35A*, 251–278. Original work presented at the meeting of the Experimental Psychology Society, Oxford, England, July 1963.

Dodge, R., & Cline, T. S. (1901). The angular velocity of eye movements. *Psychological Review*, *8*, 145–157.

Drury, C. G. (1975). Application of Fitts' Law to foot-pedal design. *Human Factors*, *17*, 368–373.

Ellson, D. G. (1949). The application of operational analysis to human motor behavior. *Psychological Review*, *56*, 9–17.

Engen, T. (1971a). Psychophysics. I. Discrimination and detection. In J.W. Kling & L. A. Riggs (Eds.), *Woodworth and Schlosberg's experimental psychology* (3rd ed., pp. 11–46). New York: Holt, Rinehart, Winston.

Engen, T. (1971b). Psychophysics. II. Scaling methods. In J.W. Kling & L. A. Riggs (Eds.), *Woodworth and Schlosberg's experimental psychology* (Third Edition, pp. 47–86). New York: Holt, Rinehart, Winston.

Fel'dman, A. G. (1966). Functional maintenance of a steady posture—II. Controllable parameters of the muscles. *Biophysics*, *11*, 565–578.

Fenn, W. O. (1938). Mechanics of muscular contraction in man. *Journal of Applied Physics*, *9*, 165–177.

Ferrell, W. R. (1965). Remote manipulation with transmission delay. *IEEE Transactions on Human Factors in Electronics*, *6*, 24–32.

Fitts, P. M. (1954). The information capacity of the human motor system in controlling the amplitude of movement. *Journal of Experimental Psychology*, *47*, 381–391.

Fitts, P. M., & Peterson, J. R. (1964). Information capacity of discrete motor responses. *Journal of Experimental Psychology*, *67*, 103–112.

Fitts, P. M., & Radford, B. K. (1966). Information capacity of discrete motor responses under different cognitive sets. *Journal of Experimental Psychology*, *71*, 475–482.

Flowers, K. (1975). Handedness and controlled movement. *British Journal of Psychology*, *66*, 39–52.

Freeman, F. N. (1914). Experimental analysis of the writing movement. *Psychological Monographs*, *17* (Whole No. 75), 1–46.

Freund, H. J., & Büdingen, H. J. (1978). The relationship between speed and amplitude of the fastest voluntary contractions of human arm muscles. *Experimental Brain Research*, *31*, 1–12.

Fullerton, G. S., & Cattell, J. McK. (1892). On the perception of small differences. *University of Pennsylvania Philosophical Series*, No. 2.

Gan, K., & Hoffman, E. R. (1988). Geometrical conditions for ballistic and visually controlled movements. *Ergonomics*, *31*, 829–839.

Garrett, H. E. (1922). A study of the relation of accuracy to speed. *Archives of Psychology*, *8* (Whole No. 56), 1–104.

Georgopoulos, A. P. (1986). On reaching. *Annual Review of Neuroscience*, *9*, 147–170.

Ghez, C. (1979). Contributions of central programs to rapid limb movement in the cat. In H. Asanuma & V. J. Wilson (Eds.), *Integration in the nervous system* (pp. 305–319). Tokyo: Igaku-Shoin.

Ghez, C., & Gordon, J. (1987). Trajectory control in targeted force impulses. I. Role of opposing muscles. *Experimental Brain Research*, *67*, 225–240.

Ghez, C., & Vicario, D. (1978). The control of rapid limb movement in the cat. II. Scaling of isometric force adjustments. *Experimental Brain Research*, *33*, 191–202.

Gordon, J., & Ghez, C. (1987). Trajectory control in targeted force pulses. II. Pulse height control. *Experimental Brain Research*, *67*, 241–252.

Green, D. M., & Swets, J. A. (1966). *Signal detection theory and psychophysics*. New York: Wiley.

Hallett, M., Shahani, B. T., & Young, R. R. (1975). EMG analysis of stereotyped voluntary movements in man. *Journal of Neurology, Neurosurgery and Psychiatry*, *38*, 1154–1162.

Hancock, P. A., & Newell, K. M. (1985). The movement speed–accuracy relationship in space–time. In H. Heuer, U. Kleinbeck, & K. H. Schmidt (Eds.), *Motor behavior: Programming, control, and acquisition* (pp. 153–188). Berlin: Springer-Verlag.

Hancock, W. M., Langolf, G., & Clark, D. O. (1973). Development of standard data for stereoscopic microscopic work. *AIIE Transactions, 5*, 113–118.

Hartson, L. D. (1939). Contrasting approaches to the analysis of movement. *Journal of General Psychology, 20*, 263–293.

Hartzell, E. J., Dunbar, S., Beveridge, R., & Cortilla, R. (1982). Helicopter pilot response latency as a function of the spatial arrangement of instruments and controls. *Proceedings of the Eighteenth Annual Conference on Manual Control* (pp. 345–364). Dayton, OH.

Hogan, N. (1984). An organizing principle for a class of voluntary movements. *The Journal of Neuroscience*, *4*, 2745–2754.

Hollerbach, J. M. (1982). Computers, brains and the control of movement. *Trends in Neuroscience*, *5*, 189–192.

Hollingworth, H. L. (1909). The inaccuracy of movement. *Archives of Psychology, 2* (Whole No. 13), 1–87.

Howarth, C. I., Beggs, W. D. A., & Bowden, J. M. (1971). The relationship between speed and accuracy of movement aimed at a target. *Acta Psychologica, 35*, 207–218.

Jagacinski, R. J. (1989). Target acquisition: Performance measures, process models, and design implications. In G. R. McMillan (Ed.), *Applications of human performance models to system design* (pp. 135–149). New York: Plenum Press.

Jagacinski, R. J., & Monk, D. L. (1985). Fitts' Law in two dimensions with hand and head movements. *Journal of Motor Behavior, 17*, 77–95.

Jagacinski, R. J., Repperger, D. W., Moran, M. S., Ward, S. L., & Glass, B. (1980). Fitts' law and the microstructure of rapid discrete movements. *Journal of Experimental Psychology: Human Perception and Performance, 6*, 309–320.

Jeannerod, M. (1988). *The neural and behavioural organization of goal-directed movements.* Oxford, England: Clarendon Press.

Karger, D. W., & Hancock, W. M. (1982). *Advanced work measurement.* New York: Industrial Press.

Keele, S. W. (1968). Movement control in skilled motor performance. *Psychological Bulletin, 70*, 387–403.

Keele, S. W. (1981). Behavioral analysis of movement. In V. Brooks (Ed.), *Handbook of physiology: The nervous system: Vol. 2. Motor control* (pp. 1391–1414). Baltimore, MD: American Physiological Society.

Keele, S. W. (1986). Motor control. In J. K. Boff, L. Kaufman, & J. P. Thomas (Eds.), *Handbook of human perception and performance*, Vol. II (pp. 1–60). New York: Wiley.

Keele, S. W., & Posner, M. I. (1968). Processing of visual feedback in rapid movements. *Journal of Experimental Psychology, 77*, 155–158.

Kerr, B. A. (1976). Motor control in aimed movements. In D. W. Landers & R. W. Christina (Eds.), *The psychology of motor behavior and sport* (pp. 269–276). Urbana, IL: Human Kinetics.

Kerr, B. A., & Langolf, G. D. (1977). Speed of aiming movements. *Quarterly Journal of Experimental Psychology, 29*, 475–481.

Kerr, R. (1973). Movement time in an underwater environment. *Journal of Motor Behavior, 5*, 175–178.

Kerr, R. (1978). Diving, adaptation, and Fitts' Law. *Journal of Motor Behavior, 10*, 255–260.

Klapp, S. T. (1975). Feedback versus motor programming in the control of aimed movements. *Journal of Experimental Psychology: Human Perception and Performance, 104*, 147–153.

Kleinman, D. L., Baron, S., & Levison, W. H. (1970). An optimal control model of human response. Part I: Theory and validation. *Automatica, 6*, 357–369.

Knight, A. A., & Dagnall, P. R. (1967). Precision in movements. *Ergonomics, 10*, 321–330.

Kvålseth, T. O. (1977). Effects of marijuana on human reaction time and motor control. *Perceptual and Motor Skills, 45*, 935–939.

Kvålseth, T. O. (1979). Note on information capacity of discrete motor responses. *Perceptual and Motor Skills, 49*, 291–296.

Kvålseth, T. O. (1980). An alternative to Fitts' law. *Bulletin of the Psychonomic Society, 16*, 371–373.

Langfield, H. S. (1915). Facilitation and inhibition of motor impulses: A study in simultaneous and alternating finger movements. *Psychological Review, 22*, 253–478.

Langolf, G. D., Chaffin, D. B., & Foulke, J. A. (1976). An investigation of Fitts' law using a wide range of movement amplitudes. *Journal of Motor Behavior, 8*, 113–128.

Lestienne, F. (1979). Effects of inertial load and velocity on the braking process of voluntary limb movements. *Experimental Brain Research, 35*, 407–418.

MacKenzie, C. L., Marteniuk, R. G., Dugas, C., Liske, D., & Eickmeier, B. (1987). Three-dimensional movement trajectories in Fitts' task: Implications for control. *Quarterly Journal of Experimental Psychology, 39A*, 629–647.

Marteniuk, R. G., & MacKenzie, C. L. (1980). Information processing in movement organization and execution. In R.S. Nickerson (Ed.), *Attention and performance VIII* (pp. 29–57). Hillsdale, NJ: Lawrence Erlbaum Associates.

Megaw, E.D. (1975). Fitts' tapping revisited. *Journal of Human Movement Studies, 1*, 163–171.

Meyer, D. E., Abrams, R. A., Kornblum, S., Wright, C. E., & Smith, J. E. K. (1988). Optimality in human motor performance: Ideal control of rapid aimed movements. *Psychological Review, 95*, 340–370.

Meyer, D. E., Smith, J. E. K., & Wright, C. E. (1982). Models for the speed and accuracy of aimed movements. *Psychological Review, 89*, 449–482.

Miller, G. A. (1953). What is information measurement? *American Psychologist, 8*, 3–11.

Nelson, W. (1983). Physical principles for economics of skilled movements. *Biological Cybernetics, 46*, 135–147.

Newell, K. M. (1980). The speed–accuracy paradox in movement control: Errors of time and space. In G. E. Stelmach & J. Requin (Eds.), *Tutorials in motor behavior* (pp. 501–510). Amsterdam: North Holland.

Newell, K. M., & Carlton, L. G. (1988). Force variability in isometric responses. *Journal of Experimental Psychology: Human Perception and Performance, 14*, 37–44.

Newell, K. M., & Hancock, P. A. (1984). Forgotten moments: Skewness and kurtosis as influential factors in inferences extrapolated from response distributions. *Journal of Motor Behavior, 16*, 320–335.

Niebel, B. W. (1962). *Motion and time study* (3rd Ed.). Homewood, IL: Irwin.

Peters, W., & Wenborn, A. A. (1936). The time pattern of voluntary movements. *British Journal of Psychology, 26*, 388–406; *27*, 60–73.

Pew, R. W. (1974). Human perceptual-motor performance. In B. H. Kantowitz (Ed.), *Human information processing: Tutorials in performance and cognition* (pp. 1–39). Hillsdale, NJ: Lawrence Erlbaum Associates.

Phillip, B. R. (1936). The relationship between speed and accuracy in a motor task. *Journal of Experimental Psychology, 19*, 24–50.

Polit, A., & Bizzi, E. (1979). Characteristics of motor programs underlying arm movements in monkeys. *Journal of Neurophysiology, 42*, 183–194.

Poulton, E. C. (1974). *Tracking skill and manual performance*. New York: Academic Press.

Prablanc, C., Echallier, J. F., Komilis, E., & Jeannerod, M. (1979). Optimal response of eye and hand motor systems in pointing at a visual target: I. *Biological Cybernetics, 35*, 113–124.

Sakitt, B. (1980). A spring model and equivalent neural network for arm posture control. *Biological Cybernetics, 37*, 227–234.

Sanes, J. N., & Evarts, E. V. (1984). Motor psychophysics. *Human Neurobiology, 2*, 217–225.

Schmidt, R. A., Sherwood, D., Zelaznik, H. N., & Leikind, B. (1985). Speed–accuracy tradeoffs in motor behavior: Theories of impulse variability. In H. Heuer, U. Kleinbeck, & K. H. Schmidt (Eds.), *Motor behavior: Programming, control, and acquisition* (pp. 79–123). Berlin: Springer-Verlag.

Schmidt, R. A., Zelaznik, H., Hawkins, B., Frank, J. S., & Quinn, J. T., Jr. (1979). Motor-output variability: A theory for the accuracy of rapid motor acts. *Psychological Review, 86*, 415–451.

Schutz, R. W., & Roy, E. A. (1973). Absolute error: The devil in disguise. *Journal of Motor Behavior, 5*, 141–153.

Searle, L. V., & Taylor, F. V. (1948). Studies of tracking behavior. I. Rate and time characteristics of simple corrective movements. *Journal of Experimental Psychology, 38*, 615–631.

Shannon, C. E. (1948). A mathematical theory of communication. *Bell System Technical Journal, 27*, 379–423, 623–656.

Shannon, C., & Weaver, W. (1949). *The mathematical theory of communication*. Urbana, IL: University of Illinois Press.

Sheridan, M. R. (1979). A reappraisal of Fitts' law. *Journal of Motor Behavior, 11*, 179–188.

Sheridan, T. B., & Ferrell, W. R. (1963). Remote manipulative control with transmission delay. *IEEE Transactions on Human Factors in Electronics, 4*, 25–29.

Sherwood, D. E., & Schmidt, R. A. (1980). The relationship between force and force variability in minimal and near-maximal static and dynamic contractions. *Journal of Motor Behavior, 12*, 75–89.

Smith, J. E. K. (1988, June). *Rapid aimed movements and the speed–accuracy trade-off.* Paper presented at the Clyde Coombs Memorial Symposium, Ann Arbor, MI.

Stein, R. B., & Lee, R. G. (1981). Tremor and clonus. In V. Brooks (Ed.), *Handbook of physiology: The nervous system: Vol. 2. Motor control* (pp. 325–344). Baltimore, MD: American Physiological Society.

Stetson, R. S. (1905). A motor theory of rhythm and discrete succession. *Psychological Review, 12*, 250–270.

Stetson, R. H., & McDill, J. A. (1923). Mechanims of the different types of movements. *Psychological Monographs, 32* (Whole No. 145), 18–40.

Travis, L.E. (1929). The relation of voluntary movement to tremors. *Journal of Experimental Psychology, 12*, 515–524.

Tsiboulevsky, I. E. (1981). Notes on the theory of the accuracy of rapid movements proposed by R. Schmidt and co-authors. *Voprosy Psikhologii* (Moscow), No. 3, May/June, 127–131. (R.Browning, Trans.).

Vince, M.A. (1948). Corrective movements in a pursuit task. *Quarterly Journal of Experimental Psychology, 1*, 85–103.

Wade, M.G., Newell, K. M., & Wallace, S. A. (1978). Decision time and movement time as a function of response complexity in retarded persons. *American Journal of Mental Deficiency, 83*, 135–144.

Wadman, W. J., Denier van der Gon, J. J., Geuze, R. H., & Mol, C. R. (1979). Control of fast goal-directed arm movements. *Journal of Human Movement Studies, 6*, 19–37.

Wallace, S. A. (1981). An impulse-timing theory for reciprocal control of muscular activity in rapid, discrete movements. *Journal of Motor Behavior, 13*, 144–160.

Wallace, S. A., & Newell, K. M. (1983). Visual control of discrete aiming movements. *Quarterly Journal of Experimental Psychology, 35A*, 311–321.

Wallace, S. A., Newell, K. M., & Wade, M. G. (1978). Decision and response times as a function of movement difficulty in preschool children. *Child Development, 49*, 509–512.

Welford, A. T. (1968). *Fundamentals of Skill*, London: Methuen.

Welford, A. T., Norris, A. H., & Shock, N. W. (1969). Speed and accuracy of movement and their changes with age. *Acta Psychologica, 30*, 3–15.

Woodworth, R. S. (1899). The accuracy of voluntary movement. *Psychological Review, 3* (Whole No. 13), pp. 1–114.

Woodworth, R. S. (1938). *Experimental psychology*. New York: Holt.

Wright, C. E. (1983). *Spatial and temporal variability of aimed movements with three contrasting goal points*. Unpublished doctoral dissertation, University of Michigan.

Wright, C. E., & Meyer, D.E. (1983). Conditions for a linear speed–accuracy trade-off in aimed movements. *Quarterly Journal of Experimental Psychology, 35A*, 279–296.

Zelaznik, H. N., Hawkins, B., & Kisselburgh, L. (1983). Rapid visual feedback processing in single-aiming movements. *Journal of Motor Behavior, 15*, 217–236.

Zelaznik, H. N., Mone, S., McCabe, G. P., & Thaman, C. (1988). Role of temporal and spatial precision in determining the nature of the speed–accuracy trade-off in aimed-hand movements. *Journal of Experimental Psychology: Human Perception and Performance, 14*, 221–230.

Zelaznik, H. N., Schmidt, R. A., & Gielen, S. C. A. M. (1986). Kinematic properties of aimed hand movements. *Journal of Motor Behavior, 18*, 353–372.

Zelaznik, H. N., Shapiro, D.C., & McClosky, D. (1981). Effects of a secondary task on the accuracy of single aiming movements. *Journal of Experimental Psychology: Human Perception and Performance, 7*, 1007–1018.

# 7 Neurophysiology of Reaching

Apostolos P. Georgopoulos
*The Johns Hopkins University School of Medicine*

## ABSTRACT

Neurophysiological studies of reaching are reviewed with emphasis on recordings from behaving monkeys. The relations between reaching parameters (direction and amplitude) and single-cell activity in the motor and parietal cortices are described. Finally, the role of motor cortex in the specifications of reaching and recent advances in spinal circuits that may subserve reaching movements are discussed.

Reaching serves to position the hand at desired points in immediate extrapersonal space. It is accomplished by well-coordinated and tightly coupled motions at the shoulder and elbow joints (Soechting & Lacquaniti, 1981), and is usually directed to visual targets. During reaching, the hand is preshaped according to the characteristics of the object that is to be manipulated (Jeannerod & Biguer, 1982). Thus, behaviorally, reaching is a complex act that involves sensory–motor integration and the concurrent preparation for handling the object of interest. I discuss in this chapter the neurophysiological studies that pertain to reaching itself. Other reviews have dealt with general aspects of reaching (Georgopoulos, 1986), the anatomical pathways involved in reaching (Humphrey, 1979), the effects on reaching of brain lesions (Jeannerod, 1986), and the visuomotor coordination underlying reaching to visual targets (Georgopoulos, 1989).

## THE COMBINED BEHAVIORAL–NEUROPHYSIOLOGICAL EXPERIMENT

The methodological milestone in studying the brain mechanisms underlying reaching was the development of the technique of recording the activity of single cells in the brain of behaving animals (Lemon, 1984). Briefly, animals, usually monkeys, are trained to sit on a primate chair and perform in various tasks. During electrophysiological recording sessions, microelectrodes are inserted into the brain area of interest through the dura and the activity of single cells recorded while the animal performs in the task. Thus, data are accumulated over many recording sessions, usually comprising several animals. This combined behavioral–neurophysiological experiment provides a direct tool and one with fine grain by which the brain mechanisms underlying performance can be studied. Most of the results discussed in this chapter come from such experiments.

FIG. 7.1. Drawing of a monkey reaching towards a lighted target in an apparatus used by Mountcastle et al. (1975) for studies of the posterior parietal cortex. The animal has just released the key and reached out to touch the lighted switch mounted on the moving carriage. The head fixation apparatus, implanted microdrive, cathode follower, and reward tube are also shown. (From Mountcastle et al., 1975; reproduced with permission).

## REACHING NEURONS IN POSTERIOR PARIETAL CORTEX (AREAS 5 AND 7)

The posterior parietal cortex was one of the first brain regions to be studied during reaching by behaving monkeys. The interest in that region apparently stemmed from the defects in reaching that are observed in human patients and in monkeys following lesions of the posterior parietal cortex (see Lynch, 1980 for a review). Indeed, cells were found in both area 5 (Mountcastle et al., 1975) and area 7 (Hyvärinen & Poramen, 1974; Mountcastle et al., 1975), which were activated during reaching in the absence of passive somesthetic driving. The activity of these "reaching" neurons was studied quantitatively using a behavioral apparatus that allowed reaching to either a stationary or a moving visual target (Mountcastle et al., 1975). A schematic drawing of the experimental setup is shown in Fig. 7.1. Figure 7.2 illustrates the impulse activity of a neuron recorded in area 5 during repeated reaching trials. This cell could not be activated by passive

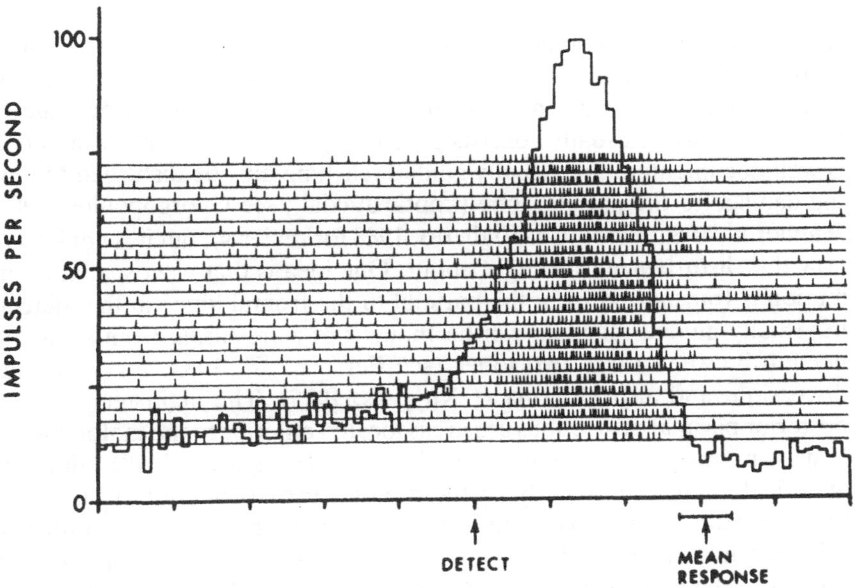


FIG. 7.2. Impulse activity of an area 5 neuron during repeated reaching movements. The cell never responded to any passively delivered mechanical stimulus to the arm, to visual or auditory stimuli, or during passive–aversive movements of the contralateral arm. *Detect* indicates key release (see Fig. 7.1), and *response* indicates contact with the lighted switch. The arrow is mean response time ± 1 SD. (From Mountcastle et al., 1975; reproduced with permission.)

manipulation of the arm and did not respond to visual or auditory stimuli. However, its activity increased several times above the background discharge when the monkey reached to pull a lever placed in front of the animal at shoulder height.

Although reaching neurons in area 7 could be activated with reaching in the absence of visual guidance (Hyvärinen & Poramen, 1974; Mountcastle et al., 1980), in other cases, neuronal activation was more intense when the animal reached with the eyes open. A particular class of neurons showed properties appropriate for an eye–hand coordination function. These neurons were somewhat activated with visual stimuli but usually not with reaching; however, they were intensely activated when the monkey tracked with the hand a visual object of interest (Mountcastle et al., 1975). Unfortunately, these neurons were not studied quantitatively, and the precise interactions of the effects of reaching and visual stimulation and/or eye fixation on single-cell activity in area 7 remain to be elucidated.

## STUDIES OF REACHING IN PREMOTOR CORTEX

It is reasonable to assume that the planning and execution of reaching involves the concurrent or sequential activation of several brain regions. In particular, reaching can be regarded as an example of sensory–motor integration, for it usually consists projecting the hand toward a target. It is well documented that visually guided reaching may be disturbed following damage to the parietal lobe ("optic ataxia"; see Georgopoulos, 1989; Jeannerod, 1986; Perenin & Vighetto, 1988 for reviews), but it is not known how this disturbance is brought about. One idea is that visuospatial information is transmitted from the parietal to the frontal lobe, so that defects in visually guided reaching can be thought of as resulting from a disconnection between these two lobes (see Ferro et al., 1983, for a succinct discussion, and also Haaxma & Kuypers, 1975). The main recipients of posterior parietal projections are areas located anterior to the motor cortex, that is, premotor cortical areas (reviewed in Humphrey, 1979), although area 5 also projects directly to the motor cortex (Caminiti et al., 1985; Strick & Kim, 1978). The premotor areas comprise a number of different subareas in both the lateral and medial surface of the hemisphere and project to the motor cortex (Barbas & Pandya, 1987; Muakkassa & Strick, 1979). In fact, these areas seem to project directly to subcortical structures, including the red nucleus (Humphrey et al., 1984) and the spinal cord (Hutchins & Strick, 1987), which means that their influence on the motor function need not be exerted exclusively through the motor cortex. Recordings of single-cell activity in these premotor areas in behaving monkeys have provided valuable information concerning their possible role in sensory–motor integration, including preparation for and anticipation of movement (see reviews by Tanji, 1984; Wise, 1985). The motor functions of premotor areas are reviewed in this volume by Wiesendanger. It is unknown how the neuronal populations in these areas cooperate to complete the preparation and planning of the upcoming movement. It is possible that these various areas ultimately converge on the motor cortex, which is then the summing point of all premotor processing. However, in view of the separate projections of the premotor areas to brain stem and spinal motor structures, it is reasonable to suppose that the process of specification of the parameters of reaching is reflected in several areas that may communicate, interact, and, possibly, gate each other both at cortical and subcortical levels.

## STUDIES OF REACHING IN THE MOTOR CORTEX

Porter and Lewis (1975) studied the changes in activity of motor cortical cells while monkeys reached out and manipulated a handle in front of them. It was found that single cells changed their activity during the task and that the latency of activation of different cells shifted to later times as more distal parts of the limb became involved in the motor act. This question of the sequential activation of motor cortical populations in reaching and grasping was investigated in more detail by Murphy et al. (1985). The activity of single cells in the forelimb area of the motor cortex was recorded in a task in which monkeys pointed to targets in front of them. The functional relation between the recording locus and the joint of the arm was determined by intracortical microstimulation. Thus, a cell could be classified as relating mainly to movements at the shoulder, elbow, hand, or fingers. It was found that in the pointing task cells were generally activated sequentially, from proximal to distal, reflecting the sequential engagement of successively more distal parts of the arm.

Murphy et al. (1982) also investigated the possible relations between motor cortical cell activity and joint motion and electromyographic (EMG) activity in muscles of the forelimb during reaching. There were three main findings of this study. First, no simple relation was observed between single-cell activity and the EMG, even when the muscle from which the EMG was recorded was activated by intracortical microstimulation. Second, single cells related to motion about the shoulder or elbow joints behaved similarly in the task, although the motions produced about these joints could be quite different. And third, the discharge of shoulder-related cells seemed to vary systematically with the movement trajectory. These results indicate that the relations between single-cell activity in the motor cortex and components (joint rotation, EMG activity) of reaching are complex.

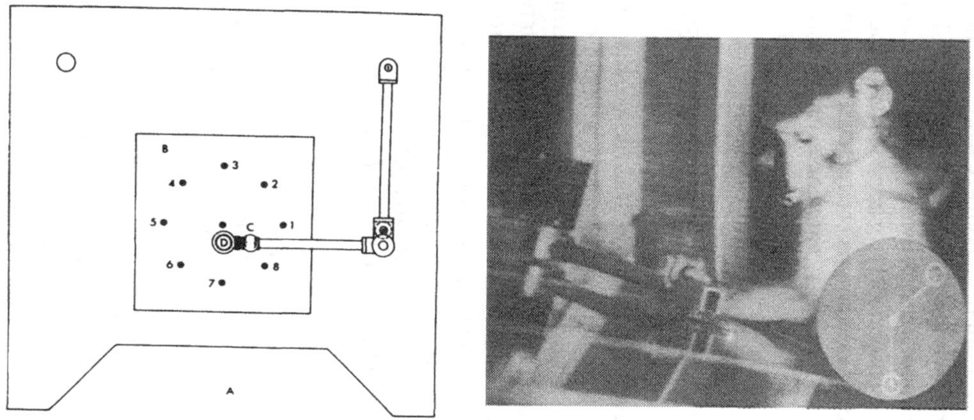


FIG. 7.3. Left: schematic drawing of the 2-D apparatus used to study "drawing" movements of monkeys. The monkey sat at A, in front of the working surface, B. The numbered light emitting diodes (LED) were placed on a circle of 8 cm radius. The monkey held the articulated manipulandum at its distal end (C) and captured a lighted LED within a 10 mm diameter transparent plexiglass circle (D). X-Y motion of the center of that circle was monitored every 10 ms with a resolution of 0.125 mm. Right: a monkey performing the task (side view). Insert shows two trajectories. (From Georgopoulos et al., 1981; reproduced with permission.)

## PARAMETRIC STUDIES OF REACHING IN MOTOR CORTEX AND AREA 5

The spatial parameters of a reaching movement are its direction and amplitude. Soechting (this volume) has presented psychophysical evidence that these two parameters are meaningful with respect to reaching performance by human subjects, and that, indeed, performance is better in the direction than in the amplitude domain. It is possible that motor cortical cell activity might indeed relate to these parameters of the reaching movement. We examined this idea and extended it to a study of area 5 of the posterior parietal cortex because this area relates to reaching (Mountcastle et al., 1975) and is reciprocally connected with motor cortex (Strick & Kim, 1978; Caminiti et al., 1985).

### Relations to the Direction of Movement

We studied the relations between the direction of pointing movements in 2-dimensional (2-D) and 3-D space and the activity of cells in the motor cortex and area 5 of the parietal cortex of rhesus monkeys. The apparatus used in the 2-D studies is shown in Fig. 7.3 and has been described in detail elsewhere (Georgopoulos et al., 1981; Georgopoulos & Massey, 1987). It consisted of a planar working surface over which an articulated manipulandum could move freely. The monkey was required to capture within the distal end of the manipulandum lighted targets that were turned on on the plane using light-emitting diodes (Georgopoulos et al., 1981) or a He-Ne laser beam backprojected onto the plane (Georgopoulos & Massey, 1987). In these studies, the amplitude of the movement was kept constant. The major finding was that the activity of single cells in the arm area of the motor cortex changed in an orderly fashion with the direction of movement in 2-D space (Georgopoulos et al., 1982): for a given cell it was highest with movements in a particular direction (the cell's preferred direction) and decreased gradually with movements made in directions farther away from the preferred one. Quantitatively, the frequency of cell discharge was a sinusoidal function of movement direction, or, equivalently, a linear function of the cosine of the angle formed between the direction of the movement and the cell's preferred direction (see Georgopoulos et al., 1982, for details of the analysis). An example is shown in Fig. 7.4A and B. In this case 8 movement directions (i.e., 45° apart) were used. The gradual changes in cell activity, especially around the peak of activation, are illustrated in Fig. 7.5, in which case 20 movement directions (i.e., 18° apart) were used. It is noteworthy that the preferred directions of cells recorded in penetrations made in the exposed part of the motor cortex tended to be similar throughout cortical depth, whereas they tended

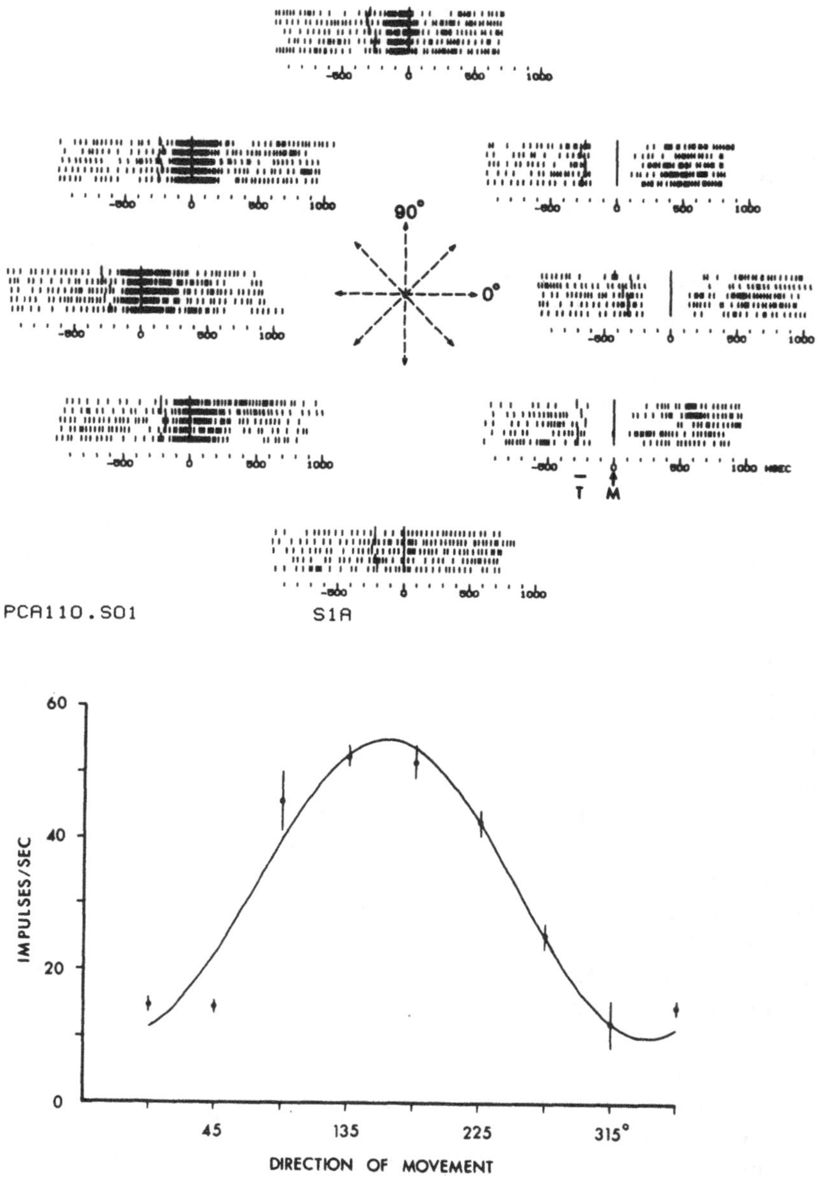


FIG. 7.4. Broad directional tuning in 2-D space of a cell recorded in the arm area of the motor cortex. A, impulse activity during five trials with movements in the directions indicated in the drawing at the center. Short vertical bars indicate the occurrence of an action potential. Rasters are aligned to the onset of movement (M). Longer vertical bars preceding the onset of movement indicate the onset of the target (T); those following

to change *en block* in penetrations made near the anterior bank of the central sulcus. These observations are compatible with the hypothesis for a columnar organization of the preferred direction (Georgopoulos et al., 1984a).

The directional properties of single cells in parietal area 5 are similar to those observed in the motor cortex (Kalaska et al., 1983; Kalaska, 1988); an example is shown in Fig. 7.6. This similarity is important because these two areas may relate to different aspects of the movement, namely the motor cortex to its initiation and area 5 to its monitoring (Kalaska et al., 1983; Kalaska, 1988). Therefore, the similar relations observed between cell activity in these structures and the direction of movement reflect a common language of communication in the spatial domain, given that these areas are anatomically interconnected (Strick & Kim, 1978; Caminiti et al., 1985). The main difference found between the two areas related to onset times of the changes in cell activity, with motor cortical cells being engaged approximately 60 ms before those of area 5 (Fig. 7.7). This finding is in accordance with the different roles postulated above for these two areas.

A different question is whether the changes in cell activity relate to the direction of the movement or the location of its target. These two variables were dissociated in a task in which the monkey moved the manipulandum to the same point starting from various positions. In both the motor cortex and area 5, the changes in cell activity related to the direction of the movement and not its target (Georgopoulos et al., 1984a, b). This is illustrated in Figs. 7.8 and 7.9 for the motor cortex and area 5, respectively.

The results described here were obtained using a manipulandum and pointing movements constrained in 2-D space. Would the same relations hold for free reaching movements in 3-D space? We investigated this question by training monkeys to point to visual targets in 3-D space in front of them. A schematic drawing of the apparatus is shown in Fig. 7.10. It allowed movements to be made in 8 directions at approximate equal angular intervals in 3-D space. The results obtained in the motor cortex under these conditions were very similar to those obtained using 2-D movements. Thus the activity of single cells were broadly tuned with respect to the direction

---

the movement indicate the entrance to the target window (see Fig. 7.2) and the delivery of reward. B, average frequency of discharge (± SEM) from the onset of the stimulus until the entry to the target window are plotted against the direction of movement. Continuous curve is a cosine function fitted to the data using multiple regression analysis; the function is of the form; $D = b + k \cos(\theta - \theta_0)$, where D is the discharge rate, b and k are constants, $\theta$ is the direction of the movement, and $\theta_0$ is the cell's preferred direction. (From Georgopoulos et al., 1982; reproduced with permission. Copyright by Society for Neuroscience.)

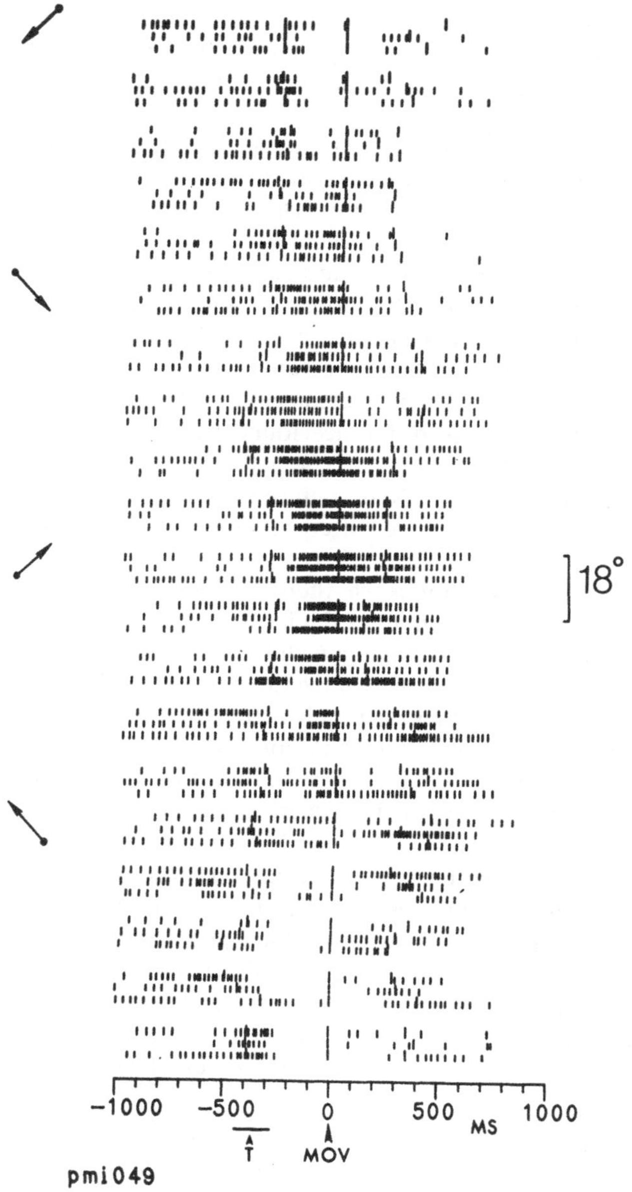


FIG. 7.5. Broad directional tuning of a motor cortical cell studied using 20 movement directions in the 2-D apparatus. (Schwartz & Georgopoulos, unpublished observations.)

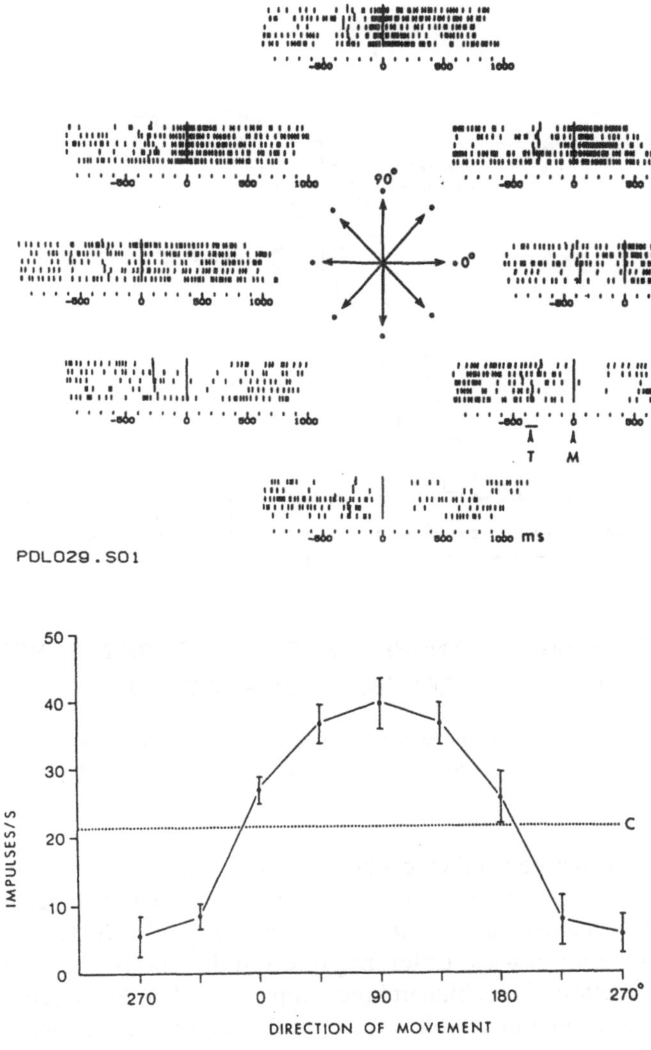


FIG. 7.6. Impulse activity of a cell from area 5 studied in the 2-D task. Conventions as in Fig. 7.4 (From Kalaska et al., 1983; reproduced with permission.)

of reaching (Fig. 7.11 A), and the frequency of cell discharge was in many cases a linear function of the cosine of the angle, θ, formed between the direction of the movement and the cell's preferred direction (Fig. 7.11 B) (Georgopoulos et al., 1986; Schwartz et al., 1988). This is illustrated in Fig.7.12. Finally, the preferred directions differed among different cells and were distributed throughout 3-D space. This is illustrated in Fig. 7.13 using an equal area projection plot.

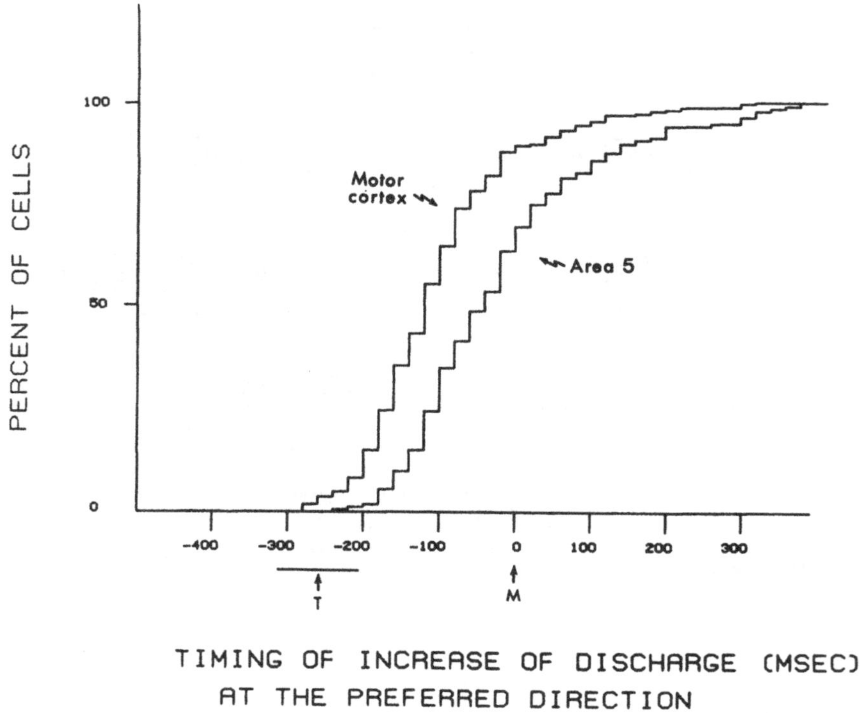


FIG. 7.7. Cumulative frequency distributions of latency of change in cell activity in motor cortex and area 5. (From Kalaska et al., 1983; reproduced with permission.)

These results described indicate that *a given cell participates in movements of various directions and that, conversely, a movement in a particular direction involves the activation of a whole population of cells*. Given that cells in the population possess different preferred directions, the question is, how is the direction of reaching uniquely represented in the directionally heterogeneous neuronal ensemble? We hypothesized that, for this representation, the motor command for the direction of movement can be regarded as an ensemble of vectors, each of which represents that contribution ("vote") of a particular cell along the axis of its preferred direction. The length of these vectors is proportional to the change in activity for the movement of a given direction. Then, the directional outcome of the operation of the neuronal population (the "population vector") is the weighted vector sum of these single-cell contributions. It was found that the population vector predicts well the direction of the movement both in the motor cortex (Georgopoulos et al., 1983, 1984a, 1986) and in area 5 (Kalaska et al., 1983). This is illustrated in Fig 7.14 for 8 movement directions studied

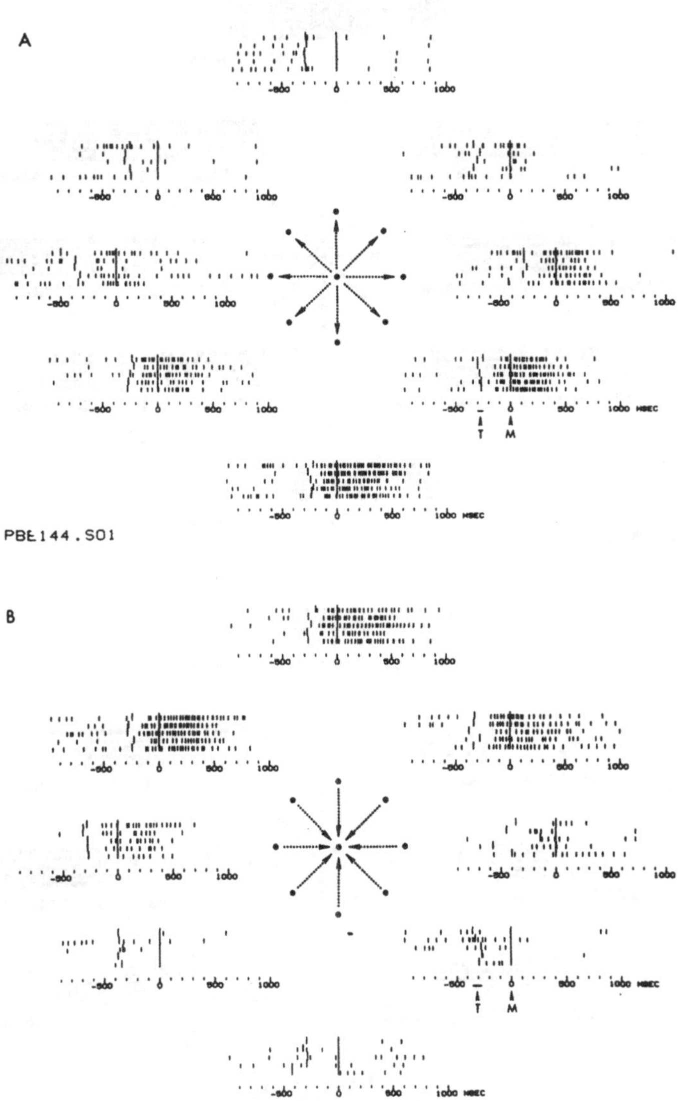


FIG. 7.8. Impulse activity of a motor cortical cell studied in the 2-D apparatus during center → out (A) and out → center (B) movements, as indicated in the center diagrams. Conventions as in Fig. 7.4. (From Georgopoulos et al., 1984a; reproduced with permission.)

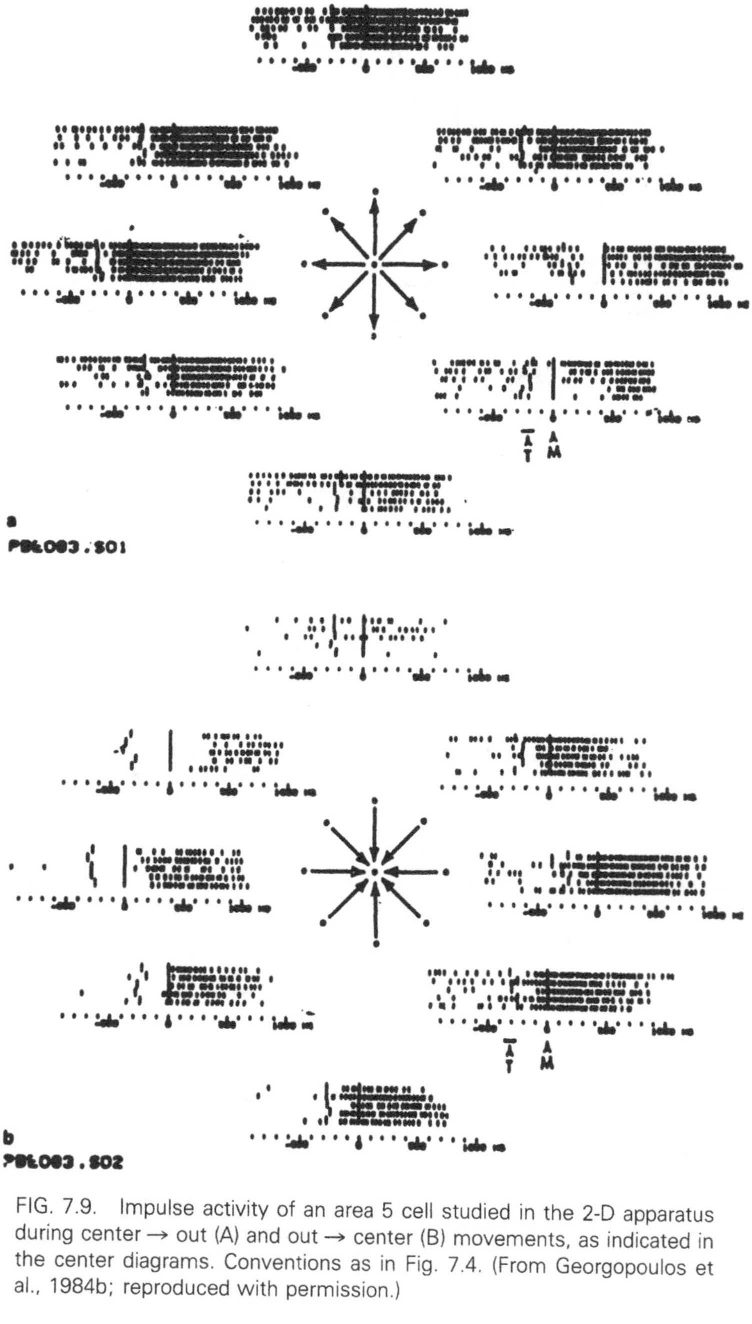


FIG. 7.9. Impulse activity of an area 5 cell studied in the 2-D apparatus during center → out (A) and out → center (B) movements, as indicated in the center diagrams. Conventions as in Fig. 7.4. (From Georgopoulos et al., 1984b; reproduced with permission.)

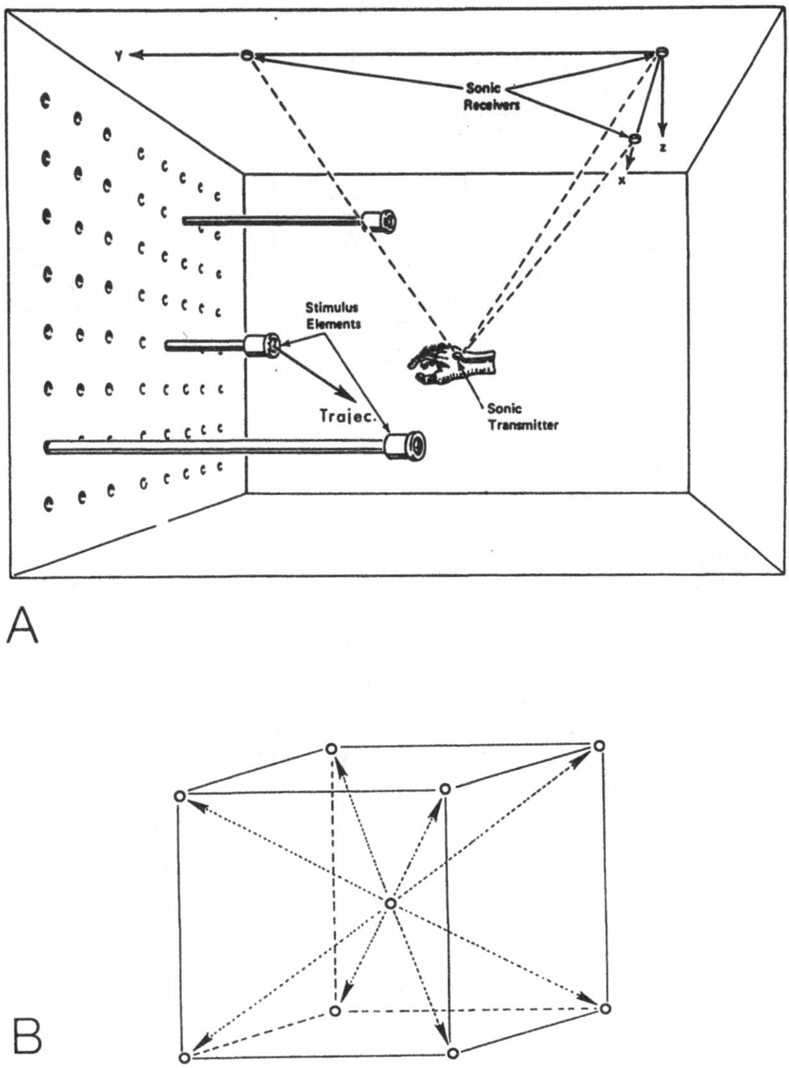


FIG. 7.10. Schematic drawing of the apparatus used to study free reaching movements in 3-D space. A, monkeys reached towards and pushed lighted buttons mounted at the end of metal rods threaded through a heavy metal plate. The movement trajectory was monitored using an ultrasonic system. B, Schematic diagram of the location of the 9 buttons used. Dotted lines indicate directions of movements. (From Schwartz et al., 1988; reproduced with permission. Copyright by Society for Neuroscience.)

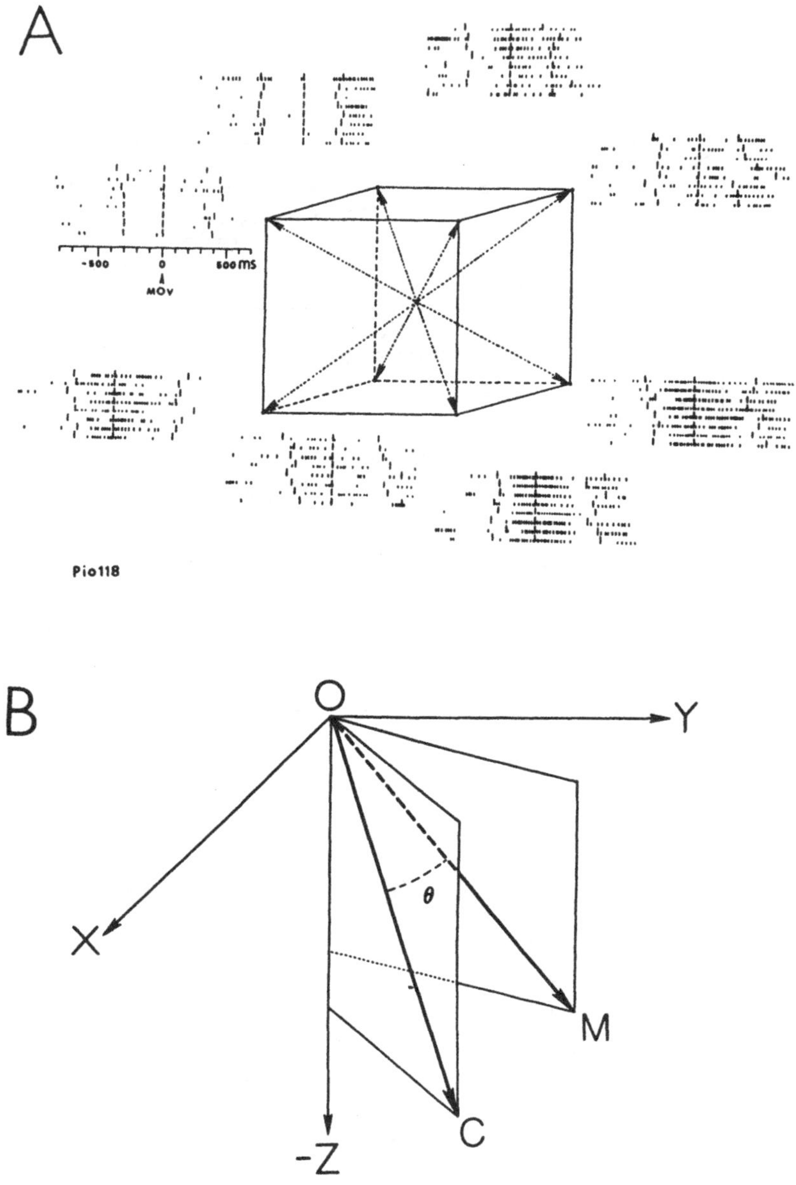


FIG. 7.11. Broad directional tuning in 3-D space of a motor cortical cell. A, impulse activity is shown in raster form for eight trials in eight movement directions indicated in the drawing at center. MOV, onset of movement. B, principle of directional tuning. **C** is the preferred direction of the cell whose rasters are shown in A; **M** is the direction of a movement; θ is the angle between **C** and **M**. The cell activity varies in a linear fashion with the cos (θ): it is maximum when cos (θ) is zero, that is, when **C** and **M** coincide; minimum when cos θ is −1, that is when **C** and **M** are in opposite directions; and in between when 1 > cos (θ) > −1, that is, when **M** is between the direction of **C** and its opposite. (Adapted from Georgopoulos et al., 1986 and reproduced with permission. Copyright by AAAS.)

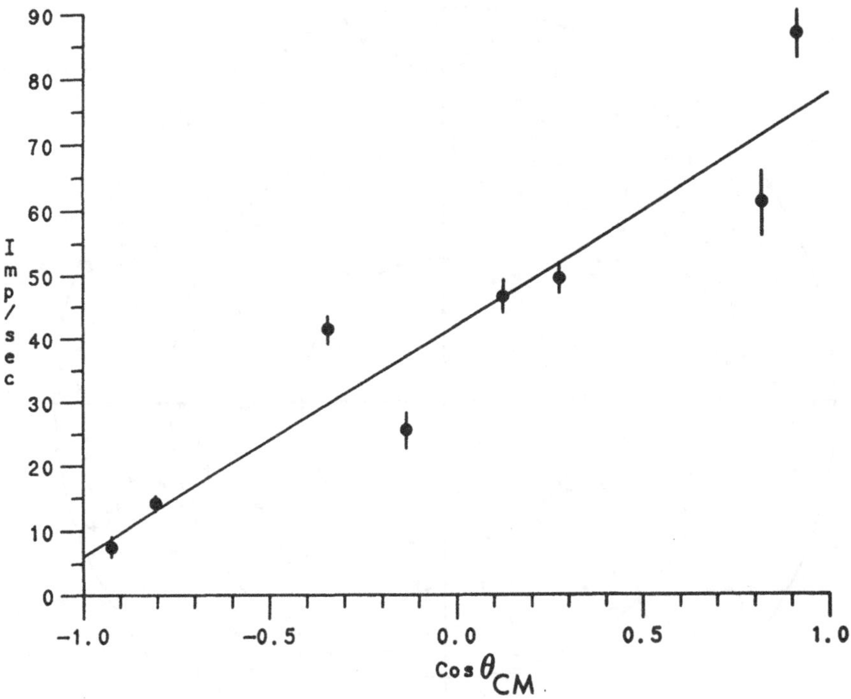


FIG. 7.12. Mean discharge rate (± SD) of a different cell during total time (from onset of target to end of movement) is plotted versus the cosine of the angle θ formed between the direction of the movement and the cell's preferred direction. (From Schwartz et al., 1988; reproduced with permission. Copyright by Society for Neuroscience.)

in the 2-D task. Similar results were obtained in the 3-D case; an example is shown in Fig. 7.15.

The data shown in Figs. 7.14 and 7.15 refer to changes in cell activity averaged over a period of time, from the onset of the target until the end of the movement, it is important to know whether the population vector can predict the direction of movement before the beginning of movement, that is, during the reaction time. For that purpose we calculated the population vector in successive 20-ms bins; the results are shown in Figs. 7.16 and 7.17 for the 2-D case and 3-D case, respectively. It can be seen that the population vector predicts the direction of the upcoming movement well in advance of the onset of the movement. In fact, this was found to be the case even when a delay was imposed (Fig. 7.18), that is, during a period of time in which the animal did not move but waited for a signal to trigger the movement (Georgopoulos et al. 1989).

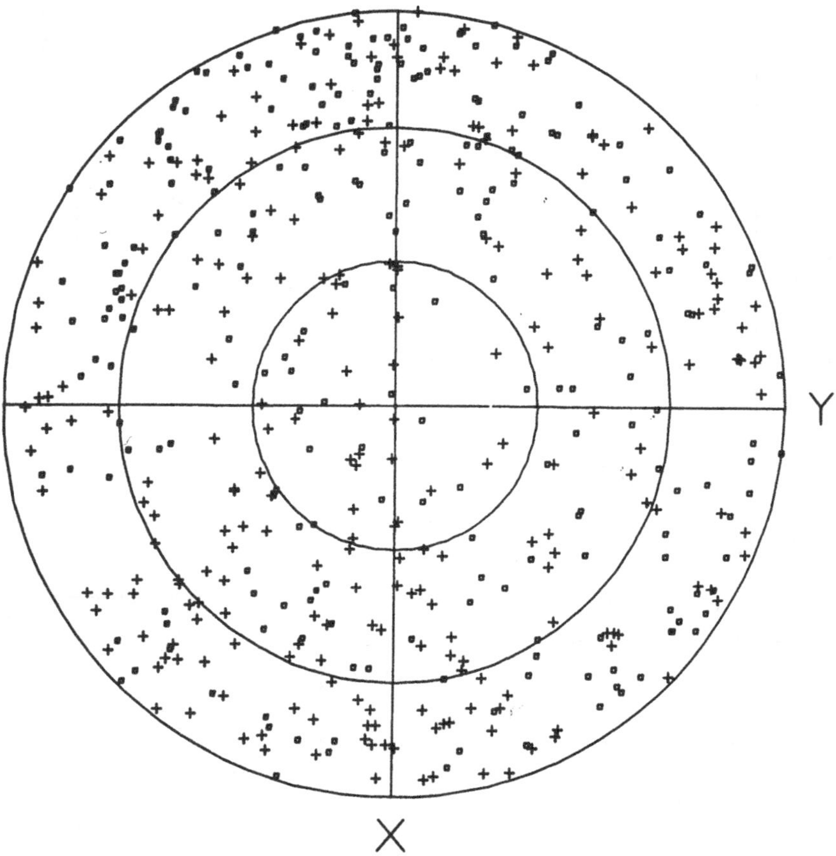


FIG. 7.13. Preferred directions of 475 cells studied in the 3-D apparatus. Equal area projection plot (Watson, 1983). The points of preferred directions on the two (upper and lower) hemispheres of the unit sphere are plotted onto a circular plane so that areas on the hemisphere are projected to regions of equal area on the plane. Preferred directions above or below the equatorial plane are indicated by squares and crosses, respectively. Points near the center indicate directions toward the poles, whereas near the outermost circle they indicate directions toward the equator. (From Schwartz et al., 1988; reproduced with permission. Copyright by Society for Neuroscience.)

In all of these cases, the movement started from the same point. It is important to know whether the population vector predicts the direction of movement when the origin of the movement changes in space. Indeed, this was found to be the case (Kettner et al., 1988). The ongoing discharge rate for a certain proportion of cells varies with the posture of the arm attained indifferent starting positions (Georgopoulos et al., 1984c; Kettner

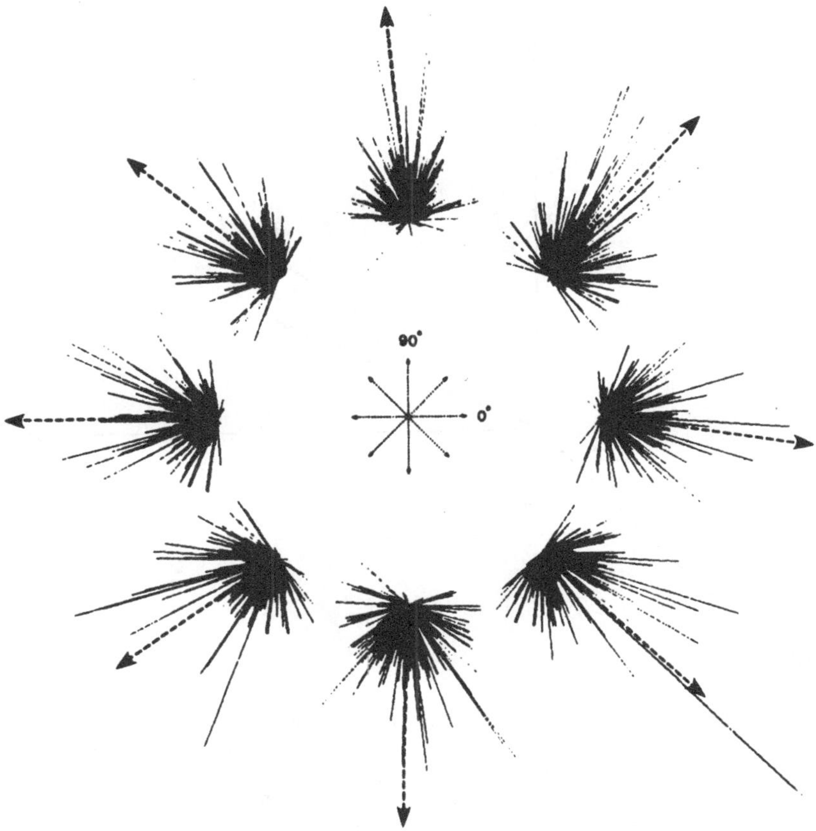


FIG. 7.14. Neuronal population vector (heavy dashed lines with arrow) calculated for 8 movement directions. All clusters represent the same neuronal population composed of individual cell vectors (thin lines, $N = 241$ cells). The dotted lines in the center indicate the direction of movement. (From Georgopoulos et al., 1983; reproduced with permission.)

et al,. 1988), but the changes in cell activity from that ongoing discharge provide, in the population analysis, an accurate signal for the direction of the movement (Kettner et al., 1988).

## Relations to the Amplitude of Movement

We studied the relations between the amplitude of pointing movements and the frequency of discharge of cells in the motor cortex and area 5 in two sets of studies using the 2-D apparatus (Fig. 7.3). In the first series of experiments (Georgopoulos, Kalaska, & Caminiti unpublished observations) movements of two amplitudes (8 cm and 16 cm) were used in the

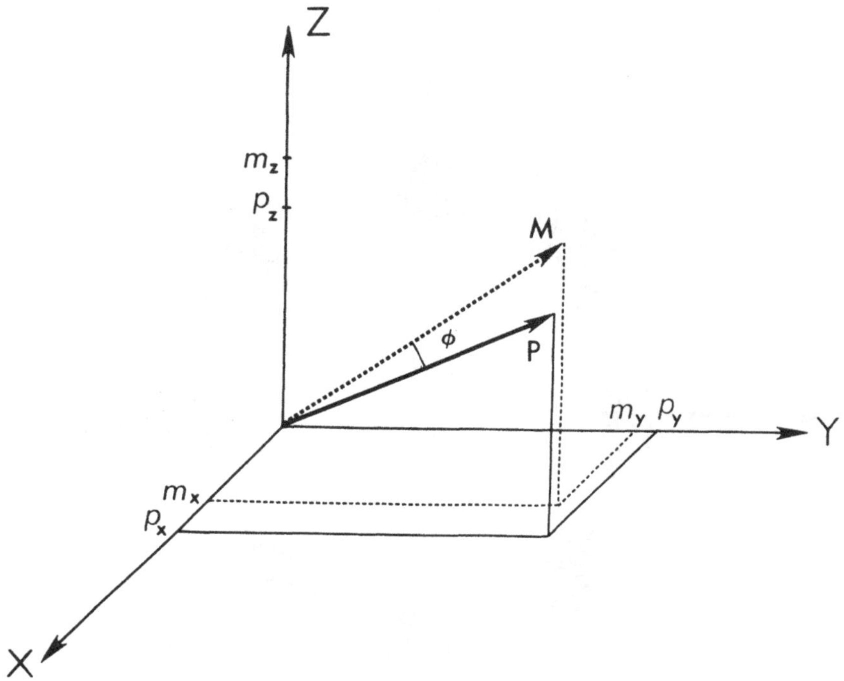


FIG. 7.15. Coding of direction of 3-D reaching, **M**, by a neuronal population whose population vector, **P**, points approximately in the direction of the movement. For illustration purposes, the two vectors are normalized to unit length. The projections of the vectors **M** and **P** onto coordinate axes *x*, *y*, and *z* are shown as $m_x$, $m_y$, $m_z$, and $p_x$, $p_y$, $p_z$, respectively. The angle formed between the two vectors is also shown. (From Georgopoulos et al., 1986; reproduced with permission. Copyright AAAS.)

same direction and along the same path. The experimental design included 8-cm movements from two different points of origin: one from the same origin as the 16-cm movement and the other from the middle of the 16-cm movement path, so that both of the latter movements ended at the same point. We were interested to find out whether the amplitude of the movement could be predicted from the changes in discharge rate observed during the reaction time (RT). Because amplitude-related changes in cell activity could differ depending upon the direction of the movement, the data were aligned to the preferred direction and analyzed for directional intervals (30°) at different angles from the preferred direction. The results obtained are shown in Figs. 7.19 and 7.20. It can be seen that the results were similar for both the motor cortex (Fig. 7.19 and area 5 (Fig. 7.20) in that the average changes in cell activity differed for movements of different directions but not for movements of different amplitudes. These findings

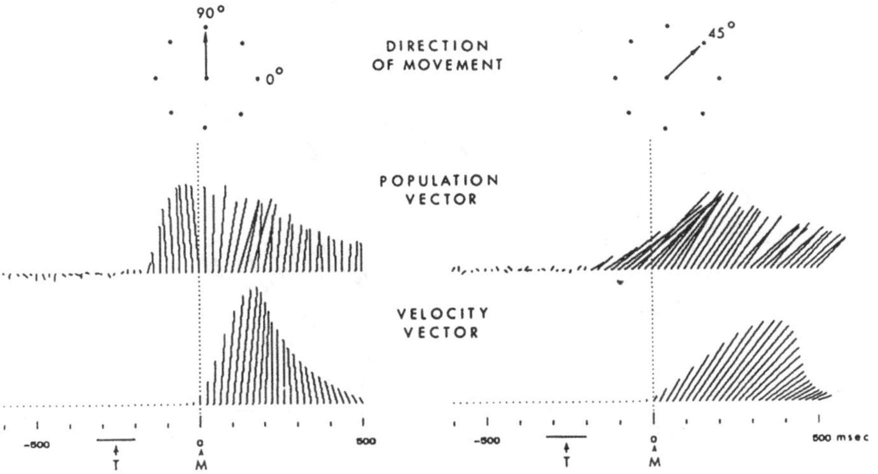


FIG. 7.16. Population vectors computed every 20 ms for movements in two different directions. Instantaneous velocity vectors are also shown. Notice that the population vector lengthens well before the movement begins and points in the approximate direction of the upcoming movement. (From Georgopoulos et al., 1984a; reproduced with permission.)

suggest that overall changes in cell activity preceding the movement, that is, during the RT, do not differ for movements of twice the amplitude but do so for movements of different directions. It is possible that the movement amplitudes of 8 cm and 16 cm used in the experiments just described did not provide enough range for an amplitude effect to be manifested. For that purpose we used a greater range of movement amplitudes (1.5, 3, 6, and 12 cm) in a second series of recordings in the motor cortex (Schwartz & Georgopoulos, 1987). All of these movements were made along the cell's preferred direction because an amplitude effect on cell activity may be most likely to be observed for movements in the cell's preferred direction. Under those conditions weak amplitude effects were observed (Schwartz & Georgopoulos. 1987). It is possible that the amplitude of movement is coded at the population level, for example, as the length of the population vector. This remains to be investigated.

## ROLE OF MOTOR CORTEX IN REACHING

The results discussed in the preceding sections show that cells in several cortical regions are engaged with reaching, and it is reasonable to suppose that other cortical areas as well as subcortical structures are involved in planning, specifying, and controlling reaching. An insight into the aspects

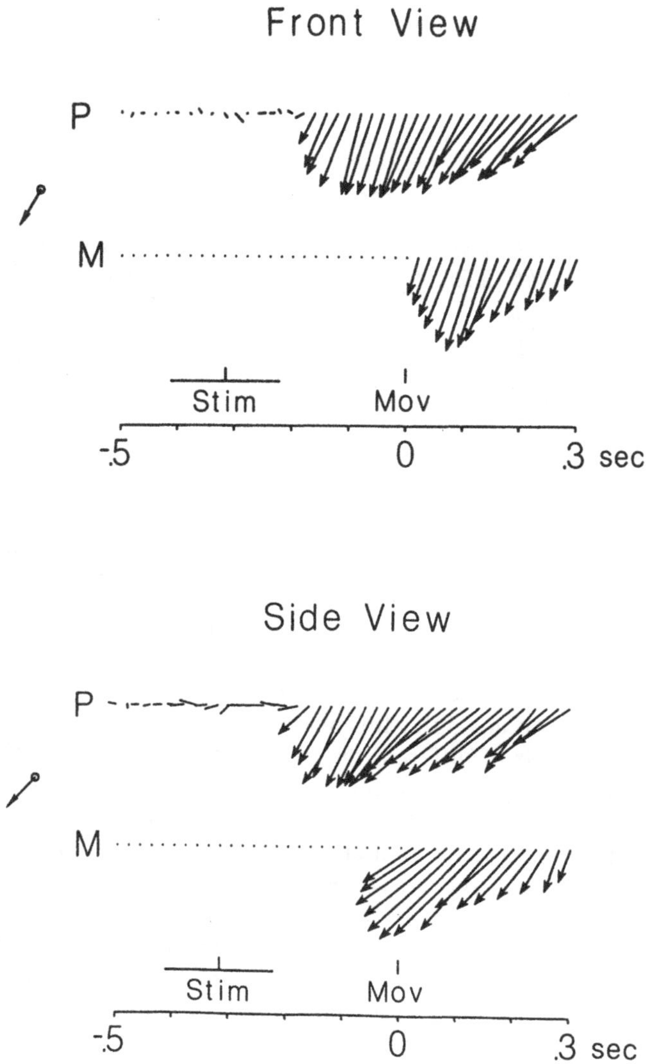


FIG. 7.17. Population vectors (P) computed every 20 ms for 3-D movements (M) in a particular direction. Two views are shown. Stim, stimulus onset (± SD); Mov, movement onset. (From Georgopoulos et al., 1988; reproduced with permission. Copyright by Society for Neuroscience.)

of reaching that the motor cortex might be involved with can be gained from parametric studies of motor cortical activity and reaching. The studies summarized above indicate that the activity of motor cortical cells varies in an orderly fashion with the direction and to a smaller degree with the amplitude of pointing movements. I discuss in the following section several

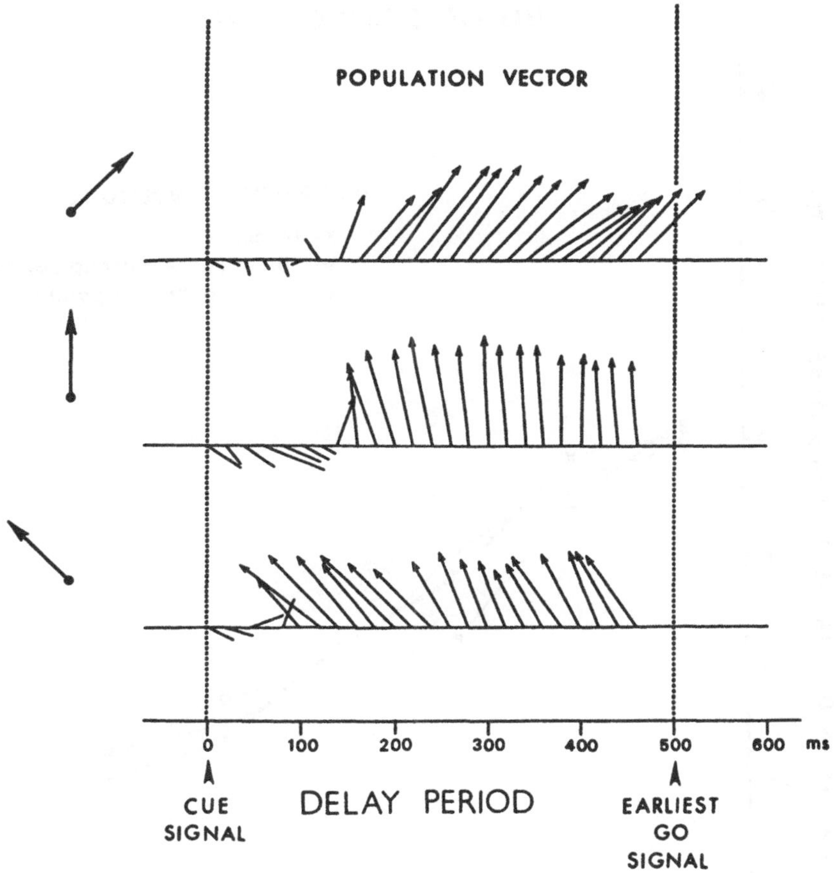


FIG. 7.18. Population vectors computed every 20 ms during a delay period point in the direction of the upcoming 2-D movement. During the delay period the animals did not move until a *go* signal was given. (From Georgopoulos et al., 1989; reproduced with permission.)

aspects of these findings that relate to the motor cortex and the movement parameters themselves.

## Intracortical Microstimulation

One question concerns the functional properties of the cells studied. The objective in our studies was to investigate quantitatively the relations between the activity of cells related to proximal joints and the parameters of reaching. Therefore, only those cells were selected for study that were active with movements of the proximal arm outside the behavioral task (Georgopoulos et al., 1982, 1984a); such cells responded sometimes to

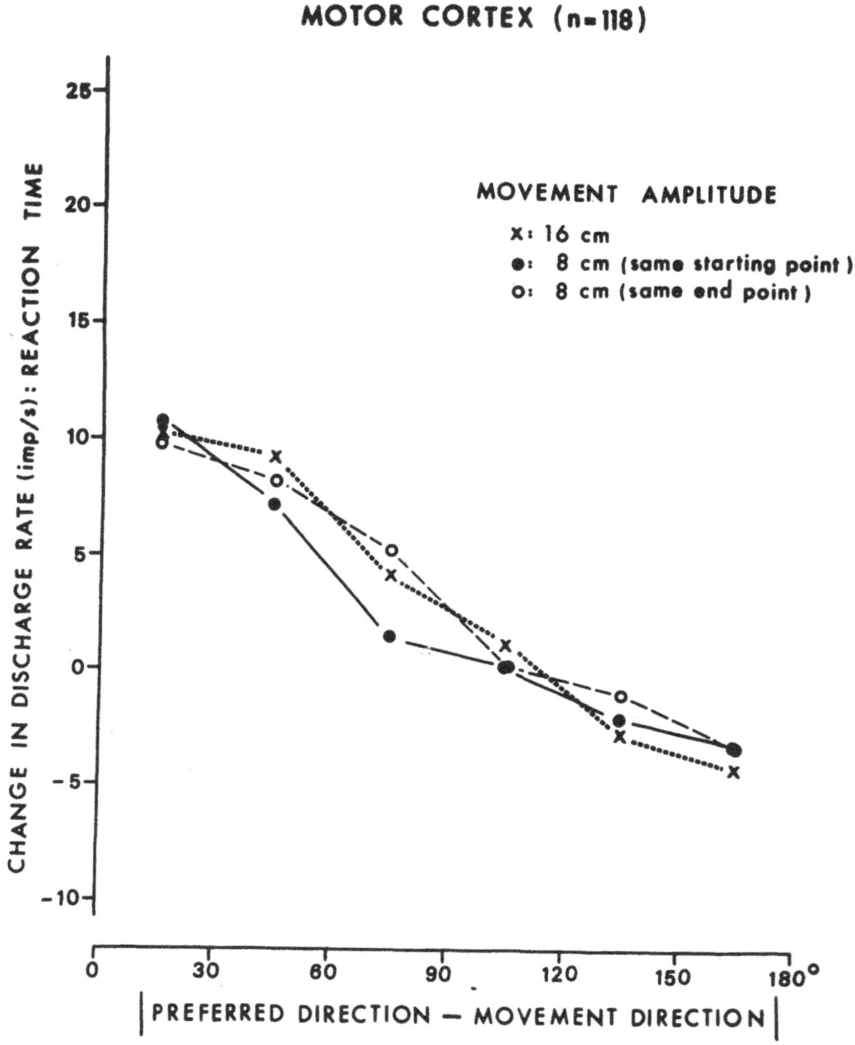


FIG. 7.19. Motor cortex: Mean discharge rate during the reaction time is plotted against movement amplitude for movement directions at different angles from the preferred direction. Clockwise and counterclockwise departures have been collapsed to one curve. (A. P. Georgopoulos, J. F. Kalaska and R. Caminiti, unpublished observations.)

passive rotations at the elbow and/or shoulder joints. Cells that were activated with movements of the hand or of the fingers in the absence of any proximal component (i.e., from the elbow and/or shoulder) were excluded from the study. The distinction between *proximal* and *distal* activation of a cell was not difficult to make following a careful examination of the

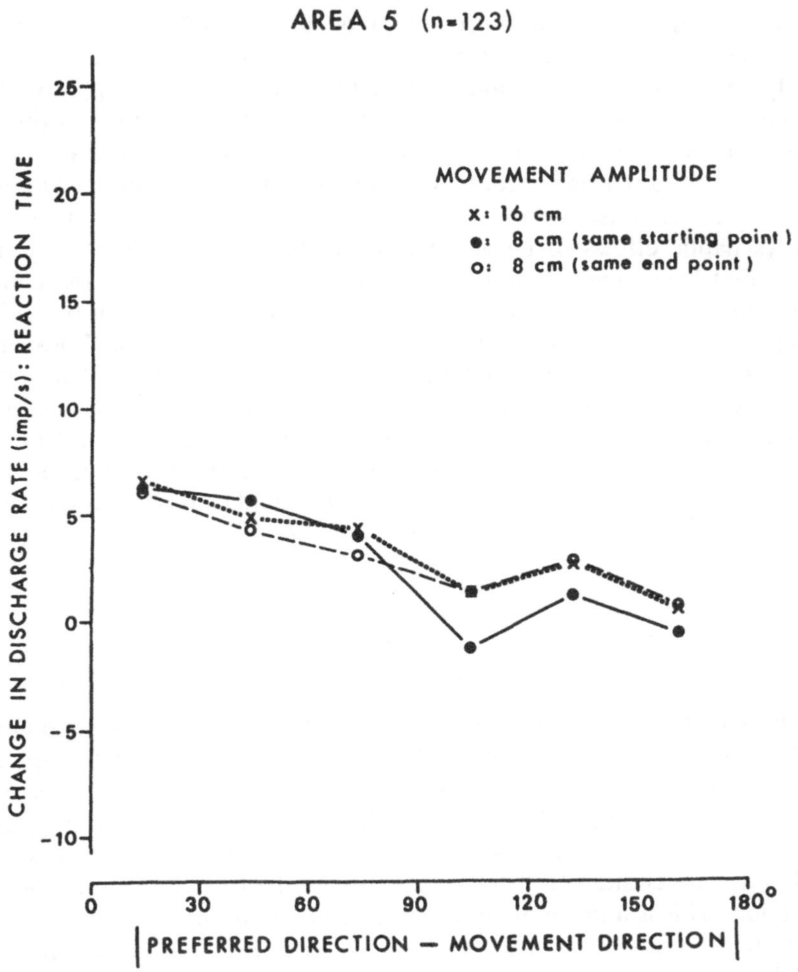


FIG. 7.20. Area 5: Mean discharge rate during the reaction time is plotted against movement amplitude for movement directions at different angles from the preferred direction. Clockwise and counterclockwise departures have been collapsed to one curve. (A. P. Georgopoulos, J. F. Kalaska and R. Caminiti, unpublished observations.)

animal, for it was common that only one of the two components was present. We did not use electrical microstimulation in these studies for three reasons. First, there is a good correspondence between the motor responses to microstimulation and the manipulations (e.g., passive motion of the limb) that are associated with changes in cell activity (Murphy et al., 1978). The second reason concerns the uncertainty of localizing the origin of the microstimulation effects in the cortex. It has been shown (Asanuma & Rosen, 1973) that intracortical microstimulation even at very low intensities (e.g., 4 uA) produces monosynaptic effects on cells within a radius of approximately 0.5 mm from the point of stimulation, and polysynaptic effects extending up to 1 mm. Moreover, the application of trains of stimuli enhances the effect of microstimulation seemingly through indirect, synaptic activation of pyramidal tract cells (Jankowska et al., 1975). A third reason concerns the unknown effects of potential alteration of cortical circuits by local electrical stimulation, given the mono- and polysynaptic effects of intracortical microstimulation. In summary, it is true that if in some experiments it is impossible to examine the animal, then microstimulation offers a localizing technique; but when the animal can be examined, then direct observation of the functional cell properties seems to be more useful. The latter technique has the additional advantage that it can be applied to other areas in which microstimulation is not useful, which is the case for area 5. In fact, in our studies, cells in both the motor cortex and area 5 were selected for quantitative studies using the same criteria mentioned above, so that the results could be directly comparable.

## Cell Identification: Antidromic Stimulation and Comparison of Functional Properties between Upper and Lower Cortical Layers

Another question concerns the "identification" of a cell being studied. There are several ways that a cell can be "identified"; for example, its shape (pyramidal, stellate, and so on), location in a particular layer, neurotransmitter used, and so forth. In the case of sensory cortices (see for example, Dow et al., 1984; Poggio, 1984) or association cortices, (see Mountcastle et al., 1975) the layer has been the main label of cell identification: This is a meaningful measure given the differential projection of cells in different layers (e.g., see Jones & Wise, 1977). It is interesting that in the case of the motor cortex identifying a cell has almost come to mean knowing the destination of the axon of a projection cell, and the technique commonly used for that purpose is electrical stimulation of other brain regions. Thus, electrical stimulation of the pyramidal tract has been used to identify motor cortical cells projecting to the spinal cord, and, in fact, several studies of the motor cortex have been focused exclusively on pyramidal tract neurons (PTNs). Although this has certainly yielded useful information on PTNs, studying cells in all cortical layers is important, and identifying cells projecting to other cortical and subcortical regions is also important but more difficult than identifying PTNs. Unfortunately, electrical stimulation of axons of motor cortical cells is not without adverse effects, including potential damage in and around the region in which the stimulating electrodes are placed. Potentially more disturbing are the possible effects of stimulating axons of pyramidal cells on the area of the cortex from which the axon is coming. These effects can be profound and comprise mainly recurrent inhibitory but also recurrent excitatory effects (Brooks,1965; Brooks & Asanuma, 1965; Stefanis, 1969; Stefanis & Jasper, 1964a, 1964b). Of course, there are additional effects on the motor system from the stimulation of the area itself. The problem is more serious if indeed every cell whose activity is recorded in a penetration is to be identified that way, in which case the effects of stimulation of several structures (e.g., pyramid, red nucleus, putamen, cerebral cortical areas) are compounded on the effects of backfiring a number of pyramidal cell axons. In summary, it seems that identification of the output projection of cells recorded in the motor cortex is appropriate for *ad hoc* studies but need not be required for all studies of the motor cortex. On the other hand, identification of the cortical layer in which a cell has been recorded is in line with studies of other cortical areas and useful and adequate for gaining an insight into motor cortical physiology.

Indeed, in our studies we strived for sampling cells from all cortical layers. In a series of experiments in which microelectrode penetrations into the motor cortex were histologically identified along their depth using small electrolytic lesions (Georgopoulos, Caminiti, & Kalaska, unpublished observations), 61 cells were located in layers II, III and IV (predominately III), and 97 cells in layers V and VI (predominantly V). The functional properties of these two populations were compared in the context of studies using the 2-D apparatus previously described with respect to the latency of cell activation, the ongoing steady-state activity during holding the hand in the same posture, the distributions of preferred directions, and the prediction of the movement direction by the neuronal population vector.

1. The distributions of the latencies of cell activation are shown in Fig. 7.21. It can be seen that the distribution from the upper and lower layers overlap extensively: statistically, they do not differ significantly (Kolmogorov-Smirnov test, two-tailed). This finding is in accord with the findings of anatomcial studies that afferent fibers to the motor cortex terminate in both the upper and lower layers (Sloper & Powell, 1979b).

2. The ongoing, steady-state cell activity during holding of the manipulandum at the center of the working surface (Fig. 7.3) differed significantly between cells in the upper and lower layers ($p<0.05$, Kolmogorov-Smirnov test, two-tailed); the median discharge rates were 3.2 impulses/sec in the upper layers versus 7 impulses/sec in the lower layers. The lower activity in the upper layers may relate to the finding that the density of symmetrical (presumably inhibitory) synapses on initial segments of pyramidal cells in the upper layers is three times higher than in the lower layers (Sloper & Powell, 1979a).

3. The distributions of the preferred directions did not differ significantly between the upper and lower layers (Kolmogorov-Smirnov test, two-tailed; data not shown). This means that preferred directions are not represented differentially in the upper or lower layers. This finding is consistent with the hypothesis of columnar organization of the preferred direction discussed above (Georgopoulos et al., 1984a).

4. Finally, the population vectors calculated from the upper and lower layers were close to the direction of movement and between themselves. The average absolute angle between the population vector calculated from cells in the upper layers and the direction of movement was 4.31° ± 2.98° (mean ± SD, $n$ = 8 movement directions), compared to 2.32° ± 2.06° for the lower layers. This finding suggests that the ensemble operation of the population vector can be realized separately in the upper and lower layers. This is important because that information can then be distributed to different structures according to the differential projections from the

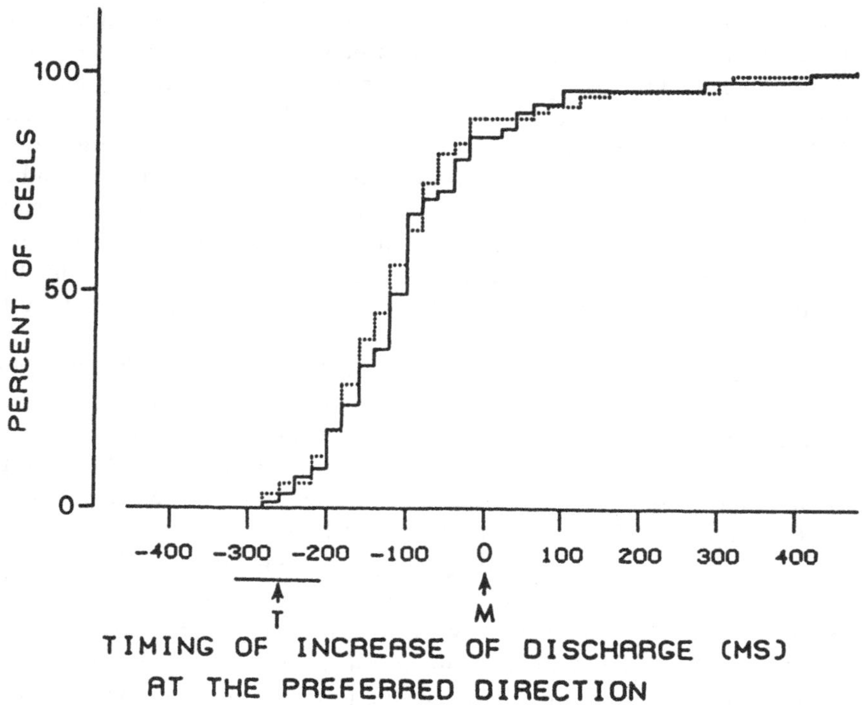


FIG. 7.21. Distributions of latencies of increase in cell activity for motor cortical cells recorded in upper layers (II, III; interrupted line, $n$ = 61) and in lower layers (V, VI; solid lines, $n$ = 97). The distributions do not differ significantly (see text).

upper and lower layers (Jones & Wise, 1977). Finally, given the probable columnar organization of the preferred direction, the population coding appears to be a tangential operation realized at different cortical layers.

## Parameters of Reaching

Direction and amplitude are the two spatial parameters that define the vector of a pointing movement. The results described above indicate that there are meaningful relations between the frequency of discharge of single cells in the motor cortex and area 5 and these movement parameters. Now, these parameters were used in our studies as operationally useful variables, and no attempts were made to dissociate several interwoven aspects of them that may place a particular brain area at a particular stage of processing. For example, the direction of a pointing movement comprises several aspects, including visual definition of the direction, concurrent contractions of various muscles that implement the movement, concurrent changes in the angles of usually two joints, and abstract components of motion of the hand in space. Our experiments were not designed to dissociate the effects of these various components on cell activity; this is the goal of ongoing studies. Instead, our studies up to now were aimed at finding out those general variables for which cells "care," and at describing the quantitative relations that hold between them and frequency of cell discharge. An additional objective was to vary systematically the origin of the movement and thus dissociate the effects of movement direction as such from the origin and endpoint of the reaching movement. The picture that has emerged from these studies is relatively simple: (a) the starting point of the movement is reflected in the ongoing, steady-state activity of single cells both in the motor cortex (Georgopoulos et al., 1984c; Kettner et al., 1988) and in area 5 (Georgopoulos et al., 1984c); (b) the changes in cell activity from this ongoing, background discharge relate to the direction of the movement, both for movements made along the same line (Georgopoulos et al., 1984b) and for movements made in directions parallel in space (Kettner et al., 1988), although interactions exist between the direction of movement and the spatial location of the trajectory (Kettner et al., 1988); and (c) the neuronal population vector predicts well the direction of movement in space even for parallel movement trajectories in extrapersonal space (Kettner et al., 1988). These findings indicate that the direction of movement is an important determinant of single-cell and population activity in the motor cortex and area 5.

The effects of movement amplitude on single-cell activity are less obvious. These amplitude effects, when present, are coexistent with directional effects (Schwartz & Georgopoulos, 1987). It is noteworthy that only weak, if any, effects of amplitude were observed for movements about a single joint (Hamada & Kubota, 1979). It is possible that the amplitude of large movements is reflected in the discharge of the neuronal ensemble if, for example, additional cells are recruited in the active population; this remains to be elucidated.

### Effects of Loads

One of the earliest findings of motor cortical recordings in behaving monkeys was that the steady-state cell activity of PTNs related to the magnitude of static force exerted by the animal (Evarts, 1969), and that changes in cell activity seemed to relate to changes in force irrespective of position along the flexion–extension axis at the wrist joint (Evarts, 1968). The relation of cell activity to the magnitude of static loads was also observed in subsequent studies (Cheney & Fetz, 1980; Hepp-Reymond et al., 1978; Schmidt et al., 1975; Thach, 1978; Evarts et al., 1983) that clarified some aspects of it. For example, this relation seems to hold for PTNs that probably make monosynaptic connections to spinal motoneurons (Cheney & Fetz, 1980). The relations between cell activity and the direction of 2-D loads of constant magnitude were studied by Kalaska and Hyde (1985). It was found that the steady-state activity of motor cortical cells in the shoulder area was a sinusoidal function of the direction of the load in 2-D space. Moreover, it was found that the neuronal population vector calculated under these static conditions pointed in a direction opposite to that of the load, that is, in the direction of the static force exerted by the animal (Kalaska et al., 1989).

A different question concerns the relations of cell activity to dynamic changes in force, that is, changes in force during the movement. To answer this question, the ongoing changes in force during the initiation and execution of the movement have to be calculated and correlated with the corresponding changes in single-cell and population activity. This problem remains to be elucidated. In this context, it is interesting that the time-series analysis of the population vector performed in our studies (Fig. 7.16) shows no reversal in the direction of the population vector, whereas the direction of force reverses due to the braking of the movement (data not shown).

In a different experimental design, Kalaska et al. (1989) studied the effects on cell activity of the direction of movement when movements of different directions were made under conditions of constant loads of various directions. It was found that the effects of the direction of the movement (reflected in the changes in cell activity) were superimposed in an additive fashion on the effects of the direction of the load (reflected in the tonic activity of the cell). Thus, signals concerning the direction of the static force and the direction of the movement can coexist and influence independently the cell activity in the motor cortex.

## MOTOR CORTICAL PROJECTIONS TO SPINAL CORD

The motor cortical influence on motor function is ultimately effected on the motoneuronal pools either directly or via intercalated spinal interneurons and/or neurons in the brainstem. There is an appreciable divergence in the projection of single pyramidal tract axons to various spinal segments (Shinoda et al., 1981). Moreover, electrophysiological studies using the technique of spike-triggered averaging have shown that motor cortical cells impinging monosynaptically on motoneurons innervating distal muscles acting on the wrist (Fetz & Cheney, 1980) and/or fingers (Lemon et al., 1986) may contact more than one motor nucleus. Conversely, other studies have shown that there is an appreciable convergence onto individual spinal motoneurons from several loci on the motor cortex (Landgren et al., 1962). Thus, even in the case of the motor cortical control of distal muscles, single cells relate, as a rule, to groups of muscles. The case for multimuscle control from the motor cortex is stronger for proximal muscles that act on the shoulder and elbow. In this case, all motor cortical control seems to be exerted through interneurons; thus, in the rhesus monkey corticospinal axons from the proximal arm area do not terminate within motor nuclei but within interneuronal zones (Kuypers, 1981), as is the case in the cat. The general pattern of connectivity between motor cortex and spinal cord is illustrated in Fig. 7.22. The interneurons involved in mediating the motor cortical command for reaching have been studied extensively in the cat using physiological, anatomical, and behavioral methods; we summarize these results in the next section.

## REACHING CIRCUITS IN THE SPINAL CORD

A major aspect of the organization of the spinal cord with respect to motor function is the existence of neural circuits specialized for a particular function. The elucidation of the spinal circuits subserving locomotion (Grillner, 1981) has provided an excellent example of the efficacy of spinal mechanisms to implement a very complicated biomechanical sequence of limb movements. Significant advances have been accomplished during the past 20-odd years concerning the spinal circuits subserving reaching movements. A major circuit comprises a set of "propriospinal" interneurons located at the C3–C4 spinal segments (Lundberg, 1979). These interneurons receive monosynaptic inputs from several supraspinal sources (Illert et al., 1978), including the pyramidal (i.e., corticospinal), rubrospinal, reticulospinal, and tectospinal tracts and distribute their axons to several proximal motoneuronal pools (Alstermark, et al., 1987); they also send an ascending collateral to the lateral reticular nucleus which projects to the cerebellum.

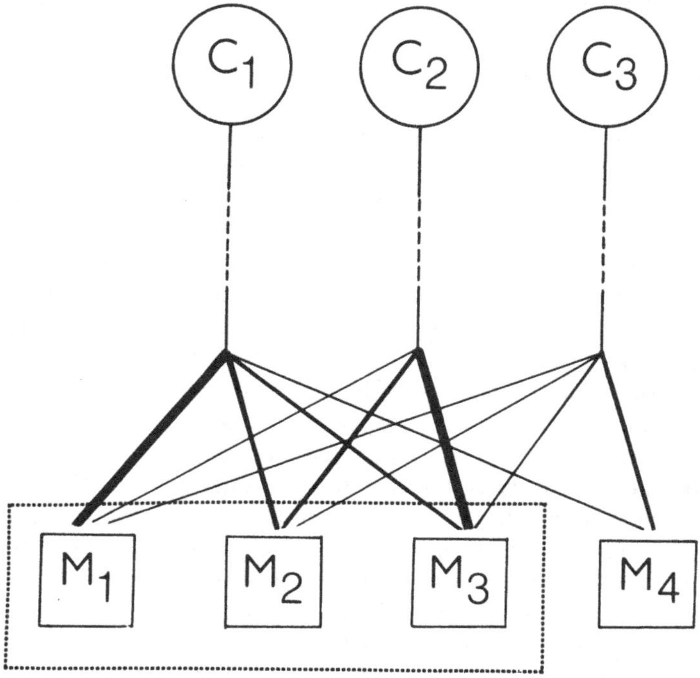


FIG. 7.22. Schematic diagram to illustrate the postulated pattern of weighted divergent and convergent influences exerted on motoneuronal pools ($M_1$, $M_2$, $M_3$, $M_4$, . . . ) by single motor cortical cells ($C_1$, $C_2$, $C_3$, . . . ). The dashed lines indicate intercalated neurons (e.g., C3–C4 propriospinal neurons). The line thickness indicates the weighted influence. Motor cortical cell $C_2$ influences a weighted combination of motoneuronal pools ($M_1$, $M_2$, $M_3$), enclosed within the interrupted parallelogram. (From Georgopoulos et al., 1988; reproduced with permission.)

(Alstermark et al., 1981a). Selective section of the output from these propriospinal neurons to their target motoneurons results in abnormal reaching with normal grasping, and similar effects are observed when the corticospinal input to the propriospinal neurons is removed (Alstermark et al., 1981b). Moreover, propriospinal neurons seem to be selectively engaged during reaching movements (Alstermark & Kummel, 1986). These results indicate that the C3–C4 propriospinal system is concerned with the neural integration of the reaching movement at the spinal level, and that the motor cortex controls reaching, most probably through that system. This motor cortical control is also exerted at other levels within the propriospinal system; for example, there is direct corticospinal input on a key inhibitory interneuron that mediates inhibition from afferent fibers to propriospinal neurons (Alstermark et al., 1984). This peripherally initiated inhibition of the propriospinal neurons is important in limiting the reaching movement because lack of peripheral input results in consistent hypermetria in reaching (Alstermark et al., 1986).

The results summarized here indicate that a large part of neural integration of the reaching movement may be accomplished in the spinal cord. Therefore, the organization and the neural integration of the reaching movement need not be the concern, or the burden, of the motor cortex. The initiation of reaching in the appropriate direction could be accomplished by the activation of neuronal populations in the motor cortex that, in turn, would engage the spinal "reaching" circuits. Some ideas concerning the translation of the motor cortical command for movement direction into levels of activation of limb muscles have been discussed elsewhere (Georgopoulos, 1988).

## ACKNOWLEDGMENTS

This work was supported by U.S. Public Health Service Grant NS17413 and ONR contract N00014-88-K-0751.

## REFERENCES

Alstermark, B., Gorska, T., Johannisson, T., & Lundberg, A. (1986). Hypermetria in forelimb target-reaching after interruption of the inhibitory pathway from forelimb afferents to C3-C4 propriospinal neurones. *Neuroscience Research, 3*, 457-461.

Alstermark, B., & Kummel, H. (1986). Transneuronal labelling of neurones projecting to forelimb motoneurones in cats performing different movements. *Brain Research, 376*, 387-391.

Alstermark, B., Kummel, H., Pinter, M. J. & Tantisira, B. (1987). Branching and termination of C3-C4 propriospinal neurons in the cervical spinal cord of the cat. *Neuroscience Letters, 74*, 291-296.

Alstermark, B., Lindstrom, S., Lundberg, A., & Sybirska, E. (1981a). Integration in descending motor pathways controlling the forelimb in the cat. 8. Ascending projection to the lateral reticular nucleus from C3-C4 propriospinal neurones also projecting to forelimb motoneurones. *Experimental Brain Research, 42*, 282-298.

Alstermark, B., Lundberg, A., Norrsell, U. & Sybirska, E. (1981b). Integration in descending motor pathways controlling the forelimb in the cat. 9. Differential behavioral defects after spinal cord lesions interrupting defined pathways from higher centres to motoneurones. *Experimental Brain Research, 42*, 299-318.

Alstermark, B., Lundberg, A., & Sasaki, S. (1984). Integration in descending motor pathways controlling the forelimb in the cat. 11. Inhibitory pathways from higher motor centres and forelimb afferents to C3-C4 propriospinal neurones. *Experimental Brain Research, 56*, 293-307.

Asanuma, H., & Brooks, V. B. (1965). Recurrent cortical effects following stimulation of internal capsule. *Archives Italliennes de Biologie, 103*, 220-246.

Asanuma, H. & Rosen, I. (1973). Spread of mono- and polysnaptic connections within cat's motor cortex. *Experimental Brain Research, 16*, 507-520.

Barbas, H., & Pandya, D. N. (1987). Architecture and frontal cortical connections of the premotor cortex (area 6) in the rhesus monkey. *Journal of Comparative Neurology, 256*, 211–228.

Brooks, V. B., & Asanuma, H. (1965). Recurrent cortical effects following stimulation of medullary pyramid. *Archives Itallíennes de Biologie, 103*, 247–278.

Caminiti, R., Zeger, S., Johnson, P. B., Urbano, A., & Georgopoulos, A. P. (1985). Cortico-cortical efferent systems in the monkey: A quantitative spatial analysis of the tangential distribution of cells of origin. *Journal of Comparative Neurology, 241*, 405–419.

Cheney, P. D., & Fetz, E. E. (1980). Functional classes of primate corticomotoneuronal cells and their relation to active force. *Journal of Neurophysiology, 44*, 773–791.

Dow, B. M., Bauer, R., Snyder, A. Z., & Vautin, R. G. (1984). Receptive fields and orientation shifts in the foveal striate cortex of the awake macaque monkey. In G. M. Edelman, W. E. Gall, & W. M. Cowan (Eds.), *Dynamic aspects of neocortical function* (pp. 41–65). New York: Wiley.

Evarts, E. V. (1968). Relation of pyramidal tract activity to force exerted during voluntary movements. *Journal of Neurophysiology, 31*, 14–27.

Evarts, E. V. (1969). Activity of pyramidal tract neurons during postural fixation. *Journal of Neurophysiology, 32*, 375–385.

Evarts, E. V., Fromm, C., Kroller, J., & Jennings, V. A. (1983). Motor cortex of finely graded forces. *Journal of Neurophysiology, 49*, 1199–1215.

Ferro, J. M., Bravo-Marques, J. M., Castro-Caldas, A., & Antunes, L. (1983). Crossed optic ataxia: Possible role of the dorsal splenium. *Journal of Neurology, Neurosurgery, and Psychiatry, 46*, 533–539.

Fetz, E. E., & Cheney, P. D. (1980). Postspike facilitation of forelimb muscle activity by primate corticomotoneuronal cells. *Journal of Neurophysiology, 44*, 751–772.

Georgopoulos, A. P. (1986). On reaching. *Annual Review of Neuroscience, 9*, 147–170.

Georgopoulos, A. P. (1988). Neural integration of movement: Role of motor cortex in reaching. *The FASEB Journal, 2*, 2849–2857.

Georgopoulos, A. P. (1989). Visual control of reaching. In G. M. Edelman, W. E. Gall, & W. M. Cowan (Eds.), *Signal and sense: Local and global order in perceptual maps*. New York: Wiley. *In press*.

Georgopoulos, A. P., Caminiti, R., Kalaska, J. F., & Massey, J. T. (1983). Spatial coding of movement: A hypothesis concerning the coding of movement direction by motor cortical populations. *Experimental Brain Research Supplement, 7*, 327–336.

Georgopoulos, A. P., Kalaska, J. F., & Massey, J. T. (1981). Spatial trajectories and reaction times of aimed movements: effects of practice, uncertainty, and change in target location. *Journal of Neurophysiology, 46*, 725–743.

Georgopoulos, A. P., Kalaska, J. F., Caminiti, R., & Massey, J. T. (1982). On the relations between the direction of two-dimensional arm movements and cell discharge in primate motor cortex. *Journal of Neuroscience, 2*, 1527–1537.

Georgopoulos, A. P., Caminiti, R., Kalaska, J. F., & Massey, J. T. (1983). Spatial coding of movement: A hypothesis concerning the coding of movement direction by motor cortical populations. *Experimental Brain Research Supplement, 7*, 327–336.

Georgopoulos, A. P., Crutcher, M. D., & Schwartz, A. B. (1989). Cognitive spatial–motor processes. 3. Motor cortical prediction of movement direction during an instructed delay period. *Experimental Brain Research, 75*, 183–194.

Georgopoulos, A. P., Kalaska, J. F., Crutcher, M. D., Caminiti, R., & Massey, J. T. (1984a). The representation of movement direction in the motor cortex: Single cell and population studies. In G. M. Edelman, W. E. Gall, & W. M. Cowan (Eds.), *Dynamic aspects of neocortical function* (pp. 501–524). New York: Wiley.

Georgopoulos, A. P., Kalaska, J. F., & Caminiti, R. (1984b). Relations between two-dimensional arm movements and single cell discharge in motor cortex and area 5: movement direction versus movement endpoint. *Experimental Brain Research Supplement, 10*, 176–183.

Georgopoulos, A. P., Caminiti, R., & Kalaska, J. F. (1984c). Static spatial effects in motor cortex and area 5: quantitative relations in a two-dimensional space. Experimental Brain Research, 54, 446–454.

Georgopoulos, A. P., Kettner, R. E., & Schwartz, A. B. (1988). Primate motor cortex and free arm movements to visual targets in three-dimensional space. II. Coding of the direction of movement by a neuronal population. *Journal of Neuroscience, 8*, 2928–2947.

Georgopoulos, A. P., & Massey, J. T. (1987). Cognitive spatial-motor processes. 1. The making of movements at various angles from a stimulus direction. *Experimental Brain Research, 65*, 361–370.

Georgopoulos, A. P., Schwartz, A. B., & Kettner, R. E. (1986). Neuronal population coding of movement direction. *Science, 233*, 1416–1419.

Grillner, S. (1981). Control of locomotion in bipeds, tetrapods, and fish. In J. M. Brookhart, & V. B. Mountcastle (Eds.), *Handbook of physiology. The nervous system II* (pp. 1179–1236). Bethesda, MD: American Physiological Society.

Haaxma, R., & Kuypers, H. G. J. M. (1975). Intrahemispheric cortical connexions and visual guidance of hand and finger movements in the rhesus monkey. *Brain, 98*, 239–260.

Hamada, I. and Kubota, K. (1979). Monkey pyramidal tract neurons and changes of movement parameters in visual tracking. *Brain Research Bulletin, 4*, 249–257.

Hepp-Reymond, M. C., Wyss, U. R., & Anner, R. (1978). Neuronal coding of static force in the primate motor cortex. *Journal de Physiologie, 74*, 287–291.

Humphrey, D. R. (1979). On the cortical control of visually directed reaching: Contributions by nonprecentral motor areas. In F. E. Talbott, & D. R. Humphrey (Eds.), *Posture and movement* (pp. 51–112). New York: Raven Press.

Humphrey, D. R. Gold, R., & Reed, D. J. (1984). Size, laminar and topographic origin of cortical projections to the major divisions of the red nucleus in the monkey. *Journal of Comparative Neurology, 225*, 75–94.

Hutchins, K. D., & Strick, P. L. (1987). Origin of corticospinal projections to the ipsilateral spinal cord. *Society for Neuroscience Abstracts, 13*, 243.

Hyvärinen, J., & Poramen, A. (1974). Function of the parietal associative area 7 as revealed from cellular discharges in alert monkeys. *Brain, 97*, 673–692.

Illert, M., Lundberg, A., Padel, Y., & Tanaka, R. (1978). Integration in descending motor pathways controlling the forelimb in the cat. 5. Properties of and monosynaptic excitatory convergence on C3–C4 propriospinal neurones. *Experimental Brain Research, 33*, 101–130.

Jankowska, E., Padel, Y., & Tanaka, R. (1975). The mode of activation of pyramidal tract cells by intracortical stimuli. *Journal of Physiology (London), 249*, 617–636.

Jeannerod, M. (1986). Mechanisms of visuomotor coordination: A study in normal and brain-damaged subjects. *Neuropsychologia, 24*, 41–78.

Jeannerod, M., & B. Biguer (1982). Visuomotor mechanisms in reaching within extrapersonal space. In D. J. Ingle, M. A. Goodale, & R. J. W. Mansfield (Eds.), *Analysis of visual behavior* (pp. 387–409). Cambridge, MA: MIT Press.

Jones, E. G. & Wise, S. P. (1977). Size, laminar and columnar distribution of efferent cells in the sensory-motor cortex of monkeys. *Journal of Comparative Neurology, 175*, 391–438.

Kalaska, J. F. (1988). The representation of arm movements in postcentral and parietal cortex. *Canadian Journal of Physiology and Pharmacology, 66*, 455–463.

Kalaska, J. F., Caminiti, R., & Georgopoulos, A. P. (1983). Cortical mechanisms related to the direction of two-dimensional arm movements; Relations in parietal area 5 and comparison with motor cortex. *Experimental Brain Research, 51*, 247–260.

Kalaska, J. F., Cohen, D. A. D., Hyde, M. L., & Prud'homme, M. (1989). A comparison of movement direction-related vs. load direction-related activity in primate motor cortex, using a two-dimensional reaching task. *Journal of Neuroscience, in press.*

Kalaska, J. F., & Hyde, M. L. (1985). Area 4 and area 5: Differences between the load direction-dependent discharge variability of cells during active postural fixation. *Experimental Brain Research, 59*, 197–202.

Kettner, R. E., Schwartz, A. B. & Georgopoulos, A. P. (1988). Primate motor cortex and free arm movements to visual targets in three-dimensional space. III. Positional gradients and population coding of movement direction from various movement origins. *Journal of Neuroscience, 8*, 2938–2947.

Kuypers, H. G. J. M. (1981). Anatomy of the descending pathways. In J. M. Brookhart, V. B. Mountcastle (Eds.), *Handbook of physiology. The nervous system II* (pp. 597–666). Bethesda, MD: American Physiological Society.

Landgren, S., Phillips, C. G. & Porter, R. (1962). Cortical fields of origin of the monosynaptic pyramidal pathways to some alpha motoneurones of the baboon's hand and forearm. *Journal of Physiology (London), 161*, 112–125.

Lemon, R. N. (1984). *Methods for neuronal recording in conscious animals.* Chisester: Wiley.

Lemon, R. N., Mantel, G. W. H., & Muir, R. B. (1986). Corticospinal facilitation of hand muscles during voluntary movement in the conscious monkey. *Journal of Physiology (London), 381*, 497–527.

Lundberg, A., (1979). Integration in a propriospinal motor centre controlling the forelimb in the cat. In H. Asanuma, & V. J. Wilson (Eds.), *Integration in the nervous system* (pp. 47–69). Tokyo: Igaku-Shoin.

Lynch, J. C. (1980). The functional organization of posterior parietal association cortex. *Behavioral and Brain Sciences, 3*, 485–534.

Mountcastle, V. B., Lynch, J. C., Georgopoulos, A. P., Sakata, H., & Acuna, C. (1975). Posterior parietal association cortex of the monkey: Command functions for operations within extrapersonal space. *Journal of Neurophysiology, 38*, 871–908.

Mountcastle, V. B., Motter, B. C., & Andersen, R. A. (1980). Some further observations on the functional properties of neurons in the parietal lobe of the waking monkey. *Behavioral and Brain Sciences, 3*, 520–522.

Muakkassa, K. F., & Strick, P. L. (1979). Frontal lobe inputs to primate motor cortex: Evidence for four somatotopically organized "premotor" areas. *Brain Research, 177*, 176–182.

Murphy, J. T., Kwan, H. C., MacKay, W. A. & Wong, Y.C. (1978). Spatial organization of precentral cortex in awake primates. III. Input-output coupling. *Journal of Neurophysiology, 41*, 1132–1139.

Murphy, J. T., Kwan, H. C., MacKay, W. A. & Wong, Y. C. (1982). Precentral unit activity correlated with angular components of a compound arm movement. *Brain Research, 246*, 141–145.

Murphy, J. T., Wong, Y. C., & Kwan, A. C. (1985). Sequential activation of neurons in primate motor cortex unrestrained forelimb movement. *Journal of Neurophysiology, 53*, 435–445.

Perenin, M. T., & Vighetto, A. (1988). Optic ataxia: A specific disruption in visuomotor mechanisms. *Brain, 111*, 643–674.

Poggio, G. F. (1984). Processing of steroscopic information in primate visual cortex. In G. M. Edelman, W. E. Gall, & W. M. Cowan (Eds.), *Dynamic aspects of neocortical function* (pp. 613–635). New York: Wiley.

Porter, R., & Lewis, M. Mc. (1975). Relationship of neuronal discharges in the precentral gyrus of monkeys to the performance of arm movements. *Brain Research, 98*, 21–36.

Schmidt, E. M., Jost, R. G., & Davis, K. K. (1975). Reexamination of the force relationship of cortical cell discharge patterns with conditioned wrist movements. *Brain Research, 83*, 213–223.

Schwartz, A. B., & Georgopoulos, A. B. (1987). Relations between the amplitude of 2-dimensional arm movements and single cell discharge in primate motor cortex. *Society for Neuroscience Abstracts, 13*, 244.

Schwartz, A. B., Kettner, R. E., & Georgopoulos, A. P. (1988). Primate motor cortex and free arm movements to visual targets in three-dimensional space. I. Relations between single cell discharge and direction of movement. *Journal of Neuroscience, 8*, 2913–2927.

Shinoda, Y., Yokota, J. I., & Futami, T. (1981). Divergent projections of individual cortico-spinal axons to motoneurons of multiple muscles in the monkey. *Neuroscience Letters, 23*, 7–12.

Sloper, J. J., & Powell, T. P. S. (1979a). A study of the axon initial segment and proximal axon of neurons in the primate motor and somatic cortices. *Philosophical Transactions of the Royal Society of London, B. Biological Sciences, 285*, 173–197.

Sloper, J. J. & Powell, T. P. S. (1979b). An experimental electron microscopic study of afferent connections to the primate motor and somatic sensory cortices. *Philosophical Transactions of the Royal Society of London, B. Biological Sciences, 285*, 199–226.

Soechting, J. F., & Lacquaniti, F. (1981). Invariant characteristics of a pointing movement in man. *Journal of Neuroscience, 1*, 710–720.

Stefanis, C. (1969). Interneuronal mechanisms in the cortex. In M. A. B. Brazier (Ed.), *The interneuron* (pp. 497–526). Berkeley, CA: University of California Press.

Stefanis, C., & Jasper, H. (1964a). Intracellular microelectrode studies of antidromic responses in cortical pyramidal tract neurons. *Journal of Neurophysiology, 27*, 828–854.

Stefanis, C., & Jasper, H. (1964b). Recurrent inhibition in pyramidal tract neurons. *Journal of Neurophysiology, 27*, 855–877.

Strick, P. L., & Kim, C. C. (1978). Input to primate motor cortex from posterior parietal cortex (area 5). I. Demonstration by retrograde transport. *Brain Research, 157*, 325–330.

Tanji, J. (1984). The neuronal activity in the supplementary motor area of primates. *Trends in Neuroscience, 7*, 282–285.

Thach, W. T. (1978). Correlation of neural discharge with pattern and force of muscular activity, joint position, and direction of intended next movement in motor cortex and cerebellum. *Journal of Neurophysiology, 41*, 654–676.

Watson, G. S. *Statistics on spheres*. New York: Wiley.

Wise, S. P. (1985). The primate premotor cortex: past, present, and preparatory. *Annual Review of Neuroscience, 8*, 1–19.

# 8 Parallel Interacting Channels in the Initiation and Specification of Motor Response Features

Claude Ghez
Wayne Hening
Marco Favilla
*Center for Neurobiology and Behavior*
*New York State Psychiatric Institute*
*College of Physicians and Surgeons, Columbia University*

ABSTRACT

The present report summarizes a series of experiments conducted in normal human subjects to characterize the processes by which information derived from a visual target is used to initiate a simple motor response and to specify the features of it's trajectory. We sought to determine which aspects of response trajectories subjects prepare in advance of target presentation, and the rate at which subjects specify amplitude and direction. We also wished to determine the relationship between the processes leading to response initiation and those that specify response features.

Subjects were required to produce isometric flexion and extension impulses at the elbow aimed to predictable and unpredictable targets of three different amplitudes in each of the two directions. To control response initiation, we used both traditional reaction time methods and another procedure, the *timed response paradigm*, that allowed us to systematically vary the time between stimulus presentation and response initiation.

Under all initiation contingencies, subjects were able to prepare accurate responses to predictable targets. However, when subjects were required to aim a response to an unpredictable target within a known range, they prepared responses with a default amplitude prior to target presentation. This default amplitude was close to the center of the target range. Over an interval of 200 to 300 msec after target presentation, response amplitude was progressively specified to conform with the target amplitude both by progressive improvement in the scaling of the initial phase of the trajectory and by corrections made to the ongoing response. When both the amplitude and direction of the required response were unpredictable, subjects prepared a single default amplitude, but selected a direction at random. After target presentation, specification of amplitude and direction occurred concurrently. However, the simultaneous need to specify direction reduced the rate of amplitude specification. In addition, the need to utilize stimulus information from the target to trigger response initiation, as occurs in the normal reaction time paradigm, also reduced the rate of amplitude specification.

We conclude that accuracy in these responses is achieved by multiple mechanisms that include both response preparation before target presentation and response specification thereafter. The processes triggering response initiation and those that specify response amplitude and direction are organized in parallel in this task, but interactions among these channels occurs.

## INTRODUCTION

A major issue concerning goal-directed motor performance relates to how the nervous system uses information derived from a target to specify accurate responses. Important questions to be answered include the rate at which such information is processed and how the nervous system breaks down complex processing into component tasks. Other questions include the extent to which responses can be prepared before their target is presented and the degree to which features of trajectories can be modified during their course. The present report addresses these questions using a simple isometric task.

### Organization and Duration of Sensorimotor Processing

Since the pioneering work of Donders (1869), one approach to assessing the rate at which the nervous system processes information derived from a target has been to determine the latency to respond to a target stimulus known as the reaction time (RT). This measure has frequently been used to estimate the duration of central information processing by contrasting RTs obtained in tasks that present different processing demands (Meyer, Osman, Irwin, & Yantis, 1988). A fundamental finding has been that when the target is unpredictable (referred to as a *choice* condition) subjects respond with a longer RT then when the target is predictable (referred to as a *simple* condition). To interpret such RT differences, many experimenters have adopted a serial model of sensorimotor processing (Luce, 1986; Welford, 1980), that postulates that component processes are organized into obligatory, sequential stages (Gottsdanker & Shragg, 1985; Sternberg, 1969; Sanders, 1980; Taylor, 1976; Welford, 1952, 1980). In this model, response initiation, which determines RT, occurs only when the stages have been completed. The increased RT in the *choice* condition (Welford, 1980) has been considered to result from the added time required to categorize the target and select the proper response, processes that should not be required in the *simple* condition.

A more recent experimental procedure, pre-cueing, has been devised to examine how information about different response features is processed. Subjects are given advance information about some but not all aspects of the required response (Rosenbaum, 1980), and the resulting differences in RT are then used to uncover the organization underlying the specification of different response features. The results of such studies have been interpreted to support serial specification of response direction and amplitude (Bonnet, Requin, & Stelmach, 1982; Rosenbaum, 1980; Zelaznik, 1981 cf, however, Goodman & Kelso, 1980).

One important difficulty with the use of RT to measure the rate of processing under differing task conditions is that it requires that the responses being compared must be initiated either after processing has been completed or after a similar degree of processing. However, such equivalence may not always apply. For example, when subjects respond at the shortest possible latencies to unpredictable stimuli, they make more errors (Baron, Menich & Perone, 1983; Hicks, 1952; Howell & Kriedler, 1963; Pachella, Fischer & Karsh, 1969; Schouten & Bekker, 1967). This inverse relationship between latency and errors has been conceptualized as a speed–accuracy tradeoff (Pachella, 1974; Pew, 1969; Swensson, 1972; Wickelgren, 1977). Because individual responses are rarely examined in detail, however, these studies tell us little about the nature of erroneous responses produced at brief RTs: It is possible that such inaccurate responses differ in some fundamental way from accurate ones.

## Information Processing and Trajectory Control

In an alternate approach to analysing how subjects use target information to specify responses, we have sought to determine how different processing demands influence trajectory control itself. We examine how response trajectories are altered during specification by performing parametric analyses of trajectory features while controlling both the information presented to the subject and the time available to process this information. To optimize our analysis of trajectory variability, we have studied a simple isometric force response. In contrast to limb movements, isometric responses do not depend on the complex relationships that exist between changes in muscle force and length and are relatively unaffected by the reflex effects of movement. We have further simplified the task by requiring subjects to only produce single uncorrected impulses of flexor or extensor force at the elbow. They were to match these peaks to a prior change in the level of a visual target. We chose to focus initially on these transient force impulses in order to minimize the subjects' ability to use visual feedback to correct their responses. Such impulses would be free of secondary submovements that can also be used to achieve accuracy (Meyer, Abrams, Kornblum, Wright, & Smith, 1988).

In our initial studies, we began by examining optimally prepared responses. We removed response urgency so that subjects could respond when all sensorimotor processing was completed. The trajectories of the force impulses were highly stereotyped in all subjects. Variations in response amplitude were produced primarily by scaling the rate of rise of force to the target force while the duration of the force rise was maintained nearly invariant (Ghez & Gordon, 1987; Gordon & Ghez, 1987a; Gordon & Ghez, 1987b). We called this pattern of amplitude modulation, *pulse-height control*. It was most evident when subjects were instructed to achieve the greatest possible accuracy rather than when they were to achieve the briefest rise time. We concluded that in order to achieve accurate matching of force impulses to their targets, subjects used target information to scale a common trajectory waveform according to a previously learned sensory-motor transform. This strategy appears to constitute a critical component of the central program for accurate specification of force impulse amplitude. In addition, a fine-grained analysis of trajectory variability revealed that small adjustments to the force rise time that correct initial trajectory errors also contribute to accuracy (Gordon & Ghez, 1987b).

In our next series of experiments (Favilla, Hening, & Ghez, 1989, Favilla, Hening, Gordon, & Ghez, 1986; Hening, Favilla, & Ghez, 1988; Hening, Vicario, & Ghez, 1988) summarized in this report, we sought to determine the time course of the processes that implement these trajectory control mechanisms. We also sought to examine how the specification of multiple response features is accomplished. To do this, we varied processing demands by presenting subjects with visual targets whose amplitudes and/or directions were either predictable or unpredictable and by varying the time available for processing target information. We planned in these experiments to use our simple, isometric task to address four key questions about trajectory formation. *First*, what aspects of trajectory control are impaired when subjects are required to respond without adequate time to accurately specify response features? *Second*, what is the time course over which the amplitude and direction of a response are specified? *Third*, are the direction and amplitude of a response specified serially or in parallel? *Fourth*, to what degree are the processes responsible for triggering the initiation of the response and those responsible for specifying its features independent from each other, and to what extent do they interact?

## METHODS

Six normal adults aged 25 to 45 participated in all experiments except for the bidirectional experiment described in a later section where only five subjects were examined. Subjects were seated with their right arm abducted and with their elbow flexed to 90°. Shoulder, arm, and forearm were immobilized using a system of padded but rigid restraints. A strain gauge measured the flexion and extension torques developed isometrically at the elbow joint. Subjects faced an oscilloscope screen where the vertical positions of two horizontal lines represented the amount of force they exerted on the strain gauge and a target level, respectively.

Testing sessions were composed of approximately 300 discrete trials that each lasted 6 to 10 sec. At the beginning of each trial, the target line was set to a zero force level. The subject aligned the force line with the target line, then, at an unpredictable time, the target was stepped to a new level on the screen. Following this visual stimulus, subjects were to produce a single impulse of force whose peak was to match the new level of the target. They were then to relax, allowing force to return passively to baseline. Subjects were instructed and trained not to correct their responses after initiation if they noted an error in amplitude or direction. Target steps had one of three possible amplitudes, in ratios of 1:3:5, with the largest force target set at approximately 40% of the maximal isometric force the subject could produce. Trials were given in blocks in which successive target steps were either predictable (*simple* condition) or unpredictable (*choice* condition).

We varied the time available to the subjects for processing target information in two ways. First (Fig. 8.1A), we used a conventional reaction time paradigm and compared the effect of two different instructions. Subjects were either to respond *as soon as possible* (*ASAP* condition) or *when ready* (*WR* condition), that is, when they felt fully prepared to respond accurately. As might be expected, reaction times were systematically prolonged in the *WR* as compared to the *ASAP* condition. Second (Fig. 8.1B), we developed an experimental paradigm to control more precisely the time elapsing between target presentation and response onset, the Stimulus–Response or S–R interval. In this *timed response paradigm*, subjects were still required to produce uncorrected force impulses to match the different visual targets; however, now they were also presented with a predictable auditory stimulus with which they were to synchronize the initiation of their responses. As illustrated in Fig. 8.1B, subjects heard a series of four tones separated by intervals of 500 ms. The interval between presentation of the visual stimulus (the target step) and the fourth tone (with which responses were to be synchronized) was systematically varied from 0 to 500 ms. As a result, we were able to examine responses made at S–R intervals both shorter and longer than the minimum latency needed by the subjects to trigger a response to the visual target.

In this paradigm, we varied the range and probabilities of targets presented to subjects in order to examine different aspects of response specification. We examined the time course of amplitude specification by presenting three targets requiring responses of different magnitudes in a single

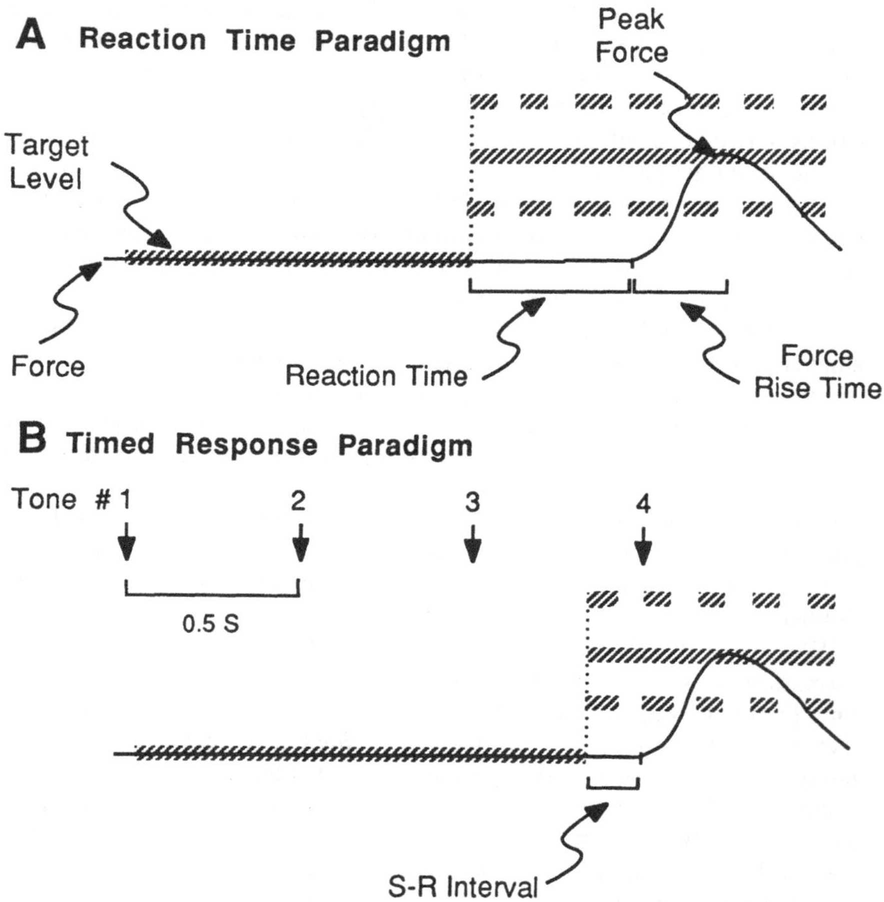


FIG. 8.1. *Reaction Time and Timed Responses Paradigms.* A. Reaction Time Paradigm. Target level shown as hatched line; subject's force output shown as single solid line. Target line shifts upwards to begin trial. Reaction time is interval from target shift to force onset (first consistent change in dF/dt, indicated by downward tick on force traces. Force rise time is interval from force onset to peak force. B. Timed Response Paradigm. Four successive tones of increasing pitch (above) allow subject to synchronize response initiation with the last tone. S–R (Stimulus–Response) interval is the interval between target shift and the time of force onset, indicated by downward tick on force line.

direction (flexion or extension). To examine the relation between specification of amplitude and of direction, we presented six targets, three each in flexion and extension, with a symmetrical distribution of amplitudes. Then, in order to examine the influence of target probability on the process of specification, we presented subjects with two targets in each direction whose probability was biased in a single direction. Finally, we compared the rate of response specification in the two paradigms. In the timed response paradigm, the target stimulus only indicates required response features, but in the reaction time paradigm, it is also the signal to evoke response initiation.

Experiments were controlled by a computer that also sampled target, force and EMG data every 2 ms. Onsets, maxima and minima in the force trajectories and in their two time deviatives, dF/dt and $d^2F/dt^2$ were marked using automatic computer programs and verified by visual inspection of the trajectories of individual responses. Response onset was operationally defined as the time of the first change in dF/dt of the force response.

To assess the accuracy of responses, we calculated the means and standard deviations of the peak forces of responses to the individual targets. From these, we derived the constant error (difference between mean value and target value) and variable error (standard deviation and coefficient of variation). In addition, we used simple and multiple regression analysis to assess the degree to which the populations of responses were scaled to the overall range of targets. The percentages of variance of peak $d^2F/dt^2$ and peak force accounted for by variation in the target were used to assess the degree to which the initial phases and the completed responses were scaled to the range of targets. These values represent the squared correlation coefficients of the relationships between target amplitude and the respective trajectory peaks. We estimated the amount of response updating that occurs between the time of the peak $d^2F/dt^2$ and the peak force by the difference between these values (variance accounted for in peak force minus that accounted for in peak $d^2F/dt^2$) (Hening, Favilla, & Ghez, 1988). Variation in this difference closely parallels the more complex index of response updating based on multiple regression analysis that we have recently reported (Hening, Favilla, & Ghez, 1988; Gordon & Ghez, 1987b). In order to examine the time course of specification, we produced scatterplots of peak force versus S–R interval. The clouds of points for each target were fitted with lines using LOWESS (locally weighted scattergram smoother, Cleveland, 1979), which does a quadratic moving average and includes a robustness criterion giving less weight to outlying data points.

## RESULTS

### Influences of Target Predictability and Response Urgency on Trajectory and Latency

Our first series of experiments was designed to identify the impairments in trajectory control that arise when subjects need to make urgent responses to unpredictable targets. These impairments, in turn, would help demonstrate the control mechanisms implemented in the course of response specification as greater accuracy is achieved. We hypothesized that decreased accuracy might result from one of several impairments. First, subjects might not be able to utilize the normal pulse–height strategy to modulate response amplitude to achieve accuracy. They might vary response amplitude by varying response duration rather than by varying rate of force rise. Second, subjects might make incorrect guesses about the required response. Third, if subjects could initiate their responses before the correct response was fully prepared, they might produce partially specified responses. In these experiments, we used a reaction time paradigm where subjects were presented with targets steps of either predictable (*simple* condition) or unpredictable (*choice* condition) amplitudes. They were instructed to initiate rapidly rising force impulses *ASAP* after the presentation of the target step and to match the peak force of their response to its amplitude. The trajectories and latencies of these urgent responses were compared to those of responses produced when subjects responded *WR*.

As expected, choice responses initiated *ASAP* to unpredictable targets were significantly less accurate then either *choice* responses initiated *WR* or responses to predictable targets. This inaccuracy had two components. The first was a modest increase in variable error. The second, and more important, was a systematic distortion in scaling of the entire set of responses to the different targets: The range of peak forces was markedly constricted around a modal value near that of the middle-sized target. Responses to the small target were grossly overshot whereas responses to the large target were undershot. This is illustrated by ensemble averages of the trajectories of choice responses obtained in one subject in Fig. 8.2A (left traces). This severe constriction of range is not seen in responses to predictable targets (Fig. 8.2A, left, filled circles) and was abolished when subjects were permitted to respond *when ready* (Fig. 8.2A, Right traces).

---

FIG. 8.2. (facing page). *Influence of Urgency and Information Processing on Scaling and Latency.* A. Average trajectories in the *choice* condition are illustrated for responses to the three targets (dashed lines) in Subject S5. Responses initiated *as soon as possible* are on the left; those initiated *when ready*, on the right. The times and amplitudes of the peak forces of comparable averages in the *simple* condition indicated by filled circles. (N's for the averages are: choice 23–28; simple, 21–25). B. Distributions of peak forces in the *choice* condition for the three target sizes (small, dotted line; middle, solid line; large, dashed line). The middle-target distribution is shaded to facilitate comparisons. Numbers below abscissa indicate the three target values. C. Relationship of accuracy, expressed as percent of variance in the peak force accounted for by variation in target force, to mean reaction time. Horizontal bars show standard errors. Data from another session of the same subject. N's for four points: 90–104.

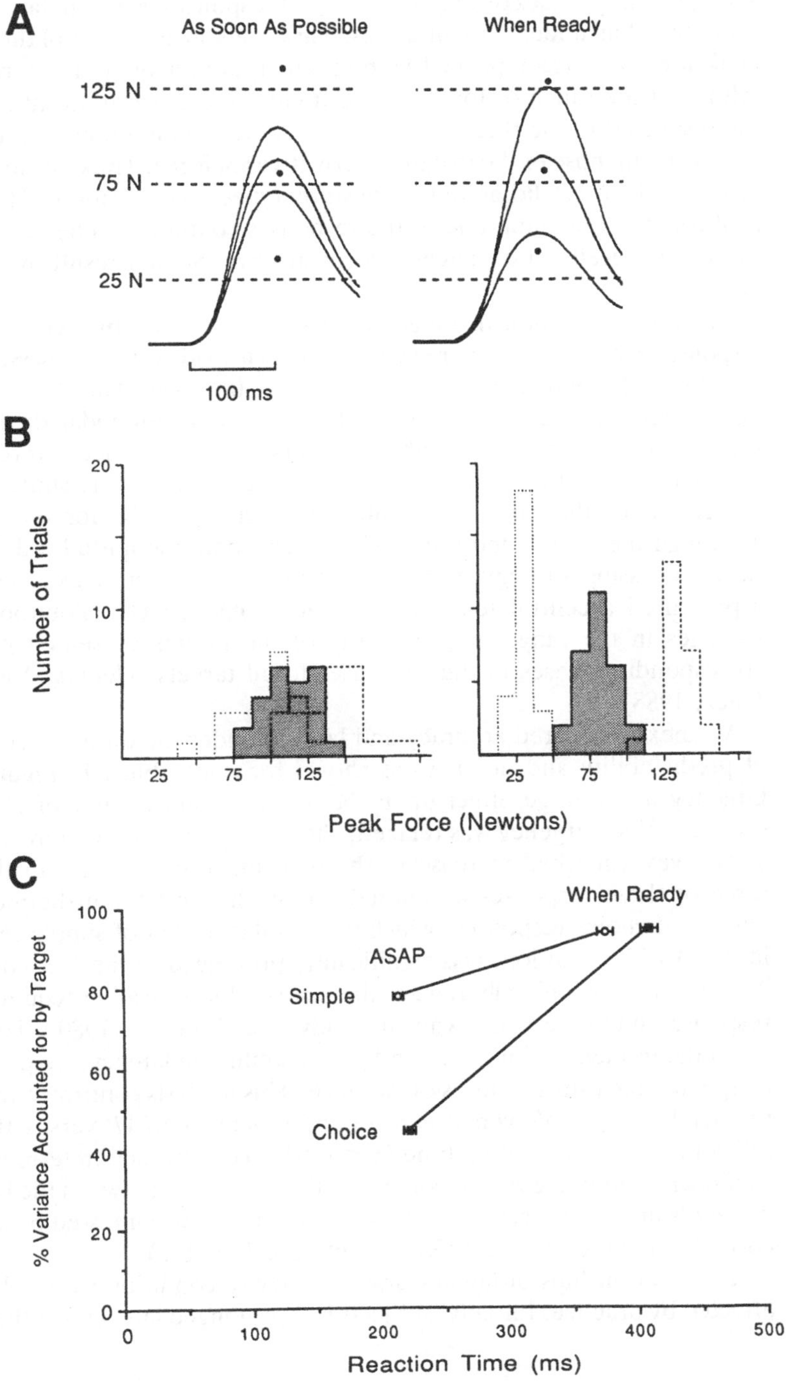

A
As Soon As Possible
When Ready
125 N
75 N
25 N
100 ms
B
Number of Trials
20
10
25
75
125
25
75
125
Peak Force (Newtons)
C
% Variance Accounted for by Target
When Ready
ASAP
Simple
Choice
100
80
60
40
20
0
0
100
200
300
400
500
Reaction Time (ms)

Then, the amplitudes of *simple* and *choice* responses were similar. It should be noted that under all conditions, variations in mean amplitude of the peak force are accomplished without changes in average force rise time. Although the range of forces is constricted, there is significant scaling of the responses to the three targets in the choice condition both in the force traces and in those of the two time derivatives of force. Thus, the instruction to respond *ASAP* did not alter the stereotyped configuration of trajectory profiles, and the amplitude of the responses continued to be governed by the same overall pulse–height control strategy. Similar results were found in all subjects.

Such a constriction in range could have also been produced if, when response amplitude was unpredictable, subjects made random guesses among the three alternative targets. This was not the case: The distribution of peak forces to individual targets always showed unimodal distributions rather than the trimodal distribution expected if the subjects were making random guesses. In the *choice* condition, all the modes were shifted toward the center of the range, as is illustrated in Fig. 8.2B for one subject. Although the constricted range of *choice* response amplitudes did not reflect a guessing strategy, it did reflect the subjects' prior experience and represented a central tendency bias or "range effect" (Poulton, 1981): Changes in the range or probability of the targets presented produced corresponding biases in the responses to all targets (Hening, Vicario, & Ghez, 1988).

We next compared accuracy and latency under the different conditions of predictability and urgency, as shown for one subject in Figure 8.2C. Urgency had a large effect upon the accuracy and latency of choice responses. When urgency was relaxed, choice responses were more accurate: There was a marked increase in the percentage of variance in the peak force of these responses accounted for by the target. Furthermore, the latency of choice responses, which was similar to that of simple responses in the *ASAP* condition, was significantly prolonged in the *WR* condition. This increase undoubtedly reflects the increased specification required when response amplitude is not known in advance (Welford, 1980). To understand the influence of urgency and predictability on latency, we performed a repeated measures analysis of variance. This analysis confirmed that both predictability (*simple* versus *choice*) and urgency (*ASAP* versus *WR*) significantly affected reaction time ($p < 0.02$). The interaction term was also significant, but in the *ASAP* condition, the difference between the latencies of *simple* and *choice* responses was not significant when tested by a simple corrected main effects test (Kirk, 1968, pp. 179–182).

The relationships of latency and accuracy to condition were selectively affected by practice. Initially, subjects had prolonged latencies in the choice condition even when required to respond urgently. However, with a few days of practice (200–400 trials per day), the difference between latencies in the simple and choice conditions was either substantially reduced or abolished. In contrast, the central tendency bias of choice responses produced *ASAP* was seen in all subjects and did not decrease even after extensive practice.

*Conclusions.* The results of this experiment indicate that, in our task, the inaccuracy of urgently produced responses to unpredictable targets results neither from alterations in the normal pulse–height strategy nor from random guesses among alternative possibilities. Rather, our finding that the trajectories of *choice* responses produced at short latencies (*ASAP*) are less well scaled than ones produced at longer latencies (*WR*) suggests that subjects can initiate a response before completely specifying its amplitude in accord with the target. The distinctive error seen with short latency responses, namely, the constriction of range of response amplitudes around the middle target, further suggests that these incompletely scaled responses might derive from a common default, that the subject prepares prior to presentation of each target.

### Gradual Specification of Response Amplitude.

In order to determine the time course of amplitude specification, we used the *time response paradigm* described earlier (Fig. 8.1B). Subjects synchronize their responses with a tone occurring at a predictable time, while the shift in the visual target signals only the required amplitude. Thus the paradigm dissociates the stimuli responsible for triggering response execution from those responsible for indicating the desired response. We also used this paradigm to determine which aspects of the response are prepared prior to target presentation and to determine how trajectory corrections contribute to increasing accuracy in the course of specification.

After one or two sessions of practice, subjects were able to synchronize their responses with the predictable tone. When the target amplitudes were predictable, response amplitude and variability were independent of S–R interval. Therefore, at the shortest S–R intervals, subjects must have prepared the amplitude of the response in advance of target presentation. When targets were unpredictable, however, response scaling depended critically on the S–R interval. Figs. 8.3 and 8.4 illustrate this dependency using data obtained from subject S2 in three consecutive sessions. Fig. 8.3A shows the amplitudes of individual responses to the three targets plotted against each response's S–R interval, together with lines fitted by LOWESS. Responses to each target are shown separately. When the S–

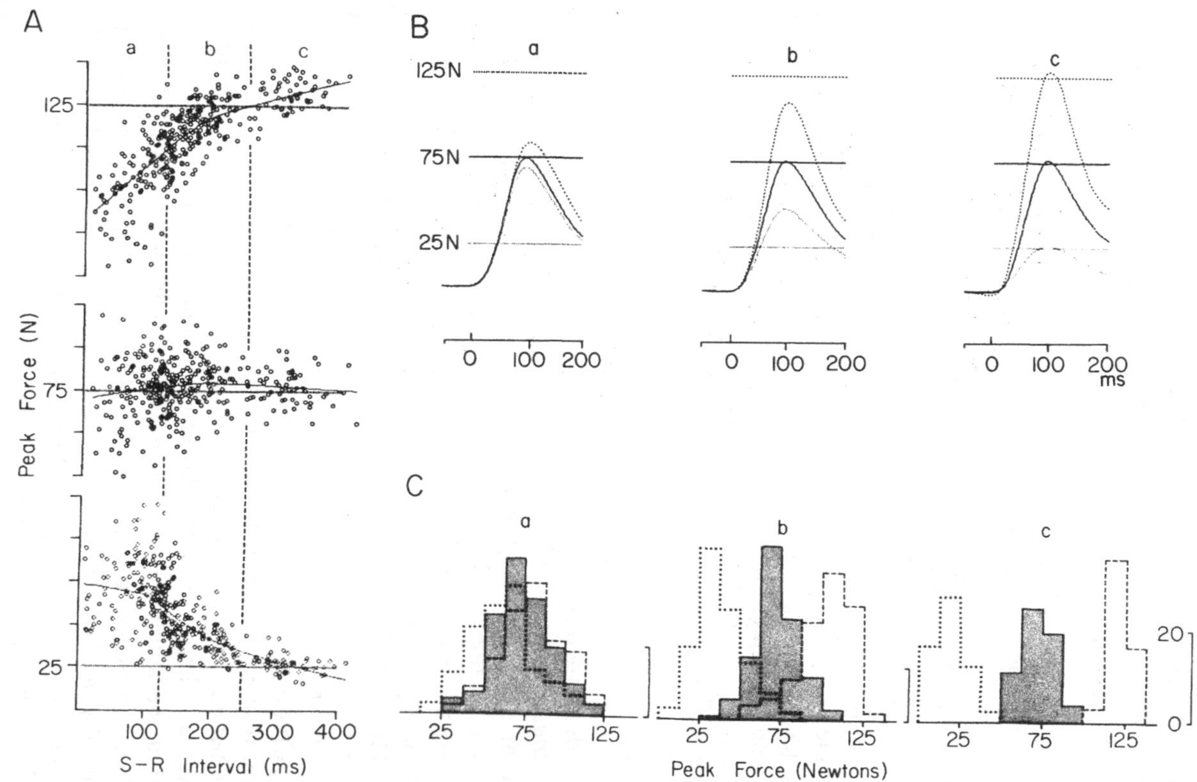
A
a
b
c
125
75
25
Peak Force (N)
100 200 300 400
S-R Interval (ms)
B
a
b
c
125N
75N
25N
0 100 200
0 100 200
0 100 200
ms
C
a
b
c
25 75 125
25 75 125
25 75 125
Peak Force (Newtons)
20
0

R interval is too short for the target to influence response amplitude (region *a*, <125ms), response amplitudes are clustered around a modal value that is the same for all three targets and near that of the middle-sized target. Ensemble averages of the trajectories made to the three targets are indistinguishable (*a*, in Fig. 8.3B). The three sets of trajectories begin to separate at S–R intervals of about 125 ms and their amplitudes gradually approach those of the targets (horizontal lines in Fig. 8.3A). Averaged force trajectories initiated between 125 and 250 ms show some degree of scaling to the targets (*b* in Fig. 8.3B) while still showing a marked constriction of range. Responses appear fully specified and show accurate scaling at S–R intervals greater than 250 ms (region *c* in Fig. 8.3A and B).

Although the amplitude distributions of responses initiated at the shortest S–R intervals showed somewhat more dispersion than those of fully specified responses, these distributions were always unimodal with a mean near the midpoint of the range (Fig. 8.3C, part *a*). With increasing S–R intervals, the means of the distributions gradually approach the targets and the dispersion of response amplitudes decreases. As in the reaction time paradigm, where we examined *choice* responses produced *ASAP*, we never observed a trimodal distribution. That distribution would have been expected had subjects, in a state of uncertainty, randomly guessed among the three responses. Based on this data, we conclude that, prior to target presentation, subjects prepared a single default amplitude near that of the middle target.

The distributions of peak forces at all S–R intervals also indicate that the specification of response amplitude is a gradual process. Specification begins with a default value and progresses until a fully specified state is achieved (see Fig. 8.3A and C). There are no sudden increases in specification indicative of an abrupt shift from default to fully specified responses. Even if the timing of the shift from default to fully specified

---

FIG. 8.3. (facing page). *Specification of Amplitude: Responses to Unidirectional Targets in the Choice Condition.* A. Single trial peak forces of responses to the small (below), medium (middle), and large (above) targets plotted against S-R interval. Distributions have been fit by LOWESS lines (fine solid lines, see Methods) while target amplitudes are indicated by dotted horizontal lines. N's for responses to 3 targets 364, 369, 364. All parts of this figure are from three consecutive sessions of subject S2. B. Average trajectories of responses to the three targets made within S–R intervals: less than 125 ms (a), 125 to 250 ms (b); more than 250 ms (c). Targets in same line type as averaged responses: small target, small dots; middle target, solid line; large target, large dots. All responses synchronized on force onset (time = 0 ms). N's of individual trajectory averages, 52–175. C. Distribution of peak forces of responses to the three targets in the three S–R intervals a–c. Histogram of the amplitudes of responses to small target is dotted; middle target, solid; and large target, dashed. Histogram for middle target is shaded. Target amplitudes in Newtons indicated on abscissa. Calibration: number of trials = 20.

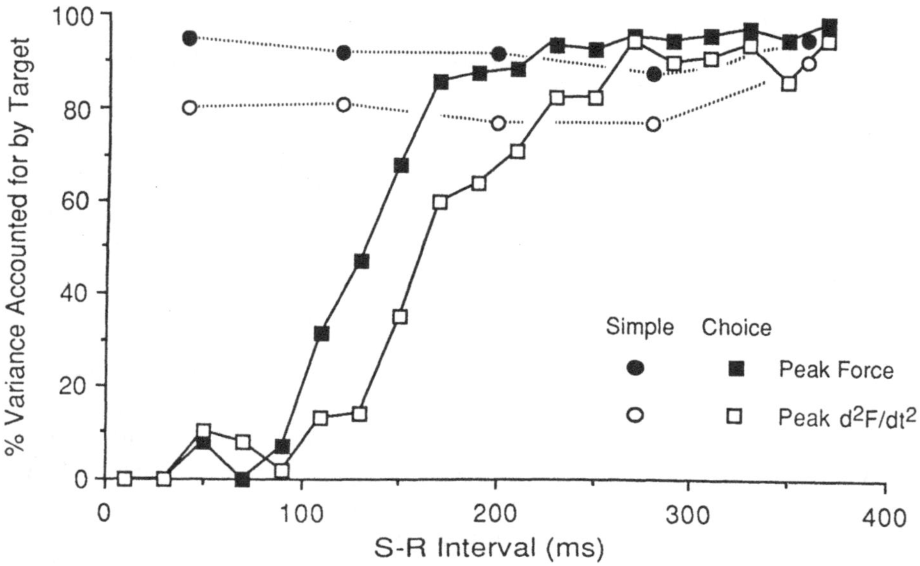


FIG. 8.4. *Time Course of Amplitude Specification of Simple and Choice responses in Unidirectional Timed Response Paradigm.* All trials are grouped by latency into S–R intervals. Percentage of variance accounted for by target force is plotted for each interval against the interval midpoint. Intervals of 80 ms in *simple* condition (circles), and 20 ms in *choice* condition (squares). Variance accounted for in peak force is shown by closed symbols and in $d^2F/dt^2$, by open symbols. Total N's: *simple*, 406; *choice*, 1097. N's for individual intervals: *simple*, 34–151; *choice* (in range of 80 to 300 ms), 54–158. Same subject and *choice* data as Figure 8.3.

responses varied from trial-to-trial, a bimodal distribution of peak forces should be evident during the course of specification, one mode for default and the other for fully specified responses. Such bimodal distributions were never seen (See Fig. 8.3C, part *b*).

The degree of specification can be expressed quantitatively as the percentage of variance in response amplitude that can be accounted for by variance in the targets. To assess the rate of specification, responses were grouped into successive 20 ms intervals. We then calculated the percent of variance in peak force and in peak $d^2F/dt^2$ accounted for by target amplitude within each of the intervals. Figure 8.4 illustrates our findings in the same subject and shows that the accurate specification of peak force occurs gradually and progressively. This result does not tell us, however, how subjects achieve greater specification. For example, they might have modified responses that shared a common initial trajectory (Cordo, 1987). Progressive specification, then, would result from continuing corrective modifications of the initial trajectory. One way to implement such a strategy would be for subjects to control response amplitude by adjustments to force rise time. Our subjects, however, did not use such a strategy. The gradual specification of peak force reflected a progressive improvement in the scaling to the targets of such initial trajectory parameters as peak $d^2F/dt^2$. The open squares in Fig. 8.4 show that when the S–R interval increases beyond 120 ms, progressively more variance in the peak $d^2F/dt^2$ of individual responses is accounted for by the targets. However, the data in this and all other subjects showed that specification of this early feature of the responses (the peak $d^2F/dt^2$ typically occurs 10–50 ms after response onset) was delayed relative to the peak force by about 20 ms. This lag in scaling of the initial phase of the trajectory indicates that subjects were able to use additional target information to produce corrective modifications of the trajectory that were implemented between the time of peak $d^2F/dt^2$ and that of peak force. By using a multiple regression analysis, we were able to determine the respective and independent contributions of initial trajectory parameters and the target itself on the peak force. This analysis was applied to all subjects, and the results confirmed that trajectories were indeed updated to improve the accuracy of responses (Favilla et al., 1989). Other subjects showed the same progressive changes in amplitude specification from a single default value. Similarly, all started specifying at S–R intervals of about 100–150 ms. However, full specification of the responses to the targets was achieved at times varying from 250 to 400 ms in different subjects.

*Conclusions.* Incompletely specified responses in the timed response paradigm share a number of distinctive trajectory characteristics with those produced *ASAP* in the reaction time paradigm. In all subjects, both sets of responses showed a strong central tendency bias and the stereotyped trajectory configuration, indicative of a pulse height control strategy, was preserved in both. Distributions of responses to single targets were always unimodal. It is therefore reasonable to believe that the processes that act to specify response amplitude are the same under the two conditions. Four conclusions can, therefore, be drawn. First, prior to target presentation, subjects prepare an adaptive or default amplitude that is near the center of the range of target amplitudes previously experienced. Second, the specification of response amplitude according to the target is a gradual process that begins earlier and ends later than a minimal RT. Third, specification is achieved both through a progressive improvement in the central program for the trajectory available prior to response initiation and by updates to the trajectory made during its course. Fourth, the mechanisms responsible for response initiation are independent of those that assure the accuracy of responses. As a result, the process of specification can be interrupted when task conditions impose temporal constraints on the time of response initiation. In addition, the finding that neither the mean amplitude nor the dispersion of simple responses showed any systematic dependence on S–R interval suggests that they did not depend on the current target for specification. It may therefore be suggested that in the reaction time paradigm, as well, the amplitudes of responses to predictable targets could be prepared in advance of the triggering stimulus.

### Parallel Specification of Response Direction and Amplitude

The timed response paradigm makes it possible to determine whether the specification of different response features follows an obligatory sequential order or whether different features can be specified concurrently, in parallel. We have addressed this question by presenting subjects with targets requiring responses whose amplitude and direction are both unpredictable. We reasoned that, if the specification of response amplitude to the target required the prior specification of response direction, amplitude specification would be delayed until direction specification were complete. On the other hand, if amplitude specification could proceed without direction specification, its time course would be little affected by the concurrent need to specify direction. Moreover, in individual responses, the degree of specification of the two features should be independent. Thus, if specification of amplitude and direction occur in parallel, right and wrong direction responses should show the same degree of specification of amplitude at a given S–R interval.

To address this question, we examined the time course over which amplitude and direction are specified in the six normal subjects. We used the timed response paradigm described in the preceding section. Now, however, upward deflections of the target line on the oscilloscope required isometric flexions, whereas downward deflections required extensions. Extension and flexion target forces had the same three amplitudes. These six target steps were again presented either in unpredictable (*choice* condition) or predictable (*simple* condition) order.

As in the unidirectional condition described earlier, responses to predictable targets were well scaled at all S–R intervals. Only a few errors in direction occurred (<2%). However, when the targets were unpredictable, both the correct choice of direction and the scaling of responses to their targets were dependent on the S–R interval. At S–R intervals of less than 100 ms, that is, before the target just presented could influence the response, about half the responses were in the wrong direction. The amplitudes of these responses had a bimodal distribution with one mode close to the middle target in each direction. Thereafter, the proportion of directional errors decreased progressively to zero and the amplitudes of the responses gradually came to approximate those of their targets. Interestingly, the progressive specification of amplitude occurred similarly among right and wrong direction responses. These findings are illustrated using the data obtained in one subject in Fig. 8.5. In these figures, right and wrong direction responses have been separated while flexion and extension responses have been grouped. (We considered this grouping acceptable because the shape of the trajectories and the duration of the force rise time were quite similar in flexion and extension.) At short S–R intervals ($< 100$ ms) response amplitudes were clustered around the same mean value. Fig. 8.5A, (part *a*) shows that the average trajectories of responses to all three target amplitudes are superimposed whereas right (directed upward) and wrong (directed downward) direction responses are mirror images of each other. Starting shortly after 100 ms, the amplitudes of the responses start to diverge (Fig. 8.5A, part *b*) and approximate those of their respective targets at S–R intervals of 200 to 300 ms (Fig. 8.5A, part *c*). After gradually decreasing in frequency, directional errors are no longer made at S–R intervals greater than 300 ms. The time course of amplitude specification for right and wrong direction responses is presented more quantitatively in Fig. 8.5C which has the same format as Fig. 8.4. It can be seen that in this subject right and wrong direction responses become specified at a similar rate. One seemingly paradoxical result of the amplitude specification occurring in wrong direction responses is that the mean absolute error of responses to the large amplitude target (Fig. 8.5A, part *b*) actually becomes greater as the S–R interval becomes longer. Although all subjects showed progressive specification of wrong direction responses, the rate and degree of specification in wrong direction responses was always somewhat less than in responses made in the right direction.

*Conclusions.* Three conclusions can be drawn from these results. First, at latencies too brief to be influenced by the most recent target, subjects produce flexion and extension responses with a common default amplitude. Second, gradual specification of amplitude is not altered by the need to select response direction. Third, the amplitude and direction of motor responses can be specified in parallel. Unfortunately, in our paradigm, the reduction of direction to a discrete, binary choice between flexion and extension, which does not allow any intermediate values, prevents us from determining whether direction can also be gradually specified from a common default. Our results do, however, suggest that correct specification of direction occurs with approximately the same time course as accurate specification of amplitude.

These results raise two issues that we address in the next sections. First, the symmetric targets used in the preceding experiment do not allow us to state whether the value of the default amplitude was prepared independent of its direction. Second, our finding that the specification of response

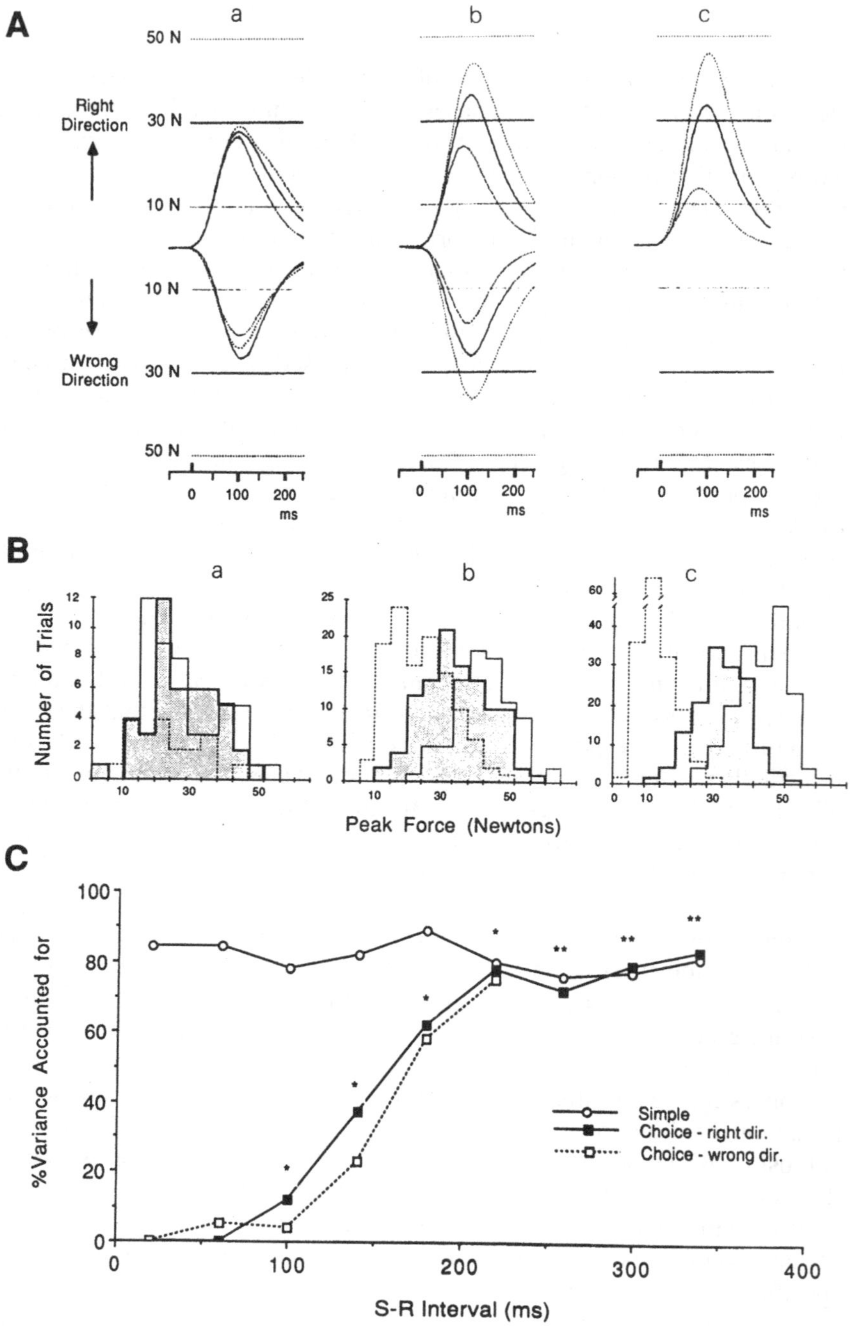


FIG. 8.5. *Specification of Amplitude and Direction: Responses to Bidirectional Targets.* A. Average trajectories in the *choice* condition. Responses to the three targets grouped by S–R intervals: less than 100 ms (a); 100

amplitude and direction occurs in parallel does not preclude these processes interacting in some way.

### Parallel Preparation of Amplitude and Direction to Unpredictable Targets

As mentioned earlier, the preceding results do not establish how the amplitudes of the flexion and extension defaults are assigned. Two possibilities are evident. On the one hand, each default amplitude may be determined only by the targets in the same direction. In this case, the default amplitude of flexion and extension responses would be computed separately. On the other hand, amplitude could be prepared independent of direction, so that there is only one default amplitude common to both directions and based on target amplitudes in both directions. The following experiment was designed to distinguish between these possibilities. We modified the timed response paradigm to selectively bias the probability of the targets in only one of the two directions, and we focused our attention on responses produced before substantial specification to the target was achieved. We reasoned that, if the default amplitude were set independently of direction, biases experienced in a single direction would be applied to responses made in both directions. If a different default amplitude were prepared for each direction, biased responses should be confined to the direction in which the probability bias was present.

In a modification of our timed response paradigm, the target step was presented at a constant time interval before the last tone and had one of

---

to 200 ms (b); more than 200 ms (c). Absolute values of trajectory have been averaged so that flexion and extension responses are combined. Correct direction responses are plotted upward and wrong direction, downward. Responses synchronized on response onset (time = 0 ms). Absolute target amplitudes are given by solid horizontal lines. N's for averages: right direction, 6–170; wrong direction, 10–27. B. Histograms of distributions of absolute values of peak forces of responses to three targets. Same time intervals (a–c) as A. Small target, dotted line; middle target, solid line; large target, dashed line.The histogram for responses to the middle target is shaded. N's for histograms: a, 50, 46, 20; b, 84, 106, 116; c, 173, 145, 161. C. Accuracy of specification of amplitude. Accuracy of amplitude is plotted as per cent of variance in peak force accounted for by target force for responses grouped in S–R intervals of 40 ms. Flexion and extension responses have been combined and absolute values of peak force and target force entered into correlation. *Simple* responses, open circles (N = 313); correct direction *choice* responses, filled box (N = 776); wrong direction *choice* responses, open box (N = 106). N's by S–R interval: *Simple*, 21–49; *Choice* correct direction, 8–151; wrong direction, 8–25. Single asterisk indicates that the probability of the response being in the correct direction was above chance levels; the double asterisk indicates that no more than one response was made in the wrong direction.

two possible amplitudes in either direction. This interval was chosen so that the mean S–R interval would be about one standard deviation shorter than the subjects' mean reaction time in a comparable reaction time task with *ASAP* instructions. This was done to elicit responses whose features were minimally specified to the target. In order to bias the default amplitude in one of the two directions, either the small or the large target in that direction was presented 4 times as often as the other (40% versus 10% of all trials). The two targets in the other direction, although also unpredictable, were equiprobable (each amplitude was presented on 25% of the trials). Responses produced under the different biases were compared with each other, with responses to unpredictable but equiprobable targets, and with responses to equiprobable and predictable targets.

Our findings are typified by the results shown in Fig. 8.6 for one subject and one direction of bias. In Fig. 8.6.A, the targets were predictable and the responses are all well scaled. In Fig. 8.6B, the targets were unpredictable and the responses are distributed around the means of the two target ranges. In Fig. 8.C, and D, the probabilities of the flexion targets were biased, whereas the probabilities of the extension targets were not. On top, the large flexion target was the most probable; on bottom, the small flexion target. When the large target is the most probable (C), all responses made in both directions are larger than those of the control with equiprobable and unpredictable targets (B). All responses are smaller when the small target is the most probable (D). Similar results were obtained in all subjects for both directions of bias (flexion and extension targets biased).

*Conclusions.* This last set of experiments shows first, that subjects prepare a default amplitude according to the probability of the targets previously experienced. The default, therefore reflects the subjects' expectation. Second, to a large degree, subjects apply the default amplitude without regard to response direction. This implies that direction and amplitude are treated as distinct features of motor set in the preparation of these responses just as they are specified independently after target presentation.

## Interactions Between Channels Specifying Direction and Amplitude

In order to assess the possibility that the processing of one response dimension interfered with processing of the other, we compared the time course of amplitude specification under conditions of unidirectional and bidirectional uncertainty. In one case targets varied in amplitude, but were all in the same direction; in the other, both amplitude and direction were unpredictable. The results of this comparison in one subject are presented in Fig. 8.7. It can be seen that both the rate of specification and the degree

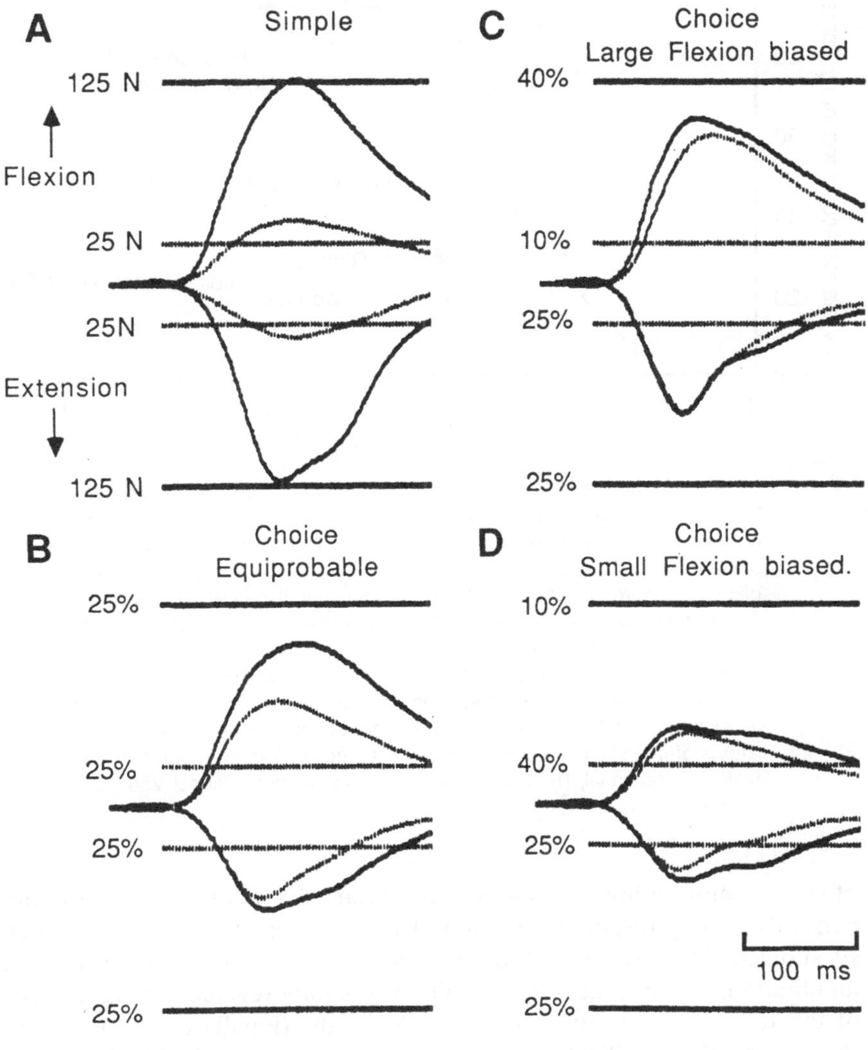


FIG. 8.6. *Independent Specification of Default Amplitude: Biased Probability Experiment.* Controls (left) and Flexion Biases (right). In each part of the figure there are four traces for averages of correct direction responses to large (solid line) and small (hatched line) targets. Target amplitudes are indicated in A and probability of each target, in three *choice* conditions (B–D), to left of target lines (line type same as averages). All averages synchronized on force onset. N's for averages: simple, 10–14; unbiased choice, 17–40; large flexion bias, 8–32; small flexion bias, 7–41.

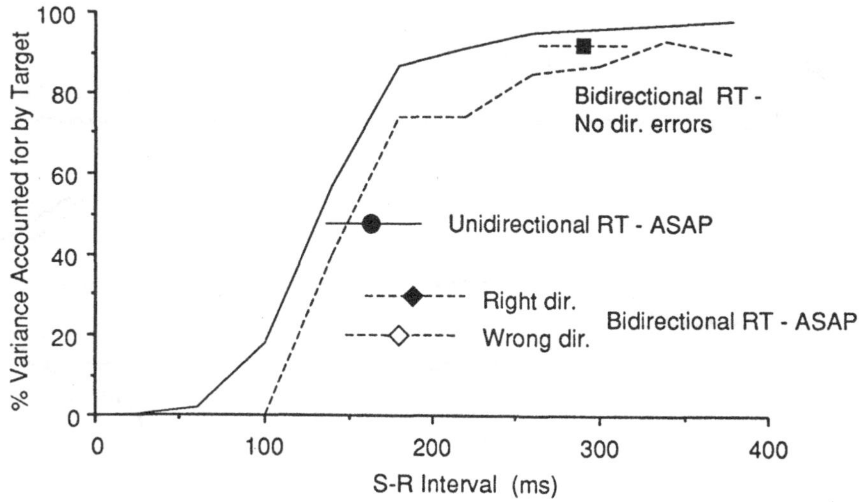


FIG. 8.7. *Interactions between Response Initiation and Response Specification.* Accuracy, measured as percent of variance in peak force accounted for by target force, is plotted against response latency for responses in *choice* condition. Lines are derived from timed response paradigm: solid line, unidirectional; dashed line, bidirectional. (Lines were created by joining values for 40 ms intervals plotted at their midpoints.) Individual points are from reaction time paradigm: unidirectional targets, *ASAP* condition, circle; bidirectional targets, *ASAP* condition: correct direction, filled diamond, and wrong direction, open diamond; and bidirectional targets, after direction specified to target, square. Horizontal bars indicate standard deviations of latencies. N's for: intervals of timed response—unidirectional, 18-281, bidirectional, 18–138; for RT points; *ASAP* unidirectional, 256; right direction, 82; wrong direction, 91; direction specified, 174.

of specification ultimately achieved are reduced when subjects must concurrently specify response direction. This same result was observed in all six subjects. However, in half of the subjects, this reduced specification of amplitude in the bidirectional condition was only revealed by an analysis of the rate of specification of the peak $d^2F/dt^2$ (Favilla et al., 1989). In these cases, the specification of the peak force was more highly dependent on compensatory adjustments made to the trajectories in the time interval between the peak $d^2F/dt^2$ and the peak force (Gordon & Ghez, 1987b).

*Conclusion.* Our results indicate that the need to specify direction prolongs the process of amplitude specification. This interaction occurs even though neither process is dependent on completion of the other. Therefore, although amplitude and direction specification are parallel processes, they are not fully independent.

## Interactions Between Response Triggering and Specification

Our next step was to determine if the need to use the same target information to trigger response initiation delayed specification of response features. In order to address this question, we compared the latencies and trajectories of isometric force impulses aimed to unpredictable visual targets of six possible sizes, three in flexion and three in extension. We required subjects to respond ASAP and urged them not to correct directional errors. We then compared amplitude specification under these bidirectional conditions to the unidirectional condition where responses were only to be made in one direction. The data plotted in Fig. 8.7 were characteristic of all subjects. When subjects were instructed not to correct directional errors, reaction times were similar for unidirectional and bidirectional conditions and many responses (30%–53%) were indeed in the wrong direction. There was a marked constriction in the range of response amplitudes for responses in both the right and wrong directions. Reaction times in the bidirectional condition (diamonds) are similar to those in the unidirectional condition (filled circle). Amplitude specification of both right and wrong direction responses is lower in the bidirectional condition. Moreover, the degree of specification achieved was always lower in the reaction time paradigm, when subjects had to use information from the visual stimulus to trigger as well as to specify response features. In Fig. 8.7, it can be seen that the degree of amplitude specification by the subject is lower in the unidirectional reaction time condition (filled circle) than that achieved at the same S–R interval in the timed response condition (solid line). Similarly, right and wrong direction reaction time responses were also less well specified than the corresponding responses in the bidirectional timed response paradigm.

In order to determine how changes in the subjects' response criteria might influence the latency of their responses and the degree of specification achieved, we studied a second bidirectional task. In this task subjects were also to produce an uncorrected impulse of force matching the target. However, subjects were only to respond once they could correctly match the direction of the target. When subjects avoided making directional errors, their reaction times were prolonged. In addition, their amplitude specification also improved such that the constriction in the range of response amplitudes was virtually eliminated. This is shown as the filled square for the subject illustrated in Fig. 8.7.

*Conclusions.* The finding that the amplitudes of responses made in the wrong direction show significant scaling under reaction time conditions confirms our hypothesis that response direction and amplitude can be specified in parallel. However, as we noted earlier in the timed response paradigm with equiprobable targets, the decrease in rate of specification when the two features are specified concurrently indicates that these two processes interact. Furthermore, the need to use information from the target stimulus to initiate responses delays amplitude scaling, indicating that the processes subserving response initiation and the specification of response features also interact.

## DISCUSSION

In this report, we have used two tasks, the timed response paradigm that independently controls the timing of response initiation and a more traditional reaction time task, to examine how target information is processed by the central nervous system to specify the features of a simple motor response. Our results provide some preliminary answers to the questions posed in the introduction. First, the degradation in accuracy that occurs when responses are produced at brief latencies takes the characteristic form of a central tendency bias. Second, response amplitude is specified by an extended process through which a default value, dependent on the range of targets previously experienced by the subject, is gradually converted into accurate amplitudes specific to each target. Third, the specification of response amplitude to the target does not require the prior specification of direction to match that of the target or vice versa. Amplitude specification, however, is prolonged by the concurrent need to specify direction. Fourth, response initiation does not depend on prior specification of response features; however, the requirement to use target information as a cue for both initiation and specification delays specification.

Our observations indicate that two phases must be distinguished when considering the information processing leading to accurately targeted motor responses. The first represents an initial preparatory phase that leads to the setting of what we have termed *default parameters* that characterize responses produced at latencies too brief to be influenced by the most recent target. The second phase is one of progressive specification during which responses converge on the appropriate target as S–R interval increases.

Four features of the response appear to be established prior to the specific stimulus eliciting a response and reflect key features of motor set (Evarts, Shinoda, & Wise, 1984): a trajectory waveform, a consistent transformational rule for scaling trajectory waveforms to a range of specific targets, a default amplitude, and a direction that is selected at random. We infer that the trajectory contour and the scaling function by which it would be matched to the target are set prior to the specific triggering stimulus because responses that were not specified to the target showed the same trajectory contour and pulse–height control as fully specified responses. The selection of the default amplitude was adaptively adjusted to reflect the subjects' prior experience in the task, and was influenced by the target on the immediately preceding trial (Hening, Favilla, & Ghez, 1988) and by the range and probability of the targets presented in the course of a block of trials (Favilla et al., 1986; Hening, Vicario, & Ghez, 1988). These influences suggest that the setting of a default amplitude has the adaptive value of reducing error in the face of uncertainty. The central tendency bias of responses produced at brief latencies in the reaction time task also reflected the default amplitude. Therefore, errors of responses produced at short intervals after target presentation in both our tasks result from incomplete specification of amplitude in relation to the current target. In the present experimental paradigm, where response directions were confined to flexion and extension, direction was randomly assigned.

In the phase of progressive specification, as S–R interval increases, the mean responses to the different targets gradually approach their respective amplitudes and response variability decreases. There were no abrupt or discontinuous changes in trajectories or accuracy, as might be expected if specification depended on the completion of a discrete processing stage. Nor did response distributions indicate the presence of two populations that might represent distinct degrees of specification. This finding indicates that information derived from the target can be continuously provided to the mechanisms setting response amplitude (Miller, 1982). This process begins well before a response can be triggered by the stimulus (i.e., at the RT) and continues for 100 ms or more longer than the RT of urgently produced responses. Similar gradual processes may occur when subjects are required to alter responses because of a second, unexpected target or when they are presented with two potential targets, one the true target and the other supernumerary, that cannot be rapidly discriminated. As well demonstrated in studies of pursuit eye movements and saccades in the oculomotor system, the transition from an initially incorrect or "averaged" response to one matched to the final target is a gradual process similar to the gradual specification we have seen in these isometric limb responses (Becker & Jurgens, 1979; Kowler, Martins, & Pavel, 1984; Ottes, Van Gisbergen, & Eggermont, 1985; Van Asten, Gielen, & Winkel, 1988). Those processes can also be influenced by such contextual information as target probability (He & Kolwer, 1987).

When subjects needed to specify both direction and amplitude, the specification of amplitude did not depend upon completing specification of direction and vice versa. This suggests that the underlying processes for specifying these two features are organized in parallel. Their relative independence is underlined by our finding that wrong direction responses show significant amplitude specification and that amplitude biases developed in one direction are applied to responses made in the unbiased direction. This indicates that parallel treatment of amplitude and direction occurs during both response preparation and specification to the target. These results, however, should not be taken to suggest that serial specification of response features could not occur. Instead, such seriality may represent an optional strategy that subjects may select when confronted with particular performance criteria.

Unfortunately, our data do not yet allow a strict comparison of the rate at which direction and amplitude are specified. Because direction in this task can only be either correct or incorrect on each trial, it cannot provide information about intermediate degrees of specification of individual responses, and no single default value of direction is possible. Despite this difference, however, the time courses of amplitude and direction specification appear to be quite similar. Experiments with different response directions are currently being conducted to determine the conditions under which a single default direction can be established, and to permit a more rigorous comparison of the time course of direction and amplitude specification. It should be emphasized that our results do not identify the physiological nature of sensorimotor processes or the "stage" that might be a rate limiting for specification. Specification may well reflect the operation of physiological processes that do not conveniently fit conventional metaphors by which processes are ordered along a sensory to motor continuum. For example, it is entirely possible that the setting of default parameters could involve the adaptive filtering of sensory input itself.

The decreased rate at which amplitude is specified when additional processing demands are imposed indicates that requirements for concurrent processing reduce the rate and efficiency of amplitude specification. This was true both of the need to specify direction, and of the need to use stimulus information to trigger the response. The interactions we observed did not arise, however, from delays imposed by the need for sequential activation of processing stages. The progressive specification of wrong direction responses shows that the specification of amplitude was not predicated in an obligatory fashion on the prior processing of the direction information. Similarly, the production of incompletely specified responses when urgency is stressed or when a short S–R interval is required indicates that initiation need not be predicated on the completion of specification according to the target. A more likely rate-limiting factor is a neural resource that is shared by the different processes. Attentional mechanisms, which have been widely viewed as having limited capacity (Duncan, 1980; Sperling & Dosher, 1986), could be responsible for the interactions we observed.

The major benefit of the timed response paradigm is that by using a predictable tone to determine the timing of response initiation we can make latency or S–R interval an independent variable under experimental control. Then, by simplifying the required response to a single uncorrected impulse of force, we can use the latency dependent variations in trajectory to assess the processes leading to accurately specified responses. In contrast, when RT alone is used as a measure of processing, the degree of processing is left uncontrolled. Our subjects, when asked both to respond ASAP and to be accurate, responded at a latency at which they achieved an intermediate degree of specification. It appears that they set an internal criterion for response initiation, which may be subject to many uncontrolled factors. The common finding that reaction time is prolonged to a greater degree by directional uncertainty than by amplitude uncertainty can, at least in part, be explained by differences in criteria that subjects spontaneously impose on response initiation. With both direction and amplitude uncertainty, subjects normally delay response initiation to avoid making directional errors. This delay allows the amplitude of responses in this bidirectional condition to be better specified than in the unidirectional condition. Reaction time to bidirectional targets can, however, be reduced by specifically urging subjects to respond ASAP and not to suppress directional errors.

Although determining the "speed–accuracy trade-off" function is a powerful means of assessing the degree of processing achieved in reaction time tasks (Sperling & Dosher, 1986; Meyer, Osman, Irwin & Kounios, 1988), the resulting assessment can be incomplete when the sole measure of response specification is the discrete determination of whether responses are correct. Our studies provide an elaboration of that methodology that permits greater control of response latency and a quantitative analysis of multiple trajectory parameters. Although it is unclear how far our conclusions can be extrapolated, the present analysis of how the trajectory of a brief and apparently simple isometric response is specified helps us understand in general how a motor response is accurately aimed to a target. Despite the apparent simplicity of the response we have studied, a careful analysis of its trajectory has revealed a multiplicity of mechanisms by which accuracy is achieved.

## ACKNOWLEDGMENTS

Supported by NIH grant NS 22715-03. Wayne Hening supported in part by the Veteran's Administration Medical Research Service.

## REFERENCES

Baron, S. R., Menich, S. R., & Perone, M. (1983). Reaction times of younger and older men and temporal contingencies of reinforcement. *Journal of Experimental Anatomy & Behav.*, *40*, 275–287.

Becker, W., & Jurgens, R. (1979). An analysis of the saccadic system by means of double step stimuli. *Vision Research, 19*, 967–983.

Bonnet, M., Requin, J., & Stelmach, G. E. (1982). Specification of direction and extent in motor programming. *Bulletin Psychonomic Science, 19*, 31–34.

Cleveland, W. S. (1979). Robust locally weighted regression and smoothing scatterplots. *Journal of American Statistical Association, 74*, 829–836.

Cordo, P. F. (1987). Mechanisms controlling accurate changes in elbow torque in humans. *Journal of Neuroscience, 7*, 432–442.

Donders, F. C. (1869). On the speed of mental processes. Translated and reprinted in *Acta Psychologica, 30*, 412–431.

Duncan, J. (1980). The demonstration of capacity limitation. *Cognitive Psychology, 12*, 75–96.

Evarts, E. V., Shinoda, Y., & Wise, S. P. (1984). *Neurophysiological approaches to higher brain functions*. New York: Wiley.

Favilla, M., Hening, W., & Ghez, C. (1989). Trajectory control in targeted force impulses: VI. Independent specification of response amplitude and direction. *Experimental Brain Research, 75*, 280–294.

Favilla, M., Hening, W., Gordon, J., & Ghez, C. (1986). Preparatory set: Independent specification of direction and size in the preparation of responses to a range of targets. *Society of Neuroscience Abstracts, 12*, 972.

Ghez, C., & Gordon, J. (1987). Trajectory control in targeted force impulses: I. Roles of opposing muscles. *Experimental Brain Research, 67*, 225–240.

Goodman, D., & Kelso, J. A. S. (1980). Are movements prepared in parts? Not under compatible (naturalized) conditions. *Journal of Experimental Psychology (General), 109*, 475–498.

Gordon, J., & Ghez, C. (1987a). Trajectory control in targeted force impulses: II. Pulse height control. *Experimental Brain Research, 67*, 241–252.

Gordon, J., & Ghez, C. (1987b). Trajectory control in targeted force impulses: III. Compensatory adjustments for initial errors. *Experimental Brain Research, 67*, 253–269.

Gottsdanker, R., & Schragg, G. T. (1985). Verification of Donders subtraction method. *Journal of Experimental Psychology: Human Perception Performance, 6*, 765–776.

He, P., & Kowler, (1987). Motor preparation as a determinant of the accuracy of short-latency saccades. *Investigative Ophthalmology and Visual Science Supplement (ARVO abstract), 28*, 332.

Hening, W., Favilla, M., & Ghez, C. (1988). Trajectory control in targeted force impulses: V. Gradual specification of response amplitude. *Experimental Brain Research, 71*, 116–128.

Hening, W., Vicario, D., & Ghez, C. (1988). Trajectory control in targeted force impulses: IV. Influences of choice, prior experience and urgency. *Experimental Brain Research, 71*, 103–115.

Hicks, W. E. (1952). On the rate of gain of information. *Quarterly Journal of Experimental Psychology, 4*, 11–26.

Howell, W. C., & Kreidler, D. L. (1963). Information processing under contradictory instructional sets. *Journal of Experimental Psychology, 65*, 39–46.

Kirk, R. E. (1968). *Experimental design: Procedures for the behavioral sciences*. Belmont, CA: Brooks/Cole.

Kowler, E., Martins, A. J., & Pavel, M. (1984). The effect of expectation on slow oculomotor control-IV. Anticipatory smooth eye movements depend on prior target motions. *Vision Research, 19*, 619–632.

Luce, R. D. (1986). *Response times. Their role in inferring elementary mental organization*. New York: Oxford University Press.

Meyer, D. E., Abrams, R. A., L. Kornblum, S., Wright, C. E., & Smith J. E. K. (1988). Optimality in human motor performance: ideal control of rapid aimed movements. *Psychological Review, 95*, 340–370.

Meyer, D. E., Osman, A. M., Irwin, D. E., & Kounios, R. (1988). The dynamics of cognition and action: Mental processes inferred from speed-accuracy decomposition. *Psychological Review, 95*, 183–237.

Meyer, D. E., Osman, A. M., Irwin, D. E., & Yantis, S. J. (1988). Modern mental chronometry. *Biological Psychology, 26*, 3–67.

Miller, J. (1982). Discrete versus continuous stage models of human information processing: In search of partial output. *Journal of Experimental Psychology, (Human Perception Performance), 8*, 273–296.

Ottes, F. P., Van Gisbergen, J. A. M., & Eggermont, J. J. (1985). Latency dependency of colour-based target vs nontarget discrimination by the saccadic system. *Vision Research, 25*, 849–862.

Pachella, R. G., Fischer, D. F., & Karsh, R. (1969). Absolute judgments in speeded tasks: Quantification of the tradeoff between speed and accuracy. *Psychonomic Science, 12*, 225–226.

Pachella, R. G. (1974). The interpretation of reaction time in information processing research. In B. Kantowitz, B. (Ed.), *Human Information Processing* (pp. 41–82). Hillsdale, NJ: Lawrence Erlbaum Associates.

Pew, R. W. (1969). The speed–accuracy operating characteristic. *Acta Psychologica, 30*, 16–26.

Poulton, E. C. (1981). Human manual control. In V. B. Brooks (Ed.), *Handbook of Physiology, Sec 1: The nervous system, Vol. 2: Motor Control* (pp. 1337–1389). Bethesda: American Physiological Society.

Rosenbaum, D. A. (1980). Human movement initiation: Specification of arm, distance, and extent. *Journal of Experimental Psychology (General), 109*, 444–474.

Sanders, A. F. (1980). Stage analysis of reaction processes. In *Tutorials in motor behavior*. G. E. Stelmach & J. Requin (Eds.), North-Holland, Amsterdam, pp. 331–354.

Schouten, J. F., & Bekker, J. A. M. (1967). Reaction time and accuracy. *Acta Psychologica, 27*, 143–153.

Sperling, G., & Dosher, B. A. (1986). Strategy and optimization in human information processing. In K. Boff, L. Kaufman, & J. Thomas, (Eds.), *Handbook of Perception and Performance*, Vol I. New York: Wiley.

Sternberg, S. (1969). The discovery of processing stages: Extensions of Donder's method. *Acta Psycholgica, 30*, 276–315.

Swensson, R. G. (1972). The elusive tradeoff: Speed versus accuracy in visual discrimination tasks. *Perception Psychodynamics, 12*, 16–32.

Taylor, D. A. (1976). Stage analysis of reaction time. *Psychology Bulletin, 83*, 161–191.

van Asten, W. N. J. C., Gielen, C. C. A. M., & Winkel, M. E. M. (1988). The effect of isoluminant and isochromatic stimuli on latency and amplitude of saccades. *Vision Research, 28*, 827–840.

Welford, A. T. (1952). The "psychological refractory period" and the timing of high speed performance. *British Journal of Psychology, 43*, 2–19.

Welford, A. T. (1980). *Reaction times*. New York, Academic Press.

Wickelgren, W. A. (1977). Speed–accuracy tradeoff and information processing dynamics. *Acta Psychologica, 41*, 67–85.

Zelasnik, H. N. (1981). The effects of force and direction uncertainty on choice reaction time in an isometric force production task. *Journal of Motor Behavior, 13*, 18–32.

# 9 Generalized Motor Programs: Reexamining Claims of Effector Independence in Writing

Charles E. Wright
*Columbia University*

## ABSTRACT

This chapter reports detailed analyses of writing data designed to provide a clearer idea of the generality and limits of effector independence in writing. Confirming previous observations of overall shape similarity across effectors, these analyses also reveal systematic changes in details of shape, kinematics, stroke decomposition, and fluency occurring with changes in effector. These results serve as the basis for reconsidering the implications of empirical effector independence in writing for the hypothesis that motor programs can be generalized to operate across effectors. Although strong assertions are premature, the results reported are consistent with the conclusion that writing by the dominant and the nondominant hands is carried out using different strategies that share only the very highest and most abstract spatial representations of what the shape of the writing should be. By contrast, writing with the dominant hand and the dominant arm appears to be controlled by mechanisms that share a common representation much lower in the hierarchy of increasing motor specificity particularly for highly overlearned productions such as one's name.

## INTRODUCTION

### Background

In the last 20 years, the nature of cognitive representations used to plan and carry out movements and movement sequences has been a major focus of research on movement control (e.g., Gentner, 1983; Keele, 1981; Kelso, 1977; Meyer, Smith, & Wright, 1982; Meyer, Abrahms, Kornblum, Wright, & Smith, 1988; Rosenbaum, 1980; Rumelhart & Norman, 1982; Saltzman & Kelso, 1987; Schmidt, 1975; Schmidt, Zelaznik, Hawkins, Frank, & Quinn, 1979; Sternberg, Monsell, Knoll, & Wright 1978). The concept of the *motor program* has played a key theoretical role in this work. Although its role has remained central, the concept itself has hardly remained static throughout this period.

In their early usage of the term, authors such as Henry and Rogers (1960), Keele (1968), and Adams (1971), for example, conceived of a motor program as a movement-specific, temporally structured collection of parameter values—that is, force, velocity, duration, muscle sequencing—that, when transmitted directly to the motor system, would initiate and carry through a specific action. This conception was proposed largely to explain the development of coordination, as skill in complex movements increases. In less skilled performance we usually observe corrective processes operating: The performer makes part of the overall movement (hereafter, a *submovement*), evaluates the result, makes another submovement, reevaluates, and so on. The result is jerky, temporally unstructured, and depends heavily on concurrent visual feedback (Pew, 1966). In skilled movements, the submovements are coordinated with one another and seem to anticipate or coincide with the events in the environment. Because the time between submovements, in a skilled movement, is often less than estimates of the time it takes to process feedback and initiate new commands, a reasonable inference is that the submovements making up the full movement have been stored in a memory structure so that they can be sequenced and executed without stimulus–response chaining or extensive feedback processing. (See Keele & Summers, 1976, for an excellent review of the evidence for and against this inference.)

More recently, a motor program has come to be thought of as a collection of generalized instructions representing not one, but a class of related movements. Before being transmitted to the muscles, these generalized motor programs must undergo a transformation: Variables contained in the generalized instructions, which allow delayed specification of at least some movement parameters, are replaced by specific values appropriate for the particular movement to be made. Bernstein (1967; but more usefully see Whiting, 1984) is probably the most influential early contributor to these ideas. The article of Schmidt (1975) had a major contemporary role in popularizing this conception among psychologists, which has been adopted and elaborated by others (e.g., Raibert, 1977; Schmidt, et al., 1979; Rosenbaum, 1980; Viviani & Terzuolo, 1980; Rosenbaum & Kornblum, 1982; Meyer, et al., 1982; Stelmach, Mullins, & Teulings, 1984; Meyer, et al., 1988).

One strong appeal of the generalized motor program hypothesis is the explanation it provides for how we produce novel variants of previously learned movements. By this hypothesis, the novelty problem is solved simply by changing the value of variables in the program. If we are limited to specific motor programs, however, we must use trial and error to create a new program for each new element in a class of "similar" movements. Empirically, the challenge posed by these alternatives is to establish criteria that can distinguish the direct transfer of learning, made possible by the existence of a generalized motor program, from the presumably slower process of generating new, specific motor programs (while perhaps benefiting from the information contained in the previously existing instances). Theoretically, the challenge of the generalized motor program hypothesis is to create models of the computations necessary for generalization of particular movement parameters that are consistent with other known regularities of motor performance.

## Effector–Independent, Generalized Motor Programs

Within this framework, I focus on the specific claim, made by some as part of the generalized motor program hypothesis, that motor programs can be *effector independent* (e.g., Schmidt, 1975, 1988; Raibert, 1977; Rosenbaum, 1980; Stelmach, et al., 1984). In the broadest interpretation of this claim, an effector-independent, generalized motor program, once established for a particular movement, can be used to control performance of that movement by any of a number of muscle–joint systems (*effectors*) in the body, although perhaps with some loss of precision. This claim evolves from the idea that specification of particular muscles and joints is not a necessary part of a motor program, but rather, as with other aspects of generalized motor programs such as amplitude and timing, the choice of effector may be represented by variables that are replaced by specific values only at the time of the movement. Considering the radically different geometries of the various muscle–joint systems in the body, this claim, at least in its most radical form, is quite extraordinary.

Take, as an example, the joints and activating muscles used for movements made with a combination of fingers and the wrist. This muscle–joint system differs radically in geometry and in the implementation of degrees of freedom from the combination of the elbow and shoulder often used for similar movements on a larger scale. These differences in muscle–joint organization are so large that it is hard to see how detailed, low-level plans for movements made by one of these combinations could be easily transformed so that they could be used to control the same movement with the other combination. And yet, the finger–wrist combination is used for normal handwriting, whereas the elbow–shoulder combination is primarily responsible for larger writing, such as on a blackboard, two related activities often suggested as candidates for control by a single, effector-independent motor program.

It is difficult for me to reconcile the two conceptions of the motor program. On the one hand, there is the idea of a motor program as a repository for the information, presumably fairly detailed and low level, acquired as skill is developed. On the other hand, there is the conception, implicit in the effector-independence hypothesis, of a representation containing no specific information about muscles and joints. My dilemma centers on how the information necessary to support skilled performance with particular effectors is stored in or integrated with the abstract information in an effector-independent representation.

### Effector Independence in Writing

Most authors when considering effector independence (e.g., Keele, 1981; Schmidt, 1988; Stelmach, et al., 1984) cite two sources for empirical support, Merton (1972) and Raibert (1977). Both of these works display samples of a phrase written with different muscle–joint effector systems. Merton (1972) compared an example produced by the fingers and wrist of the dominant hand with an example, roughly ten times larger, produced by the elbow and shoulder of the dominant limb. Raibert's (1977) palindromic demonstration compared one sample of the phrase "Able was I ere I saw Elba" produced by each of the dominant hand, the dominant arm with the wrist immobilized, the nondominant hand, the head with the pen gripped in the teeth, and the leg with the pen taped to the foot. Both Merton (1972) and Raibert (1977) emphasize striking similarities of shape between performances with different effectors.

So far as it exists, this evidence of effector independence in writing is unambiguous. My concern is with the fact that these demonstrations do little to establish the limits of this phenomenon. If this empirical similarity is interpreted only as evidence for *some* common underlying representation across effectors, then there is little problem. But, in most cases, the empirical observations are interpreted more strongly, as specific evidence that the underlying representations are effector-independent, generalized motor programs. The evaluation of this claim requires both that the claim be elaborated more fully and that we examine the empirical phenomenon more carefully.

A major obstacle to evaluating the claim for the existence of effector-independent, generalized, motor programs in writing is knowing the level of representation that is common or generalized across the same movement made with different effectors. Saltzman (1979), for example, has proposed a specific sequence of transformations and refinements between the task-level representation of an activity and the specific sequence of muscle activations needed to carry it out. (Arbib's chapter, in this book, discusses another such scheme.) Adopting this framework, the question becomes less one of whether there is effector independence and more one of how up in the representation hierarchy one must go to find the representation common across effectors. Presumably, for instance, all *nominally* identical movements—for example, "Press the button"—are related at the highest, intentional level of representation. But if this is the lowest level of representation common to two instances of that nominal movement task, I would not wish to label this an effector-independent, generalized motor program.

Although the existing data addressing effector independence are suggestive, their usefulness for inferring the existence of or theorizing about the structure of effector-independent, generalized motor programs is limited because these data lack sufficient detail to constrain the space of possible hypotheses. Consider the task of someone attempting to forge your signature. Forgery is possible because we usually only evaluate the gross outlines of a signature, and, relative to the forger's ability to mimic, this evaluation is relatively easy to satisfy. But if, in addition, we examine closely the structural details of a signature, as a handwriting expert might do, the job of the forger is much harder. If, as in done in systems using signature analysis to authenticate identities, we examine information about the kinematics of the signature, then the job of the forger is harder still. The point of this example is simply, what if the motor system is, in some sense, capable of forgery? What I mean by this is, what if, in addition to motor programs that allow us to produce overlearned actions skillfully and fluently with preferred effectors, the motor system is also able to use other effectors on an *ad hoc* basis to simulate some of the salient aspects of such skilled performance. If in evaluating empirical data supporting effector independence we restrict our examination only to those salient characteristics, then we may well be taken in by the "forgery."

This reasoning suggests that along with the aspects of the performance that constitute the primary goal for the performer—for instance, overall trajectory profiles in writing—we should also be looking at other aspects of the performance. One possibility is simply to look for systematic differences in the fine-grained details of nominally identical trajectories produced with different effectors. For example, a close examination of the samples displayed by both Merton (1972) and Raibert (1977) reveals "minor" differences from sample to sample. To some degree, these differences are certainly due simply to variability across individual performances. Unfortunately, without several examples of nominally identical performance from each effector system to compare, we cannot evaluate the contribution of this variability to these differences.

Similarly, we should look at kinematic aspects of performance. As in the case of the forger, these may reveal additional differences across effectors even when similarity of overall trajectory shape is maintained. One possible "forgery" mechanism that might pass tests of shape similarity but that would be revealed by differences in kinematics is the visual-feedback-based, closed-loop strategy typical of unskilled behavior. These unprogrammed performances are, presumably, guided by visual feedback and a nonmotoric conception of what one's writing should look like—in particular, its distinctive features. Using this strategy, unprogrammed performances are controlled by repeatedly deciding on a stroke to make, based on the conceptual representation, making the stroke, and then, after visually comparing the current output with our conception, deciding on the next stroke. The resulting writing performance, although potentially similar in shape to that of the dominant hand, would be typical of unskilled behavior: slow, jerky, uncoordinated, and imprecise.

## Overview of a New Experiment

The rest of this chapter describes a new experiment following those of Merton (1972) and Raibert (1977). Although intended as a replication and extension of the previous work, this experiment differs from those on which it is modeled in a number of important ways.

1. This experiment does not include the same effector conditions as either previous experiment. Three effector systems were studied:

   i. the dominant (right) hand writing small as on lined paper (Condition DS);
   ii. the dominant hand writing large as on a blackboard (but note that the graphics tablet on which subjects wrote remained flat on the table), with the subjects instructed to move the stylus primarily with their arm rather than with their wrist and fingers (Condition DL);
   iii. the nondominant (left) hand writing small as on lined paper (Condition NS).

The intent, in these choices, was for writing in the DL condition to be four times larger than that in the DS or NS condition. The obtained ratio was smaller, however.

The comparison of the DS and DL conditions is interesting because these conditions were used in previous studies and this comparison seemed likely to provide as good an example of effector independence as any comparable contrast. The comparison between the DS and NS conditions is also potentially quite interesting. Given the differences in orientation of the dominant and nondominant hands, their laterality differences, and the lack of experience subjects have had writing with their nondominant hand, we might expect that this comparison would not show effector independence. Yet, this comparison was included in Raibert's (1977) demonstration of effector independence, along with others that seem *a priori* even more extreme.

2. This experiment included the manipulation of concurrent visual feedback. As detailed in the Method section, on half of the trials subjects were able to see their complete trajectory on a display screen as they were writing (Condition + V) and on half of the trials they only saw the complete trajectory after the movement was complete (Condition −V). The −V condition was included to disrupt and thus to detect if visual-feedback-based, closed-loop control strategies were being used.

3. In the experiment all subjects wrote two different patterns. In the *NAME*-pattern condition, subjects cursively wrote their common or "short" name: for example, *Craig*, *Michelle*, and so on. This condition was chosen because the writing of one's name in this form is typically (a) short, (b) overlearned, (c) legible (often much more than a full signature), but (d) stylized. Although we do not know with certainty the units of planning and performance in writing, the overlearned and stylized nature of this material may have allowed subjects to use a single, preexisting optimized motor program for the entire string, at least in some conditions, if they ever do such a thing. Finally, although the confounding of material with subjects makes detailed comparison of the writing performances across subjects impossible, it does allow summary comparisons that are directly generalizable across both subjects and stimuli.

In the *EQUATION*-pattern condition, subjects wrote the expression $X + Y = Z$. This expression provides a useful contrast to the material used in the *NAME*-pattern condition in several ways. First, although the expression was undoubtedly familiar to our subjects and probably had been written previously by all of them, it was, presumably, not nearly as overlearned as that in the *NAME*-pattern condition. Probably partially as a result, these performances were not nearly so stylized or distinctive across subjects (although each subjects's hand was clearly distinguishable). In addition, the nature of this material suggested that subjects' performances might be controlled by sequencing subprograms for stroke or letter-sized segments rather than by using a single, highly optimized motor program. Finally, detailed comparisons across subjects are facilitated because the elements of this expression are typically, and always by these subjects, written distinctly in a block writing style. In this style, the strokes making up each element are all, nominally at least, straight lines.

4. This experiment included multiple (20), nominally identical performances for each of the three effector systems studied with each of six subjects. This richer data set allows evaluation of whether observed differences between instances represent systematic or random variation.

5. As detailed in the Method section, this experiment used different technology to collect and store the data. This change allows more and better data to be collected but also may have affected subjects' writing strategies.

## METHOD

### Subjects

The data reported here are from the first of two hour-long sessions for 6 right-handed Columbia University undergraduates. They received $5.00 per hour for their participation.

### Design

In each session, a subject produced 130 repetitions in one of the two pattern conditions: *NAME*-pattern or *EQUATION*-pattern. The *NAME*-pattern condition was always run first.

The 130 repetitions (hereafter *trials*) of each pattern, were divided into 13 blocks of ten trials each. Subjects were instructed that the ten trials in each block were to be as close to exact repetitions as possible. The blocks differed in the effector and manner with which subjects were instructed to write (DS, DL, and NS effector conditions) and the presence or absence of concurrent visual feedback (+V, −V feedback conditions). The combination of the three effector systems and two types of concurrent visual feedback produced six conditions, each of which was studied in two of the 12 blocks of ten test trials. The first block of trials was considered practice and was always done in the DS+V condition. The order of conditions within a subject was balanced so that linear practice effects could be averaged out. The within-subject orders were further balanced across subjects using several Latin squares.

### Apparatus

The data were collected using an IBM-AT compatible computer with a Summagraphics Model 1812 digitizing tablet (resolution of 1000 points per in. on each dimension of its *X*, *Y* coordinate system). The inductive sensing system of this digitizer allows position information to be collected whenever the stylus is within approximately 1.25 cm of the digitizer surface (there is some loss of accuracy when the stylus is off of the tablet surface since stylus orientation also affects the position sensed). X-Y coordinate pairs were sampled and stored 110 times per sec by the computer along with the sample time, measured to millisecond accuracy using a Metrabyte clock card. The computer presented visual information to the subject on a 25-cm by 18-cm NEC color graphics display driven by a QuadRam EGA + color display board (running in mode 16 with 640 × 350 resolution and a 60 Hz noninterlaced refresh rate).

Subjects wrote using a stylus with a nylon tip on the surface of the digitizer. Subjects were able to see the current position of the stylus and their hands. However, because the stylus left no marks on the digitizer surface, the trajectory up to the current point could only be seen if the computer presented it on the display screen. This display sat directly behind the digitizer, with its screen almost perpendicular to the digitizer surface. Although this is not an ideal arrangement for concurrent visual feedback manipulation because subjects had to look up from their hand to see the screen, it was possible for subjects to view the screen without moving their heads.

For each effector system there was a rectangle visible on the surface of the digitizer: For the DS and NS conditions the rectangle was 1.91 cm high by 7.62 cm wide and for the DL condition it was scaled proportionately, 7.62 cm high by 30.48 cm wide. Subjects were instructed that their writing should stay within the rectangle, approximately filling it in one dimension. Displays of writing performance were scaled for each effector condition, to maintain the aspect ratio of the actual performance. The box visible to the subject on the digitizer was mapped on a similar outline displayed on the screen, which was approximately 5.72 cm high and 22.86 cm wide.

### Procedure

The experimenter remained with the subject throughout each session. At the beginning of each block, a display informed the subject about the visual feedback and effector conditions for that block. The experimenter made sure that the subject understood these instructions and then pressed a key on the keyboard connected to the computer to begin the sequence of trials.

Data recording began when the subject touched the stylus down on the surface of the digitizer. Once data collection began, the subject had 15 sec to finish writing. Data collection ended after the subject lifted the stylus off the surface of the digitizer for more than 750 ms. After each trial, the computer displayed what the subject had written (this amounted to simply leaving up the display drawn during the trial in blocks with concurrent visual feedback). The data points recorded with the stylus off the surface of the digitizer were excluded from these feedback displays. When this display indicated a systematic problem or confusion on the part of the subject, the experimenter pointed this out; otherwise the experimenter simply waited to initiate the next trial when the subject indicated a readiness to proceed. Because there was no norm against which to compare each performance, no scores or other systematic feedback about the performance were given to the subjects.

### Data Reduction

To aid in further analyses, a semiautomated procedure was developed to identify analogous stroke endpoints across each of a subject's repetitions of a response. This was done after each subject had finished the experiment. Underlying this procedure is the assumption, which I simply accept from the work of others (Denier van der Gon & Thuring, 1965; Hollerbach, 1981; Morasso, Mussa-Ivaldi, & Ruggiero, 1982; Viviani & Cenzato, 1985), that the stroke constitutes a basic unit of analysis and performance in handwriting.

To begin the process of segmentation, possible stroke endpoints were identified automatically using an algorithm that identified local minima of sufficient depth in the tangential velocities. The experimenter then worked interactively with the computer to select which of these potential endpoints matched the stroke endpoints in a previously segmented prototype movement. Prototype trials were chosen based on their representativeness and fluency from the set of trials made using the DS effector with concurrent visual feedback. All 130 trials for each subject were then marked to match this prototype. The coordinates of the resulting marked points as well as characterizations of the strokes between the marks were the basis of the remaining analyses.

The decision embedded in this procedure to force the marking of each performance to match that of the prototype is based on the premise that all 130 trials represent a single underlying target pattern realized through a noisy transmission channel. With this procedure, I hoped to maximize the possibility that effector independence would be revealed if it exists.

The experimenter-directed part of this marking process. although usually straightforward, did occasionally require substantial judgment. Among the more describable of the problems encountered were: (a) cases where two usually distinct strokes were coalesced into a single, looping stroke, (b) cases where some feature was omitted, added, or transformed, usually throughout a condition, and (c) cases where the subject simply appeared to be confused about the task. In addition, the automatic stroke-identification algorithm often ran into problems when subjects wrote slowly and deliberately in the NS condition. I never excluded trials from analysis because of these marking problems because that might have introduced a poorly understood sampling bias. Occasionally, however, not all marks could be placed for every trial, and later analyses had to be tailored to deal with the resulting uncertainty in the data.

## RESULTS

### Comparison of Single Movements

At the level of analysis previously used, the data from this experiment replicate and extend the studies of Merton (1972) and Raibert (1977) in that the overall shape of the written strings displayed marked similarities across effectors for both pattern types. Figures 9.1 and 9.2 give some idea of both this similarity in shape and the variability across instances for replications by a single subject. Figure 9.1 shows 10 repetitions by one subject in the *NAME*-pattern DS + V and NS + V conditions. Figure 9.2 allows the same type of comparison for a different subject in the *EQUATION*-pattern DS − V and NS − V conditions. Although these figures provide a useful "feel" for the data, to get a clearer impression of what subjects were trying to accomplish in each condition, it is more useful to compare average performances rather than the individual instances shown in these figures.

### Comparison of Averaged Movements

Averages were computed for the "markable" trials that each subject produced for each patten (there were between 18 and the maximum of 20). These averages were obtained through an iterative process. This involved first rotating, translating, and scaling all of the instances produced by a subject for a pattern. These transformations were done so that the marked points in each instance made the best possible match to their analogues in the current best estimate of the "average" performance. These transformed instances were then summed to create a new estimate of the average across conditions and the entire process was repeated. Once the overall averages produced by this process stabilized (usually after only two or three iterations), the transformed instances were averaged separately for each of the six combinations of effector and concurrent visual feedback. These averages by condition include not only the marked points but ten interpolated points spaced evenly in time between each pair of marked points. These interpolated points allow the average trajectories to be visualized and compared in more detail.

Examples of the results from this averaging process can be seen in Figs.

FIG. 9.1. Comparison of ten trials (one block) in the *NAME*-pattern DS + V and NS + V conditions for one subject. Solid lines were made with the stylus on the digitizer surface; dashed lines were recorded with the stylus off the digitizer surface.

9.3 and 9.4. These figures display comparisons of the average performance across trials in all six conditions. The crosses at the bottom left corner of each frame represent a pooled estimate of ± 2 standard errors of the mean (each line has a total length of four standard errors of the mean) for each point in these average trajectories.[1]

---

[1]Except for a few isolated cases, typically occurring near the ends of the trajectories or where the stylus was off the surface of the digitizer, the standard deviations at the individual points were reasonably homogeneous across a trajectory.

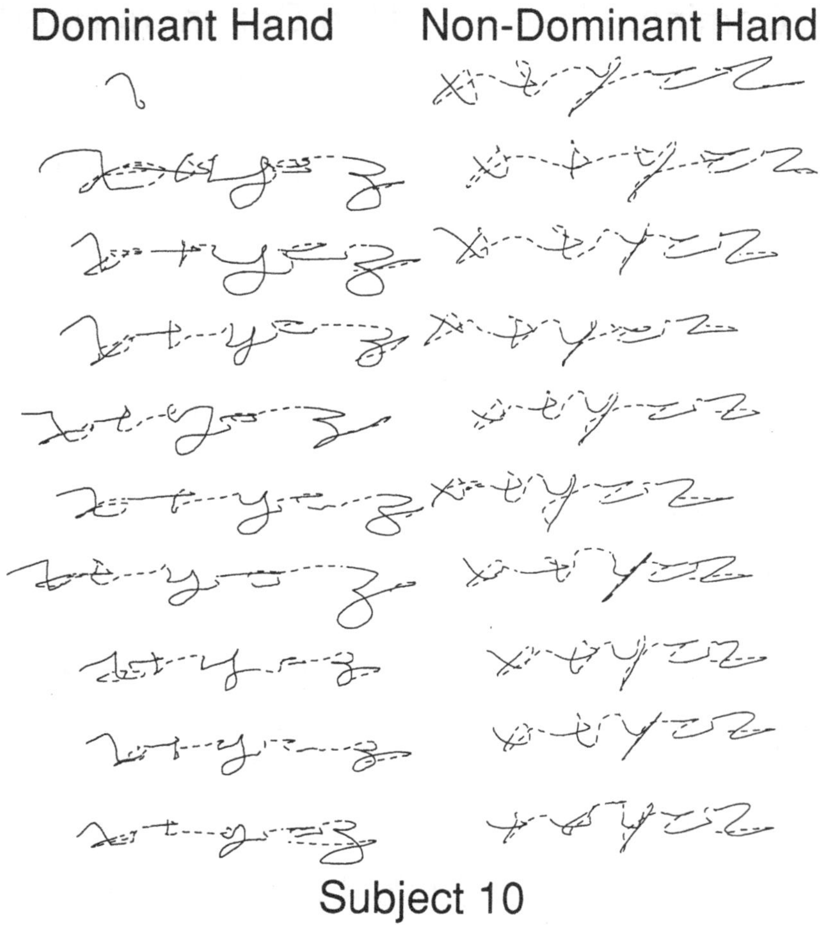


FIG. 9.2. Comparison of ten trials (one block) in the *EQUATION*-pattern DS–V and NS–V conditions for one subject. Solid lines were made with the stylus on the digitizer surface; dashed lines were recorded with the stylus off the digitizer surface.

Like the individual movements displayed in Figs. 9.1 and 9.2, the averages in Figs. 9.3 and 9.4 show clear similarities across conditions. What these averages reveal just as clearly is that there are differences between the average performance in different conditions, and these differences are systematic, not merely trial-to-trial variation. These differences tend to be most striking for comparisons involving conditions with the nondominant hand (NS) and one of the other two effector conditions (DS and DL). Although they are usually less dramatic, there also are systematic differences between the two conditions performed with the dominant hand (DS versus DL) and the conditions performed with and without concurrent visual feedback (+V and −V).

Notice the formation of the letter *K* in Fig. 9.3, for example. Typical of the first letter in many signatures, the subject forms this letter from four strokes in a fairly ornate style. The first stroke is a vertical stroke starting at the top of the letter. The second stroke, which is off the surface of the digitizer, returns from the bottom of the letter to its upper right-hand corner. The third stroke is the diagonal from the upper right back to the midpoint of the vertical line. This stroke starts out as a wide curve that

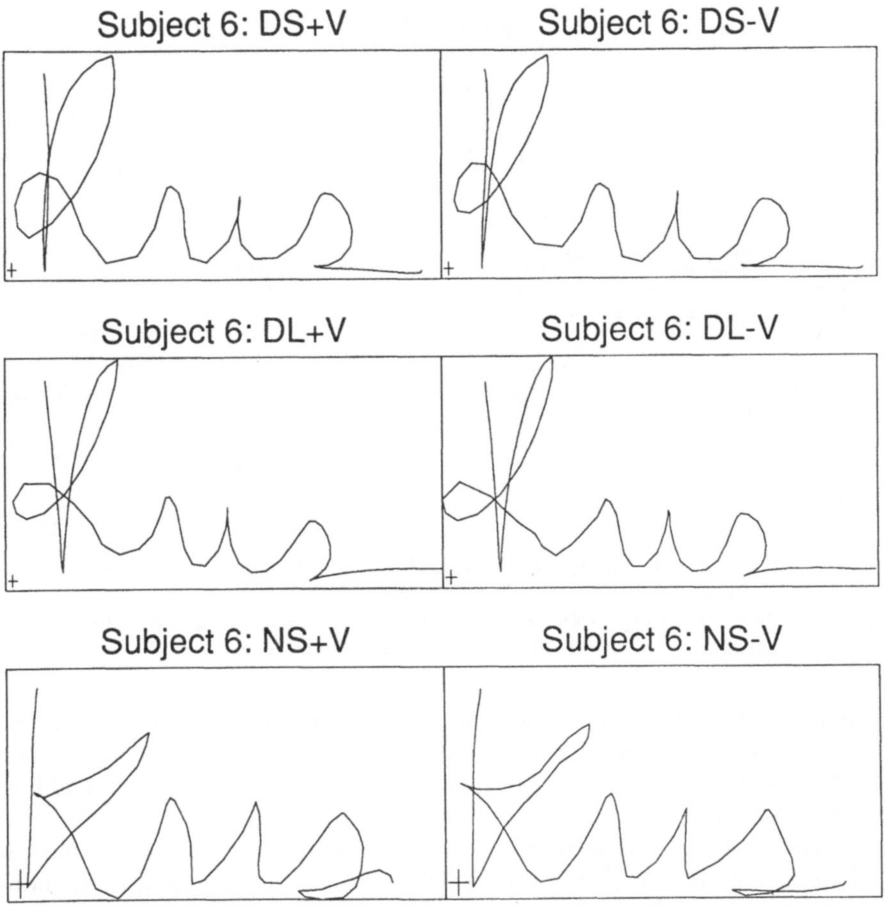


FIG. 9.3. The average performance of one subject writing the *NAME*-pattern is each of the six conditions. Note that the averaging process did not distinguish strokes made with the stylus on the surface of the digitizer and those made with the stylus off the digitizer surface.

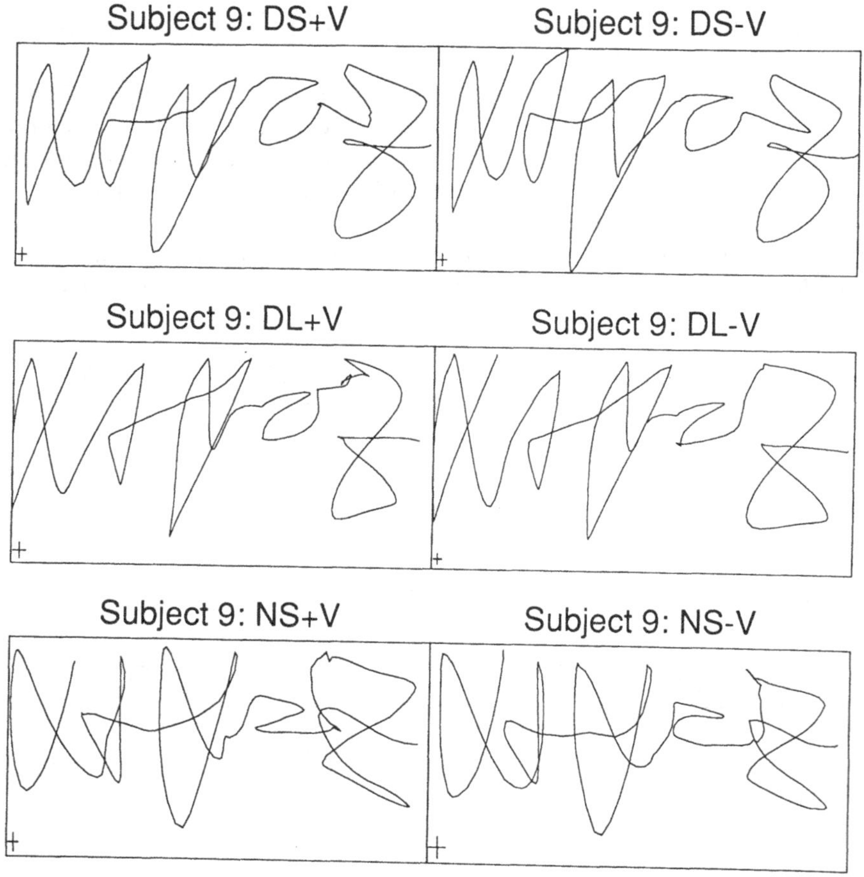


FIG. 9.4. The average performance of one subject writing the *EQUATION*-pattern in each of the six conditions. Note that the averaging process did not distinguish strokes made with the stylus on the surface of the digitizer and those made with the stylus off the digitizer surface.

tightens to form the lower half of a loop at the vertical line, the characteristic embellishment of this subject. The fourth stroke begins at the back of this loop, forms its top, and then continues down and right to lead into the next letter.

This description is accurate for all four conditions in which the dominant hand is the effector (DS and DL). When this subject writes with the nondominant hand (the NS conditions), however, the characteristic loop is largely eliminated. And, although the subject still maintains the four stroke composition of this letter, the relationship between these stokes also changes somewhat. In both NS conditions, stroke two, instead of starting by retracing the path of the first (vertical) stroke, immediately heads off to the right side of the character. In the process, this stroke fails to reach the top of the character, causing stroke three to be closer to horizontal than normal.[2]

The same modifications already noted for movements made with the nondominant hand also occur, although less dramatically, for movements made with the dominant arm (DL conditions) and movements made without the aid of concurrent visual feedback (−V conditions). Thus the area of the loop bounded by strokes three and four is larger for the DS+V condition than in these other three conditions [for the comparison with DS − V, $F(1,5) = 6.95$, $p = .06$; for the comparison with DL+V, $F(1,5) = 20.03$, $p < .01$; for the comparison with DL − V, $F(1,5) = 18.27$, $p < .01$].

Similarly, the loop bounded by strokes two and three is larger in the DS+V condition than in these other three conditions [for the comparison with DS − V, $F(1,5) = 16.35$, $p < .01$; for the comparison with DL + V, $F(1,5) = 25.83$, $p < .005$; for the comparison with DL − V, $F(1,5) = 63.18$, $p < .001$]. Along the same lines, note that the point where strokes three and four cross over is out to the right farther from the first vertical stroke in the DS conditions than in the DL conditions, whereas stroke two is back closer to vertical stroke one in the DS conditions than the DL conditions.

Interestingly, the presence or absence of visual feedback appears to have had no effect for this last pair of comparisons. Similarly, although the absence of concurrent visual feedback reduced the area of the loops in the DS conditions, there was no equivalent effect for the comparisons between the DL conditions [for the loop formed by strokes three and four, $F(1,5) = 0.82$; for the loop formed by strokes two and three, $F(1,5) = 2.13$, $p > .25$]. Visual inspection of the NS conditions suggests that there was also little systematic effect of concurrent visual feedback on this effector.

Although the exact nature of these differences across effectors and the presence/absence of concurrent visual feedback depend strongly on the subject and the pattern, it is a reasonable summary that similar patterns of systematic differences occur for all six subjects and both patterns. These analyses have concentrated on changes in large features. Figure 9.5 provides examples of comparisons that reveal differences at a subtler level.

Figure 9.5 shows, superimposed, the average trajectories for one subject (named Thomas) writing the *NAME* pattern in the NS+V condition and

[2]Note that the tendencies discussed here are not artifacts of averaging, but, instead, are typical of most of the individual trials of each type.

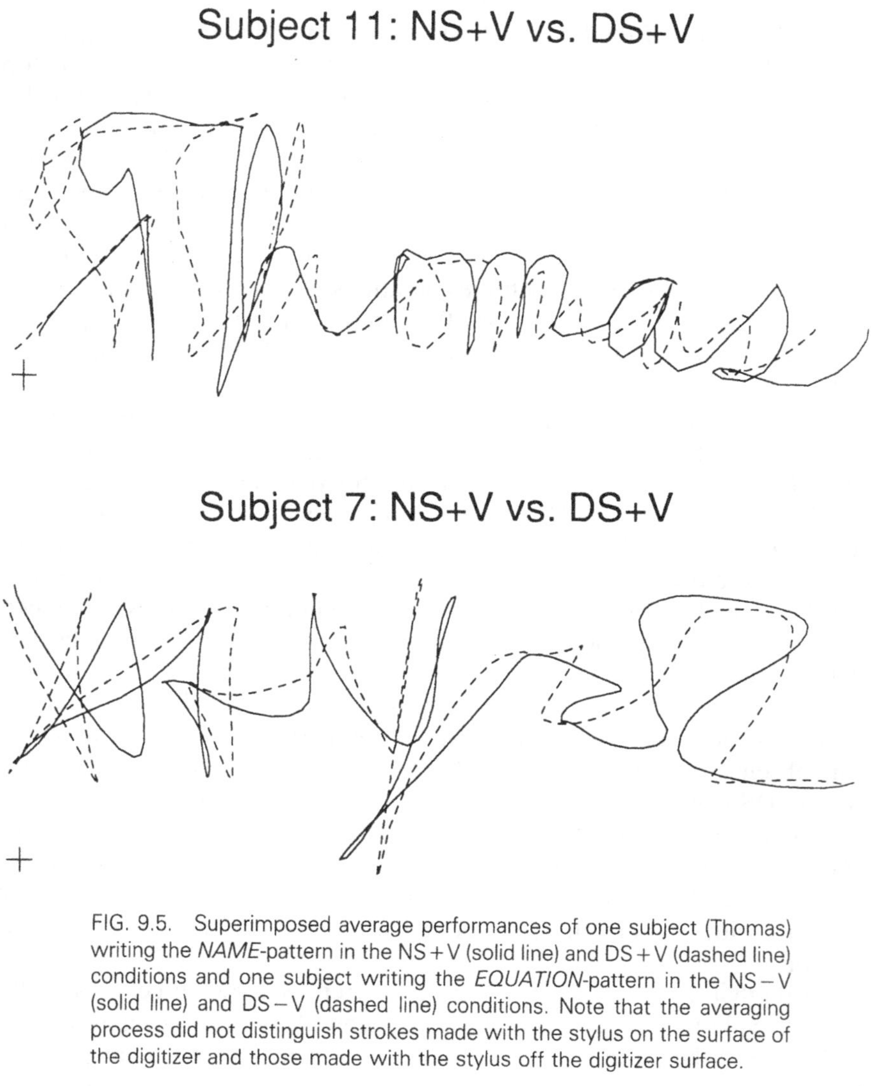


FIG. 9.5. Superimposed average performances of one subject (Thomas) writing the *NAME*-pattern in the NS + V (solid line) and DS + V (dashed line) conditions and one subject writing the *EQUATION*-pattern in the NS − V (solid line) and DS − V (dashed line) conditions. Note that the averaging process did not distinguish strokes made with the stylus on the surface of the digitizer and those made with the stylus off the digitizer surface.

the DS + V condition. Figure 9.5 also shows a similar comparison for a different subject in the *EQUATION*-pattern condition. The cross at the lower left corners represent a pooled estimate of ± 2 standard errors of the mean difference between corresponding points (each line has a total length of four standard errors of the mean) for each point in these average trajectories. The two average trajectories have been scaled and oriented to maximize their overlap. Despite this, it is clear that there are changes in the relative size of parts of these averages as well as changes in the way the space within them is used.

### Comparison of Kinematic and Other Summary Data

I now turn from descriptions that depend on the shape of the trajectory to other aspects of the writing performance: stroke composition, duration, and peak velocity. These aspects of the writing behavior tell a consistent story across subjects and the two pattern conditions. To summarize, the dominant hand writing small (DS condition) was fluent and quick. There is some suggestion that writing large letters with the dominant hand (DL condition) was somewhat less fluent, although it was as stable across repetitions. Finally, writing with the nondominant hand (NS condition) was clearly less fluent, slower, more jerky, and more variable across repetitions.

The data supporting this summary are in Table 9.1. Within this table, each line breaks out one aspect of performance by the pattern and effector conditions. Although there were systematic effects of the visual feedback condition within the data of individual subjects, the pattern of these effects was not consistent across subjects. The data in Table 9.1 are, therefore, averaged across visual feedback condition. Thus, each cell in Table 9.1 is the average of roughly 40 replications for each of 6 subjects. For each pattern condition there is also a 10-degree of freedom, pooled estimate of the repeated-measures, standard error for each of the three means.

Line 1 of Table 9.1 shows the total trajectory length measured from the beginning of the first marked stroke to the end of the last marked stroke, including strokes made both on and off the digitizer surface. There is no reliable difference in this length between the DS and NS conditions. The length in the DL conditions is roughly 2.5 times that in the DS conditions. This ratio is less than the target of 4.0 that I had hoped the stimulus conditions would induce.

Line 2 of Table 9.1 shows the time, in seconds, from the beginning of the first marked stroke to the end of the last marked stroke. In both pattern conditions, this time increases systematically from the DS to the DL effector conditions and additionally from the DL to the NS effector conditions. Although an increase in time by a factor of 1.25 between the DS and DL conditions may seem plausible given an increase of 2.5 in the trajectory length, this finding was somewhat unexpected in light of previous research involving size changes in handwriting. For example, Hollerbach (1981, p. 153) states: "Handwriting frequency is independent of writing size. This independence is an accepted observation in the handwriting literature (Denier van der Gon & Thuring, 1965; Yasuhara, 1975) and is an observation substantiated in my own measurements." I can only spec-

TABLE 9.1.
Summaries of writing behavior by pattern type and effector condition averaged across visual-feedback conditions and subjects (occasionally also segments).

| | | *NAME-Pattern* | | | *Pooled* | *EQUATION-Pattern* | | | *Pooled* |
|---|---|---|---|---|---|---|---|---|---|
| *Measure* | *Units* | *DS* | *DL* | *NS* | *StdErr* | *DS* | *DL* | *NS* | *StdErr* |
| 1. Trajectory Length | cm | 50 | 112 | 44 | 10 | 31 | 84 | 34 | 2.0 |
| 2. Duration | s | 2.21 | 2.78 | 4.01 | 0.11 | 2.74 | 3.41 | 4.35 | 0.14 |
| 3. Average Tangential Velocity | cm/s | 24 | 43 | 12 | 4.6 | 11.7 | 25.3 | 8.1 | 0.66 |
| 4. Peak Tangential Velocity | cm/s | 52 | 91 | 35 | 5.0 | 49 | 94 | 35 | 4.1 |
| 5. Number of Extra Segments | | 0.66 | 0.89 | 1.98 | 0.15 | 0.45 | 0.71 | 1.36 | 0.08 |
| 6. Std-Dev X Coordinates | * | 11.0 | 11.9 | 20.6 | 1.4 | 14.1 | 15.1 | 21.6 | 1.1 |
| 7. Std-Dev Y Coordinates | * | 6.7 | 6.9 | 11.5 | 0.5 | 8.1 | 8.8 | 12.3 | 0.5 |

*The standard deviations were computed after size normalization and so have arbitrary units of length.

ulate on the source of this discrepancy without further investigation; however, a critical difference appears to be that subjects in the current experiment were explicitly instructed to form their larger trajectories using their arms instead of their hands, whereas subjects in previous experiments may have produced instances of writing at different sizes using the same effectors.

If, as previous theories suggest (Denier van der Gon & Thuring, 1965; Hollerbach, 1981), invariant movement time is a critical element of the motor representation underlying handwriting, then this difference between the DS and DL effector conditions indicates different motor programs may have controlled the performances in these conditions. This indication receives some further support from the fact that the size of this increase is remarkably stable across the pattern conditions (1.26 in the *NAME*-pattern condition and 1.24 in the *EQUATION*-pattern condition; a 95% confidence interval for the difference of these values is ± 0.06). The much larger difference in writing time between the DS and NS effector conditions (1.7 to 1) supports this indication even more strongly because this difference cannot be attributed to any differences in the size of the trajectories.

Line 3 of Table 9.1 shows the average tangential velocity in each segment pooled across the segments. In these data there are reliable differences between all the effector conditions for both pattern conditions.

Line 4 of Table 9.1 shows the peak tangential velocity in each segment pooled across the segments. Once again, there are reliable differences among all the effector conditions for both pattern conditions: Peak velocity is higher in the DS than in the NS condition and still higher in the DL condition. These data also allow an intriguing comparison between the DS and DL effectors across the two pattern conditions. For the *NAME*-pattern condition, there is an increase in movement duration going from the DS to the DL effector of 1.26 and an increase in peak velocity of 1.75. The product of these two increases, 2.20, almost perfectly matches the actual increase in the trajectory length, 2.24. This outcome is consistent with models of movement control in which changes of distance and/or time are achieved by rescaling a velocity profile whose shape is invariant (e.g., Meyer, Smith, & Wright, 1982). For the *EQUATION*-pattern condition, there is an increase in movement duration going from the DS to the DL effector of 1.24 and an increase in peak velocity of 1.92. This increase in peak velocity is not, however, sufficient to allow the product of these two increases, 2.38, to match the actual increase in the trajectory length, 2.71. This discrepancy means that the shape of the velocity profile must be different for these two effectors in the *EQUATION*-pattern condition. Confirmation and an explanation of this discrepancy must await further detailed analyses.

Line 5 of Table 9.1 shows the number of "extra" potential segment boundaries within each marked segment. The Methods section describes how each instance of a written pattern was segmented into a fixed set of strokes by marking stroke endpoints. The stroke endpoints for this analysis were determined by an algorithm that identified local minima in the tangential-velocity profile. Although all marks were constrained to fall at stroke endpoints, not all of the potential stroke endpoints found by this algorithm were used as the basis of marks. The stroke-composition measure, then, is simply the count of potential stroke endpoints occurring between each pair of marks. Whereas the absolute number of the endpoints determined by the algorithm depends on details of the algorithm itself, the relative number of these endpoints found across conditions should indicate the fluency of the writing activity.

As Line 5 shows, even the DS condition reliably contained such extra potential stroke endpoints. In this condition, most of the extra endpoints appear, however, to be due to positioning movements made with the stylus off the surface of the digitizer. The number of extra endpoints increased somewhat in the DL condition: The 34% increase for the *NAME*-pattern condition is not reliable across subjects, whereas the 73% increase for the *EQUATION*-pattern condition is reliable. The increase in the number of extra endpoints between the DS and NS effector conditions is more substantial and consistent across the pattern conditions: There is a 258% increase for the *NAME*-pattern condition and 231% increase for the *EQUATION*-pattern condition, both reliable across subjects. These large increases suggest the writing was decomposed into smaller strokes in the NS condition. Presumably, concatenating many smaller strokes either requires less coordination or reduces the possibility of large, unacceptable spatial errors.

Lines 6 and 7 of Table 9.1 show the average standard deviations of the $x,y$ coordinates for the marked points, computed after the transformations to align the movement instances. Because these standard deviations were computed after rescaling, any tendency for endpoint variability to vary proportionally with movement size is eliminated. In fact, these standard deviation estimates are virtually identical across the DS and DL effector conditions; the small increase in Table 9.1 is not reliable across subjects. These variability measures do, however, increase substantially and reliably between the DS and NS effector conditions, consistent with the suggestion that movements in the NS condition are less precise and thus less fluent.

## DISCUSSION

This research begins an attempt to use the phenomenon of effector independence in handwriting as a tool to gain a better understanding of the motor representations used to control complex movement sequences and, in particular, to better understand how motor programs can be generalized. To make this phenomenon more useful as a tool, it is necessary to begin by identifying some of its limits. Just as the linguist's search for language universals starts by identifying features that do and do not change between languages, so, I hope, a better understanding of how motor programs are generalized can emerge by studying the features of writing movements that are and are not invariant across effectors.

## General Summary

Extending the previous work of Merton (1972) and Raibert (1977), the data reported here demonstrate that, at the level of overall shape, the trajectories of highly overlearned and less-well-practiced writing movements are similar when performed, with or without concurrent visual feedback, by any of the three effector systems studied. Perhaps more importantly, these data also demonstrate that changing effector system or removing concurrent visual feedback both lead to changes in the details of the written trajectories. These changes in detail are systematic across repetitions, and, thus, they are not simply due to the variability of particular instances or the lack of skill in the use of particular effector systems. The data reported here also reveal reliable differences across effectors in measures of kinematics, fluency, and repeatability of the writing movements. These differences need not be interpreted as violations of the principle of effector independence; rather, their importance lies in the limits these differences circumscribe for the generality of this observation.

Along more theoretical lines, we can ask, to continue the analogy drawn in the Introduction, whether the differences observed for writing with different effectors constitute "forgery." Alternatively, these differences could be interpreted as predictable changes occurring when a single, effector-independent, generalized motor program is carried out by effector systems with somewhat different properties. If we consider these differences to be consistent with the hypothesis of a single, effector-independent, generalized motor program, then we must also address the question of the level of specification common to the performance across effectors. Of course, the second version of this question might be considered logically to subsume the first.

## Dominant Versus Non-Dominant Hand

Evaluating the comparison between the dominant and the nondominant hand for evidence of a forgery, we might conclude that, as such, it is fairly inept. Although the overall shape of the written trajectory is preserved, many salient details of that trajectory are noticeably altered. In addition, writing with the nondominant hand is less fluent. Simply because subjects are less practiced writing with their nondominant hand, we might expect this hand to be slower and to have lower spatial accuracy across repetitions. More telling is the observation that writing with the nondominant hand uses a different decomposition of letters into strokes, involving many more, smaller strokes. If, as several authors suggest (e.g., Edelman & Flash, 1987; Hollerbach, 1981; Viviani & Cenzato, 1985), strokes constitute the primary level of organization for the planning of writing movements, the difference in stroke composition for two versions of the same movement suggests that these movements were produced according to different plans or motor programs. And, if we consider these movements to have been produced under the control of a single program, that program must be quite abstract, incorporating little more than overall shape information and almost nothing strictly motoric.

To my mind, a more plausible explanation for these observed differences is that writing with the nondominant hand is controlled in a different way than writing with the dominant hand. One obvious candidate for an alternative mechanism is the process of successive approximation based on feedback typical of unskilled activities. This is not, of course, to imply that writing with both hands could not, perhaps after sufficient practice, be controlled by the same motor program but that for the tasks, subjects, and level of practice involved here, this was not the case. One possible problem with this interpretation is the expectation, outlined in the Introduction, that writing produced using successive approximation based on feedback might use a feedback loop involving, specifically, visual feedback. In this case, I would have expected performance to depend critically on the availability of concurrent visual feedback. The results are inconsistent with that suggestion. If the suggestion had been correct, I would have expected removal of concurrent visual feedback to have had dramatic negative effects on performance with the nondominant hand. Although removing concurrent visual feedback had a measurable impact, this manipulation did not cause major disruptions. However, as noted in the Method, the procedure used to manipulate the presence of concurrent, visual feedback in this experiment was, for practical reasons, perhaps less than ideal. It is possible that a more natural manipulation of concurrent visual feedback would lead to larger effects. This is a possibility that has to be explored in future research.

## Small Versus Large Writing with the Dominant Hand

The differences observed in the comparison between the DS and DL conditions, the dominant hand writing small as on a piece of paper versus the dominant hand and arm writing large as on a blackboard, are more subtle because the similarities are much stronger. Evaluating this comparison for evidence of "forgery" would be much more difficult because, although there are systematic differences in performance at various levels of analysis, no difference itself is compelling. Once again, the overall shape of the writing is similar, and, in fact, it is necessary to make a careful examination of the fine details of each stroke before any differences become apparent.

Without having a formal analysis to support this supposition, it is plausible to me that movements made with different effectors controlled by a single effector-independent, generalized motor program would result in systematic, if subtle, differences in trajectory. Consider, for example, the model for handwriting proposed by Hollerbach (1981). In this model, cursive writing movements are represented as changes in the parameters of two coupled oscillators: one operating in the direction of a superimposed, constant rightward movement, and the other operating in a direction roughly orthogonal to the first. The assumption of effector independence comes in the supposition that the degrees of freedom of the effector system used for writing can be organized into functional synergies that implement the required, orthogonally coupled oscillators. If, however, the *virtual effector* created by these synergies differs, even slightly, when implemented using the arm as opposed to the hand, then we would expect minor, systematic differences in the observed trajectories.

The observed difference in movement times between the DS and DL conditions is suggestive of a change in control strategy or motor program, but far from conclusive. These results suggest this because it is well accepted that the writing control system acts to keep movement times constant across variations in movement size and stylus friction (Denier van der Gon & Thuring, 1965; Hollerbach, 1981; Yasuhara, 1975). These results are not conclusive because it is unclear if the earlier results on which this conclusion is based involved comparisons across effectors. It is plausible that an effector-independent, generalized motor program for writing would include relative, not absolute, timing information that could then be transformed differently by the functional synergies used to implement this program for different effectors. This possibility suggests the desirability of a controlled comparison of timing invariance in writing within and across effectors.

Finally, two differences between the DS and DL effectors that occur only in the *EQUATION*-pattern condition might be taken to suggest that an effector-independent, generalized motor program was operating in the *NAME*-pattern but not the *EQUATION*-pattern conditions. First, as in the comparison of the DS and NS effectors, there is an increase in the number of potential stroke endpoints with the DL effector. As before, this can be taken to suggest a different decomposition of letters into strokes. However, this conclusion is less persuasive than in the previous comparison both because the increase is smaller and because I cannot be certain that this increase is not an artifact due to an interaction between the larger stroke sizes with the DL effector and the algorithm used to identify potential stroke endpoints.

The second of these differences has to do with the relative size of the increases in trajectory length, duration, and peak tangential velocity. As outlined in the Results, the size of the increase in trajectory length going from the DS to the DL effectors can be accounted for in the *NAME*-pattern condition by the simultaneous increases in duration and peak tangential velocity according to the principles of velocity-profile rescaling. A similar accounting does not work, however, in the *EQUATION*-pattern condition. Because invariance of velocity profile across effectors is a stronger constraint than invariance of trajectory shape, this result might also be interpreted as evidence for the operation of an effector-independent, generalized motor program in the *NAME*-pattern but not in the *EQUATION*-pattern condition. However, before giving credence to this possibility, the underlying assertion about the velocity profiles can and should be tested directly.

## CONCLUSIONS

This chapter reports the application of several detailed analyses of writing data designed to provide a clearer idea of the generality and limits of effector independence in writing. Confirming the previous observations that overall shape is strikingly similar across effectors, these analyses also confirm that changing effectors results in systematic changes in details of shape, kinematics, stroke decomposition, and fluency. Clearly, more work needs to be done refining and extending these analyses before we have a complete picture of what stays the same and what differs when the effectors used for writing are changed.

I have used the preliminary results reported here to reopen the discussion of the implications of effector independence in writing for the hypothesis that motor programs can be generalized to operate across effectors. I believe this discussion should be reopened because the previously existing evidence of shape similarity across effectors only weakly constrains the set of possible explanations. Although it is early to make strong assertions, I believe the results reported are consistent with the conclusion that writing by the dominant and the nondominant hands is carried out using different strategies that share only the very highest and most abstract spatial representations of what the shape of the writing should be. By contrast, writing with the dominant hand and the dominant arm appear to be controlled by mechanisms that share a common representation much lower in the hierarchy of increasing motor specificity particulary for highly overlearned productions such as one's name.

## ACKNOWLEDGMENTS

Preparation of this chapter was supported by National Science Foundation Grant BNS-87-11273 to the author. I thank Barbara Landau, Nina Macdonald, David E. Meyer, and an anonymous reviewer for their useful comments on earlier drafts of this chapter.

## REFERENCES

Adams, J. A. (1971). A closed-loop theory of motor learning. *Journal of Motor Behavior, 3*, 111–150.

Bernstein, N. (1967). *The co-ordination and regulation of movements*. Oxford: Pergamon.

Denier van der Gon, J. J., & Thuring, J. Ph. (1965). The guiding of human writing movements. *Kybernetik, 2*, 145–148.

Edelman, S., & Flash, T. (1987). A model of handwriting. *Biological Cybernetics, 57*, 25–36.

Genter, D. R. (1983). Keystroke timing in transcription typing. In W. E. Cooper (Ed.), *Cognitive aspects of skilled typewriting* (pp. 95–120). New York: Springer-Verlag.

Henry, F. M., & Rogers, D. E. (1960). Increased response latency for complicated movements and a "memory drum" theory of neuromotor reaction. *Research Quarterly, 31*, 448–458.

Hollerbach, J. M. (1981). An Oscillation Theory of Handwriting. *Biological Cybernetics, 39*, 139–156.

Keele, S. W. (1968). Movement control in skilled motor performance. *Psychological Bulletin, 70*, 387–403.

Keele, S. W. (1981). Behavioral analysis of movement. In V. B. Brooks (Ed.), *Handbook of physiology. Vol. II: Motor control, part 2* (pp. 1391–1414). Baltimore, Maryland: American Physiological Society.

Keele, S. W., & Summers, J. J. (1976). The structure of motor programs. In G. E. Stelmach (Ed.), *Motor control: Issues and trends*. New York: Academic Press.

Kelso, J. A. S. (1977). Motor control mechanisms underlying human movement reproduction. *Journal of Experimental Psychology: Human Perception and Performance, 3*, 529–543.

Merton, P. A. (1972). How we control the contraction of our muscles. *Scientific American, 226*, 30–37.

Meyer, D. E., Abrahms, R. A., Kornblum, S., Wright, C. E., & Smith, J. E. K. (1988). Optimality in human motor performance: Ideal control of rapid aimed movements. *Psychological Review, 95*, 340–370.

Meyer, D. E., Smith, J. E. K., & Wright, C. E. (1982). Relations between the speed and accuracy of aimed limb movements. *Psychological Review, 82*, 449–482.

Morasso, P., Mussa-Ivaldi, F. A., & Ruggiero, C. (1982). Modelling the generation of handwriting. *Behavioral Brain Research, 5*, 112.

Pew, R. W. (1966), Acquisition of hierarchical control over the temporal organization of a skill. *Journal of Experimental Psychology, 71*, 764–771.

Raibert, M. H. (1977). *Motor control and learning by the state space model.* Technical Report AI-M-351, Massachusetts Institute of Technology. NTIS AD-A026-960.

Rosenbaum, D. A. (1980). Human movement initiation: Specification of arm, direction, and extent. *Journal of Experimental Psychology: General, 109*, 444–474.

Rosenbaum, D. A., & Kornblum, S. (1982). A priming method for investigating the selection of motor responses. *Acta Psychologica, 51*, 223–243.

Rumelhart, D. E., & Norman, D. A. (1982). Simulating a skilled typist: A study of skilled cognitive-motor performance. *Cognitive Science, 6*, 1–36.

Saltzman, E. (1979). Levels of sensorimotor representation. *Journal of Mathematical Psychology, 20*, 91–163.

Saltzman, E., & Kelso, J. A. S. (1987). Skilled actions: A task dynamic approach. *Psychological Review, 94*, 84–106.

Schmidt, R. A. (1975). A schema theory of discrete motor skill learning. *Psychological Review, 82*, 225–260.

Schmidt, R. A. (1988). *Motor control and learning: A behavioral emphasis.* Champaign, IL: Human Kinetics Publishers.

Schmidt, R. A., Zelaznik, H. N., Hawkins, B., Frank, J. S., & Quinn, J. T. (1979). Motor output variability: A theory for the accuracy of rapid motor acts. *Psychological Review, 86*, 415–441.

Stelmach, G. E., Mullins, P. A., & Teulings, H.-L. (1984). Motor programming and temporal patterns in handwriting. In J. Gibbon & L. Allan (Eds.), *Timing and time perception,* Vol. 423 of the *Annals of the New York Academy of Sciences* (pp. 144–157). New York: New York Academy of Sciences.

Sternberg, S., Monsell, S., Knoll, R. L., & Wright, C. E. (1978). The latency and duration of rapid movement sequences: Comparisons of speech and typing. In G. E. Stelmach (Ed.), *Information processing in motor control and learning* (pp. 118–152). New York: Academic Press. Reprinted in R. A. Cole (Ed.), *Perception and production of fluent speech.* Hillsdale, NJ: Lawrence Erlbaum Associates, 1980.

Viviani, P., & Cenzato, M. (1985). Segmentation and coupling in complex movements. *Journal of Experimental Psychology: Human Perception and Performance, 11*, 828–845.

Viviani, P., & Terzuolo, C. (1980). Space–time invariance in learned motor skills. In G. E. Stelmach & J. Requin (Ed.), *Tutorials in motor behavior.* Amsterdam: North-Holland.

Whiting, H. T. A. (1984). *Human motor actions: Bernstein reassessed.* Amsterdam: North-Holland.

Yasuhara, M. (1975). *Experimental studies of handwriting process.* Rep. Univ. ElectroComm. 25-2 (Sci. and Tech. Sect.), 233–254. Cited in Hollerbach, 1981.

# 10 Constraints for Action Selection: Overhand Versus Underhand Grips

David A. Rosenbaum, Frank Marchak, Heather Jane Barnes, Jonathan Vaughan, James D. Slotta, and Matthew J. Jorgensen
*Department of Psychology University of Massachusetts*

## ABSTRACT

A fundamental problem in action research is to determine how the motor system selects among alternative methods for achieving specific goals. We suggest that a useful approach to this problem is to identify optimization constraints. One constraint, introduced here, is to minimize times in joint angles far from resting positions. This constraint is shown to account for data from tasks in which adult human subjects pick up a cylindrical bar and transport it to various locations. The data take the form of overhand versus underhand grip choices. As assessed by awkwardness ratings of these two grip postures, grip posture choices are best predicted by assuming that subjects attempt to minimize awkwardness at the end of the first bar transport move they perform. We show that this outcome is consistent with the Extreme Joint Angle Constraint and is optimal in the sense that awkward joint angles, when they are adopted, allow for passive rotation of the arm into desired final positions.

## INTRODUCTION

The question of how the motor system chooses among action alternatives has often been discussed in the context of the degrees-of-freedom problem, introduced by Bernstein (1967). The main approach taken to the problem is one that Bernstein advocated—to identify dependencies within the motor system that effectively constrain or reduce the degrees of freedom that need to be independently controlled. These dependencies have been called *synergies* (Bernstein, 1967), and much recent work has been aimed at identifying how they are organized (Kelso, Holt, Kugler, & Turvey, 1980).

Although synergies play a role in coping with the degrees-of-freedom problem, they are not the only means by which the problem is solved. The motor system also makes use of *optimization constraints* to help determine which actions should be performed in which contexts. An optimization constraint is a stipulation, represented explicitly or implicitly, that a function take on one or more prescribed values (usually a minimum or maximum). An example of an optimization constraint is minimization of the mean squared magnitude of jerk (the third time derivative of position); this constraint helps account for the bell-shaped velocity profiles that are often observed in studies of aimed hand movements (Hogan, 1984). Other optimization constraints that have been proposed are minimization of distance, time, peak velocity, energy, and peak acceleration (Nelson, 1983).

Most previous studies that have pursued the constraint approach to motor control have focused on relatively microscopic features of movements performed in restricted laboratory settings. Our aim in this chapter is to consider large-scale characteristics of spontaneously chosen actions. To give an example of the sort of issue we wish to study, consider the following observation made by one of the authors (Rosenbaum). While entertaining a one-year-old infant, he held a cylindrical rattle in front of the baby so she could grasp the rattle by lifting her hand (which was in a palm down orientation) and closing her fingers around it. When the baby appeared to be interested in the rattle but before moving her hand toward it, the rattle was shifted past her hand to a more eccentric position. The baby's response was to lower her hand and come around the rattle from above, taking hold of it with an overhand grip. She did not perform another movement that, conceivably, she could have—twisting her arm around to grab the rattle with an underhand grip. The decision to take the longer path and grab the rattle with an overhand posture illustrates the use of planning about a macroscopic feature of manual control, whether to turn the hand counterclockwise and grab the rattle with an underhand posture, or turn the hand clockwise and grab the rattle with an overhand posture. The main question addressed in the present research is how such decision making occurs.

Another naturalistic observation shows that people occasionally adopt awkward postures to end up in comfortable ones. While dining at a restaurant, Rosenbaum noticed a waiter reaching for a glass that was standing upside down on a table. The waiter reached for the glass with an underhand posture, that is, with his thumb pointing down; after picking up the glass, he turned it over to fill it with water. In principle, the waiter could have reached for the glass with an overhand grip, but this would have put him in the awkward position of having to hold the glass with his thumb pointing down during the pouring operation. To avoid ending in this awkward position, the waiter initially grabbed the glass in a manner that was relatively uncomfortable.

These examples illustrate the importance of planning for everyday actions. However, very little research has been done on the way such planning occurs. (Most research falling under the rubric of motor control has been concerned with the question of how already-chosen movements are carried out.) If planning is omitted from motor-control research, we may end up understanding how movements are *executed* with little or no idea of how they are *chosen*. What we believe is needed, therefore, is a *grammar* of action as well as a *mechanics* of action.

What form should the grammar take? One possibility is that the actor has stored in memory lists of condition–action pairs. For example, the baby may learn: If a rattle is at such and such a location, grab it with an overhand grip; if it is at some other location, grab it up with an underhand grip, and so on. The problem with this approach is that it may not be helpful in novel situations. For example, it may never explain how someone whose arms are filled with packages determines that she can close a light switch with her chin. Another disadvantage of the approach is that it may load memory unnecessarily.

Another possibility is that the planning system does not specify in advance which actions *are* allowed in which situations, but rather specifies which actions, or properties of actions, are *not* allowed. This proscriptive method has been adopted by Chomsky (1986) in his analysis of linguistic grammar. If a similar approach can be pursued for a grammar of action, the challenge is to identify the constraints that the system uses to rule out actions in particular contexts. Ultimately, the grammar should have a minimal set of constraints that maximally predict performance.

One constraint that can by hypothesized at this point comes from the examples mentioned above. When the baby brought her hand around the rattle to grab it with an overhand grip, she avoided twisting her arm. Similarly, when the waiter picked up the glass with an underhand grip, he apparently tried to minimize the time his hand would be in an awkward posture. Hence, a constraint underlying the selection of manual action (and perhaps other sorts of actions) may be: *Minimize times in extreme joint angles*. By extreme joint angles, we mean joint angles that bring the joint far from its resting position. Bringing the joint to an extreme angle requires extreme force, given the length–tension properties of muscle (cf. Berkenblit, Feldman, & Fucson, 1986; Houk & Rymer, 1981). The fact that muscle and tendon stretch receptors tend to fire most vigorously at extreme joint angles, which in turn gives rise to corrective responses (Houk & Rymer, 1981), accords with the assumption that extreme joint angles are generally avoided.

The experiments reported here were designed to test the Extreme Joint Angle Constraint using an ecologically valid task modeled after the naturalistic observations described above. The first experiment addressed the question of whether subjects would pick up a horizontally positioned cylindrical bar with an overhand or underhand grip when the left or right end of the bar was to be brought in contact with a horizontal surface. The second and third experiments tested alternative interpretations of the results of Experiment 1.

## EXPERIMENT 1

### Method

Figure 10.1A shows the apparatus. A bar, 32 cm long, 2 cm in diameter, weighing approximately 43 gm, with a black tape on one end and gray tape on the other, rested on two supports, 26 cm apart and fastened to a table 76 cm high. The bar lay on the supports, with the black end on the left support and the gray end on the right support. When the bar lay in this position, its bottom edge was about 5 cm above the table, and so it could be picked up with the fingers fitting beneath it. To the left of the bar lay a blue disk, and to the right lay a red disk. Each disk was 12 cm in diameter with its midpoint about 16 cm from the nearest support. Subjects stood about 140 cm from the bar and were instructed to stand with the midsagittal plane of the body aligned with the bar's midpoint. On each trial, subjects were asked to step toward the bar, pick it up, and place either the black or gray end squarely on the red or blue disk, depending on the instruction, and then to return the bar to its start position. Performance was videotaped through a one-way mirror with the subject's prior consent. There were four conditions achieved by combining the tasks of bringing the left or right end of the bar to the left or right disk. The original design called for the four conditions to be run in a completely balanced order, which would have required 24 subjects, but because the first twelve subjects performed identically, the experiment was terminated after this group was tested. Subjects were not told in advance which hand was to be used to complete the task. If a subject used one hand for some translations and the other hand for others, that subject was asked, after completing all four tasks, to use only the preferred hand from then on; the four tasks were then retested. The experiment took about 5 minutes per subject and was conducted without speed pressure. No specific instructions were given to the subjects about how to grab the bar. The subjects were all neurologically normal university students and staff. They participated on a volunteer basis.

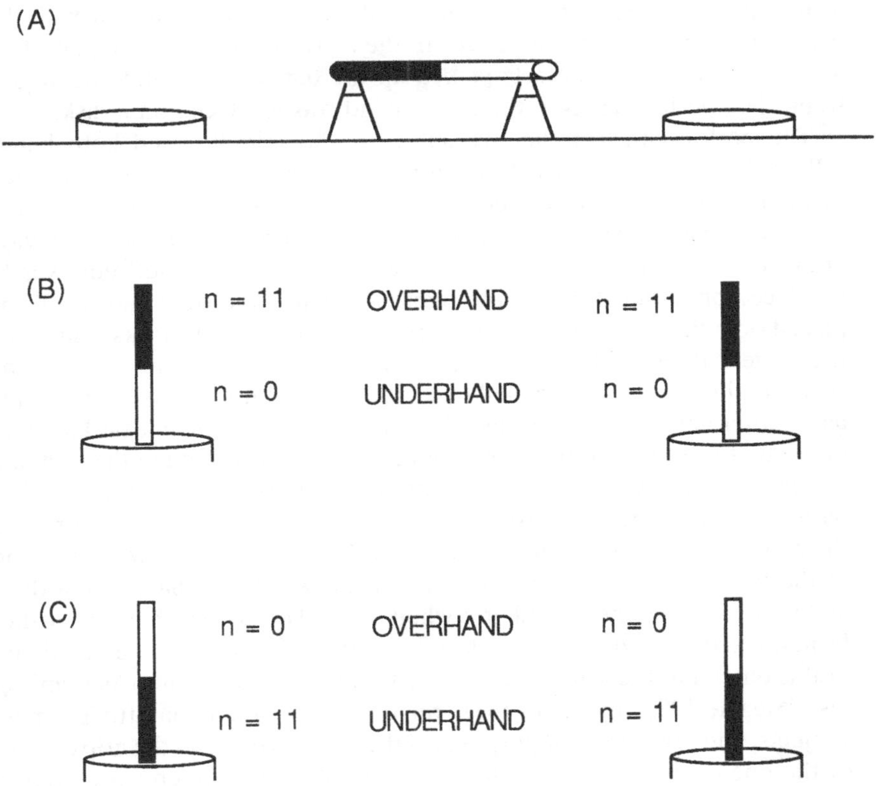


FIG. 10.1 Apparatus used in Experiment 1 (*A*). Of the subjects who used the right hand exclusively by the second cycle of conditions (11 of 12 subjects), all used overhand grasps to bring the white (right) end of the bar to both target disks (*B*), and all these subjects used underhand grasps to bring the black (left) end of the bar to both target disks (*C*). The one subject who used the left hand behaved in the mirror image fashion.

## Results

Of the twelve subjects, seven used only one hand, whereas the other five subjects used both hands. Of the seven subjects who used one hand, six used the right hand only. All the subjects in the one-hand group who used the right hand used an overhand grip when the right end of the bar was to be placed on the left or right disk and an underhand grip when the left end of the bar was to be placed on the left or right disk (see Fig. 10.1B). For the one subject who always used the left hand, performance mirrored that of the right-hand-only subjects; this subject used the underhand grip when placing the right end of the bar on the left or right disk and the overhand grip when placing the left end of the bar on the left or right disk.

Thus, all the subjects who used one hand only picked up the bar with the overhand grip when the end closer to the active hand was to be placed on a disk, and all these subjects picked up the bar with the underhand grip when the end farther from the active hand was to placed on a disk.

The performance of the five subjects who initially used both hands followed the same pattern. (All subjects in this "switching" group opted to use the right hand in the second run through the tasks when they were instructed to use the preferred hand only.) One subject in this group always grabbed the bar overhand, using the left hand when the left end was to be placed on either disk and the right hand when the right end was to be placed on either disk. When this subject repeated the four tasks using only the preferred (right) hand, she performed in exactly the same way as the right-hand subjects described above. The other four subjects who initially used either hand made hand selections in an apparently haphazard manner in the first run through the tasks, though every grasp, both in the first and second run, followed the generalization presented at the end of the last paragraph. Thus, without exception, subjects in the switching group, like the subjects in the one-hand group, used the overhand grip when the end of the bar that was closer to the active hand was to be placed on a disk, and they used the underhand grip when the end of the bar that was farther from the active hand was to be placed on a disk. To evaluate the likelihood of this outcome, the grip preference data were analyzed with Cochran's $Q$ test (Siegel, 1956, pp. 161–165), relying on the final four trials for all subjects, and reversing, for purposes of the analysis only, all the grip choices of the one subject who used the left hand. The grip preferences differed significantly from chance: $Q = 33.00$, $p < .001$, ($df = 3$).

## Discussion

What accounts for the interaction between the hand grips that subjects used (overhand versus underhand) and the end of the bar that was supposed to be placed on the target disks? Our hypothesis is that subjects tried to avoid extreme joint angles. Consider a subject who used the right hand only. If the subject used the overhand grip when the left end of the bar was to be placed on a disk, this would have resulted in an awkward final position because the arm would have adopted extreme joint angles when the end of the bar was brought to the disk. Similarly, if the subject used the underhand grip when the right end of the bar was to be placed on a disk, this would have resulted in an awkward final posture. The results of this study can be taken to suggest, then, that subjects planned the initial grip so as to minimize the awkwardness of the final position. This outcome is all the more remarkable when it is recalled that the planning was done with respect to a final stimulus position that could not be seen and a final arm position that could not be felt. Thus, the results extend the findings of others (Arbib, Iberall, & Lyons, 1985; Jeannerod, 1981) who have reported that people adjust their grip patterns depending on the properties of the objects for which they are reaching. The present results indicate that subjects adjust their grip patterns depending on what they plan to do with the objects being grabbed. A similar conclusion has been reached by Klatzky, McCloskey, Doherty, Pellegrino, and Smith (1987) in a study of haptic exploration and by Marteniuk, MacKenzie, Jeannerod, Athenes, and Dugas (1987) in a study of grip aperture control.

## EXPERIMENT 2

The next experiment was designed to test our interpretation of the results of Experiment 1. The experiment had three main objectives. One was to determine how decisions about awkwardness actually determine grip choice. A second objective was to study grip choices in more situations than the ones used in Experiment 1. A third objective was to see what would happen if subjects had to move the bar to a series of targets rather than one. Would the initial choice of grip be affected by the task to be performed after the first one? If so, this would provide evidence for extensive pre-planning.

### Apparatus

Fig. 10.2 shows the apparatus. Instead of having a disk to the right and left of the cradle, as in Experiment 1, both disks were on the right side of the cradle, with one disk about 65 cm above the other. The disks now had holes in them, and subjects were asked to insert one or the other end of the bar into the hole of the disk named in the instruction. The task was either to bring the bar down into the hole of the bottom disk until it contacted the table on which that disk rested, or to bring the bar into the hole of the upper disk until it contacted the frame which supported that disk. The bar used in this experiment was 3.5 cm in diameter, 24 cm long, weighed 71 gm, and was painted half black and half white in the same manner as the bar used in Experiment 1. The bases of the supports were color coded black and white to indicate how the bar was to be positioned. The two disks were 1.75 cm thick and 15 cm in diameter with a 4-cm hole in the center. The disk that rested on the table (14 cm from the edge of the table closest to the subject and with its center 18 cm from the right support) was red. The disk in the top position was blue.

### Rating Study

Prior to the main experiment, we conducted a rating study to obtain estimates of the perceived awkwardness of holding the bar in every static

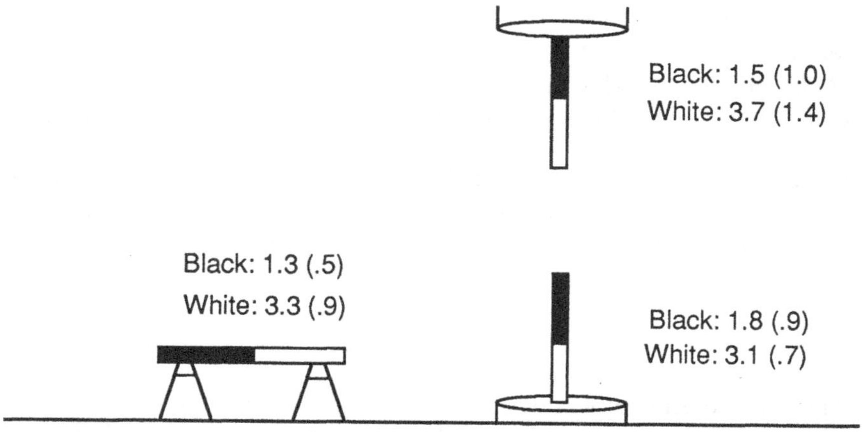


FIG. 10.2 Apparatus used in Experiment 2. A single bar was used in the experiment. It is shown here in each of the three positions it could occupy (though the opposite black–white orientations were also possible in the two vertical positions during the main experiment but not during the rating study). The sizes of the cradle, bar, and disk, and the distances between the disks are not shown drawn to scale (see text for exact measurements). Next to each bar position is the mean (and standard deviation) of the awkwardness ratings given by the independent raters ($n$ = 12) when they were asked to hold the bar with the base of the thumb toward the black or white end. *1* = Least Awkward; *5* = Most Awkward.

position that was possible in the main experiment. (Ratings of awkwardness of static postures have also been collected by Parsons, 1987). The ratings were used to test predictions about alternative methods that subjects used in determining grip preferences.

Twelve right-handed subjects participated in the rating study. They were asked to hold the bar with the right hand, with the thumb wrapped around the bar as subjects in Experiment 1 had done, but with the base of the thumb toward the black or white end. Ratings were obtained with the bar in each of three positions; resting horizontally in the cradle (with the black end to the left), resting vertically in the hole of the lower disk (with the black end up), and resting vertically in the hole of the upper disk (with the black end up). The raters were simply asked to hold the bar and evaluate how awkward the static position felt, where *1* was least awkward and *5* was most awkward. The order in which the positions were tested was randomized for each subject, and each subject went through the six conditions (three locations × two hand postures) twice, the first cycle being used to familiarize subjects with the rating scale. The means and standard deviations of the second set of ratings are shown in Figure 10.2, where it is seen that rated awkwardness was less when the thumb pointed to the left than to the right when the bar was horizontal, and was less when the thumb pointed up than down when the bar was vertical.

## Method

The experimental trials began with the subject standing 75 cm from the table. The tasks involved taking one step forward, grabbing the bar, performing the manipulation designated in the instructions, and returning the bar to its original position.

As seen in Table 10.1, there were two types of bar transport moves—single and double—as well as two start positions—horizontal (in the cradle) and vertical (in the red disk). Single moves (Tasks 1–4) involved placing the end of the bar in the hole of one disk and returning it to the start position. In Task 2, for example, where the instruction was *Black to Red*, the subject was supposed to remove the bar from the cradle and put the black end in the hole of the red disk, then return the bar to the cradle with the black end positioned over the black support. Double moves (Tasks 5–8) required a series of two bar transports. In Task 7, for example, where the instruction was *Black to Red, White to Blue*, subjects were supposed to put the black end of the bar in the hole of the red disk, then put the white end in the hole of the blue disk, and then return the bar to the cradle. There was also a single move condition in which the subject was supposed to flip the bar so its white end went into the red disk after the bar stood in the red disk on its black end (Task 9).

For the horizontal starting position, half the subjects had the black end on the left, and the other half had the white end on the left. For the vertical starting position, half the subjects had the black end on the bottom, and half had the white end on the bottom. The manipulations were tested in a balanced order. All subjects were told to manipulate the bar with the right hand only.

Twelve right-handed subjects participated for payment or course credit. As in Experiment 1, their performance was videotaped with prior consent. The experiment took about 20 minutes. Subjects were paid or received course credit.

## Results and Discussion

Inspection of the videotapes revealed a striking individual difference. Six out of twelve subjects picked up the bar with the thumb on one side and the opposing fingers on the other side and occasionally rotated the bar within the hand by momentarily holding the bar between the index and middle fingers. We call these subjects *twirlers*. Having twirling as a behavioral option reduces the constraints on the grip patterns that may be

TABLE 10.1
Tasks Awkwardness Ratings, and Observed Grips in Experiment 2

| *Task* | *Start* | *Action* | *Thumb Direction* | *Awkwardness Ratings* *P1* | *P2* | *P3* | *Mean* | *Observed* |
|---|---|---|---|---|---|---|---|---|
| 1 | Cradle | White to Red | Black | 1.3 | 1.8 | – | 1.6 | 6 |
| | | | White | 3.3 | 3.1 | – | 3.2 | 0 |
| 2 | Cradle | Black to Red | Black | 1.3 | 3.1 | – | 2.2 | 1 |
| | | | White | 3.3 | 1.8 | – | 2.6 | 5 |
| 3 | Red | White to Blue | Black | 3.1 | 3.7 | – | 3.4 | 0 |
| | | | White | 1.8 | 1.5 | – | 1.7 | 6 |
| 4 | Red | Black to Blue | Black | 3.1 | 1.5 | – | 2.3 | 5 |
| | | | White | 1.8 | 3.7 | – | 2.8 | 1 |
| 5 | Cradle | Black to Red, Black to Blue | Black | 1.3 | 3.1 | 1.5 | 2.0 | 0 |
| | | | White | 3.3 | 1.8 | 3.7 | 2.9 | 6 |
| 6 | Cradle | White to Red, White to Blue | Black | 1.3 | 1.8 | 3.7 | 2.3 | 5 |
| | | | White | 3.3 | 3.1 | 1.5 | 2.6 | 1 |
| 7 | Cradle | Black to Red, White to Blue | Black | 1.3 | 3.1 | 3.7 | 2.7 | 1 |
| | | | White | 3.3 | 1.8 | 1.5 | 2.2 | 5 |
| 8 | Cradle | White to Red, Black to Blue | Black | 1.3 | 1.8 | 1.5 | 1.5 | 6 |
| | | | White | 3.3 | 3.1 | 3.7 | 3.4 | 0 |
| 9 | Red | White to Red | Black | 3.1 | 1.8 | – | 2.5 | 2 |
| | | | White | 1.8 | 3.1 | – | 2.5 | 4 |

Note. All task descriptions assume starting positions with the black end of the bar in the left end of the cradle or in the red (bottom) disk. P1, P2, P3 denote positions 1, 2, 3, respectively.

initially adopted in this task because rotating the bar within the hand makes it possible always to avoid awkward hand postures. It is interesting that some subjects twirled the bar whereas others never did. Nothing in our instructions prevented subjects from twirling. In fact, we purposely said nothing about this aspect of performance because we wanted to observe the entire range of behavior that subject would spontaneously adopt. Twirling the bar was an efficient behavior, given that the bar was light and short. Five of the six subjects who twirled the bar did so on more than one occasion, suggesting that the twirling strategy was consistently available to them. There was no obvious physical difference between the twirling and nontwirling subjects such as gender, size, or apparent strength. Our conjecture is that the difference between the two subject groups arose from a difference in the way they cognitively represented the task. In the traditional problem-solving literature, it is well known that problems can be represented differently, and in fact may be solved or not solved depending on the availability of the right kind of representation (Wickelgren, 1974). It may be that the twirlers and nontwirlers represented the bar-transport problem in fundamentally different ways. If this is so, problem representation is a critical feature of motor tasks, even those as mundane as picking up a bar and bringing it to a new location. This conclusion accords with the view that cognitive abilities may originate in motor abilities (Keele, 1987; Kimura, 1979; Lieberman, 1984; Piaget, 1952).

Since the way one initially grasps an object is more critical if the grasp cannot change than if it can, initial grasps were more diagnostic of future planning in nontwirling than in twirling subjects. Accordingly, we restricted our analysis of initial grip patterns to the six subjects who never twirled the bar. For purposes of analyzing their grip pattern, we departed from the categories *overhand* and *underhand* and instead described grip patterns in terms of the end of the bar toward which the thumb, or the base of the thumb, pointed. The reason for the new method of description was to allow the named grip to remain constant as long as the bar's orientation within the hand remained fixed, regardless of the hand's orientation in space. Referring to the end of the bar to which the thumb points has one disadvantage, however: Because the bar started with the black end to the left for half the subjects and the white end to the left for the other half of the subjects (and similarly for the vertical start position), referring to a particular color without regard to the starting orientation could obscure the consistency of subject's grip preferences. However, color had no apparent effect on the grip preferences that were exhibited. Therefore, it was possible to adopt the convention that the bar always started with the black end on the left cradle support (for the horizontal starting position) and with the black end in the red disk (for the vertical starting position).

The data are shown in Table 10.1. To analyze the data using Cochran's Q test, we separated conditions in which the bar started in the vertical and horizontal orientations because differences in grip preferences for the two starting positions could have been due to the immediate physical demands of picking up the bar rather than to the expected demands of bringing the bar to its target position(s). The grip preference data differed significantly from chance, both in the conditions where the bar initially lay in the cradle ($Q = 23.78$, $p < .001$, $df = 5$) and in the conditions where the bar initially stood in the red disk ($Q = 7.00$, $p < .05$, $df = 2$).

Let us review the data from each task. In Task 1, lifting the bar from the cradle and placing the white end in the red disk, all six subjects picked up the bar with the thumb pointing to the black (left) end. As seen in

Table 10.1, the mean awkwardness ratings were lower for the thumb-to-black grip than for the thumb-to-white grip, both for the initial *and* final positions. Thus, subjects performed optimally in Task 1, at least judging from the awkwardness ratings.

In Task 2, lifting the bar from the cradle and placing the black end in the red disk, five out of six subjects picked up the bar with the thumb pointing to the white (right) end. This outcome replicates Experiment 1. With respect to the awkwardness ratings, the majority of subjects seemed to minimize the awkwardness of the *final* position. That is, five out of six subjects used the thumb-to-white posture even though the awkwardness rating for this grip in the initial position was *higher* than for the overhand grip. Also, the *mean* awkwardness rating was higher for the thumb-to-white posture than for the thumb-to-black posture, suggesting that subjects were not trying to minimize *mean* awkwardness. The tentative conclusion from these results is that subjects attempted to minimize the awkwardness of the *end* of the bar transport move. Note that this conclusion is consistent with the data from Task 1.

In Task 3, the bar initially stood in the red disk with the black end pointing down. Subjects were instructed to bring the white end up into the hole of the blue disk. All subjects performed this task with the thumb toward the white end. This was the optimal grip, judging from the awkwardness ratings, both for the initial and final positions.

In Task 4, as in Task 3, the bar initially stood in the red disk with the black end down, but the task was to place the *bottom* end (the black end) into the hole of the blue disk. Under this condition, five out of six subjects picked up the bar with the thumb pointing down (to the black end). Presumably, they adopted this initial grip to end with the thumb pointing up. This is the first instance in which subjects used the thumb down grip when holding the bar in the red disk. Recall that subjects *avoided* this posture in the other conditions we described. This finding is consistent, however, with the view that subjects tried to minimize the awkwardness of the *final* position, a view that accords with the grip preference data from all the tasks discussed so far.

Was it the end of the *first* bar transport move or the end of the *final* bar transport move that determined subjects' choices? In the tasks considered so far, we cannot decide between these alternatives because the first bar transport move was also the last. To decide between these alternatives, we needed tasks in which subjects made *series* of transport moves rather than one. The theoretically most important of these conditions was Task 5. Here subjects took the bar from the horizontal position, placed the left end in the red (bottom) disk and then placed the lower end of the bar into the hole of the blue (top) disk. Suppose subjects tried to minimize the awkwardness of the final position. It would be better for them to grab the bar with the thumb toward the left end than with the thumb toward the right end because the left-pointing posture would result in a more comfortable final position than would the right-pointing posture. However, as we have seen, when subjects transported the left end of the bar from the cradle to the bottom disk (which is also the first move in the two-move sequence required in Task 5), they preferred to pick up the bar with the thumb toward the *right* end rather than the left. Hence, if subjects assigned priority to the final position in this two-move sequence, they should have changed from the thumb-to-right grip to the thumb-to-left grip.

As seen in Table 10.1, subjects did *not* make this change. Every subject grabbed the bar with the thumb pointing to the white end of the bar. Examining the awkwardness ratings for the initial, intermediate, and final positions, it is seen that the grip preference data were not predicted by the awkwardness of the final position, the awkwardness of the initial position, or the mean awkwardness. Rather, the grip preference data were accounted for the awkwardness of the end of the first bar transport move.

The data from the other two-move conditions support this interpretation. When subjects were given the instruction *White to red, white to blue* (Task 6), five out of six subjects picked up the bar with the thumb toward the black end. (Note that this outcome is *inconsistent* with the hypothesis that *final* awkwardness was the critical determinant of initial grip choice.) When subjects were given the instruction *Black to red, white to blue* (Task 7), five out of six subjects picked up the bar with the thumb toward the white end, again consistent with the notion that subjects tried to minimize awkwardness at the end of the first bar transport move. When subjects were given the instruction *White to red, black to blue* (Task 8), all six subjects picked up the bar with the thumb toward the black end, again consistent with the hypothesis that minimizing the awkwardness of the end of the first bar-transport move was the main optimization constraint.

There is one condition where minimizing the awkwardness of the end of the first bar transport move did not account for the grip preference data. This was Task 9, which entailed flipping the bar when it rested in the red disk. As seen in Table 10.1, two out of six subjects in this condition grabbed the bar with the thumb pointing to the black end (enabling them to end with the less awkward thumb up posture), but four of the six subjects grabbed the bar with the thumb pointing to the white end (causing them to end with the more awkward thumb up posture). This was the only condition in which fewer than five out of six subjects adopted a single grip pattern, and it is the only condition in which subject's grip preferences did not accord with the strategy of minimizing awkwardness at Position 2.

## EXPERIMENT 3

Give the small number of subjects in Experiment 2 whose data were used, we felt it was important to replicate the results with a larger group. To increase the number of usable subjects, participants in the third experiment were told not to twirl the bar. Instead they were urged to "grab the bar in the middle, holding it like a tennis racket." Several new tasks were also introduced (see Table 10.2). Besides including all the conditions from Experiment 2, we added tasks in which the bar was to be brought from the cradle directly to the blue (top) disk, either from the cradle or from the red (bottom) disk, and where the move to the blue disk was either the last move to be preformed (before returning the bar to the original location) or intermediate between two other transport moves. There was also a single move condition (Task 16) in which the bar initially stood in the red disk (with the black end down) and subjects were asked to bring the bar to the cradle, placing the black end on the cradle's left side. Because this task could be regarded as the opposite of Task 2 (placing the black end of the bar into the red disk after removing it from the cradle), we expected that subjects would show different grip preferences in Tasks 2 and 16. Specifically, we expected that in Task 16 subjects would grab the bar with the thumb toward the black end, whereas in Task 2 they would grab the bar with the thumb toward the white end.

## Method

For all subjects the bar started with the black end on the left side of the cradle or with the black end in the red disk. The apparatus was the same as in Experiment 2. Twelve new subjects participated for pay or course credit. The order of conditions was randomized for each subject. The experiment lasted about 30 min.

## Results and Discussion

The results are shown in Table 10.2. The data were analyzed as in Experiment 2. Grip preferences differed significantly from chance both in the conditions where the bar initially lay in the cradle ($Q = 38.58, p < .001, df = 11$) and in the conditions where the bar initially stood in the red disk ($Q = 17.32, p < .001, df = 3$).

In terms of performance within tasks, the results of Experiment 3 essentially replicate the findings of the previous experiments. As in Experiment 2, grip preferences were accounted for most effectively by assuming that subjects tried to minimize the awkwardness of the end of the first bar-transport move. Fourteen of the sixteen tasks followed this generalization; only Tasks 9 and 15 yielded data inconsistent with it. Task 15 was the task in which subjects flipped the bar; this was also the condition in Experiment 2 that yielded data inconsistent with the end-of-first-move principle. Task 9 was comparable to the other two-move sequences that were studied. In

TABLE 10.2
Tasks Awkwardness Ratings, and Observed Grips in Experiment 3

| Task | Start | Action | Thumb Direction | Awkwardenss Ratings P1 | P2 | P3 | Mean | Observed |
|---|---|---|---|---|---|---|---|---|
| 1 | Cradle | White to Red | Black | 1.3 | 1.8 | – | 1.6 | 11 |
| | | | White | 3.3 | 3.1 | – | 3.2 | 1 |
| 2 | Cradle | Black to Red | Black | 1.3 | 3.1 | – | 2.2 | 3 |
| | | | White | 3.3 | 1.8 | – | 2.6 | 9 |
| 3 | Red | White to Blue | Black | 3.1 | 3.7 | – | 3.4 | 0 |
| | | | White | 1.8 | 1.5 | – | 1.7 | 12 |
| 4 | Red | Black to Blue | Black | 3.1 | 1.5 | – | 2.3 | 8 |
| | | | White | 1.8 | 3.7 | – | 2.8 | 4 |
| 5 | Cradle | White to Blue | Black | 1.3 | 3.7 | – | 2.5 | 1 |
| | | | White | 3.3 | 1.5 | – | 2.4 | 11 |
| 6 | Cradle | Black to Blue | Black | 1.3 | 1.5 | – | 2.3 | 12 |
| | | | White | 3.3 | 3.7 | – | 2.8 | 0 |
| 7 | Cradle | Black to Red, Black to Blue | Black | 1.3 | 3.1 | 1.5 | 2.0 | 4 |
| | | | White | 3.3 | 1.8 | 3.7 | 2.9 | 8 |
| 8 | Cradle | White to Red, White to Blue | Black | 1.3 | 1.8 | 3.7 | 2.3 | 10 |
| | | | White | 3.3 | 3.1 | 1.5 | 2.6 | 2 |
| 9 | Cradle | Black to Red, White to Blue | Black | 1.3 | 3.1 | 3.7 | 2.7 | 6 |
| | | | White | 3.3 | 1.8 | 1.5 | 2.2 | 6 |
| 10 | Cradle | White to Red, Black to Blue | Black | 1.3 | 1.8 | 1.5 | 1.5 | 11 |
| | | | White | 3.3 | 3.1 | 3.7 | 3.4 | 1 |
| 11 | Cradle | White to Blue, Black to Red | Black | 1.3 | 3.7 | 3.1 | 2.7 | 1 |
| | | | White | 3.3 | 1.5 | 1.8 | 2.2 | 11 |
| 12 | Cradle | White to Blue, White to Red | Black | 1.3 | 3.7 | 1.8 | 2.3 | 3 |
| | | | White | 3.3 | 1.5 | 3.1 | 2.6 | 9 |
| 13 | Cradle | Black to Blue, Black to Red | Black | 1.3 | 1.5 | 3.1 | 2.0 | 12 |
| | | | White | 3.3 | 3.7 | 1.8 | 2.9 | 0 |
| 14 | Cradle | Black to Blue, White to Red | Black | 1.3 | 1.5 | 1.8 | 1.5 | 11 |
| | | | White | 3.3 | 3.7 | 3.1 | 3.4 | 1 |

TABLE 10.2
*(Continued)*

| | | | | *Awkwardenss Ratings* | | | | |
|---|---|---|---|---|---|---|---|---|
| *Task* | *Start* | *Action* | *Thumb Direction* | *P1* | *P2* | *P3* | *Mean* | *Observed* |
| 15* | Red | White to Red | Black | 3.1 | 1.8 | – | 2.5 | 5 |
| | | | White | 1.8 | 3.1 | – | 2.5 | 6 |
| 16 | Red | Black to Cradle | Black | 3.1 | 1.3 | – | 2.2 | 8 |
| | | | White | 1.8 | 3.3 | – | 2.6 | 4 |

Note. Data from eleven rather twelve subjects is included because one of the twelve subjects misunderstood the instructions and did not perform this task.

this task subjects brought the black end of the bar from the left end of the cradle to the red disk and then brought the white end of the bar to the blue disk. Ideally, all the subjects in this condition should have picked up the bar with the thumb pointing to the white end, but only half the subjects did so. Inspection of the videotapes of subjects' performance in this task did not reveal anything unusual about their performance, nor did examination of the sequential position of Task 9 vs à vis the other tasks; for example, it was not the case that, by chance alone, Task 9 was tested after an unusually strong bias was created for the thumb-to-black grip. The results of Task 9 cannot be explained by saying that subjects were simply insensitive to differences in the awkwardness of Positions 1, 2, or 3, because in other comparable conditions the same subjects exhibited clear grip preferences (for example, in Task 11, where the awkwardness ratings of Position 1 as well as the mean awkwardness ratings matched those of Task 9). For these reasons, we are forced to conclude that the results of Task 9 were simply attributable to chance variation.

Returning to the flipping task (Task 15), the other condition that yielded results inconsistent with the end-of-first-move hypothesis, let us consider why subjects divided themselves approximately equally between starting with thumb up and thumb down postures. One possibility is that subjects based their grip choices on the mean awkwardness of positions 1 and 2. In the flipping task, the mean awkwardness of positions 1 and 2 was of course the same regardless of whether the thumb up or thumb down position was initially adopted. However, this averaging hypothesis is vitiated by the results of two other tasks in this experiment—Tasks 2 and 16. These tasks were opposites, and so their mean awkwardness ratings had to be the same, yet subjects made opposite grip choices in them. In Task 2, three times as many subjects grabbed the bar with the thumb toward the white end as toward the black end, but in Task 16 twice as many subjects grabbed the bar with the thumb toward the black end as toward the white end. These results are inconsistent with the averaging hypothesis but consistent with the end-of-first-move principle. Because the averaging hypothesis was violated here and in other conditions of the present and preceding experiment, we hesitate to accept it as an explanation of the results from the flipping conditions.

## HYPOTHESES

How then can we account for the results of all the conditions? Can we do so with a single model? This section reviews several possibilities we have considered.

### Working Backwards

One hypothesis is that subjects worked backward from the end of the first move, imagining or otherwise representing the optimal posture for the *end* of the first move before considering the *start* of the first move. If subjects planned backwards, and if they felt compelled to start moving without too much delay after hearing the instructions, they would have been more likely to curtail planning of the *initial* hand state than the *final* hand state, and so the initial hand state would have been more likely to be awkward than the final hand state was. There are two arguments against this hypothesis. First, if subjects worked backwards, they should have done so from the *final* end position, yet, as we have seen, they minimized the awkwardness of the end of the *first* move even if another move was coming. A second argument is that when we watched the videotapes of subjects' performance, we observed that subjects often switched from a preshaped overhand posture to a preshaped underhand posture, or vice versa, just before grabbing the bar. This outcome provides support for Jeannerod's (1981) findings about hand shaping occurring in the late phases of reaching, except that here subjects altered their hand shape in anticipation of a *later* task, not the one to be performed immediately. This late change in the orientation of the hand is damaging to the working-backwards hypothesis because working backwards predicts that subjects should select a hand shape relevant to the end of the first move *before* initiating the first move.

### Fatigue

A second hypothesis is that the planning system anticipates the effects of fatigue and tries to get awkward postures out of the way as quickly as possible. One argument against this hypothesis is that subjects assigned priority to the end of the first move even when there was another move to come. If the fatigue hypothesis were correct, the ultimate final position should have been assigned highest priority. Another argument comes from the data in the flipping conditions. If subjects consistently chose easy grip postures first, they should have done so in the flipping conditions as well as the other conditions, yet, as we have seen, this was not the case. In the flipping conditions, subjects were about equally likely to adopt the easy or difficult grip pattern. Given this outcome, neither the fatigue hypothesis nor the working-backwards hypothesis can be accepted as a generally correct explanation of subjects' grip preferences.

### Minimize Time in Awkward Postures

A third hypothesis is that subjects only planned one move at a time (perhaps because they were not highly practiced in this task) and tried to minimize time in awkward postures. Suppose that when subjects planned to transport the bar from one location to another they expected the *ends* of the moves to be more time consuming than the *starts*. This seems reasonable because initiating the moves simply required grabbing the bar, whereas terminating the moves required aiming to place the bar in the cradle or hole it was supposed to occupy. If subjects sought to minimize the amount of time in awkward postures, it would have been adaptive for them to avoid awkwardness during the final aiming phase. In the conditions that required flipping the rod, the actual or perceived difficulty of aiming relative to grabbing may have been less than in the other conditions.

We have tested this hypothesis in another experiment in which we varied the difficulty of aiming. In one condition, the bar stood on a very large disk, and in another condition, the bar stood on a very small disk. There was no hole in either disk, and the bar was a bit wider and shorter than the bar used in the present experiments so it could easily be stood on end. In both the large- and small-disk conditions, subjects were asked to flip the rod, making sure it ended up within the confines of the disk. Because aiming was more difficult for the small disk, we predicted that when subjects flipped the bar on the small disk they would be more likely to grab the bar with the thumb pointing down than with the thumb pointing up. By contrast, when subjects flipped the bar on the large disk, we predicted that they would be equally likely to grab the bar with the thumb toward the top or bottom end. These predictions were *not* confirmed. Subjects were equally likely to grab the bar with the thumb pointing up or down regardless of whether the disk was large or small. Similarly, when the bar lay on the cradle and had to be transported to the large or small disk, the size of the disk had no effect on the types of grips that were adopted. Thus the data from this experiment do not support the aiming hypothesis. Although it is possible that the hypothesis might have been confirmed if the range of aiming difficulty were larger, we are doubtful that this is the case because we believe there is an explanation of the results of this experiment as well as the other experiments we have reported that is both simpler than the aiming hypothesis and more powerful.

## Exploiting Awkwardness

The key to the more powerful hypothesis is that subjects may have not only tried to minimize awkwardness as Position 2 but may have tried to *exploit* awkwardness at Position 1. The rationale behind this proposal can be appreciated by noting that almost all the tasks we studied required translation as well as *rotation* of the hand. It is a property of the right arm, attributable to the elastic properties of its musculature, that when the thumb points toward 12, 1, 2, 3, 4, or 5 o'clock, the arm's tendency is to turn in the *counterclockwise* direction, back to the resting position at around 11 o'clock, but when the thumb points toward 5, 6, 7, 8, 9, or 10 o'clock, its tendency is to turn in the *clockwise* direction, back again to the 11 o'clock resting position. The tendency of the hand to turn clockwise or counterclockwise depending on its initial position, coupled with the possibility that the planning system has access to this information, can explain our results. When subjects had to transport the *right* end of the bar from the cradle to the red disk, they picked up the bar with the thumb pointing to the left; this grip facilitated the necessary clockwise rotation of the bar. However, when subjects had to transport the *left* end of the bar from the cradle to the red disk, they picked up the bar with the thumb pointing to the right; this grip facilitated the necessary *counterclockwise* rotation. When subjects had to *flip* the bar, they had the option of turning it clockwise or counterclockwise, and so they could pick up the bar either with the thumb up or down, which is what they did. All the main features of our results, therefore, can be explained with this simple model, provided we also assume that subjects planned no further than Position 2 (or were unaffected if their early grasping behavior if they did plan beyond Position 2).

A prediction that arises from this model is that directions of hand rotation should depend on initial hand orientation. We have studied videotapes of subjects' performance and found that this prediction is beautifully confirmed: Virtually without exception when subjects rotated the bar after picking it up with the thumb pointing down or to the left, they rotated the bar in the clockwise direction, and virtually without exception when subjects rotated the bar after picking it up with the thumb pointing up or to the right, they rotated the bar in the counterclockwise direction. Thus the observed directions of rotation were as predicted. Given this outcome, we are reasonably confident that our explanation of the grip preference data is correct. In the next section, we consider the implications of this conclusion for our analysis of action planning and the implications of our research strategy for ongoing research in the field of motor control.

## CONCLUSIONS

We began this chapter by saying that students of action should pay more attention to the selection of macroscopic features of motor behavior–that they should develop a *grammar* of action as well as *mechanics* of action. We suggested that a potentially profitable way to approach this problem is to identify constraints that the planning system uses to rule out possible movements. We hypothesized that one important constraint might be avoidance of extreme joint angles. Our research has shown that the planning system follows this constraint, although it also *exploits* extreme joint angles when they cannot be avoided.

It is doubtful that the only constraint used for action selection concerns extreme joint angles. As mentioned earlier, Hogan (1984) and Nelson (1983) suggested kinematic constraints, and Meyer et al. (this volume) adopted a constraint perspective to account for movement timing in the Fitts aiming tasks. Another constraint that is likely to play a role in selecting macroscopic action features concerns balance. Presumably, we usually select actions that prevent us from toppling over. This constraint, like the Extreme Joint Angle Constraint, can be experimentally investigated in the manner pursued here: determining the conditions under which actors perform in one way or another depending on features of the task to be performed. The same empirical approach can be used to address a variety of straightforward behavioral questions: When does one use one hand or the other to perform a task? When does one use two hands? When does one take a step toward an object to be grasped, and so on? Addressing a range of questions like these may permit us to discern the constraints that underlie action selection in a broad range of tasks.

A final remark about our study concerns the implications of our work for the rift in motor control between those who focus on the *mechanical* properties of the motor system and those who focus on the *cognitive* properties of the motor system. It has been said that neither side pays adequate attention to the concerns of the other. We have seen that the two approaches cannot be divorced from one another. The planning system has access to biomechanical limitations, but those limitations alone cannot account for the planful behaviors that are displayed in tasks like those studied here and, presumably in other tasks as well.

## ACKNOWLEDGMENTS

Supported in part by National Science Foundation grant BNS-8710933, a National Science Foundation Research Opportunity Award (to DAR and JV), and a Research Career Development Award from the National Institutes of Health (to DAR). We thank Dan Anderson, Andrew Gordon, Judith F. Kroll, and Nora and Sarah Kroll-Rosenbaum for useful contributions to various aspects of the research.

Jonathan Vaughan's primary affiliation is with the Department of Psychology, Hamilton College, Clinton, New York 13323.

Correspondence should be sent to David A. Rosenbaum, Department of Psychology, Tobin Hall, University of Massachusetts, Amherst, Massachusetts 01003.

## REFERENCES

Arbib, M. A., Iberall, T., & Lyons, D. (1985). Coordinated control programs for movements of the hand. In A. W. Goodwin & I. Darian-Smith (Eds.), *Hand function and the neocortex* (pp. 111–129). Berlin: Springer-Verlag.

Berkenblit, M. B., Feldman, A. G., & Fucson, O. I. (1986). Adaptability of innate motor patterns and motor control. *Behavioral and Brain Sciences, 9*, 585–638.

Bernstein, N. (1967). *The coordination and regulation of movements*. London: Pergamon.

Chomsky, N. (1986). *Knowledge of language*. New York: Praeger.

Hogan, N. (1984). An organizing principle for a class of voluntary movements. *The Journal of Neuroscience, 4*, 2745–2754.

Houk, J. C., & Rymer, J. C. (1981). Neural control of muscle length and tension. In V. B. Brooks (Ed.), *Handbook of physiology: Motor control (Section 1, Vol. 2)* (pp. 257–323). Bethesda, MD: American Physiological Society.

Jeannerod, M. (1981). Intersegmental coordination during reaching at natural objects. In. J. Long & A. Baddeley (Eds.), *Attention and performance IX* (pp. 153–169). Hillsdale, NJ: Lawrence Erlbaum Associates.

Keele, S. W. (1987). Sequencing and timing in skilled perception and action: An overview. In A. Allport, D. MacKay, W. Prinz, & E. Scheerer (Eds.), *Language Perception and Production* (pp. 463–487). London: Academic Press.

Kelso, J. A. S., Holt, K. G., Kugler, P. N., & Turvey, M. T. (1980). On the concept of coordinative structures as dissipative structures: II. Empirical lines of convergency. In G. E. Stelmach & J. Requin (Eds.), *Tutorials in motor behavior* (pp. 49–70). Amsterdam: North-Holland.

Kimura, D. (1979). Neuromotor mechanisms in the evolution of human communication. In H. D. Steklis & M. J. Raleigh (Eds.), *Neurobiology of social communication in primates*. New York: Academic.

Klatzky, R. L., McCloskey, B., Doherty, S., Pellegrino, J. & Smith, T. (1987). Knowledge about hand shaping and knowledge about objects. *Journal of Motor Behavior, 19*. 187–213.

Lieberman, P. (1984). *The biology and evolution of language*. Cambridge, MA: Harvard.

Marteniuk, R. G., MacKenzie, C. L., Jeannerod, M., Athenes, S., & Dugas, C. (1987).

Constraints on human arm movement trajectories. *Canadian Journal of Psychology, 41*, 365–378.

Nelson, W. L. (1983). Physical principles for economics of skilled movements. *Biological Cybernetics, 46*, 135–147.

Parsons, L. (1987). Imagined spatial transformations of one's hands and feet. *Cognitive Psychology, 19*, 178–241.

Piaget, J. (1952). *The origins of intelligence in children.* (2nd ed.). (M. Cook, Translator). New York: International Universities Press.

Siegel, S. (1956). *Nonparametric statistics for the behavioral sciences.* New York: McGraw-Hill.

Wickelgren, W. A. (1974). *How to solve problems.* New York: W. H. Freeman.

# IV THE STRUCTURE OF MOTOR PATTERNS IN LEARNED MOVEMENTS AND SPEECH

# 11 Common Factors in the Control of Free and Constrained Movements

Paolo Viviani
*University of Geneva*

ABSTRACT

It is argued that both spontaneous movements and movements under continuous visual monitoring share more common factors than hitherto acknowledged. A brief review is presented of some behavioral data supporting the notion that human gestures have unique features that set them apart from artificial movements. I try to show that these features correspond to specific rules that are obeyed in the planning and execution phases of the movements. Three experiments on visuo-manual pursuit are presented to demonstrate that, even when the geometry and kinematics of the motor response are in principle dictated from without, these rules cannot be waived. A compatibility problem may then arise whenever the input in a visuo-manual coordination task does not comply with the intrinsic properties of the human motor control system. The traditional linear-system approach to the study of pursuit tracking seems insufficient to deal with this issue.

## THE SYSTEM ANALYSIS APPROACH TO THE STUDY OF TRACKING

Tracking movements are one of the many adaptive responses with which we react to changes in the environment. They are strongly characterized by the fact that the response is effective only if it mimics accurately some measurable parameters of the world. In most cases, of course, both the environment and the response change in time, and the required matching of corresponding parameters involves on line control. Thus, the overall quality of the response—which is related to some average measure of the instantaneous mismatch—depends critically on the dynamic properties of the entire perceptuo-motor loop. The importance of these adaptive responses for everyday life need not be stressed. It is also clear that the study of these responses can provide valuable information for understanding the various components of the perceptuo-motor loop involved in their generation. Not surprisingly, then, tracking has been one of the earliest motor skills to be investigated in a laboratory setting. However, interest in the tracking skill really took off after World War II as a result of the conjunction of two related circumstances. On the one hand, the war-related development of sophisticated weapons required an increasing integration of the human operator within complex systems of measuring devices and actuators. Therefore, the operator's capacities of information processing had to be assessed with an unprecedented level of accuracy. On the other hand, applied psychologists got acquainted with the analytical techniques developed by electrical engineers for the design of servomechanisms. Together, these two circumstances are largely responsible for the fact that, since the late 1940s, the basic conceptual framework for studying the tracking skill has been provided by the mathematical theory of system control (see Wickens, 1986). Within this framework, the transduction from the sensory message to the motor response in a tracking situation is construed as a real-time functional transformation (filtering) from an input into an output.

To the extent that one is interested mostly in the ergonomic, human-engineering aspects of the tracking skill, the system analysis approach offers several advantages. Foremost among them is the fact that whenever the transduction is linear, all the dynamic aspects of the performance can be summarized in the very compact form of transfer functions. It is then easy to obtain a global description of systems comprising both human operators and man-made devices that are also characterized by transfer functions (see Frost, 1972). Moreover, the transfer function representation of the perceptuo-motor loop affords a most convenient way of dealing with the man–machine interaction problem, that is, with the problem of matching the dynamic response of the controlled devices to the properties of the human controller.

## Some Reasons for Concern

The purpose of this chapter is to argue that, despite the advantages mentioned above, there are reasons to be concerned with an unqualified adoption of the system analysis approach. In the first place, there is evidence that the assumptions that one has to make in order to apply the linear system analysis approach are too stringent to encompass the wide variety of operating conditions in which the tracking skill is called upon in real life. It is well known that a certain chain of processes may display an approximately linear behavior even though some of the components are themselves irreducibly nonlinear. Sometimes this is the result of nonlinearities cancelling each other. In some other cases the apparent linearity is the result of testing the system under very constrained laboratory conditions. In either case, the description of the system as a linear operator is then bound to be fragile. The model derived in one condition may cease to be valid if the relative weights of the component processes are altered.

The second reason for concern is related to the growing importance that the study of adaptive responses is taking as a tool for investigating the perceptuo-motor system. In fact, the feature that makes the system approach so appealing in some applied contexts is the possibility of neglecting the detailed structure of the processes responsible for the input–output transformation. If tracking behavior is not considered globally, as the applied psychologist would do, but rather as the result of the coordinated operation of a series of identifiable nervous mechanisms—perceptual, motor as well as cognitive—each one of which has to be elucidated independently, this intrinsic opacity of the system analysis approach represents a serious methodological drawback. The situation bears some resemblance to what happens when system analysis is applied to specific motor mechanisms. Just consider the example of the myotatic reflex. It is well known that if muscle stretch is taken to be the input to the system and the shortening force its output, the reflex behaves quite linearly. Moreover, over most of the biologically relevant range of frequencies, the gain and phase of the reflex transfer function are flat. Thus, as long as reflex actions are to be integrated into a larger scale representation, it may be legitimate to represent them globally by springs with controllable stiffness. However, if the main focus of interest is the actual operation of the reflex, the overall transfer function is not very informative. To achieve a true understanding of this operation, it is necessary to consider the behavior of each subsystem of the reflex in its own right—and some turn out to be far less linear than the reflex itself—and then verify that the coordinated action of all the subsystems results in the observed global behavior (Rosenthal, Roberts, McKean, & Terzuolo, 1970).

By analogy with this physiological example, it appears that the validity of the system analysis approach to the perceptuo-motor system involved in tracking is contingent upon the possibility of breaking up the loop into identifiable components, and then verifying for each component the criteria of applicability of the analysis. In this respect the results to be presented in later sections are not very encouraging inasmuch as they document certain types of behavior that cannot be accommodated within the classical framework.

### Visuo-manual Pursuit

Visuo-manual pursuit is but one of the many forms that the tracking skill can take. It is characterized jointly by three conditions: (a) the target is provided visually, (b) the response involves one or more segments of the arm, and (c) the response must be a faithful copy of the target (this last stipulation distinguishes pursuit tracking from error tracking, an other common form of the skill (see Poulton, 1974). The results presented here were all obtained under these conditions, and some of the conclusions cannot be generalized to other forms of tracking. In particular, eye tracking raises problems and issues that are specifically related to the unique characteristics of the oculomotor system: the coexistence of two sharply different modes of control—smooth pursuit and saccades—and the fact that the eye is used both for seeing the input target and for producing the holding response. These questions can only be addressed within the context of oculomotor studies (see Zuber, 1981). Nevertheless, certain issues raised in connection with visuo-manual pursuit tracking appear to be relevant for most forms of tracking behavior and can be discussed even on the basis of the results obtained in a specific condition.

For the purpose of the presentation I assume that the mechanisms underlying the tracking performance can be partitioned into two broad classes: those involved in the central representation of the target, and those that pertain to the planning of the motor response. Within the second class, I further distinguish between the case of simple targets that can be reproduced on the basis of a single, completely specified plan, and the case of targets that are either unpredictable or too complex for such a unique plan to exist. The organization of the chapter derives from these distinctions. In each case I have selected a few experimental facts that cannot be reconciled with the basic assumptions of the system analysis approach. Because the goal of the chapter is simply to substantiate the concerns expressed above, the selection is somewhat biased. Moreover, it should be obvious that the results are much too fragmentary to provide the basis for a coherent alternative to system analysis.

## THE ROLE OF COGNITIVE FACTORS

The influence of cognition on many aspects of visual perception is so pervasive that talking of perception without considering the higher psychological processes may in fact be meaningless. Yet, it is not common to see this influence discussed within the context of visuo-manual tracking studies. Indeed, the only cognitive process that has been considered in any detail is forecasting, that is, the almost compulsive tendency by the pursuer to entertain expectations about the future course of the target (e.g., Gottsdanker, 1952; Pew, 1974; Rouse, 1976). When the target is following a relatively simple trajectory with a well-behaved law of motion, forecasting is indeed useful inasmuch as the motor response can be tuned not to the actual target but to its assumed future position (Runeson, 1975). However, it is also well known that this tendency can result in large tracking errors whenever the target undergoes some unexpected changes in direction. One contingent reason why forecasting is about the only cognitive component of the performance that has been analyzed and modeled is the role, already remembered before, that the engineering approach has traditionally played in tracking studies. In fact, although actual forecasting is certainly mediated by exceedingly complex and so far unknown mental processes, its effects on the performance can be simulated approximately by a simple mathematical device used extensively in the engineering literature (see Wickens, 1986). More sophisticated schemes of linear forecasting are also known in the technical literature (e.g., the Kalman filters) but, as far as I am aware, their application to the modeling of tracking has not been attempted yet (see, however, Wewerinke, 1981).

Another, more profound reason for the neglect of the cognitive factors is the relative poverty of test stimuli. To begin with, the majority of laboratory experiments have considered one-dimensional stimuli that, by definition, do not define a shape.[1] Thus, the stimuli lack one of the most important prerequisites for engaging perceptual and cognitive processes of some complexity. This lack of shape has also other implications that are discussed in a later section. Second, for historical reasons the subjects tested in tracking experiments have almost always been adults. Quite independently from their degree of skill, normal adults master completely the conceptual categories of causality, permanence, grouping, and identity that are among the basic cognitive tools involved in the perception of complex dynamic displays. Therefore, it should not be surprising that, especially in handling simple visual configurations, visuo-motor performances bear little evidence of the involvement of these cognitive categories. Only by having recourse to sensitive psychophysical techniques (e.g., chronography) can one hope to assess the computational load associated with the use of these categories, but I know only one tracking study that has attempted to use these techniques to measure interference effects (McLeod, 1977).

In general, the closer the degree of effectiveness of a cognitive procedure to its optimum final-state value, the harder it is to disentangle its specific reflections on the total performance from those of all other concurrent

[1]In the context of avionics complex displays are often used to simulate the actual array of gauges in the cockpit. However, the utilizations of icons is not to be confused with the dimensionality of the stimuli. Only sophisticated flight simulators provide realistic two-dimensional dynamic displays.

processes. In particular, this is certainly true for the specific process whereby the various dimensions of a dynamic event are made to coalesce into a perceptual or motor unit. Once again, the difficulty of discriminating the components that concur in a perceptuo-motor performance may not be a major concern as long as one is only interested in a global assessment of skill proficiency. It becomes so, however, when a deeper understanding of the underlying mechanisms is pursued. Here lies the considerable importance of developmental studies. As argued cogently by the Piagetian school (see, for instance, Mounoud, 1986; 1988), the analysis of the performances of young children affords the opportunity of testing the perceptuo-motor system before the attainment of final state, when there is still a chance of detecting behavioral evidence of the various processes that cooperate in a complex task. The next section provides some evidence that indeed this opportunity is real.

## A Developmental Study

The relevance of the developmental approach for investigating the role of cognitive factors in visuo-manual tracking is exemplified by the results of an experiment performed on 5- to 9-year old children (Mounoud, Viviani, Hauert, & Guyon, 1985). In order to dramatize the main conceptual point of the study, the task and the stimuli for the experiment were kept extremely simple (Fig. 11.1). A galvanometric mirror was used to displace sinusoidally a point light source on a translucent screen. Subjects were sitting on the opposite side of the screen. By rotatory movements of the right forearm in the horizontal plane, they could displace another light source on the screen. The assignment was phrased as a skill game and required to keep the controlled spot as close as possible to the target spot throughout a fixed number of movement cycles. Two frequencies—0.2 Hz and 0.8 Hz—were tested on 50 grade-school male children and ten adults. A nonmonotonic evolution of the performances with age can be demonstrated with the traditional spectral techniques. However, for the purpose of this discussion it is sufficient to consider the raw data.

A majority of children within all age groups managed to track the target with reasonable accuracy. A performance was judged qualitatively successful (Fig. 11.2) if—over and above the age-specific distortions present in the pursuit track—the three basic parameters of the target—frequency, amplitude, and phase—were restituted correctly. Successful performances differ sharply with those, judged unsuccessful, in which at least one of these parameters was grossly distorted (Fig. 11.3). Let us consider first track C produced by a 6-year-old child. Only occasionally does this child succeed in generating an oscillatory motion. During the best part of the trial, the child hovers around the middle of the screen. Such a dramatic

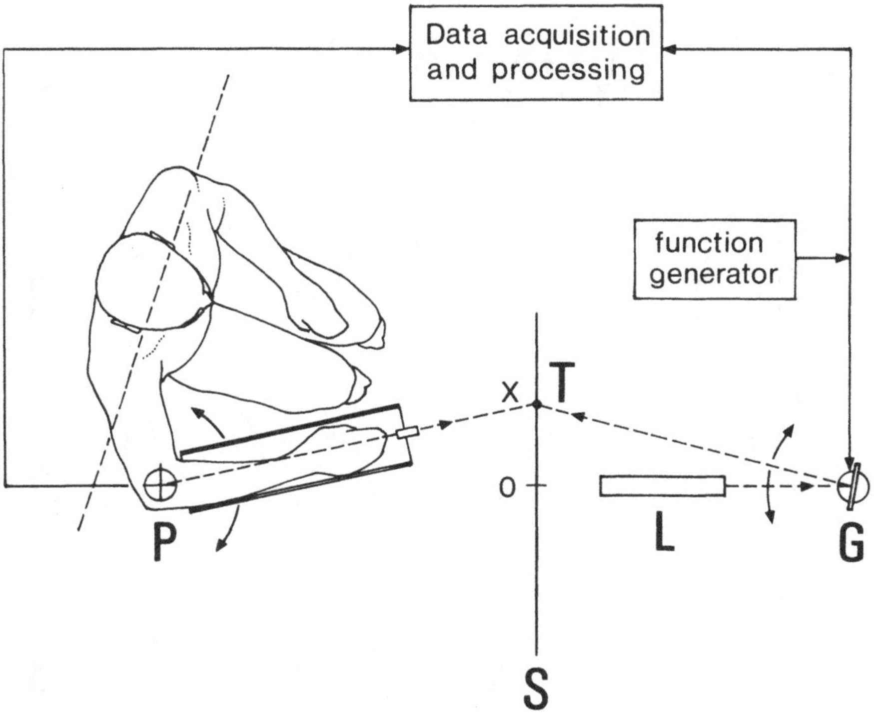


FIG. 11.1 Schematic view of the experimental set-up used for the tracking experiment described in the text. (From Mounoud et al., 1985).

failure may have several explanations, ranging from a deficiency in motor coordination to attentional deficits that prevent the child from anchoring the response to the target. The design of the experiment does not allow one to discriminate among alternative hypotheses but, at least in principle, one could represent the deficit in terms of real-time filtering with a low, irregular gain. Although representational factors cannot be ruled out, there is no unequivocal indication from the data that they indeed play any role. Track D, again from a 6-year-old child, is more puzzling in that both frequency and amplitude of the pursuit match those of the target. Thus, the motor coordinations necessary for generating a smooth oscillatory track are obviously in place. However, the phase is reversed during most of the trial. One may suspect that for some reason the phase parameter is not correctly represented, but the failure of an unspecified synchronization mechanism may also be at the origin of the mismatch. In any event, the system analysis representation must be strained almost beyond reason (time-variable phase with constant gain) to accommodate a performance as the one in D. Finally, let us consider tracks A and B in which the gain is

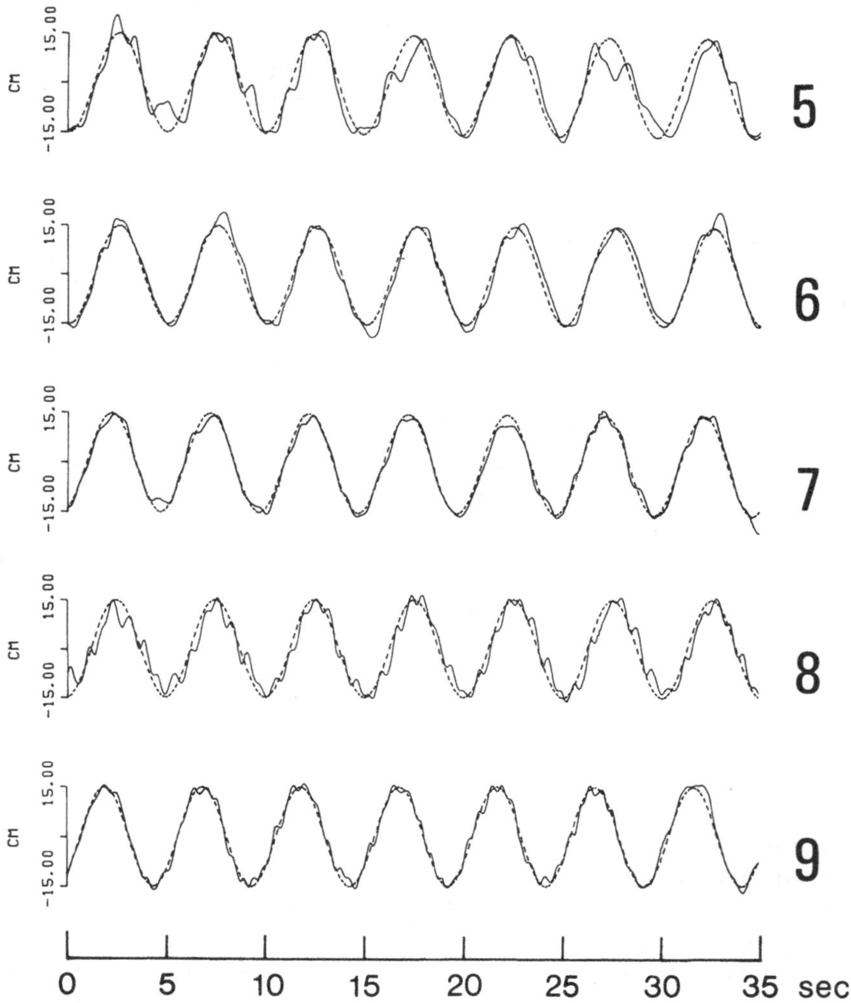


FIG. 11.2 Examples of successful performances in the sinusoidal tracking experiment (0.2 Hz). Each record shows a typical record of the pursuit movement (continuous lines) at the indicate age. Dashed lines indicate the targets. Notice the presence of high frequency corrections in the records from 8- and 9-year-old boys. (From Mounoud et al., 1985)

approximately correct but the pursuit frequency is respectively double and half of that of the target. Because frequency is not correctly restituted, a phase cannot even be defined. Here the departure from the real-time, filtering behavior is dramatic: There is just no way of making any sense of performances like these within the framework of system analysis theory. Because, as noted above, motor deficits seem to be ruled out by the smoothness and stability of the movements, and synchronization cannot be an issue, an explanation must sought in terms of representational factors.

A possible line of reasoning for interpreting these results may be the following. In classical feedback models the response is driven by the instantaneous distance between itself and the target. Instead, let us assume that, although the tracking assignment is always understood as the necessity of keeping the response "close" to the target, the metric by which distances are measured is not applied to the instantaneous parameters of the movements (position, velocity, acceleration) but to their abstract internal representation. Specifically, let us qualify the General Motor Program hypothesis (see Rosenbaum, 1980; Schmidt, 1975) by assuming that regular and learned movements are represented internally by a combination of a "type" label and a set of characterizing parameters. To each type is associated a multidimensional space whose axes code the values of these

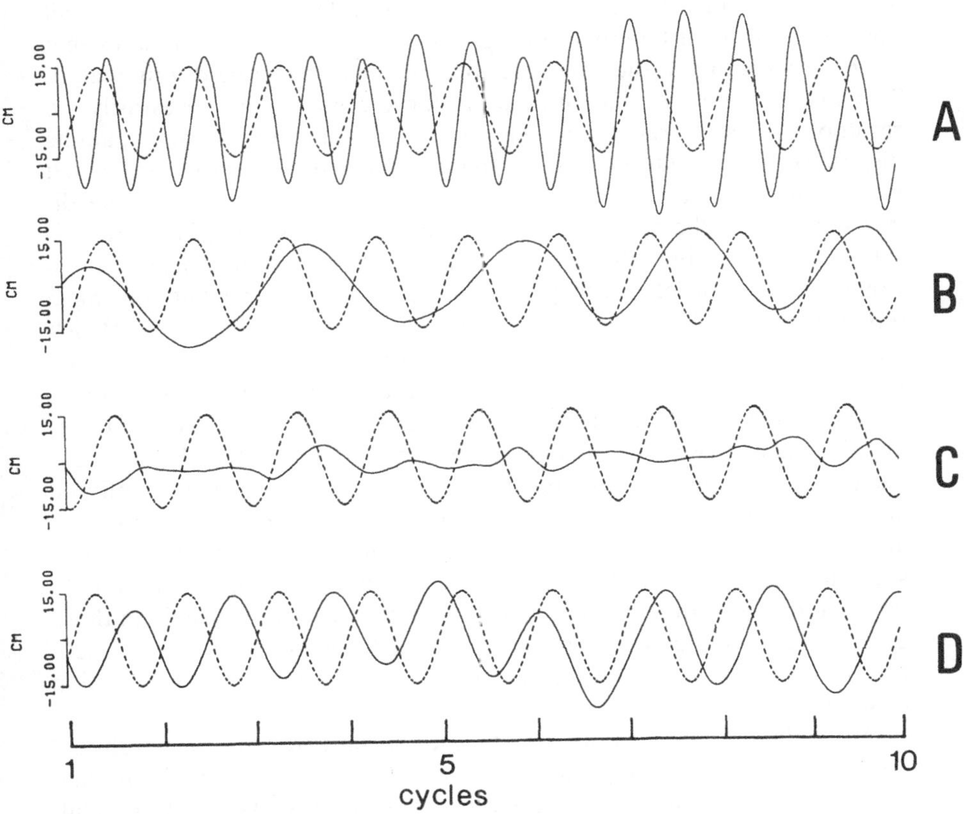


FIG. 11.3 Examples of unsuccessful performances at 0.8 Hz. Fully discussed in the text. (From Mounoud et al., 1985)

parameters. For instance, a specific sinewave would be coded as a point in the two-dimensional space associated with the type "sinewave" with one axis for frequency and one for amplitude (phase is a relational parameter that cannot be defined independently). More complex movements would of course necessitate spaces of higher dimensionality. In all cases, however, reproducing an external template requires at least three clearly distinct steps: the identification of the correct type "label," the matching of the representative points in the parameter space, and the actual production of the response. If these three steps are carried out satisfactorily, proximity between target and response would then be maintained by on line corrections based on the instantaneous parameters of the movements.

The general scheme outlined above can accommodate performances like those shown in Figs. 11.3A and 11.3B. If for any reason one of the parameters that characterizes uniquely a movement template is not reckoned as pertinent, the representational space collapses over the corresponding axis, and this in turn creates a class of equivalence over the set of all possible movement templates of a given type. As a result, templates that are clearly distinct in the complete space will not in fact be discriminated. To be sure, the establishment of equivalence classes is not necessarily the result of a failure to reckon the relevance of a parameter. Normally, adults dispose of representational spaces that are far richer than what is needed in any one specific circumstance, but they know how to reduce the dimensionality of the representation to suit the necessities. Thus, for instance, in a tracking situation like the one described previously, adults would have no problem accepting that the target and pursuit spot have different shapes. The trouble comes, however, when the space is not rich enough to cope with the assignment, and this is precisely what might be happening with some of the younger children. Apparently, they possess the type "sinewave" (or at least the more general type "thru-and-for movement") and can discriminate stimuli along the amplitude axis. By contrast, frequency is not represented and all movement tokens of a given amplitude are considered as identical. The frequency of the responses need not bear any relation to that of the stimuli (it seems, however, that integer ratios seem to be preferred). Notice that these children make no attempt to improve the performance. Such attempts are instead quite visible in those cases when the deficit concerns the use of feedback information (cf. tracks for 5- and 8-year-old children in Fig. 11.2) or temporal synchronization (trace D in Fig. 11.3).

According to this viewpoint, the failures documented in A and B should not be attributed to carelessness or to lack of comprehension about what "staying close" means. Within their system of representation, those children are tracking perfectly the target stimulus. The failures originate instead from the fact that their cognitive system has not learned yet how to attach discriminal relevance to certain parameters. As this happens, and the child evolves toward the skill mastery typical of adult subjects, it may become difficult to find behavioral evidence of the underlying coding of both stimulus and response. However, there is no reason to believe that whatever structure this coding might have in children, it changes drastically with age. In summary, the point I have tried to make is that at all ages successful tracking is contingent on cognitive and representational factors that cannot, even remotely, be taken into account by the classical system analysis approach. With the next section I turn to other aspects of this skill that again point to the basic inadequacy of such an approach.

## FACTORS RELATED TO THE IMPLEMENTATION OF THE RESPONSE

The large number of articulations in the arm-hand-fingers chain, the great variety of muscles types, and the equally great variety of muscle insertions subserving this chain afford us with an astonishingly large number of dynamic configurations. Indeed, watching someone who performs highly skillful and well-learned movements such as those involved in playing a musical instrument, we may get the impression that the limits to our motor performances are only set by trivial biomechanical constraints, and that very few (if any) limitations are specifically related to the controlling motor system. In other words, one could be led to believe that within certain ranges of force, frequency, and velocity, the arm-hand-finger system has unlimited flexibility. Superficially, the mode of production of this system shows little evidence of an intrinsic "texture" and, in this respect, it seems drastically different from, say, the oculomotor system mentioned earlier. However, indications of such a texture emerged as early as the turn of the century from the analysis of long sequences of simple movements, such as in Morse code tapping (Bryan & Harter, 1897; 1899) and typing (Book, 1908). These authors showed conclusively that the timing of the movements is far from uniform and undifferentiated. A temporal structure exists in all sequences that carries the idiosyncratic imprint of their central coding. Also, it was soon recognized that velocity and extent of hand movements are mutually constrained in a way that cannot be ascribed to biomechanical factors alone (Freeman, 1914). Later, all these observations were confirmed and extended both in the case of discontinuous temporal sequences and the case of continuous gestures (see Viviani, & Terzuolo, 1983). Moreover, as already noted over a century ago (Müller, 1840), certain combinations of movements are awkward or downright impossible even though the single elements of the combination are well within the permissible biomechanical range. Recent demonstrations of these constraints have been provided both for bimanual coordination (Kelso, 1984) and the activation of ipsilateral upper and lower limbs (Baldissera, Cavallari, & Civaschi, 1982).

If indeed these constraints originate from the way motor sequences are represented centrally—as some authors maintain—they might apply only to the production of voluntary movements (both learned and extemporaneous). In fact, despite the results of the previous section, one could still argue that movements guided from without—as tracking movements are—need not be represented centrally. Consequently, pursuit motor actions would be more flexible then their voluntary counterpart and limited only by the frequency response of the neuromuscular plant. The aim of this section is to provide evidence against this point of view. I argue that in addition to the constraints already mentioned above, human movements are further constrained by regularities that reflect both planning and the implementation stage of the motor plan. To the extent that these regularities concern the execution of movements, they must be satisfied in all cases, irrespective of whether the action is guided from within or without. Thus, a conflict may arise whenever the intrinsic properties of the response do not match those of the external template to be followed. The results to be discussed were obtained from experiments with predictable and unpredictable targets. In both cases I first document a structural property common to all spontaneous productions of a certain class of movements. Then I show the impact of this property on the performance in a tracking situation where the same movement must be reproduced from an external model.

### The Case of Simple Learned Movements

If the arm-hand system were designed as most man-made mechanical manipulators, the properties of 2- and 3-dimensional movements would be entirely deducible from those of 1-dimensional movements. Moreover, the biological system would share the same flexibility of its mechanical counterpart. In particular, any spatial trajectory could be traced using infinitely many different sets of 1-dimensional components. Let us consider the case of a familiar 2-dimensional trajectory, the ellipse. Mathematically, the most straightforward way of tracing an ellipse is by composing vectorially two harmonic functions (Lissajous mode). It is then convenient to use this mode of generation as a reference for studying the way humans go about tracing the same figure. Indeed, sinewaves are among the simplest predictable motions and have been used extensively in tracking experiments. Both adults and children above the age 8 or 9 years are able to produce spontaneously very good approximations to this waveform (Viviani & Zanone, 1988). Moreover, even as an input target, sinewaves are far too simple to pose to adult subjects the problems evoked previously. Whether we consider wrist or forearm movements, normal adults can pursue a 1-dimensional sinewave very accurately up to frequencies of about 1 or 2 Hz, depending on the amplitude (see Poulton, 1974).

Standard calculations show that in the Lissajous mode the tangential velocity of the motion is proportional to the cubic root of the instantaneous radius of curvature. This we can write as:

$$\log V = \log K + 1/3 \log R \qquad (1)$$

In an experiment on the development of motor skills (Viviani, & Schneider, in preparation) we asked subjects to trace elliptical forms of various sizes. The size of the required movement was suggested by a printed template, but it was made clear that execution should be as free and spontaneous as possible. Movements were recorded with the help of a standard high-resolution digitizing table. Both adults and children aged between 5 and 12 years were tested, each subject executing twice a series of ten figures in a different randomized order. The perimeters of the templates were spaced logarithmically within the range [2.343–53.025] cm. Figure 11.4 shows typical performances of an 8-year-old child (A) and of an adult (B). As discussed at length in previous articles (Lacquaniti, Terzuolo, & Viviani, 1983; Viviani, & Cenzato, 1985) a simple way of using the Lissajous model of generation as a reference for characterizing human performance is to compare these performances with the predictions of equation (1). In the present case this was done by performing a linear regression analysis of the logarithm of the tangential velocity V versus the logarithm of the radius of curvature:

$$\log V = \alpha + \beta \log R \qquad (2)$$

for every subject and every size. Under the condition that the resulting coefficient of correlation turn out to be high, the performance is compared to the Lissajous model simply by looking at the slope parameter $\beta$ in the regression equation. The closer this parameter is to the value $\frac{1}{3}$, the better the agreement with the model. The meaning of the intercept parameter $\alpha$ is discussed later.

The data in Table 11.1 show that the conditions for carrying out this type of analysis are fully satisfied at all ages, because this group average of the correlation coefficient is never lower that .95. The results are clear: The Lissajous mode is followed almost perfectly by the adults; by contrast, significant differences with the theoretical value $\beta = \frac{1}{3}$ emerge from the children data. The discrepancy is larger for the younger age groups and disappears progressively. However, in all cases the tangential velocity is a power law of the radius of curvature. Thus, the experiment confirms and qualifies the general conclusion, already arrived at in previous work, that

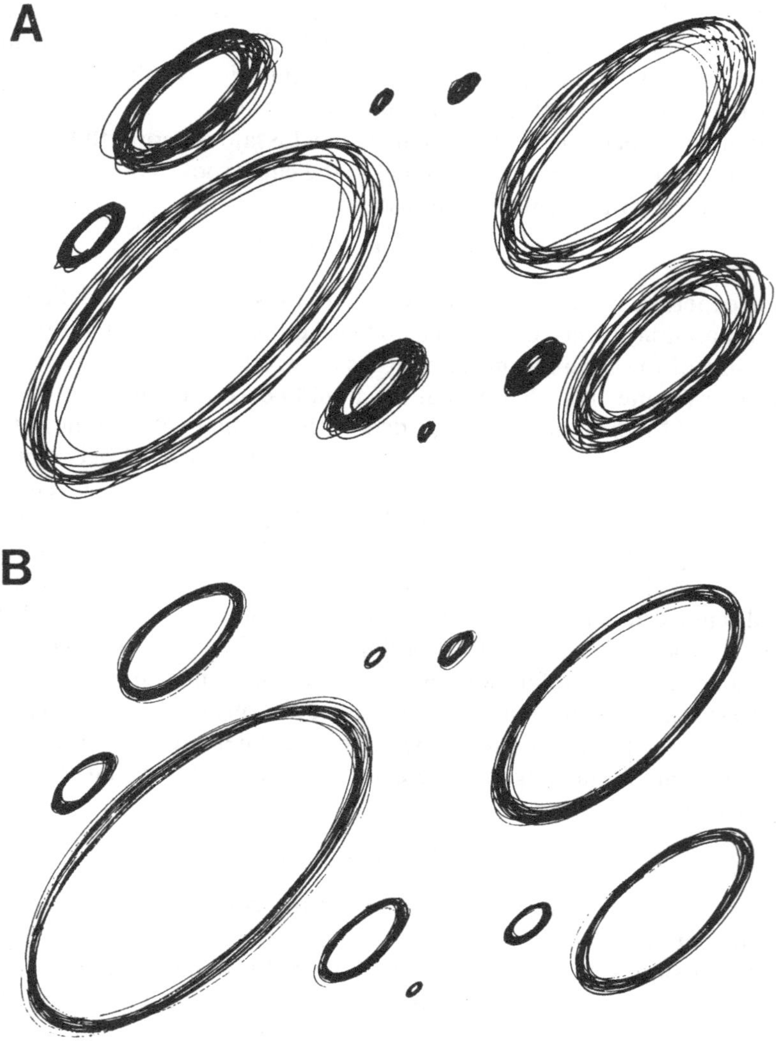


FIG. 11.4 Typical examples of free drawing of ellipses of different size. A: Data from one 8-year-old boy. B: data from one adult subject.

human movements as simple as those considered here are characterized by a specific covariation of their kinematic and geometrical variables. Two points deserve to be emphasized about these results. First, the power law:

$$V = KR^{\beta} \tag{3}$$

that corresponds to the linear regression (2) is a significant generalization of the Lissajous model. The structure of the relation between V and R is preserved. However, in order to encompass in a single formulation the age development of the motor behavior and its final state, it is necessary to relax the constraint on the value of the exponent. As shown in Fig. 11.5, when the coefficient β is either higher or lower than ⅓, the horizontal and vertical velocity components of an elliptic movement that satisfies Equation (3) depart noticeably from the sine and cosine functions. Consequently, also the displacement components are no longer harmonic functions. In other words, if development is taken into account, we can no longer equate the functional relation that exists between tangential velocity and radius of curvature with the simple recipe: To take an ellipse, just oscillate sinusoidally along two orthogonal axes (Hollerbach, 1981). As a second point, it should be noted that although this recipe is mathematically straightforward there is nothing compelling about it. Any trajectory can be traced out using an infinite number of different components, and to each of them corresponds a different time course of the tangential velocity. Typically, in designing the control of mechanical manipulators, one is either concerned with the minimization of some integral index of performance or with the path of the endpoint (see Brady, Hollerbach, Johnson, Lozano-Pérez, & Mason, 1984; and in particular chapter 4). In neither case does the design include explicitly the specification of the velocity profile at the endpoint, and, in fact, many profiles have been envisaged for the same basic trajectories. Thus, the fact that in generating ellipses adults follow exactly the Lissajous mode represents a significant peculiarity of the mature human motor control system. Preliminary results (T. Flash and P. Viviani, unpublished observations) seem to indicate that this mode is associated with the minimization of the derivative of the acceleration (jerk) but this point is not pursued here.

## A Tracking Experiment

We can now move to consider the implications for tracking behavior of the fact that geometry and kinematics of spontaneous movements are not independent. As mentioned before, it is conceivable that such a relation comes into being only for spontaneous movements. If so, it should matter little in a tracking context whether or not the movement to be imitated

TABLE 11.1
Spontaneous generation of elliptic trajectories

| *Age* | *5* | *6* | *7* | *8* | *9* | *10* | *11* | *12* | *Adults* |
|---|---|---|---|---|---|---|---|---|---|
| r | .958 | .960 | .960 | .968 | .975 | .969 | .968 | .986 | .992 |
| β | .259 | .230 | .241 | .267 | .280 | .289 | .325 | .317 | .335 |
| sd | .063 | .060 | .069 | .061 | .074 | .061 | .040 | .038 | .016 |

Analysis of the relation between the tangential velocity and the radius of curvature (see equation (2) in the text). For each age group are indicated the correlation coefficient r of the linear regression, the slope β of the regression line and its standard deviation sd.

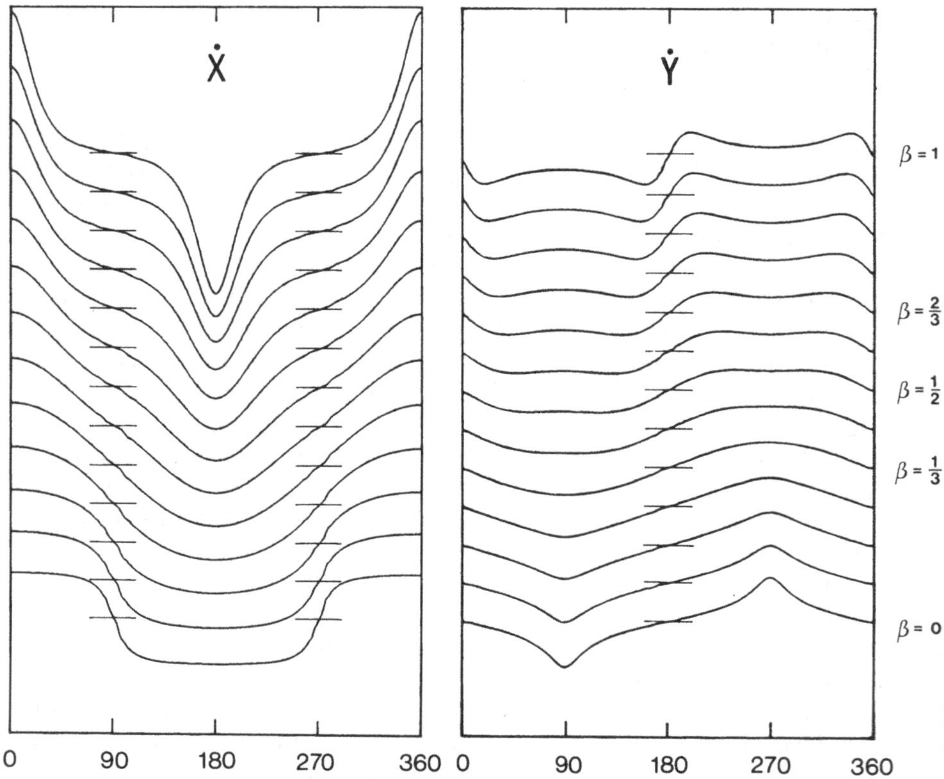


FIG. 11.5 Simulated data. For each of the indicated values of β are shown the horizontal and vertical velocity components of an elliptical movement satisfying equation (3) in the text. When the exponent $\beta = \frac{1}{3}$ the movement is generated according to the Lissajous model and the two components are harmonic functions. As the value of β moves away from the value $\frac{1}{3}$, the components are no longer sine and cosine functions.

complies itself with the condition expressed by Equations (1) or (3). Otherwise, we should expect a compatibility problem to arise whenever the target does not share the structural properties of the pursuit response.

A tracking experiment was designed to investigate this point (Viviani & Mounoud, in press). The gist of the experiment is to contrast the performance in two conditions that differ only in the nature of the relation between geometry and kinematics. In all cases, subjects were to track elliptical targets of constant size (Major axis A = 12.57 cm; minor axis B = 5.48 cm; eccentricity = 0.9). The average velocity across one movement cycle could take the following seven values: 6.10, 12.20, 18.31, 24.42, 30.52, 36.62, 42.73 cm/sec. The orientation of the main axis of the trajectory could be either vertical or horizontal. In the control condition (*N*) the target was generated according to the Lissajous model which, as we have just seen, is also followed by adult humans. In Lissajous targets of period $2\pi/\omega$ the tangential velocity oscillates regularly along the trajectory between a maximum of $\omega A$ attained where the trajectory is flattest, and a minimum of $\omega B$ at the two poles of the ellipse where the curvature is maximum. In the control condition (*T*) the velocity oscillates between the same extremes, but the correspondence with the minima and maxima of curvature is inverted: The movement accelerates at the poles and decelerates along the sides of the ellipse. It is important to stress that, from the point of view of system analysis, test targets are just as acceptable as the controls. In fact, although the cartesian components of the test targets are no longer harmonic functions (Fig. 11.5), their spectral content is well within the dynamic range of the motor system (Viviani, 1988). As targets for a 1-dimensional task, each component independently could be tracked quite accurately.

Figures 11.6 and 11.7 summarize with examples in two subjects the finding of the study that is relevant here. Panels marked *N* and *T* are relative to the corresponding conditions. The lower diagrams in each panel show the average over several cycles of movement of the pursuit trajectory together with their least square elliptical approximation. The upper diagrams are polar plots of the instantaneous delay between target and pursuit. The instantaneous delay at each point along the trajectory is proportional to the radial segment comprised between the plot and the elliptic zero reference. Whenever the plot lays inside the reference line (as in panels *T* of both figures) the pursuit is in advance with respect to the target. These examples are typical of what was observed in all subjects: The performance in condition *T* is dramatically worse than in condition *N*. This is already evident in the spatial domain because the trajectories are significantly more distorted in *T* trials then in *N* trials. However, the degradation of the performance is especially serious in the time domain: Lissajous targets are pursued effectively with a time lag that may vary from subject to subject but remains almost constant throughout the entire trial. By contrast, the instantaneous delay in *T* trials is longer and varies systematically over the cycle, which indicates that subjects can no longer reproduce the very same velocity profile that they were able to cope with in condition *N*. The fact that minima and maxima of curvature and tangential velocity are not matched as in spontaneous movements is directly responsible for this failure. The harder subjects strive to reproduce accurately the shape of the target, the closer the velocity will be to the theoretical value given by Equation (1) which, of course, is not that of the target. Dynamic accuracy must be traded for spatial accuracy.

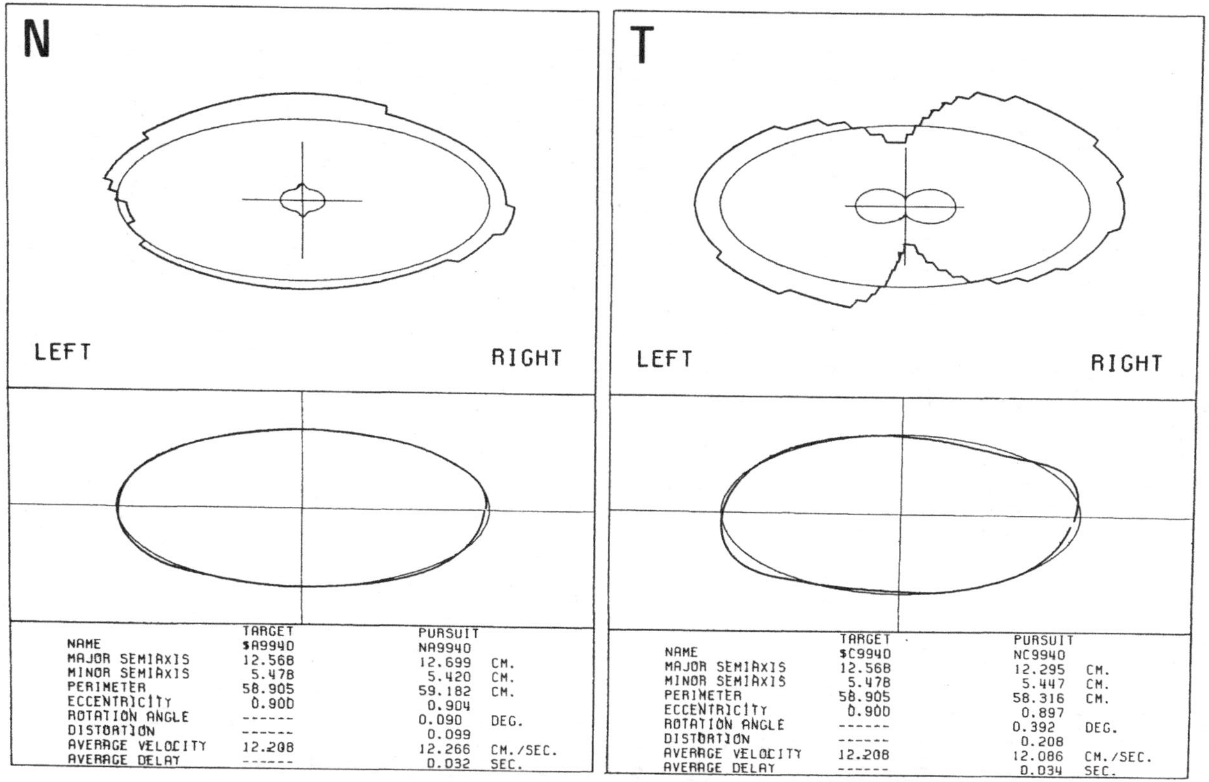


FIG. 11.6 Results of the elliptic tracking experiment. Averages computed from ten repetitions in one subject. The target has an average velocity of 12.80 cm/sec and its major axis is horizontal. The upper diagrams are polar representations of the instantaneous delay. The plots at the center of the ellipse indicate the angular velocity of the target for all radial directions. The lower diagrams show the average pursuit track (thick line) and its least square elliptical approximation (thin line).

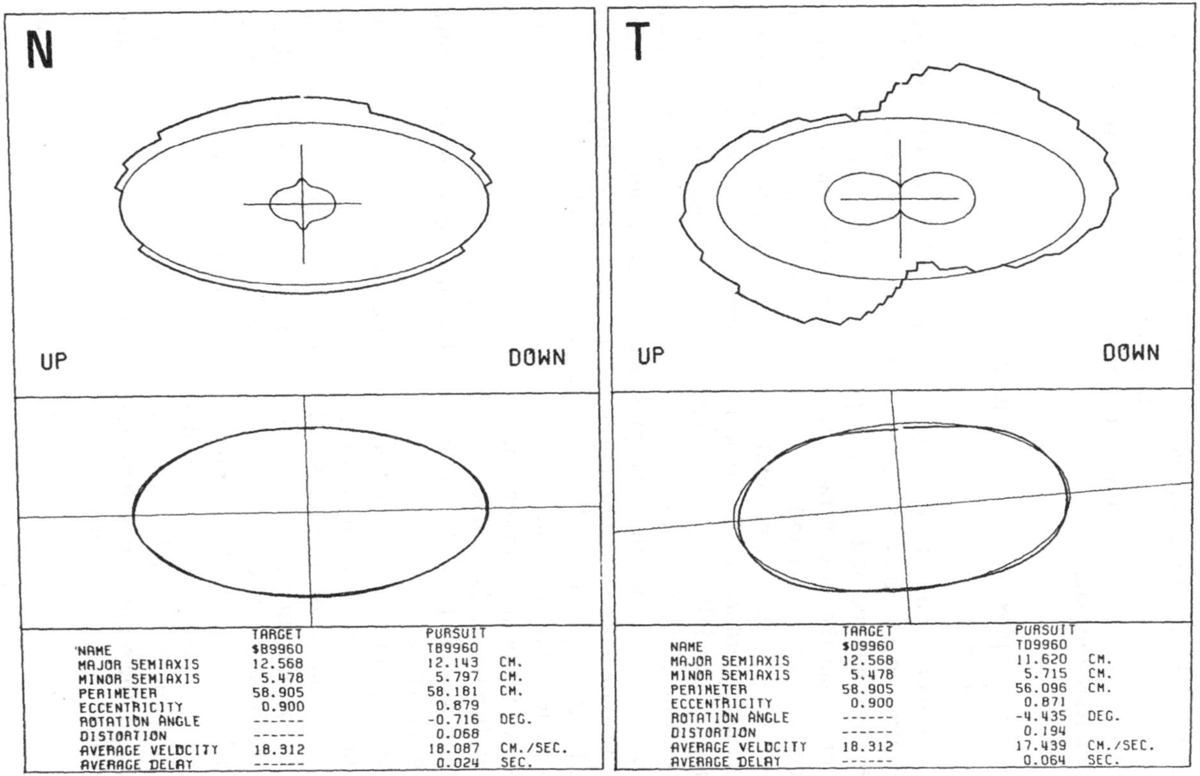


FIG. 11.7 Results of the elliptic tracking experiment. Data in a different subject. The target has an average velocity of 18.31 cm/sec and its major axis is vertical. Same conventions as in Figure 11.6.

In conclusion, the study provides an unequivocal answer to the question raised at the beginning of the section. Even when targets have very simple shapes, they will not pursued effectively unless their kinematic and geometrical parameters covary as in natural movement. There is indeed a compatibility issue that can only be exposed by abandoning the classical system analysis approach.

## The Case of Complex Movements: Hierarchical Structures and Their Behavioral Manifestations

Having considered one of the simplest geometrical forms that one can learn, I pursue now the main argument of the chapter by moving, so to speak, to the other end of the range of possible 2-dimensional movements. I consider in fact the case of random movements (scribbles) that cannot be learned or memorized and that can be made as complex as we wish. As before, let me begin by presenting a few selected facts concerning the spontaneous generation of this type of movement. An experiment was conducted on 12 adults who were asked to trace random continuous trajectories within circular frames of various sizes. No constraint was imposed on the movements other than the presence of well-defined boundaries. Ten frame radii spaced logarithmically in the interval [1–10] cm were selected for the experiment. The entire sequence was administered three times in a single session. Frames were drawn on separate sheets and displayed sequentially in a different random order for each subject and each repetition. In all cases the analysis was performed on a sample of movement of constant duration ($T = 10$ sec) extracted from a longer uninterrupted period of recording.

Behaviorally, the most immediate conclusion from the analysis of the data is the amazingly fine tuning of the velocity across frame sizes. Specifically, let $r$ and $V_{av}$ be the radius of the frame and the average tangential velocity of the movement, respectively. Confirming previous results (Lacquaniti, Terzuolo, & Viviani, 1984), we found that these two quantities are strongly dependent: The average for all subjects of the coefficient of correlation between the logarithm of $V_{av}$ and the logarithm of $r$ is .976 with a minimum of .957 and a maximum of .996. Averaging the individual parameters of the regression, and expressing the result as a power function, one finds that the expression:

$$V_{av} = 5.2\, r^{.783} \tag{4}$$

accounts for more than 95% of the total variance. Of course, the average velocity across the entire trajectory is a post hoc abstract concept that seems to have no relevance to the on line control of the movement. In fact, the instantaneous velocity varies considerably during the execution of the movement (see Lacquaniti et al., 1983). How then are we going to explain the fact that these variations cancel each other over the entire movement with such an extravagant accuracy? In order to answer this question it is necessary to hark back again to the relation between geometry and kinematics expressed by Equation (3). It was shown (Lacquaniti et al., 1984) that at all points of the trajectory sufficiently removed from the points of inflection the constant of proportionality K can be interpreted as a velocity gain factor. This important descriptor of the movement can be expressed in terms of the derivatives of the cartesian components:

$$K = (\dot{x}^2 + \dot{y}^2)^{(.5+1.5\beta)} (\dot{x}\ddot{y} - \ddot{x}\dot{y})^{\beta} \quad (5)$$

and summarizes the presetting by which the overall timing of the movement is controlled. In the particular case of simple closed movements it was found empirically that K is a power function of the perimeter of the trajectory (Viviani, & McCollum, 1983). When the movement results from the sequencing of two closed shapes, the velocity gain factor depends jointly on the linear extent of the movement components being executed and on the total linear extent (Viviani, & Cenzato, 1985). A connection between the velocity gain factor and the size of the movement emerges also in the case of open, extemporaneous movements. The median of $K$ over the entire movement ($K_m$) correlates with the frame size $r$ just as well as the average velocity (mean correlation between log $K_m$ and log $r$ for all subjects: .953). Moreover, $K_m$ is also highly correlated with the total linear extent $L$ of the trajectory. The empirical expression:

$$K = K_0 L^{\alpha} \quad (6)$$

accounts in fact for an astonishing 99 per cent of the total variance. As discussed elsewhere (Viviani, 1986), it is implausible that complex and/or extemporaneous movements be represented and planned as a whole. A more sensible assumption is that these movements are in fact composed of a sequence of smaller units of motor action. In an attempt to apply this notion to the scribble data, we considered the decomposition of the movement that results by identifying the end points of the units with the points of inflection of the trajectory. The rationale for this choice is discussed elsewhere (Schneider, 1987). Here, it is sufficient to show that such a decomposition permits us to expose the internally consistent structure of the movement that underlies the scaling of average velocity with frame size.

Because of the nature of the task, the total length of the trajectory cannot be available to the motor control system at any time during the execution of the movement. Thus, the relation expressed by Equation (6) must result from more local interactions between the geometry of the trajectory and the setting of the kinematic variables. There are many ways in which the concept of unit of motor action can be brought to bear on the problem of identifying these interactions. Possibly the simplest hypothesis that one can entertain in this respect is that the mean of the velocity gain factor over each unit of motor action ($K_s$) depends on the length of the corresponding stretch of trajectory ($L_s$) in a manner that is similar to the power law expressed by Equation (6). The hypothesis can be tested by performing a linear regression analysis on the logarithms of the quantities $L_s$ and $K_s$. For reasons that need not to be detailed here (Schneider, 1987) we actually carried out the statistical analysis on the logarithms of the normalized gain $K_s/K_m$ and of the normalized length $L_s/L_{av}$. Once again, the average coefficient of correlation turned out to be high (.830). Thus, within each stretch of trajectory that we have identified with the expression of one unit of motor action, we observe the same phenomenon already verified for the entire movement, namely that the gain factor increases (non-linearly) with the linear extent of the trajectory. In the case of closed trajectories, this relation has been identified with one component of the so-called Isochrony Principle (Viviani, & Terzuolo, 1983) that states that the duration of execution of human movements tends to be independent of their size. From these results it follows that a similar principle is at work also in the case of open, extemporaneous movements.

The fact that the whole and its parts are tied together by one and the same principle may suggest that some kind of group principle is at work. However, a word of caution is in order. The data point clearly to the rather paradoxical fact that the length–gain relation is stronger at the global level than at the level of the constituent units of motor action. Consequently, global properties cannot be deduced entirely from local ones: Some form of internal feedback must be present that tightens up the length–gain dependency as the movements unfold. We have no clue yet as to the nature of this feedback, but the very fact that such a mechanism is implied by the data can only bring additional support to the thesis that natural movements are organized and executed as a whole.

### Local Control of Velocity

Finally, we come to consider the instantaneous modulations of velocity that are present within each unit of motor action. On the basis of previous work, also in the case of scribbling movements we identify the source of these modulations with the relation between curvature and tangential velocity expressed by Equation (3). A robust procedure for validating this hypothesis consists in carrying out a regressing analysis between the values of these two parameters of the movement at isolated points of the trajectory. Limiting the analysis to fewer points has an advantage over the more conventional technique used in other reports (Lacquaniti et al, 1983; Viviani, & Cenzato, 1985; Viviani, & Schneider, in preparation) whenever it is important to eliminate the spurious correlation that always exists between successive samples of continuous movements. Specifically, let $t_{jm}$ be the temporal midpoint of the *j-th* segment of trajectory. Moreover, let us express Equation (3) in the equivalent form:

$$A(t)/K(t) = C(t)^{1-\beta} \tag{7}$$

where $A$(t) stands for the angular velocity and $C$(t) for the curvature. When $1 - \beta = \frac{2}{3}$ Equation (7) represents the so-called Two-Third Power Law (Lacquaniti et al., 1983). However, for the purpose of this analysis the value of the exponent β has not been fixed. If Equation (7) holds within each unit of motor action, we expect that the two quantities $\log A(t_{jm})/K(t_{jm})$ and $\log C(t_{jm})$ be highly correlated with a regression slope of $1-\beta$ and an intercept close to zero. The analysis was carried out for each subject, pooling the mid-point data for all segments and all frame sizes. The average across subjects of the regression parameters were found to be: slope = .672, intercept = −.002, coefficient of correlation = .908. Considering that both the mean length and the mean radius of curvature of the segments included in this analysis range between a few millimeters and several centimeters, we can conclude that the Two-Third Power Law ($1 - \beta = \frac{2}{3}$) predicts accurately the within-segment modulations of velocity for almost all kinds of trajectories produced spontaneously.

In summary, the absolute tangential velocity during scribbling movements is specified by the cooperation of two mechanisms. The first one sets the (almost constant) velocity gain over the segment as a function of the segment length. The second one multiplies this gain by a geometrical factor related to curvature. An unspecified mechanism is responsible for the fact that the piece-wise assembly of the units of motor action into a complete gesture displays a higher degree of consistency than each one of the components.

## A Tracking Experiment

What are the consequences of this complex hierarchical organization of the movement on the tracking performance? Some data relevant to this question come from an experiment in which subjects were asked to pursue as closely as possible a random, 2-dimensional target (Viviani, Campadelli, & Mounoud, 1987). Targets for this study were created from actual scribbling movements recorded from someone who did not participate in the experiments. Two conditions were tested. In both of them the trajectory of the target was the one produced spontaneously. However, whereas in the first condition ($N$) also the original velocity profile was preserved, in the second one ($T$) the tangential velocity was made constant throughout

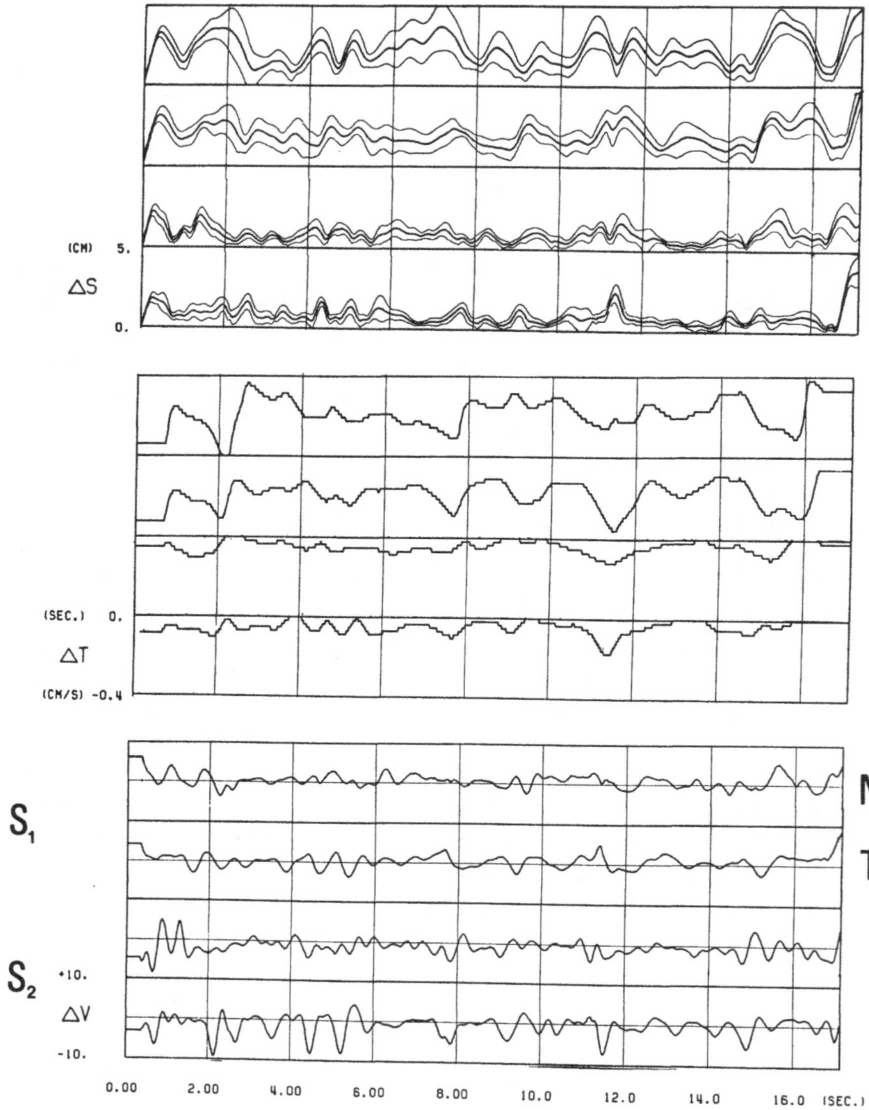


FIG. 11.8 Tracking an unpredictable movement. Data in two subjects. Each of the three panel compares the time course of the indicated performance parameter (ΔS: spatial error; ΔT: temporal delay; ΔV: velocity error) between the two subjects ($S_1$ and $S_2$) and the two conditions (N and T). Thick lines are averages over ten repetitions. Thin lines above and below the average spatial error encompass 2 standard deviations. Notice that because of the experimental design (see text) the same time instant correspond to different locations along the pursuit trajectories in condition N and T. (Unpublished data from Viviani et al., 1987)

the movement. We reasoned that, although a constant velocity movement might appear easier to track than a variable velocity one, in fact, it forces the subjects to depart from the mode of movement production that they would follow if they had to trace the same trajectory spontaneously. Consequently, we expected a significant deterioration of the tracking performance in the second test condition. In particular, it was predicted that both the average temporal delay and the average spatial error between target and pursuit would have been larger.

The expectations were not borne out by the results. Significant variations of these two performance parameters between the two conditions were indeed detected by statistical analysis, but they remained qualitatively marginal. The reason is quite simple: No subject ever succeeded in condition T to depart from the natural mode of movement as the task required. This is shown in Figs. 11.8 and 11.9 that summarize the relevant results in two subjects. Each panel in these figures contrasts the time course of one performance parameter in the two subjects (S1 and S2) and the two conditions (N and T). Figure 11.8 shows from top to bottom the time course of the spatial error ($\Delta S$), the temporal delay ($\Delta T$), and the velocity error ($\Delta V$). Figure 11.9 shows the tangential velocity (V) and the curvature (C) of both target (thick line) and pursuit (light line) tracks. The performances of the two subjects differ in several respects. Among other things, subject $S_2$ has small spatial errors, short delays, and little geometrical distortion in both conditions, whereas subject $S_1$ is far less accurate from the geometrical point of view. However, over and above individual differences, it is clear that neither subjects succeeded in reproducing the constant target velocity of condition T. Actually, the one who is more accurate as far as time and geometric errors are concerned is much worse in terms of velocity tracking. From the inspection of the traces in Fig. 11.9 it can be sent that the unintentional modulation of the velocity in condition T correlates with the simultaneous changes of the trajectory curvature in a way that is reminiscent of the phenomenon documented above in spontaneous movements. In conclusion, faced with an input target that violates a basic regularity found in all spontaneous movements, even experienced subjects cannot satisfy simultaneously both the geometrical and kinematical constraints imposed by the tracking task. In particular, the velocity profile continues to bear the imprint of the general principles of organization for spontaneous movements, even though this is in contrast with the specifications of the target.

## SUMMARY AND CONCLUSIONS

I have tried to argue that tracking movements exhibit a considerable degree of commonality with learned and extemporaneous movements. Despite the fact that the blueprint for tracking movement is dictated by an external

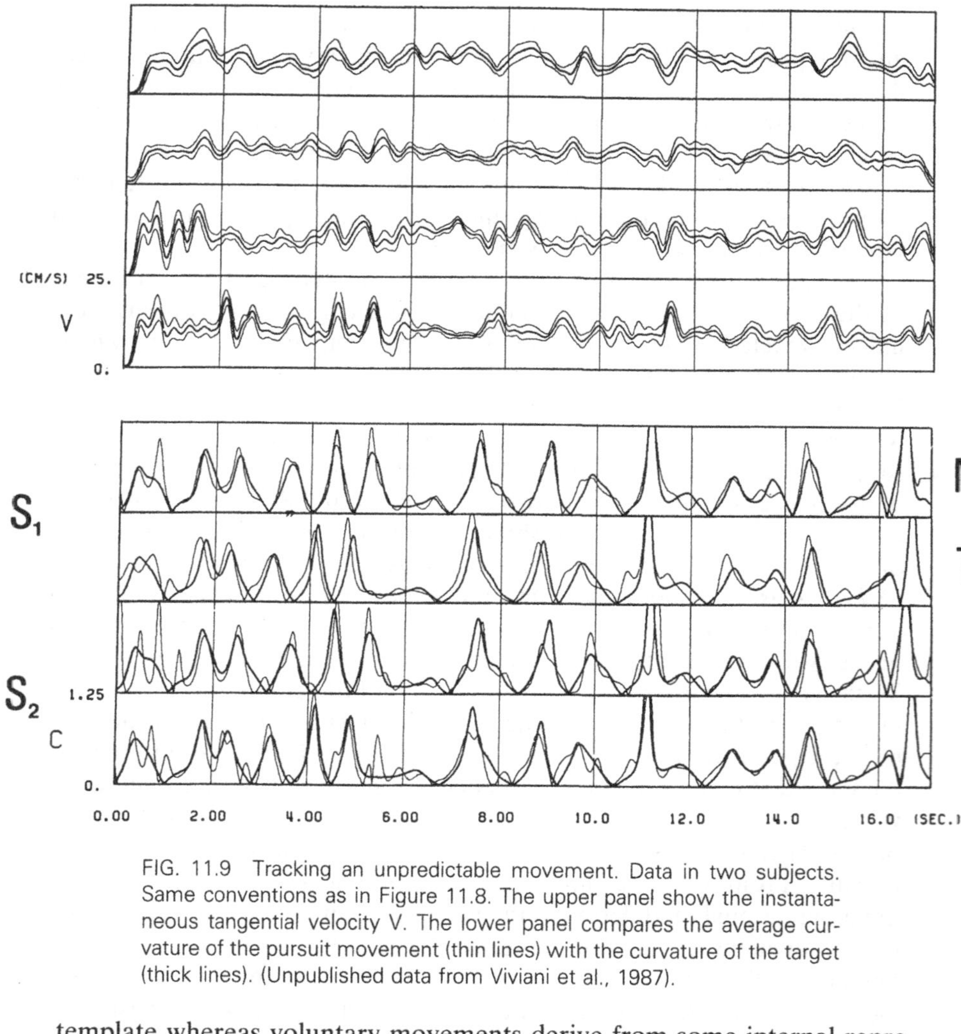


FIG. 11.9 Tracking an unpredictable movement. Data in two subjects. Same conventions as in Figure 11.8. The upper panel show the instantaneous tangential velocity V. The lower panel compares the average curvature of the pursuit movement (thin lines) with the curvature of the target (thick lines). (Unpublished data from Viviani et al., 1987).

template whereas voluntary movements derive from some internal representation, both share a common set of principles of organization. It appears as though, independently from where the specification of the movement comes, to produce a gesture always require going through a similar sequence of stages. If this premise is valid, two related consequences follow necessarily. The first one is that the framework for investigating tracking behavior must be at least as powerful and articulated as the one required for the study of general motor behavior. This is true not only when the tracking task is used as a tool for probing the organization of motor control, but also in the context of human engineering studies. The second consequence is that the system analysis approach, which has been so much privileged in the past, cannot alone provide such an adequate framework.

The main point of the chapter was to show that both inferences are actually supported by specific observations. Although this is a gross oversimplification of reality, I have drawn a distinction between factors relative to the central representation of the motor plan and factors relative to the actual implementation. In both cases, only very limited experimental evidence was presented whose sole purpose was to demonstrate the reasonableness of the argument. This is particularly true as far as the problem of the representation is concerned, which certainly involves much more than the coding of relevant parameters discussed here. The implementation process was touched on in greater detail, but also in this case the approach was basically descriptive.

Selected features of several types of voluntary movements were contrasted with those observed during the pursuit of comparable trajectories. In all cases we observed certain types of failures in the performance that could not be explained by limitations in the dynamic range of the perceptuo-motor loop. The origin of the failures was traced instead either to an inadequacy of the central representation or to a conflict between the relational properties of the target and those of the motor system. A mathematical analogy might help clarify the relation that seems to exist between the spontaneous productions of the motor system and the effectiveness with which an external dynamic template can be reproduced. It is well known that a linear operator is a mapping of a function space into another. Many "black boxes" that appear in functional diagrams stand in fact for a linear operator of some sort. In general, sending a function through a linear operator modifies its shape quite dramatically. However, for a given operator there are certain functions (called eigenfunctions) that enjoy a peculiar privilege: They can go through the operator and come out (almost) unchanged. Besides turning input functions into outputs, linear operators are also capable of generating "spontaneously" (really, when kicked appropriately) a sterotyped output (called impulse response) that carries a very detailed imprint of the type of transformations that are normally applied to the inputs. A deep result of linear algebra relates the spontaneous outputs of an operator with its response to an eigenfunction (Zadeh & Desoer, 1963). In other words, there are close connections between what the operator produces by itself and what it lets pass through with minimum distortion. Let us now equate the entire perceptuo-motor loop with just one big operator. Then, targets that can be pursued effectively may be taken to represent the eigenfunctions of the operator and spontaneous productions would represent impulse response.

It should be obvious that the analogy is inaccurate on many counts and should not be construed as a way of rescuing system analysis. In particular, the bewildering variety of voluntary movements cannot be subsumed under one mathematical construct. Moreover, the presence of internal states that are crucial for determining the response to a given input calls for much more general concepts than the linear operators evoked above. The analogy does place, however, the problem of studying tracking behavior in what I take to be the appropriate framework, one in which pursuit responses and voluntary movements are no longer separated conceptually.

## ACKNOWLEDGMENTS

The preparation of this chapter was supported by FNRS Research Grant 1.150.0.85. Many of the results reported here were obtained in collaboration with other researchers. I am glad to express my gratitude to them all.

## REFERENCES

Baldissera, F., Cavallari, P., & Civaschi, P. (1982). Preferential coupling between voluntary movements of ipsilateral limbs. *Neuroscience Letters, 34*, 95–100.

Book, W. F. (1908). *The psychology of skill.* Missoula: Montana Press.

Brady, M., Hollerbach, J. M., Johnson, T. L., Lozano-Pérez, T. & Mason, M. T., (1984). (Eds.), *Robot Motion: Planning and Control.* Cambridge, Mass.: MIT Press.

Bryan, W. L., & Harter, N. (1897). Studies in the physiology and psychology of the telegraphic language. *Psychological Review, 4*, 27–53.

Bryan, W. L., & Harter, N. (1899). Studies on the telegraphic language; the acquisition of a hierarchy of habits. *Psychology Review, 6*, 345–375.

Freeman, F. N. (1914). Experimental analysis of the writing movement. *Psychological Review Monograph supplement, 17*, 1–46.

Frost, G. (1972). Man–Machine systems. In: H. P. Vancott & R. G. Kinkade (Eds.), *Human engineering guide to equipment design.* Washington, D.C.: U.S. Government Printing Office.

Gottsdanker, R. M. (1952). Prediction-motion with and without vision. *American Journal of Psychology, 65*, 533–543.

Hollerbach, J. M. (1981). An oscillatory theory of handwriting. *Biological Cybernetics, 39*, 139–156.

Kelso, J. A. S. (1984). Phase transitions and critical behavior in human bimanual coordination. *American Journal of Physiology, 15*, R1000–R1004.

Lacquaniti, F., Terzuolo C. A., & Viviani, P. (1983). The law relating kinematic and figural aspects of drawing movements. *Acta Psychologia, 54*, 115–130.

Lacquaniti, F., Terzuolo C. A., & Viviani P. (1984). Global metric properties and preparatory processes in drawing movements. In S. Kornblum & J. Requin (Eds.), *Preparatory states and processes* (pp. 357–370). Hillsdale, NJ: Lawrence Erlbaum Associates.

McLeod, P. A. (1977). Dual task response modality effect: Support for multiprocessors models of attention. *Quarterly Journal of Experimental Psychology, 29*, 651–667.

Mounoud, P. (1986). Action and Cognition. In M. G. Wade & H. T. A. Whiting (Eds.), *Motor development in children: Aspects of coordination and motor control* (pp. 373–390). Dordrecht: Martinus Nijhoff Publishers.

Mounoud, P. (1988). The ontogenesis of different types of thought: Language and motor behavior as non-specific manifestations. In L. Weiskrantz (Ed.), *Thought without language* (pp. 25–45). Oxford: Clarendon Press.

Mounoud, P., Viviani, P., Hauert, C. A., & Guyon, J. (1985). Development of visuo-manual tracking in 5- to 9-year-old boys. *Journal of Experimental Child Psychology, 40*, 115–132.
Müller, J. (1840). *Handbuch der Physiologie des Menschen*, Koblenz: Hölscher.
Pew, R. W. (1974). Levels of analysis in motor control. *Brain Research, 71*, 393–400.
Poulton, E. C. (1974). *Tracking skill and manual control*. New York: Academic Press.
Rosenbaum, D. A. (1980). Human movement initiation: Specification of arm, direction, and extent. *Journal of Experimental Psychology: General, 109*, 444–474.
Rosenthal, N. P., Roberts, W. J., McKean, T. A., & Terzuolo, C. A. (1970). Frequency analysis of stretch reflex and its main subsystems in tricepts surae of the cat. *Journal of Neurophysiology, 33*, 713–749.
Rouse, W. B. (1976). A model of the human in a cognitive prediction task. *IEEE Transaction on Systems, Man, and Cybernetics, SMC-6*, 337–343.
Runeson, S. (1975). Visual prediction of collision with natural and nonnatural motion functions. *Perception and Psychophysics, 18*, 261–266.
Schmidt, R. A. (1975). A schema theory of discrete motor skill learning. *Psychological Review, 82*, 225–260.
Schmidt, R. A. (1988). *Motor control and learning: A behavioral emphasis*. Champaign, IL: Human Kinetics Publishers.
Schneider, R. (1988). *Le concept d'unite d'action motrice et son role dans l'organisation spatio-temporelle des gestes graphiques*. Doctoral Dissertation, University of Geneva.
Viviani, P. (1986). Do units of motor action really exist? In H. Heuer & C. Fromm (Eds.), *Generation and modulation of Action Patterns*. Berlin: Springer-Verlag, (pp. 201–216).
Viviani, P. (1988). Two-dimensional visuo-manual tracking: beyond the system analysis approach. Proceedings of the 1988 IEEE International Conference on Systems, Man, and Cybernetics, (pp. 404–412). Beijing: International Academic Publishers.
Viviani, P., & Cenzato, M. (1985). Segmentation and coupling in complex movements. *Journal of Experimental Psychology: Human Perception and Performance, 11*, 828–845.
Viviani, P., & McCollum, G. (1983). The relation between linear extent and velocity in drawing movements. *Neuroscience, 10*, 211–218.
Viviani, P., Campadelli, P., Mounoud, P., (1987). Visuo-manual pursuit-tracking of human two-dimensional movements. *Journal of Experimental Psychology: Human Perception and Performance, 13*, 62–78.
Viviani, P., & Mounoud, P. (in press). Perceptuo-motor compatibility in pursuit tracking of two-dimensional movements. *Journal of Motor Behavior*.
Viviani, P., & Schneider, R. (in preparation). Age-related changes in the organization of two-dimensional periodic movements.
Viviani, P., & Terzuolo, C. A. (1983). The organization of movement in handwriting and typing. In B. Butterworth (Ed.), *Language production: Vol II. Development, writing and other language processes* (pp. 103–146). New York: Academic Press.
Viviani, P., & Zanone, P. G. (1988). Spontaneous covariations of movement parameters in 5- to 7-year-old boys. *Journal of Motor Behavior, 20*, 5–16.
Wewerinke, P. A. (1981). A model of the human decision maker observing a dynamic system. (Tech. Rep. NLR TR 81062L). The Netherlands: National Lucht-en Ruimtevaartlaboratorium.
Wickens, C. D. (1986). The effects of control dynamics on performance. In K. R. Boff, L. Kaufman, & J. P. Thomas (Eds.), *Handbook of perception and human performance, Vol II*, (pp. 39.1–39.60). New York: John Wiley.
Zadeh, L. A., & Desoer, C. A. (1963). *Linear system theory: The state space approach*. New York: McGraw Hill.
Zuber, B. L (Ed.). (1981). *Models of oculomotor control*. Boca Ration, FL: CRC Press.

# 12 Rhythmic Precision in the Performance of Piano Scales: Motor Psychophysics and Motor Programming

C. L. MacKenzie
D. L. Van Eerd
*Music Performance Laboratory, Department of Kinesiology*
*University of Waterloo, Waterloo, Ontario*

## ABSTRACT

An important music performance skill is to play scales, triads, and arpeggios with rhythmic precision. Rhythmic precision implies regularity, smoothness, and crispness in the timing and loudness of notes. The present study investigated rhythmic precision in the performance of piano scales. We had two related goals: first, to provide motor psychophysical functions for scales performance, and second, to make inferences about motor programming. We asked how tempo and unimanual or bimanual playing conditions would affect piano performance. Seven highly skilled pianists performed two C-major scales with the left and right hands alone, in contrary motion (anatomically similar) or in similar motion (melodically similar) at the following tempos: 60, 90, 120, and 180 metronome beats/min (corresponding to 4, 6, 8, and 12 notes/sec) and as fast as possible. An on line computer system monitored keyboard motion. Position-time data for each key were analyzed to derive: keypress velocities, internote intervals, note durations, and gap between notes. Results indicated that variability of internote intervals (SD and CV across notes in one trial) increased with increases in playing speed, as did the variability (SD) of velocity of keypresses. Thus, tempo affected rhythmic precision. These variability measures were not affected by playing condition or hand. Left and right hands did not differ in internote intervals, but they did differ in keypress velocities, note durations and overlap between consecutive notes as a function of unimanual and bimanual playing conditions. The results do not support a phonograph analogy of generalized motor programs, and are discussed in terms of anatomical, motor, and musical constraints on the programming and control of rhythm in piano performance.

## MOTOR PSYCHOPHYSICS AND MOTOR PROGRAMMING

The purpose of the present chapter is to examine rhythmic precision in the performance of scales on the piano. An important music performance skill is to play scales, triads, and arpeggios with rhythmic precision. In examining the scales performance, we have two related goals. First, we provide motor psychophysical functions to describe the effects of tempo and playing conditions on the performance of piano scales. Second, we make inferences about motor programming and provide a specific test of the phonograph analogy of generalized motor programs. The performance of highly skilled pianists performing such a practised task provides an excellent opportunity to evaluate how the central nervous system programs the activity. Such programming should show itself clearly in the rapid, "automatized" performance of scales by highly skilled pianists.

Piano performance has been measured by either electromechanical (Hartmann, 1932; Sears, 1902; Seashore, 1938; Shaffer, 1980; Wagner, 1971; Wills, MacKenzie, Harrison, Topper & Walker, 1985) or acoustical (Bengtsson & Gabrielsson, 1980, 1983; Povel, 1977) systems since the turn of the century. There are advantages and disadvantages to both approaches. Measurement of the acoustical characteristics is certainly closer to the music, but overlapping information in spectrographic records presents problems. Physical characteristics of the instrument, combined with the performer's actions and the acoustics of the room produce the acoustical signals. On the other hand, measurement of the keyboard or hammer motion is certainly closer to the functional output of the motor system, but not all of this motion is musically meaningful (e.g., the keys on the keyboard bounce) and the music itself, the musical goal, is not measured. Relationships among touch, the keyboard motion, hammer motion, and string motion and sound have been investigated recently on the grand piano by Askenfelt (1988) and Askenfelt and Jansson (1988). Advances in computer and microprocessing technology enable more detailed and extensive measurements of piano performance today than before (but see Seashore, 1938); however, the management, reduction, analyses, and interpretations of these data still present a major challenge to researchers who study piano performance.

In our view, the piano performance output and psychophysical functions resulting from measurement of piano performance are reflecting two main things: multiple levels of constraints and the central representations involved in planning and programming the movements. There must be mappings among multiple representations (musical, spatial, auditory, and motor) in the central nervous system. The nature of these representations has been the subject of much recent research (e.g., Clarke, 1982; Gabrielsson,

1987; Shaffer, 1981; Sloboda, 1983). Sloboda (1985) has emphasized that the emotional and expressive aspects of music are dependent on such central representations. Rather than a complete separation of the knowledge-based, cognitive representations from the representations necessary for implementation by the motor system, we suggest that all of these representations interact in a distributed system and are characterized by extensive sensorimotor integration, feedforward, and feedback processes (see also Arbib, 1985; MacKenzie & Marteniuk, 1985).

Further, there are many levels of constraints on observable piano performance. The term *constraint* is frequently used, and we think it was introduced to the motor control literature by Turvey (1977). We use it to refer to some factor that limits motion in some particular manner or direction (see MacKenzie, 1985; MacKenzie & Marteniuk; 1985; Marteniuk, MacKenzie, Jeannerod, Athenes, & Dugas, 1987). Table 12.1 shows low-level constraints on performance, such as anatomical and biomechanical limitations (e.g., we only have two hands, with a given span or maximum distance between the thumb and little finger), and high-level constraints on piano performance, such as knowledge of musical structure (e.g., harmonic relations) and expressive intentions (e.g., emotional aspects). We view these low-level and high-level constraints as being *within* the human performer. As well, there are *external*, physical constraints on performance, such as the mechanical characteristics of the piano action and the acoustical environment, that limit the piano performance. These constraints are not necessarily limiting in a negative sense; indeed, a highly skilled performer learns to exploit the biomechanical and physical characteristics of systems in order to achieve the functional goals of music performance.

With respect to the first goal of describing motor psychophysics in piano performance, there has been little systematic study. Our starting point is the empirical rhythm research of Gabrielsson and colleagues (Bengtsson

TABLE 12.1
Levels of Constraints on Piano Performance

| | |
|---|---|
| High Level | •musical structure<br>•interpretation<br>•expression<br>•action |
| Low level | •sensorimotor loop times<br>•biomechanical constraints<br>•perceptual and motor hardware<br>•anatomy |
| Physical | •dyamics of piano action<br>•piano acoustics<br>•environmental acoustics |

& Gabrielsson, 1977; Gabrielsson, 1979). Figure 12.1 shows that a musical performance (by one or more musicians playing on their instruments) gives rise to sound sequences (the music) that in turn gives rise to rhythm responses in the listener. The performances and rhythm responses include experiential, behavioral, and physiological components. There is feedback to modify performances, and it is important to note that the performer is a listener as well.

To begin, our focus is on the arrow describing psychophysical relations resulting from the sound sequence. Traditional sensory psychophysical experiments present these sound sequences or musical stimuli to the listener and measure the perception or experience. The sensory psychophysical results present the subjective perception of the musical stimuli as a function of the objective, physical characteristics of the sound sequence. Now we shift our focus to the performance box. In contrast, motor psychophysical experiments present objective goals in the musical task to the performer (e.g., play at a specified tempo) and measure the produced motor behavior. Motor psychophysical results present the produced piano performance as a function of the objectively defined musical task or stimulus characteristics.

From a psychophysical perspective, rhythm is a multidimensional concept because it is more than simply timing and relative timing (see Fraisse, 1974; Gabrielsson, 1979, 1986; Seashore, 1938, 1947). Rhythm is defined by the Oxford English Dictionary (1982) as, "movement marked by the regulated succession of strong and weak elements or of opposite or different conditions." *Accent*, a related term, is defined as ". . . stress recurring at intervals of time which are generally fixed, but may be varied by syncopation and cross accentuation." Thus, rhythm and accent are intimately related aspects of performance. There are temporal and intensive dimensions measurable in rhythm, and rhythm has both individual (at the level of single movements or notes) and collective (at the level of groups of movements or notes) aspects. In a rhythmic tapping series, Semjen and Garcia-Colera (1986) and Keele, Ivry, and Pokorny (1987) have examined the covariation of the forces and timing of accented taps, suggesting that force and timing are not independent, but separate ways of accenting a given tap.

Related findings in piano performance have been presented by Gabrielsson (1987) and Shaffer (1981), although the functions describing force and timing in performance have yet to be elucidated. In piano performance, pitch, timbre, and intensity can vary. Besides the note played, the pianist is able to vary the timing and intensity. In our research we examine both temporal and intensive (dynamic) aspects of accenting in rhythm, and their relationship. The temporal measures in our research include: note duration, internote interval, and gap between consecutive notes. The dynamic or

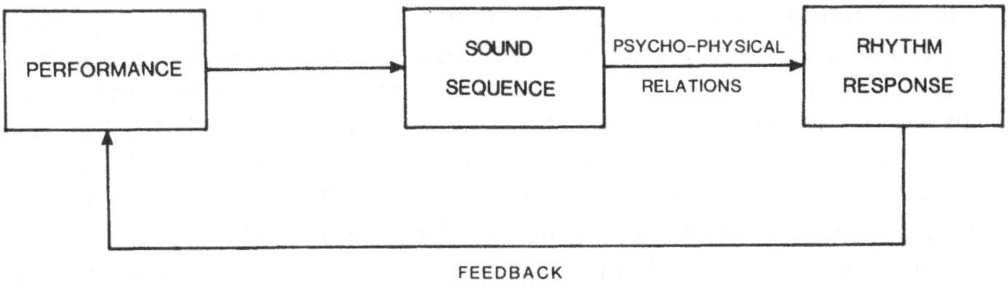


FIG. 12.1. A general framework for empirical rhythm research (adapted from Gabrielsson, 1979). Sensory psychophysics refers to perceived rhythms as a function of objective characteristics of rhythmic stimuli. Motor psychophysics refers to performed rhythms as a function of objective characteristics of the rhythmic task or stimuli.

intensive measure is keypress velocity: The intensity of loudness of a note is a function of the velocity of the hammer when it hits the string, determined by the keypress velocity at the time of escapement. At this time, we provide psychophysical functions for each of these measures. Eventually, we wish to examine their multivariate contributions in performance.

With respect to the second goal of how pianists program the movements in scales performance, an important starting point is the classic address on serial order in behavior by Lashley (1951). He noted that, like speech, there is the potential for up to 15 movements/sec in piano performance. He suggested that with such fast movements, feedback could not be used, at least not in a serial chaining way, where the sensory feedback from one element in the sequence triggers the motor execution of the next element. Hence, there must be central motor programs used by the nervous system. Following Lashley's lead, Keele (1968) defined a motor program as a central representation for a movement sequence that allowed the movement to be executed in the absence of peripheral feedback.

Examining the concept of motor programming in piano performance, Shaffer (1980, 1981) showed that concert level pianists played repetitions of the same pieces of music very consistently. Further, there were deviations in the timing of the notes in local groups or phrases (rubato), and these expressive deviations in the timing were consistent in repeat performances. In an earlier study, we showed that, although all pianists were very consistent in their repetitions, highly skilled pianists played with greater consistency in repeat performances than less skilled pianists (MacKenzie, Nelson-Schultz, & Wills, 1983).

As Shaffer (1981) noted in his investigations of motor programming, performers deviate consistently from the score-specified timing of notes; that is, they do not play with mechanical regularity in the timing. Similar findings of deviations in note timings were obtained by Sloboda (1983) who manipulated the meter of a piece through changing bar line placement.

He also noted that more experienced performers showed more expressive variation than less experienced performers, and that performers showed agreement in where the expressive deviations in performance took place. In a recent report, we showed how manipulations of tonal structure also affected the systematic variation of timing in performance (MacKenzie, Van Eerd, Graham, Huron, & Wills, 1986). There was less systematic variation in the timing when the tonal structure conformed to rules of traditional Western music than under atonal transformation conditions. In performances of Erik Satie's "Vexations," Clarke (1982) found that the tempo of playing affected the relative patterns of timing in rhythmic performance. In experimental studies of tapping tasks as well, there are deviations from regularity (Povel & Collard, 1982; Rosenbaum, Inhoff, & Gordon, 1984; Semjen & Garcia-Colera, 1986, Semjen, et al., this volume; Wing, 1980).

Systematic variations from regularity in the performance of musical rhythms have been extensively investigated by Gabrielsson and colleagues (Gabrielsson, 1974, 1987; Gabrielsson, Bengtsson, & Gabrielsson, 1983). They have examined deviations from regularity in the timing and intensities of notes in many types of music and on different instruments, concluding that such systematic variations are influenced by metric, phrasing, melodic, and harmonic aspects of musical structure in addition to the notated rhythm in the music. We agree, and we suggest further that the systematic deviations are reflecting central representations as well as constraints.

In his pioneering work, Seashore (1947) argued that deviations from regularity in timing and pitch provide the basis for esthetics in music. "Beauty in the rendition of a composer's design lies most frequently in the artistic play with deviations from the regular-true pitch, even loudness, metronomic time or pure tone, or any of their combinations or derivatives" . . . it is through the "principle of artistic deviation that artists find unlimited opportunities for variation and interpretation, for the expression of their creative power and artistic sensitivity, and for differentiating their individuality as artists" (Seashore, 1947, p. 47). He notes further, however, that skilled artists tend to use common principles of deviation from regularity in their performance, consistent with the empirical results reported above.

In the performance of scales, however, the goal is to play with rhythmic precision. To play music with rhythmic precision implies regularity, smoothness, and crispness in the timing and loudness of notes. One might expect equal timing and loudness for all notes in the scale; systematic deviations from regularity in performance would be unwanted. However, in practice, it is know that there is variation in the timing. Wagner (1971) had skilled pianists perform scales at different tempos and suggested that there was an optimum speed (about 6–9 notes/sec) for minimum variability in the timing. Our earlier study (Van Eerd, MacKenzie, Wilcox, Graham,

Huron, & Wills, 1986) replicated this finding, though our minimum variability in the internote intervals occurred at about 4–6 notes/sec. It seems, then, that the programming of piano performance includes deviations from regularity, and that these deviations are consistently reproduced over repetitions of the performances, by they sonatas, fugues, or technical exercises.

The hypothesis of generalized motor programs (Schmidt, 1975, 1982) makes clear predictions about the performance of piano scales. It implies that if the tempo of performance is varied, the absolute variability of timed intervals should decrease as tempo increases (i.e., the standard deviations should decrease as the timed intervals decrease). The variability would be a constant proportion of the mean time interval, e.g., 10% or 15% (i.e., the coefficient of variation should be constant). Further, the performance pattern of note timing or intensities across notes in a scale is predicted to change as a proportional function of playing tempo or speed. For example, in playing a scale twice as fast, the pattern of timing across notes should be the same, but with half the timing of each note. This view of a generalized motor program with a multiplicative rate parameter has received support (Carter & Shapiro, 1984; Keele & Summers, 1976; Shapiro, 1976; Shapiro, Zernicke, Gregor, & Diestel, 1981; Terzuolo & Viviani, 1979, Viviani & Terzuolo, 1980) and nonsupport (Gentner, 1982, 1987; Heuer & Schmidt, 1988, Winter, 1983; Zelaznik, Schmidt, & Gielen, 1986) from studies of movement in locomotion, typewriting, handwriting, and limb movement sequences.

Another approach to understanding how the nervous system programs movements in piano performance is to vary the playing hand (left or right), and then playing conditions for the scales (alone or together). In bimanual conditions, the task for the two hands can be anatomically the same (playing in contrary motion; the right hand ascends while the left hand descends the scale) or the bimanual scales task can be melodically the same (playing in similar motion: both hands ascend and descend the scale simultaneously). If the timing pattern is centrally generated by a generalized motor program or by a single module (see Keele, this volume), then one would expect to obtain similar patterns of results for left and right hand performance, regardless of whether the scales are being played with one hand alone, in contrary, or in similar motion.

## METHOD

### Subjects

Seven right-handed pianists served as subjects. Both males and females, between 20 and 30 years of age, participated. All were trained through standardized conservatory methods and certified by the Royal Conserva-

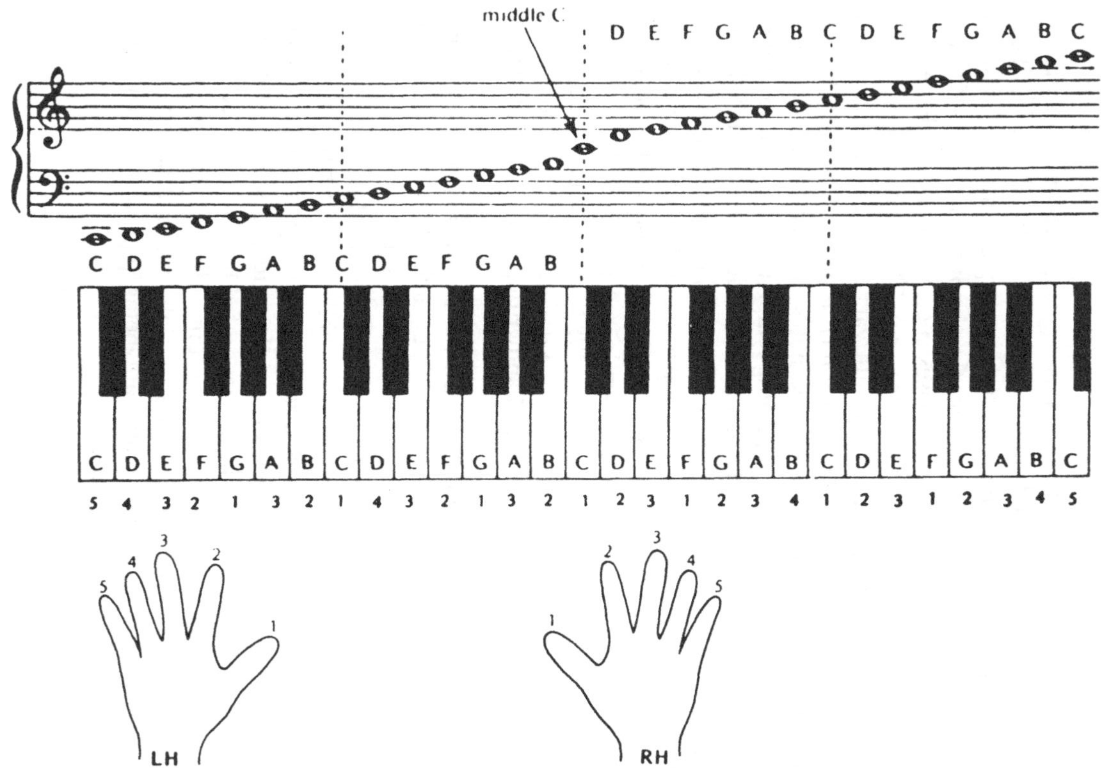


FIG. 12.2. The music, keyboard locations and fingering used. Pianists performed two octaves of a C major scale with left and right hands in the following conditions: alone; together in contrary motion (same fingering); together in similar motion (same notes).

tory of Music, Toronto. Two were Associates of the Conservatory (ARCT, the highest performance level), and were actively practicing for competitions. Three had Grade 10, one had Grade 8 and one had Grade 6 performance level certification. The pianists were naive as to the purpose of the experiment. They provided informed consent for the audio and piano monitoring of their performances, and were assured of anonymity.

## Musical Task

The pianists were asked to perform two-octave C-major scales, ascending and descending. They were instructed to play the scales as evenly as possible, with a forte or loud dynamic level. The music, keyboard locations, and fingering are provided in Fig. 12.2 (these were not shown to the pianists). The right hand always started ascending from middle C and the left hand played the two octaves below middle C. The scales were performed in the following playing conditions, with left and right hands: alone; together in similar motion (left and right hands ascend and descend the scales together, requiring simultaneously different fingering); and together in contrary motion (left descends while right ascends the scale, then left ascends while right descends the scale, requiring simultaneously identical fingering). The order of the playing conditions was randomized for each subject. Within each playing condition, subjects played at the following tempos: 60, 90, 120, and 180 metronome beats/min, and as fast as possible (with no metronome). The metronome beats corresponded to quarter notes and the scales were to be played in sixteenths. The metronome tempos thus corresponded to 4, 6, 8, and 12 notes or keypresses/sec. The order of the 5 tempos was also randomized within each playing condition.

For metronome-paced tempos, each trial was preceded by 8 beats of the metronome. Scales were performed with no metronome beats. Subjects were free to start anytime after the 8 beats of the metronome, and almost always played the first note of the scale in time with the next (nonexistent) metronome beat. For the condition of playing the scales at fast as possible, subjects were free to start anytime after the experimenter said, "READY." For each tempo, subjects performed 5 consecutive trials that were monitored with a tape recorder and the piano monitoring system. Subjects rarely made errors; however only the first 3 trials with "clean" data after preprocessing (described later) were used for subsequent analyses.

In the testing session, pianists practiced scales or other exercises of their choice, to adjust to the piano and studio. They then performed the scales in the randomized order determined a priori. The testing session lasted 45–50 min.

## Piano Monitoring System (PMS)

An on line computer system monitored keyboard action. A preliminary report of the system can be found in Wills, MacKenzie, Harrison, Topper, & Walker (1985); the planning and hardware details of implementation, correspondence of key action with hammer action, and reliability and validity tests are available in Topper (1984). The PMS consists of a grand piano, a hardware interface, and a microcomputer. Infrared sensors are mounted below each of the 88 keys of a Kawai acoustic grand piano, on a circuit board running the length of the keybed. A graded translucent flag mounted to the side of each key passes between the infrared emitter and detector and conveys position changes as a given key on the keyboard is depressed. The voltage output from the sensor is linearly proportional to the displacement of the piano key. A hardware interface performs analog-to-digital conversion and specifies time to the nearest ms to accompany key number and position information. The instrumentation is unobtrusive; there appears to be no effect on the feel of the keyboard or the sound of the piano.

Prior to each session, the system was calibrated by recording key *up* and *down* positions for the entire keyboard. These calibration files were used with monitored performance data files to produce the position-time data files, which include key numbers, positions, and times.

## Data Analyses

Software has been developed for preprocessing and pattern recognition on the position-time data files (Graham, 1984, 1985). For our purposes here, the position-time data were reduced to keypresses; each keypress was represented by the four position-time points of a trapezoid (see Fig. 12.3). From these data, we defined as *clean* those trials with complete data for keypresses for the scales pattern defined for each hand.

The keypress trapezoids were used to define the following dependent measures. *Note duration*, defined as the time from key down to key up for a given note, $(t3_n - t1_n)$, represents the length of time a note was played. The other temporal measures are relational and perhaps more important in accenting: Note gap and internote interval are computed from consecutive keypresses in a pattern. *Note gap*, defined as the time from key down of $note_{n+1}$, minus the key up time of $note_n$, $(t1_{n+1} - t3_n)$, represents the gap or overlap between consecutive notes. A measure of articulation, positive values indicate gaps, or staccato performance, and negative values indicate overlaps or legato performance. *Internote interval* is defined as the time between consecutive key downs, $(t1_{n1} - t1_n)$ and represents the time between consecutive note onsets. *Keypress velocity*, the average velocity from key up to key down $[(p1 - p0)/(t1 - t0)]$, reflects the intensity

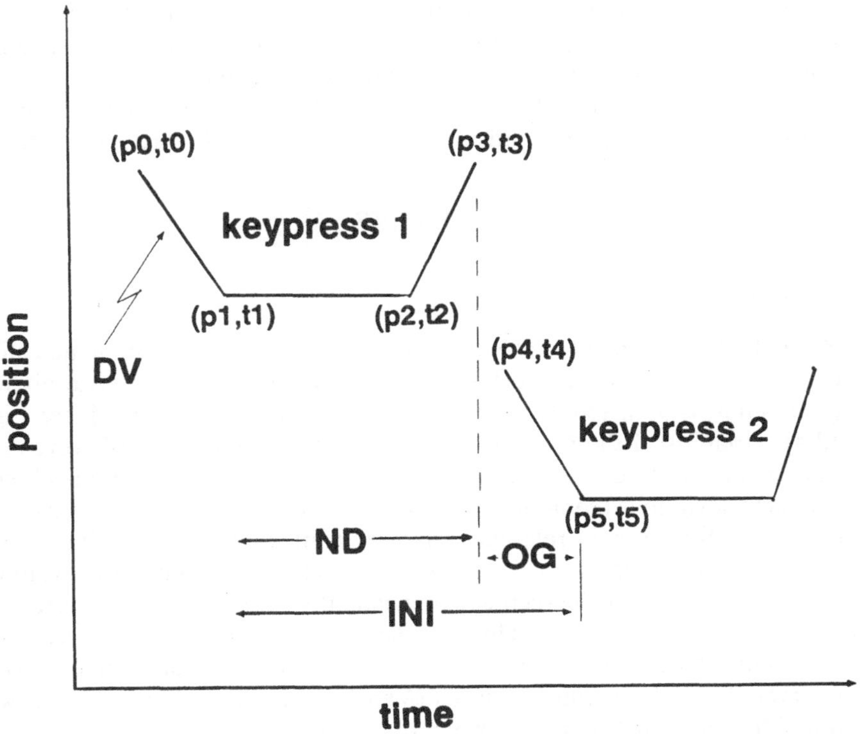


FIG. 12.3. Preprocessing and pattern recognition software were used to reduce position-time data to keypresses and musical patterns. Each keypress is represented by the 4 position-time points of a trapezoid. The following dependent measures were used: note duration (ND), note gap (OG), internote interval (INI), and keypress velocity (DV). See text for details.

of the note, or dynamic accenting. We recognize that these variables do not correspond exactly with the sound events (see Askenfelt & Jansson, 1988); however they were our best estimates of the keyboard measures relevant for the corresponding sound events. These measures were computed for all keypresses.

Consistent with our earlier work (MacKenzie et al., 1986; Van Eerd et al., 1986), our first step in analysis was to compute means, standard deviations, and coefficients of variation for internote intervals, note durations, and keypress velocities (over the notes in each trial). We intended to use the variability of timing (internote intervals) as an initial indicator of rhythmic precision in performance. These descriptive statistics were analyzed by separate repeated measures analyses of variance for each dependent measure, using as factors subjects, playing condition, tempo, trials, and hand ($7 \times 3 \times 5 \times 3 \times 2$).

Then, for each dependent measure, the data were analyzed separately for right ascending, right descending, left ascending and left descending notes. We used subjects × playing condition × tempo × trial × note (7 × 3 × 5 × 3 × 14) repeated measures ANOVAs. Given the large number of significance tests, only statistics significant at $p < .01$ are reported.

## RESULTS

### Variability of Internote Intervals

Recall that we intended to use the variability of timing as an initial indicator of rhythmic precision in performance. Results indicated that variability (across notes) of internote intervals increased with playing speed (see Fig. 12.4). Analysis of the standard deviations of internote intervals revealed significant effects of tempo, $F(4,24) = 6.02$, $p < .0017$. Within-subject standard deviations (computed over notes) tended to increase as a function of tempo. Increases in relative variability (reflected by the coefficient of variation) with tempo were even more pronounced. The standard deviation becomes an increasing percentage of mean internote interval with increases in tempo or playing speed. This was the case for all pianists. This extends the results of Wagner (1971) and our earlier study of variability of internote intervals for right ascending scales (Van Eerd et al., 1986), in which we found that standard deviations were a U-shaped function of tempo. In Van Eerd et al. (1986), the standard deviation of INI increased for tempos slower than the 60 beats/min or 4 notes/sec used in the present study (consistent with the expected finding that standard deviations increase with the magnitude of the mean internote interval). In the present study, the finding of increasing variability in internote intervals with playing speed is a robust one across playing conditions, that is, it does not vary as a function of playing hand or playing condition.

### Timing Measures

*Internote intervals (INI).* The analysis of mean internote intervals (collapsed over notes) revealed significant effects of tempo, $F(4,24) = 91.97$, $p < .0001$, and a playing condition by tempo interaction, $F(8,48) = 5.17$, $p < .0001$. Figure 12.5 shows that, although INI decreased with tempo (as it should), there was a trend for playing faster than specified at slow tempos and slower than specified at faster tempos. The condition by tempo interaction in Fig. 12.5 shows that at "slow" tempos (60 beats/min or 4 notes/sec), subjects played longest INIs in alone conditions, then in contrary motion, and shortest INIs in similar motion. At "faster" tempos (120 and

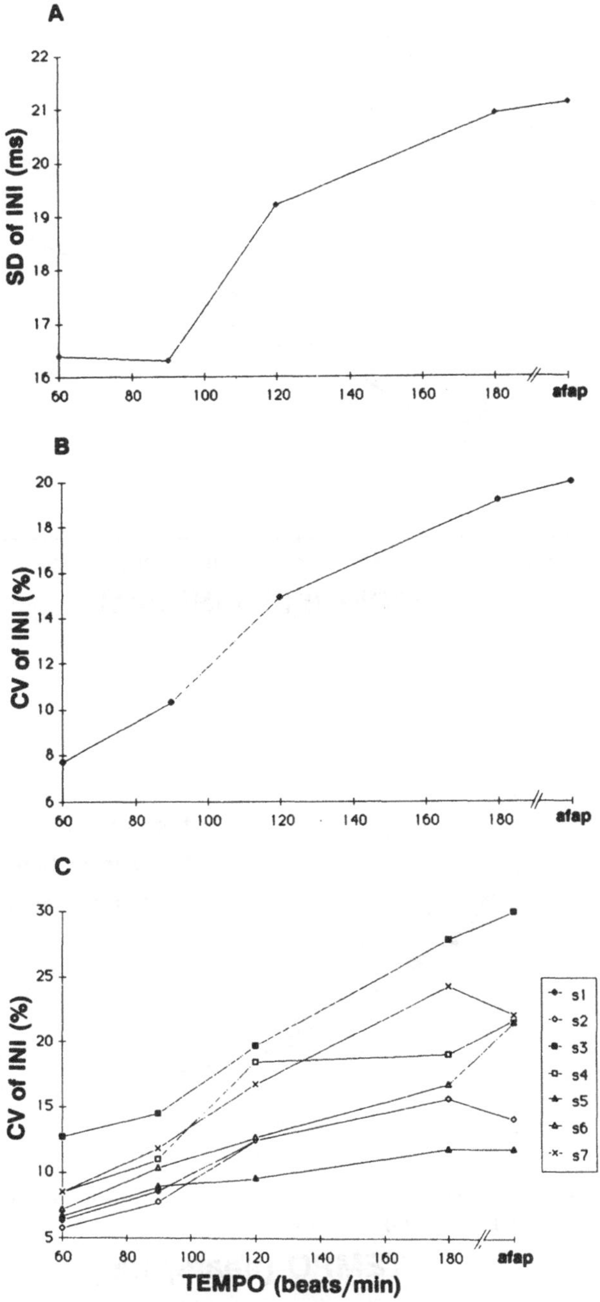


FIG. 12.4. Variability of internote intervals in the scale as a function of tempo (AFAP = as fast as possible). The variability measures are computed over notes for single trials, then averaged over all conditions of the experiment. (a) Mean within subject standard deviations of internote intervals in the scale (ms) (b) Coefficients of variation of internote intervals in the scale (%) (c) Coefficients of variation (%) for each pianist.

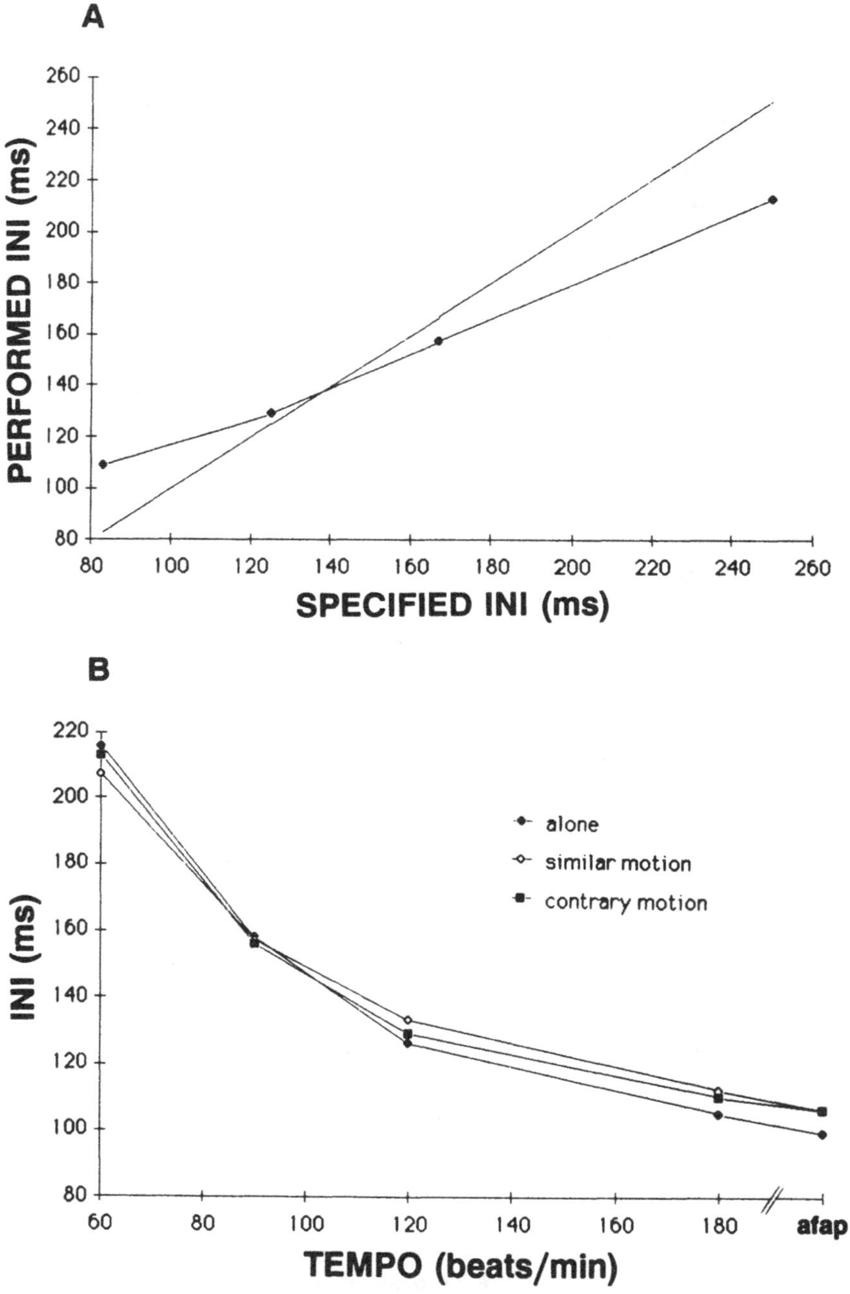


FIG. 12.5. (a) Performed internote interval (msec) as a function of internote interval specified by the metronome. Subjects played slower than specified at the fastest tempos and faster than specified at the slowest tempos. (b) Performed internote interval by playing condition. Pianists played truest to metronome specified tempos when playing alone, and with greatest deviation from metronome specified tempos when playing with the two hands in similar motion.

180 beats/min or 8 and 12 notes/sec), subjects played shorter INIs in alone conditions and longest INIs in similar conditions. That is, subjects played "truest" to the metronome specified tempos in the hands alone condition, then in contrary motion, then similar motion. When playing as fast as possible, INIs are shorter when playing alone than in either of the hands together conditions.

Subsequent analyses were done including notes as a factor separately for right ascending, right descending, left ascending, and left descending notes. The interaction of playing condition with tempo in the analysis of internote intervals was significant in the analyses of right ascending and right descending notes, with the same effects as described above for the mean INI data. Playing condition and tempo did not interact in the separate analyses of left ascending and left descending scales, suggesting that the effects are more pronounced in right hand than in left hand data.

Internote intervals also changed significantly as a function of note position in the scale, even though the scales were to be played evenly in sixteenths. This is consistent with other reports of departures from metronomic or mechanical regularity in the expressive performance of music (e.g., Gabrielsson, 1974; Seashore, 1947). There were main effects of note position in right ascending, $F(13,78) = 7.92, p < .0001$, right descending. $F(13,78) = 3.01, p < .0013$, left ascending, $F(13,78) = 3.12, p < .0009$, and left descending, $F(13,78) = 5.04, p < .0001$, internote intervals. The pattern of longer and shorter internote intervals appears to reflect, in part, the biomechanical aspects of crossovers and crossunders in fingering. For example, in right ascending data, as the pianist plays CDEFGAB . . . with fingering 1231234 . . . , the internote interval is slightly shorter as the thumb crosses under to play F, slightly longer as the index finger crosses over to play G and slightly shorter as the middle finger plays A (refer to Fig. 12.6).

This pattern of internote intervals as a function of note/fingering is modified by playing tempo. There was a significant tempo by note interaction in the analysis of INI for right ascending, $F(52,312) = 3.19, p < .0001$, right descending, $F(52,312) = 2.63, p < .0001$, left ascending, $F(52,312) = 3.67, p < .0001$, and left descending, $F(52,312) = 2.42, p < .0001$, scales. Shown in Fig. 12.6, the local increases and decreases in INI (or zig zag pattern of long–short–long–short) at the slowest tempo (60 beats/min) become more pronounced in magnitude (both in a relative and absolute way), but less frequently occurring at the faster playing speeds. That is, at the higher speeds, subjects do not always alternate with long and short INIs. This is true of individual subjects as well. The timing pattern across internote intervals changes systematically as a function of tempo, rather than retaining the proportional durations predicted by a motor program with a multiplicative rate parameter.

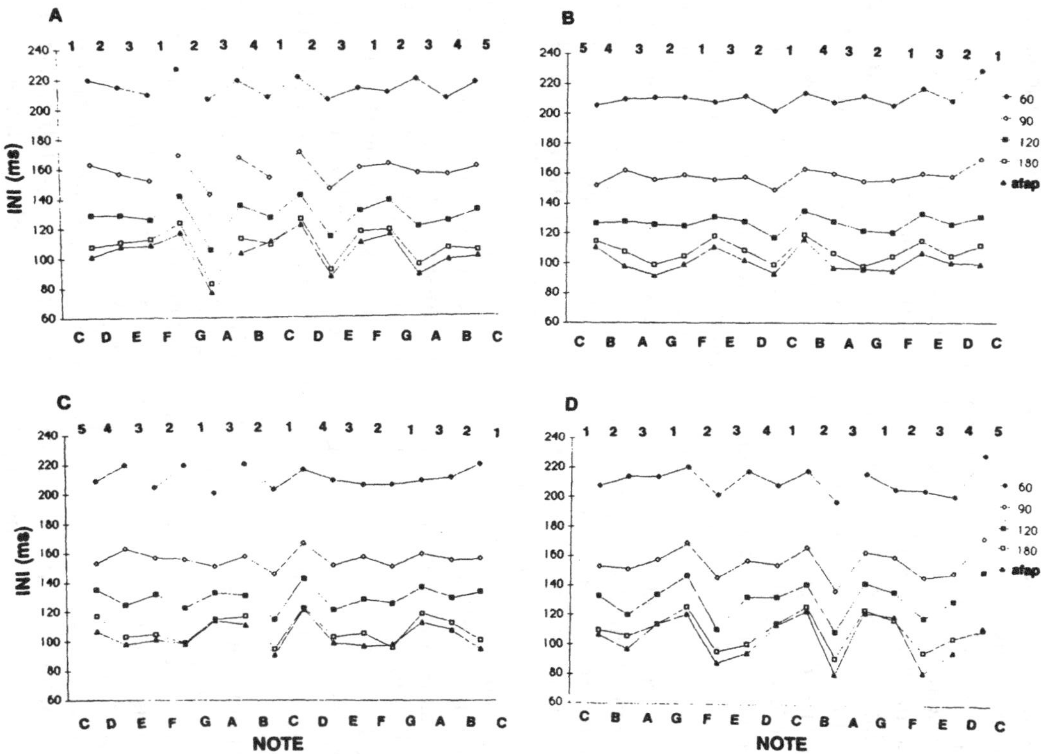


FIG. 12.6. Mean internote interval (msec) as a function of tempo and note position for (a) right ascending, (b) right descending, (c) left ascending, and (d) left descending scales. Fingering as indicated at the top: 1 = thumb, 2 = index, 3 = middle, 4 = ring and 5 = little fingers. Tempos are beats/min specified by the metronome and as fast as possible (AFAP).

This pattern of internote intervals as a function of note/fingering is also modified by the task requirements for the other voice/hand, as revealed by the significant playing condition by note interaction in the analysis of internote intervals in the right descending, $F(26,156) = 3.85, p < .0001$, and left descending scales, $F(26,156) = 2.13, p < .0025$. Figure 12.7 shows the effects. Our interpretation of the pattern for left descending data is that the variations in INI due to crossovers and crossunders in performance are magnified by the anatomically mirrored crossovers and crossunders required by the right hand in contrary motion conditions relative to the other playing conditions. Such is not the case, however, for the right descending data; here it appears that the same right hand fingering gives rise to greater variation in INIs and longer INIs when playing in similar motion than in the other two playing conditions.

In no case was there a second order note by tempo by playing condition interaction in the analysis of INI. Internote intervals were modified by tempo, and internote intervals were modified by playing condition, but internote intervals across note positions were never modified by both playing condition and tempo.

*Note durations.* The analyses of note duration data revealed the same significant effects as the internote interval analyses, that is, significant tempo, tempo by playing condition, note, tempo by note, and condition by note effects. The timing patterns were graphically similar in note duration and internote interval data, and are therefore not detailed again here. There was however, one important exception. In the mean note duration data, there was a significant tempo by playing hand interaction, $F(4,24) = 11.82, p < .0001$. The interaction indicated that the right hand played with longer durations than the left hand at the two slowest tempos (i.e., closer to metronome specified time) and with shorter durations than the left hand when playing as fast as possible.

*Note gap.* The hand effects, already discussed, in timing were also apparent in the mean gap overlap measure (averaged over notes). Figure 12.8 illustrates that the articulation or degree of legato performance changed significantly as a function of tempo, $F(4,24) = 6.75, p < .0009$, with the greatest overlap at the slow tempos. As tempo increased, the legato or overlap between consecutive notes decreased. Although there were individual differences in the measure, the trend for increasing separation between the notes as tempo increased held even for pianist number 5 who played with gaps between notes at all tempos. (As an aside, it may be of interest that subject number 5 was one of the most skilled pianists). This trend for increasing separation between notes as tempo increases is small but reliable.

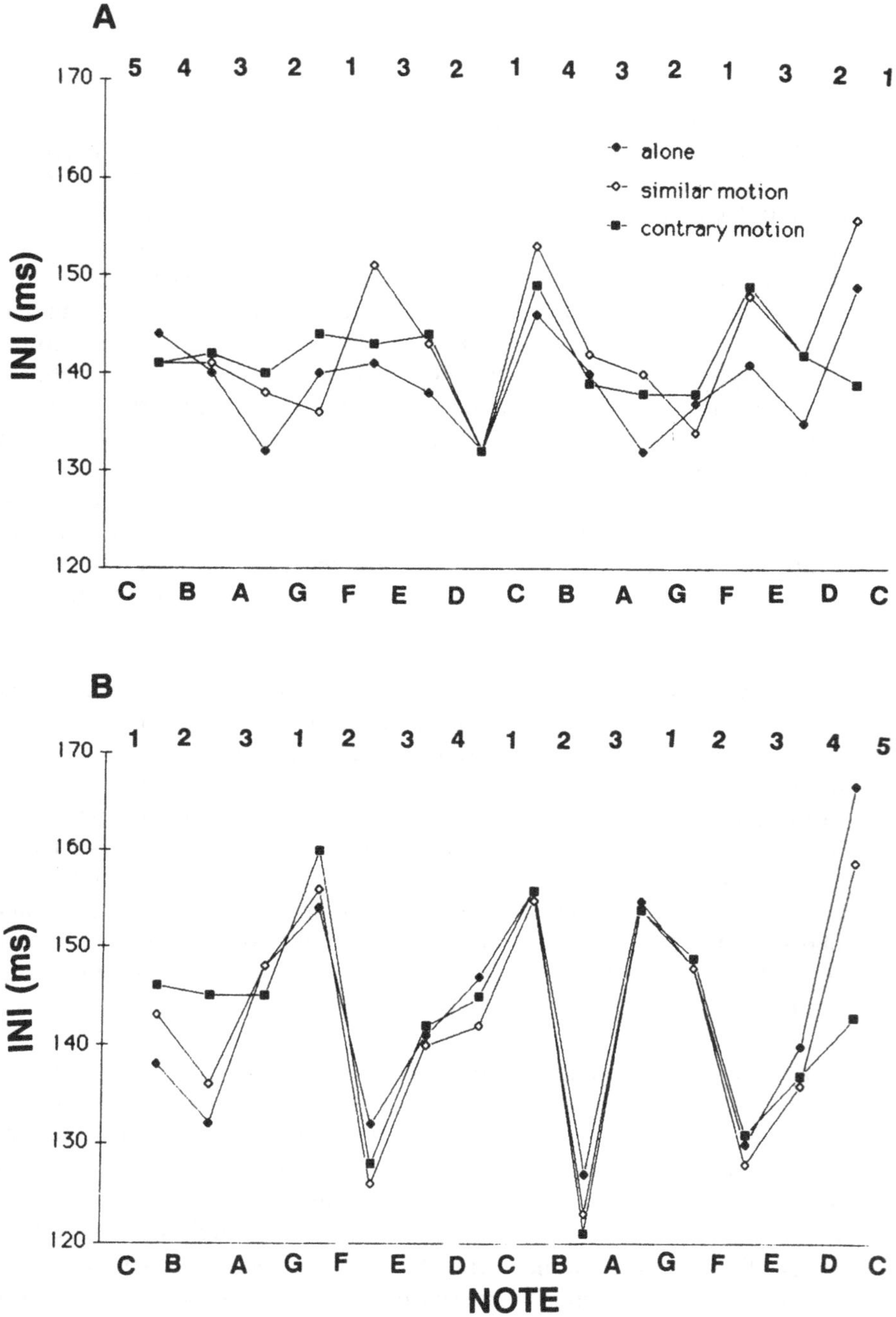


FIG. 12.7. Mean internote interval (msec) as a function of playing condition and note position for (a) right descending, and (b) left descending scales. Fingering is indicated at the top.

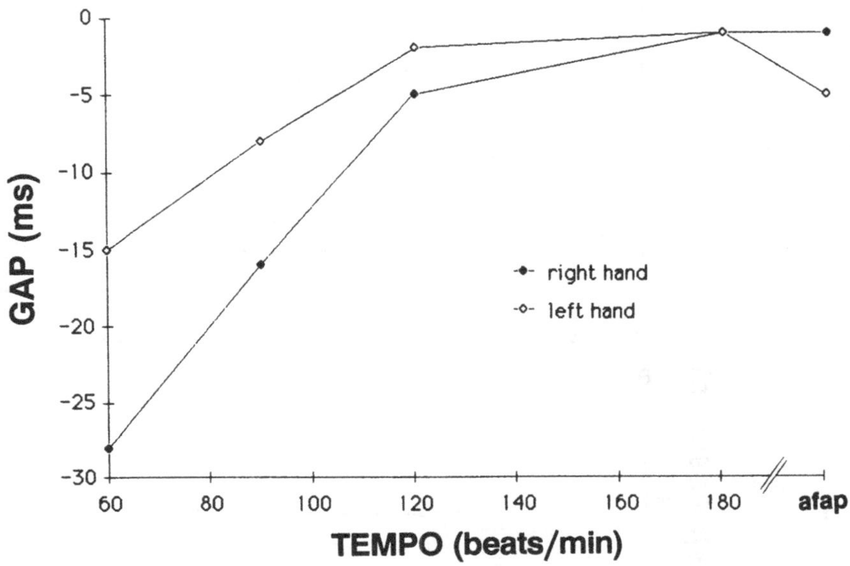


FIG. 12.8. Gap (msec) between consecutive notes as a function of tempo and playing hand. Negative gap values indicate legato playing, with overlap between consecutive notes.

In addition, the significant hand by tempo interaction, $F(4,24) = 12.59$, $p < .0001$, illustrated in Fig. 12.8, shows that although both hands showed a decreasing amount of overlap as tempo increased, the right hand had a steeper slope. Also, the right hand played the notes with more overlap (legato) than the left hand at all metronome specified tempos but the fastest. When playing as fast as possible, the left hand had more overlap than the right. Thus, the right hand appears to be playing more legato than the left hand at all tempos except as fast as possible. These results parallel the hand by tempo interaction described above for note durations.

## Keypress Velocity

Analysis of mean velocity of keypresses (averaged over notes) indicated that as speed of playing increased, so did speed of keypressing, $F(4,24) = 13.67$, $p < .0001$. Figure 12.9 illustrates this. It is not surprising that increases in overall speed of playing gives rise to increases in speed of individual keypresses. However, for a given key, an increase in keypress speed gives rise to an increase in the dynamics or loudness. Theoretically, it should be possible to alter the dynamics independently of the playing speed, within limits. Our pianists were instructed to maintain the same dynamic level (forte) at all speeds.

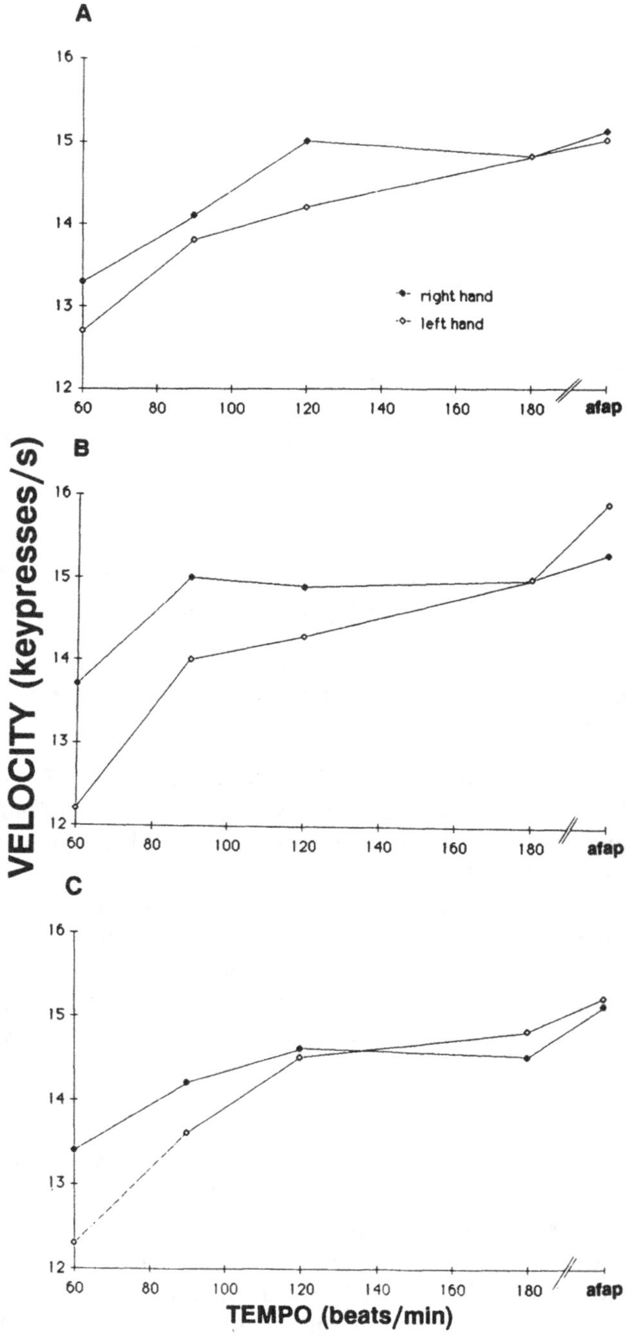


FIG. 12.9. Keypress velocity (keypresses/sec) as a function of tempo for left and right hands when playing: (a) alone, (b) in similar motion, (c) in contrary motion. The keypress velocity increases with increases in playing speed, and is always higher for the right hand in metronome specified tempos (AFAP = as fast as possible).

As well, there was a tempo by hand interaction, $F(4,24) = 11.04$, $p < .0001$. Figure 12.9 shows that the right hand has a higher keypress velocity than the left hand at all metronome-paced tempos but the fastest (180 beats/min). When playing as fast as possible, the left hand has a slightly higher keypress velocity than the right hand. Thus, in addition to the increased legato of the right hand, the right hand also has a higher keypress velocity than the left.

Mean keypress velocities (averaged over notes for each trial) also revealed a significant tempo by hand by playing condition interaction, $F(8,48) = 2.97$, $p < .01$. Graphed in Fig. 12.9, it appears that in the contrary motion condition, the two hands have more similar keypress velocities at a lower tempo (120 beats/min) than in similar motion and hands alone playing conditions.

There were also significant note effects in the separate analyses of keypress velocities for right ascending, $F(13,78) = 4.86$, $p < .0001$, right descending, $F(13,78) = 47.9$, $p < .0001$, left ascending, $F(13,78) = 8.07$, $p < .0001$, and left descending scales, $F(13,78) = 9.55$, $p < .0001$. Notes were accented as a function of fingering: for example, in right ascending scales there were often higher keypress velocities for the thumb (playing C and F) and middle finger (playing A). These higher velocities were for the notes that usually preceded the longer internote intervals. Thus, the dynamic accenting through increased loudness was followed by a temporal accenting through increased internote interval. However, notes were also accented as a function of scale position. There was a contrast in the dynamic accenting of the first and second notes in the right and left descending data. The loud high C was followed by a quieter B as the scale was descended, perhaps to emphasize the octave.

However, like the internote intervals, the significant changes in keypress velocity as a function of note were modified by tempo and playing condition. That is, velocity for a given note changed with playing speed and bimanual playing conditions. There was a significant tempo by note interaction, graphed in Fig. 12.10, for right ascending, $F(52,312) = 1.99$, $p < .0002$, and the right descending scales, $F(52,312) = 2.24$, $p < .0001$. The differences among the tempos appeared most pronounced for those notes that had the highest and lowest velocities.

There was a significant playing condition by note interaction on keypress velocity for right descending, $F(26,156)=2.71$, $p<.0001$, left ascending, $F(26,156)=3.78$, $p<.0001$ and left descending scales, $F(26,156)=4.37$, $p<.0001$. The interaction of playing condition and note did not reach significance in right hand ascending data. Figure 12.11 shows that for left descending data velocity of keypress is lower in contrary motion than in other playing conditions for all notes but the first (when the left and right thumbs must play middle C simultaneously). For left ascending scales,

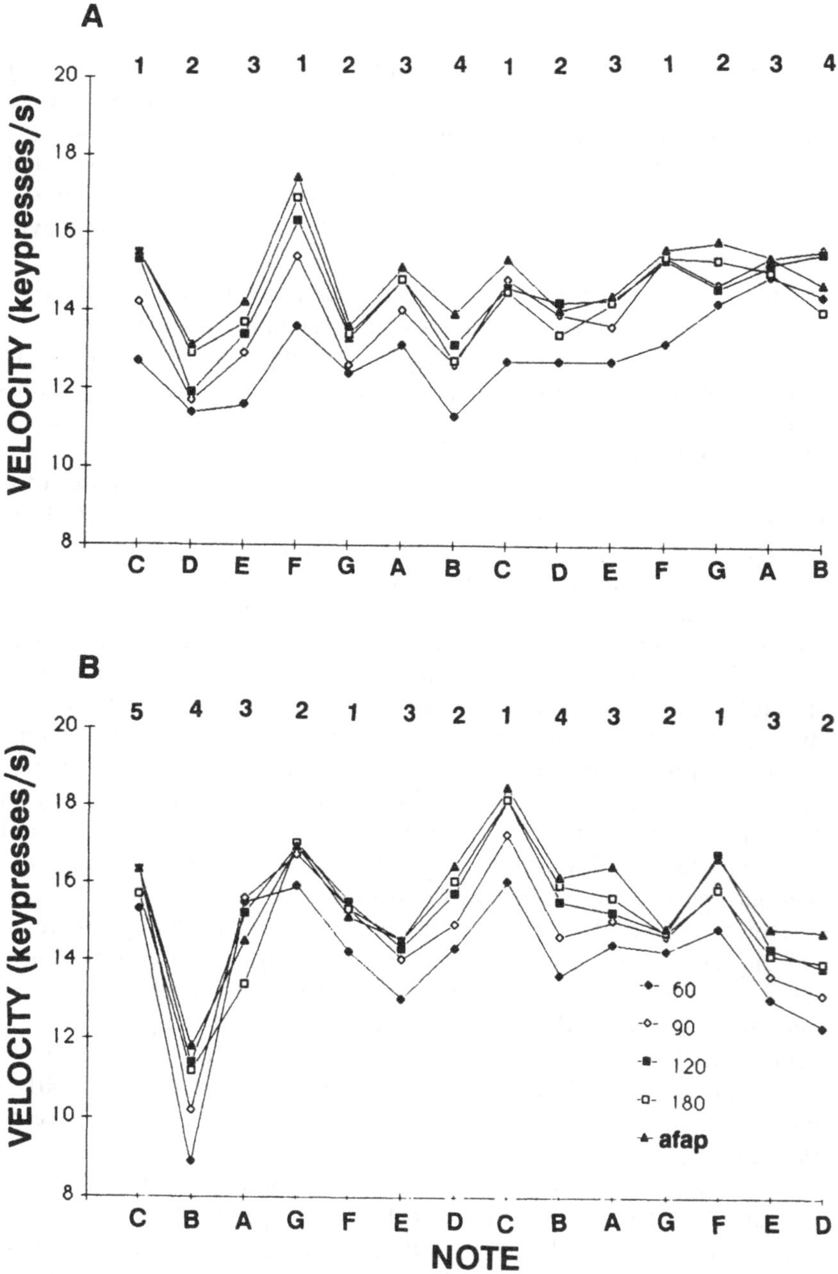


FIG. 12.10. Keypress velocity (keypresses/sec) as a function of tempo for (a) right ascending and (b) right descending scales. Fingering is indicated at the top and is defined in Fig. 12.6 which shows corresponding internote intervals.

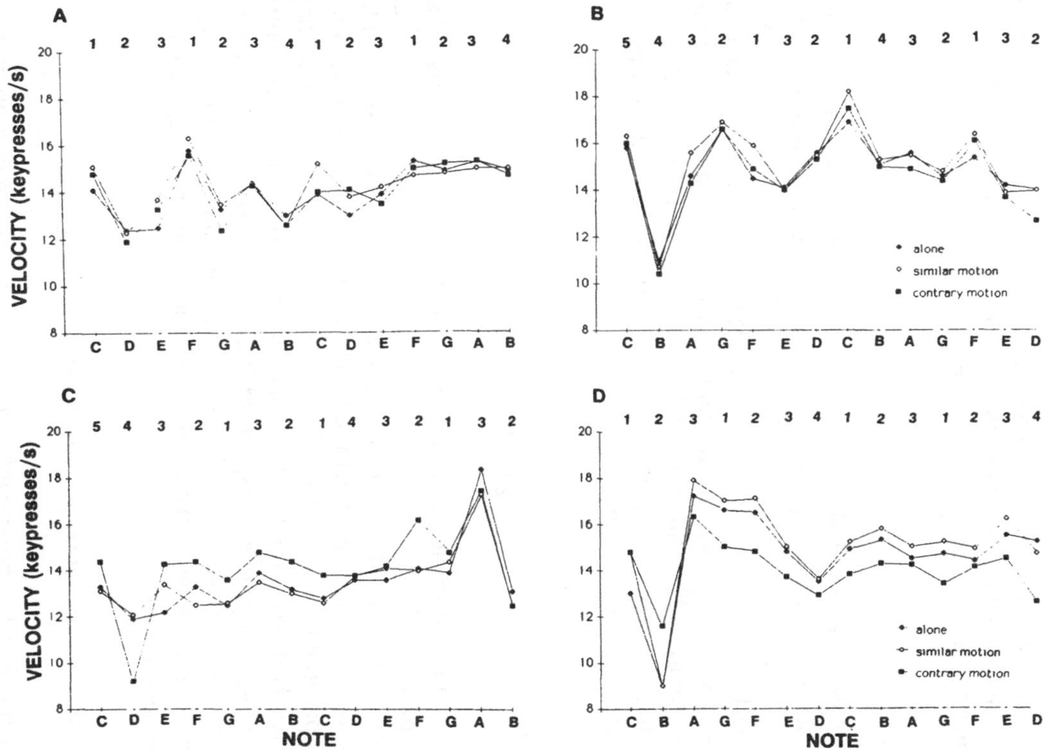


FIG. 12.11. Keypress velocity (keypresses/sec) as a function of playing condition for (a) right ascending, (b) right descending, (c) left ascending, and (d) left descending scales. Fingering is indicated at the top.

velocity of keypresses are higher in contrary motion than in other playing conditions for all notes but the second note (D). This is when the right hand is making a contrasting low-velocity keypress (B) to the previous high C mentioned above. As well, Fig. 12.11 shows that the note by playing condition interaction is more pronounced for left hand than right hand keypress velocity data, suggesting that the left hand is more affected by the playing conditions than the right hand.

### Trial Effects

In no case was trials a significant factor in the above analyses. In only one or two cases were higher order interactions involving trials significant, but with very small *F* ratios (less than 2). We think this is a reasonable indication that the performances were highly consistent (repeated by each pianist in almost exactly the same way from trial to trial), in agreement with earlier work of Seashore (1938), Shaffer (1981), Gabrielsson (1974) and Sloboda, (1983). Thus, the reported pattern of results are reliably indicating the pianists' rhythmic performances.

## DISCUSSION

The performance of piano scales by highly skilled pianists is a task that has been preceded by many practice hours, with the goal being to play with rhythmic precision. From an empirical perspective, the present results are interesting in that they reflect the multiple levels of representation for planning, programming, and execution of the task. They are also reflecting multiple levels of constraints on the movement, from musical structure through to the biomechanical aspects of fingering transitions. In all cases, the performer is effectively achieving the goal of rhythmic precision, though it is clear from listening to the performers that some are doing this with more crispness or clarity.

### Motor Programs and Sensory Information Processing

*Evidence against the phonograph analogy.* In terms of motor programs, the results do not support the predictions of a phonograph analogy of a central representation of relative timing (Schmidt, 1975; 1982). A constant pattern of timing and velocities over the notes in a scale, adjusted proportionally for playing speed, was predicted. The variability in the timing and loudness of the notes increased as tempo or playing speed increased. The INI values changed over notes in the scales performance. Neither the timing nor keypress velocities showed a proportional adjustments with changes in tempo. Consistent with the work of Gentner (1982, 1987), the timing in the scales performance does not support proportional duration models. Figure 12.6 shows that for right ascending, right descending, left ascending, and left descending scales, there were differences in internote interval patterns as a function of note position with changes in tempo. It appeared that at the faster tempos, the long–short alternating pattern of INIs was less frequently occurring, but the differences between long and short INIs when they did occur were greater. Thus, we think it is of great significance that the variability of INIs increased with increases in tempo.

It is important to emphasize the range of tempos used in the piano performance—the "slowest" tempo was 60 beats/min or 4 notes/sec, giving rise to specified INIs of 250 msec, and the "fastest" metronome specified tempo was 180 beats/min or 12 notes/sec, giving rise to specified INIs of 83 msec. When playing as fast as possible, subjects played even faster than this, getting close to Lashley's limit, where he suggested motor programs must be operating.

Traditionally, one of the arguments in favor of motor programs is a default one: For movements less than 200 msec in duration, a motor program must be operating because there is not enough time for feedback to alter the ongoing movement. Estimates of visual feedback processing time are classically of 200 msec (Keele & Posner, 1968). More recent research has indicated shorter times for visual feedback to modify ongoing, discrete movement (e.g., Carlton, 1981; Zelaznik, Hawkins & Kisselburgh, 1983), of less than 150 msec. Auditory and kinesthetic reaction times are known to be faster than visual ones.

More important, modification of an ongoing sequence of movements is different from modification of a discrete aiming movement. Sensory information may be nonautogenic, feeding forward to other movement components in the sequence. Abbs, Gracco, and Cole (1984) have argued for latencies as short as 40–65 msec for modification of articulatory movements to achieve the speech goal, based on perturbations of the lower lip. In the present study, for "slower" tempos (e.g., 4 or notes/sec), auditory and proprioceptive information could be making on line modifications to the next note in the scales performance. As tempo increases, the possibility of using sensory information from a given keypress/note to modify the following note in the movement sequence decreases. At the fastest performances (12–15 notes/sec), the possibility for sensorimotor modifications on a note by note basis becomes increasingly unlikely.

We would like to suggest that sensory information is indeed being used at the slower tempos, and that as playing speed increases, sensory information is still being used, but less and less to make modifications on a note by note basis. Alternatively, one might say that there is an increased reliance on programming without rapid use of sensory information as the playing speed increases. At the fastest playing speeds, the performers are indeed near the limit of sensorimotor processing loop times for modifications of keypresses, at least at the level of one or two elements hence. At the faster speeds, subjects are using sensory information as well as they can. Sensory information is modifying the performance less frequently (in larger groups, not alternate notes), but with larger adjustments when modifications are made. These suggestions are consistent with the less frequent but larger modifications in internote intervals seen in the data.

The idea of increased variability being evidence *for* programmed performance and against sensory modifications is contrary to the view presented in the introduction that motor programs should be characterized by low variability. A similar argument has been presented by Wing, Turton, and Fraser (1986) for prehension. In examining the spatial variability of the thumb position in grasping, they found greater spatial variability in the time interval immediately prior to object contact when subjects were grasping under speeded conditions than when subjects were grasping under normal conditions. According to models in which grasping is composed of ballistic and sensory modification phases, for very fast movements, adjustments cannot be made as well to "groove" the movement during the sensory modification phase prior to object contact. Thus, there is increased variability for the very fast movements. Likewise, in very fast piano playing, adjustments cannot be made on every note to correct for deviations in the desired pattern of timing. Thus, the pattern of variability in timing increases with speed. If this explanation of our results holds, then the sensory based performance modifications are not an all-or-none phenomenon, based on a certain feedback loop time of *x* msec, but a matter of degree, with the frequency of occurrence of modifications decreasing, and the magnitude of the modifications increasing with increases in playing speed. In summary, we suggest that subjects are programming the scales at all tempos, and that sensory information is being used when possible to achieve the goal of rhythmic precision.

*Playing with one or two hands.* Another important finding in the data was that playing with one hand alone, or with both hands together in similar motion or contrary motion did not affect the variability but did affect the mean internote intervals. In all playing conditions, subjects played faster than required at the slower tempos and slower than required at the fastest tempos. However, subjects played closer to the metronome-specified tempos when playing alone than in bimanual conditions. It is as if the coordination of the two hands in the bimanual conditions takes some attention, detracting from the temporal precision in performance (see Peters, 1981). There was greatest deviation from the metronome-specified tempos when playing in similar motion. In this sense, playing in similar motion (different fingering) might be interpreted as more difficult than playing in contrary motion. Often, scales in similar motion are practiced more frequently by pianists than scales in contrary motion.

The pattern of timing over notes also changed as a function of playing condition. This might be interpreted as evidence against a central module giving rise to a pattern of timing that is constant regardless of whether scales are being played with one or two hands, in contrary or similar motion. The timing pattern in performance is thus reflective not only of the fingering and tonal relations for the playing hand of interest, but also of the fingering and tonal relations of the other playing hand. The right hand appeared to play with less variability over notes when playing alone than when playing in similar and contrary motion. This was not the case for the left hand (see Fig. 12.11a and 12.11b). The right hand seemed more affected in similar than contrary motion, whereas, the left hand timing seems more variable over notes in contrary than in similar motion. These asymmetrical effects are reminiscent of the "cross talk" between left and right hands seen in earlier studies of bimanual coordination in pointing (Kelso, Southard, & Goodman, 1979; Marteniuk, MacKenzie, & Baba, 1984). As well, there were significant differences between the two hands, and we now focus on these effects.

*Hand differences in performance.* There were some compelling results on differences between left and right hands in performance. These results must be interpreted while keeping in mind that there were no significant differences between the two hands in the mean INI analysis. That is, the time intervals between consecutive tones did not, on average, differ between right and left hand scales performance.

Although the mean internote intervals did not differ between right and left hands, the other temporal measures of mean note durations and overlap between consecutive notes were significantly different between the two hands. Likewise, the right and left hands differed significantly in their keypress velocities. These findings argue against a central representation for the rhythmic performance that is constant regardless of effector.

These results must be interpreted with caution. At face value, the results suggest the right hand treble scales are played with greater overlap (legato) and greater intensity (loudness) than the left hand bass scales. Thus it would appear that the pianists are emphasizing the right hand in the scales performance; it is louder, more continuous, and more prominent than the performance of the left hand.

We are unaware of literature formally documenting psychophysical or acoustic reasons for the hand differences. One possible explanation for the overlap effects is due to sound decay: Bass strings on the piano are reported to have a longer decay time than the treble strings (A. Askenfelt & J. Sundberg, personal communication, May 26, 1988); therefore, the larger overlap with the right hand notes (treble strings) could serve to perceptually equate bass and treble notes in the bimanual conditions. With respect to hand differences in keypress velocities, the finding corresponds to one of Sundberg's Rules for Music Performance Synthesis (Sundberg, 1988)—increase the loudness as a function of pitch. It may be that a given keypress velocity gives rise to different dynamics (loudness), depending on the key location on the keyboard. The same keypress velocity on bass notes is reported to produce a louder sound than on treble notes; in fact there may be a continuum along the keyboard, (A. Askenfelt, J. Sundberg, and G. Weinreich, personal communication, May 26, 1988). However, such a continuum is not obviously evident in the velocities of Figs. 12.10 and 12.11. Again, the greater keypress velocities for the right hand notes could serve to perceptually equate treble and bass notes in bimanual playing conditions. However, the overlap and velocity differences between left and right hands occur in the unimanual playing conditions as well as the bimanual scales.

The differences between hands may also reflect differences in articulatory aspects of the motor control between right and left hands of our right-handed pianists (Kimura, 1979; MacKenzie, 1980; Peters, 1981; Roy, 1985). Our legato and keypress velocity results suggest that the available range of articulatory control is greater for the right hand. This is consistent with views of the right hand as more dexterous and specialized for sequencing and articulatory aspects of communication (Kimura, 1979). However, for highly skilled pianists, we certainly did not expect to find differences between the two hands in the highly practiced task of playing scales. One final observation is important for these considerations—when playing as fast as possible, in the absence of the external pacing stimulus (8-beat lead in by the metronome), the situation reverses; here the left hand has a slightly longer duration, more overlap, and a higher keypress velocity than the right hand. Regardless of the relative acoustic or motor reasons for the observed hand differences in performance, at this time we believe that the apparent right hand emphasis is musically intended by the highly skilled pianists in order to convey the smoothness, evenness, and synchrony required for the goal of rhythmic precision in scales performance.

## Motor Psychophysics of Rhythm

One goal of this research was to provide motor psychophysical functions for rhythmic precision in the performance of piano scales. As shown in Fig. 12.5, pianists had a significant tendency to shorten the long internote intervals required in slow tempos and to lengthen the short internote intervals required in fast tempos. This bias is consistent with sensory psychophysical results in time and distance estimation, labeled by Poulton (1979a, 1979b), *range effects* or *contraction bias*. Pianists played truest to metronome-specified tempos when playing alone, and with greatest deviation from metronome-specified tempos when playing with the two hands in similar motion.

With an objectively defined tempo for performance, variability of performance increased with tempo (Fig. 12.4). This was true for velocity of keypresses as well as the internote intervals shown in Fig. 12.4. The specific patterns over notes of internote intervals and velocity giving rise to this variability are shown in Fig. 12.6 and 12.10 respectively. Just as performing faster affects spatial precision in aiming movements, so does performing scales faster affect temporal precision in rhythmic movement sequences. Our results do not address the question of whether the increases in performance variability with tempo are perceptible.

Rhythm is a multidimensional concept in many senses. It is multidimensional in that many dependent measures may be used to examine rhythmic aspects of performance. In our analyses, we have considered the timing (internote intervals), articulation (overlap between consecutive notes), and dynamics (keypress velocities) as important measures of rhythm and rhythmic precision. How are these measures interrelated?

At one level, it appears that the gap and keypress velocities are showing similar results, reflecting common programming and control processes. For example, the hand by tempo interactions are very similar for overlap and keypress velocity data (graphed as Figs. 12.8 and 12.9 respectively). Such hand by tempo effects did not exist for the internote interval data. On a note by note basis in performance, comparison of internote intervals and keypress velocities showed that sometimes they appeared to covary. For example, looking at the right ascending data for internote intervals (in Fig. 12.6) and keypress velocity (Fig. 12.10), there were corresponding changes in the two measures due to the biomechanical aspects of crossovers and crossunders in fingering. When playing the notes, EFG . . . with fingering 312 . . . , there was lower keypress velocity for E, a shorter internote interval as the thumb crosses under, a higher keypress velocity for the thumb playing F, a longer internote interval as the index crosses over, and a lower keypress velocity for G, and so on. This type of pattern (the correspondence of keypress velocity with preceding and succeeding internote intervals) was seen for most crossovers and crossunders. In contrast, the same figures demonstrate that sometimes the internote intervals and keypress velocities seemed quite independent. The first few notes (CDE) had almost no variation in internote intervals, but systematic variation in keypress velocities, such as dynamic accenting of the first C in the scale.

Motor psychophysical functions are a necessary foundation to further our understanding of the underlying representations and constraints in music performance. They may be used to generate hypotheses about the nature of specific constraints, testable in subsequent studies. For example, it may be that the pattern of crossovers and crossunders described above are constrained not by the shortest interval (for a given tempo), but by the *longest* internote interval (J. McClelland, personal communication, June 30, 1988). The performer thus "averages" internote intervals accordingly to achieve the specified tempo.

One goal of future research in motor psychophysics might be to quantify the interrelations among these timing, articulation, and keypress velocity variables from a multivariate perspective (such as multidimensional scaling). For example, just as the present chapter has examined how timing, articulation, and keypress velocity performance variables are used to convey rhythmic information, research is needed to describe how these variables are combined to communicate musically dynamic information, like forte and crescendo–diminuendo.

Rhythm is multidimensional in other aspects as well. Returning to Fig. 12.1, Gabrielsson (1979, 1986) has referred to the experience of rhythm (including perceived grouping, pulse, tempo, experienced motions and emotions), the overt behaviour of rhythm (as a listener or performer), and the physiological aspects of rhythm (respiratory, neural, and electromyographic expressions of rhythmic activity). The integration of these multiple facets of rhythm also present a challenge to researchers of music and human performance. After Harold Seashore, Gabrielsson (1979) suggested that "the psychophysical relations between the performer and the listener must be worked out in detail." The present study can be regarded as one of many small steps in this direction.

## ACKNOWLEDGMENTS

The authors wish to thank: our pianists for playing scales; David Huron, Lisa Wilcox and Barry Wills for initial discussions; Jude Charlton for her assistance in data collection; and Evan Graham for the hardware, preprocessing, and pattern recognition software he developed. Thanks to Anders Askenfelt, Alf Gabrielsson, Marc Jeannerod, Graham Naylor, Marie Therese Perenin, Henry Shaffer, and Johan Sundberg for comments on an earlier draft.

The grand piano and studio facilities for the Music Performance Laboratory are provided by Conrad Grebel College at the University of Waterloo. The Piano Monitoring System was developed with the aid of NSERC and the University of Waterloo Research Grants Program. This research was supported by grants from the Natural Sciences and Engineering Research Council of Canada to the first author. Preparation of the manuscript was done while on leave at the Experimental Neuropsychology Laboratory, INSERM U94, Bron, France with support from NRC France/Canada Science and Technology Research Exchange, and NSERC International Collaborative Research Grants. Correspondence should be addressed to Christine L. MacKenzie, Department of Kinesiology, University of Waterloo, Waterloo, Ontario, N2L 3G1 Canada.

## REFERENCES

Abbs, J. H., Gracco, V. L., & Cole, K. J. (1984). Control of multimovement coordination: Sensorimotor mechanisms in speech programming. *Journal of Motor Behavior, 16,* 195–231.

Arbib, M. A. (1985). Schemas for the temporal control of behavior. *Human Neurobiology, 4,* 63–72.

Askenfelt, A. (1988). *Analyzing acoustic characteristics of melody, instruments and voice.* Ph.D. dissertation. Department of Speech Communications and Musical Acoustics, Royal Institute of Technology, Stockholm.

Askenfelt, A., & Jansson, E. (1988, May 27). *The Acoustics of the Piano.* Proceedings of a seminar arranged by the Royal Swedish Academy of Music, Stockholm, Sweden.

Bengtsson, I., & Garbrielsson, A. (1977). Rhythm research in Uppsala. *Music Room, and Acoustics*, Publications issued by the Royal Swedish Academy of Music. Stockholm, *17*, 19–56.

Bengtsson, I., & Gabrielsson, A. (1980). Methods of analyzing performance of musical rhythm. *Scandinavian Journal of Psychology, 21*, 257–268.

Bengtsson, I., & Gabrielsson, A. (1983). Analysis and synthesis of musical rhythm. In J. Sundberg (Ed.), *Studies of Music Performance*, Publications issued by the Royal Swedish Academy of Music. Stockholm: *31*, 27–60.

Carlton, L. G. (1981). Visual information: The control of aiming movements. *Quarterly Journal of Experimental Psychology, 33*, 87–93.

Carter, M. C., & Shaprio, D. C. (1984). Control of sequential movements: Evidence for generalized motor programs. *Journal of Neurophysiology, 52*, 787–796.

Clarke, E. F. (1982). Timing in the performance of Erik Satie's 'Vexations'. *Acta Psychologica, 50*, 1–19.

Fraisse, P. (1974). *Psychologie du rythme.* Paris: Presses Universitaires de France.

Gabrielsson, A. (1974). Performance of rhythm patterns. *Scandinavian Journal of Psychology, 15*, 63–72.

Gabrielsson, A. (1979). Experimental research on rhythm. *Humanities Association Review, 30*, 69–92.

Gabrielsson, A. (1986). Rhythm in music. In J. R. Evans & M. Clynes (Eds.), *Rhythm in psychological, linguistic and musical processes*. Springfield, IL: Charles C. Thomas.

Gabrielsson, A. (1987). Once again: The theme from Mozart's Piano Sonata in A Major (K.331). In A. Gabrielsson (Ed.), *Action and Perception in Rhythm and Music*. Publications issued by the Royal Swedish Academy of Music No. 55.

Gabrielsson, A., Bengtsson, I., & Gabrielsson, B. (1983). Performance of musical rhythm in $\frac{3}{4}$ and $\frac{6}{8}$ meter. *Scandinavian Journal of Psychology, 24*, 193–213.

Gentner, D. R. (1982). Evidence a central control model of timing in typing. *Journal of Experimental Psychology: Human Perception and Performance, 8*, 793–810.

Gentner, D. R. (1987). Timing of skilled motor performance: Tests of the proportional duration model. *Psychological Review, 94*, 255–276.

Graham, E. D. (1984). *Data acquisition for the piano monitoring system*. Unpublished manuscript. University of Waterloo.

Graham, E. D. (1985). *Development of a Piano Monitoring System (PMS) as an experimental tool*. Unpublished manuscript. University of Waterloo.

Hartmann, A. (1932). Untersuchungen über metrisches Verhalten in musikalischen Interpretationsvarianten. *Archiv für die gesamte Psychologie, 84*, 103–192.

Heuer, H., & Schmidt, R. A. (1988). Transfer of learning among motor patterns with different relative timing. *Journal of Experimental Psychology: Human Perception and Performance, 14*, 241–252.

Keele, S. W. (1968). Movement control in skilled motor performance. *Psychological Bulletin, 70*, 387–403.

Keele, S. W., Ivry, R. I., & Pokorny, R. A. (1987). Force control and its relation to timing. *Journal of Motor Behavior, 19*, 96–114.

Keele, S.W. and Posner, M. I. (1968). Processing of visual feedback in rapid movements. *Journal of Experimental Psychology, 77*, 155–158.

Keele, S. W., & Summers, J. J. (1976). The structure of motor programs. In G. E. Stelmach (Ed.), *Motor control: Issues and trends*. New York: Academic Press.

Kelso, J. A. S., Southard, D. L., & Goodman, D. (1979). On the coordination of two-handed movements. *Journal of Experimental Psychology: Human Perception and Performance, 5*, 229–238.

Kimura, D. (1979). Neuromotor mechanisms in the evolution of human communication. In H. D. Steklis & M. J. Raleigh (Eds.), *Neurobiology of social communication in primates: An evolutionary perspective*. New York: Academic Press.

Lashley, K. S. (1951). The problem of serial order in behavior. In L. A. Jeffress, (Ed.), *Cerebral mechanisms in behavior*. New York: John Wiley. 112–136.

MacKenzie, C. L. (1980). *Structural constraints on response selection and organization in finger sequencing*. Unpublished doctoral dissertation, University of Waterloo, Ontario.

MacKenzie, C. L. (1985). Structural constraints on timing of human finger movements. In D. Goodman, R. B. Wilberg & I. M. Franks (Eds.), *Differing perspectives in motor learning, memory and control*. Amsterdam: North-Holland.

MacKenzie, C. L., & Marteniuk, R. G. (1985). Motor skill: Feedback, knowledge and structural issues. *Canadian Journal of Psychology, 39*, 313–327.

MacKenzie, C. L., Nelson-Schultz, J. A., & Wills, B. L. (1983). A preliminary investigation of motor programming in piano performance as a function of skill level. In D. R. Rogers & J. A. Sloboda (Eds.), The acquisition of symbolic skills. New York: Plenum Press.

MacKenzie, C. L., Van Eerd, D. L., Graham, E. D., Huron, D. B., & Wills, B. L. (1986). The effect of tonal structure on rhythm in piano performance. *Music Perception, 4*, 215–226.

Marteniuk, R. G., MacKenzie, C. L., & Baba, D. M. (1984). Bimanual movement control: Information processing and interaction effects. *Quarterly Journal of Experimental Psychology, 36A*, 335–365.

Marteniuk, R. G., MacKenzie, C. L., Jeannerod, M., Athenes, S., & Dugas, C. (1987). Constraints on human arm movement trajectories. *Canadian Journal of Psychology, 41*, 365–373.

Peters, M. (1981). Attentional asymmetries during concurrent bimanual performance. *Quarterly Journal of Experimental Psychology, 33A*, 95–103.

Poulton, E. C. (1979a). Models for biases in judging sensory magnitude. *Psychological Bulletin, 86*, 777–803.

Poulton, E. C. (1979b). Range effects and asymmetric transfer in studies of motor skill. In C. H. Nadeau, W. R. Halliwell, K. M. Newell & G. C. Roberts (Eds.), *Psychology of Motor Behavior and Sport*. Champaign: Human Kinetics Publishing, 339–359.

Povel, D. J. (1977). Temporal structure of performed music. Some preliminary observations *Acta Psychologica, 41*, 309–320.

Povel, D. J., & Collard, R. F. A. (1982). Structural factors in patterned finger tapping. *Acta Psychologica, 52*, 107–124.

Rosenbaum, D. A., Inhoff, A. W., & Gordon, A. M. (1984). Choosing between movement sequences: a hierarchical editor model. *Journal of Experimental Psychology: General, 113*, 372–393.

Roy, E. A. (1985). *Neuropsychological studies of apraxia and related disorders*. Amsterdam: North-Holland Publishing.

Schmidt, R. A. (1975). A schema theory of discrete motor skill learning. *Psychological Review, 82*, 225–260.

Schmidt, R. A. (1982). *Motor control and learning*. Champaign, IL: Human Kinetics Publishers.

Sears, C. H. (1902). A contribution to the psychology of rhythm. *American Journal of Psychology, 13*, 28–61.

Seashore, C. A. (1938). *Psychology of music*. New York: McGraw-Hill.

Seashore, C. A. (1947). *In search of beauty in music*. Westport, CT: Greenwood Press. Reprinted in 1981 from 1947 edition published by Ronald Press, New York.

Semjen, A., & Garcia-Colera, A. (1986). Planning and timing of finger-tapping sequences with a stressed element. *Journal of Motor Behavior, 18*, 287–322.

Shaffer, L. H. (1980). Analyzing piano performance: A study of concert pianists. In G. E. Stelmach & J. Requin (Eds.), *Tutorials in Motor Behavior*. Amsterdam: North-Holland.

Shaffer, L. H. (1981). Performances of Chopin, Bach and Bartok: Studies in motor programming. *Cognitive Psychology, 13*, 326–376.

Shaffer, L. H., Clarke, E. F., & Todd, N. P. (1985). Metre and rhythm in piano playing. *Cognition, 20*, 61–77.

Shapiro, D. C. (1976). A preliminary attempt to determine the duration of a motor program. In D. M. Landers & R. W. Christina (Eds.), *Psychology of sport and motor behavior, 1*, Champaign, IL: Human Kinetics.

Shapiro, D. C., Zernicke, R. F., Gregor, R. J., & Diestel, J. D. (1981). Evidence for generalized motor programming using gait-pattern analysis. *Journal of Motor Behavior, 13*, 33–47.

Sloboda, J. A. (1983). The communication of musical metre in piano performance. *Quarterly Journal of Experimental Psychology, 35A*, 377–396.

Sloboda, J. A. (1985). *The musical mind. The cognitive psychology of music*. New York: Oxford University Press.

Sundberg, J. (1988). Computer synthesis of music performance. In J. A. Sloboda (Ed.), *Generative processes in music. The psychology of performance, improvisation and composition*. London: Clarendon Press.

Terzuolo, C. A., & Viviani, P. (1979). The central representation of learned motor patterns. In R. E. Talbott & D. R. Humphrey (Eds.), *Posture and movement*. New York: Raven Press.

Topper, T. N. (1984). *A piano monitoring system*. Unpublished thesis. Department of Systems Design Engineering, University of Waterloo.

Turvey, M. T. (1977). Preliminaries to a theory of action with reference to vision. In R. Shaw & J. Bransford (Eds.), *Perceiving, acting and knowing: Toward an ecological psychology*. Hillsdale, NJ: Lawrence Erlbaum Associates.

Viviani, P., & Terzuolo, C. A. (1980). Space–time invariance in learned motor skills. In G. E. Stelmach & J. Requin (Eds.), *Tutorials in motor behavior*. Amsterdam: North-Holland.

Wagner, C. (1971). The influence of the tempo of playing on the rhythmic structure studied at pianists playing scales. *Medicine and Sport, 6: Biomechanics II*, 129–132.

Wills, B. L., MacKenzie, C. L., Harrison, D. B., Topper, T. N., & Walker, G. A. (1985). On the measurement of pianists' keyboard performance. In D. A. Winter, R. W. Norman, R. P. Wells, K. C. Hayes & A. E. Patla (Eds.), *Biomechanics IX-B: International series on biomechanics*. Champaign, Ill: Human Kinetics.

Wing, A. M. (1980). The long and short of timing in response sequences. In G. E. Stelmach & J. Requin (Eds.), *Tutorials in motor behavior*. Amsterdam: North-Holland.

Wing, A. M., Turton, A., & Fraser, C. (1986). Grasp size and accuracy of approach in reaching. *Journal of Motor Behavior, 18*, 245–260.

Winter, D. A. (1983). Biomechanical motor patterns in normal walking. *Journal of Motor Behavior, 15*, 302–330.

Zelaznik, H. N., Hawkins, B., & Kisselburgh, L. (1983). Rapid feedback: processing in single aiming movements. *Journal of Motor Behavior, 15*, 217–236.

Zelaznik, H. N., Schmidt, R. A., & Gielen, C. C. A. M. (1986). Kinematic properties of aimed hand movements. *Journal of Motor Behavior, 18*, 353–372.

# 13 Rapid Serial Movements: Relation Between the Planning of Sequential Structure and Effector Selection

Andras Semjen
*Centre National de la Recherche Scientifique*
*Marseille, France*

Robert Gottsdanker
*University of California, Santa Barbara*

ABSTRACT

This chapter reports three experiments that examined the interrelations of temporal planning and effector specification for trains of movements through reaction time procedures. More specifically, the question addressed was whether sequential plan and effector are selected one at a time and, if so, whether order of selection is fixed. In all experiments, the task of the subjects consisted of making one of four finger-tapping sequences that differed in the location of an emphasized element and hand employed. Subjects could receive advance information regarding one or another of these attributes either long-in-advance (> 2000 ms) or briefly-in-advance (0, 30, 60, 90 ms) of the reaction signal that specified the remaining response attribute. The lack of consistent headstart effect except for advance hand information when initial emphasis was known argues against serial parameterization of the hypothesized "sequential plan" and "effector" dimensions, but suggests serial programming of the constituents of the upcoming movement sequence. A serial-programs model is presented in which specification of effector is integrated with specification of its mode of action to program the initial response. Foreknowledge of the sequential plan facilitates the programming of the sequence elements, but such a plan can be flexibly updated with new information when programming or execution is already in progress.

It has long been known that the reaction time (RT) to initiate rapid movement sequences depends on certain characteristics of the sequence as a whole, one of which is the number of elements (e.g., Klapp, Abbott, Coffmann, Greim, Snider, and Young, 1979; Sternberg, Monsell, Knoll, & Wright, 1978). Structural features of the upcoming movement sequence also affect RT. For example, patterned sequences of heterogeneous elements have longer RT than sequences of identical elements (Klapp & Wyatt, 1976; Semjen & Garcia-Colera, 1986). Moreover, choice RT increases with the number of different orderings of the sequential elements (Semjen, Garcia-Colera & Requin, 1984). It also depends on the structural similarity of the alternative sequences (Cooper & Ehrlich, 1981; Rosenbaum, Inhoff, & Gordon, 1984). These and other studies on sequence choice lead to the conclusion that sequence planning contributes in a substantial way to RT, and, therefore, the plan for sequential organization is selected before execution starts. However, is the plan selected before other attributes of the movement sequence? A fixed order of programming with sequential plan selected first seems appealing for a number of reasons. The representation of the sequential structure of a skill is often considered the core aspect of the motor program (Summers, 1981). Such a representation grasps invariances across different realizations of a motor task: it confers consistency to motor actions when repeated, especially to learned ones; it allows for generalization and transfer. Furthermore, motor activities very frequently differ from each other only in the order of a few basic elements. In such activities, the main goal of an upcoming action is to produce, or to reproduce, a particular serial order. Because selection of the goal of an upcoming action is the first decision made before acting, one would expect programming of the sequence to begin with the selection of a plan for sequential organization. The experiments to be reported here tested this hypothesis through reaction time procedures.

## EXPERIMENT 1

The subjects were required to produce movement sequences that differed in sequential structure and effector. They received in advance either full information (that is, sequential structure and effector), partial information (that is, sequential structure or effector), or no information regarding the movement sequence they were to perform. Under such a procedure, analogous to the movement precuing method of Rosenbaum (1983), the effects of partial advance information upon choice RT (CRT) appear critical for establishing constraints in the specification of the precued attributes. Suppose, for example, that information about effector proves useless when information about the sequence structure is lacking, whereas the latter information, when delivered alone, reduces the CRT in comparison to the no-information control condition. Such a pattern of results would suggest that the attributes of the movement sequence are selected one at a time, and that sequence specification has primacy over effector selection. The alternative outcome, namely a reduction in CRT whatever the precued movement attribute, would be considered opposing the notion of fixed order programming (see Miller, 1982; Rosenbaum, 1983).

The movement sequence studied consisted of 8 repetitive finger taps performed at a high rate. Sequential organization was introduced by the use of a single distinguishing element, that is, an accented tap. The sequence was performed with either the left or the right hand. In the 4-choice (4C) condition, subjects chose between sequences that differed in the location of the accent and required hand (no advance information). In the location choice (L) condition, subjects chose between sequences that differed in the location of the accent only while effector hand was prespecified. In the hand choice (H) condition, subjects chose between hands while accent location remained the same. In the ambiguous choice (A) condition, a different accent location was uniquely assigned to each hand. This condition was introduced in order to control for the effects of advance information that did reduce the number of stimulus–response alternatives but did not allow for preselection of either response attribute. Finally, in the simple reaction (SR) condition subjects produced the sequence with the same accent location and with the same hand for a series of trials.

In the conditions where subjects chose between sequences that differed in accent location (4C, A, and L) the alternative accent locations were either 2 and 3, or 5 and 6. Use of these particular pairings was justified by earlier studies (Garcia-Colera & Semjen, 1988) that showed a decrease in choice reaction time (CRT) as the alternative accent locations moved from the initial part of the sequence toward a later part of the sequence. Such a result can be accounted for in terms of a shift from plan preselection to concurrent planning. It appears that when the alternative accent locations occur near the beginning of the sequence, a plan for its organization must be selected before execution begins and this time-consuming process lengthens the CRT. However, when the alternative accent locations occur later in the sequence, execution may start with a program for the initial elements known before the reaction stimulus was presented. Planning of the later part of the sequence, as specified by the reaction stimulus, can be performed concurrently with execution (Garcia-Colera & Semjen, 1988). In the case where the alternative movement sequences differ soon after outset, the hypothesis of primacy of sequential planning leads to the expectation that advance information about sequential structure would shorten the CRT by a larger amount than advance information about the hand to be used. In contrast, in the case where the alternative sequences share a longer initial run and differ in their terminal part, because of the availability of concurrent planning, advance information about the hand to be used should reduce the CRT far more than advance information about the (terminal) sequence structure.

## Method

*Procedure.* Subjects produced the movement sequences by tapping on one of two circular metal plates, 2 cm in diameter, housed 15 cm apart in a wooden support. Located beneath each metal plate was a piezoelectric shock sensor to convert the energy of impact into a proportional electric signal. The subject was free to use either the index or the middle finger for tapping, but was required to use the same finger throughout the experiment.

A trial began with the presentation of a temporal model delivered to the subject via headphones. It consisted of 8 clicks of equal intensity separated by 150-ms intervals. The last click of the model initiated a preparatory period at the end of which the reaction signal (RS) appeared. The preparatory period lasted, with equal probability, for 700, 800, 900, or 1000 ms. The RS was a visually presented digit that indicated the serial position (2, 3, 5, or 6) of the tap to be accented. It appeared on one of two digital displays mounted on a 18 × 14 cm panel placed in front of the subject. On each trial, the location of the display (left or right) on which the digit was presented indicated to the subject the hand for response. The RS remained present until the subject had executed the entire sequence. A trial was aborted if any interval between successive taps was over 300 ms. The intertrial interval was about 6 s.

The subject's task was (a) to produce, by tapping on one of the two metal plates, a sequence of taps in which the number of taps and intertap intervals reproduced the temporal model; (b) to accentuate the particular tap indicated by the RS; (c) to initiate the sequence as quickly as possible after the presentation of the RS. After each trial, the RT was displayed on a television screen placed above the stimulus panel. Subjects were asked to read these RT values and to try to maintain performance close to their best achievable level. Information about errors was given to the subjects after a trial block had been completed.

*Measurements.* The experimental events and measurements were controlled by a PDP-12 computer. Any contact between the finger and a response plate triggered an electronic circuit whose output allowed for identification of the onset of a tap. For each response sequence we recorded: (a) the RT, that is, the delay between the RS onset and the onset of the first tap; (b) the interval between successive taps, defined as the difference between their respective onset times; and (c) the force of each tap as the peak output voltage of the piezoelectric sensor. No particular force values were required; the measures were used only in a relative way to ensure that the accent location was correct.

## Subjects and Design

Eight paid volunteers performed in the experiment. Six of them had previous experience with this type of task and two were new subject. The two new subjects had to be replaced, one because of failure to learn the task, the other because of an excessively high error rate. The replacement subjects were, of course, given the standard initial practice.

Data were collected for each subject in two experimental sessions that differed in the accent locations used (2 and 3, or 5 and 6). Half of the subjects started with the 2–3 pairing, and the other half with the 5–6 pairing. Each session began with the performing of the SR condition. This condition involved four blocks of 16 trials, one block for each accent location and hand. The SR condition was followed by the L, H, and A, and 4C conditions presented in a counterbalanced order. For each the L and H conditions two block of 32 trials were run, one at each level of the other variable. For the A condition, two blocks of 64 trials were run, one for each set of hand-location pairings. For the 4C condition a single block of 64 trials was run. Within every L, H, A, or 4C trial block, there was an equal probability of the alternative presentations. Information regarding hand, location, or hand-to-location assignment was given orally before each block of trials and was also available in printed form on the display itself.

## Results

*Errors.* A sequence was classified as erroneous if the number of taps differed from 8 and/or the accent was not placed on the required tap. The overall mean error rate was 18.1% for sequences with accent locations 2 or 3, and 17.5% with accent locations 5 or 6. When examined as a function of precue condition, the error scores did not show significant variations, $F(4, 28) = 1.32$. The only sizeable difference was linked to side of responding: left-hand sequences showed more error (23%) than right-hand sequences (13.3%). The results reported in the following paragraphs are based on correct sequences only.

*Sequence timing.* The successive intertap intervals of the sequence are presented in Fig. 13.1 as a function of accent location and hand used to execute the sequence. A 7 × 5 × 2 analysis of variance (ANOVA) was conducted for each accent location to test the effects of interval's serial position, precue condition and hand. The effect of interval's serial position was significant in all four ANOVAs (at $p < .05$ or better), confirming the previously reported finding that accent production lengthens the intertap intervals that immediately precedes and follows the accented tap (Semjen et al., 1984). Sequences tapped with the right hand were faster than sequences tapped with the left hand, but the difference proved significant only when tap 2 or 3 was accented. The effect of precue condition did not reach the significance level; the ANOVAs yielded $F(4, 28)$ ratios of 2.02 (when sequences with the accent on tap 2 were considered), 1.60 (accent on tap 3), 0.81 (accent on tap 5), and 0.96 (accent on tap 6). The Precue × Interval's serial position interaction, which reflects variations in timing

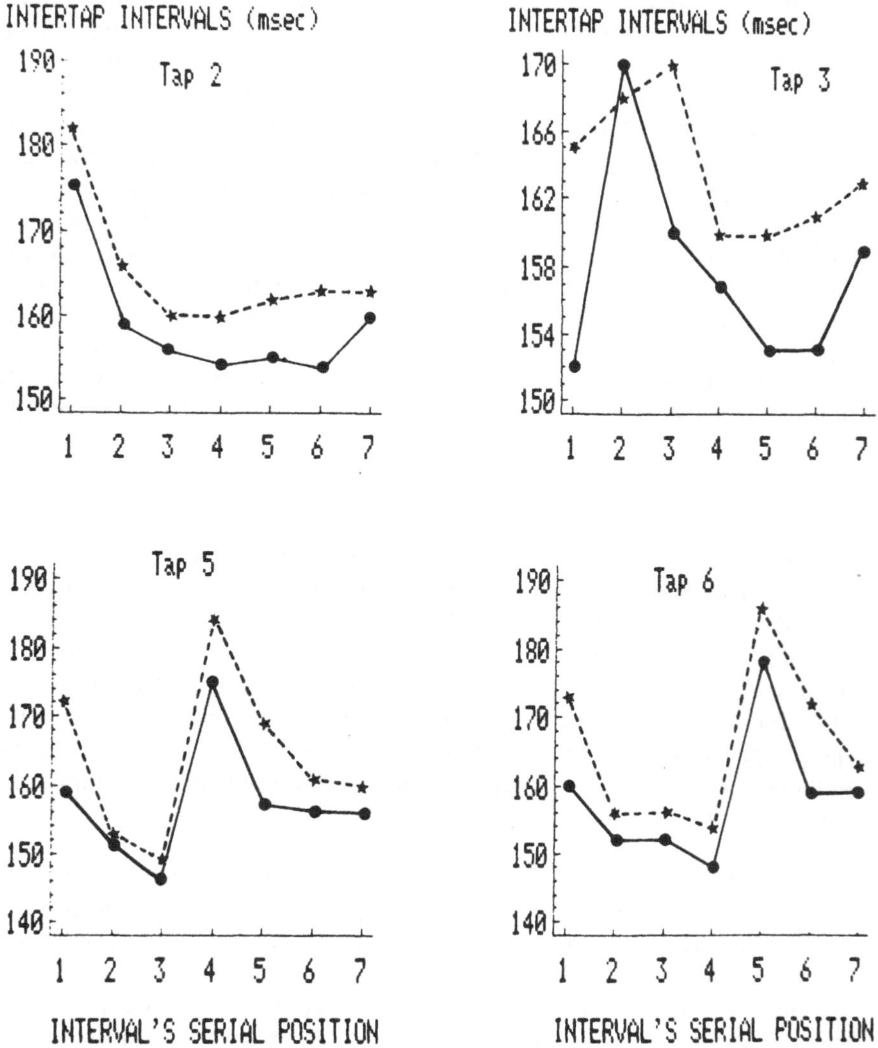


FIG. 13.1. Mean duration of the successive intertap intervals in a sequence as a function of accent location (tap 2, 3, 5, or 6) and hand (filled circles: right hand; stars: left hand). Averaged over precue conditions.

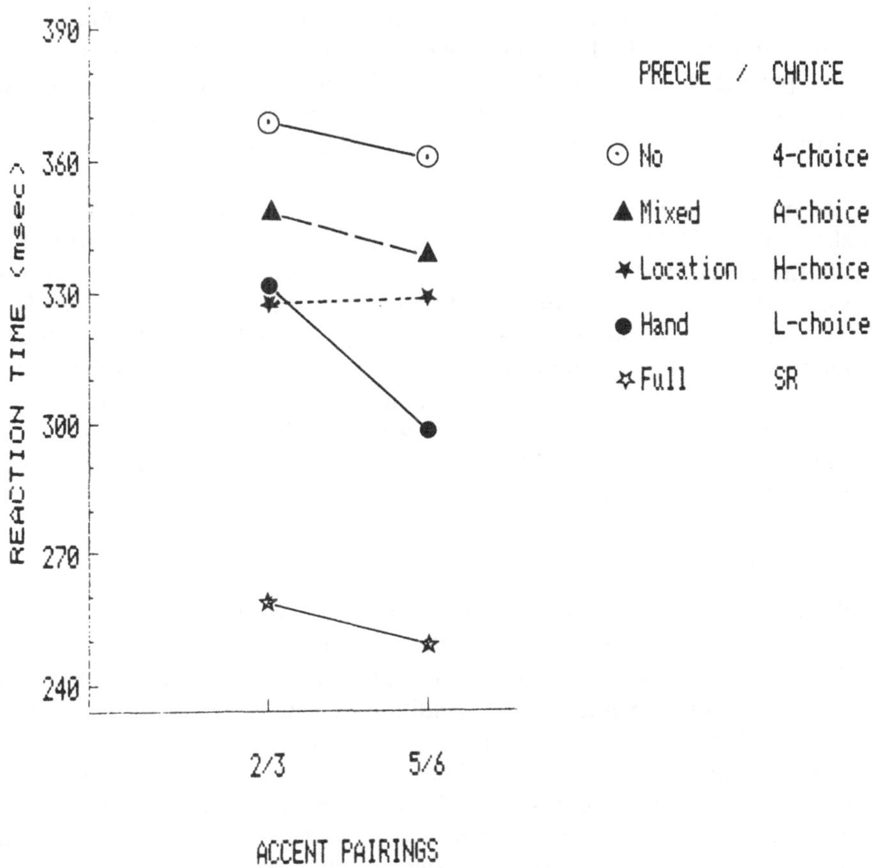


FIG. 13.2. Mean RT as a function of precue and choice condition, and alternative accent locations (pairings).

across precue conditions, reached the significance level only when tap 6 was accented, $F(24, 168) = 1.70, p < .05$. It remained below the significance level when accent location 2, 3, or 5 was considered (here the F ratios were, respectively, 1.51, 1.06, and 0.81). Given such consistency of sequence execution across Precue conditions, it seems justified to compare the RTs observed for each of them.

*Reaction times.* Fig. 13.2 shows the mean RTs calculated from individual medians, as a function of Precue (choice) condition and Accent pairing. RT varied as a function of amount of advance information, because it was shortest when full advance information was given (SR), intermediate when partial advance information was given (H, L, and A), and longest when no advance information was given (4C). Accent pairing had a strong effect only under the L condition where the RT was 33 ms shorter for the 5–6 pairing than for the 2–3 pairing. The effects of giving no or partial advance information was tested in a 4 × 2 (Precue condition × Accent pairing) ANOVA that yielded a significant main effect for Precue condition, $F(3, 21) = 20.49$, $p < .001$, and a significant Precue condition × Accent pairing interaction, $F(3, 21) = 4.35$, $p < .05$. Pairwise comparisons showed that RT for the 4C condition was significantly different from RT for the L, H, and A conditions. RT differences were tested among the three two-choice conditions (H, L, and A) in a second ANOVA. This 3 × 2 (Precue condition × Accent pairing) ANOVA yielded a significant main effect for Precue condition, $F(2, 14) = 9.22$, $p < .01$, and a significant interaction between Precue condition and Accent pairing, $F(2,14) = 5.32$, $p < .05$. Additional comparisons showed the RT in the A condition was significantly longer than RT in both the H condition, $F(1, 7) = 11.05$, $p < .05$, and L condition, $F(1, 7) = 17.23$, $p < .01$.

## Discussion

Partial advance information that prespecified an attribute (hand or sequential organization) common to the remaining response alternatives reduced the RT to a larger extent than prespecifying two sequences that did not share a common attribute (A condition), despite the possible advantage of redundant stimulus information in the latter. The greater effect of precuing a common attribute is usually accounted for in terms of preprogramming: the precued aspect of the response is programmed before the RS, whereas the remaining (unspecified) aspect of the response is programmed during the RT period (Larish, & Frekany, 1985; Rosenbaum, 1980).

In this experiment, our main concern was in finding out whether selection of sequential plan and hand occurred in a fixed order, and this could be tested only in the case where the alternative sequences differed soon after outset. Fixed order of programming would have been indicated if use of one type of information (for instance, *hand*) had depended on existing knowledge of another attribute (for example, sequential order). The data did not support such a hierarchy because in the 2–3 pairing condition precuing of accent location or hand reduced RT substantially with respect to the no-precue (4C) condition.

When the alternative accent positions were 5 or 6, advance information about the hand to be used reduced the CRT far more (62 ms) than advance information about the accent position (32 ms). The very great effect of advance hand information is consistent with the notion that in the case where the sequences differ in their terminal part only, (a) hand information is sufficient to plan the initial run of non-accented taps, and (b) the terminal sequence structure is planned concurrent with execution. However, in view of the sizeable difference between the RT for 5–6 location choice and the RT in the SR condition one must consider the possibility that the discrimination and/or planning processes involved in the 5–6 choice were initiated (although not completed) during the RT period.

## EXPERIMENT 2

Because knowing either attribute of the upcoming tapping sequence is beneficial, it might be reasoned that the selection of sequence structure and hand is flexible in that either may be selected first. However, a more precise test of this hypothesis may be made using the method of Brief Precue-to-Stimulus Intervals (PSIs). This was done in the next experiment. It was reasoned that if sequence structure may be selected first, providing the location information briefly (for instance, 20 ms) in advance of hand information should reduce RT be exactly that amount by allowing a head start in processing. In general a −1 slope between RT and PSI would hold over PSIs that are shorter than the shortest duration of the location selection operation. Likewise, in the flexible model, precuing of hand of response would also provide a −1 slope. In a previous experiment, when hand of response was precued in a choice reaction, a −1 slope was found between RT and PSI, and was interpreted as showing the strict seriality between the discrimination–selection operations and response actualization (Gottsdanker & Shragg, 1985.) It should be noted that advance information that reduces the number of alternative responses does not necessarily produce a −1 slope. Leonard (1958) found an overall reduction of only 4 ms for information given 40 ms in advance of the RS that reduced the number of stimulus–response alternatives from 6 to 3. The clearest set of results that would argue against the flexibility hypothesis and support a fixed serial order of the sequence and hand selection operations would be a −1 slope over brief PSIs for either location or hand, and a 0 slope for the complementary attribute. For that attribute, the reduced RT would appear only with PSIs sufficiently long for a cognitive reorganization (for instance, over 300 ms). This latter kind of result was obtained for precues that indicated an increase in event probability on a simple reaction (Gottsdanker, 1975). In Experiment 2, then, brief PSIs, ranging from 0 to 90 ms were used to precue either accent location or hand of the response. The long PSI was represented by a value of 2100 ms.

### Method

*Procedure.* A digitial display was mounted midline between two light-emitting diodes (LEDs) that were displaced horizontally, 26 mm between centers. The appearance of a digit indicated required accent location, and the illumination of the left or right LED indicated required hand. As in Experiment 1, a trial begun with presentation of a temporal model. However, here, the last click of the model initiated a countdown interval in order to reduce time uncertainty, subjective as well as objective. This interval consisted of three epochs of 1050 ms duration each. The end of the first and second epochs was marked by simultaneous illumination for 10 ms of all three parts of the display. The reaction signal (RS) was presented at the end of the third epoch. It was preceded 0, 30, 60, or 90 ms by a precue stimulus. The precue and RS provided complementary information: If the precue indicated accent location, the RS indicated required hand, and vice versa. The response of the subject extinguished both the precue stimululs and the RS. In control–trial blocks, the precue stimulus was presented for 30 ms at the end of the first and second epochs. In such trial blocks the PSI was thus 2100 ms.

## Subjects and Design

Six paid volunteers performed in the experiment. Four of them had extensive previous experience with this type of task. The two new subjects learned to produce the tapping sequences during a preliminary session in which no short PSI was used.

Half of the subjects first performed all the trial blocks where accent locations 2 and 3 were used, and continued the experiment by performing the trial blocks where accent locations 5 and 6 were used. The other subjects received these conditions in the reverse order. The order in which subjects received the Precue conditions (for instance, location versus hand) was counterbalanced in a similar way. A trial block consisted of 32 trials equally distributed between four movements sequences (2 Hands × 2 Locations × 8 Repetitions). In every Accent pairing and Precue condition subjects performed on groups of 4 trial blocks. In the first 3 blocks of a group, the same brief PSI was used (for example, 30 ms), whereas in the last block the PSI was 2100 ms. Block 1 was considered as practice, block 2 to 4 as experimental. The brief PSIs were presented to each subject in a different order, two PSIs per experimental sessions. A total of 8 sessions were needed to run a subject in the 4 PSIs × 2 Precue × 2 Accent pairing conditions.

## Results

*Errors.* The mean error rate was 14.6% when hand to be used was precued, and 14.7% when accent location was precued. It showed weak and inconsistent variations across PSIs. When hand was precued, the mean percentage error rates were 16.8%, 13.3%, 14.4%, 12.2%, and 16.5%, from the shortest to the longest PSI. When accent location was precued, the corresponding values were 12.7%, 17.1%, 15.8%, and 12.4%. However, sequences tapped with the left hand showed more errors (18.8%) than those tapped with the right hand (10.6%).

*Sequence timing.* The timing of the tapping sequences was almost identical to that observed in Experiment 1. The effects of brief PSIs on the intertap intervals were first tested separately for right-hand and left-hand sequences, for each accent location and precue condition. A total of 16 ANOVAs were thus performed with Interval's Serial Position (7) and PSI (4) as factors. The effect of PSI reached significance ($p < .05$) in only one of these analyses. Even then, the variations in mean intertap interval remained within a range of 4.5 ms. Furthermore, the interaction between Interval's Serial Position and PSI did not reach statistical significance in any of the 16 ANOVAS. Given such an outcome of the first analysis, the intertap interval values were collapsed over PSIs, and the effects of Precue condition, Interval's serial position and Hand were tested in four 2 × 7 × 2 ANOVAs, one for each accent location. The $F(1, 5)$ ratios found for Precue condition were 2.49 (sequences with accent on tap 2), 2.32 (accent on tap 3), 1.00 (accent on tap 5), and 0.75 (accent on tap 6), all $ps > .05$. Moreover, the interaction between Interval's serial position and Precue condition was nonsignificant in all four ANOVAs (the $F(6, 30)$ values ranged from 0.30 to 1.21). Over all, the PSI manipulation had no effect on the speed and the time profiles of the tapping sequences.

*Reaction times.* Fig. 13.3 shows the mean RTs calculated from individual medians, as a function of PSI, Precue condition, and Accent pairing. The effects of brief PSIs differed according to whether hand or accent location was precued. For accent location precues, the slope between RT and PSI was consistently shallower than −1 (about half that amount) through the 90 ms PSI for both accent pairings. For required-hand precues, the slope was very slightly steeper than −1 through the 30 ms PSI for both accent pairings. For the 2–3 case, the rate of decrement leveled off beyond that point, whereas for the 5–6 case the −1 slope persisted through the 60 ms PSI.

Because only the 30-ms PSI was within the range of potential RT reduction in all conditions, the individual median RTs observed at PSIs of 0 and 30 ms were subjected to a PSI (2) × Precued Dimension (2) × Accent Pairing (2) ANOVA that yielded a significant effect for PSI [$F(1, 5) = 12.54, p < .05$], whereas the effect of Precued Dimension fell short of the significance level, $F(1, 5) = 6.46, p < .10$, and accent pairing was nonsignificant, $F(1, 5) = 0.69$. However, the difference in slope of the RT reduction was supported by a highly significant interaction between PSI and Precued dimension [$F(1, 5) = 51.29, p < .001$].

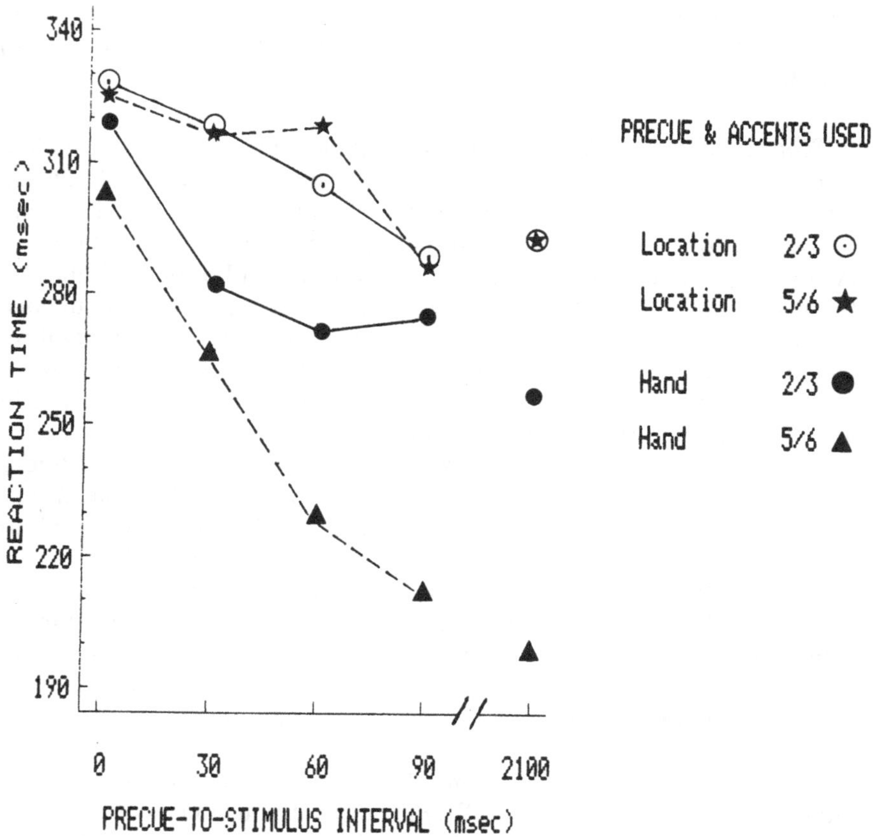


FIG. 13.3. Mean RT as a function of PSI, precue condition, and alternative accent locations (pairings).

The effects of advance information delivered with a long PSI differed on two points from those found in Experiment 1. Whereas in that experiment, L-choice and H-choice RT were equal for the 2–3 pairing, in this experiment, choosing between locations 2 or 3 (with hand known in advance) took less time than choosing between hands (with accent location known in advance), $F(1, 5) = 16.58, p < .01$. Moreover, precuing of hand produced a larger additional decrease in RT for the 5–6 pairing, making the RT in this accent-pairing condition close to the value of a simple reaction time.

## Discussion

The hypothesis of flexible seriality, in which either sequential structure or hand of response may be selected first, was not supported. RT did not drop with a $-1$ slope over brief PSIs when accent location was precued.

In fact, the initial decline was about one-third of that amount. This outcome may be contrasted with the −1 slope found when hand was precued, both for the 2–3 and 5–6 cases. The difference between the two outcomes cannot be attributed to a possibly more difficult discrimination or to reduced S-R compatibility with location precued. Such factors would simply raise the level of RT at each value of brief PSI; it would not diminish the slope. A head start would produce a −1 slope regardless.

Thus, the results indicate that when hand was precued, processing began immediately, but when location was precued, there was only some facilitation of response, the nature of which is considered later. The most obvious general conclusion would seem to be that effector (here, hand) is selected before sequential structure (here, location of accented and non-accented taps). However, it somewhat belatedly occurred to the authors that hand and location precuing are not strictly comparable for either of the two cases studied, 2–3 or 5–6. With hand precued, there is no uncertainty regarding the required force of the first tap; it is always weak. Thus, given hand information, the first response may be entirely programmed before location information is received. However, with location precued, the first response cannot be entirely programmed as long as hand remains unknown. Consequently, a third experiment was planned in which a location choice would be used in which certainty of the force required on the initial tap was eliminated.

## EXPERIMENT 3

The inclusion of uncertainty in the force required on the initial tap was provided by use of a sequence in which either first or second tap was to be accented, that is, the 1–2 case. If, indeed, hand is selected before serial structure, a −1 slope should be found over brief PSIs for hand precued in this case, just as we found in Experiment 2 for the 2–3 and 5–6 cases.

### Method and Design

The same procedure was used and the same subjects participated as in Experiment 2, but the experimental design was slightly modified. An experimental session comprised 8 blocks of 32 trials. The first 2 blocks served as a warm-up. Here hand or accent location was precued with a PSI of 2100 ms. The remaining blocks were divided into three groups of 2 blocks and a different PSI was used in each of them. The subjects received these PSIs in a variable order. In one group of blocks the long (2100 ms) PSI was used, in the other two, different brief PSIs. The first block in each group was considered as practice. The first two experimental sessions were devoted to the precuing of either hand or location; these sessions differed only in the values of brief PSIs. Session 3 and 4 differed from the first two sessions only in the response attribute precued. The order of precued response attributes was counterbalanced over subjects.

## Results

In order to obtain data sets as comparable as possible to the previous experiments in which the tapping sequences always started with a weak tap, response sequences with accent on tap 2 were only considered for establishing individual error scores, intertap intervals, and RTs.

*Errors.* When hand was precued, the mean percentage error rates were 10.4%, 9.3%, 8.3%, 9.3%, and 10.9% from the shortest to the longest PSI. With accent location precued, the corresponding values were 8.3%, 17.7%, 15.6%, 11.4%, and 10.9%.

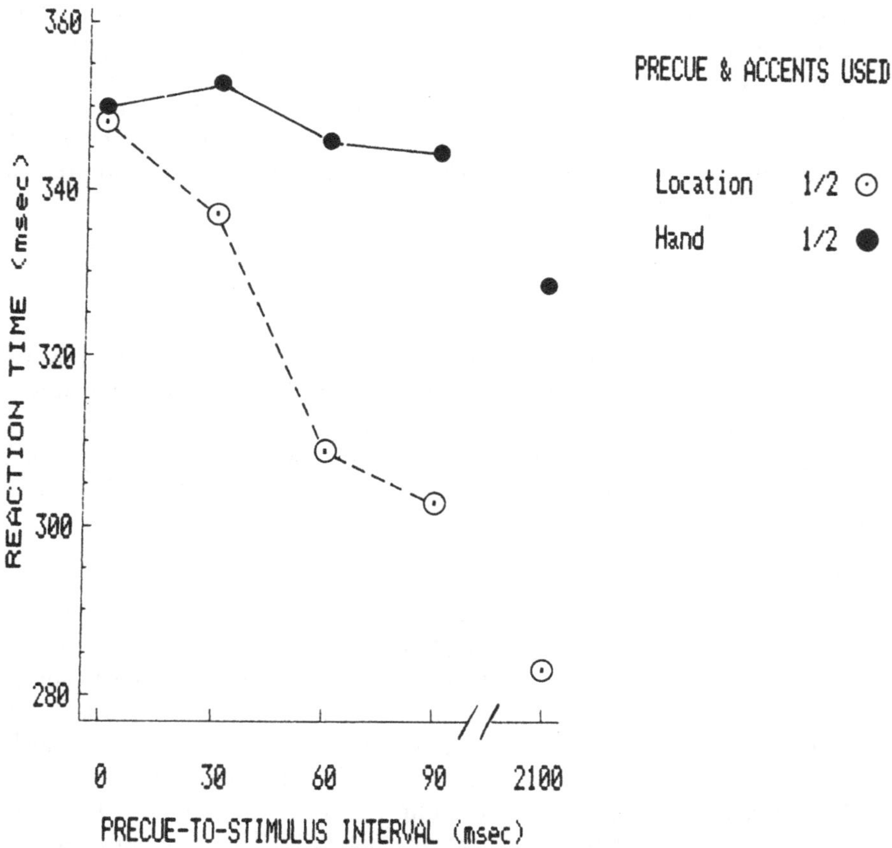


FIG. 13.4. Mean RT as a function of the PSI and precue condition.

*Reaction times.* The effects of precuing hand or accent location with brief PSIs are shown in Fig. 13.4. The individual median RTs were subjected to a PSI (4) × Precue Condition (2) ANOVA that yielded significant effects for Precue Condition, $F(1, 5) = 6.99$, PSI, $F(3, 15) = 5.35$, and their interaction, $F(3, 15) = 3.45$, all $ps < .05$. As can be seen in Figure 13.4, precuing of hand of response with brief PSIs did not induce sizeable changes in RT. Within this condition, the effect of PSI and the linear trend of the RT change over PSIs were nonsignificant ($F < .30$). In contrast, precuing of accent location with increasing PSIs resulted in a gradual decrease in RT. Within this condition, the effect of PSI was significant, as was the linear trend of RT decrease, $F(1, 5) = 17.37$, $p < .01$. It must be noted, however, that the decrease in RT did not follow a $-1$ slope, and that the first 30-ms increment in PSI resulted only in 11 ms shortening of the RT, $F(1, 5) = 0.16$.

The comparison between the RTs observed for PSIs of 0 or 2100 ms showed a significant decrease in both precue conditions. However, required location precue induced a larger effect than required hand precue as shown by the significant PSI × Precue Condition interaction, $F(1, 5) = 16.34$, $p < .01$. In the present experiment, then, the benefit from knowing accent location was greater than the benefit that resulted from knowing hand, whereas the opposite held true in the 2–3 and 5–6 accent pairing conditions of the previous experiment.

## GENERAL DISCUSSION

Taken together, the data do not support serial specification, either flexible or rigid, of hypothesized sequential plan and effector dimensions in programming a rapid train of responses. When the alternative accent locations were in the initial part of the sequence, preknowledge of the hand reduced RT very little, and not at all when the precue only briefly preceded the stimulus indicating location of the accent. With location of the accent precued briefly before the hand choice stimulus, the drop in RT with PSI was far shallower than the $-1$ slope indicative of a head start in programming.

A more plausible account of how information on sequential structure and required hand is utilized is provided by a model that treats the programming of each response in the series concretely, rather than as a consequence of overall specification of two dimensions. This serial-programs model is defined through the following set of propositions:

1. A rapid sequence of responses is programmed in a strictly serial order.

2. Programming a response consists of impressing a mode of action (here, strong or weak force) on an effector unit.
3. Knowing the mode in advance (here, accent at a location) reduces the required time for programming a response.
4. Actualization of a rapid response sequence begins when two responses have been programmed.

Proposition 3 implies that there is a sequential plan in existence and that it is modified by the precuing of location. Proposition 4 refers to an optimal solution, as later responses are programmed concurrently with later processing stages or even responses.

The programming operations that precede overt execution of a rapid response sequence according to the serial-programs model are indicated in Table 13.1. Presentation of the reaction stimulus is indicated by *S*. Events to the left of the stimulus line occur before the reaction stimulus has occurred, and those to the right follow the stimulus occurrence. *R* is the time needed to program a response with no advance information regarding its mode (that is, whether accented or nonaccented), while *r* is the time needed to program a response, the mode of which is known in advance ($R > r$). Response actualization (*A*) refers to the processes, central as well as peripheral, that precede overt execution of an already programmed response. Preparation for the programming of responses in location 1 and 2 is indicated by L1 and L2 to the left of *S*.

The table shows that in the 4-choice (no precue) condition 2, 1, or 0 response(s) must be programmed without previous preparation, and 0, 1, or 2 response(s) can be programmed in a prepared state, depending on the alternative accent locations, that is 1–2, 2–3, or 5–6. Because $R > r$, the programming time should decrease over accent pairings.

TABLE 13.1
Hypothesized programming operations leading to response sequence actualization

| | *Accent Pairings* | | |
|---|---|---|---|
| *Choice* | 1/2 | 2/3 | 5/6 |
| 4C | S R1 R2 A | L1 S r1 R2 A | L1 S r1 r2 A |
| | * | * | L2 * |
| | * | * | * |
| | * | * | * |
| HAND | L1 S r1 r2 A | L1 S r1 r2 A | L1 S r1 r2 A |
| | L2 * | L2 * | L2 * |
| | * | * | * |
| | * | * | * |
| LOC. | S R1 R2 A | R1 S R2 A | R1 R2 S A |

4C: 4-choice; HAND: hand-choice; LOC.: location-choice

In the hand choice (location precued) condition, 2 responses can be programmed with foreknowledge of the response mode, that is, in a prepared state. Here, the programming time should be the same in all accent-pairing conditions.

In the location choice (hand precued) condition 2, 1, or 0 response(s) must be programmed after the reaction stimulus has occurred, and 0, 1, or 2 response(s) can be programmed before the occurrence of that stimulus. This account provides a sharp decrease in programming time over accent pairings, and a head start in programming of the initial response(s) when hand information is delivered shortly before location information 2–3 or 5–6.

Expectations derived from the model may be evaluated against the RTs obtained in Experiments 2 and 3 at the 0 and long (2100 ms) PSIs, in Fig. 13.5. The expected RT ranks from the shortest (1) to the longest (5) are shown above the points for the corresponding means. The expectations are based on the number and preparation status of responses programmed after the appearance of the RS. In the evaluation, one adjustment was made on the data, and one assumption was made in deriving the expected ranks. The adjustment was to subtract 20 ms from the three 4-choice means; the assumption was that the required time for programming an unprepared response is less than twice the time for programming a prepared response.[1] The rank–order fit of model to corresponding means is seen to be close, with no reversals.

The serial-programs model presents similarities with the hierarchical editor (HED) model (Rosenbaum et al., 1984) and its revised version (Rosenbaum, Hindroff, and Munro, 1987) in that both suggest serial determination of the subprograms for the constituents of the sequence, and both admit temporal overlap between programming and execution. However, in the present model, specification of hand is made in the context of the required mode of the initial response to produce a program for that response. Whether specification of hand, in this sense, is then immediately distributed over all locations in the sequence (as is the abstract hand attribute in the HED model), or is transmitted chainlike with the programming of each response (as the present model suggests), remains in question.

---

[1]The 20 ms adjustment is equal to the difference found between 4-choice RTs and the two 2-choice cases that are indicated as having identical operations. It is suggested that articulation of the indicators of the two attributes is impeded by their simultaneous presentation and that temporal separation can thus reduce RT up to 20 ms. The assumption $R < 2r$ is conservative and is supported by subtractions that may be made in the model between instances that differ in only one respect. For example, the difference between the $\frac{1}{2}$ and $\frac{2}{3}$ pairings on location choice gives an estimate of 72 ms for the duration of R1, and the 4-choice difference provides an R1 minus r1 estimate of 25 ms.

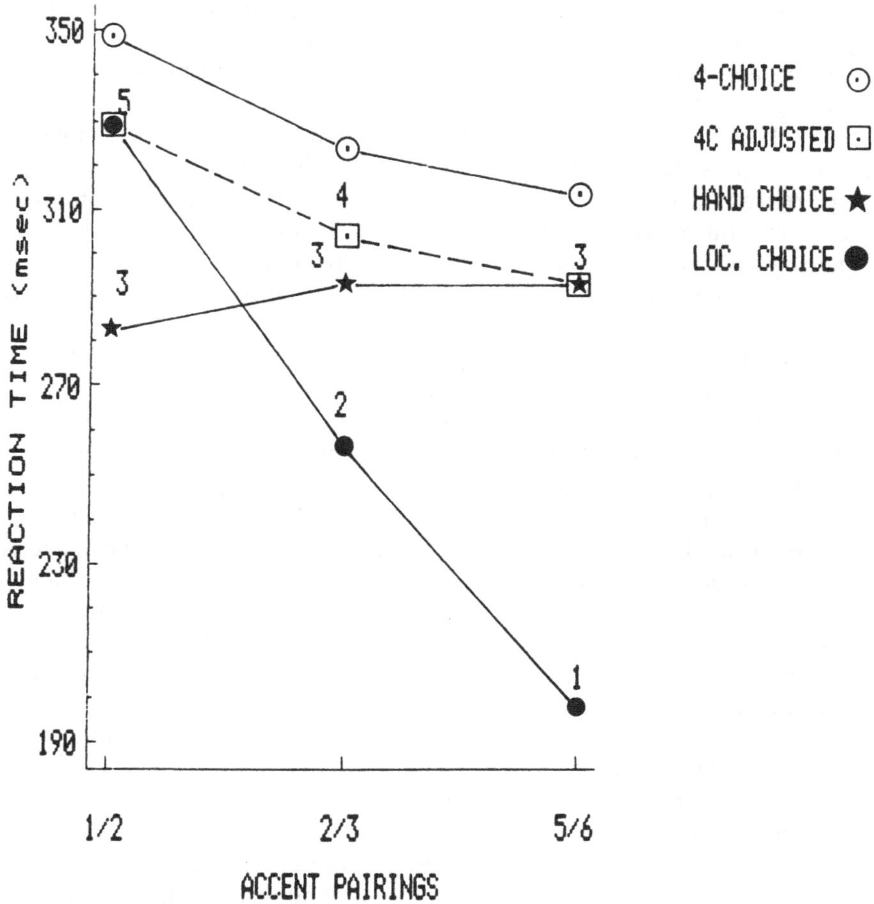


FIG. 13.5. 4-choice (0 ms PSI) and 2-choice (2100 ms PSI) RT in experiments 2 & 3. The broken line arise from the shifting of the 4-choice data points by 20 ms (see text).

## ACKNOWLEDGMENTS

Preparation of this chapter was supported in part by a grant for international collaboration in research from the NATO Scientific Affairs Division (reference 0135/87) to Robert Gottsdanker.

## REFERENCES

Cooper, W. E., & Ehrlich, S. F. (1981). Planning speech: Studies in choice reaction time. In J. Long & A. Baddely (Eds.), *Attention and Performance IX* (pp. 537–552). Hillsdale, NJ: Lawrence Erlbaum Associates.

Garcia-Colera, A., & Semjen, A. (1988). Distributed planning of movement sequences. *Journal of Motor Behavior, 20*, 341–367.

Gottsdanker, R. (1975). The attaining and maintaining of preparation. In P. Rabbitt and S. Dornic (Eds.), *Attention and Performance V* (pp. 33–49). New York: Academic Press.

Gottsdanker, R., & Shragg, G. P. (1985). Verification of Donders' subtraction method. *Journal of Experimental Psychology: Human Perception and Performance, 11*, 765–776.

Klapp, S. T., & Wyatt, E. P. (1976). Motor programming within a sequence of responses. *Journal of Motor Behavior, 8*, 19–26.

Klapp, S., Abbott, J., Coffmann, K., Greim, D., Snider, R., & Young, F. (1979). Simple and choice reaction time methods in the study of motor programming. *Journal of Motor Behavior, 11*, 91–101.

Larish, D. D., & Frekany, G. A. (1985). Planning and preparing expected and unexpected movements. Reexamining the relationship of arm, direction and extent of movement. *Journal of Motor Behavior, 17*, 168–189.

Leonard, J. A. (1958). Partial advance information in a choice reaction task. *British Journal of Psychology, 49*, 89–96.

Miller, J. (1982). Discrete versus continuous stage models of human information processing: In search for partial output. *Journal of Experimental Psychology: Human Perception and Performance, 9*, 273–296.

Rosenbaum, D. A. (1980). Human movement initiation: Specification of arm, direction and extent. *Journal of Experimental Psychology: General, 109*, 444–474.

Rosenbaum, D. A. (1983). The movement precuing technique: Assumptions, applications and extensions. In R. A. Magill (Ed.), *Memory and control of action* (pp. 231–274). Amsterdam: North-Holland.

Rosenbaum, D. A., Inhoff, A. W., & Gordon, A. M. (1984). Choosing between movement sequences: A hierarchical editor model. *Journal of Experimental Psychology: General, 113*, 372–393.

Rosenbaum, D. A., Hindroff, V., & Munro, E. M. (1987). Scheduling and programming of rapid finger sequences: Tests and elaborations of the hierarchical editor model. *Journal of Experimental Psychology: Human Perception and Performance, 13*, 193–203.

Semjen, A., & Garcia-Colera, A. (1986). Planning and timing of finger tapping sequences with a stressed element. *Journal of Motor Behavior, 18*, 287–322.

Semjen, A., Garcia-Colera, A., & Requin, J. (1984). On controlling force and time in rhythmic movement sequences: The effect of stress location. *Annals of the New York Academy of Sciences, 423*, 168–182.

Sternberg, S., Monsell, S., Knoll, R. L., & Wright, C. E. (1978). The latency and duration of rapid movement sequences: Comparisons of speech and typewriting. In G. E. Stelmach (Ed.), *Information processing in motor control and learning* (pp. 117–152). New York: Academic Press.

Summers, J. J. (1981). Motor programming. In D. H. Holding (Ed.), *Motor skills* (pp. 42–64). Chichester: Wiley.

# 14 Phase Transitions in Speech Production and Their Perceptual Consequences

Betty Tuller
J. A. Scott Kelso
*Center for Complex Systems and Department of Psychology*
*Florida Atlantic University, Boca Raton*

## ABSTRACT

A central issue in speech has involved uncovering aspects of articulation that remain invariant in the face of suprasegmental transformations. Much previous experimental work has shown that such invariance is unlikely to be exhibited at the level of individual muscles or articulator movements. Here we introduce the concept of stability of a collective macroscopic variable—the relative phasing of articulatory events—that has been observed to change little across rather large variations in individual articulators' kinematics. To establish the veracity of this concept, we study how articulatory patterns change, with the aim of understanding both stability and change of coordinated speech patterns as two manifestations of the same underlying dynamics. To this end, subjects were asked to say /ip/ or /pi/ repetitively, at increasing speaking rates, while their glottal and lip movements were monitored. In general, the observed interarticulator phase relationships did not depend on the rate-scaling parameter. However, at critical values of rate the relative phase of articulator movements changed markedly. Experiment 2 found that these transition points correspond precisely to perceptual changes in syllable affiliation of the consonant. More generally, the work is directed at reorienting speech research to the identification and analysis of articulatory and perceptual *stability* instead of its present search for invariants.

When a speaker produces a word, the acoustic instantiation of the word varies widely with many factors. Among these are the rate at which the utterance is produced, the word's emphatic or syllabic stress level, and the identity of surrounding speech segments. Nonlinguistic factors that are also potentially disruptive include a noisy environment or the frequency filtering introduced by the telephone system. Luckily, the perception of linguistic content is remarkably insensitive to these acoustic variations.

The robustness of speech perception sparked a long and continuing search for articulatory invariants that underlie individual speech segments. Despite its aim, the search served mostly to document the highly context-dependent nature of speech production. One early hypothesis was that the motor commands sent to the muscles would stand in one-to-one identity with an individual speech segment, and that this invariance would be observable in electromyograms (Liberman, Cooper, Shankweiler, & Studdert-Kennedy, 1967). However, it was quickly apparent that electromyographic activity was itself strongly context-dependent (e.g., MacNeilage & DeClerk, 1969). Shifting the focus of analysis to articulatory gestures also failed to reveal an invariant relationship between motor variables (kinematics) and speech segments (e.g., Butcher & Weiher, 1976; Gay, Ushijima, Hirose, & Cooper, 1974; Kozhevnikov & Chistovich, 1965; Kuehn & Moll, 1976; Liberman et al., 1967; Lindblom, 1963). In fact, the variability of individual articulatory gestures as a function of stress, rate, and contextual change is one of the few aspects of speech production on which most investigators agree (but see, for example, Kelso, Bateson, Saltzman, & Kay, 1985, for evidence of stable kinematic *relationships* in individual articulators across rate and stress; also Vatikiotis-Bateson, 1988).

A more controversial notion—that the relative timing among speech articulator movements is less variable than the absolute timing of individual articulators—is slowly gaining ground (e.g., Kent & Moll, 1975; Kent & Netsell, 1971; Kozhevnikov & Chistovich, 1965; Löfqvist & Yoshioka, 1981). Our own work has focussed on the relative timing of vowel-related (V) and consonant-related (C) events across changes in speech rate and syllable stress. Basically, we reported that across these suprasegmental changes the timing of medial consonant production in CV1#CV2C utterances (e.g., *ba wab*) remains stable relative to the interval between events associated with the flanking vowels. The relationship between these articulatory event durations (the interval from V1 to C and from V1 to V2) was described well by a linear function with a positive slope and nonzero intercept. This occurred whether "segment-associated activity" was observed at the electromyographic (e.g., Tuller, Kelso, & Harris, 1982; 1983) or kinematic levels (e.g., Tuller & Kelso, 1984; Tuller, Kelso, & Harris, 1983). One reason these "relative-timing" results may be important is that they hint at the presence of a functional *constraint* on movement patterns whereby individual system elements are constrained within a particular temporal relationship, whereas metrical adjustment of the components affords flexibility.

The foregoing experiments motivated a flurry of activity among speech scientists, many of whom subsequently corroborated the basic result (Bladon & Al-Bamerni, personal communication; Gentil, Harris, Horiguchi, & Honda, 1984; Linville, 1982; Lubker, 1983). On the other hand, the data are open to two central criticisms, one statistical and one theoretical. The statistical issue concerns the use of Pearson's product–moment correlations when the variables correlated represent overlapping intervals (so-called part–whole correlations). Barry (1983) rightly expressed reservations to using part–whole correlations because any high correlations obtained could be simply statistical artifact (see also Gentner, 1987). Löfqvist (1986), Munhall (1985), and Tuller et al. (1983) used various approaches in attempts to avoid the statistical artifact inherent in part–whole correlations, but all were shown by Benoit (1986) to be mathematically deficient (but see Kelso & Tuller, 1987, for a different approach).

The second, and perhaps more basic problem with the experiment on relative timing concerns the issue of *invariance*. It is now commonplace to assume that if a variable does not change much over a range of parameter values, it is an invariant. Schmidt (1982), for example argues that the invariance of proportions of time spent in various phases of the locomotion step cycle provides evidence for walking and jogging programs. Criteria for invariance are seldom specified, however, leading to much confusion (cf. Sternberg, this volume). One may ask: invariant with respect to what? Or, how much can the mean value of a variable change before one decides that the variable is not invariant? The aim of the present approach is to show that a collective variable for coordinated speech (relative phase) remains *stable* over a range of parameter values (speaking rates), then shifts to a new relative phase state as a consequence of *loss of stability*. Thus, pattern stability and change are deemed to arise from the same underlying nonlinear dynamics. We use observations of phase transitions in speech movement patterns as a means of demonstrating that the dynamical concept of stability is fundamental and is tightly coupled to experimental measures. In this sense, we believe the concept of stability (which incorporates stochastic features) better serves the understanding of speech and other forms of motor control, than the rather vague term *invariance*.

Our approach originated from the study of bimanual movement coordination and is based on theories of self organization and pattern formation in nonequilibrium systems, particularly Haken's (1975; 1983) synergetics (see Kelso, Del Colle, & Schöner, this volume). The aim is to understand coordinated movement as a pattern formation process. In synergetics, when a certain parameter or combination of parameters (referred to as *controls*) reach a critical point, structure or change in structure occurs spontaneously. At so-called nonequilibrium phase transitions, the emerging patterns may be characterized by *collective variables* whose dynamics (equations of motion) are low dimensional. Synergetics thus promotes a search for these collective variables (also called order parameters) in experimental model systems and the determination of control parameters that move the system through collective states.

The dynamics of collective variables may be determined by studying the stability and loss of stability of movement patterns. Specific empirical phenomena are predicted to occur jointly at transition points. One is the presence of *enhanced fluctuations*, in which the values of the collective variable undergo large fluctuations with no change in the mean state (cf. earlier remarks on invariance). Another is called *critical slowing down*, which simply means that when a system is close to a transition point, it reacts more and more slowly to external perturbations than in other situations when it is far from criticality. Both features are characteristic of self-organization, and both have been observed experimentally in movement systems (for recent reviews see Kelso, & Schöner, 1988; Kelso, Schöner, Scholz, & Haken, 1987; Schöner & Kelso, 1988).

What are the collective variables and control parameters in the speech production system? There are many possible modifiers of articulation that might lead the system through different coordinative patterns and hence reveal their differential stability. These include, but are not limited to, stress, loudness, and speaking rate (which is manipulated in the present experiments). If the nonspecific influence of a continuously varying control parameter results in a change of pattern, then one can identify collective variables that correspond to the observed patterns and determine their dynamics through an analysis of stability. There is already a hint in the literature that certain kinds of phonetic and syllabic change correspond directly to phase transitions among articulatory gestures (see Kelso, Saltzman, & Tuller, 1986a) and hence may enable us to specify relevant collective variables. Consider an example, discussed in some detail by Stetson (1951; also Kelso & Munhall, 1988). A subject produces the syllable *at* repetitively. As speaking rate is gradually increased, the syllable affiliation of the consonant changes, producing the series *ta, ta*. . . . Stetson thought that the need to simplify coordination caused the syllable-final consonant to become syllable–initial because the final consonant was "off-phase, out of step with the syllable movement (p.96)." Our interpretation of Stetson is that the phasing among component gestures is the collective variable underlying both stability and change in syllabic structure (Kelso, Saltzman, & Tuller, 1986a). The discontinuities created by rate scaling are at least consistent with a nonequilibrium phase transition explanation of articulatory change.

The present experiments explore this proposal explicitly. In Experiment 1, we ask whether the relative phase of laryngeal and oral events may be considered a suitable collective variable to characterize the coordinative patterns corresponding to the specific utterances investigated. Experiment

2 explores whether the measured collective variable adequately predicts listeners' perceptions of segment identity and syllable type.

## EXPERIMENT 1

### Subjects

The subjects were two adult males and two adult females. One female and one male subject (SN, ES) were born in America, one female subject (SH) was a native of England, and one male subject (AL) was a native of Sweden.

### Material and Procedure

Subjects were asked to produce the utterance /ip/ repetitively at a slow speaking rate. They were told that while they were speaking, the experimenter would signal them (using a distinct hand gesture) to speed up their rate of production in a steplike manner. Subjects were asked to speed up approximately 4 to 6 times within each trial, depending on their initial speech rate and the step size they elected. Typically, the experimenter would signal the subject to increase speech rate approximately every 3 to 5s. Subjects were also requested not to breathe within a trial. An identical procedure was used for the utterance /pi/.

### Data Recording

Articulatory movement in the up–down direction was monitored using an optoelectronic tracking device (a modified SELSPOT system). In this system modulated light from lightweight, infrared, light-emitting diodes (LEDs) is captured by a camera equipped with a Schottky planar diode located in its focal plane. The output of the photodiode is fed to associated electronics that decode the signals and compute pairs of $x$ and $y$ coordinates. Calibration is achieved by moving an LED through a known distance in the field of view. In this experiment, the relevant LEDs were attached to the subject's upper lip, lower lip, and nose bridge. A headrest was used to minimize head movements during the experiment.

A measure of glottal motion was obtained using an updated version of the transillumination technique (also referred to as photoglottography) originally developed by Lisker, Abramson, Cooper, and Schvey (1969). A fiberoptic laryngoscope was used to aim light toward the glottis; the light transmitted through the glottis was measured by a phototransistor that is coupled by a light-tight enclosure to the pretracheal wall, where it is translucent. Previous studies have shown a close correspondence between the transillumination signal and glottal width (Baer, Löfqvist, & McGarr, 1983), particularly with regard to the temporal course of glottal area changes (Löfqvist & Yoshioka, 1979; Yoshioka, Löfqvist, & Hirose, 1981).

Intraoral air pressure for three of the four speakers was monitored by a plastic tube inserted into the pharyngeal cavity through the nose and connected to a pressure transducer.

The acoustic speech signal was monitored simultaneously with the glottal and oral movements, and all were recorded onto FM tape for off line computer analysis. Analysis of the acoustic signal is described in Experiment 2. The movement and air pressure records were digitized using a sampling rate of 200 Hz, and smoothed with a 25 ms triangular window. To correct for up–down head movements, output of the nose LED was software subtracted from output of the LEDs attached to the lips. Similarly, a tract variable, lip aperture, was software derived and defined as the vertical distance between upper and lower lips.

A peak-picking routine was used to determine the time of (a) peak glottal opening and (b) minimum lip aperture. These values were used to calculate the phasing between glottal and lip movements for each utterance, arbitrarily defining the successive occurrences of minimum lip aperture as the reference cycle and peak glottal opening as the event whose phase was determined relative to the reference cycle. When intraoral air pressure recordings were available, they were software differentiated using a two-point central difference algorithm. The sharp rise in intraoral air pressure as the lips come into contact is easily discernible as a large peak in the velocity trace. The phasing between peak glottal opening and the abrupt increase in intraoral air pressure was calculated, defining successive occurrences of lip implosion as the reference cycle and peak glottal opening as the event whose phase was determined relative to the reference cycle. The articulatory events chosen—peak glottal opening and minimum lip aperture, or lip implosion, during production of /ip/ and /pi/—seem reasonable given Löfqvist and Yoshioka's (1981) demonstration that the interval from peak glottal opening to lip implosion varies linearly as a function of closure duration, at least for bilabial stop-vowel productions.

## Results

First, we focus on the hypothesized collective variable, relative phase. For all speakers, the initial estimate of relative phase was based simply on the phase of peak glottal opening (PGO) relative to successive occurrences of minimum lip aperture. Examples from one /pi/ trial and one /ip/ trial are presented in Fig. 14.1 and Fig. 14.2, respectively, which show trajectories over time of lip aperture (the dotted line) and glottal aperture (the solid line) and the corresponding relative phase values for subject AL. The vertical ticks denote the onsets of minimum lip aperture and peak glottal opening.[1] In the case of /pi/ it is obvious that the phase relation between minimum lip aperture and PGO is quite steady. Over the range of speaking rates examined, PGO lags minimum lip aperture by roughly 40–50 degrees. In the case of /ip/, the phase relation is also steady, although different from that observed for /pi/, for speaking rates of approximately 1–4 syllables/s. Then a clear and abrupt jump in phase occurs to the phasing pattern for /pi/. Note that this abrupt change in interarticulator patterning occurs well below maximum syllable rates.

In order to facilitate comparison among trials, frequency plateaus within each trial were identified, and mean relative phase and SD were determined over all cycles within each plateau. Each subject's data were then examined individually, using ANOVA, for the effects of required utterance (/ip/ or /pi/), speaking rate, and their interaction on the mean relative phase of PGO and minimum lip aperture and its variability. In discussing the quantitative results we also refer to qualitative aspects of the data, as shown in Fig. 14.3, which plots plateau means of relative phase as a function of speaking rate (syllables/s). In this figure and those following, trials in which the subject was instructed to produce /pi/ are denoted by filled squares; /ip/ trials are denoted by open squares.

Mean relative phase was significantly influenced by required utterance and speaking rate for three of the four subjects (all but ES). Individual *F*-values ranged from 4.83 to 83.14 for utterance ($ps < .05$) and from 4.05 to 11.55 for rate ($ps < .01$). For AL, the interaction of utterance and rate also affected mean relative phase, $F(5,31) = 8.32$, $p < .001$. Simple main effects analysis showed mean relative phase to be smaller for /ip/ than /pi/ at speaking rates up to 6 syllables/s ($p < .01$). At that rate and faster, relative phase for /ip/ and /pi/ were equivalent ($ps > .1$). Note, however, that the individual trial means shown in Fig. 14.3 do not all follow the overall pattern indicated by ANOVA. For one /ip/ trial, the relative phase value calculated did not change as a function of rate. Subject SH also showed a significant interactive effect of utterance and rate, $F(5,39) = 3.02$, $p < .05$. Simple main effects analysis showed that this subject produced equivalent phase values for both /ip/ and /pi/ at all but the slowest speaking rate used in each trial. At the slowest rate, $F(1,39) = 14.51$, $p < .001$; thereafter, $ps > .1$.

A similar qualitative pattern for /pi/ trials and a subset of /ip/ trials was observed for speaker SN. Namely, consistent but different relative phase values for /ip/ and /pi/ at slow speaking rates with an abrupt increase in phase value of /ip/ trials at faster speaking rates, equalling relative phase

[1]A similar figure, published in Kelso, Saltzman, & Tuller (1986a), mistakenly identified the ticks as denoting lip opening (consonant release) rather than lip closure.

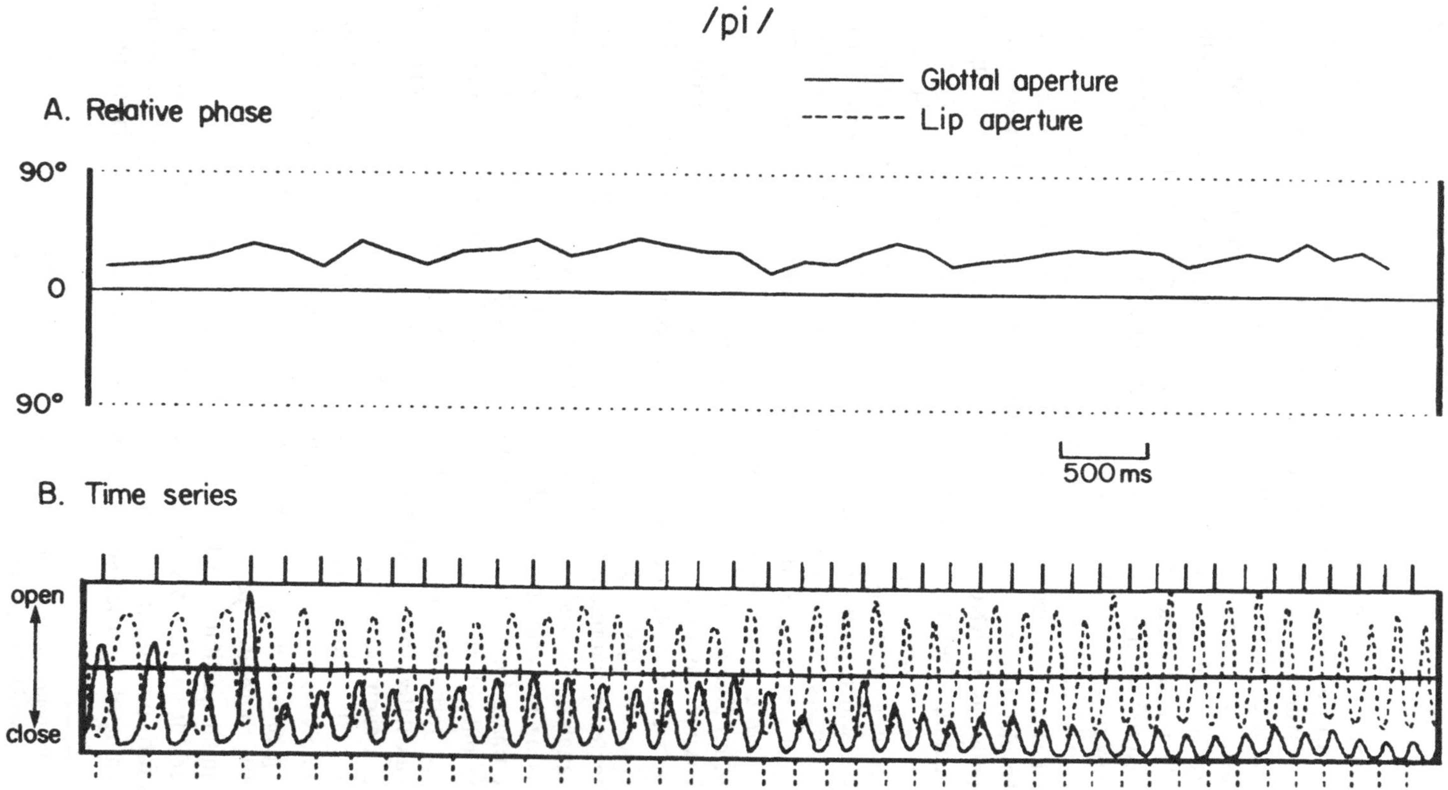


FIG. 14.1. For the required utterance /pi/, (A) phase relation between peak glottal aperture and minimum lip aperture and (B) trajectories of each variable over time as the subject speeds up rate of production.

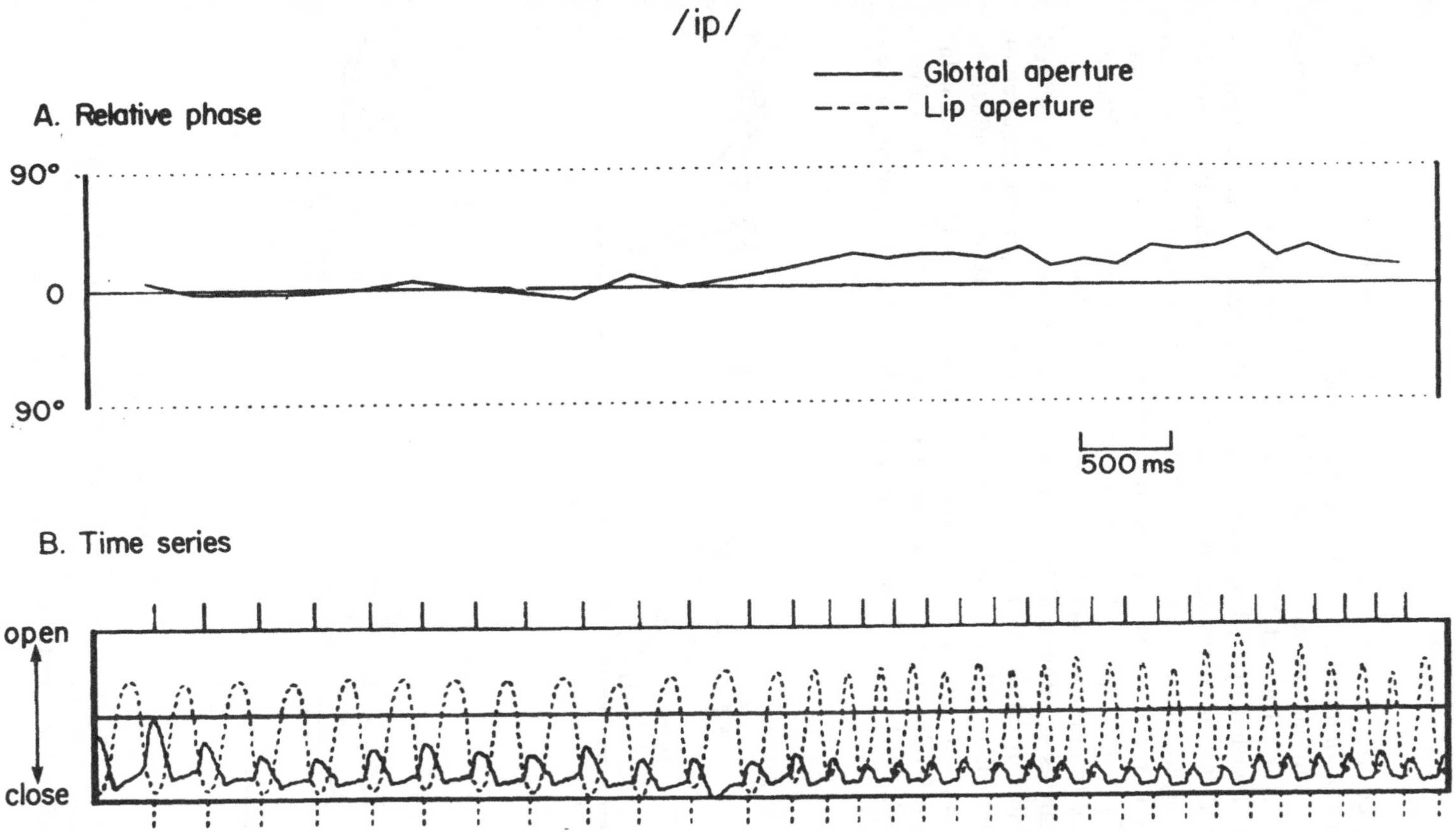


FIG. 14.2. For the required utterance /ip/, (A) phase relation between peak glottal aperture and minimum lip aperture and (B) trajectories of each variable over time as the subject speeds up rate of production.

observed for /pi/ at the same frequency. However, for this subject, the interaction of utterance and rate did not reach significance, $F(5,30) = 1.9$, $p > .1$, and one can see from Fig. 14.3 that in two /ip/ trials, the relative phase value calculated did not change as a function of rate.

The fourth subject, ES, never established a steady phase relation between PGO and minimum lip aperture for either utterance type, and the effect of utterance was not significant ($p > .1$). Both /ip/ and /pi/ trials tended to show increased relative phase values as speaking rate increased, $F(5,22) = 6.39$, $p < .001$. This subject's mean relative phase data are omitted from further analysis but are discussed below.

Analyses of within trial SD as a function of utterance and rate showed variability of relative phase to be unaffected by the utterance required or its interaction with rate (*ps* > .1). However, speaking rate influenced relative phase variability for three of the four speakers (all but AL). Individual *F*-values ranged from 2.76 to 9.67, *ps* < .05. Subsequent analyses (Tukey's HSD) showed that rate affected relative phase variability differ-

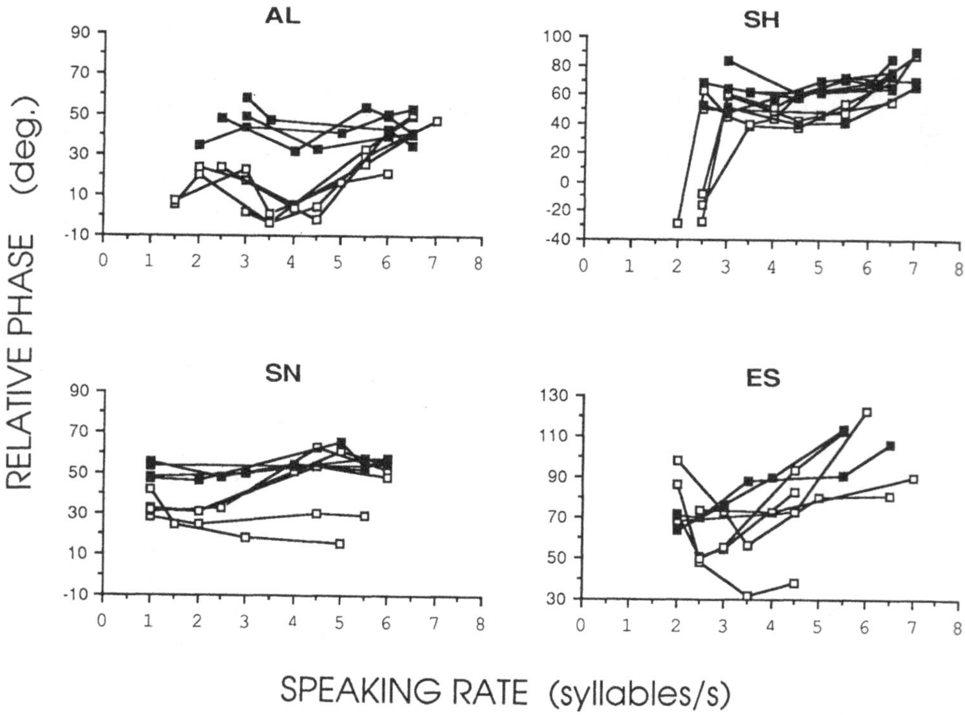


FIG. 14.3. Mean relative phase of peak glottal opening and maximum lip closure as a function of speaking rate. Individual trials are shown for each of the four subjects for required utterance /pi/ (■) and required utterance /ip/ (□).

ently for the three speakers. None of the subsequent comparisons for SH reached significance; qualitatively, SD showed a local peak at rates of 3–4 syllables/s, then increased again at the fastest rate. For SN, variability was large at both the slowest and the fastest rates produced ($ps < .05$). SD of the three slower rates that ES elected to produce were less variable than the three faster rates ($ps < .05$).

In order to facilitate comparison among subjects, /ip/ trials were examined for evidence of a change in relative phase as speaking rate increased. If such a change occurred, the rate plateau on which it occurred was identified. This plateau was designated the *transition* and was averaged across trials and speakers. The plateau just before the transition (t – 1) was also averaged across subjects and trials. Two plateaus before the transition, and two after the transition (if present), were handled similarly. Trials in which the subjects produced /pi/, and /ip/ trials that did not show a clear shift in relative phase, were also divided into rate plateaus. The plateau showing the closest frequency value to the /ip/ transition plateau was determined on a trial-by-trial basis for each speaker. These were averaged across trials and speakers separately for /pi/ and /ip/, and are plotted in Figure 14.4 under the *transition* designation on the abscissa. Across subjects[2], speech frequency ranged from 3.32 to 6.80 Hz at the transition plateau, from .77 to 4.10 Hz at $t-2$, and from 4.00 to 6.25 Hz at $t+2$. Mean and SD for the two pretransition plateaus, and two posttransition plateaus were also determined. Note that no posttransition values appear for /ip/ trials showing no transition because these trials tended to be produced at slower peak frequencies than /ip/ trials that did show a transition.

The data in Fig. 14.4 are quite clear. As speaking rate increases, the /pi/ trials maintain the relative phase between PGO and minimum lip aperture. Relative phase observed for /ip/ trials with no transition is also maintained, but at a different value than that observed for /pi/. In contrast, when a transition is observed during /ip/ trials, relative phase begins at about the same value as for /ip/ trials that do not include a transition. As the control parameter increases, relative phase then shifts to the level observed for /pi/ trials. Further increases in speaking rate produce no further changes in relative phasing between glottal and lip apertures.

Figure 14.4 also shows the pattern of fluctuations in mean relative phase calculated across trials and subjects as SD of the plateau means (Fig. 14.4b), and mean SD of the plateaus (Fig. 14.4c). SD of the means is relatively low and steady in all but one case, namely, during /ip/ trials in the plateau

[2]The absolute value of the control parameter at which a transition occurs is not critical and may vary markedly across speakers and across tasks. For example, explicitly instructing subjects to inhibit phonetic alterations may well increase the speaking rate at which transitions occur, or inhibit those transitions altogether.

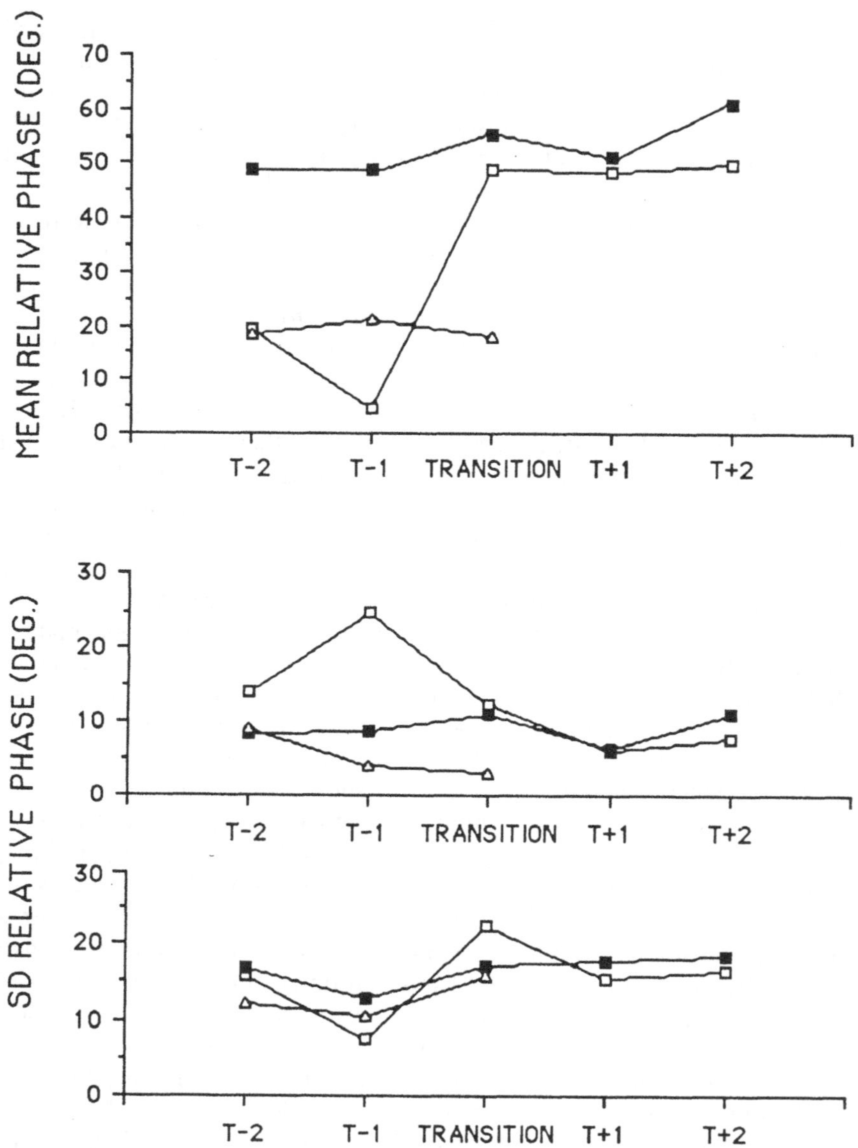


FIG. 14.4. Top: mean relative phase across subjects and trials, aligned with respect to transition plateau. Middle: SD of the means of relative phase. Bottom: Mean within trial SD of relative phase. Required utterance /pi/ (■); Required /ip/ trials with a transition (□); Required /ip/ trials without a transition (△).

immediately preceding the transition. Although the SD change is in the direction predicted, we do not consider it strong evidence for enhanced fluctuations prior to a transition because the mean SD *within* each trial changed little across plateaus ($p > .1$; Fig. 14.4c).

Intraoral air pressure was monitored for three subjects (AL, SH and SN), and relative phase between velocity peaks in intraoral air pressure (lip implosion) and PGO were calculated. This measure is not necessarily identical to minimum lip aperture versus PGO. The lips are compliant masses, allowing some compression after the point of contact. Thus, unless the lips are extremely stiff and do not compress noticeably after contact, our measure of minimum lip aperture would occur a short time after lip contact has actually occurred. Consequently, a sharp rise in intraoral air pressure would precede minimum lip aperture, as measured, and the calculated phase relation of PGO and lip implosion would be greater than that of PGO and lip aperture. Estimated lip stiffness during closing gestures has been shown to increase with speaking rate (Kelso, Bateson, Saltzman, & Kay, 1985), so that the difference in relative phase of PGO with lip aperture versus lip implosion, may itself vary as a function of rate.

The phasing of PGO relative to successive velocity peaks in intraoral air pressure is shown in Fig. 14.5. Individual trial means are shown separately for the three speakers. For AL and SH, the absolute phase values are different from those shown in Fig. 14.3 but the patterns are identical. For both subjects, the main effects and their interaction once again significantly influence mean relative phase ($F$-values ranged from 4.76 to 72.38, $ps < .01$). Qualitatively, there is a clear separation between /ip/ and /pi/ trials until a critical point, at which /pi/ trials maintain the relative phase between PGO and lip implosion but /ip/ trials tend to shift in relative phase to that observed for /pi/. The quantitative analyses are also consistent regarding when the shift in relative phase occurs. Simple main effects analysis shows significant differences between relative phase for /ip/ and that for /pi/ only at rates slower than 6 syllables/s for AL ($F$-values ranged from 4.70 to 41.48, $ps < .05$) and only at the slowest rate for SH, $F(1,39) = 4.3$, $p < .01$. Relative phase values for faster /ip/ productions equal those for /pi/ ($ps > .1$). Thus, the shift in relative phase occurs at the same time whether indexed by the relation between PGO and lip implosion, or PGO and minimum lip aperture. Note, however, that one /ip/ trial produced by AL still shows no change in relative phase as a function of rate. SN again shows main effects of utterance and rate, $F(1,30) = 44.86$ and $F(5,30) = 14.01$, respectively, $ps < .001$. In contrast to the earlier analysis, however, the two /ip/ trials that maintained relative phasing between PGO and minimum lip aperture despite increases in speaking rate, now show a phase shift to the value appropriate for /pi/. This is

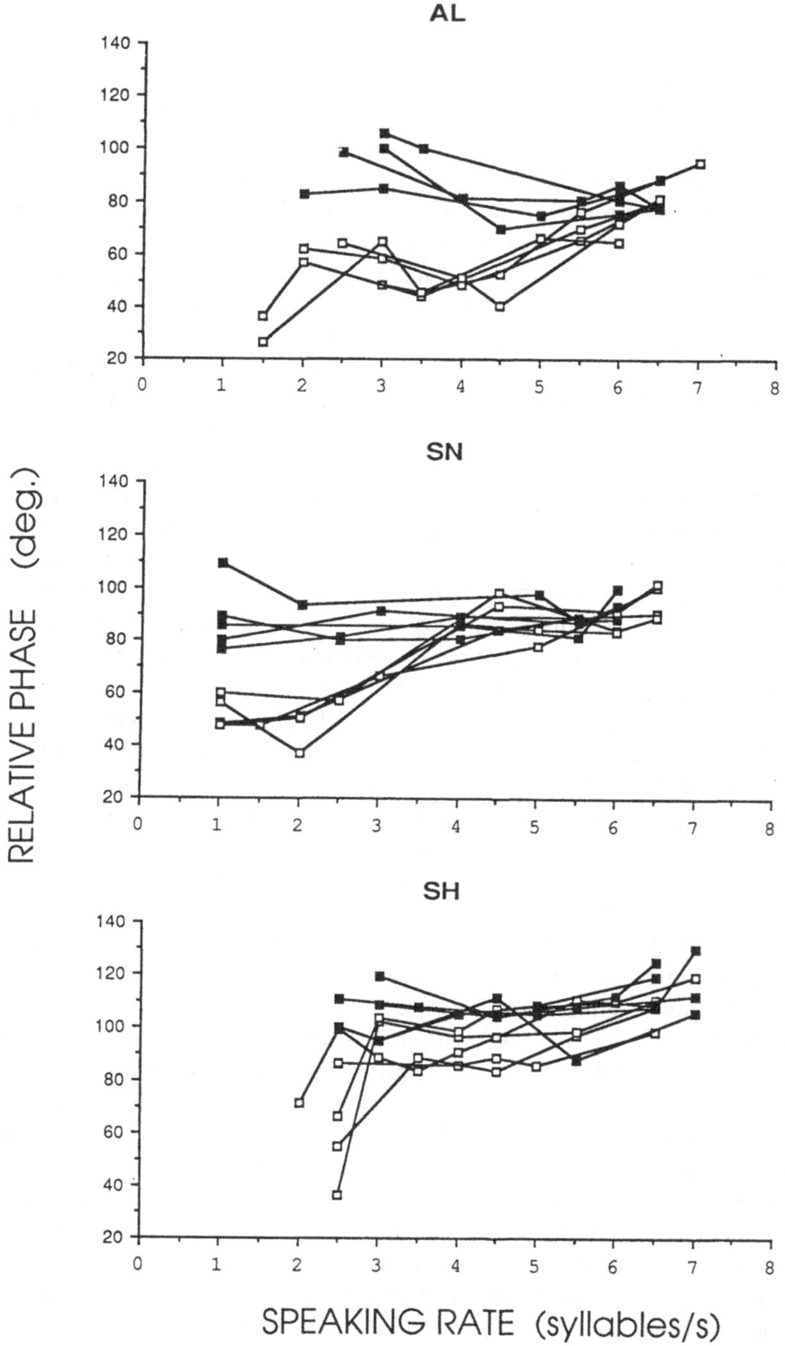


FIG. 14.5. Mean relative phase of peak glottal opening and intraoral air pressure as a function of speaking rate. Individual trials are shown for each of the four subjects. Required utterance /pi/ (■); Required utterance /ip/ (□).

quantified in the now significant interaction of utterance and rate, $F(5,30) = 11.41$, $p < .001$. Simple main effects analysis shows that the difference between relative phase of /ip/ and /pi/ ceased being significant at rates greater than 3.5 syllables/s. $F$-values ranged from 5.86 to 71.13 at rates slower than 3.5 syllables/s ($ps < .05$); thereafter $ps > .1$. This suggests that for SN, lip stiffness increased nonlinearly with speaking rate.

## Discussion

Investigation of the mean phase relation between lip and glottal events results, by and large, in the expected pattern. For three of the four speakers, the phase relation for /ip/ and /pi/ is consistent, but different, at slow speaking rates. At higher rates, /pi/ maintains the same phase relation as at slow rates. In contrast, as speaking rate increases, the phase relation for /ip/ tends to show a clear jump in phase, usually within a single cycle, to the phasing pattern for /pi/. This phase shift occurs at very different speaking rates across the three subjects, and can occur at well below maximum syllable rates (see particularly Fig. 14.3, SH). In the case of both /ip/ and /pi/ it might be tempting to describe the relative phasing between glottis and lips as *invariant* over a range of parameter values (speaking rates). Clearly there are statistics, first and second moments, corresponding to the two patterns. The fact that (a) an abrupt transition from one pattern to the other occurs at a critical rate, and (b) beyond the transition only the /pi/ pattern is stable suggests an alternative formulation. Namely, relative phase is a collective variable, or order parameter, whose dynamics are bistable in one parameter regime and monostable in another. The shift in pattern is a result of loss of stability of the /ip/ pattern and may be understood as a phase transition. The mean speech data thus are qualitatively like previous experiments on bimanual coordination.

Several questions remain, however. First, we need to explore why the change in relative phase between lip and glottal events did not occur on every /ip/ trial, and why ES failed to show any consistency on this articulatory measure. To this end, it is important to know whether the jump in relative articulatory phase has any perceptual consequences. Specifically, is a shift in relative phase for /ip/ to the relative phase of /pi/ *perceived* as a change from /ip/ to /pi/? Do trials that fail to show a shift in relative phase maintain their perceptual identity throughout the range of speaking rates produced? Obviously, a jump in relative articulatory phasing might well have acoustic consequences including, but not limited to, differences in the amount of aspiration, fundamental frequency at consonant release, or timing from lip implosion or release to voicing onset. Rather than ask what acoustic variable, or variables, signals a perceptual change from /ip/ to /pi/, in Experiment 2 we ask simply whether the distributed acoustic effects of a change in relative articulatory phase are sufficient to yield a corresponding perceptual change.

## EXPERIMENT 2

### Subjects

Subjects were 12 undergraduate students at Florida Atlantic University who participated in order to receive extra credit in a statistics course. All were naive to the experiment's purpose and all reported having normal hearing in both ears.

### Materials and Procedure

Sixteen source trials were chosen for perceptual analysis. These included two /ip/ trials and two /pi/ trials from each speaker recorded in Experiment 1. For both AL and SN, /ip/ trials chosen included one in which a shift in relative phase of PGO and minimum lip aperture was observed as speaking rate increased and one in which relative phase did not change as a function of frequency.

The acoustic signal was digitized using a sampling rate of 10 kHz, preemphasis, and 5 kHz low-pass filtering. Within each trial, 3 sec of silence were inserted at the onset of voicing for every seventh vowel using an interactive wave-editing program (ILS). Perceptually, this transforms a trial of, for example, 48 continuous syllables into a trial containing 8 sequences of 6 syllables each. Note that inserting the silence at voicing onset may introduce a bias toward perception of the vowel–consonant form. A stimulus tape was prepared that contained 6 instances of the 16 source trials separated into sequences of 6 syllables (total sequences = 780). The test order was randomized across speakers ($N$ = 4) and utterance types (/ip/ and /pi/), but the order of sequences within a trial was not altered. The stimulus tape was presented to listeners binaurally through headphones. Listeners were given an answer sheet with five response columns, labelled *eep, pee, eeb, bee,* and *other*. Subjects were required to indicate what they heard by checking the corresponding column on the answer sheet during each 3s silent period. If the utterance seemed to change within a sequence, subjects were asked to indicate which utterance changed, and to what.

### Results

Results of the listening test are shown in Fig. 14.6, which plots the proportion of responses that placed the consonant in syllable-initial position

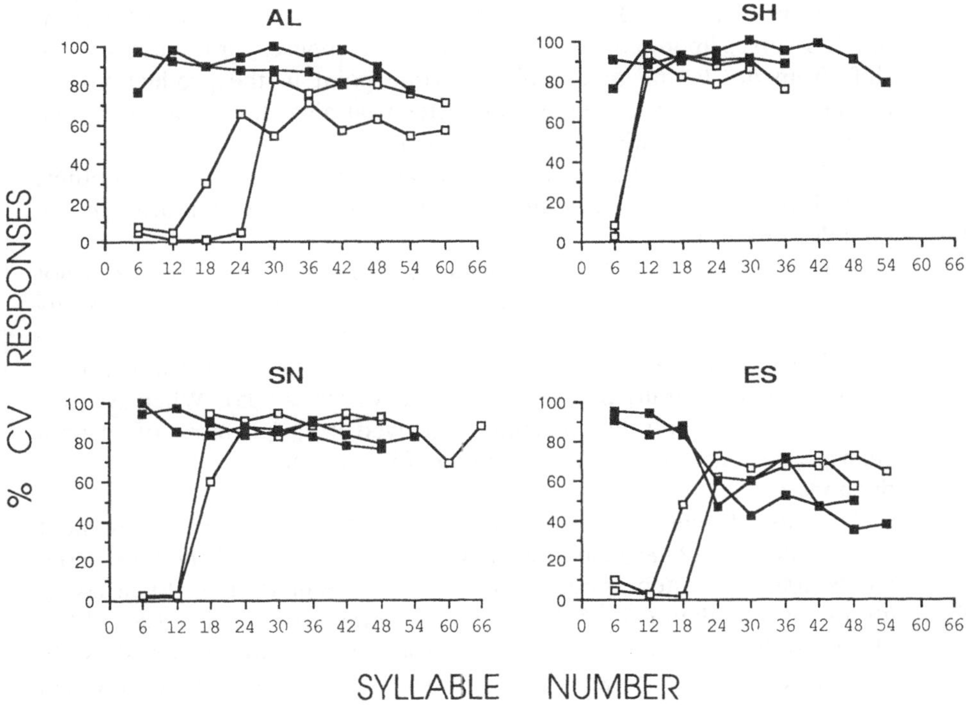


FIG. 14.6. Identification functions for two /pi/ trials (■) and two /ip/ trials (□) per speaker.

as a function of sequence presented, for each speaker. First, notice that /pi/ trials from three of the four speakers were perceived correctly by listeners as having a syllable-initial consonant at least 80% of the time. In contrast, /pi/ trials produced by ES are perceived correctly at slow rates of speech, but accuracy drops to chance level as rate increases.

Listeners also judged the identity of syllables in two /ip/ trials per speaker. For AL, one /ip/ trial showed a shift in relative phase to that observed for /pi/. This is mirrored in listeners' responses. Listeners perceived the trial to begin as /ip/ but, beginning with sequence 5 (syllable 30) perceived the speaker to be saying /pi/. This is *exactly* where the shift in relative phase occurred in production of this trial. The second /ip/ trial included for this speaker did not show a reliable shift in relative articulatory phase. Listeners were correspondingly less sure of the syllable affiliation of the consonant, hearing the first 12 syllables as /ip/, but responding at chance level for 5 of the next 8 sequences ($p > .05$).

The /ip/ trials produced by SN also included one trial in which relative phase showed an abrupt shift (on syllable 13), and one in which a shift was absent on the PGO versus minimum lip aperture analysis, but present in PGO versus lip implosion. The perceptual responses for the former trial shift from /ip/ to /pi/ at exactly the same point in the production and perception tasks. Interestingly, the latter trial also showed a perceptual shift in syllable affiliation of the consonant.

The perception and production results for SH fit well with each other. In both measures, /ip/ trials undergo a shift from /ip/ to /pi/ after only 6 syllables in the /ip/ mode.

The fourth speaker, ES, produced relative phase values that did not differentiate well between /ip/ and /pi/ trials, and both changed markedly as speaking rate increased. This subject's production of /ip/ and /pi/ were reliably perceived as such only at slow speaking rates. Thereafter, identification was generally no better than chance ($ps > .05$). When ES produced /ip/ at faster rates, listeners perceived it as /pi/ more often than chance ($p < .05$) but not as reliably as for the other speakers' /ip/ productions.

The judged segmental identity of the syllables was also very consistent across listeners. *Other* responses accounted for only 13 of the 9360 judgments (.001%). Listeners perceived three of the four speakers to be saying /ib/ or /bi/ less than 6% of the time. In contrast, for speaker ES, 12.6% of /ip/ productions, and 32.5% of /pi/ productions, were perceived as /ib/ or /bi/. As one would expect, the proportion of /b/ responses increased with speaking rate.

## Discussion

The results of Experiment 2 demonstrate a remarkable fit between the articulatory measures and perception of syllable form. The three speakers who showed a consistent phase relation for production of /pi/ across speaking rates were also perceived as saying /pi/ across all rates. Speakers /ip/ trials in which relative phase jumped to the /pi/ mode were perceived as /ip/ during the pretransition plateaus; perception switched to /pi/ on exactly the same cycle as the jump in relative phase. Inconsistent relative phase values were perceived inconsistently (ES). Interestingly, perception of SN's productions followed the relationship between PGO and lip implosion more closely than that between PGO and minimum lip aperture.

We feel confident in concluding that a change in articulatory coordination as indexed by relative phase has as its counterpart a perceptual change in the production's syllabic form. Moreover, dynamic pattern theory predicts that more than one type of transition is possible because the system has additional degrees of freedom available (see Buchanan, de Guzman, & Kelso, 1988; Kelso & Scholz, 1985). The perceptual results suggest that an additional transition might be found in the distinction between voiced and voiceless bilabial stop consonants, as the system is scaled.

## GENERAL DISCUSSION

The experiments reported here support the notion that the articulatory system undergoes a phase transition when speaking rate is varied across a wide range of values. When an articulatory phase transition was observed, it corresponded precisely to a perceived change in syllable structure. This supports Stetson's (1951) observation that articulatory phasing changes across speaking rates, and rationalizes *why* phase shifts occur, according to the differential stability of syllable types.

The results also shed light on discrepancies in the "relative timing" literature, concerning whether relative timing, or relative phase, of articulatory gestures is maintained within a given phonetic string as speaking rate or stress is varied (cf. Gentil et al., 1984; Kelso et al., 1986a; 1986b; Linville, 1982; Löfqvist, 1986; Lubker, 1983; Munhall, 1985; Nittrouer, Munhall, Kelso, Tuller, & Harris, 1988; Tuller et al., 1983). The results presented here argue that the interpretation of relative phase information depends crucially on the range of control parameter values selected for study. As an example, consider the following scenario: While studying the relative phasing between labial and laryngeal gestures during production of /ip/ across speaking rates, we arbitrarily choose (or allow our subjects to choose) rates between 2.5 and 6.7 Hz (i.e., the range adopted by 3 of the 4 speakers reported in Nittrouer et al., 1988). Inspection of Fig. 14.2 reveals that under these conditions, none of our subjects shows invariant mean relative phasing of lip and larynx gestures for /ip/. However, if we had chosen a narrower range of control parameter values, from 3.8 to 6.2 Hz (the range adopted by the fourth speaker reported in Nittrouer et al., 1988), SH and SN would appear to maintain invariant mean relative phasing of lip and larynx gestures. Obviously, the interpretation of relative phase information depends crucially on the range of control parameters selected for study. Relative phase may change little across some range of control parameter values, but the fact that it undergoes an abrupt change may be more fundamental, undercutting the invariance concept and instead promoting the dynamical concept of stability.

Although our observations were generally supportive of a phase transition interpretation of speech patterns, there were anomalies in the data that need further explanation. Specifically, why did ES show a lack of stationarity at almost all speaking rates, paralleled by listeners' perceptions that his speech was variable and ambiguous? ES was a native speaker of English and did not have particularly small or noisy transillumination or lip movement recordings. One possible confounding is the accuracy with which the syllable strings were produced. Although no formal evaluation of accuracy was performed, only ES produced the syllables with a strong alternating stress pattern.

A more likely explanation is that the lack of stationarity in ES's data and a second apparent anomaly—the lack of evidence across subjects for enhancement of relative phase fluctuations as the transition region was approached—are both related to shortcomings in the experimental design. In particular, the relationship between the system's *time scales* plays a decisive role in determining the nature of the phase transition, that is, what is observed (Schöner, Haken, & Kelso, 1986; Kelso et al., 1987). To characterize coordinative movement patterns within a dynamical description, three types of time scales are relevant. The first is the typical time scale on which the system is observed, $\tau_{obs}$ (i.e., how long the experimenter observes the system in a given initial condition). The second is the local relaxation time, $\tau_{rel}$, namely, the time it takes for the system to relax into a local stationary state following a small perturbation. The third time scale is the equilibrium time (or global relaxation time, $\tau_{equ}$) which is the time it takes for the system to achieve the lowest potential minimum (the most stable state available). If the time scales fulfill the following relation

$$\tau_{rel} \ll \tau_{obs} \ll \tau_{equ} \qquad (1)$$

then the interpretation of observed states as attractors is consistent. When coordinative patterns in an experiment are referred to as stable, stationary states, times scales relation (Equation 1) is obeyed. However, at a transition point, time scales relation (Equation 1) breaks down. That is, $\tau_{rel}$ and $\tau_{equ}$ become roughly equivalent and a transition is observed—the system switches to the most stable state.

Whether features of critical phenomena (fluctuation enhancement, slowing down, etc.) are observed depends on how quickly the control parameter changes ($\tau_{par}$, the time scale for parameter change). As long as the relation

$$\tau_{rel} \ll \tau_{par} \ll \tau_{equ} \qquad (2)$$

holds, the system changes state only as the old state becomes unstable, and critical phenomena are observed. If, however,

$$\tau_{rel} \ll \tau_{equ} \ll \tau_{par} \qquad (3)$$

the system switches state (to the lowest potential minimum available) before the old state becomes unstable, and none of the features of critical phenomena would be seen.

For example, in a study of human bimanual coordination, Scholz, Kelso, & Schöner (1987) found no evidence for critical fluctuations on the pre-transition frequency plateau despite earlier observations of fluctuation enhancement before the critical point (Kelso & Scholz, 1985; Kelso, Scholz, & Schöner, 1986). Using a previously derived theoretical model for the stochastic dynamics of relative phase (Haken, Kelso, & Bunz; 1985; Schöner, Haken, & Kelso, 1986), the actual values of the model parameters were determined from the empirical data. When fluctuation enhancement *was* observed (Kelso, Scholz, & Schöner, 1986), $\tau_{par}$ was much less that $\tau_{equ}$, thus allowing the system to come closer to the critical point before time scales relation (Equation 2) was violated. When fluctuation enhancement was *not* observed, the calculated $\tau_{equ}$ was much shorter than $\tau_{par}$, again entirely consistent with theory and time scales relation (Equation 3).

Let us bring the discussion back to the present experimental observations on speech. The basic point is that time scales relations play an important, but seldom recognized role in the design of experiments. In particular, in the present Experiment 1, rate of speaking and the time spent at each rate were not precisely controlled. Subjects were allowed to self-select their speaking rates and they typically produced very large increases in frequency when told to speak faster (see Fig. 14.3). Although this may reflect the speaker's desire to get the experiment over with quickly, it may also signify that subjects do not elect to speak at rates at which their productions would be unstable (see, for example, Hoyt & Taylor, 1981). Experiments that manipulate the speaking rate and the length of time spent at each rate are necessary to test the foregoing time scales assumptions explicitly, and to enable enhanced fluctuations to be observed as the critical region is approached. Such work is in progress. In addition, modeling of the stochastic dynamics of relative phase for the utterances we have investigated, and determination of the time scales relations, is crucial to our understanding and interpretation of speech patterns, stability, and change as phase transitions.

Although the understanding of speech production in terms of dynamic pattern theory is still in its infancy, the nature of the limited pattern change that has been observed is still informative. In particular, in the small set of utterances observed, vowel–consonant utterances shifted to the consonant–vowel form, but when the consonant–vowel was initially produced, no shift to vowel–consonant occurred. This suggests that the consonant–vowel form is more intrinsically stable. The fundamental nature of consonant–vowel syllables has long been noted in the linguistics literature. Jakobson (1966) pointed out that the consonant–vowel syllable is the only syllable type present in all languages (cf. Abercrombie, 1967; Clements & Keyser, 1983). It is also the predominant syllable type in infant babbling (Locke, 1983). Previous empirical work attempting to provide a biological basis for phonemic coding assumed a similar syllable organization in that the trajectories used as primitives moved from closed to open vocal tract positions (Lindblom, 1986; Lindblom, MacNeilage, & Studdert-Kennedy, 1983). It should be noted, however, that the literature on speech errors presents a different view. In spontaneous speech errors, the most frequent single-segment errors involve initial consonants of syllables and words; errors in final consonants are much less common. Similarly, when speakers are required to exchange phonological segments that occur in like positions in a pair of words, they make initial exchanges more rapidly and accurately than final-consonant exchanges, suggesting that initial consonants cohere less strongly with their vowel than do final consonants (Fowler, 1987). Studies such as those presented here are intended to reveal which forms of coordination are more stable, and thus may rationalize linguists' and psychologists' observations (cf. Kelso & Tuller, 1987).

In summary, the approach developed here is strongly biased toward substituting an analysis of articulatory and perceptual *stability* for the search for invariants. Fluctuations are viewed not as unwanted noise, but as an essential source of discovery for new stable patterns. The spontaneous emergence of constraints in articulatory patterns observed here offers preliminary insights into the basis of perceptual stability and may ultimately rationalize more formal descriptions of language.

## ACKNOWLEDGMENTS

This work was supported by NIDCD grant NS-24771 and NIMH grant MH-42900, both to the Center for Complex Systems, Florida Atlantic University. Additional support was provided by NINCDS grant NS-13617 to Haskins Laboratories. The authors thank Linda Fishman, Kevin Munhall, and Eric Vatikiotis-Bateson for their help with data collection and analysis.

## REFERENCES

Abercrombie, D. (1967). *Elements of general phonetics*. Chicago: Aldine.

Baer, T., Löfqvist, A., & McGarr, N. S. (1983). Laryngeal vibrations: A comparison between high-speed filming and glottographic techniques. *Journal of the Acoustical Society of America, 73*, 1304–1308.

Barry, W. (1983). Some problems of interarticulator phasing as an index of temporal regularity in speech. *Journal of Experimental Psychology: Human Perception and Performance, 9*, 826–828.

Benoit, C. (1986). Note on the use of correlations in speech timing. *Journal of the Acoustical Society of America, 80*, 1846–1848.

Buchanan, J. J., deGuzman, G. C., & Kelso, J. A. S. (1988). Further phase transitions in biological coordination: Rhythmic movement in three dimensions. In J. A. S. Kelso, A. Mandell, M. F. Schlesinger (Eds.), *Dynamic Patterns in Complex Systems* (p. 397). Singapore: World Scientific.

Butcher, A., & Weiher, E. (1976). An electropalatographic investigation of coarticulation in VCV sequences. *Journal of Phonetics, 4*, 59–74.

Clements, G. N., & Keyser, S. J. (1983). *CV phonology: A generative theory of the syllable.* Cambridge, MA: M.I.T. Press.

Fowler, C. A. (1987). Consonant–vowel cohesiveness in speech production as revealed by initial and final consonant exchanges. *Speech Communication, 6*, 231–244.

Gay, T., Ushijima, T., Hirose, H., & Cooper, F. S. (1974). Effect of speaking rate on labial consonant–vowel articulation. *Journal of Phonetics, 2*, 46–63.
Gentil, M., Harris, K. S., Horiguchi, S., & Honda, K. (1984). Temporal organization of muscle activity in simple disyllables. *Journal of the Acoustical Society of America, 75*, S23.
Gentner, D. R. (1987). Timing of skilled motor performance: Tests of the proportional duration model. *Psychological Review, 94*, 225–276.
Haken, H. (1975). Cooperative phenomena in systems far from thermal equilibrium and in nonphysical systems. *Rev. Mod. Phys., 47*, 67–121.
Haken, H. (1983). *Advanced synergetics*. Heidelberg: Springer-Verlag.
Haken, H., Kelso, J. A. S., & Bunz, H. (1985). A theoretical model of phase transitions in human hand movements. *Biological Cybernetics, 51*, 374–356.
Hoyt, D. F., & Taylor, C. R. (1981). Gait and the energetics of locomotion in horses. *Nature, 292*, 239–240.
Jakobson, R. (1966). Implications of language universals for linguistics. In J. H. Greenberg (Ed.), *Universals of language* (pp. 263–278). Cambridge, MA: M.I.T. Press.
Kelso, J. A. S., Bateson, E., Saltzman, E. L., & Kay, B. (1985). A qualitative dynamic analysis of reiterant speech production: phase portraits, kinematics, and dynamic modeling. *Journal of the Acoustical Society of America, 77*, 266–280.
Kelso, J. A. S., Del Colle, J., & Schöner, G. This volume.
Kelso, J. A. S., & Munhall, K. G. (1988). *R. H. Stetson's motor phonetics: A retrospective edition*. San Diego: College-Hill.
Kelso, J. A. S., Saltzman, E. L., & Tuller, B. (1986a). The dynamical perspective in speech production: Data and theory. *Journal of Phonetics, 14*, 29–60.
Kelso, J. A. S., Saltzman, E. L., & Tuller, B. (1986b). Intentional contents, communicative context, and task dynamics: A reply to the commentators. *Journal of Phonetics, 14*, 171–196.
Kelso, J. A. S., & Scholz, J. P. (1985). Cooperative phenomena in biological motion. In H. Haken (Ed.), *Complex systems: Operational approaches in neurobiology, physical systems and computers* (pp. 124–129). Berlin: Springer-Verlag.
Kelso, J. A. S., Scholz, J. P., & Schöner, G. S. (1986). Non-equilibrium phase transitions in coordinated biological motion: Critical fluctuations. *Physics Letters A, A118*, 279–284.
Kelso, J. A. S., & Schöner, G. (1988). Self-organization of coordinative movement patterns. *Human Movement Science, 7*, 27–46.
Kelso, J. A. S., Schöner, G., Scholz, J. P., & Haken, H. (1987). Phase-locked modes, phase transitions and component oscillators in biological motion. *Physica Scripta, 35*, 79–87.
Kelso, J. A. S., & Tuller, B. (1987). Intrinsic time in speech production: Theory, methodology and preliminary observations. In E. Keller and M. Gopnik (Eds.), *Motor and sensory processes of language* (pp. 203–222). Hillsdale, NJ: Lawrence Erlbaum Associates.
Kent, R. D., & Moll, K. L. (1975). Articulatory timing in selected consonant sequences. *Brain and Language, 2*, 304–323.
Kent, R. D., & Netsell, R. (1971). Effects of stress contrasts on certain articulatory parameters. *Phonetica, 24*, 23–44.
Kozhevnikov, V. A., & Chistovich, L. A. (1965). *Speech: Articulation and perception*. (U. S. Department of Commerce, Trans.). Clearinghouse for Federal Scientific and Technical Information, Washington, DC.
Kuehn, D. P., & Moll, K. L. (1976). A cineradiographic study of VC and CV articulatory velocities. *Journal of Phonetics, 4*, 303–320.
Liberman, A. M., Cooper, F. S., Shankweiler, D. P., & Studdert-Kennedy, M. (1967). Perception of the speech code. *Psychological Reviews, 74*, 431–461.
Lindblom, B. (1963). Spectographic study of vowel reduction. *Journal of the Acoustical Society of America, 35*, 1773–1781.

Lindblom, B. (1986). On the origin and purpose of discreteness and invariance in sound patterns. In J. S. Perkell and D. H. Klatt, (Eds.), *Invariance and variability in speech processes*, (pp. 493–523). Hillsdale, NJ: Lawrence Erlbaum Associates.

Lindblom, B., MacNeilage, P., & Studdert-Kennedy, M. (1983). Self-organizing processes and the explanation of phonological universals. *Linguistics, 21*, 181–203.

Linville, R. (1982). *Temporal aspects of articulation: Some implications for speech motor control of stereotyped productions*. Unpublished doctoral dissertation, University of Iowa, Iowa City.

Lisker, L., Abramson, A. S., Cooper, F. S., & Schvey, M. H. (1969). Transillumination of the larynx in running speech. *Journal of the Acoustical Society of America, 45*, 1544–1546.

Locke, J. (1983). *Phonological acquisition and change*. New York: Academic Press.

Lubker, J. (1983, October). *Comment on "Temporal invariance in the production of speech," by Harris, Tuller and Kelso*. Paper presented at Conference on invariance and variability in speech processes, Cambridge, MA.

Löfqvist, A. (1986). Stability and change. *Journal of Phonetics, 14*, 139–144.

Löfqvist, A., & Yoshioka, H. (1979). Laryngeal articulation in Swedish obstruent clusters. *Haskins Laboratories Status Reports on Speech Research, SR-59/60*, 103–125.

Löfqvist, A., & Yoshioka, H. (1981). Interarticulator programming in obstruent production. *Phonetica, 38*, 21–34.

MacNeilage, P. F., & DeClerk, J. (1969). On the motor control of coarticulation in CVC monosyllables. *Journal of the Acoustical Society of America, 45*, 1217–1233.

Munhall, K. G. (1985). An examination of intra-articulator relative timing. *Journal of the Acoustical Society of America, 78*, 1548–1553.

Nittrouer, S., Munhall, K., Kelso, J. A. S., Tuller, B., & Harris, K. S. (1988). Patterns of interarticulator phasing and their relation to linguistic structure. *Journal of the Acoustical Society of America, 84*, 1653–1661.

Schmidt, R. A. (1982). *Motor control and learning: a behavioral emphasis*. Champaign, IL: Human Kinetics.

Scholz, J. P., Kelso, J. A. S., & Schöner, G. (1987). Nonequilibrium phase transitions in coordinated biological motion: Critical slowing down and switching time. *Physics Letters A, 123*, 390–394.

Schöner, G., Haken, H., & Kelso, J. A. S. (1986). A stochastic theory of phase transitions in human hand movement. *Biological Cybernetics, 53*, 247–257.

Schöner, G., & Kelso, J. A. S. (1988). Dynamic pattern generation in behavioral and neural systems. *Science, 239*, 1513–1520.

Sternberg, S. This volume.

Stetson, R. H. (1951). *Motor Phonetics: A study of speech movements in action*. Amsterdam: North Holland.

Tuller, B., & Kelso, J. A. S. (1984). The timing of articulatory gestures: Evidence for relational invariants. *Journal of the Acoustical Society of America, 76*, 1030–1036.

Tuller, B., Kelso, J. A. S., & Harris, K. S. (1982). Interarticulator phasing as an index of temporal regularity in speech. *Journal of Experimental Psychology: Human Perception and Performance, 8*, 460–472.

Tuller, B., Kelso, J. A. S., & Harris, K. S. (1983). Converging evidence for the role of relative timing in speech. *Journal of Experimental Psychology: Human Perception and Performance, 9*, 829–833.

Vatikiotis-Bateson, E. (1988). *Linguistic structure and articulatory dynamics: A cross-language study*. Unpublished doctoral dissertation, Indiana University.

Yoshioka, H., Löfqvist, A., & Hirose, H. (1981). Laryngeal adjustments in the production of consonant clusters and geminates in American English. *Journal of the Acoustical Society of America, 70*, 1615–1623.

# 15 Acquisition of Speech Production: Frames, Then Content

Peter F. MacNeilage
Barbara Davis
*The University of Texas at Austin*

## ABSTRACT

Serial ordering errors in adult speech reveal a frame/content mode of representation whereby content elements—consonants and vowels—are placed into syllable structure frames. It is suggested that the initial stages of reduplicative canonical babbling in infants can be characterized as primarily involving production of Pure Frames, in that control signals specific to single segments within the successive syllablelike babbling envelopes are virtually absent. Subsequent variegated babbling, and first words, which closely conform to babbling patterns, can be characterized primarily in terms of a process of frame differentiation by local modifications—a process proceeding toward independent representation of segments, but, for the most part, not closely approaching this result by the age of two.

## INTRODUCTION

The acquisition of motor representation and control for speech has received relatively little attention considering the privileged status of speech as probably the most complex naturally occurring function in living organisms. This is probably partly due to the complexity itself and to the difficulty of making direct measurements, some of which involve invasive techniques, on infants (MacNeilage, 1980). It is also because most interest in early speech production has been about the way sounds form a system of *linguistic contrasts*, a question that involves the *message* level but not the level of *signal* generation. The field has been dominated by the formalistic approach to linguistic theory—by the influential views of Jakobson (1968) and by Generative Phonology (Chomsky & Halle, 1968). (For a brief historical account see Menn, 1980).

Because of the relative lack of attention to the acquisition of speech motor control, there is at present no unified view of how motor control develops (though Menn, 1980, has provided a list of problems for the construction of such a view). In this chapter we suggest a unified view and use it in an attempt to provide a coherent approach to a large body of findings scattered among the literature of different subdisciplines. In brief, we single out a putative property that we consider central to adult speech motor organization and consider evidence as to how this property emerges during development. The property is what has been described as the frame/content mode of organization (MacNeilage, Studdert-Kennedy & Lindblom, 1985). The associated claim is that speech production includes a stage in which content elements—consonants and vowels—are "placed" into frames determined by syllable structure constraints. In the following section we present the frame/content hypothesis and review the evidence for it.

## ADULT SPEECH: THE FRAME/CONTENT HYPOTHESIS

The evidence for the frame/content mode of organization in adult speech comes primarily from serial ordering errors in the speech of normal speakers. The phenomenon on which the frame/content mode is predicated is the position-in-syllable constraint: When speech segments (consonants and vowels) occur in incorrect locations, these locations are almost invariably in the same position in syllable structure that they occupied in the correct word. The existence of this phenomenon is universally acknowledged in the literature on speech errors. (Representative papers in which the phenomenon is discussed are Dell, 1986; MacKay, 1972; Shattuck-Hufnagel, 1979). The constraint is most easily illustrated by spoonerisms—errors in which two units exchange positions with each other in an otherwise correct utterance. The error possibilities are illustrated in Fig. 15.1. using the English word *catnip* (slightly misspelled in the figure to facilitate the illustration). It can be seen that the errors most likely to occur in a word of this type involve units erroneously occurring in the same position in syllable structure that they presumably came out of. The same statement would be true if the syllables were, or were in, different words. Syllable-initial consonants exchange with syllable-initial consonants, vowels exchange with vowels, and final consonants exchange with final consonants. Exchanges of initial and final consonants are rare. Other logically possible exchanges do not occur, even though the result would be a word that had

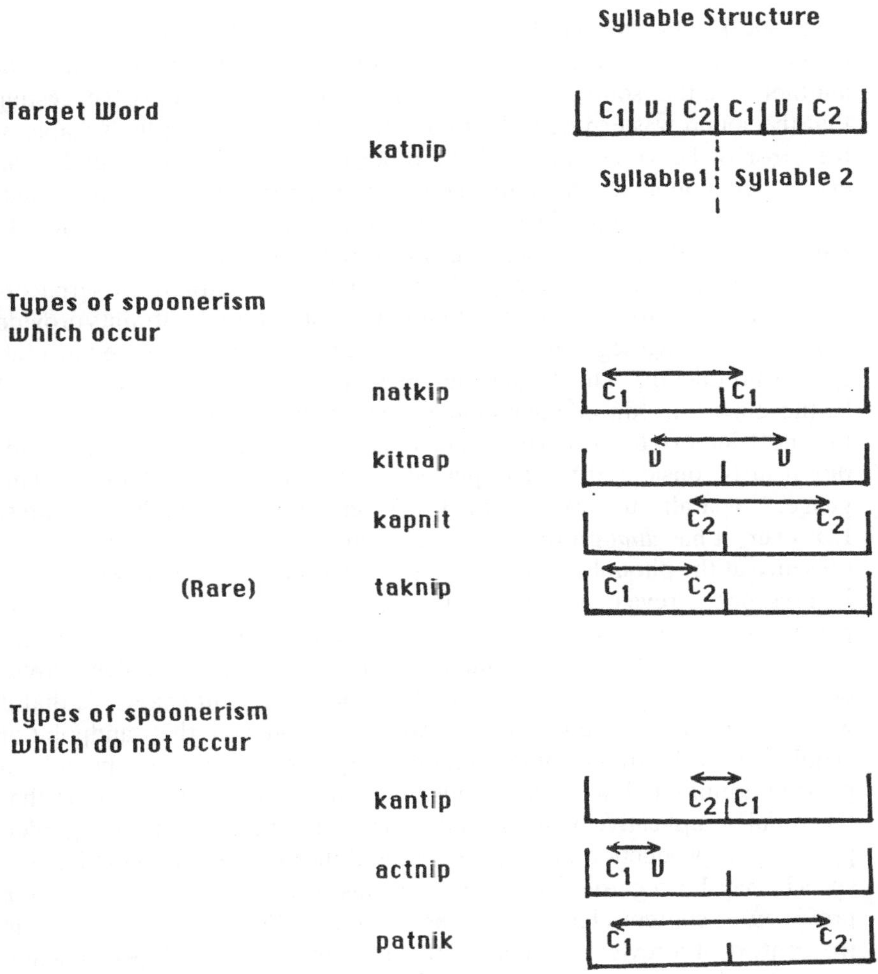


FIG. 15.1. Schematic illustration cf syllable structure characteristics associated with occurring and nonoccurring types of spoonerism.

a sequence of sounds that is permissible in the language. It is particulary notable that two of the illustrated nonoccurring outcomes of exchanges of segments in nonequivalent syllable positions (*kantip* and *actnip*) involve *adjacent* sounds. One might have expected, a priori, that adjacent sounds would be most likely to exchange with each other as they do in typing (MacNeilage, 1985).

Though the position-in-syllable constraint is most easily illustrated in spoonerisms, because we know where the misplaced sounds "came from," it is also applicable to other types of error. When a single sound *shifts* from one location to another, it moves to the same position in syllable structure that it came out of. When a sound is *added* or *substituted* for another, and that sound occurs again in the immediate vicinity (suggesting that the sound was erroneously used twice), the position in the syllable of the error is the same as that of the other occurrence of the sound. And even when a putative source of a substitution error cannot be determined, vowels and consonants do not substitute for each other. See Shattuck-Hufnagel (1979) for examples relevant to these points.

The dominant role of the position-in-syllable constraint on serial ordering errors led to the metaphor that syllable structure constraints provide a *frame* into which segmental *content* elements are normally placed in adult speech production. This frame/content mode of organization is regarded by those who introduced the metaphor (MacNeilage et al, 1985) as perhaps the most important property of the serial organization of speech. In addition, it is considered to have played a key role in the evolution of language. No parallel to it exists in the vocal communication of other primates. However, it has *dual* representation in human language. In addition to its presence at the phonological level, it also exists at the syntactic level where its presence is revealed by errors in which content words are incorrectly placed in syntactic frames (Garrett, 1975). (A detailed exposition of this view is given in a forthcoming monograph, Lindblom, MacNeilage, Studdert-Kennedy, in preparation). The thesis of the present chapter is that if we regard the acquisition of speech motor function from the standpoint of acquisition of the frame/content mode of organization, we may be able to provide a unified view of the acquisition process. In brief, we argue that the initial reduplicative babbling of the 7-month-old-infant may be regarded primarily as production of *pure frames*, and that subsequent babbling and speech may be regarded primarily in terms of a progression toward the provision of content elements for these frames. We develop this thesis in the context of a review of many of the main phenomena of speech acquisition, beginning with vocalization patterns that precede babbling.

## EARLY VOCAL COMMUNICATION

Vocal capacity is, of course, present in a baby at birth. Infant cry involves quasiperiodic vibration of the vocal folds concurrent with the expiratory phase of respiration and with mouth opening achieved by depression of the mandible. The absence of aperiodic waveforms (when the child is not *extremely* disturbed) makes this initial cry phonation similar to that produced in voiced sounds of subsequent speech. The accompanying mandibular depression is analogous to that produced for vowels during speech.

In addition to this distress cry, a baby, in the first 2 months of life, also makes what have been described as "comfort sounds" (Stark, 1980). One of these sounds is cooing. Like crying, cooing involves quasiperiodic vocal fold vibration with the mouth open. However, coos are produced with a much more uniform, fundamental frequency contour than cries, and occur in quiet resting states where the level of affect is more similar to that typical in the subsequent speech stage. Coos have been considered to be protovowels. *Gooing* has also been described (Oller, 1980). This has a similar phonatory component to cooing, and an open mouth, but apparently involves constriction or closure of the vocal tract by means of juxtaposition of the tongue body with the soft palate. Stark (1980) also describes various other consonantal elements: stops, clicks, friction noises, trills, and a few nasal sounds and liquids.

The foregoing comfort state vocalizations are not necessarily produced in company. However, the work of Trevarthen (e.g., Trevarthen & Marwick, 1986) shows that systematic communicative routines, including vocalization, are present in the interaction between mother and infant from birth. In face-to-face mother–infant interactions, infants can produce coos or goos in turn, and also produce what Trevarthen describes as *prespeech* mouth and tongue movements unaccompanied by phonation. These vocal tract events often occur in integrated packages that include hand movements (more often of the right hand) and whole-body movements. They occur in relation to particular patterns of activity in the mother, for example, a smile with eyebrows raised accompanied by a vocalization with a progressive increase in fundamental frequency. Presumably, the motor patterns produced by the infant at this stage are minimally dependent on experience, though Meltzoff and colleagues (Meltzoff, 1986) have shown convincingly that infants, virtually at birth, are capable of both manual and oral imitation. From an early age, infants can control the communicative interchange by turning away at specific junctures, initiating smiling, or making a silly face (clowning).

At about 3 months of age, infants seem to lose interest in the mother in favor of more attention to the environment, and the mother responds with more forceful attention-getting attempts, including teasing, more roughhousing, and the playing of communicative games with predictable outcome and terminal body contact, for example, "This Little Piggy." Intonation and rhythmic form of these games are probably universal, and at this stage adult and child have "protoconversations" involving turn taking within particular rhythmic time frames (Trevarthen & Marwick, 1986).

From 4 to 7 months of age, vocalization goes through an *expansion* stage (Oiler, 1980). The infant gives the impression of purposefully expanding its repertoire. Without obvious regard for a recipient, the infant produces a series of examples of yells, screams, roars, trills (raspberries, or "Bronx cheers") and clicks. During this time, English-speaking adults use fundamental frequency variations to achieve joint attention to objects or events (e.g., "Look at *this*") and to indicae social control categories—rising intonation for approval, low, falling intonation for disapproval (Fernald, 1984).

## CANONICAL BABBLING

At about 7 months, children begin canonical babbling. This is defined as a rhythmic alternation between an open and closed configuration of the vocal tract accompanied by phonation. The opening and closing is achieved primarily by an alternation between an elevation and a depression of the mandible, the elevation resulting in a consonantlike sound, and the depression in a vowel-like sound. A typical utterance would be "babababa." Now for the first time the infant's vocalizations have a truly speechlike quality. The time-locked quality of successive mandibular cycles the impression of a regular sequence of syllables.

A very salient characteristic of canonical babbling is its reduplicative quality. The same consonant and vowel sequence tends to be repeated over and over again. Oller (1980, p. 98) has noted the "lack of substantial variation within utterances in terms of consonantal and vocalic units occurring in a sequence." He points out that "Often infants seem to focus on a particular syllable type to the exclusion of all others during concentrated periods."

Up until recently, in accordance with the highly influential views of Jakobson (1968), it was thought that a child, in babbling, produces all of the sounds of the world's languages without obvious favorites. However, this now seems to be clearly not the case (Locke, 1983). Certain consonant sounds, such as ejectives, implosives, liquids, fricatives and affricates, are extremely rare if they occur at all, and high and back vowels are also rare. Even clicks and trills, which occur alone during the expansion phase, are not produced in alternation with vowels in babbling. Instead, the sound repetoire at the initial stages of babbling is severely limited. Complete closures of the vocal tract occur predominantly at the lips and in the tongue tip blade region giving listeners the impression that voiceless, unaspirated labial or dental-alveolar stop consonants are being produced (sounds such as *b* and *d* in the words *ball* and *doll* in English—phonetically [*p*] and [*t*]). Less frequently, the nasal consonants made with the same vocal tract closures as *b* and *d* (*m* and *n* are observed. A less than total closure of the vocal tract often gives rise to the impression that semivowels such as *w* and *y* as in *we* and *you* in English are being produced. The open phase of the vocal tract most often gives the impression that mid- to low-vowels and front to central vowels are being produced—vowels such as *e* [ɛ] in *bet*, *a* [æ] in *bat*, *u* [ʌ] in *but*, *a* [a] in *father* and *e* [ə] in *the* when it is produced as an unstressed syllable (Kent & Murray, 1982). Consonant like sounds, often in alternation with vowel sounds, are also produced by changes in vocal fold configuration. Closure of the folds gives rise to a glottal stop, and abduction of the folds to an *h*-like sound. Although children have so far been studied in only a limited number of language communities (Locke, 1983, summarizes results for consonants from 15 language communities), it is presently considered that the patterns described are universal. No significant biases in preference toward the sounds of the child's specific language have been described in initial babbling.

Although the hierarchy of sound preferences in initial babbling may not be biased toward those of the ambient language, the presence of that language is necessary for normal babbling to be initiated on schedule. Oller, Eilers, Bull, and Carney (1985) have shown that a deaf child did not begin babbling until several months after the normal onset time, and even then the sounds produced were in some respects atypical. It is perhaps not surprising that auditory exposure to adult language is important in vocal development in the light of the sustained and intense nature of communication between the infant and the adult, alluded to earlier, in the first few months of life. However, if the specific properties of the adult target language are not influential in this development, one might expect the earliest sounds to be universal in languages. With respect to sounds associated with the closing phase, the following facts can be noted (based on Maddieson's 1984 survey of 317 languages). All languages have at least one series of stop consonants, and closures in the labial and dento-alveolar regions are highly favored by languages as well as babblers. In addition, almost all languages (7%) have at least one nasal consonant. Semivowels (called vocoid approximants by Maddieson) are the least favored in languages of all the consonantal sounds produced by articulatory manuevers in babbling, although even here, more than 90% of the languages sampled by Maddieson have at least one semivowel.

In contrast to the almost universal use of the vocal tract constrictions made in babbling for consonant production in languages, the vocal *fold* maneuvers used to produce consonant-like sounds in babbling are a good deal less than universal in languages. Less than 66% of languages use *h* [h], and only 45% of languages have the glottal stop [ʔ].

The common factor underlying the entire initial babbling repertoire may be the induction of what could be called "dummy speech." The child may have developed motor capabilities sufficient to "capture" the fact that speech tends to alternate between louder sounds (the vowels) and sounds that are less loud (the consonants). A useful analogy may be made with the child who captures the fact that reading aloud involves holding an open book in front of one and talking. It is basically correct from the standpoint of the appearance of the overall performance, but it does not yet include the appropriate (in this case visually based) control of the fine structure of the spoken level of language. However, in the course of developing the capacity for simulating the alternation of sound levels that is characteristic of speech, the child develops *one* behavior that is a universal motor base for subsequent speech—the open–close alternation of the mandible. In contrast, the *glottal* alternations, though contributory to the initial simulation of the sound-level alternations, turn out to be an ontogenetic blind alley, just as doing without visual input does, in learning to read.

## CANONICAL BABBLING AS PURE FRAMES

It is a central claim of this chapter that the initial phases of canonical babbling, at least those that involve articulatory alternations, can be regarded primarily as the production of *pure frames*. The dominant articulatory event is the oscillation of the mandible between a closed (elevated) and open (depressed) position. The reduplicative quality of babbling would then be produced primarily by successive *cycles* of the same oscillatory event. These would not be frames of the kind postulated earlier on the basis of adult speech errors. The adult error patterns suggested that syllable frames had a *premotor* representation that constrained the positioning of individual segments, which also had a premotor representation. Instead, the infant's cycles of mandibular oscillation may be dynamic prototypes from which a structurally homomorphic premotor representational level of syllable frames eventually develops in adults. The infant's frames may be considered pure in the sense that they do not include constraints on segmental positioning. In fact, there may be little analogy with segmental content units (consonants and vowels) in adult speech that not only have discrete premotor representations but also tend to be accompanied by motor events specific to them. (See MacNeilage & DeClerk, 1969, for data on this point. Their electromyographic study shows that a number of muscles can be active for a particular single segment in production of a syllable.) Contrarily, motor events confined to only a single apparent vowel and consonant (in addition to those controlling the mandible) may be very rare in initial babbling.

How can this claim be reconciled with the fact that even though the variety of consonant and vowel sounds produced in early babbling is not great, there is *some* variety? We suggest three factors may be responsible for this. First, differences in vowel height (mid versus low) in different utterances, may simply be due to differences in the amount of depression of the mandible without specific positioning of the tongue for each individual open phase. This quantitative dimension of mandibular action may be free to vary to a certain extent from utterance to utterance. Lindblom and Sundberg (1969) have shown, using an articulatory model of the vocal tract, that a range of front vowels varying in height can be produced simply by variations in mandibular depression with the tongue position fixed in relation to the mandible. Second, babbling episodes with labial stop consonants [p] or [m] may be produced without any specific positioning of the lips themselves. The fact that the lower lip "rides" on the mandible may be sufficient to produce lip closure when the mandible is elevated. Third, for dental-alveolar consonants and variations in vowels along the front-central dimension, the tongue may be set in a configuration that eventually produces these sounds, *before* mandibular oscillation begins, and consequently there may be no activity specific to the tongue for the individual consonantlike or vowel-like sounds that are heard during the utterance. Another possibility for vowels that differ in the front–back dimension is that there may simply be a range of naturally occurring, tongue-resting positions along the front–back dimension of the vocal tract, and the front–back position of a vowel in babbling might simply be determined by what the preceeding rest position was. In the case where later sounds differ from earlier ones within a single utterance (except for vowel height variations that might be ascribed to variations in the amplitude of the mandibular cycle), one must postulate the activity of some articulator other than the mandible for the production of the later sounds. Such utterances would therefore depart from the pure frame form of organization.

The pure frame hypothesis is testable. Frontal videotapes of infants would show the extent to which vowel height, determined acoustically by the frequency of the first format, is correlated with mandibular depression. There is no a priori reason why this correlation should be very high (as the pure frame hypothesis would predict) because the tongue and the mandible can, in principle, exercise considerable independence from each other, as Lindblom (1983) has pointed out. Videotapes would also show whether there is any active change in lip configuration for the production of labial consonants. The hypothesis would predict that utterances with nasal consonants would show acoustic signs of heavy nasalization of the intervening vowels, because separate palatal elevation gestures for each vowel (to make them nonnasal, like adult vowels) would not be predicted. It might also be expected that if tongue front elevation for dento-alveolar consonants is present throughout utterances with such consonants, vowels in such utterances might, in general, be higher and more front than in utterances without such consonant-related positioning. Such a trend has been predicted by Kent and Murray (1982) for speech acquisition and observed in early words by Stoel-Gammon (1983) and by us (Davis & MacNeilage, 1990) as we discuss later, but it has not been investigated in early babbling.

## VARIEGATED BABBLING

There is agreement in the literature that initial babbling, which, as we have noted, is primarily reduplicative, is followed at about 10 months by a stage called variegated babbling, which lasts beyond the occurrence of the first words. In this stage, according to Oller (1980, p. 99) "the restriction against varied consonantal and vocalic types within a multisyllabic string seems to be at least partially overcome." In fact, Oller claims that "infants systematically produce utterances with differing consonantal and vocalic elements." (In contrast, there appears to be little progress toward mastering new sounds during the period.)

From the perspective of the hypothesis that babbling begins with pure frames, the variegated babbling stage must be regarded as a stage of internal modification of frame structures. In terms of the frame/content hypothesis, applied to adult speech, variegated babbling may be considered to be the first step in the development of content elements. To the extent that presetting operations are responsible for particular consonantlike and vowel-like properties in reduplicated multisyllabic babbling, the occurrence of these qualities as *changes* in consonantlike or vowel-like entities *within* nonreduplicative utterances must mean that the operations are now being applied more locally. In other words, the operations may be an initial step in moving from a purely cyclical utterance level *toward* the segmental level. The child seems to be primarily involved in a process of *differentiation* of the superordinate structure provided by mandibular oscillations. This process can be contrasted with a process of putting hitherto separate entities together—a process of integration. Most early babblers can readily produce instances of the monosyllables *ba* and *da*. If speech acquisition were to occur by integration, there might then be a stage in which the child would produce *ba* and *da* close together in time but with a silence between them. Finally, they would be combined in a continuous utterance. This sequence of events has not been reported, to our knowledge, which tells us that although the syllable, usually the consonant-vowel syllable, can occur alone, entire individual syllables are probably not separate functional units in multisyllable utterances that have different sounds in successive syllables.

Thus, our conception of the variegated babbling stage is of an infant able to vary the number of cycles of a mandibular oscillation in an utterance, and involved in the task of making local modulations in this reduplicated oscillatory envelope. The rhythmic constancy of individual syllables in multisyllable utterances during the variegated babbling stage is evidence for the continuing primacy of the oscillatory cycle. Describing the individual utterances of a group of five 13-month-old-babblers, Kent and Bauer (1985) stated that "once a rhythm was established, it usually was maintained during the syllable train." They maintain that "—the actual rhythm or period of these repetitive utterances is variable across utterances but tends to be held constant within an utterance, as though a preset periodicity is followed throughout a syllable train—" (p. 509). This uniform rhythm was observed in utterances in which the internal phonetic content was variegated as well as in reduplicated babbling sequences. The authors present an example of an 8-syllable sequence beginning with 4 consonant-vowel syllables and terminating with 4 consonant-vowel-consonant syllables. They conclude that, "Despite the change in syllable shape the basic periodic structure of the utterance was preserved" (p. 508).

Another finding of Kent and Bauer contributes further to the picture of an interactive relation between two distinct subprocesses: a basic, uniform, oscillatory sequence, and a process of local modifications of these frames by internal adjustments. The term *harmony* is traditionally used in child phonology to denote cases in which the same vowel or consonant is used throughout a multisyllabic utterance. Kent and Bauer observed a direct relation between the tendency to produce disyllabic utterances and the relative frequency of consonant and/or vowel harmony. The relation was statistically significant for consonant harmony (rho = .98 $p < .05$). This result suggests that the production of variegation (nonharmonic sequences) has a cost in terms of the number of syllables that can be produced in a single utterance. The cost of variegation is further indicated by the overall frequencies with which vowel and consonant harmony were being produced in the children studied by Kent and Bauer—children who had presumably been babbling for an average of 6 months. About 30% of their two-syllable utterances had both consonant and vowel harmony (were reduplicative) and an additional 44% had either consonant harmony or vowel harmony. Thus, after 6 months of babbling, about 75% of the disyllabic utterances of these five children had harmony in at least one of the two major types of sound.

Of course the achievement of variegation in babbling moves vocal output patterns in the direction of production of the adult language, in which vowel or consonant harmony in multisyllabic words is typically rare. In terms of the specificity with which babbling patterns approach those of the target language, the current belief has been that infants show no approaches to language-specific aspects of sounds or sound patterns of the ambient language during the babbling period. In contrast to this view, de Boysson Bardies and her colleagues have recently presented evidence that by 10 months of age both consonants (de Boysson Bardies, Sagart, Halle, & Durand, 1986) and vowels (de Boysson Bardies, Halle, Sagart, & Durand, 1989) show a bias toward the ambient adult language.

## THE FIRST WORDS

At about 1 year of age, an infant usually begins to produce its first single words. The resulting stage, which typically lasts at least 6 months, has been designated, from a phonological perspective, the stage of the first 50 words (Ferguson & Farwell, 1975). In addition to being wrong about the versatility of babbling, Jakobson has also turned out to be wrong in his influential claim that there is no relation between the sound preferences of babbling and the sounds of the first words. It was Jakobson's belief that the choice of sounds in the first words was determined by an innate universal mechanism maximizing the perceptual distances between sounds that applies in the same way to each child. But recent work, particularly by Locke, (1983) and by Vihman, Macken, Miller, Simmons, and Miller, (1985) has shown that rather than being unrelated to babbling preferences, the sounds of the first words are for the most part exactly what would be predicted from babbling preferences. Most first words have the form CV or CVCV, consonants are composed primarily of labial and dentoalveolar stops with some nasals, and favored vowels range from mid to low and front to central. Thus, it might be expected that any control capabilities that had been developed for achieving variegation in babbling would be used in production of the first words. According to the hypothesis presented here, the main control process that is being developed in the first words is a continuation of the local modification of basic mandibular oscillatory frame structures that began during the variegated babbling period.

This view is contradictory to a hypothesis originating in the work of Ferguson and Farwell (1975) that the *whole-word* is a unit in early speech. Menn (1980) has pointed out that the whole-word hypothesis initially arose in opposition to the views of Jakobson and others that sublexical units developed in early words as units *in contrast* with other units. Ferguson (1986) reflects this particular opposition in a recent summary of the whole-word hypothesis in which he stated that, "the unit of *contrast* [underlining ours] in production—is predominantly the whole word rather than the syllable or the segment" (p. 37). However, later in the paper, Ferguson omits the reference to the issue of contrast and concludes as follows: "In summary, the basic phonological production unit in the period of early phonology is the whole word shape. The phenomena that point to the existence of smaller (and larger) phonological units than the word are either incipient or nonexistent at this stage" (p. 52).

The basic problem with the whole-word hypothesis has been stated by Menn (1980). Hypotheses about contrast do not necessarily translate in a straightforward way into hypotheses about the motor control of production. This can be shown in this particular case by considering the implications for motor control of the three main lines of evidence that Ferguson cites for the whole-word hypothesis (Ferguson, 1986, pp. 39–40): "(a) Variable stability. The consonantal onset may be stable in one word and variable in another even though the two words seem comparable in articulatory difficulty or begin with the same consonant in the adult source word." The implication is that these onset differences occur because the onsets belong to different whole-words. Later in this chapter we emphasize that there has been virtually no consideration of the role of vowels in early speech and present evidence that different vowels can differentially affect the production of particular consonants. These considerations lead to an alternative *sub*lexical account of phenomena of this type in terms of coarticulatory problems in combining consonants with vowels.

"(b). Segmental mismatching. Consonantal onsets of the child's words may show splits and combinations with respect to the adult initial consonants. For example, K had an "m" phone class of words starting with *m* but occasionally with *n* in adult words, a "b" class of words starting with *b* or occasionally with *m* in the adult models, and a small "n" class of words beginning with adult *n*." A similar explanation of the preceding one can be suggested for this second class of phenomena, namely, that vowels can influence consonants. For example, Stoel-Gammon (1983) has compiled, from the literature, numerous instances of alveolar consonants (of which *n* is an example) substituting for labial consonants in the environment of high front vowels.

"(c) Prosodic variability. A child's word shows such variation in articulatory features and their position in the word that the child seems to have located such features as somewhere in the word rather than in particular segments in a particular order." Although this phenomenon shows that the word may have holistic status in that it is a semantic unit that, in some sense "dominates" a particular family of motor attributes, it also shows that from the motor control point of view these sublexical units may be independently controllable. Insofar as this is true, the whole-word may be a unit of lexical access but not a unit of motor control.

We conclude from this analysis that evidence cited by Ferguson for the whole-word production hypothesis may be, for the most part, reinterpreted in motor control terms as involving processes of local modification of syllable frame structures. Elsewhere, Ferguson himself emphasizes the importance of sublexical, though suprasegmental factors (for us, frame variables) by noting that partial reduplication is "very widespread in early words," and states that it "can be seen as an acquisitional strategy that assists in several aspects of phonological development" (Ferguson, 1986, p. 42). Among the authors Ferguson cites in favor of this conclusion are Fee and Ingram (1982, p. 91) who state that reduplication is "a general pattern during the earliest stages of phonological development." In addition, Shibamoto and Olmstead (1978) have noted evidence for reduplication in the data of Ferguson and Farwell (1975) and have argued for the existence of phonological (sublexical) processes in those data. We believe there is insufficient evidence that the word is a *production* unit in early speech acquisition in the sense that there is a single coherent motor package associated with it. If variegated babbling primarily involves local modification of frame structures, and if the first words are constructed from the babbling repertoire, then the first words also involve local modification of frame structures.

## BEYOND THE FIRST FIFTY WORDS

Just as there has been some consensus that production of the first 50 words is holistic, there is also a consensus that beyond this stage the infant begins to achieve independent control of speech segments. Much of the impression that this is so probably comes from the fact that in a child's words at this time some new sounds (which *appear* to develop independently) become more prominent. But perhaps more importantly, children typically "substitute" sounds for others that they cannot produce, giving the impression that the substituted sounds at least are under independent control. (e.g., when *yip* is produced for *lip*, one gets the impression that *y* is under independent control). However, as intimated earlier, from a motor control standpoint no satisfactory evidence has been provided that the sounds substituted (usually consonants) are controlled separately from the vowels that they usually abut. In fact, a great difficulty in evaluating the motor control of early words in general results from the fact that hardly anything is known about the acquisition of vowels. There is not a single, comprehensive case study on the course of acquisition of correct vowel production in various consonantal contexts in the literature. Almost all work on the phonology of early words has been devoted to consonants. Most concern has been with the development of consonants in the three main positions in words—initial, medial and final—in relation to the requirements imposed by adult words. This focus tends to give an *artefactual* impression that consonants are produced independently, because the problem of how consonants are coarticulated with the vowels that they invariably abut is finessed.

Part of the reason for the neglect of vowels is that they are usually considered easy to produce. However, this connection has never been documented in quantitative terms, though it needs to be, because the possibility that vowels are easy to acquire whereas consonants are difficult is of considerable theoretical interest. It is also of interest that, judging by the results of babbling studies reviewed earlier, some vowels may be easier than others. A small-scale spectral study by Bickley (1983) has shown that it takes a number of months for acoustic patterns for high vowels (which are not favored in early babbling) to cluster in distinct spectral regions. This development presumably allows them to eventually be perceived correctly, although that particular question was not examined in the study. There is also the possibility that vowels may be easier to produce in some positions in words than others. For example, in a CVCV word, the vowel between two consonants might be more difficult than the terminal vowel. One question that arises, in English at least, for vowels but not for consonants is how the correct use of vowels to signal the distinction between stressed and unstressed syllables is acquired. Finally, it might be asked whether there are any mutual coarticulatory constraints between consonants and vowels as suggested by the study of Stoel-Gammon (1983) cited earlier. She found evidence of a special articulatory compatibility between high front vowels and alveolar consonants.

Information about these questions regarding vowels in early words is necessary to evaluate the hypothesis that early words, like variegated babbling, can be viewed as primarily a process of internal modification of frame structures. Consequently, we have performed a comprehensive case study of the acquisition of correct vowel production in an English-speaking child (Davis & MacNeilage, 1990). We now summarize aspects of the study that bear on the questions that have just been raised.

## ACQUISITION OF CORRECT VOWEL PRODUCTION: A CASE STUDY

A female English-speaking child was studied from the age of 14 months to 20 months during which time the child's vocabulary grew from about 25 words to over 700 words. Approximately 40 hours of tape recordings were made in the subject's natural home environment, spaced at once- or twice-weekly intervals during the period. Phonetic transcriptions were made of over 25,000 speech and babbling utterances. A computer program was used for tabulation of data and extraction of statistical regularities. Analysis was restricted to CV and CVCV sequences in monosyllables and disyllables in which consonants had labial, alveolar, or velar places of articulation.

Vowels were *not* found to be particularly easy to produce by this child. Of a set of over 11,000 vowels in words of the forms just mentioned, only about 66% were judged by the transcriber to be correctly produced, and this percentage stayed approximately constant for the course of the study. This low percentage was particularly notable because this child's vocabulary increase over the time period studied was exceptional. A typical vocabulary for a 20-month-old-child would be 1–300 words. However, although this child's vocabulary development was exceptional, analysis of 20 other English-speaking children suggests that the figure of 66% correct vowel production in this period may be typical (Paschall, 1983).

Position of the vowel in the word was found to be a major influence on its correctness. Rank correlation coefficients between correctness levels of the set of 13 vowels studied, in the three word positions studied, (C*V*, C*V*CV, and CVC*V*) were not statistically significant. It might seem particularly surprising that the correctness of vowels in C*V* and CVC*V* positions was not significantly correlated because they share a word-final post-consonantal context. However, most of the two-syllable words produced by the subject required the second syllable to be unstressed. And overall correctness levels of vowels in this position were significantly lower than in the other two contexts. Many of the errors in the unstressed syllables took the form of the production of the neutral vowel [ə] instead of other vowels. Thus, in this subject at least, vowels were not sufficiently well mastered to be uniformly correctly produced in all environments. They exhibited very little context independence. Furthermore, the need to control vowel quality in relation to stress was found to be a major problem.

Evidence of coarticulatory constraints between alveolar consonants and high front vowels were found, consistent with the findings summarized by Stoel-Gammon (1983). The subject favored a co-occurrence of alveolar consonants and high front vowels. An analysis of two syllable words in the *Daniel Jones Pronouncing Dictionary of the English Language* (Jones, 1917) shows that this co-occurrence constraint is also present in the language. An interesting question that could be answered by cross-language research is whether the constraint is present in the child because it is present in the particular adult language, or whether it is universally present in both children and languages because it results from a natural propensity. It should be remembered that such a co-occurrence constraint was suggested earlier in the context of the hypothesis regarding pure frames as a possible concomitant of a presetting of tongue front elevation, preparatory to the production of early babbling episodes with alveolar consonants. The results of the case study suggest that this question can be placed in the broader perspective of the role of language specific versus language universal constraints on speech acquisition.

Two additional findings are directly relevant to the hypothesis that speech acquisition involves local modifications in frame structures. First, it was shown that even at the age of 20 months, the subject's two-syllable words still showed a good deal more consonant and vowel reduplication than is called for in the adult language. Consonants with the same place of articulation (labial, alveolar, or velar) were reduplicated in 67% of the subject's words, whereas 12% of the subject's two-vowel sequences were reduplicative. An estimate based on the two syllable words in the *Daniel Jones Pronouncing Dictionary* is that 18% of consonant sequences reduplicate place of articulation and only 3% of vowel sequences are reduplicative. This suggests that strong traces of the basic reduplicative frames, presumably first used in this child's initial babbling, are still present at 20 months of age.

Another finding that bears directly on the hypothesis of local modification of frames was an inverse relation between vowel reduplication and consonant reduplication within utterances. It was found that more consonant reduplication occurred in disyllables in which the vowels were not reduplicated (68.3%) than in disyllables in which they were (58.0%). This, of course, also means that more vowel reduplication occurred in disyllables in which the consonants were not reduplicated (13.2%) than in disyllables in which they were (8.8%). Menn (1978) has also analyzed sound patterns of early words in terms of a limited capacity for modification of basic forms, though with more emphasis on consonants alone. The present result suggests that the constraint on modification may apply to vowels as well as consonants, and that the constraints on vowels and consonants are sensitive to each other.

## PROCESSES OF LOCAL MODIFICATION

In the following paragraphs we present some suggestions as to how the local modifications, which we believe to be central to variegated babbling and the first words, might occur. Some evidence is available for some of these suggestions, but others arise more indirectly from considerations of anatomy and the articulatory demands of later developing sounds and sound patterns relative to earlier ones from which they may be derived.

One well-documented example of speech sound acquisition that could be described as local modification comes from a study by Macken and Barton (1980) of voicing of word-initial stop consonants in English. They showed that voiceless aspirated stop consonants, which involve voicing onset 50–70 msec after release of the stop, develop from voiceless unaspirated stops, which involve voicing beginning almost immediately after stop release by a gradual process of increasing the time lag between stop release and voice onset. In motor control terms, this requires delaying the onset of vocal fold adduction relative to the onset time that seems to occur naturally in initial babbling. This development could be described as a modification in *relative timing*.

We have noted that vowel acquisition involves expansion of use of the vowel space, particularly in the direction of high vowels. This may involve different processes for high front and high back vowels. Insofar as the front–back dimension of the tongue is under active control in initial babbling, then the fronting movement may be under the control of the genioglossus, which runs from the tongue body to the anterior mandible. Because of the incompressibility of the tongue, it also becomes higher as it moved frontward. Consequently, in order to produce high front vowels such as [I] and [i], we suggest, *force increase* applied to contraction of the genioglossus is necessary. In this case, more contraction than typically accompanies the front vowels of early babbling is necessary to produce high front vowels.

In contrast to the effect of fronting of the tongue, a backing movement, other things being equal, should lower it. Thus, production of *high* back vowels presumably requires a *new movement.* Anatomical considerations suggest that this new movement is achieved by the cooperation of two muscles inserting on the styloid process of the cranium, the styloglossus and stylohyoid muscles. (In addition, it is interesting to note that a study of adults by MacNeilage and Scholes, 1964, suggests that genioglossus muscle activity is also necessary to keep the base of the tongue in a forward position for high *back* vowels.) Another new movement, that of lip rounding, is also necessary for high back vowels. It is also of interest to note that high back vowels, which, according to our analysis, make more demand on *new* movements, are acquired later than high front vowels.

Of course, in most languages including English, a large number of sounds not favored in babbling must be added to the infant's repertoire, and complexity of syllable organization must increase beyond the CV stage by acquisition of word-final consonants, and consonant clusters. We suggest that at least some of these new sounds and combinations might be made by *Temporospatial Variation* in sounds already available. One notable fact about the subject we studied was that from one week to the next, a word that was pronounced [ʌbi] became pronounced [ʌbis] (for *open*). We suggest the possibility that the movement for making the high front vowel [i] was temporally and spatially extended until the opening of the vocal tract in the alveolar region became small enough to create the turbulent airflow heard as the fricative [s]. A prediction from this suggestion is that later occurring consonants, most of which require articulation of the tongue with the roof of the mouth, might develop first in the context of high vowels.

A final suggestion about the form of local modification related to individual consonants and vowels concerns the sounds present in initial babbling. Insofar as any of these sounds resulted from a presetting of the articulatory apparatus that was then maintained throughout a multisyllabic utterance, the presetting gestures must become subject to *Contextual Restriction* in order to confine them only to the segments for which they provide a major contribution to the acoustic substance of communication. For example, the elevated tongue front for the alveolar consonant would no longer also be maintained during surrounding vowels. The decrease in amount of coarticulation, as Nittrouer, Studdert-Kennedy, and McGowan (1988, p. 89) have noted in older children, might in part be related to this type of development.

## SUMMARY AND IMPLICATIONS

We have presented a hypothesis about the general form of speech acquisition, namely that initial babbling is composed primarily of pure frames, and that the subsequent development of babbling and early speech is primarily a matter of frame differentiation by means of local modifications. The hypothesis can perhaps be tested by determining from video and audio recordings how much of the child's initial babbling can be accounted for by a process of mandibular oscillation plus a small number of presetting conditions, and one of which can determine acoustic properties of consonantlike or vowel-like sounds that appear repeatedly in each cycle of oscillation occurring in reduplicative babbling.

Evidence has been reviewed that favors a second stage of frame differentiation by local modification rather than a stage of integration of individual syllables that can already occur in isolation. Gradual development of aspirated stop consonants from unaspirated stop consonants and gradual expansion of the vowel space were given as examples of local modification and five processes were suggested for local modification based on the spatial and temporal relations between existing events and new developments. That such developments are position-specific was already clear for consonants, but the case study we reported showed that it is also clear for vowels.

One thing that repeatedly emerged from our review is the lack of data presently available to test as general a hypothesis as the one we have presented. More detailed studies of the complete repertoire of individual children and particularly of the organization of that repertoire from the onset of initial babbling to acquisition of the entire adult sound system are necessary to test the present hypothesis or any other general view. In word production, the problem is particularly acute for vowels. Although there have been several intensive studies on the acquisition of consonant systems in individual children, the scarcity of information on vowels makes it impossible to derive conclusions regarding the organization of entire sound systems from these studies. We have made an initial step toward remedying this problem with an intensive case study of vowel acquisition in a single child.

The degree to which pure frames are playing a role in the development of speech at any time can be estimated quantitatively by the percentage of reduplication of consonants, vowels, or syllables in the infant's repertoire. The high frequency with which consonant harmony (e.g., *goggy* for *doggy*) has been reported as a "strategy" for word acquisition for particular subsets of words is additional evidence for the continuing importance of reduplication in early words. (Fee & Ingram, 1982). Exactly how to parcel out frame and content when a typical child is in the course of acquiring an increasingly diverse total repertoire is not clear at the moment. However, it can be addressed with some success in a number of children who seem to create *special purpose, nonreduplicative frames* as part of their attempt to produce an increasing repertoire. For example, Macken (1978) reports a Spanish child who favored the sequence of word-initial labial consonant and then alveolar consonant in words so strongly that a number of very differently formed adult words were subject to this constraint. Examples:

*sopa* [*p*o*t*a]

*telefono* [*f*a*nt*ino]

*zapato* [*p*a*t*o]

(The labial and alveolar dental consonants in the output are underlined.) Priestly (1977) reports a child who favored a CVyVC sequence. ([Y] is the semivowel or glide in words such as *yet* and *you*.) Examples:

*sucker* [fayak]

*hanger* [hayaŋ]

*elephant* [eyat]

*lizard* [ziyan]

These results provide one important piece of evidence for the view being advocated in this chapter. Parts of the restrictions in these forms are abstract in the sense that the *particular* vowel or consonant produced is not uniquely specified by the control program. This suggests, in accordance with the concept of frames, that the alternation between vowels and consonants has some representational independence from the phonetic details of the utterance.

Why have we not used serial ordering errors as clues to the development of the frame/content mode of organization just as they are used as clues to adult organization? Informal observation suggests that such errors are relatively rare in young children, which probably in itself means that the segmental organization of the adult has not yet been acquired. In addition, young children make many errors because of their inability to make correct sounds, and the variability that often accompanies unsuccessful attempts makes it difficult to unequivocally establish instances of serial ordering errors. Corpora of serial ordering errors in infants might be very valuable, but collection would need to be done by a person with intensive exposure to the particular child's developing system, without which interpretation of errors would not be optimal. Even under these conditions, interpretation necessarily includes a subjective component.

At this point we can attempt to explicitly address the following question: To what degree do infants at the age at which our case study terminated (20 months) have the frame/content mode of organization that adults reveal by their serial organization errors? We have described developments in variegated babbling and early speech in terms of local modification but with relatively little evidence as to *exactly* what these modifications are. Consequently, we are not in a good position to say exactly how closely these local events relate to the segments of adult speech. In this regard, it should be noted that adult speech errors usually involve *entire* segments, not just features or other subsegmental components (Shattuck-Hufnagel & Klatt, 1979). We know of no evidence that entire segments are separately represented in the speech of 20-month-old-children, despite the apparent consensus in the literature that such a capability exists. The evidence is much more consistent with the claim that local modification involves subsegmental components. Neither do we know, in the absence of serial ordering errors, whether local modifications can be accidentally displaced to a corresponding position in another syllable, as *segments* can be in adult speech. Similarly, it should be acknowledged that the pure frames we postulate for initial babbling are not the same as the frames hypothesized for adults on the basis of the serial ordering error data. In the former case, the frames are inferred directly from the rhythmic oscillatory movements, and the exact sense in which these movements themselves might have an underlying representation is not at all clear. In the latter case, the frames are clearly premotor representations, because the movements necessary to interface misplaced segments with their neighbors must be programmed after the misplacement occurs in order for the erroneous productions to be as fluent as they typically seem to be. However, we assume that the frames that appear to underlie the position-in-syllable constraint on adult serial ordering errors might be related to the pure frame of infants, in that the adult frames might develop from the mandibular cycle, which is so dominant in initial babbling.

There is one question that is very interesting whether or not our speech acquisition hypothesis is valid: How does speech develop to the point where it exhibits the particular set of error contingencies that are observed in adults? In adults, serial ordering errors are most likely to involve syllable-initial consonants (and to a lesser degree vowels) in stressed syllables, and their frequency seems for the most part to be a function of the frequency of the segment type in the language. However, in our conception of initial babbling we have regarded the initial-consonant and its following vowel as a cohesive unit (the cycle) and have noted that the true consonants most likely to be involved in initial babbling (voiceless, unaspirated stops and nasals) tend to be language universal—and therefore probably frequently used in a given language. The question that arises is: How does speech develop from a stage of supposedly cohesive CV syllables to a stage where the C and the V are separately represented in stressed syllables?

Finally, how should the present approach be characterized in the context of other approaches to the understanding of speech acquisition? We have placed a very strong emphasis on differentiation rather than integration as the main developmental process in speech acquisition. (For a background on these two major concepts relative to the development of motor function see Fentress, 1984). The implication of this emphasis is that we should concentrate our attention on small changes over time in production capabilities in particular segmental contexts. This contrasts with the more common view of speech acquisition at the stage beyond 50 words, which has focused on consonant segments with supposedly independent status being placed in various positions in words (initial, medial, final) without regard to the particular segmental contexts provided by adjacent vowels. In terms of the particular theoretical framework, we have offered two things that are new:

1. An explicit designation of a target for the acquisition of speech motor function in the form of a central property of adult speech motor function—the frame/content mode of organization.
2. The postulation of pure frames in reduplicative babbling as the dynamic antecedent of the adult frames, and of the local modification of these frames in variegated babbling and beyond as the first steps toward the formation of the segmental content elements of adult speech.

The approach can be regarded as the beginnings of a response to Menn's (1980, p. 30) challenge for the field of Developmental Phonology to come up with "a general adequate account—of how children end up with phonemes" that would serve as a "replacement for Jakobson" (p. 30).

## ACKNOWLEDGMENTS

We would like to thank Marlys Macken, Michael Studdert-Kennedy, and Betty Tuller for their comments on the manuscript. The analyses of sound frequencies from the *Daniel Jones Pronouncing Dictionary* were made at Haskins Laboratories with the support of NICHD Contract No. 1-HD-5-2910.

## REFERENCES


Bickley, C. (1983). Acoustic evidence for phonological development of vowels in young children. Presented at *Tenth International Congress of Phonetic Sciences*. Utrecht, Holland.

Chomsky, N., & Halle, M. (1968). *The Sound Pattern of English*. New York: Harper & Row.

Davis, B., & MacNeilage, P. F. (1990). Acquisition of correct vowel production: A quantitative case study. *Journal of Speech and Hearing Research*.

de Boysson-Bardies, B., Halle, P., Sagart, L., & Durand, C. (1989). A cross-linguistic investigation of vowel formants in babbling. *Journal of Child Language*. (In press).

de Boysson-Bardies, B., Sagart, L., Halle, P., & Durand, C. (1986). Acoustic investigations of cross-linguistic variability in babbling. In B. Lindblom & R. Zetterstrom (Eds.), *Precursors of early speech* (pp. 113–126). New York: Stockton Press.

Dell, G. (1986). A spreading activation theory of retrieval in sentence production. *Psychological Review, 93*, 283–321.

Fee, M., & Ingram, D. (1982). Reduplication as a strategy of phonological development. *Journal of Child Language, 9*, 41–54.

Fentress, J. C. (1984). The Development of coordination. *Journal of Motor Behavior, 16*, 99–136.

Ferguson, C. A., & Farwell, C. B. (1975). Words and sounds in early language acquisition: English initial consonants in the first fifty words. *Language, 51*, 419–439.

Ferguson, C. A. (1986). Discovering sound units and constructing sound systems: It's child's play. In J. S. Perkell & D. H. Klatt (Eds.), *Invariance and variability in speech processes* (pp. 36–53). Hillsdale, NJ: Lawrence Erlbaum Associates.

Fernald, A. (1984). The perceptual and affective salience of mother's speech to infants. In L. Feagans, C. Garvey, R. Golinkoff, M. Greenberg, C. Harding, & J. Bohannon (Eds.), *The origins and growth of communication* (pp. 5–29). Norwood, NJ: Ablex.

Garrett, M. (1975). The analysis of sentence production. In G. Bower (Ed.), *The psychology of learning and motivation, Vol. 9.* (pp. 133–175). New York: Academic Press.

Jakobson, R. (1968). *Child language, aphasia, and phonological universals*. The Hague: Mouton.

Jones, D. (1917). *An English Pronouncing Dictionary*. London: Dent.

Kent, R. D., & Bauer, H. R. (1985). Vocalizations of one year olds. *Journal of Child Language, 12*, 491–526.

Kent, R. D., & Murray, A. D. (1982). Acoustic features of infant vocalic utterances at 3,6, and 9 months. *Journal of the Acoustical Society of America, 72*, 353–365.

Lindblom, B. (1983). Economy of speech gestures. In P. F. MacNeilage (Ed.), *The production of speech* (pp. 217–245). New York: Springer-Verlag.

Lindblom, B., MacNeilage, P. F., & Studdert-Kennedy, M. G. (In preparation). *The evolution of spoken language*.

Lindblom, B., & Sundberg, J. (1969). A quantitative method of vowel production and the distinctive features of Swedish vowels. *Quarterly progress status report*. Speech Transmission Laboratory, Royal Institute of Technology, Stockholm, *1*, 14–32.

Locke, J. (1983). *Phonological acquisition and change*. New York: Academic Press.

MacKay, D. G. (1972). The structure of words and syllables: Evidence from errors in speech. *Cognitive Psychology, 3*, 210–227.

Macken, M. A. (1978). Permitted complexity in phonological development: One child's acquisition of Spanish consonants. *Lingua, 44*, 219–243.

Macken, M. A., & Barton, D. (1980). The acquisition of the voicing contrast in English: A study of voice onset time in word initial stop consonants. *Journal of Child Language, 7*, 41–74.

MacNeilage, P. F. (1980). The control of speech production. In G. Yeni-Komshian, J. F. Kavanagh, & C. A. Ferguson (Eds.), *Child phonology, Vol. 1: Production* (pp. 9–20). New York: Academic Press.

MacNeilage, P. F., & DeClerk, J. L. (1969). On the motor control of coarticulation in CVC monosyllables. *Journal of the Acoustical Society of America, 45*, 1217–1233.

MacNeilage, P. F., & Sholes, G. N. (1964). An electromyographic study of the tongue during vowel production. *Journal of Speech and Hearing Research, 7*, 2208–2232.

MacNeilage, P. F., Studdert-Kennedy, M. G., & Lindblom, B. (1985). Planning and production of speech: An overview. In J. Lauter (Ed.), Proceedings of the conference of planning and production of speech by normally hearing and deaf people. *American Speech and Hearing Association Reports.*

Maddieson, I. (1984). *Patterns of sounds*. Cambridge: Cambridge University Press.

Meltzoff, A. N. (1986). Imitation, intermodal representation, and the origins of mind. In B. Lindblom & R. Zetterstrom (Eds.), *Precursors of early speech* (pp. 245–265). New York: Stockton Press.

Menn, L. (1980). Phonological theory and child phonology. In G. Yeni-Komshian, J. F. Kavanagh, & C. A. Ferguson (Eds.), *Child Phonology, Vol. 1: Production* (pp. 23–41). New York: Academic Press.

Nittrouer, S., Studdert-Kennedy, M., & McGowan, R. S. (1988). The emergence of phonetic segments: Evidence from the spectral structure of fricative-vowel sequences spoken by children and adults. *Journal of Speech and Hearing Research.*

Oller, D. K. (1980). The emergence of the sounds of speech in infancy. In G. Yeni-Komshian, J. F. Kavanagh, & C. A. Ferguson (Eds.), *Child Phonology, Vol. 1: Production* (pp. 93–112). New York: Academic Press.

Oller, D. K., Eilers, R., Bull, D., Carney, A. (1985). Prespeech vocalizations of a deaf infant: A comparison with normal metaphonological development. *Journal of Speech and Hearing Research, 28*, 47–62.

Paschall, L. (1983). Development at two years. In J. V. Irwin & S. P. Wong (Eds.), *Phonological development in children: 18–72 months* (pp. 73–81). Carbondale: Southern Illinois University Press.

Priestly, T. M. S. (1977). One idiosyncratic strategy in the acquisition of phonology. *Journal of Child Language, 4*, 45–66.

Shattuck-Huffnagel, S. (1979). Speech errors as evidence for a serial ordering mechanism in speech production. In W. E. Cooper & E. C. Walker (Eds.), *Sentence processing: Psycholinguistic studies presented to Merrill Garrett* (pp. 295–342). Hillsdale, NJ: Lawrence Erlbaum Associates.

Shattuck-Huffnagel, S., & Klatt, D. H. (1979). The limited use of distinctive features and markedness in speech production: Evidence from speech error data. *Journal of Verbal Learning and Verbal Behavior, 18*, 41–55.

Shibamoto, J. S., & Olmstead, D. L. (1978). Lexical and syllabic patterns in phonological acquisition. *Journal of Child Language, 5*, 417–457.

Stark, R. (1980). Stages of speech development in the first year of life. In G. Yeni-Komshian, J. F. Kavanagh, & C. A. Ferguson (Eds.), *Child phonology, Vol. 1: Production* (pp. 73–92), New York: Academic Press.

Stoel-Gammon, C. (1983). Constraints on consonant-vowel sequences in early words. *Journal of Child Language, 10*, 455–457.

Trevarthen, C., & Marwick, H. (1986). Signs of motivation for speech in infants, and the nature of a mother's support for development of language. In B. Lindblom & R. Zetterstrom (Eds.), *Precursors of early speech* (pp. 279–308), New York: Stockton Press.

Vihman, M., Macken, M., Miller, R., Simmons, H., & Miller, J. (1985). From babbling to speech: A reassessment of the continuity issue. *Language, 60*, 397–445.

# V SENSORIMOTOR TRANSFORMATION AND THE REPRESENTATION OF ACTION COORDINATES

# 16 Sensorimotor Transformations and the Kinematics of Arm Movements in Three-dimensional Space

J. F. Soechting
C. A. Terzuolo
*Department of Physiology*
*University of Minnesota, Minneapolis*

## ABSTRACT

It is hypothesized that the organization of limb movements involves a series of sensorimotor transformations: between intrinsic and extrinsic coordinates, between the intrinsic coordinates and dynamics, and between dynamics and muscle activation. It is also hypothesized that these transformations involve approximations and the imposition of constraints that lead to uniqueness of behavior. This chapter focuses on the first of these transformations. Data from psychophysical experiments are summarized. These experiments were designed to lead to the identification of a preferred coordinate representation of the intrinsic frame of reference of the arm and the characteristics of the mapping between points in the extrinsic and intrinsic frames of reference. Results of experiments that gave insight into the algorithm whereby wrist motion in arbitrary planes in space could be produced are also summarized. The results are consistent with implementations that involve approximations in the mappings between intrinsic and extrinsic parameter spaces.

## INTRODUCTION

Goal-directed movements involve sensorimotor transformations. Although this statement may seem obvious, it nevertheless provides a good starting point for an investigation into the manner in which limb movements in three-dimensional space are organized and controlled. Consider a task in which the subject is asked to reach for and to grasp an object. The neural activity which results from the image of that object falling on a particular region of the retina must ultimately be transformed into patterns of motoneuron activity in all of the muscles of the arm, patterns appropriately coordinated so that the hand makes contact with the object.

The essential nature of the problem is that there is no one-to-one relation between the retinal image location and the motor output pattern. First of all, the position of the eyes relative to the head and of the head relative to the body can vary from one instance to another. However, even in the simpler case where the eyes, head, and trunk remain fixed, the problem remains underdetermined. As Bernstein (1967) pointed out some time ago, the central nervous system has at its disposal extra degrees of freedom in the performance of any motor task. Thus, a given position of the hand in space, relative to the trunk, can be achieved with an infinite variety of postural orientations of the arm because the arm has 7 degrees of freedom (3 at the shoulder, 1 at the elbow, and 3 at the wrist) whereas only 3 parameters are required to describe hand location.

Furthermore, a given postural orientation of the arm can in principle be maintained by an infinite variety of combinations of activations of arm muscles, even when one rules out the possibility of co-contraction of agonist and antagonist. Consider the case where shoulder and elbow flexors are required to counteract gravity, and consider for simplicity only three muscles: brachio-radialis (an elbow flexor), anterior deltoid (a shoulder flexor), and biceps brachii (an elbow and shoulder flexor). The limb posture could be maintained by activating solely brachio-radialis and deltoid. Alternatively, biceps could also be activated as long as the activation of the other two muscles is reduced commensurately.

Finally, in most motor tasks the movement of the end point (e.g. the hand) is not completely specified. For example, in the reaching task mentioned above, only the initial and the final positions of the hand are unique; there are infinitely many paths that could be taken by the hand in going from one to the other. Furthermore, for any particular path, there are infinitely many velocity profiles that could be assumed by the movement. Thus, there is the potential for spatial and temporal variability in the motion of the end point and for even greater variability in the angular motions of the limb segments and in the patterns of activation of the muscles that contribute to that motion.

Nevertheless, in the case of arm movements at the shoulder and elbow, it has been found that any given task is performed in a stereotyped manner, the variability within and between subjects being low. For point-to-point arm movements, it has been shown that there is little variability in the path taken by the wrist and in the angular motion at the shoulder and elbow (Abend, Bizzi, & Morasso, 1982; Georgopoulos, Kalaska, & Massey, 1981; Lacquaniti, Soechting, & Terzuolo, 1986; Soechting & Lacquaniti, 1981) and that the variability decreases with practice. The path is independent of the speed of the movement (Soechting & Lacquaniti, 1981), of any inertial loads on the arm (Atkeson & Hollerbach, 1985; Lacquaniti, Soechting, & Terzuolo, 1982) and of accuracy constraints placed on the movement (Soechting, 1984). There is also little variability in the temporal characteristics of the motion; the speed at the wrist typically exhibits a bell-shaped profile (Abend et al., 1982; Atkeson & Hollerbach, 1985; Morasso, 1981). For continuous movements such as the drawing of circles and ellipses or in handwriting, one finds a characteristic relationship between curvature and angular velocity (2/3 power law, Lacquaniti, Terzuolo, & Viviani, 1983; Viviani & Terzuolo, 1982). Thus, both for point-to-point and for continuous arm movements, one finds stereotypical spatial and temporal characteristics of the motion. Finally, for any particular movement performed at a given speed, one also finds stereotypical patterns of muscle activation (Gielen & van Zuylen, 1986; Wadman, Denier van der Gon, & Derksen, 1980).

Thus, despite the theoretical possibility for infinite variety, it has been possible experimentally to define invariants that hold true for entire classes of limb movements. The significance of such invariants is that they can be, and have been, used to deduce the algorithms used by the central nervous system in organizing and controlling limb movements. Before going into more detail concerning such invariant characteristics, it may be useful to discuss in general terms some of the hypotheses that have been formulated. It is also pertinent to mention one more point. The hypothesis that the expression of such invariants involves a process of learning is appealing (Soechting, Lacquaniti, & Terzuolo, 1986). Although there is little evidence in its support at present (von Hofsten, 1979), a strong argument in its favor can be made on theoretical grounds (Loeb, 1983).

One approach that has gained much attention recently is based on neural network models (Sejnowski, 1986) where the strengths of synaptic connections are modified by suitable learning rules (cf. Linsker, 1986). Such an approach has been used by Kuperstein (1988a, 1988b) to model the posture assumed by the arm in grasping an object, that is, the mapping from a retinotopic space to the space of limb muscles mentioned at the outset of this section. His approach utilizes a set of topographically ordered maps for gaze and retinotopic target location to generate a target map that in turn is used to produce the motor signals to the relevant muscles. Computer simulations have been successful in generating limb postures with average position errors of 4% of the arm length and orientation errors of 4° (Kuperstein, 1988b).

Another formulation of the problem that is similar to the approach used in robotics (Hollerbach, 1982) makes explicit a sequence of sensorimotor transformations (cf. Soechting & Terzuolo, 1986). Consider, for example, the task of pointing to a target. Target location is represented in the extrinsic coordinates of extrapersonal space (the target maps of Kuperstein, 1988b) and the first transformation involves a mapping from these extrinsic coordinates into the intrinsic coordinates describing the angular orientation of the arm. Thus, this transformation involves only kinematics. A second transformation is from kinematics to dynamics, that is, given a trajectory of angular motions at the shoulder and elbow, the torque at these joints adequate to produce the desired motion is determined. The third transformation is required to partition the torque among the different muscles of the limb.

As mentioned before, the first and third of these transformations is not unique. Therefore, to achieve a unique behavior appropriate neural constraints must be imposed. Furthermore, each of these relations is mathematically complex. In contrast to the field of robotics where these equations are solved exactly on digital computers, one must entertain the possibility that biological systems utilize simpler, approximate solutions. One consequence of this possibility is that subjects should produce predictable errors in their performance and that some movements should be difficult if not impossible to execute.

The main reason for assuming that the organization of movements involves a sequence of transformations, aside from the fact that the problem becomes more tractable conceptually, is that a number of invariants have been described at the kinematic level, implying that there is an explicit transformation between extrinsic and intrinsic coordinates. Some of these have already been mentioned, and others are described more fully later. The general topic of kinematic—as well as dynamic—invariants has also been dealt with extensively in a recent review (Soechting, 1988).

As will be shown, if one makes the assumption that there exists a mapping between extrinsic and intrinsic coordinates, one can uncover some of the elements of the algorithm used to achieve this transformation. How such an algorithm could be implemented remains an open question. Obviously one possibility is by neural network models, such as those used by Kuperstein (1988a, 1988b). In fact his model could be made congruent with the framework already outlined here if two additional layers (one kinematic, one dynamic) were interposed between his target map and his muscle map. One can expect that such network models would lead to stereotypical behavior in any one subject as a consequence of learning. Whether it would always converge to the same behavior and what, if any, additional constraints would need to be imposed to assure such convergence remain open questions.

## IDENTIFICATION OF PREFERRED COORDINATE SYSTEMS

In order to begin to identify the elements of the transformation between the extrinsic and intrinsic frames of reference, one must first choose a set of parameters that can be used to describe the location of a point in the two reference frames. In both cases, a number of alternatives is possible. For example, in the extrinsic frame of reference the position of the hand can be described in terms of the distance from the eyes in the anterior, lateral, and vertical directions, that is, in a Cartesian coordinate system. However, it could be described equally well in terms of the hand's radial distance from the shoulder and the azimuth and elevation angles, that is, in terms of a spherical coordinate system.

Similarly the orientation of the arm, in the intrinsic frame of reference, can be defined in a variety of ways. The top row in Fig. 16.1 illustrates two such possibilities. The position of the elbow relative to the shoulder is defined by two angles. One alternative is given by the angular elevation of the upper arm (i.e., the angle it makes relative to the trunk or the vertical axis, measured in the vertical plane) and the yaw angle (measured in the horizontal plane). However, one could also use the angle of forward flexion and a complementary angle ($\theta_2$ in the top right panel in Fig. 16.1) to define the orientation of the arm.

To distinguish among these and other possibilities, we (Soechting & Ross, 1984) performed some psychophysical experiments. We reasoned that if there were a preferred representation of limb orientation, then subjects' performance in a task in which they were asked to match one of these angles with both arms would be superior. Therefore, we asked subjects to point to a target with their right arm (in the dark) and then to match either the angular elevation or the angle of forward flexion by movement of their left arm. Motion of the left arm was constrained to be in the sagittal plane; in this plane the angular elevation and the angle of forward flexion are the same. Typical results from one subject are shown in the bottom panels of Fig. 16.1. All subjects were significantly better at matching the angular elevation. Similarly, they also performed better at matching angular elevation than the angle of abduction at the shoulder. Therefore, we concluded that the angular elevation and yaw angles constituted the preferred set of parameters for representing the orientation of the upper arm.

Other psychophysical experiments indicated that the same two angles (elevation and yaw) also constituted the preferred representation of forearm orientation. In particular subjects were better able to match forearm angular elevation than they were the angle of flexion–extension at the

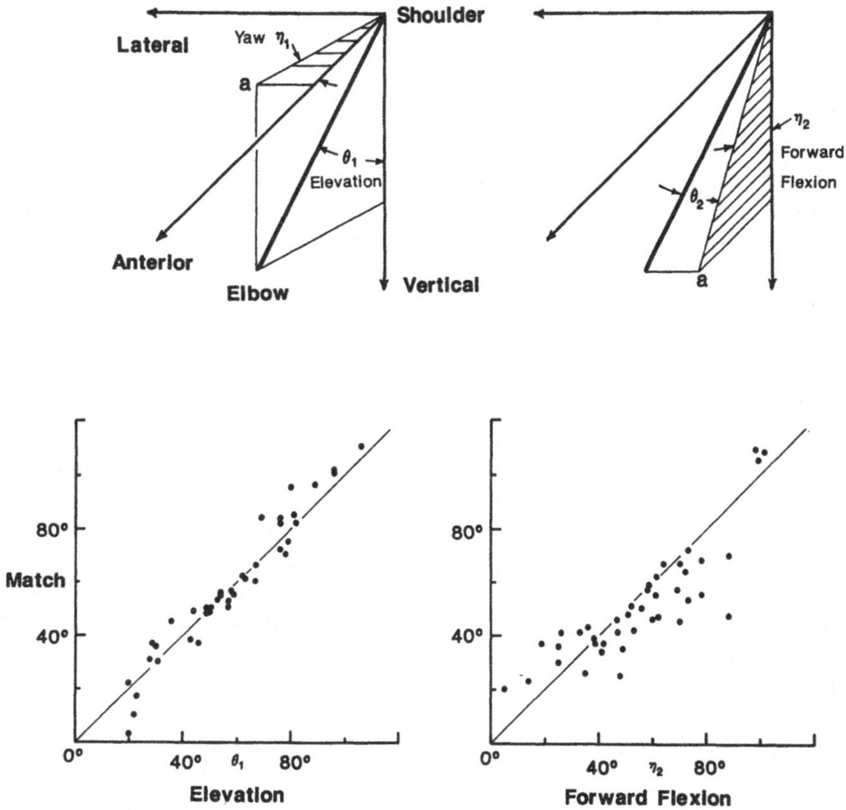


FIG. 16.1. The top row shows two possible ways to describe the orientation of the upper arm. On the left, the position of the elbow relative to the shoulder is given by the elevation, measured in the vertical plane passing through the shoulder and elbow, and by the yaw angle measured in the horizontal plane as the angle between the anterior direction and the line from the point *a* (the projection of the elbow on the horizontal plane) to the shoulder. On the right, the angle of forward flexion is measured in the sagittal plane, and a second angle ($\theta_2$) is measured in the plane of the shoulder, elbow and the elbow's projection onto the sagittal plane. Below are shown results of an experiment in which a subject was asked to match either the elevation (left) or the angle of forward flexion (right) with both arms. Modified from Soechting and Ross (1984).

elbow (Soechting, 1982). In fact, when subjects were asked to match elbow angle, they made consistent errors that were strongly biased in the direction of matching forearm elevation (Worringham, Stelmach, & Martin, 1987).

What the preferred representation of a point in the extrinsic frame of reference might be remains to be determined.

## SENSORIMOTOR TRANSFORMATIONS BETWEEN POINTS IN EXTRINSIC AND INTRINSIC FRAMES OF REFERENCE

As has already been mentioned, the mathematical relations between joint angles (intrinsic frame of reference) and a point in extrapersonal space are complicated. In particular

$$\begin{aligned} x &= \ell\,(\sin\theta\cos\eta + \cos\beta\cos\alpha) \\ y &= \ell\,(\sin\theta\sin\eta \text{ pl } \cos\beta\sin\alpha) \\ z &= \ell\,(\cos\theta - \sin\beta) \end{aligned} \tag{1}$$

where $x$, $y$, and $z$ denote the distance in the anterior, lateral, and vertical direction, $\theta$ and $\eta$ are the angular elevation and yaw of the upper arm, and $\beta$ and $\alpha$ are the corresponding angles of the forearm. For the sake of simplicity it is assumed that the lengths of the upper arm and forearm are both equal to $\ell$. The difficulty of the problem becomes apparent if the task is stated thusly: Given $x$, $y$, and $z$ calculate $\theta$, $\eta$, $\beta$, and $\alpha$. The problem becomes no simpler if one defines a point in extrinsic space in terms of its distance $R$ and direction relative to the shoulder

$$R^2 = 2\ell^2\,[1 + \sin\theta\cos\beta\cos(\eta - \alpha) - \cos\theta\sin\beta] \tag{2}$$

If this problem is solved by approximation, then one might expect subjects to make predictable errors when they are asked to point to a target, provided that visual feedback is not available for error correction. We have begun to address this question by using the following experimental paradigm: We present a target at random locations in extrapersonal space, within the subject's reach, ask the subject to remember the location, and to point to the spot in the dark. To identify the constant errors in performance, we computed least square polynomial fits between the position of the index finger and the target's actual location

$$\begin{aligned} R &= a_0 + a_1R_T + a_2\chi_T + a_3\psi_T \\ \chi &= b_0 + b_1R_T + b_2\chi_T + b_3\psi_T \\ \psi &= c_0 + c_1R_t + c_2\chi_T + c_3\psi_T \end{aligned} \tag{3}$$

where $R$ is the radial distance from the shoulder, $\chi$ the azimuth and $\psi$ the elevation. The subscripted variables (i.e. $R_T$) denote target location, the unsubscripted variables the position of the index finger. Equation (3) is a linear polynomial fit to the data; quadratic and cubic polynomials were also used.

Figures 16.2 through 16.4 show representative results from one subject. In the top panels of Fig. 16.2 are shown the least square fit of radial distance ($R$), azimuth, and elevation to the corresponding target variables. A 45° line denotes a perfect fit between each of the two variables, and the scatter of data points reflects the contribution of other terms (for example, $a_2$ and $a_3$). The lower row shows the variable error that remained after the fit. Results from the same experiment in which cubic polynomials were used to fit the data are shown in Fig. 16.3. Finally, 16.4 shows the relationship between the position the subject pointed to and the target's location. Each panel represents one vertical plane of constant azimuth, and target location is denoted by light traces, whereas finger position is indicated by the heavier lines.

It is evident in these figures that subjects pointed in the correct direction but that there were substantial errors in the radial distance from the shoulder. For targets close to the shoulder there was an overshoot; when targets were further away but still within reach there was an undershoot. On average (4 subjects) the regression coefficient $a_1$ was substantially different from unity ($a_1 = 0.61 \pm 0.08$), whereas those for azimuth ($b_2 = 0.97 \pm 0.05$) and elevation ($c_3 = 0.94 \pm 0.02$) were very close to 1.

The average root mean square total error in these subjects ranged from 6.7 to 11.7 cm. Of this total, the variable error accounted for 39% to 79% of the total (3.8 to 5.3 cm). The remainder was due to the systematic errors illustrated in Figs. 16.2 through 16.4.

When these experiments were repeated with the lights on, very similar results were obtained. In this case the subjects had visual as well as proprioceptive feedback of the position of their arm, but they were still pointing to a virtual target. The same tendency to overshoot or undershoot the distance was found ($a_1 = 0.70 \pm 0.09$) as before.

This set of results raised two possible interpretations of the data: (a) there were errors in the subjects' perception of the distance at which the target was located, or (b) there was an error in the mapping between extrinsic and intrinsic coordinates, and the subjects' use of visual feedback was ineffective in compensating for the error. To distinguish among these possibilities we performed another set of experiments. The subjects were given a pointer and asked to point to the target's location with the tip of the pointer. Visual feedback of pointer position was provided. In this experiment, the relationship between extrinsic and intrinsic coordinates is very different, most of the arm movement now occurring at the wrist. The subjects' performance on this task was substantially better than it was on the other two ($a_1 = 0.90 \pm 0.01$).

Thus, although perceptual errors in judging distance cannot be ruled out entirely, it appears that the major error in the pointing task in the absence of visual feedback is due to an error in the transformation between

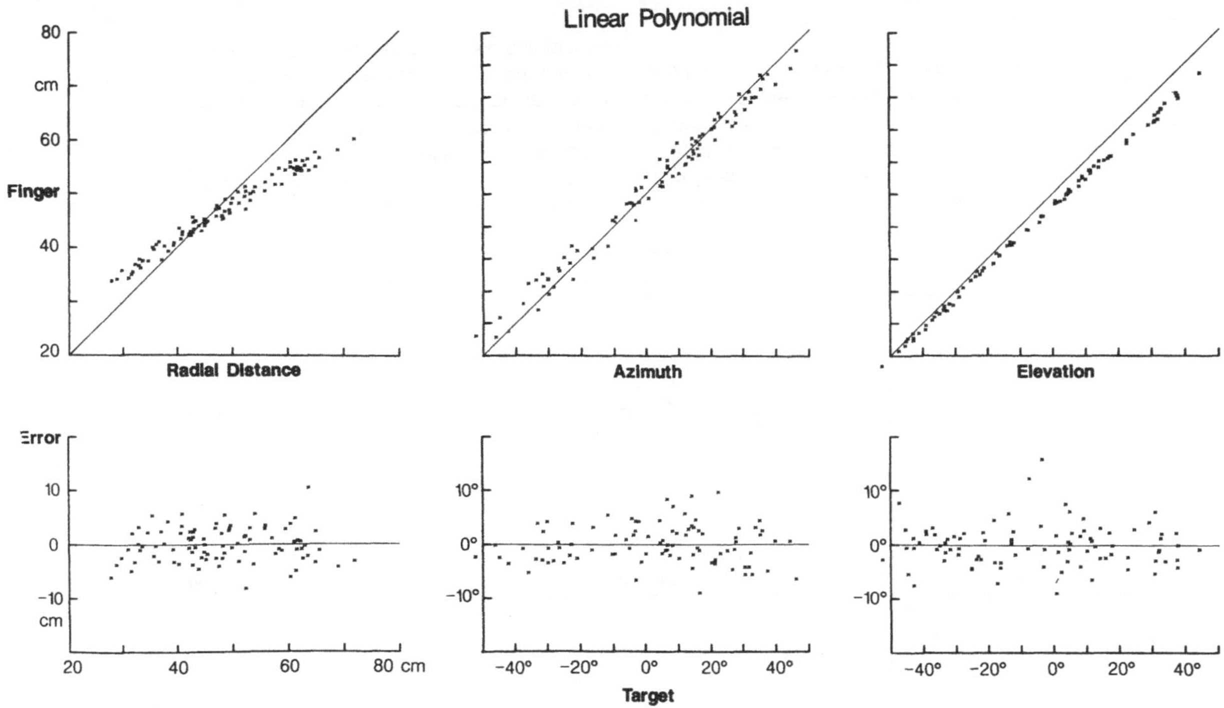


FIG. 16.2. Errors in pointing in the absence of visual feedback. The panels in the top row show the results of a least square polynomial fit between target location and finger position (Equation 3 in text). They give an indication of the constant error in the performance. For perfect performance, the data points would lie on the 45° line. The panels in the lower row indicate the error remaining after the least squares fit (variable error). Note that the subject made consistent errors in the distance to which the subject pointed (left panel) but that consistent errors in direction (azimuth and elevation) were much less.

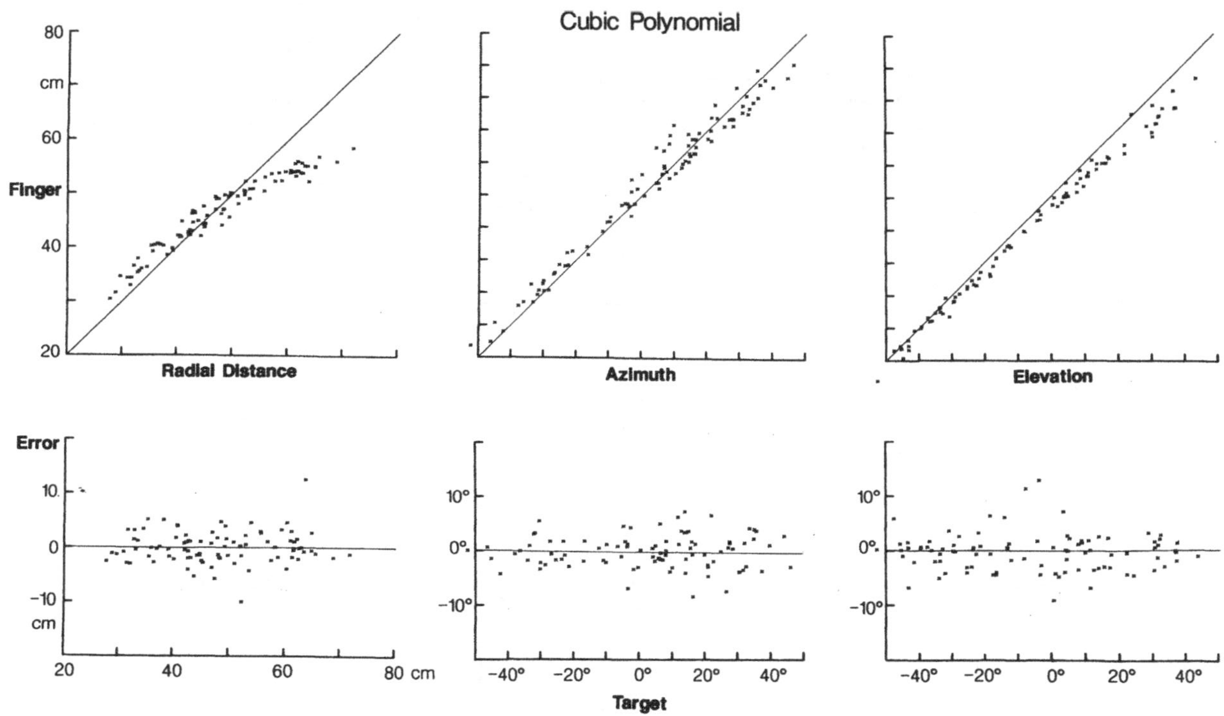


FIG. 16.3. Errors in pointing in the absence of visual feedback. The experimental data are the same as those in Fig. 16.2, but they have now been fitted with a cubic polynomial. There is very little difference in the constant error (top panels) between Figs. 16.3 and 16.2, the major difference being that the relationship between target distance and finger distance (left panel) now exhibits a slight curvature. This analysis indicates that the linear polynomial is adequate to characterize the error in the mapping between target position and arm orientation (extrinsic and intrinsic coordinates).

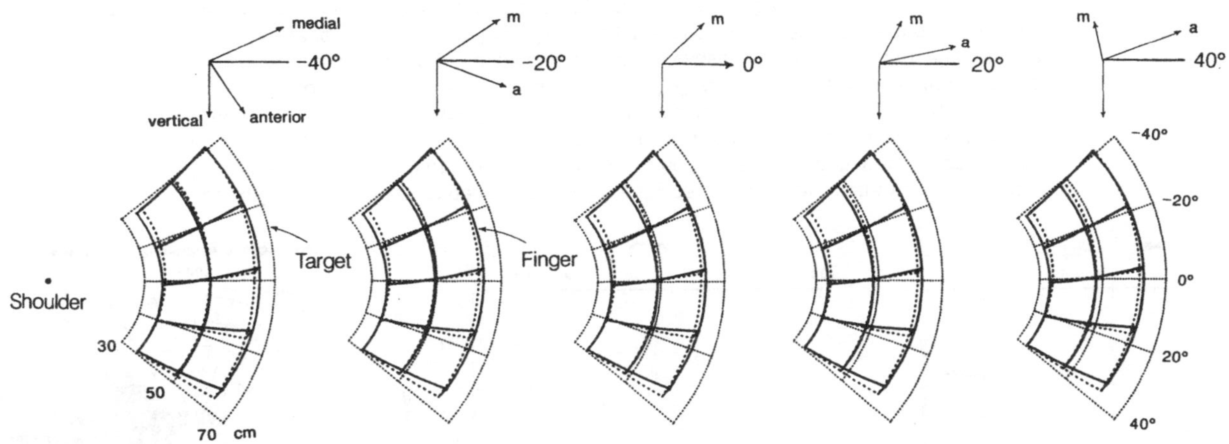


FIG. 16.4. Distortion in the mapping between target and finger location. This figure replots the results from the two previous figures. Target location is indicated by the light lines, each panel representing lines of constant radial distance and elevation in one plane of constant azimuth ranging from 40° medial to 40° lateral. The expected finger position, computed from the linear polynomial coefficients, is indicated by the heavy solid lines; its projection onto the plane of constant azimuth is given by the dashed lines.

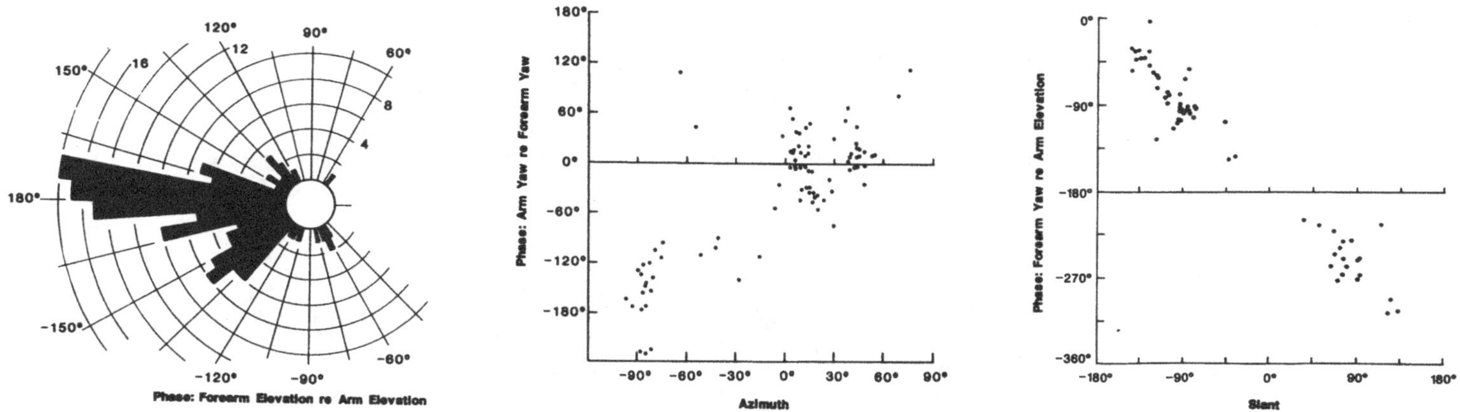


FIG. 16.5. Simple rules for generating curvilinear wrist motion in an arbitrary plane in space. The azimuth of the plane of motion is linearly related to the phase between sinusoidal yaw motion of the upper arm and forearm (middle panel), whereas the slant of ellipses is linearly related to the phase between forearm yaw and angular elevation motions (right panel). A slant of 90° corresponds to a circle, slants of 0° or 180° to straight line motion. Both of these simple relations derive from the constraint that the phase between upper arm and forearm angular elevations be close to 180°. The left panel shows a polar histogram of the experimentally measured values. From Soechting and Terzuolo (1986).

extrinsic and intrinsic coordinates. This finding suggest that Equation (2), which relates radial distance to the orientation angles of the arm, is not solved exactly but only approximately and that such approximations lead to predictable distortions.

More details concerning the algorithm that might be used in this task cannot be given at this time. However, it is noteworthy that there is independent support for the idea that distance and direction are encoded separately. Georgopoulos and his colleagues (Georgopoulos, 1986; Georgopoulos, Caminiti, Kalaska, & Massey, 1983; Georgopoulos, Schwartz, & Kettner, 1986) have described neurons in motor cortex whose frequency of discharge was tuned to the direction of the movement but only weakly tuned to the amplitude of the movement (Schwartz & Georgopoulos, 1987).

## WRIST MOTIONS IN DIFFERENT PLANES IN SPACE

In addition to the three parameters that are needed to specify the location of an object in space, three more parameters are needed to specify its orientation. For the task we have considered so far, namely grasping that object, it has been shown (Lacquaniti & Soechting, 1982; Soechting, 1984) that motion at the shoulder and elbow depends little on the object's orientation, whereas wrist motion (prono–supination and flexion–extension) does.

However, in drawing and in gestural movements, shoulder and elbow motion does depend on the orientation of the plane in which the figure is drawn. In this instance, we have been able to characterize most of the elements of the algorithm by means of which motion of the wrist can be translated from one plane to another (Soechting, Lacquaniti, & Terzuolo, 1986; Soechting & Terzuolo, 1986).

When subjects are asked to draw circles and ellipses in different planes in space, the linear motion of the wrist and the angular motion at the shoulder and elbow (defined in the preferred coordinate system described in the previous section) are close to sinusoidal. Therefore, the motion in the intrinsic coordinates could be described simply by the amplitudes and phases of the fundamental components. Three parameters were needed to describe the motion in extrinsic coordinates: the azimuth and the elevation of the plane of motion and the slant of the ellipse.

We found particularly simple relations between two of these extrinsic parameters and the intrinsic parameters of the angular motion of the arm. They are (as illustrated in Fig. 16.5):

1. The phase difference between upper arm and forearm yaw was linearly related to the azimuth of the plane of motion.

2. The phase difference between forearm yaw and angular elevation was linearly related to the slant of the ellipse.

These two simple relations came about because the phase relation between forearm and upper arm angular elevation was always close to 180° (± 25°) independently of the azimuth and slant and over a wide range of planar elevations (within 45° of the vertical). This last finding, illustrated in the left panel in Fig. 16.5, constitutes a constraint that ensures uniqueness in the behavior (by reducing the number of independent degrees of freedom of the arm from four to three). Simulation studies showed that this constraint ensured that the two relations between intrinsic and extrinsic parameters are obeyed, in an average sense, for arbitrary combinations of amplitude modulations of the orientation angles.

Although these rules were uncovered by studying periodic, elliptical wrist motions, there is a way to extend their validity to more arbitrary motions provided one makes one additional assumption: Such motions consist of segments, each segment constituting an arc of an ellipse and constructed according to the rules described above. If so, wrist motion during each segment should be confined to one plane, but there is the possibility that the plane of motion can change abruptly from one segment to the next. Furthermore, one would predict that subjects should not be able to produce wrist motion where the tangent plane changes smoothly and continuously.

We tested these predictions experimentally be asking subjects to produce more complicated drawings in free space with their arm, such as figure 8s, stars, and scribbles (Soechting & Terzuolo, 1987a). We also asked them to produce drawings where the tangent plane of the wrist motion should change smoothly, such as circles wrapped around the long axis of a cylinder or the seams on a tennis ball (Soechting & Terzuolo, 1987b). The results were in agreement with the predictions. In all instances, movements consisted of segments in which the motion at the wrist was confined to one plane, even when the task required otherwise. Furthermore, the plane of wrist motion could change abruptly and appreciably from one segment to the next. For example, when subjects were asked to draw figure 8s, the top and bottom loops of the figure each generally constituted one segment, and in about 30% of the trials planar elevations changed by more than 20° from one segment to the next.

The rules relating extrinsic and intrinsic coordinates described are approximations to the mathematically exact nonlinear relations. One consequence is that, as we have just described, some tasks are impossible for subjects to perform. Another consequence is that some movements are predictably distorted. For example, when subjects are asked to draw a circle in the sagittal plane, there is a flattening of the proximal portion of the trajectory (Soechting et al., 1986).

## CONCLUSION

We have summarized evidence that suggests that the sensorimotor transformation from the extrinsic parameters of extrapersonal space to the space of muscle activations involves at least one intermediate step—a kinematic transformation between extrinsic and intrinsic variables. Furthermore, the elements of this transformation are approximate. This hypothesis can account for a large body of experimental evidence, namely, the existence of kinematic invariants and the existence of predictable errors in performance.

## ACKNOWLEDGMENT

This work was supported by the U. S. Public Health Service (NS-15018) and by the National Science Foundation (BNS-8418539).

## REFERENCES

Abend, W., Bizzi, E., & Morasso, P. (1982). Human arm trajectory formation. *Brain, 105*, 331–348.

Atkeson, C. G., & Hollerbach, J. M. (1985). Kinematic features of unrestrained vertical arm movements. *Journal of Neuroscience, 5*, 2318–2330.

Bernstein, N. (1967). *The Coordination and Regulation of Movements*. Oxford: Pergamon.

Georgopoulos, A. P. (1986). On reaching. *Annual Review of Neuroscience, 9*, 147–170.

Georgopoulos, A. P., Caminiti, R., Kalaska, J. F., & Massey, J. T. (1983). Spatial coding of movement: A hypothesis concerning the coding of movement direction by motor cortical populations. *Experimental Brain Research Supplement, 7*, 327–336.

Georgopoulos, A. P., Kalaska, J. F., & Massey, J. T. (1981). Spatial trajectories and reaction times of aimed movements: Effects of practice, uncertainty and change in target location. *Journal of Neurophysiology, 46*, 725–743.

Georgopoulos, A. P., Schwartz, A. B., and Kettner, R. E. (1986). Neuronal coding of movement direction. *Science, 233*, 1416–1419.

Gielen, C. C. A. M., & van Zuylen, E. J. (1986). Coordination of arm muscles during flexion and supination: Application of tensor analysis approach. *Neuroscience, 17*, 527–539.

Hollerbach, J. M. (1982). Computers, brains and the control of movement. *Trends in Neuroscience, 5*, 189–192.

Kuperstein, M. (1988a). Neural model of adaptive hand–eye coordination for simple postures. *Science, 239*, 1308–1310.

Kuperstein, M. (1988b). An adaptive neural model for mapping invariant target position. *Behavioral Neuroscience, 102*, 148–159.

Lacquaniti, F., & Soechting, J. F. (1982). Coordination of arm and wrist motion during a reaching task. *Journal of Neuroscience, 2*, 399–408.

Lacquaniti, F., Soechting, J. F., & Terzuolo, C. A. (1982). Some factors pertinent to the organization and control of arm movements. *Brain Research, 252*, 394–397.

Lacquaniti, F., Soechting, J. F., & Terzuolo, C. A. (1986). Path constraints on point-to-point arm movements in three-dimensional space. *Neuroscience, 17*, 313–324.

Lacquaniti, F., Terzuolo, C., & Viviani, P. (1983). The law relating the kinematic and figural aspects of drawing movements. *Acta psychologica, 54*, 115–130.

Linsker, R. (1986). From basic network principles to neural architecture: emergence of spatial-opponent cells. *Proceedings of the National Academy of Science USA, 83*, 7508–7512.

Loeb, G. E. (1983). Finding common ground between robotics and neurophysiology. *Trends in Neuroscience, 6*, 203–204.

Morasso, P. (1981). Spatial control of arm movements. *Experimental Brain Research, 42*, 223–237.

Schwartz, A. B., & Georgopoulos, A. P. (1987). Relations between the amplitude of 2-dimensional arm movements and single cell discharge in primate motor cortex. *Society for Neuroscience Abstracts, 13*, 244.

Sejnowski, T. J. (1986). Open questions about computation in cerebral cortex. In J. L. McClelland, D. E. Rumelhart, & and PDP Research Group (Eds.), *Parallel Distributed Processing: Computational Models of Cognition and Perception* (pp. 373–392). Cambridge, MA: MIT Press.

Soechting, J. F. (1982). Does position sense at the elbow reflect a sense of elbow joint angle or one of limb orientation? *Brain Research, 248*, 392–395.

Soechting, J. F. (1984). Effect of target size on spatial and temporal characteristics of a pointing movement in man. *Experimental Brain Research, 54*, 121–132.

Soechting, J. F. (1988). Elements of coordinated arm movements in three-dimensional space. In S. A. Wallace (Ed.), *Perspectives on the Coordination of Movement*. New York: North Holland Publishing.

Soechting, J. F., & Lacquaniti, F. (1981). Invariant characteristics of a pointing movement in man. *Journal of Neuroscience, 1*, 710–720.

Soechting, J. F., Lacquaniti, F. & Terzuolo, C. A. (1986). Coordination of arm movements in three-dimensional space. Sensorimotor mapping during drawing movement. *Neuroscience, 17*, 295–311.

Soechting, J. F., & Ross, B. (1984). Psychophysical determination of coordinate representation of human arm orientation. *Neuroscience, 13*, 595–604.

Soechting, J. F., & Terzuolo, C. A. (1986). An algorithm for the generation of curvilinear wrist motion in an arbitrary plane in three-dimensional space. *Neuroscience, 19*, 1395–1405.

Soechting, J. F., & Terzuolo, C. A. (1987a). Organization of arm movements. Motion is segmented. *Neuroscience, 23*, 39–52.

Soechting, J. F., & Terzuolo, C. A. (1987b). Organization of arm movements in three-dimensional space. Wrist motion is piece-wise planar. *Neuroscience, 23*, 53–61.

Viviani, P., & Terzuolo, C. (1982). Trajectory determines movement dynamics. *Neuroscience, 7*, 431–437.

von Hofsten, C. (1979). Development of visually directed reaching: The approach phase. *Journal of Human Movement Studies, 5*, 160–178.

Wadman, W. J., Denier van der Gon, J. J., & Derksen, R. J. A. (1980). Muscle activation patterns for fast goal-directed arm movements. *Journal of Human Movement Studies, 6*, 19–37.

Worringham, C. J., Stelmach, G. E., & Martin, Z. E. (1987). Limb segment inclination sense in proprioception. *Experimental Brain Research, 66*, 653–658.

# 17 The Geometric and Dynamic Implications of the Coherence Constraints in Three-Dimensional Sensorimotor Interactions

Jacques Droulez
Christian Darlot
*Laboratoire de Physiologie Neurosensorielle*
*Centre National de la Recherche Scientifique*
*Paris, France*

ABSTRACT

In the past decade, the investigations on the role of vision and vestibular organs in eye movement control, posture control and egomotion perception have emphasized the importance of interactions between sensory signals and the necessity to take into consideration any intervening motor commands in the sensory central processing itself. Most endeavors have focused on the interactions between semicircular canals and visual information. The otolithic information processing, on the other hand, was shown to be more complex and difficult to analyze, very likely because of the ambiguous nature of the otolithic signal (both gravitational and inertial). A comprehensive model, explaining how information originating from visual, otolithic and semicircular canals systems might interact, is still lacking.

In the present work, our attempt was to develop such a model, based on a few basic assumptions. The main hypothesis was that interactions do not occur between primary sensory signals, but rather between central representations of physical variables, such as head or gaze velocity and orientation. Each central representation receives a direct message from only one specific dedicated captor (for instance, the head angular velocity estimator receives a direct input from the semicircular canals), which provides a first approximated measure of the related physical variable. The central representations are then cross-coupled through functional connections that mimic the interrelations between physical variables, thus improving the coherence of the internal estimations.

Three interdependent examples of this model are given: the head/gaze angular velocity coupling (involving the classical visual-vestibular interactions), the head linear acceleration/velocity coupling (involving the interactions between otolithic and visual informations) and the head angular velocity/orientation coupling (involving the intravestibular interactions). Quantitative simulations of the model are shown to be consistent with most experimental results obtained in a wide range of experimental conditions. A new perceptual illusion, still to be experimentally demonstrated, is also predicted by the model.

## INTRODUCTION

The central representation of selfmotion (or egomotion) and the generation of compensatory eye movements have been extensively investigated for several years as an experimental model of sensorimotor interactions (see Henn, Cohen, & Young, 1980, for a review). Most experimental paradigms used in animals or humans involve the rotation of the head and/or the rotation of the visual scene around the vertical Z axis, colinear with the gravitational acceleration. In this case, the dynamic complementarity of the visual and semicircular canal systems has been demonstrated in several species, including the goldfish (Allum, Graf, Dichgans, & Schmidt, 1976), the rabbit (Dichgans & Brandt, 1972), the cat (Keller & Precht, 1979), the monkey (Waespe & Henn, 1979) and man (Dichgans & Brandt, 1978). From a theoretical point of view, this dynamic complementarity has been explained by either a positive feedback loop (Robinson, 1975, 1977) or a velocity storage mechanism (Raphan, Cohen, & Matsuo, 1977; Raphan, Matsuo, & Cohen, 1979).

However, when a linear or off-vertical angular egomotion is evoked by actual head movement or by visual stimulation, the role of a third sensory organ—the otolithic system—has to be logically taken into account. Considering the variety of movements in which the otolithic system is involved, the vertical axis rotation could be viewed as a special case in which the otolithic information is not absent but only kept constant. Therefore, the otolith signal may be involved in the multisensory interaction process even during pure vertical rotations.

The development of modern transportation and, more specifically, space exploration technology and the problem of adaptation to weightlessness gave rise to numerous investigations on the role and the processing of the otolithic informations. The dynamic stimulation of the otolithic organ by off vertical axis rotation (OVAR) at constant velocity in the dark induces a compensatory nystagmus and an egomotion sensation that last as long as the stimulation continues (Guedry, 1965; Benson & Bodin, 1966; Stockwell, Turnipseed, & Guedry, 1970; Harris & Barnes, 1985; Darlot, Denise,

Droulez, Cohen, & Berthoz, 1988; Denise, Darlot, Droulez, Cohen, & Berthoz, 1988). The role of the primary otolithic afferences in these permanent responses has been demonstrated (Correia & Money, 1970; Goldberg & Fernandez, 1981, 1982).

Numerous experimental data have also been obtained during horizontal or vertical translations in the dark that elicit egomotion sensation (Meiry, 1965; Melvill-Jones, Rolph, & Downing, 1980) and compensatory L-nystagmus, that is, nystagmus evoked in the dark by head linear acceleration along the interaural axis (Jongkees & Philipszoon, 1962; Mac Cabe, 1964; Buizza, Léger, Droulez, Berthoz, & Schmid, 1980).

Nonlinear interactions between visual and otolithic stimulation during horizontal translation has been invoked in the modulation of the optokinetic nystagmus, smooth pursuit eye movements and perceived velocity of the visual scene in humans (Pavard & Berthoz, 1977; Buizza et al., 1980; Buizza, Schmid, & Droulez, 1981). Another example of visual–otolithic interaction is also illustrated by the perceptual responses to rotation of the visual scene about a horizontal axis. For instance, the continuous sensation of roll motion, elicited by a large dome rotating around the subjects' line of sight, is markedly reduced when this axis is horizontal as compared to the case of a vertical rotation axis. Moreover, in spite of the continuous, visually induced roll (or pitch) selfmotion, subjects experienced only limited angular displacements with respect to gravity (Dichgans & Brandt, 1978). These effects have generally been interpreted as resulting from sensory conflicts between vision and gravireceptors (otolithic and somatosensory systems).

Several attempts have been made to develop a mathematical description of these complex visual–semicircular canal–otolith interactions. Some of these models focused only on partial aspects: visual–otolith interactions (Finke & Held, 1978; Pavard & Berthoz, 1977), semicircular canal–otolith interaction (Mayne, 1974; Ormsby & Young, 1977), visual–semicircular canal interactions during smooth pursuit (Raphan et al., 1979; Robinson, 1977). Only a few models deal with the problem of the complete reconstruction of the 3D egomotion from visual, semicircular canals, and otolithic informations (Borah, Young, & Curry, 1980; Hain, 1986).

Hain's model accounts for compensatory eye movements evoked by OVAR stimulation and horizontal linear head displacement. It is conceived as a 3-Dimension (3D) extension of Robinson's models (Robinson, 1977, 1982) and includes nonlinear processing of the otolithic inputs. However, it cannot account for the deviation of perceived vertical during vertical optokinetic stimulation and more generally for the 3D selfmotion perception.

The theory of Kalman filters (Kalman, 1960), which is extensively used in artificial navigation systems (Schmidt, 1966), has also been proposed as a model of animal and human navigation function. The most interesting feature of such models is their use in providing optimal estimations of movements based on the dynamical properties and the uncertainty of the afferent signals. However, the neural implementation of Kalman filters, involving matrix inversion and manipulation, is far from evident. The biological systems may have developed simpler and more robust solutions.

The goal of this chapter is to present a new scheme of multiple information fusion, based on a few logical assumptions, that could be applied to the generalized 3D egomotion perception and compensatory movement generation. Numerical simulations of the model show that this scheme accounts for most experimental results. A new perceptual effect is also predicted by the model. This effect, yet to be experimentally demonstrated, results from complex nonlinear interactions between visual, semicircular canal, and otolithic systems.

## THE CONCEPTUAL FRAMEWORK

### The Central Representation of Movement and Orientation

The hypothesis that those physical entities that present a high biological interest may be reconstructed and coded in the central nervous system is a widely accepted assumption. It overlaps somewhat with the notion of sensorimotor integration. More precisely, we assume here that to each variable, for instance, the head angular velocity, corresponds a unique central representation or internal estimation that could be used to drive eye, head, or body movements, and on which the selfmotion perception is based. Although the existence of neurons coding head movements and receiving multiple sensory or motor signals has been evidenced in various species (review in Henn et al. 1980), the unicity of such a representation is still questionable.

Other aspects of the central representation have to be further elaborated and are discussed in the following:

1. The input and output signals of the internal estimations have yet to be defined, as well as the list of the centrally represented variables involved in egomotion and orientation perception.
2. Most of the variables are multidimensional: In what coordinate system are they coded (see for instance, Pellionisz & Llinas, 1979) and, more generally, what is the code that transcribes the internal estimation into the activity of a neural population?

3. The way the central representations are elaborated and how they interact with each other has still to be specified.
4. The anatomical localizations of the central representations is still largely unknown.

## The Captor Specificity

The hypothesis of captor specificity is different from the notion of sensory complementarity that is classically considered as the basis of sensory integration. We assume here that to each centrally represented variable corresponds, as a specific source of information, the captor whose output is the most closely correlated to this variable, that is, yields the most faithful measurement. The source of information is the set of primary afferents (or a subset of afferent fibers) originating from a sensory end-organ. By extension, we consider also as a source of information the "corollary discharge" or the "efferent copy" of a motor command.

The captor specificity is a central hypothesis in our scheme. It means that the cooperation between sensory signals is only required to refine the direct measure and to extend the working range of the dedicated captor. The retinal slip, classically considered as the visual input to head angular velocity estimation, is not directly interpretable in terms of head rotation with respect to the stationary environment. This signal must be combined with the eye velocity in the head, which is clearly not a visual information. On the contrary, the semicircular canal system provides a rather precise and direct measure of the head angular velocity in a very wide frequency range (Wilson & Melvill-Jones, 1979).

The captor specificity is also a guide to answer some of the questions left open by the central representation hypothesis. First, it helps us to draw a preliminary list of the centrally coded variables (see Table 17.1). These variables are separated into two groups: the *exocentric variables* that define

TABLE 17.1
The Captor Specificity

| *PHYSICAL VARIABLE* | *DEDICATED SOURCE OF INFORM.* | *NOT.* |
|---|---|---|
| HEAD ANGULAR VELOCITY | SEMICIRCULAR CANALS | H |
| GRAVITATIONAL ACCELER. | OTOLITHIC REGULAR UNITS | G |
| INERTIAL ACCELERATION | OTOLITHIC IRREGULAR UNITS | A |
| GAZE ANGULAR VELOCITY | RETINAL SLIP | R |
| HEAD LINEAR VELOCITY | IMAGE DEFORMATION | V |
| INTERSEGMENTAL POSITION and VELOCITY | PROPRIOCEPTIVE AFFERENTS and EFFERENT COPY | P E |

the segmental movements or orientation with respect to the environmental stationary frame of reference, and the *intersegmental variables* that define the relative movement or orientation of one body segment with respect to another adjacent segment. In the following, we focus mainly on the way the exocentric variable estimations are performed, though the role of the intersegmental variables and their internal estimations are not overlooked.

The present work does not deal with the problem of visual processing itself: how the primary visual inputs are separated into retinal orientation and movement with respect to the environment, and how the linear and angular displacements are disentangled. The problem of optic flow processing is known to involve the internal representation of the 3D environmental structure (Longuet-Higgins & Pradzny, 1980; Waxman & Ullman, 1985; Koenderink & Van Doorn, 1987). This aspect is to be further developed as an extension of the present work.

The central projections of the specific source of information provide also the best candidates for the anatomical structure involved in the central representation. Similarly, the coordinate system used by the captor yields a natural coding scheme for the internal estimations.

## The Coherence Constraint

The set of centrally represented variables listed in Table 17.1 is clearly overdetermined. Indeed, they represent physical variables that are linked together by both geometric and dynamic relationships. For instance, the two exocentric variables H and R (head and gaze angular velocities with respect to the environment) and the intersegmental variable E (eye velocity with respect to the head) are linearly dependent, that is, two of the three determine the third. Similarly, the head orientation with respect to gravity (or, more generally, with respect to the environment) and the head angular velocity are dynamically related. The coherence constraint hypothesis states that to these physical relationships correspond internal relationships between centrally coded variables. Several illustrations of the geometrical and dynamical coherence constraints follow.

In our model, the internal relationships between the estimated variables are implemented by the "coherent copies" of these variables. Let us consider, for instance, the head angular velocity estimator. As stated above, it receives a direct input from its specific captor: the semicircular canal system. But it receives also an input from the coherent copy of the head angular velocity. This coherent copy is computed from the other centrally coded variables: namely, the gaze and eye angular velocities and the head orientation with respect to gravity.

There are, then, two sources of information for each centrally represented variable: The first one is *external* and originates from the specific captor; The second one is *internal* and originates from other central representations through the implementation of the coherence constraints.

In other words, the notion of convergence of informations is replaced in our model by the notion of coherence constraint, and it is the need for coherence rather than the complementarity that imposes the sensorimotor integration.

## The Truth Constraint

The central representation must not only be coherent but also true, as far as possible. This implies that the combination of the direct external measure and the internal coherent copy takes into account the specific input/output characteristic function of the captor. Various combinations of these two signals yield to a correct central estimation. Taking into account the uncertainty attached to the measure and to the coherent copy, there exists one optimal combination that minimizes the uncertainty on the central representation. We assume that the characteristic function of the captor has been submitted to evolutionary pressure in order to equalize the uncertainty on the measure (it would not be economical to extend the working range of a captor if that had increased the uncertainty in such a way that the measure would not be reliable and therefore not usable for further processing).

In this case, the optimal estimation is rather simple. Let x be the physical variable, $y$ its central representation and $y'$ the coherent copy, then:

$$y = f(x) + y' - f(y') \tag{1}$$

where $f(.)$ denotes the input/output characteristic function of the captor. The same function is applied, but with a negative sign, to the coherent copy $y'$.

In normal situation, the coherent copy is also a true estimation of the physical variable $x$, then the equation (1) yields:

$$y = y' = x \tag{2}$$

An internal model of the characteristic function $f$ of the peripheral captor must be centrally implemented. However, it does not require this internal model to be fixed and predefined. On the contrary, the internal characteristic function that has to be applied to the coherent copy is very likely adaptable. The error signal that drives the adaptive process could be the difference between $y$ and $y'$. For instance after destruction of the captor, the internal function $f$ is progressively set to zero, and the central representation $y$ becomes equal to the coherent copy $y'$.

## The Geometric and Dynamic Coherence Constraints

In the following, the same notation used for a neural message it also used, for the sake of simplicity, for the operator that elaborates it.

### The Cross-Coupling Between Head and Gaze Angular Velocities

A first illustration of the four basic hypotheses is given by the cross-coupling between internal estimations of head and gaze angular velocities. Both are tridimensional exocentric variables. The head angular velocity vector H is directly measured by the semicircular canals in a wide frequency range. The gaze angular velocity vector R is directly measured by the angular retinal slip (as mentioned above, we bypass the problem of separation between angular and linear displacements in the visual processing of optic flow). The central representations of these two physical variables are $\underline{H}$ and $\underline{R}$, respectively. They are linked by the internal estimation of the intersegmental variable $\underline{E}$ (eye velocity with respect to the head). The two coherent copies R′ and H′ are computed by, respectively, adding $\underline{E}$ to $\underline{H}$ and substracting $\underline{E}$ from $\underline{R}$ (see Fig. 17.1). The input/output functions of the two captors are modelized by two linear transfer functions: for the semicircular canal system (output Hc), a high pass filter (time constant Tc = 5 sec):

$$Hc/H = sTc/(1 + sTc) \tag{3}$$

and for the visual system (output Rr), a low pass filter (time constant Tr = 1 sec.)

$$Rr/R = 1/(1 + sTr) \tag{4}$$

As depicted in Fig. 17.1, the same two transfer functions are applied to the coherent copies H′ and R′, respectively.

The internal representation of the intersegmental variable E is not detailed here. The eyeball has no external load and practically no inertia. Therefore the eye movements in the head can always be correctly estimated, using proprioceptive inputs and efferent copy of the motor command. In order to stabilize the retinal images, the eye movements are driven by two compensatory signals: the head velocity estimation $\underline{H}$, with a negative gain: $-$ a ($|a| < 1$) the gaze velocity estimation $\underline{R}$, with a negative gain: $-$ b.

With respect to the other models, these gains are different from a pure vestibulo-ocular reflex (VOR) gain or open loop gain (for a) and from a pure optokinetic gain or close loop gain (for b). In our model, both internal representation $\underline{H}$ and $\underline{R}$ convey visual and vestibular informations. During head movements in darkness, for instance, the gaze velocity is estimated

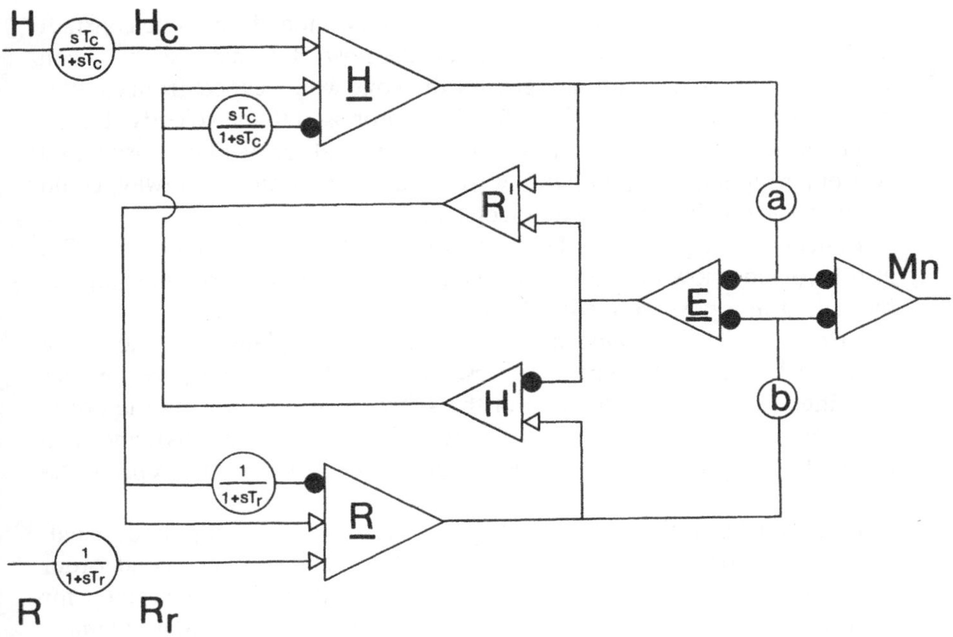


FIG. 17.1. The coherence constraints between head (H) and gaze (R) angular velocity estimations. The central representations are underscored. They receive the specific afferent input (from canals: Hc; from retinal slip: Rr) and the coherent copies H' and R', first as a direct input, secondly as an indirect input through an internal filter. Open triangles: excitatory input; full circles: inhibitory input. E is an efferent copy of the eye velocity command, which is sent also to the motoneurons. The final integrator that adds the required tonic signal to the motoneurons is not represented in the figure.

through the coherent copy R', and the resulting slow phase velocity of vestibular nystagmus is dependent upon both gains a and b:

$$-E/H = sTc.(a + (a+b) .sTr)/(1-a + sTc + (1+b)\ s^2TcTr) \quad (5)$$

During high frequency oscillations in darkness, the overall VOR gain can be approximated by:

$$-E/H = (a+b)/(1+b) \quad (6)$$

which is close to 1, even if the "direct pathway" gain, a, is small, provided that b is great compared to 1.

Because the time constant of visual inputs (Tr) is supposed to be short as compared to the time constant canal inputs (Tc), the overall VOR behavior, as predicted by the model at low-frequency oscillations, is close to that of a high-pass filter with a time constant equal to Tc/(1 − a), which is greater than the time constant of canal inputs. Indeed, the time constant of the slow phase eye velocity decay during postrotational tests has been shown to be three or four times greater than one would expect from cupular dynamics. Raphan et al. (1977, 1979) have proposed that a velocity storage mechanism, that is, a neural integrator that would receive both visual and vestibular inputs, acting in parallel with a direct vestibular pathway, could account for this increase in time constant.

Functionally equivalent, the model proposed by Robinson (1975, 1977) also involved a low-pass filter receiving both visual and vestibular inputs, but acting in a positive feedback loop.

In our model, the same dynamic behavior is explained by the cross-coupling between the central representations of head and gaze angular velocities (and also, as detailed in the subsequent sections, by the cross-coupling between head angular velocity and gravity estimations), and then results logically from the required coherence between central representations.

It should be noticed that for high values of b, this simplified visual-vestibular model presents damped oscillations due to the second order denominator in Equation (5). These damped oscillations may account for the existence of secondary and tertiary phases of postrotational nystagmus. The quantitative stimulations are detailed below.

In the light, during head movements in a stationary visual environment, the internal estimations $\underline{H}$ and $\underline{R}$ are equal to the physical variables:

$$\underline{H} = H \text{ and } \underline{R} = H + E = R \tag{7}$$

This result is independent of the values of the gains a and b, and is also true if an additional eye movement command is added (smooth pursuit for instance).

During optokinetic stimulation with head fixed, the slow phase eye velocity is proportional to the stimulation velocity S:

$$E/S = (a+b + sbTc)/(1+b)\ (1 + sTc + s^2TcTr) \tag{8}$$

that could be approximated at low frequency by:

$$E/S = (a+b)\ /\ (1+b) \tag{9}$$

In conclusion, the simple application of the four basic hypotheses to the head and gaze coupling leads to a quantitative model that explains the main dynamic characteristics of the visual-vestibular interactions during vertical axis rotation. The "velocity storage mechanism" (Raphan et al, 1977) may be considered as a consequence of the coherence constraint that links the two exocentric centrally represented variables.

## The Interaction Between Head Linear Velocity and Acceleration

In the preceding paragraph, we had to combine variables of the same kind (angular velocity) but corresponding to two different body segments (geometric coherence). We consider now the case of two central representations of the same body segment displacement (the linear head movement), but with different dynamics: the linear velocity estimation $\underline{V}$ and the linear acceleration estimation $\underline{A}$. The captors involved in their measurements are the visual system ($\underline{V}$) and the irregular and phasic afferents of the otolithic system ($\underline{A}$). As already mentioned, the separation of the primary visual inputs into angular and linear velocities is supposed to be solved. The similar problem of separation of the otolith signal into gravitational acceleration and inertial acceleration are treated next. For the time being, we should simply remind that:

1. There is no linear intersegmental displacements between eye and head, and the rotation centres of the two segments are close enough, so that the linear velocity of the retina with respect to the environment, measured by the visual system, is practically equal to the head linear velocity with respect to space.
2. If the gravity vector and the head angular velocity are known, then the gravitational contribution to the phasic discharge of the otolithic irregular units is computable and can be removed from the afferent signal.

When the head is not rotating, the linear acceleration may be obtained by temporal derivation of the linear velocity (and, reciprocally, the linear velocity is the exact integral of the acceleration). However, this is not true when the head—and by extension for the moving body segment—is submitted to angular movements. For instance, during a circular movement at constant speed (as when driving a car in a highway curve), the head linear velocity is a constant vector in the craniotopic frame of reference, but the head linear acceleration is not zero. In order to take into account the angular velocity H of the frame of reference, the linear acceleration must be computed as follows:

$$A = dV/dt + H \times V \qquad (10)$$

The Equation (10) is included in the model to form the coherent copy A′ starting from the central representation of the head angular and linear velocities ($\underline{H}$ and $\underline{V}$), and reciprocally to form the coherent copy V′ from the central representation of head angular velocity and head linear acceleration. The scheme of the model, shown in Fig. 17.2, is very similar to the previous one: Here the cross product $\underline{H} \times \underline{V}$ plays the role of the intersegmental eye velocity $\underline{E}$.

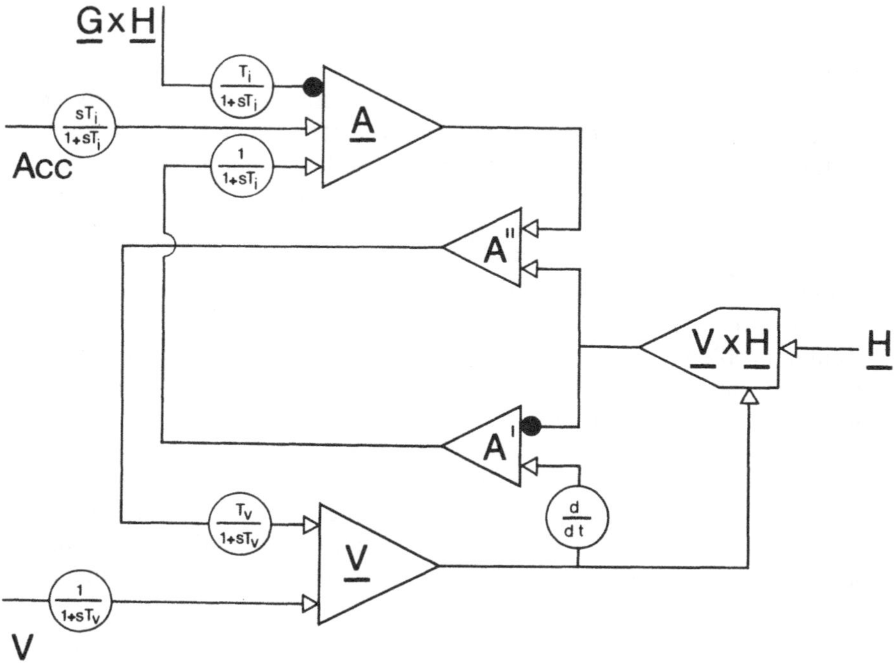


FIG. 17.2. The coherence constraints between head linear accleration (A) and velocity (V). Same notations as in Fig. 17.1. The gravitational jerk (computed by the cross-product $\underline{H} \times \underline{G}$) must be added to the primary afferences (from otolithic irregular units) dedicated to the linear acceleration estimation, through a transfer function which reproduces the input/output characteristics of the captor. The coherent copy of the linear velocity is computed by integration and filtering of its temporal derivative A', originating from the linear acceleration estimation.

As shown by Fernandez and Goldberg (1976) in the monkey, the otolithic primary afferent units can be segregated in two classes: the regular units whose discharge is closely related to linear acceleration, and the irregular—or phasic—units, that preferably respond to acceleration changes. Several authors (Mayne, 1974; Ormsby & Young, 1977) have postulated that this segregation might reflect different functions, and it is often assumed that the regular unit signals are likely used to estimate the gravity vector, whereas the irregular unit signals probably serve as an input to estimate the linear displacements. Indeed, the gravitational forces acting on the otolith end-organs are usually constant because of the head angular stabilization in normal posture, whereas the inertial forces, such as those that occur during normal locomotion, are usually transient.

The signal originating from the otolithic irregular units (Fernandez & Goldberg, 1976) is simulated by a simple high-pass filter applied to the linear acceleration (time constant Ti = 1.5 sec), whereas the visual measure of head linear velocity is simulated by a low-pass filter (time constant Tv = 1 sec.). The same transfer functions are also applied to the coherent copies, A′ and V′, respectively.

These transfer functions are oversimplified versions of the input/output characteristics of the captors: In particular, the nonlinearities have been overlooked. Nonetheless, the quantitative simulations fit quite well to the experimental data.

### The Interaction Between Head Orientation and Angular Velocity

The regular otolithic units preferably respond to a constant acceleration, which on earth is mainly due to the gravitational force. The gravity vector in head coordinates system is certainly an important information that plays a major role both in postural readjustments and in head orientation with respect to the vertical. As an angular positional vector, it must interact with its time derivative: the head angular velocity.

The gravitational acceleration is a practically constant vector with respect to the earth (there are of course very small variations; they are ignored here). By applying the equation (10) to the gravity vector in head coordinate system, we obtain

$$dG/dt + (H \times G) = 0 \tag{11}$$

The equation (11) can be used to compute the coherent copies H′ and G′: Its integration leads to an indirect estimation of G from the head angular velocity. The application of the double cross-product law leads to an indirect estimation of H from the gravity vector:

$$H' = ( d\underline{G}/dt \times \underline{G} + \langle \underline{H},\underline{G} \rangle . \underline{G}) / \langle \underline{G},\underline{G} \rangle \tag{12}$$

where <.,.> denotes the dot product between two 3D vectors.

Because the magnitude of the physical gravity vector is constant, the denominator in the equation (12) can be replaced by a constant parameter. The scheme of the model is illustrated on Fig. 17.3. Indeed, the coherent copy of the head angular velocity originates from two sources: from the gaze angular velocity R (see Fig. 17.1) and from the gravity vector G (Fig. 17.3). Any linear combination of the two coherence constraints can be used in the model. More generally, a centrally represented variable may receive multiple coherent copies from other internal estimations: These signals can be combined in different ways. The head-orientation has also to be internally defined by nonvertical vectors (probably originated from visual information). These centrally represented vectors must interact with both estimated gravity vector $\underline{G}$ and head angular velocity $\underline{H}$.

In the simplified scheme presented in Fig. 17.3, the regular otolithic

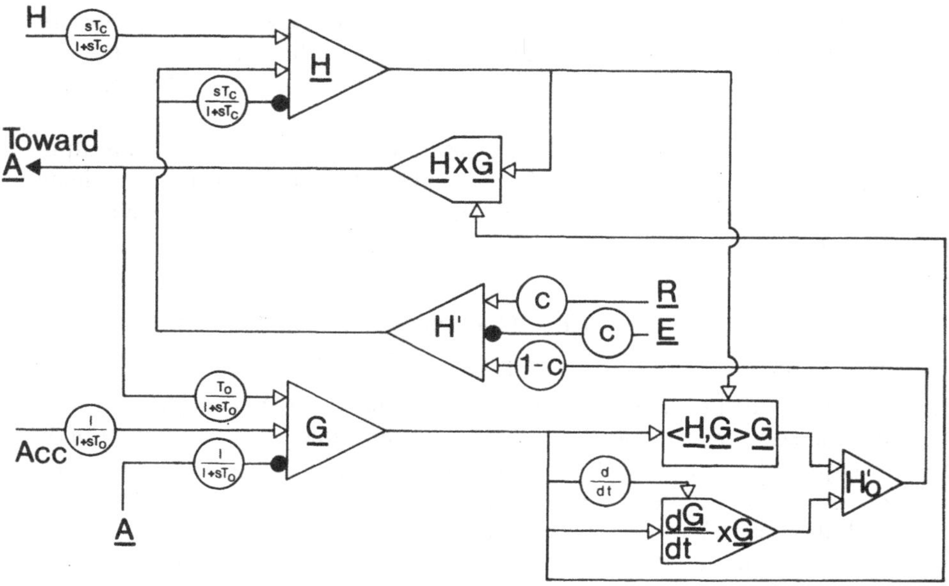


FIG. 17.3. The coherence constraints between the gravity vector (G) and the head angular velocity (H). Same notations as in Fig. 17.1. The linear acceleration is removed from the otolithic regular afferents, and the gravity vector estimation is completed in the high-frequency domain by the gravitational jerk, which is the cross-product H × G. The gravitational coherent copy of the head angular velocity (H'o) is computed according to the double cross-product law (see text). H'o is linearly combined with the coherent copy originating from the gaze and eye angular velocities.

captor is modelized by a low-pass filter (time constant To = 0.5 sec). The gravity vector estimation receives the output of this filter, the coherent copy G′ originating from the internal estimation of head angular velocity, and an input from the internal estimation of head linear acceleration A. The central representation of head angular velocity is similar to the one presented in Fig. 17.1, except that it now receives also an input from the gravity estimation. The coherent copy H′ is now formed by linear combination of a signal derivated from head orientation (H'o) and the difference between the gaze and the eye angular velocity. The linear combination is determined by a single constant parameter: c.

The separation between the gravitational and inertial components of the otolithic signal is primarily based on the dynamic characteristics of the regular and irregular units (Mayne, 1974): It relies on the gravity vector constancy on the one hand, and on the usually transient nature of the inertial acceleration on the other hand. However, this difference in dynamic properties would be true only if the two variables were measured in an earth-based frame of reference. Because the captors are moving with the head, the angular velocity of the frame of reference has to be included in the dynamic separation process: This is done by the coherence copy G′ that constrains the gravity estimation to be constant in space (through equation [11]) and by the coupling between head linear velocity and acceleration.

It should be recalled that the three schemes presented in Figs. 17.1, 17.2, 17.3 are not independent: They are three modules of the same model. In all experimental situations simulated, these three modules strongly interact, even when a single sensorial modality is solicited. For instance, a circular optokinetic stimulation about the earth-horizontal X axis induces a circular vection (roll vection), based on the central representation $\underline{H}$ of the head angular velocity. This in turn changes the orientation of the estimated gravity vector through its coherent copy G′. However, this change of orientation is limited because of the transfer function applied to G′ (this transfer function results from the input/output characteristic of the regular otolithic afferents). Finally, the gravitational component of the coherent copy H' strongly limits the circular vection itself.

## QUANTITATIVE SIMULATIONS

### Rotations Around the Vertical Z Axis (VOR Gain and Phase and Postrotational Nystagmus)

During rotations about the vertical Z axis, there is no dynamic stimulation of the otolith organ. The behavior of the model is then linear. The slow phase eye velocity may be computed as a linear combination of the physical angular velocities of the head (H) and of the visual stimulation (S):

$$E = A.H + B.S \tag{13}$$

where A and B denotes the following transfer functions:

$$A = -\ sTc\ (a + (a+b)sTr)/(c(1 - a) + sTc + (1+b)s^2TcTr) \tag{14}$$

$$B = (c(a+b) + bsTc)/(1 + b).(c + sTc + s^2TcTr) \tag{15}$$

The transfer function A corresponds to the classical VOR measured in the dark. In Fig. 17.4, we have plotted the gain and phase of the slow phase eye velocity with respect to head velocity, as predicted by the model and compared the theoretical curves to the experimental data on human VOR recently published by Wall and Black (1984). There is, of course, intersubject variability, so that we have not tried to precisely fit the model to the experimental data. For instance, other authors have reported VOR

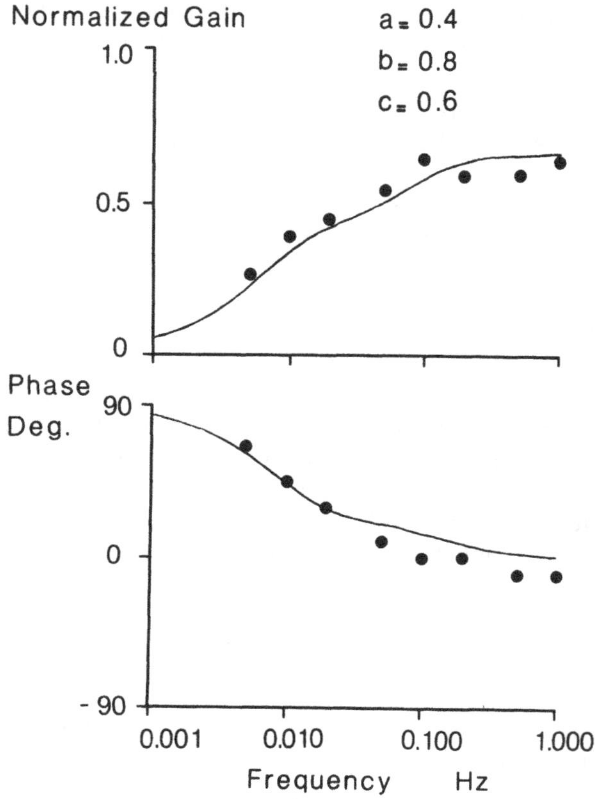


FIG. 17.4. VOR gain and phase during rotation around the vertical Z axis at different frequencies in darkness. Dots indicate mean experimental values obtained by Wall and Black (1984). The parameters a, b, and c are defined in the text.

gain close to 1 at high frequency oscillations (Tomlinson et al., 1980; Baloh et al., 1984). This could be modelized by increasing the gain b of the gaze stabilization in our model.

The transfer function B corresponds to the dynamic characteristic of the optokinetic nystagmus. The smooth pursuit system—not included in our model—is generally assumed to play a role in the fast-rising phase of the eye velocity just after the optokinetic stimulation onset. Fig. 17.5 shows a sample of the model output during and after an optokinetic stimulation. The eye velocity trace exhibits a rapid increase after the stimulation onset and a rapid decrease after the light switching off. Therefore, the fast initial response observed in man and, to a lesser extent, in monkey, can be partly explained without involving the smooth pursuit system.

The eye movements induced by postrotatory tests are illustrated in Fig.

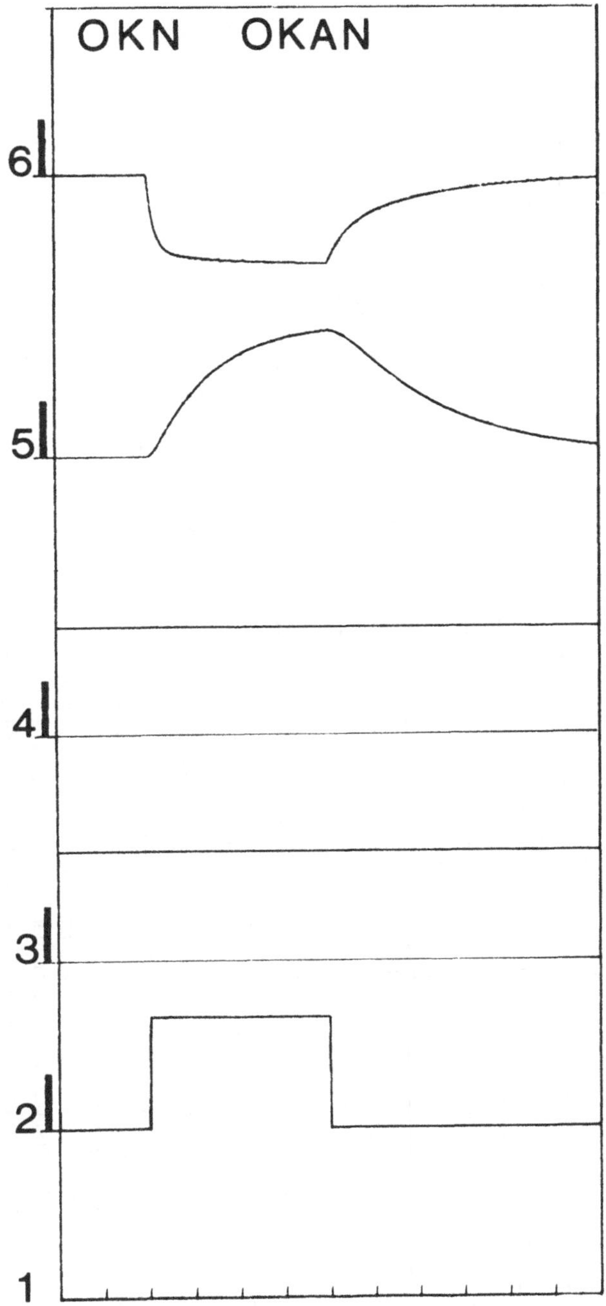


FIG. 17.5. Simulation of optokinetic nystagmus and optokinetic after-nystagmus (rotation around the vertical Z axis). Traces from bottom to top are defined as follows:

1. actual head velocity H
2. light indicator
3. total head acceleration along the Z axis (scale: 5 m/s²)
4. estimated gravity vector along the Z axis (scale: 5 m/s²)
5. estimated head angular velocity around Z (scale: 0.5 rad/sec)
6. resulting eye velocity (scale: 0.5 rad/sec)

Time scale (in abscissae): 5 sec between marks.

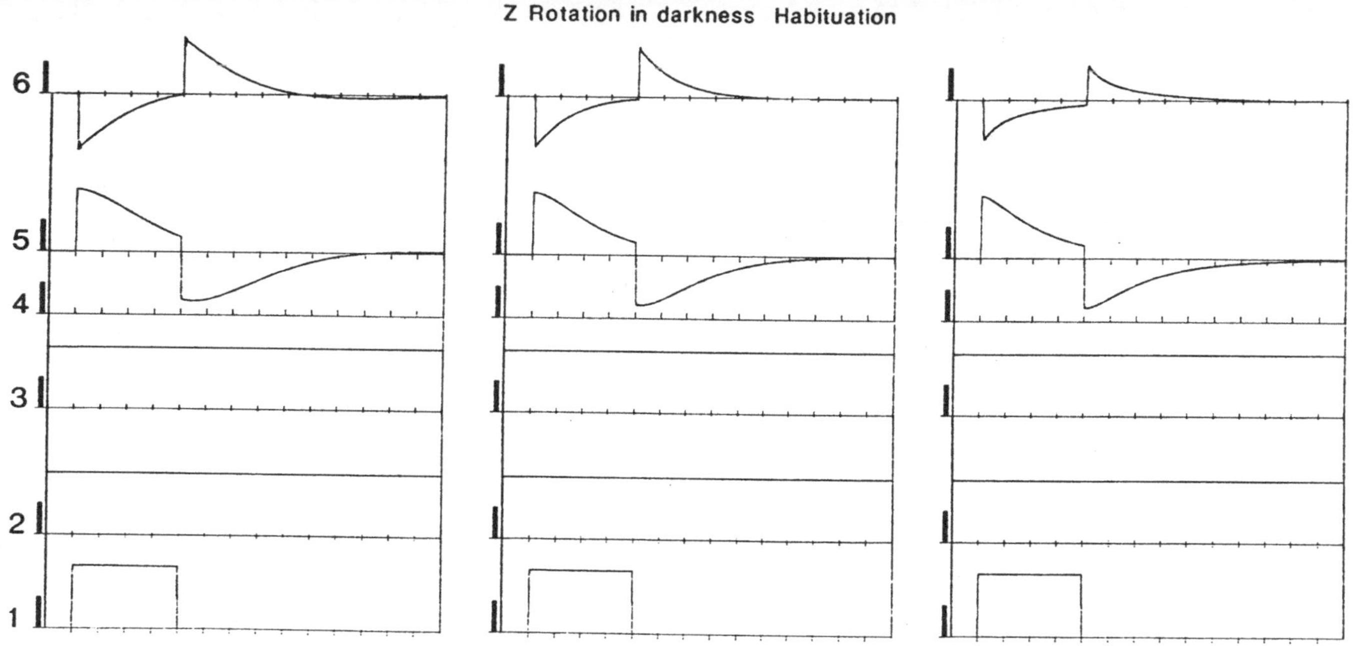


FIG. 17.6. Simulation of postrotatory nystagmus for different values of the gaze compensation gain, b. Traces from bottom to top are defined as follows:
1. actual head velocity H (scale: 0.5 rad/sec)
2. total head acceleration (Acc) along the Z axis (scale: 5 $m/s^2$)
3. estimated gravity (G) component along the Z axis (scale: 5 $m/s^2$)
4. estimated gravity (G) component along the interaural axis (same scale)
5. estimated head angular velocity around the Z axis (scale: 0.5 rad/sec)
6. resulting slow phase eye velocity around the Z axis (scale: 0.5 rad/sec).

Time scale (abscissae): 5 sec between 2 marks.

From left to right, the gaze compensation gain is decreased as follows: on the left: b = 6; in the middle: b = 3; on the right: b = 1.

The decrease of b yields a significant decrease of the peak eye velocity, time constant and a clear suppression of the secondary phase of the postrotatory nystagmus

17.6. The eye velocity decreases with a "time constant" greater than the time constant of the semicircular canals (5 to 6 sec). This effect is also explained by the internal feedback loop of Robinson (1975) or by the velocity storage mechanism (Raphan et al., 1977). More interesting is the existence of a secondary phase of the postrotatory nystagmus. As mentioned, this effect is explained in the model by the cross-coupling between the central representations of head and gaze velocities. It should be noticed that the secondary phase appears only for high values of the gain b. Repetitive postrotatory tests (habituation) have been shown to decrease the initial eye velocity peak and to decrease or even suppress the secondary phase of the nystagmus. A simple reduction of the gain b may account for these observations (Fig. 17.6).

### Combined Visual-Otolithic Stimulation During Head Translation (L-Nystagmus and Linear Vection)

The frequency characteristics of the linear vection was investigated in man by Berthoz et al. (1975) and Chu (1976). The gain and phase of the perceived linear displacement are compatible with the low-pass filter with a time constant of 1 second which is included in the model as the preprocessing of the visual signals (see Fig. 17.1 and 17.2).

Lateral linear accelerations of the head induce a small but significant nystagmus (Niven et al., 1965; Buizza et al., 1980). Recently, Viirre et al. (1986) have reformulated the L-nystagmus by computing the eye velocity E that would exactly compensate for a given head angular velocity ($\underline{H}$) and a given head linear velocity ($\underline{V}$):

$$E = -\underline{H} + (\underline{V} \times T)/\langle T,T \rangle \tag{16}$$

where T denotes the current craniotopic location of the target. Because the target is normally foveated, the vector T is colinear to the craniotopic eye position vector. Therefore, Equation (16) is equivalent to:

$$E = -\underline{H} + (\underline{V} \times \underline{P}) \tag{17}$$

where $\underline{P}$ is a vector colinear to the craniotopic eye position, and the magnitude of which is the proximity of the foveated target. We used this expression in the model to generate the compensatory eye movements induced by a central representation of the head linear velocity $\underline{V}$. We assumed that the target proximity is set in darkness to a resting value of 0.2 $m^{-1}$, whereas it is determined by the visual system in the light.

Buizza et al (1980) measured the L-nystagmus sensitivity in the light and in darkness. In Fig. 17.7 is shown a simulation of their experiments. The value of the resting proximity has been chosen to fit the experimental sensitivity of the L-nystagmus in darkness. However, the phase relationship

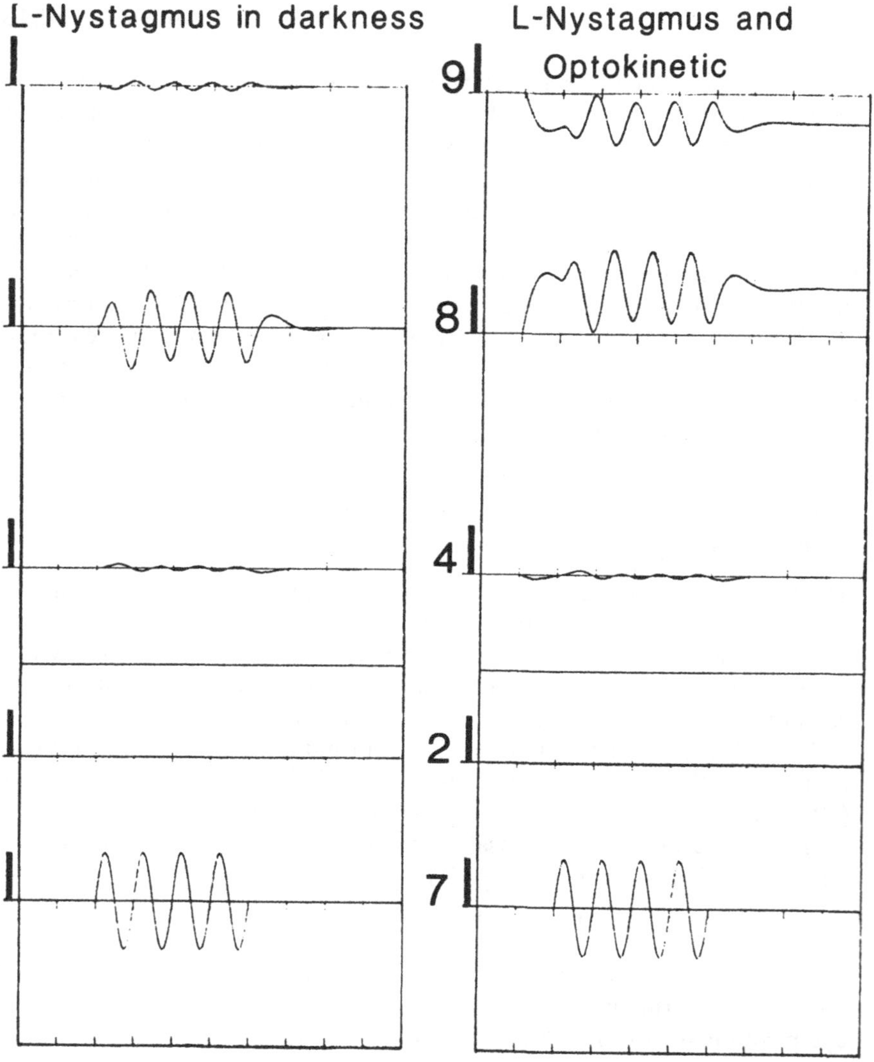


FIG. 17.7. Simulation of L-nystagmus and modulation of optokinetic nystagmus during horizontal lateral displacement of the head.
In darkness, the weakness of the L-nystagmus is due to the absence of target; in the light the important modulation of optokinetic nystagmus is due to the important value of the proximity (Buizza et al, 1980). Traces are defined as follows:

2. total linear acceleration along Z (scale 5 $m/s^2$)
4. estimated gravity component along Y (same 1 $m/s^2$)
7. total linear acceleration along Y (scale 1 $m/s^2$)
8. estimated head linear velocity along Y (scale 1 m/s)
9. resulting slow phase eye velocity (scale 1 rad/sec)

Time scale (abscissae): 5 sec between 2 marks

between eye velocity and head linear acceleration is also correctly predicted by the model. Moreover, the modulation of the optokinetic reflex induced in the light by the otolithic stimulation is also predicted by the model.

### Off-Vertical Axis Rotation in Darkness (OVAR and "Barbecue" Stimulation)

We have first simulated the "barbecue" experiments that consist in a rotation about the horizontal longitudinal axis at a constant speed. After several seconds of such a stimulation, the semicircular canal afferents cease to inform the central nervous system. However, as described by Guedry (1965) the sensation of rotation and the compensatory nystagmus last as long as the stimulation continues. Fig. 17.8 shows an example of the stimulation during OVAR: The central representation of head angular velocity as well as the mean slow phase eye velocity ("bias" component) decrease after the onset of the rotation but are maintained to a constant level even after 3 min of stimulation. The model predicts also the modulation of the slow phase eye velocity ("sinusoidal" component) and the perception of a rotating head linear velocity.

The level of constant "bias" component and the peak-to-peak modulation of the slow phase eye velocity have been computed for different angular velocities and compared to the experimental results obtained in man by Benson and Bodin (1966) and Correia and Guedry (1966) (see Fig. 17.9). The best fit was obtained with the time constants To = 0.2 sec and Ti = 0.9 sec and an overall VOR gain of 0.9.

The permanent bias component is due to the coherent copy of head angular velocity that derives from the rotating gravity vector estimation, through the computation of the cross product (Equation [12]). The low values obtained during rotations at high velocities result from a decrease in the internal estimation of the gravity vector magnitude. This decrease is due to the dynamic cross-coupling between the central representations ($\underline{H}$, $\underline{G}$, $\underline{A}$ and $\underline{V}$). Guedry (1974) reported that during high-velocity rotations in "barbecue" experiments, the subjects feel as if they are upright. This perceptual effect may be explained by the very small magnitude of the perceived gravity vector, as predicted by the model at high-velocity rotations, a situation that can be compared to free-floating in weightlessness.

Similarly, the rotations about a tilted axis (OVAR) induce a permanent "bias" component of the horizontal nystagmus and a sensation of rotating along a cone (Darlot et al., 1988; Denise et al., 1988). The main characteristics of the slow phase eye movements are predicted by the model: They do not differ qualitatively from the barbecue nystagmus. However, the existence of a constant gravitational component along the longitudinal

## Off-Vertical Axis Rotation (OVAR)

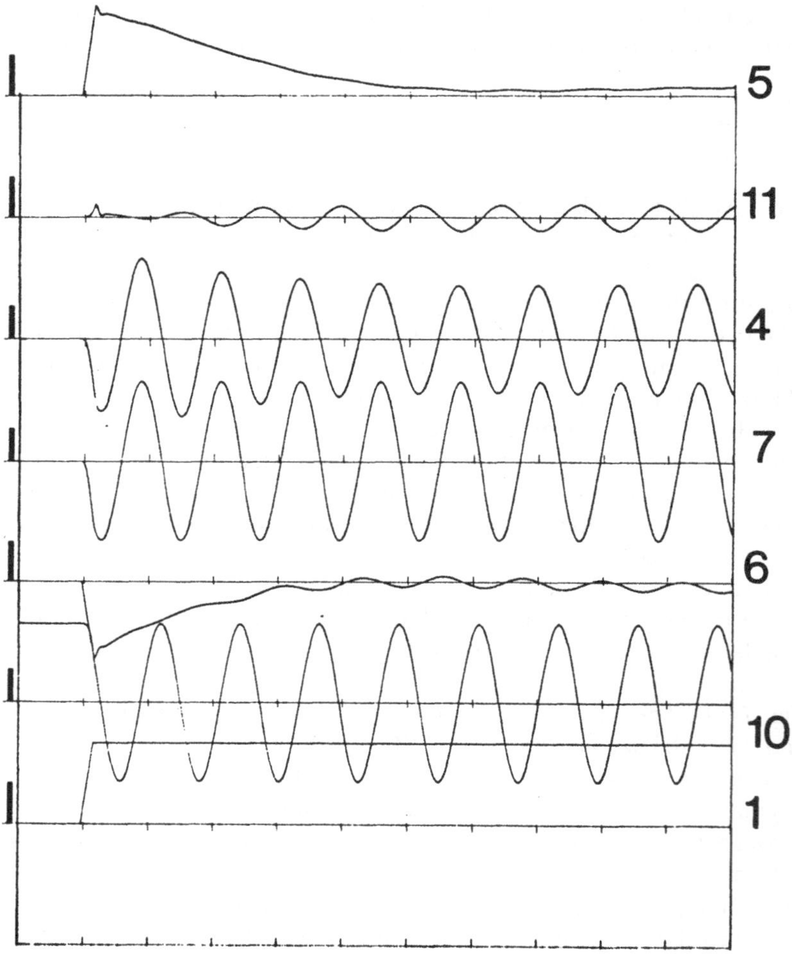


FIG. 17.8. OVAR simulation in darkness: the subject Z axis is tilted at 30 degrees with respect to the vertical. The angular velocity is 1 rad/s (57 deg/s). Traces are defined as follows:

1. head angular velocity, Z axis (scale: 0.5 rad/s)
4. estimated gravity vector (G), Y axis (scale: 2 $m/s^2$)
5. estimated head angular velocity, Z axis (scale: 0.5 rad/s)
6. resulting slow phase eye velocity, Z axis (scale: 0.5 rad/s)
7. actual acceleration (Acc), Y axis (scale: 2 $m/s^2$)
10. actual acceleration (Acc), X axis (same scale)
11. estimated head angular velocity, Y axis (scale: 0.5 rad/s)

Time scale (abscissae): 5 sec between 2 marks.

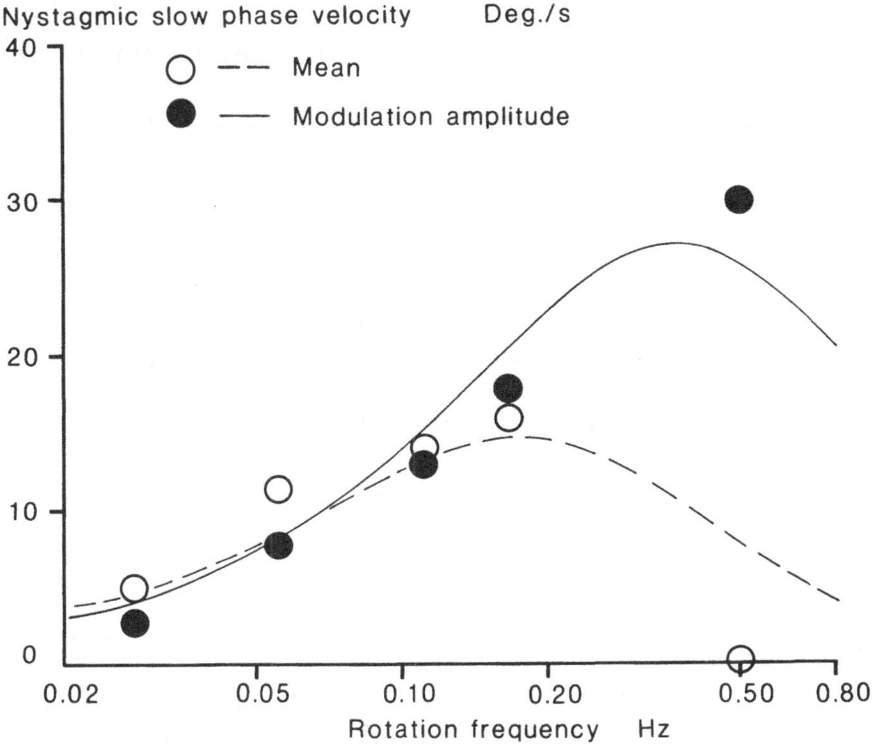


FIG. 17.9. Simulation of the barbecue experiment. In ordinate: mean or "bias" eye velocity (broken line) and sinusoidal amplitude (continuous line). In abscissae: frequency (the corresponding angular velocity is 360 deg/sec at 1 Hz). Experimental data have been reported by Benson & Bodin (1966) and Correia & Guedry (1966).

axis at small tilt angles produces a mean eye velocity greater than the value that would be expected if only the X and Y otolithic stimulations were taken into account. In Fig. 17.8 an example is shown of OVAR simulation at a tilt angle of 30° and an angular velocity of 1 radian/sec (57 deg/sec). The orientation of the centrally represented gravity vector is slightly in advance with respect to the actual G axis: This is also consistent with the sensations reported by the subjects during OVAR stimulation.

Tilting the head during postrotatory tests induces a significant decrease of the nystagmus time constant. Figure 17.10 shows a simulation of this experiment: the time constant is reduced in this case from 15 sec to about 8 sec. A similar reduction of the time constant of the perceived head angular velocity is also observed. The time constant decrease is due to the coupling between the gravity and the angular velocity estimators. The semicircular canal stimulation induced by the sudden rotation stop yields also a significant but limited rotation of the central representation of the gravity vector.

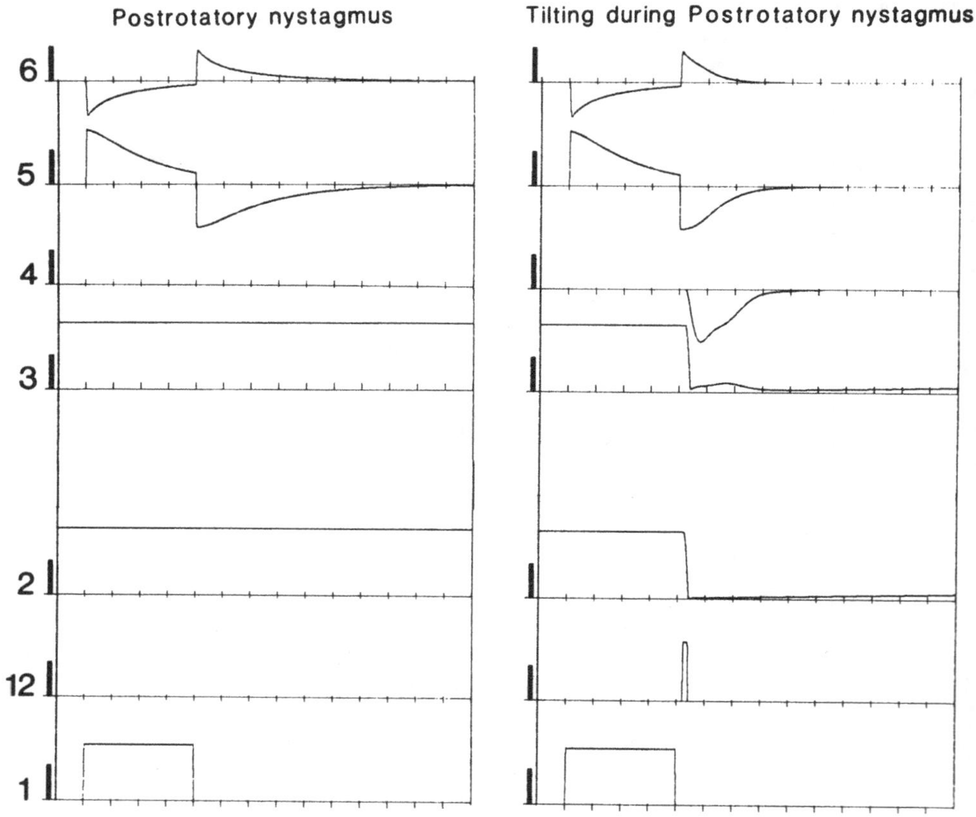


FIG. 17.10. Simulation of tilting during a postrotational test. On the left: standard postrotational test around the vertical Z axis. On the right, the subject Z axis is rapidly moved from a vertical orientation to an horizontal orientation, 1 second after the chair has stopped. Traces are defined as follows:

1. actual head angular velocity, Z axis (scale: 0.5 rad/sec)
2. total head acceleration, Z axis (scale: 5 m/s$^2$)
3. estimated gravity vector, Z axis (same scale)
4. estimated gravity vector, Y axis (same scale)
5. estimated head angular velocity, Z axis (scale: 0.5 rad/s)
6. resulting slow phase eye velocity, Z axis (same scale)
12. actual head angular velocity, Y axis (scale: 1 rad/s)

Time scale (abscissae): 5 sec between 2 marks.

### Optokinetic Stimulation Around An Off-Vertical Axis (Vertical Circular Vection)

The optokinetic stimulation around the X axis (frontal plane) elicits a circular vection much stronger when the subject lies on his back (the X axis is aligned with the gravity) than when the subject is upright (the X axis is then horizontal). This effect is explained in our model by the coherence constraint between the gravity and the head angular velocity estimators: When the rotation axis is colinear to the gravity, the perceived velocity is not affected by the gravity estimation. But, when the rotation axis is horizontal, the coherence constraint induces not only a reduction of the time constant, as discussed in the preceding paragraph, but also a marked reduction of the estimated angular velocity. The ratio of circular vection amplitudes according to the orientation of the rotation axis with respect to gravity is predicted by the model as equal to the parameter c, which is the gain of the gaze estimator input to the coherent copy H′ (see Fig. 17.3).

The model accounts also for the perceived tilting of the vertical during optokinetic stimulation around a horizontal axis. For small angular velocity of the stimulation (S), the gravity estimation deviates from the actual vertical of an angle T equal to:

$$T = \text{Arctang}\,[\,c.(To + Ti).S] \tag{18}$$

Dichgans et al (1972) reported that during circular vection, the perceived vertical deviation is proportional to the stimulation velocity but saturates at 15 degrees above 30 to 40 deg/sec. These data are consistent with the time constants used in the simulation. However, the existence of a saturation probably reflects a nonlinear characteristic of the visual signal preprocessing.

## DISCUSSION

### Functional Characteristics of the Model

The model has generally a nonlinear behaviour, except in the case of a pure rotation about the gravity axis. The nonlinearities are due to the cross-product operations between central representations. There are, of course, other sources of nonlinearities that have not been considered here for the sake of simplicity. These other nonlinearities are generally considered to originate from the input/output function of the captors: particularly in the visual system during high-stimulation velocity, and in the otolith system during high-acceleration level or for unusual head orientation. These characteristics have not yet been included in the model.

The linear behavior of the model during vertical axis rotation is not markedly different from the functional characteristic of the classical models of visual-vestibular interaction. It is not surprising because Robinson's model, for instance, is clearly embedded in ours (the internal positive feedback loop is interpreted in our model by the connection that links the gaze velocity to the head velocity). The principles on which the proposed model is based are therefore consistent with the kind of organisation proposed by Robinson (1975, 1977, 1982).

Most experimental results concerning compensatory eye movements and perceptual effects due to off-vertical rotations or head translations are qualitatively and quantitatively accounted for by the coherence constraints between central representations. Except for the gains a, b, and c, all the parameters of the model could be directly infered from the dynamical characteristics of the afferent signals.

## Connectivity

Three types of connections can be distinguished in the model. For instance, in Fig. 17.1:

1. Direct connections from one receptor to the corresponding operator, for instance, from canals to operator $\underline{H}$.
2. Indirect connections from one operator to another, which convey unprocessed central informations, for instance, from operator $\underline{H}$ to operator $\underline{R}$. The conveyed information is here the central representation of head rotational velocity, and in the connectivity pattern of Fig. 17.1 reaches $\underline{R}$ via R′. But it can be conceived that a connection from $\underline{H}$ to $\underline{R}$ could exist, by-passing R′, and complemented by an input from $\underline{E}$ to $\underline{R}$. These two connections would be parallel to a connection via R′, whereas this latter would only transmit the filtered informations. This expanded version of the connectivity pattern (not illustrated) does not change the structure nor the functioning of the model, but allows comparison with anatomical connections.
3. Indirect connections conveying processed (here filtered) informations, for instance, from operator R′ to operator $\underline{R}$ via the filter 1/(1 + s.Tr). The conveyed information is here the coherent copy of the message from the canals.

A tentative identification of these connections with connections between neural structures can be proposed. The indirect connections conveying unprocessed information would belong intrinsically to the brainstem, whereas the connections conveying processed informations would be projections from the cerebellum onto brainstem structures. Therefore, the filters would appear to stand for the processing of informations by the cerebellum. Formal similarities with connectivity of previous models and with anatomical connections are in agreement with this identification.

The model proposed by Schmid (1986) to account for habituation has two loops, both converging on a single operator, one feedforward and one feedback loop, this latter accounting for the so-called Velocity Storage Mechanism. Two variable gains, one on each loop, can change the weights of the loops and account respectively for habituation of VOR and adaptation to wearing of prisms or lenses. Similarly, Cohen et al. (1985) proposed that the output of the uvulo-nodulus can change the gain of the VSM, as represented in the model of Raphan (Waespe et al. 1983). Although the two models are not homeomorphic, it is tempting to compare the feedforward loop of Schmid's model, with that part of Raphan's that contains uvulo-nodulus, and the feedback loop with the part containing the flocculus.

In the model proposed here, the weights of the connections are fixed, because it is not devoted to account for neural plasticity. The only parameters that are not set by the constraints are a, b, and c, and are therefore the only ones that can be varied to describe the changes in VOR performances. However, a comparison can be made with the two previous models. The filter equivalent to transfer function of semicircular canals [operator (s.Tc)/(1 + s.Tc) located between H′ and H] can be compared to the uvulo-nodulus and the feedforward loop, and the filter equivalent to the transfer function of vision [operator 1/(1 + s.Tr) located between R′ and R] can be compared to the flocculus and the feedback loop. The symmetrical connectivity of the two operators would account for the reciprocal roles of the flocculus and the uvulo-nodulus, put into evidence by Waespe et al. (1985).

Connectivity of the floccular target neurons (FTN) as described by Lisberger (1988) is also consistent with this interpretation. Indeed, FTN receive indirect vestibular inputs, as do R via H, and inputs from the flocculus comparable to the indirect processed input that R receives from R′ via the filter. In addition, R receives a visual input, that the FTN either could directly receive, or that they could receive indirectly via a projection from the nucleus reticularis tegmenti pontis or pontine nuclei (Keller & Crandall, 1983). Moreover, the projections from the uvulo-nodulus to the vestibular nuclei (Precht et al. 1976) are also quite comparable to the connectivity of H in the expanded version of our model.

Several pieces of functional evidences support the proposed identification: Flocculectomy reduces accuracy of visual pursuit and efficiency of visual suppression of VOR, whereas uvulo-nodulectomy reduces tilt suppression (Cohen et al. 1985).

In addition, nodulectomy reduces motion sickness susceptibility (Bard et al. 1947), which can be simply accounted for by the present model. Motion sickness would arise from a discrepancy between the input from one captor and the prediction of this input by the internal filter. This discrepancy would disappear when the output from the filter would also disappear.

Adaptive changes could be prompted when discrepancies between normally equal messages would be detected. When a captor would be impaired or forced to work outside its normal range, adaptation would partly consist in equalizing the transfer function of the internal model of the captor to the new actual transfer function of the captor itself. As the internal model would come closer to the captor transfer function, the discrepancies would be reduced, and the functional role of the sensori–motor interaction would be recovered. In the same time, the motion sickness, frequently observed at the beginning of the adaptation process, would disappear. This could give a functional ground to the sensory conflict hypothesis, which is indeed replaced in our model by the notion of level of incoherence.

The duality of processed and unprocessed indirect connections could explain why movements are still possible after cerebellectomy, but with modified temporal pattern and reduced accuracy. Because the processed indirect connections convey the predicted dynamic responses of the captors, their removal would drastically change the dynamic pattern of any ongoing movement.

## Prediction and Extension of the Model

Most of the experimental stimulations involve a combination of only two sensory modalities. The coherence constraints can be used to predict the effect of a triple stimulation: visual, semicircular canal and otolithic systems. For instance, what would be the sensations of a subject sitting on a rotating chair (vertical axis) and submitted to a constant horizontal antero–posterior linear vection? The prediction would be the following: The vestibular stimulation induces a sensation of rotation about the vertical axis ($\underline{H}$z); the visual stimulation induces a sensation a linear vection along the antero-posterior axis ($\underline{V}$x); the cross-coupling between internal estimations of head linear velocity and acceleration yields a sensation of linear acceleration, through the computation of the cross-product $\underline{H} \times \underline{V}$, which, in turn, induces a lateral tilt of the perceived gravity $\underline{G}$. Because $\underline{G}$ and $\underline{H}$ are also coupled through the computation of their cross-product $\underline{H} \times \underline{G}$, this yields a deviation of the perceived gravity in the antero–posterior direction and consequently changes the linear vection itself.

Because the model is based on a set of very general assumptions, the same coherence scheme could be applied to other aspects of multisensory interaction and to other central representations. For instance, the coupling between eye position and eye velocity, classically considered as resulting from the final neural integrator, could be analyzed with a model very similar to the coupling scheme between gravity and head angular velocity. This question, and other aspects of the model, such as adaptation or connectivity, that have been too briefly discussed in the preceding sections, are to be detailed in the future.

## REFERENCES


Allum, J. H. J., Graf, W., Dichgans, J., & Schmidt, C. L. (1976). Visual-vestibular interactions in the vestibular nuclei of the goldfish. *Experimental Brain Research, 26*, 463–485.

Baloh, R. W., Hess, K., Honrubia, V., & Yee, R. D. (1984). Low and high frequency sinusoidal rotational testing in patients with peripheral vestibular lesions. *Acta Otolaryngologica, Suppl. 406*, 189–193.

Bard, P., Woolsey, C. N., Snider, R. S., Mountcastle, V. B., & Bromiley, R. B. (1947). Delimitation of central nervous mechanisms involved in motion sickness. *Federal Proceedings, 6*, 72.

Benson, A. J., & Bodin, M. A. (1966). Interaction of linear and angular accelerations on vestibular receptors in man. *Aerospace Medicine, 37*, 144–154.

Berthoz, A., Pavard, B., & Young, L. R. (1975). Perception of linear horizontal self-motion induced by peripheral vision (linear vection). Basic characteristics and visual vestibular interaction. *Experimental Brain Research, 23,* 471–489.

Borah, J., Young, L. R., & Curry, R. E. (1980). Optimal estimator model for human spatial orientation. *IEEE Transactions on Systems, Man and Cybernetics*.

Buizza, A., Leger, A., Droulez, J., Berthoz, A., & Schmid, R. (1980). Influence of otolithic stimulation by horizontal linear acceleration on optokinetic nystagmus and visual motion perception. *Experimental Brain Research, 39*, 165–176.

Buizza, A., Schmid, R., & Droulez, J. (1981). Influence of linear acceleration on oculomotor control. In A. Fuchs & W. Becker (Eds.), *Progress in oculomotor research* (pp. 517–524). Amsterdam: Elsevier.

Chu, W. H. N. (1976). Dynamic response of human linear vection. S. M. Thesis, MIT, Cambridge, MA.

Cohen, B., Raphan, T., & Waespe, W. (1985). Vestibulo-cerebellar control of the vestibulo-ocular reflex (VOR). In E. L. Keller, E. L., & D. S. Zee (Eds.), *Advances in the biosciences (Vol. 57), Adaptive processes in visual and oculomotor systems*. New York: Pergamon Press.

Correia, M. J., & Guedry, F. E. (1966). Modification of vestibular responses as a function of the rate of rotation about an earth horizontal axis. *Acta Otolaryngologica, 62*, 297–308.

Correia, M. J., & Money, K. E. (1970). The effect of blockage of all six semicircular canal ducts on nystagmus produced by dynamic linear acceleration in the cat. *Acta Otolaryngologica, 69*, 7–16.

Darlot, C., Cohen, B., Berthoz, A., & Denise, P. (1987). Eye movements and perceptual effects induced by Off-Vertical Axis Rotation at small angles of tilt. In M. D. Graham & J. L. Kemink (Eds.), *The vestibular system, neurophysiologic and clinical research* (pp. 271–280). New York: Raven Press.

Darlot, C., & Denise, P. (1988). Nystagmus induced by rotation about off-vertical axis in the cat. *Experimental Brain Research, 73*, 78–90.

Darlot, C., Denise, P., Droulez, J., Cohen, B., & Berthoz, A. (1988). Eye movements induced by off-vertical axis rotation (OVAR) at small angles of tilt in man. *Experimental Brain Research, 73*, 91–105.

Denise, P., Darlot, C., Droulez, J., Cohen, B., & Berthoz, A. (1988). Motion perception induced by Off-Vertical Axis Rotation (OVAR) at small angles of tilt in Man. *Experimental Brain Research, 73*, 106–114.

Dichgans, J., & Brandt, Th. (1972). Visual-vestibular interaction and motion perception. In J. Dichgans & E. Bizzi (Eds.), *Cerebral control of eye movement and motion perception* (pp. 327–338). Basel-New York: Karger.

Dichgans, J., & Brandt, Th. (1978). Visual-vestibular interaction, effects on self-motion perception and postural control. In R. Held, H. Leibowitz, & H. L. Teuber (Eds.), *Handbook of sensory physiology, Vol. VIII, Perception* (pp. 755–804). Berlin-New York: Springer-Verlag.

Fernandez, C., & Goldberg, J. (1976). Physiology of peripheral neurons innervating otolith organs of the squirrel monkey (I, II and III). *Journal of Neurophysiology, 39*, 970–1008.

Finke, R. A., & Held, R. (1978). State reversals of optically induced tilt and torsional eye movements. *Perception & Psychophysics, 23*, 337–340.

Goldberg, J. M., & Fernandez, C. (1981). Physiological mechanisms of the nystagmus produced by rotations about an earth-horizontal axis. *Annals of New York Academy of Sciences, 374*, 40–43.

Goldberg, J. M., & Fernandez, C. (1982). Eye movements and vestibular nerve responses produced in the squirrel monkey by rotation about an earth-horizontal axis. *Experimental Brain Research, 46*, 393–402.

Guedry, F. E. (1965). Orientation of the rotation axis relative to gravity, its influence on nystagmus and the sense of rotation. *Acta Otolaryngologica, 60*, 30–48.

Guedry, F. E. (1974). Psychophysics of vestibular sensations. In H. H. Kornhüber (Ed.), *Handbook of sensory physiology* (Vol. VI/2, pp. 3–154). Berlin, Heidelberg, New York: Springer-Verlag.

Hain, T. C. (1986). A model of the nystagmus induced by off vertical axis rotation. *Biological Cybernetics, 54*, 337–350.

Harris, L. R., & Barnes, G. R. (1985). The orientation of vestibular nystagmus is modified by head tilt. In M. D. Graham & J. L. Kemink (Eds.), *The vestibular system, neurophysiologic and clinical research* (pp. 260–270). New York: Raven Press.

Henn, V., Cohen, B., & Young, L. R. (1980). Visual-vestibular interaction in motion perception and the generation of nystagmus. *Neuroscience Research Program Bulletin, 18*, 4.

Jongkees, L. B. W., & Philipszoon, A. T. (1962). Nystagmus provoked by linear accelerations. *Acta Physiologica Pharmacologica Neerlandensis, 10*, 238–247.

Kalman, R. E. (1960). A new approach to linear filtering and prediction problems. *Journal of Basic Engineering, D82*, 35–45.

Keller, E. L., & Crandall, W. F. (1983). Neural responses to optokinetic stimuli in pontine nuclei of behaving monkey. *Journal of Neurophysiology, 49, 1*, 169–187.

Keller, E. L., & Precht, W. (1979). Visual-vestibular responses in vestibular nuclear neurons in intact and cerebellectomized, alert cat. *Neuroscience, 4*, 1599–1613.

Koenderink, J. J., & Van Doorn, A. J. (1987). Facts on Optic Flow. *Biological Cybernetics, 56*, 247–254.

Lisberger, S. G. (1988). The neural basis for motor learning in the vestibulo-ocular reflex in monkeys. *Trends in Neuroscience, 11, 4*, 147–152.

Longuet-Higgins, H. C., & Pradzny, K. (1980). The interpretation of moving retinal images. *Proceedings of the Royal Society, London, B208*, 385–387.

Mac Cabe, B. F. (1964). Nystagmus response of the otolith organs. *Laryngoscope, 74*, 372–381.

Mayne, R. (1974). A system concept of the vestibular organs. In H. H. Kornhüber (Ed.), *Handbook of sensory physiology*. (Vol. VI/2, pp. 493–580). Berlin, Heidelberg, New York: Springer-Verlag.

Meiry, J. L. (1965). The vestibular system and human dynamic space orientation. Washington NASA CR-628. Science Doctoral Thesis. MIT, Cambridge, MA.

Melvill-Jones, G., Rolph, R., & Downing, G. H. (1980). Comparison of human subjective and oculomotor responses to sinusoidal vertical linear acceleration. *Acta Otolaryngologica, 90,* 431–440.

Niven, J. I., Hixson, W. C., & Correia, M. J. (1965). Elicitation of horizontal nystagmus by periodic linear acceleration. *Acta Otolaryngologica, 62,* 429–441.

Ormsby, C. C., & Young, L. R. (1977). Integration of semicircular canal and otolith information for multisensory orientation stimuli. *Mathematical Biosciences, 34,* 1–21.

Pavard, B., & Berthoz, A. (1977). Linear acceleration modifies the perceived velocity of a moving visual scene. *Perception, 6,* 529–540.

Pellionisz, A., & Llinas, R. (1979). Brain modelling by tensor network theory and computer simulation. The Cerebellum, distributed processor for predictive coordination. *Neuroscience, 4,* 323–348.

Precht, W., Volkind, R., Maeda, M., & Giretti, M. L. (1976). The effects of stimulating the cerebellar nodulus in the cat on the responses of vestibular neurons. *Neuroscience, 1,* 301–3012.

Raphan, T., Cohen, B., & Matsuo, V. (1977). A velocity-storage mechanism responsible for optokinetic nystagmus (OKN), optokinetic after-nystagmus (OKAN) and vestibular nystagmus. In R. Baker & A. Berthoz (Eds.), *Control of gaze by brain stem neurons* (pp. 37–47). Amsterdam: Elsevier.

Raphan, T., Matsuo, V., & Cohen, B. (1979). Velocity storage in the vestibulo-ocular reflex arc (VOR). *Experimental Brain Research, 35,* 229–248.

Robinson, D. A. (1975). Oculomotor control signals. In G. Lennerstrand & P. Bach-y-Rita (Eds.), *Basic mechanisms of ocular motility and their clinical implication* (pp. 337–374). New York: Pergamon Press.

Robinson, D. A. (1977). Vestibular and optokinetic symbiosis: An example of explaining by modelling. In R. Baker & A. Berthoz (Eds.), *Control of gaze by brain stem neurons* (pp. 49–58). Amsterdam: Elsevier.

Robinson, D. A. (1982). The use of matrices in analyzing the three-dimensional behavior of the vestibulo-ocular reflex. *Biological Cybernetics, 46,* 53–66.

Schmid, R. (1986). Habituation of vestibulo-ocular responses to repeated vestibular stimulations in darkness. In S. Ron, R. Schmid, & M. Jeannerod, M. (Eds.), *Sensorimotor plasticity, theoretical, experimental and clinical aspects. Plasticité sensorimotrice, aspects théoriques, expérimentaux et cliniques*. Paris: Les éditions de l'I.N.S.E.R.M.

Schmidt, S. F. (1966). Application of state-space methods in navigation problems. In C. T. Leondes (Ed.), *Advances in control systems* (Vol. 3, pp. 293–340). New York: Academic Press.

Stockwell, C. W., Turnipseed, G. T., & Guedry, F. E. (1970). *Nystagmus responses during rotation about a tilted axis (NAMRL-1129, USAARL 71-15)*. Pennsacola, FL: Naval Aerospace Medical Research Laboratory and U.S. Army Aeromedical Research Laboratory.

Tomlinson, R. D., Saunders, G. E., & Schwarz, D. W. F. (1980). Analysis of human vestibulo-ocular reflex during active head movements. *Acta Otolaryngologica, 90*, 184–190.

Viire, E., Tweed, D., Milner, K., & Vilis, T. (1986). A reexamination of the gain of the vestibulo-ocular reflex. *Journal of Neurophysiology, 36, 2*, 439–450.

Waespe, W., & Henn, V. (1979). The velocity response of vestibular nucleus neurons during vestibular, visual and combined angular acceleration. *Experimental Brain Research, 37,* 337–347.

Waespe, W., Cohen, B., & Raphan, T. (1983). Role of the flocculus and paraflocculus in optokinetic nystagmus and visual-vestibular interactions, effects of lesions. *Experimental Research, 50*, 9–33.

Waespe, W., Cohen, B., & Raphan, T. (1985). Dynamic modification of the vestibulo-ocular reflex by the Nodulus and Uvula. *Science, 228*, 199–202.

Wall, C., III, & Black, F. O. (1984). Intersubject variability in VOR responses to 0.005-1.0 Hz sinusoidal rotations. *Acta Otolaryngologica, 406*, 194–198.

Waxman, A. M., & Ullman, S. (1985). Surface structure and three-dimensional motion from image flow kinematics. *The International Journal of Robotics Research, 4, 3*, 72–94.

Wilson, A. J., & Melvill-Jones, G. (1979). *Mammalian vestibular physiology*. New York: Plenum Press.

# 18 Sensory-motor Adaptation to High Force Levels in Parabolic Flight Maneuvers

James R. Lackner
*Ashton Graybiel Spatial Orientation Laboratory*
*Brandeis University*

## ABSTRACT

Human sensory-motor control is dynamically adapted and calibrated to the terrestrial force background provided by the gravitational acceleration of Earth. The existence of this calibration becomes perceptually apparent when movements are carried out during exposure to greater than 1G levels of acceleration. Under this circumstance, the surface of support no longer feels stable as one moves in relation to it, and one misperceives the extent of one's body movements.

If repeated whole-body movements are made during exposure to a high level of acceleration, the apparent stability of self and environment during voluntary movement is gradually regained, but at the cost of a loss of adaptation to the force background of Earth. Consequently, on re-exposure to 1G acceleration levels, voluntary locomotion again elicits illusory motion of self and environment, but it is now of opposite sign. Such observations provide a basis for understanding many of the postural and sensory-motor aftereffects that astronauts and cosmonauts experience on return to Earth after space flight.

## INTRODUCTION

All bodily movements and activities on Earth are carried out within the context of a 1G[1] background acceleration, the acceleration of gravity. Within this terrestrial environment, changes in activity at the body's re-

[1]G is the gravitational acceleration of Earth, $9.8 m/s^2$.

ceptor surfaces that are contingent on self-motion are distinguished from those that are contingent on motion of the environment (Holst & Mittelstaedt, 1950; Sperry, 1950). In this way, the nervous system continually monitors and takes into account the consequences of bodily movements in order to preserve a stable distinction between sensory changes owing to self-locomotion, whether achieved actively or passively, and those attributable to external causes. Such automatic monitoring of the consequences of our movements forms part of a specific adaptation of our bodies to the terrestrial force level of Earth (Lackner, 1985, 1988a; Lackner & Graybiel, 1981, 1984).

The existence of this adaptation only becomes apparent when the body is exposed to unusual background force levels, that is, abnormal levels of gravitoinertial force (Lackner & Graybiel, 1981). For example, Fig. 18.1 shows what a person experiences when making a deep knee bend at twice Earth gravity acceleration level generated in an aircraft performing para-

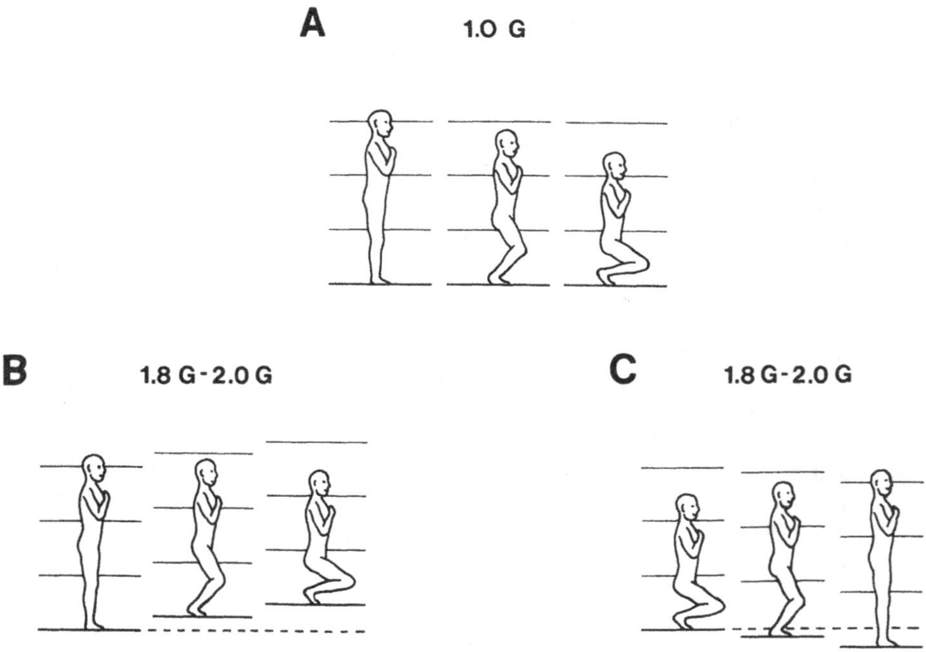


FIG. 18.1. Panel A illustrates an individual's true motion when doing deep knee bends; viewed left to right, the subject is lowering his body, right to left the subject is raising his body. Panel B depicts what the individual perceives his body and the visual world (schematically represented by the horizontal lines) to be doing as he lowers his body during exposure to a high acceleration level, 1.8G–2.0G. Panel C illustrates what he experiences in rising from the squatting position.

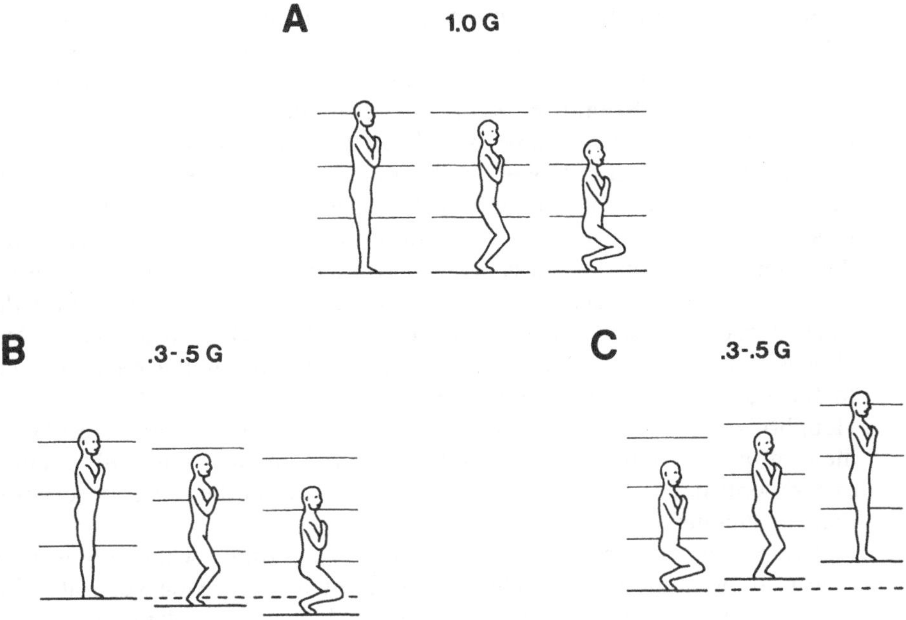


FIG. 18.2. Panel A illustrates an individual's true motion as he does a deep knee bend; viewed left to right, the subject is lowering his body, right to left, raising it. Panel B depicts what the individual perceives his body and the visual world (indicated by the horizontal lines) to be doing as he lowers himself during exposure to a low acceleration level, .3G–.5G. Panel C illustrates his apparent body motion and the apparent motion of the visual world as he rises up.

bolic maneuvers. As the person lowers the body, there is the compelling feeling that the move downward was too rapid and that the deck has risen up under the person's feet. When rising, the person feels as though he or she is moving upward too slowly and that the immense efforts in regaining an upright posture are pushing the deck of the aircraft downward. During exposure to lower than 1G acceleration levels, the inverse illusory patterns are experienced (Fig. 18.2). This G-force dependency indicates that under normal terrestrial conditions sensory-motor control is actively tuned to the 1G background force level.

Another way that the existence of this tuning can be recognized is by standing upright on both feet and then shifting stance to one foot. In this circumstance, the force on the sole of the stance foot has doubled. Nevertheless, little if any increase in force on the stance foot is perceived. By contrast, if comparable forces are passively applied to the foot when the body is horizontal, an enormous increase in force is perceived (Lackner & Graybiel, 1984).

Since the beginning of manned space travel, it has been known that astronauts and cosmonauts who return to Earth from prolonged space flight experience a variety of postural and sensory-motor disturbances (Johnston & Dietlein, 1977; Schmitt & Reid, 1985). In addition, postflight changes in various otolith-spinal reflexes have been described (Reschke, Anderson, & Homick, 1984; Watt, Money, & Tomi, 1986). These reflexes, as well as the tonic vibration reflex (Lackner & DiZio, 1989; Lackner, DiZio, & Fisk, 1989) that is an indicator of postural tonus, are also transiently modified during exposure to force variations in parabolic flight maneuvers. Such observations suggest that adaptation to the force background of Earth can be lost or degraded during exposure to non-1G background force levels. However, it may be that this loss is accompanied by a dynamic recalibration to the new force environment. For example, more than 100 years ago, Helmholtz (1866) found that visuo-motor adaptation to optical displacement of the visual array could be achieved quite rapidly. Many subsequent experiments have revealed adaptation to a variety of other types of sensory rearrangements.

The present experiments were devised to determine whether sensory-motor adaptation could be achieved to the high force levels generated in parabolic flight maneuvers. We wanted to see how rapidly adaptation might occur, how specific this adaptation would be to the particular "types" of movements made during exposure, and whether it would simultaneously entail a loss of adaptation to Earth gravity force level. We expected that if adaptation to 1G were lost during repeated exposure to high force levels, then, on return to 1G force levels subjects would experience the patterns illustrated in Fig. 18.2 when making a deep knee bend. In other words, they should have similar experiences to subjects adapted to 1G who move in a less than 1G force environment.

## METHODS

### Subjects

The subjects were 12 men; six different individuals participated in each of two experiments. All of them had already taken part in at least one test week involving four, 40-parabola flights. This experience was important so that they would already have gained some familiarity with exposure to high and low force levels and the demands of locomotion in such situations. In terms of susceptibility to motion sickness in parabolic maneuvers, all were *Category I* subjects (cf. Lackner & Graybiel, 1984), in that they were insusceptible to motion sickness during passive exposure to gravitoinertial force variations in parabolic flight. All the subjects had met the stringent medical requirements necessary for participation in parabolic maneuvers and had received a physiological training course in a high-altitude chamber.

## Procedure

Two separate forms of experimental observations were carried out in the high-force phases of parabolic flight maneuvers. These involved deep knee bends and stepping onto and off a low platform. Figure 18.3 illustrates the flight profile of the Boeing KC-135 aircraft used in our studies. Testing was initiated when the force reading on a digital accelerometer reached 1.8G during the pull-up phase of a parabola.

Each set of experimental observations involved baseline measurements made both during straight-and-level flight and during the first parabola of a flight, followed by interspersed measurements during the subsequent parabolas to track any adaptation that might occur.

*Deep knee bends.* The deep knee bends were carried out to a metronome beat and timed to achieve knee flexion or extension in approximately 1 s. The subjects practiced the movements before the flight and then again during straight-and-level flight before the onset of the parabolas. During the movements, the subjects, who were tested individually on separate test days, kept their eyes open as they moved downward 15 to 20 cm toward a squatting position and raised up again, pausing approximately 1 s after each component of the movement pattern.

The in-flight testing protocol involved ten practice movements in straight-and-level flight. Then, 2 minutes before the onset of the 1st parabola, the subjects performed knee bends after which they (a) made magnitude estimations in inches[2] of any apparent motion of the base of support and specified its direction, and (b) indicated whether their experienced bodily motion was too fast or too slow in relation to their intended bodily motion. During the 1st parabola, the subjects again made three cycles of movements and repeated the magnitude estimations and apparent bodily motion judgments; this provided a baseline rating for initial exposure to the high-force level. In each subsequent parabola, the subjects made six cycles of deep knee bends during the high-force, pull-up phases; and, in every 10th parabola they repeated the magnitude estimation and apparent motion judgments. This procedure was followed through the last parabola of the 40-parabola flight (the last measurement being on parabola 40). In straight-and-level flight after completion of the parabolas, the subjects made three

[2]When our subjects made estimates of apparent displacement, they did so in inches because they were familiar and comfortable with this measure. The estimates were converted into centimeters for presentation in tables and for statistical analysis.

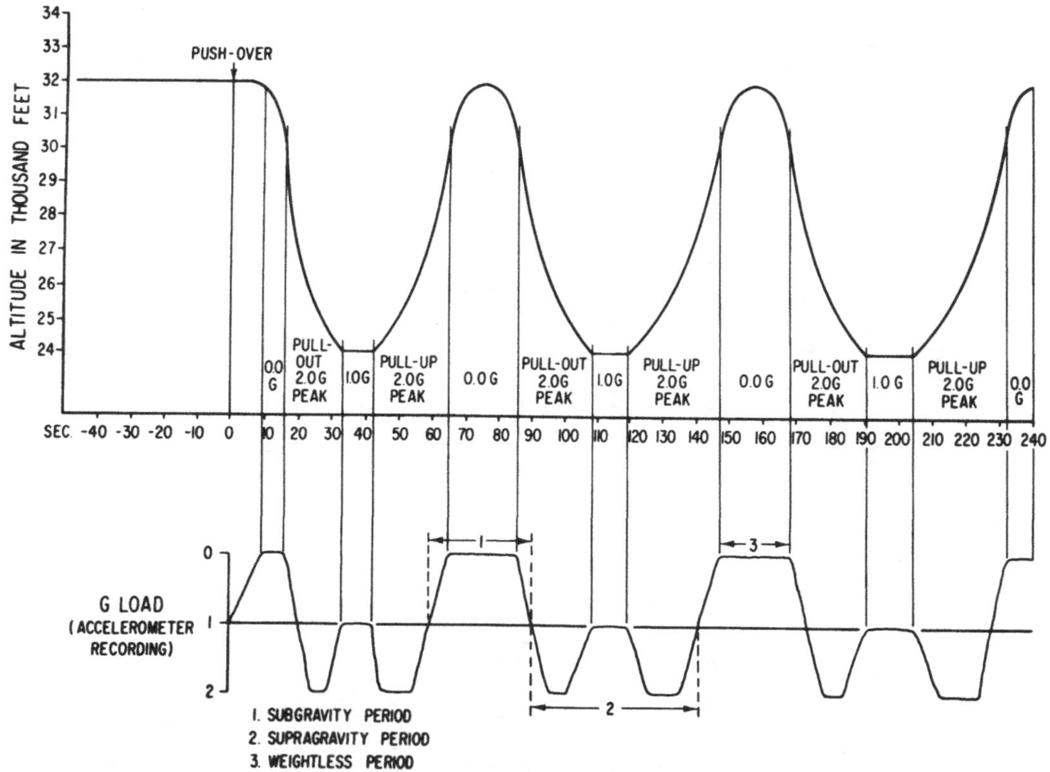


FIG. 18.3. Schematic illustration of the flight profile of the Boeing KC-135 aircraft used in our studies.

additional cycles of deep knee bends, as well as magnitude and apparent motion judgments.[3] An additional three cycles of movements followed by judgments were made postflight in a briefing room; to get to this briefing room, the subjects had to walk approximately 400 ft and climb a low flight of stairs.

The postparabola judgments provided an index of aftereffects of exposure to a high-force level. Each subject was also instructed to keep a record of any unusual sensory-motor phenomenon noticed during the week after the flight.

*Stepping up and down.* These observations concerned the patterns of motion and displacement perceived when stepping onto and off a 15 cm high platform attached to the deck of the aircraft. Subjects were required to stand on two feet in front of the platform and then to step up onto it, bringing both feet into contact and then to step back down to the deck. Subjects attempted to step in a normal way at a normal pace. Each subject repeated these movements for three cycles: (a) in straight-and-level flight, (b) in the 1st parabola, (c) in every 10th parabola, (d) in straight-and-level flight immediately after the 40th parabola,[3] and (e) in the briefing room postflight as soon as possible after landing. After each set of movements, the subjects reported the direction and magnitude in inches of any apparent motion of the platform and of the deck as they stepped up and down. They also indicated whether the apparent forces on their feet as they stepped up on, stood on, and stepped off the platform had changed in relation to the baseline forces experienced in straight-and-level flight. After the initial ratings in parabola 1, the subjects starting with parabola 2 stepped up and down from the platform six times during the high-force pull-up phase of each subsequent parabola, except parabolas 10, 20, 30, and 40 during which they only made three cycles of movements.

## RESULTS

### Deep Knee Bends

*Straight-and-level flight.* None of the subjects experienced aberrant motion of their bodies or of the aircraft when they did deep knee bends prior to the onset of the parabolas.

---

[3]Two of the subjects who took part in the deep knee bend study and two who took part in the stepping up and down study were tested for cross-over adaptation to the other series. Following their last regular tests in straight-and-level flight, the subjects from the deep knee bend study stepped onto and off the platform three times, and the other two subjects did three sets of deep knee bends. After their movements, the subjects made self-motion and environment motion judgments as described in the methods.

*First parabola.* All of the subjects experienced the airplane rise up under their feet as they lowered their bodies, they felt and saw the plane move upward and felt that they had moved downward too rapidly. As they raised their bodies, they felt and saw the plane move downward in relation to themselves and felt that they had raised up too slowly and that their strenuous efforts had pushed the plane downward. The magnitudes of the apparent displacement of the aircraft ranged from 3 in. to 12 in. for the different subjects.

*Later parabolas.* As the flight proceeded, all of the subjects experienced a progressive diminution in the amplitudes of the illusory self-motions and aircraft motions associated with their deep knee bends. By the end of 40 parabolas, an attenuation of 40% to 100% had occurred for the different subjects.

*Straight-and-level flight, postparabolas.* Five of the six subjects experienced a reversal of the illusory patterns initially felt during the high-force phases of the preceding parabolas: As they lowered their bodies, it felt as though they were moving downward too slowly, and they felt and saw the aircraft move downward; as they raised their bodies, it felt as though they rose upward too quickly, and they felt and saw the aircraft move upward as well. The remaining subject felt that he and the aircraft were less stable during the deep knee bends and that much less than normal effort was required, but the subject could not specify how things had changed.

*Postflight, briefing room.* Three of the subjects continued to experience, albeit with diminished intensity, the patterns that they had felt in straight-and-level flight after the parabolas. For two of these three subjects, these patterns recurred periodically for several days. For example, after sitting down for several hours or on getting up after sleeping, the floor would temporarily feel unstable to them as they moved about. The experimental findings are summarized in Table 18.1 and illustrated in Fig. 18.4.

## Stepping Up and Down

*Straight-and-level flight.* The subjects accurately perceived their stepping movements and felt that the platform remained stationary as they stepped onto it, and that the deck of the aircraft remained stationary as they stepped back down onto it.

TABLE 18.1

Apparent motion experienced when making deep knee bends in the increased gravitoinertial force phases of parabolic flight maneuvers and afterwards in straight-and-level flight. Self-entries specify the direction of perceived body motion up (↑) or down (↓), and whether that motion was perceived as too fast (TF), too slow (TS), or normal (N) in relation to intended motion. The aircraft entries indicate the direction, up (↑) or down (↓), of felt and seen motion of the aircraft and the magnitude of that motion in cm; the ranges are presented within parenthesis. Correlated *t*-tests were performed to compare the means of parabola 1 and parabola 40, and parabola 1 and straight-and-level flight for both sets of aircraft motion entries; all four comparisons were significantly different at the $p<.01$ level, $N=6$.

| | *Lower the Body* | | *Raise the Body* | |
|---|---|---|---|---|
| | *Self* | *Aircraft* | *Self* | *Aircraft* |
| PARABOLA 1 | ↓ TF | ↑ 15.1cm (7.60–30.4) | ↑ TS | ↓ 13.4cm (8.9–25.4) |
| PARABOLA 10 | ↓ TF | ↑ 8.6cm (5.1–21.6) | ↑ TS | ↓ 8.1cm (5.1–16.5) |
| PARABOLA 20 | ↓ TF | ↑ 5.6cm(1.3–16.5) | ↑ TS | ↓ 4.7cm (2.5–10.2) |
| PARABOLA 30 | ↓ TF | ↑ 4.3cm (0–12.7) | ↑ TS | ↓ 2.9cm (1.3–7.6) |
| PARABOLA 40 | ↓ N | ↑ 2.8cm (0–7.6) | ↑ N | ↓ 1.8cm (0–5.1) |
| LEVEL FLIGHT | ↓ TS | ↓ 11.2cm (5.1–23.2) | ↑ TF | ↑ 8.7cm (5.1–16.5) |

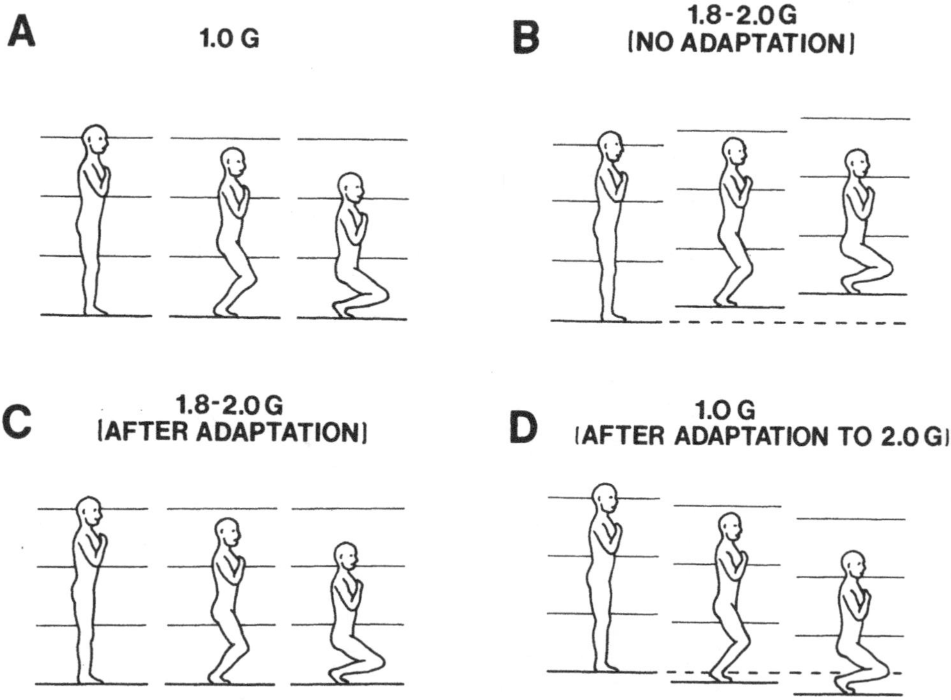


FIG. 18.4. Panel A illustrates an individual's body motion and the apparent stability of the visual world as he lowers his body in a deep knee bend; the horizontal lines represent the visual world. Panels B, C, and D illustrate what the individual perceives his body and the visual world to be doing as he does a deep knee bend during exposure to a high acceleration level initially, Panel B, and then after partial adaptation to the high acceleration, Panel C. Panel D illustrates that after adapting to a high acceleration level the individual has lost sensory-motor calibration to the acceleration background of Earth.

*First parabola.* All of the subjects found that as they stepped onto the platform it seemed to move downward under their stepping foot, and their visual surroundings appeared to move downward as well (Fig. 18.5); as they brought their other foot into contact with the platform, it felt as though the platform rebounded slightly upward under their feet. Greater than normal effort was experienced in making the movement. The inverse pattern was experienced as they stepped off the platform; it felt as though the deck rose up under their stepping foot after contact was made, and that they had moved downward too rapidly in stepping off the platform.

*Subsequent parabolas.* Every subject with continued exposure reported a progressive diminution in the magnitude of the illusory motions experienced during stepping up and down. Two of the six subjects reported on parabola 40 that the platform and deck remained stable as they stepped up and down and that their movements required normal or near normal effort. The other four subjects still experienced illusory motion, but only 20% to 40% of that initially present.

*Straight-and-level flight.* All six subjects reported that the platform seemed to rise up under their foot as they stepped onto it and that the deck of the aircraft moved downward as they stepped down onto it. This apparent motion was accompanied by perceived visual motion of the surroundings of corresponding direction.

*Postflight debriefing room.* Each of the six subjects continued to experience the same illusory patterns during stepping up and down as they had in straight-and-level flight prior to landing. The amplitudes of these aftereffects attenuated rapidly within 15 min. The experimental observations are summarized in Table 18.2.

*Long-term aftereffects.* Over the course of the next several days, three of the subjects when walking up and down stairs sometimes experienced the same types of aftereffects as they had in the debriefing room, especially after having been quiescent for some period. These effects were enhanced if they closed their eyes.

## DISCUSSION

Sensory-motor control on Earth is actively tuned to the 1G terrestrial background acceleration of gravity. Because of this calibration, during exposure to nonterrestrial force levels, we no longer correctly perceive locomotory movements of our body, and the environment no longer seems stable as we move about (Lackner & Graybiel, 1981, 1984). The present experiments demonstrate that with repeated exposure to high force levels, adaptive changes in sensory-motor control can occur to accommodate to the new force environment. In fact, partial recalibration can be initiated and accomplished quite rapidly, in the course of executing less than 100 movements in a high force field. As a consequence of this adaptation, the apparent instability of the support stratum first experienced during voluntary body movements quickly begins to abate.

The extent of adaptational specificity—the extent to which the adaptation achieved was limited to the particular types of movements executed during the adaptation period—is an important issue. Two of the subjects in each of the experiments were tested for generalization to the other type

TABLE 18.2

Apparent subjective displacement experienced when stepping onto and off a stationary platform in the high gravitoinertial force, 1.8G, phase of parabolic flight maneuvers. The arrows indicate the direction of perceived motion. Means and ranges of apparent displacements of the platform and deck are presented. Correlated *t*-tests were performed comparing the means for parabola 1 and 40, and parabola 1 and straight-and-level flight for stepping up and stepping down, $p<.01$ for all comparisons, $N=6$.

| | *Step-up* | | *Step-down* | |
|---|---|---|---|---|
| | *Self* | *Platform* | *Self* | *Deck* |
| PARABOLA 1 | ↑ TS | ↓ 9.1cm (5.1–16.5) | ↓ TF | ↑ 10.8cm (5.1–17.8) |
| PARABOLA 10 | ↑ TS | ↓ 6.3cm (2.5–11.4) | ↓ TF | ↑ 7.1cm (2.5–14.0) |
| PARABOLA 20 | ↑ TS | ↓ 5.1cm (1.3–8.9) | ↓ TF | ↑ 4.9cm (2.5–8.9) |
| PARABOLA 30 | ↑ TS | ↓ 3.6cm (1.3–7.6) | ↓ TF | ↑ 3.2cm (1.3–5.1) |
| PARABOLA 40 | ↑ TS | ↓ 2.3cm (0–6.4) | ↓ TF | ↑ 2.3cm (0–6.4) |
| LEVEL FLIGHT | ↑ TF | ↑ 8.9cm (5.1–12.7) | ↓ TS | ↓ 7.8cm (5.1–14.0) |

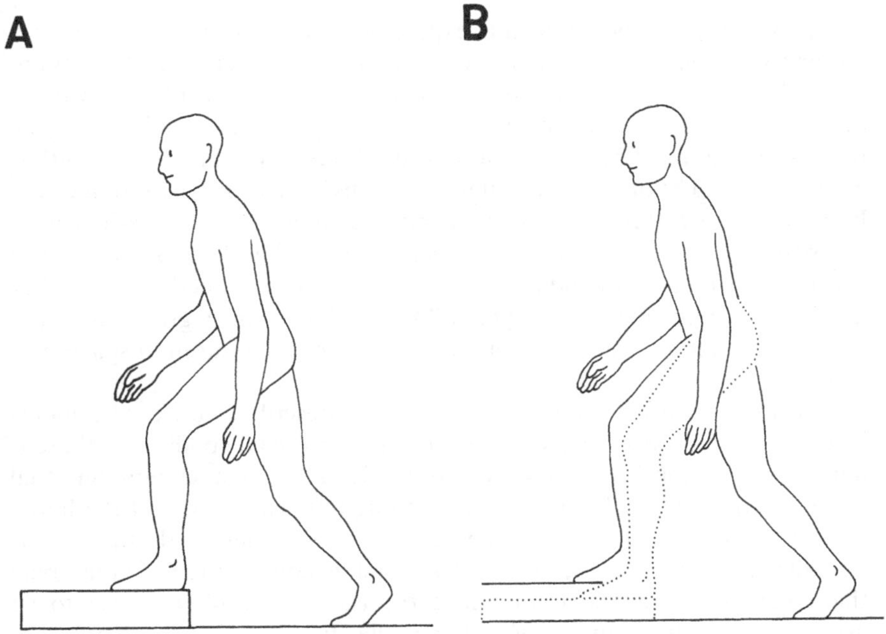


FIG. 18.5. When a subject steps up onto a low platform during exposure to a 1.8G level (A), the platform will seem to move downward under his stepping foot (B). The subject feels as if his immense efforts have pushed the platform downward and that his body has risen too slowly. After prolonged exposure, his movements will again seem normal.

of movement pattern following their last judgments in straight-and-level flight: Subjects who had made deep knee bends stepped onto and off the platform, and those who had made stepping up and down movements executed deep knee bends. Both pairs of subjects showed partial transfer of adaptation: They experienced the same types of apparent motion as the subjects initially had who had taken part in the full adaptational series with those movements, but only to about one third the extent.

On the basis of other observations, we have additional evidence for specificity of adaptive compensation to the patterns of exposure experienced. For example, individuals who no longer experience aberrant motion of the aircraft deck or of their body when moving their legs may nevertheless experience inappropriate movement of themselves and of the aircraft deck when they attempt "half push-ups." These are push-ups in which the legs remain horizontal and the torso is raised with the arms (full push-ups are too strenuous in a 1.8G force environment for most subjects). During these half push-ups, as the subjects raise their torso, the deck will seem to move downward somewhat and the torso will seem to move up too slowly; the reverse pattern is experienced on lowering the body. Accordingly, if relatively restricted movements (e.g., deep knee bends) are made during brief exposure periods, adaptation seems confined to those or closely related movements and is not a nonspecific adaptation applying to all movements executed in the unusual force environment. Put differently, the adaptation is not a generalized sensory-motor adaptation of the body to the new force environment but is confined to the muscle groups or appendages used during the exposure period. The factors that govern specific and nonspecific adaptation need to be determined because they have great significance for the possibility of using artificial gravity to maintain sensory-motor control accuracy during very long-duration space missions.

The adaptation achieved in the experiments described here can be understood in the following items. Under normal terrestrial conditions, there is a dynamic relationship among patterns of alpha and gamma motoneuronal innervation of skeletal muscles, the ensuing physical motion of the body, and the pattern of muscle spindle activity in the innervated muscles (cf. Lackner, 1985; Matthews, 1981). (The possible contribution of beta fibers that innervate both fusimotor and extrafusal muscle fibers is yet to be ascertained in the human.) Muscle spindle discharge level is highly important in the determination of limb position. Goodwin, McCloskey, and Matthews (1972a, 1972b) have shown that if the level of spindle activity within a muscle is abnormally high for the level of alpha-gamma motoneuronal activity present, then that muscle will be interpreted as longer than it actually is, and the abnormal "length" will be referred to the limb(s) that it controls. For example, abnormally high levels of spindle discharge in the biceps brachii muscle, achieved by mechanically vibrating the surface of the muscle, lead to the illusion that the forearm is more extended than it is. We (Lackner, 1985, 1988b; Lackner & Levine, 1979) have shown that by creating abnormally high levels of spindle activity in postural support muscles one can elicit illusions of body tilt or body rotation that depend on which muscles are vibrated. Such apparent motion of the body is often accompanied by "compensatory" nystagmoid eye movements, just as if it were real motion.

Spindle discharge per se can affect position sense (Matthews, 1981, 1988). McCloskey, Cross, Honner, and Potter (1983) have shown that when the extensor hallucis longus is surgically detached from the big toe, pulling on the tendon gives rise to the sensation of the toe moving downward. Electrical stimulation of the ulnar nerve at intensities insufficient to elicit muscular contraction but adequate to stimulate 1a afferent fibers from muscle spindles can evoke sensations of finger displacement (Gandevia, 1985).

Nevertheless, spindle discharge is normally interpreted within a context of information about intended and ongoing movements. This is necessary because patterns of spindle activity can only be meaningfully rendered in relation to the levels of fusimotor and alpha motor activity present (cf. Matthews, 1988). That is, the spindle discharge must be interpreted in relation to corollary discharge signals about intended movements—and possibly golgi tendon organ, joint capsular, and somatosensory signals as well—and, in this computational milieu they provide information about the progress of body movements.

With these considerations in mind, it can be assumed that when an individual who is accustomed to making movements in the normal terrestrial force field of Earth initially makes deep knee bends in a high-force field, the level of muscle spindle activity present in the antigravity muscles would be unexpectedly high throughout the lowering motion of the body for the level of alpha and gamma motoneuron innervation present. This is because body weight has increased and the spindles would be unloaded less rapidly than they normally would be for the innervation levels present (Lackner & Graybiel, 1981, 1984). We believe this abnormally high level of activity is interpreted as stretch of the antigravity muscles and "referred" to the joints about which they operate—in this case, ankle, knee, and hip joints. As a consequence, the angles of these joints are interpreted as being more flexed than they actually are during lowering of the body. The heightened spindle activity is attributed to the aircraft deck having moved upward under the feet stretching the muscles. (We have preliminary evidence under terrestrial conditions that supports this interpretation. If the quadricep brachii muscles are vibrated bilaterally at 120 Hz as the individual lowers the body in a deep knee bend, the individual experiences a similar pattern of illusory motion as when making a deep knee bend in a high force field.) The same considerations explain the patterns experienced during raising of the body from a deep knee bend and the patterns experienced during the stepping up and down movements (Lackner, 1985; Lackner & Graybiel, 1981, 1984).

As repeated deep knee bends are made in the high force phases of parabolic flight maneuvers, the relationship between patterns of alpha-gamma innervation and spindle feedback is gradually recalibrated until perceptual stability of self and environment is regained. Such adaptive recalibration is a familiar phenomenon for exposure to various other forms of sensory-motor discordance (cf. Howard & Templeton, 1966; Howard, 1982; Lackner, 1978, 1981; Wallach, 1976; Welch, 1978).

After adaptation to the high force level is achieved, return to a terrestrial force level again leads to misapprehensions of self-motion and of the apparent stability of the support surface because the new calibration is inappropriate for a 1G force environment. When the body is lowered in a deep knee bend, the level of spindle feedback present would now be inapropriately low for the level of alpha-gamma coactivation. Consequently, in keeping with the contribution of muscle spindle feedback to position sense, the antigravity muscles of the legs would be interpreted as being too short during the movement, that is, the knees seem more extended than they actually are. The body would seem to move downward too short a distance, and the deck would also seem to move downward under the feet. This is, of course, precisely the pattern we have observed during testing in 1G after prolonged exposure to the high force level.

At present, the neurophysiological basis of adaptation to high force levels is unknown. However, several possible ways amongst others include (a) alpha-gamma remapping, (b) vestibulo-spinal gain modulation, and (c) regulation of Renshaw cell activity. The alpha-gamma remapping strategy would achieve adaptation by adjusting the pattern of alpha and gamma motoneuronal firing to take into account the changed load on the body. This would be possible during exposure to both increased and decreased G-force fields if the nonterrestrial force level were maintained at a stable value for some period of time. The remapping would establish a new relationship between the physical movements of the body and the spindle discharges associated with particular levels of alpha-gamma motoneuronal activity. This could be done by altering the level of alpha motoneuronal activity in relation to gamma activation levels: increasing alpha in relation to gamma activity for adaptation to high force levels, decreasing it for low force levels. The effect would be to make the level of spindle discharge appropriate for the course of the voluntary movement and also to make the sense of effort one has with regard to his intentional movements again seem normal. Modulating the gain of the vestibulo-spinal tract would be another way of accommodating the postural muscles to altered force levels. The excitatory vestibulo-spinal influences on the alpha and gamma motoneurones of the antigravity musculature could be enhanced during exposure to a high force level to increase muscle stiffness thereby compensating in part at least for the increased weight of the body, and they could be diminished during exposure to low force levels. In fact, during brief exposure to high force levels, there is a virtually immediate increase in the intensity of tonic vibration reflexes, and during brief exposure to weightlessness an immediate intensity decrease; both are likely related to vestibulo-spinal influences on fusimotor activity thereby affecting spindle sensitivity (Lackner & DiZio, 1989; Lackner, DiZio, & Fisk, 1989). Finally, it would be possible to achieve adaptation by adjusting the activity levels of spinal cord interneurons. Renshaw cells are excited by axon collaterals from alpha motoneurones and they project back in an inhibitory fashion on the same alpha motoneurones. Pompeiano (1988) has demonstrated that the locus coeruleus can modulate the threshold of Renshaw cells. This provides a potential mechanism by which the same patterns of alpha and gamma motoneuronal activity could be associated with different degrees of muscle stiffness depending on the descending influence of the locus coeruleus on the Renshaw cells. In principle, any one or any combination of the three mechanisms described could participate in sensory-motor adaptation to altered force levels, as well as to the normal terrestrial background.

Dynamic sensory-motor tuning is, of course, also crucial developmentally because we, as all animals, in the progression from infancy to adulthood undergo great changes in body mass for which accommodation needs to be made. It is important to note that when we carry large weights and move about, we still correctly perceive the extent of our movements and perceive the environment as stationary. This means that sensory-motor calibration is to self-weight and that changes in touch, pressure, and proprioceptive activity associated with carrying objects are taken into account when interpreting the spindle feedback accompanying voluntary motor commands and movement of the body.

Given that the proprioceptive signals from muscle spindles are strongly represented in the cerebellum (Sherrington's *head ganglion* for proprioception; Sherrington, 1906) and given that the cerebellum has within it multiple representations of other sensory modalities as well, and motor commands (Ito, 1984; Marr, 1969; Thompson, 1986), it is natural to suppose that at least part of the recalibration described could reflect cerebellar processing. The specialized nature of the adaptations achieved and the cross-reference between different sensory and motor systems, however, suggest that a broader central nervous system representation of these phenomena may also be present. This view of a more distributed representation of adaptive processing would be in line with evidence that different sensory systems (Hubel, Wiesel, & LeVay, 1977; Kaas, Merzenich, & Killackey, 1983; Lisberger, 1988; Stryker & Sherk, 1975) have considerable plasticity and that different motor systems may as well (Evarts, Shimoda, & Wise, 1984; Miles & Lisberger, 1981; Lisberger, Miles, & Zee, 1984).

Evidence for this viewpoint has recently been gained using the apparatus illustrated in Fig. 18.6. This apparatus makes it possible to alter the normal relationships among an individual's voluntary stepping movements, the physical displacement of the body relative to the support surface, and the visual flow pattern he receives. Lackner and DiZio (1988) have investigated a wide range of experimental configurations, only two of which are described here. In one situation, the apparatus is adjusted so that the test subject is making forward stepping movements but is actually walking in place (because the floor is rotating to provide a circular treadmill) while the large optokinetic drum is rotating around the subject in the same

FIG. 18.6. This apparatus permits independent control of the subject's visual environment by means of the large optokinetic drum, his stepping movements by means of the floor that he stands on while holding the small arm of the L-shaped bar, the long arm of which is coaxial with the center of rotation of the drum and floor. The drum, floor, and bar can be rotated or held stationary independently of one another thereby allowing the creation of any desired configuration of visual flow, stepping movements, and actual displacement (if any) of the subject.

direction that the floor is turning, but at twice the velocity. This pattern creates a circumstance in which the individual's stepping movements do not displace him in relation to external space, but at the same time the visual flow velocity is twice what would be expected for the voluntary stepping movements being made. In this situation, subjects perceive themselves to be walking in a circular path on a stationary floor in a visually stationary environment. Simultaneously, they experience either an increase in stride length, or an increase in the rate at which they are voluntarily stepping, or both. If an increase in stride length is perceived, an increase in push-off force on the "propelling" foot is also experienced. However, because of the physical configuration of the experimental situation, the subject is actually forced to maintain near-constant step frequency and stride length.

In the other test situation, the apparatus is set so that the subject is making forward stepping movements on the rotating treadmill floor while the visual surround is rotating at the same rate as the floor but in the opposite direction. This creates a visual flow pattern that normally is associated with walking backward. Individuals exposed to this situation fall into three general categories:

1. Some feel that at the same time they are making forward stepping movements in a visually stationary environment, they are being moved backward faster by the floor than they are stepping; they also experience an increase in step force.
2. Some feel they are stepping forward on a stationary floor in a stationary visual surround, but that their steps are paradoxically pushing them backward; they also experience a decreased stride length and an increase in step force.
3. Others feel that they are voluntarily stepping backward on a stationary floor in a stationary visual surround, but at the same time they experience a decrease in stride length, step rate, and step force.

These findings provide further evidence for dynamic sensory-motor tuning to terrestrial force levels. They show that when the normal coupling among voluntary stepping movements, displacement of the body in relation to the support surface, and visual flow are altered, then a variety of illusory changes are evoked in apparent step rate, stride length, step force, and direction of stepping. In most of these perceptual remappings there is a tendency to perceive the visual environment as stationary. These remappings also highlight the complexity of the processes that must underlie the determination of various perceptual experiences that we generally take for granted, for instance, direction of voluntary stepping and stride length. They point, too, to ways in which the position sense and perceptual schema of the body can be updated over time (Lackner, 1988b).

The sets of observations described here as well as our earlier work (Lackner, 1981, 1983, 1985; Lackner & Graybiel, 1981, 1982, 1984) lead to the prediction that astronauts and cosmonauts who are exposed to prolonged space flight should lose their sensory-motor calibrations to a 1G force level and during reentry and on return to Earth should temporarily experience a variety of sensory-motor disturbances. We can expect that during reentry with the return of a significant G-load, the "feel" of the controls in a spacecraft would seem inappropriate, for example, generating the same level of force as preflight to operate foot pedals would seem to require much more "effort." On return to Earth, astronauts would initially experience instability of themselves and the ground as they walk, and if they lower their body in a deep knee bend, the ground may seem to rise under their feet. Over the years, some of these phenomena have been partly described by astronauts, but they have been considered to be a form of postflight disorientation. It is now clear that these effects should not be thought of as disorientation but rather as maladaptation resulting from a loss of sensory-motor calibration to the terrestrial force level of Earth concomitant with new adaptation to the conditions of space flight.

## ACKNOWLEDGMENTS

Support was provided by National Aeronautics and Space Administration Contract, NAS 2-295. I thank Dr. Hans Wallach and Dr. A. M. Graybiel for advice and comments on the manuscript.

## REFERENCES

Evarts, E., Shimoda, Y., & Wise, S. (1984). *Neurophysiological approaches to higher brain functions*. New York: Wiley.

Gandevia, S. C. (1985). Illusory movements produced by electrical stimulation of low-threshold muscle afferents from the hand. *Brain, 108*, 965–981.

Goodwin, G., McCloskey, D. I., & Matthews, P. B. C. (1972a). Proprioceptive illusions induced by muscle vibration. Contribution to perception by muscle spindles? *Science, 175*, 1382–1384.

Goodwin, G., McCloskey, D. I., & Matthews, P. B. C. (1972b). The contribution of muscle afferents to kinesthesia shown by vibration-induced illusions of movement and by paralyzing joint afferents. *Brain, 95*, 705–748.

Helmholtz, H. von (1866). *Treatise on Physiological Optics*. New York: Dover.

Holst, E. von, & Mittelstaedt, H. (1950). Das Reafferenzprinzip. Wechselwirkungen Zwischen Zentral nerven system und peripherie. *Naturwissenschaften, 37*, 464–476.

Howard, I. P. (1982). *Human Visual Orientation*. New York: Wiley.

Howard, I. P., & Templeton, W. B. (1966). *Human Spatial Orientation*. New York: Wiley.

Hubel, D. H., Wiesel, T. N., & LeVay, S. (1977). Plasticity of ocular dominance columns in monkey striate cortex. *Philosophical Transactions of the Royal Society of London, Series B*, *278*, 377–409.

Ito, M. (1984). *The cerebellum and neural control.* New York: Raven.

Kaas, J., Merzenich, M., & Killackey, H. (1983). The reorganization of somatosensory cortex following peripheral nerve damage in adult and developing animals. *Annual Review of Neuroscience*, *6*, 325–356.

Lackner, J. R. (1978). Some mechanisms underlying sensory and postural stability in man. In R. Held, H. W. Leibowitz, & H.-L. Teuber (Eds.), *Handbook of sensory physiology, Vol. III. Perception* (pp. 805–845). Berlin: Springer-Verlag.

Lackner, J. R. (1981). Some aspects of sensory-motor control and adaptation in man. In R. Walk & H. Pick (Eds.), *Intersensory perception and sensory integration* (pp. 143–173). New York: Plenum.

Lackner, J. R. (1983). Human spatial orientation in unusual force environments. *Proceedings of the XXIst International Congress of Aviation and Space Medicine* (pp. 35–38).

Lackner, J. R. (1985). Human sensory-motor adaptation to the terrestrial force environment. In Ingle, D., Jeannerod, M., & Lee, D. (Eds.), *Brain mechanisms and spatial vision* (pp. 175–210). Amsterdam: Nijhoff.

Lackner, J. R., & DiZio, P. (1988). Visual stimulation affects the perception of voluntary leg movements during walking. *Perception*, *17*, 71–80.

Lackner, J. R., & DiZio, P. D. (1989). Muscular tonus depends on gravitoinertial force level. Manuscript submitted for publication.

Lackner, J. R., DiZio, P. D., & Fisk, J. (1989). Postural tonus adjusts rapidly to a changing gravitoinertial force load. *Aviation, Space and Environmental Medicine.* Manuscript submitted for publication.

Lackner, J. R., & Graybiel, A. (1981). Illusions of postural, visual, and substrate motion elicited by deep knee bends in the increased gravitoinertial force phase of parabolic flight. *Experimental Brain Research*, *44*, 312–316.

Lackner, J. R., & Graybiel, A. (1982). Rapid perceptual adaptation to high gravitoinertial force levels: Evidence for context-specific adaptation. *Aviation, Space and Environmental Medicine*, *53*, 766–769.

Lackner, J. R., & Graybiel, A. (1984). Perception of body weight and body mass at twice Earth-gravity acceleration levels. *Brain*, *107*, 133–144.

Lackner, J. R., & Levine, M. S. (1979). Changes in apparent body orientation and sensory localization induced by vibration of postural muscles: Vibratory myesthetic illusions. *Aviation, Space and Environmental Medicine*, *50*, 346–354.

Lisberger, S. G. (1988). The neural basis for motor learning in the vestibuloocular reflex in monkeys. *Trends in Neuroscience*, *11*, 147–152.

Lisberger, S. G., Miles, F. A., & Zee, D. (1984). Signals used to compute errors in monkey vestibulo-ocular reflex: Possible role of flocculus. *Journal of Neurophysiology*, *52*, 1140–1153.

Marr, D. (1969). A theory for cerebellar cortex. *Journal of Physiology*, *202*, 437–470.

Matthews, P. B. C. (1981). Muscle spindles: Their messages and their fusimotor supply. In V. B. Brooks (Eds.), *Motor control, Sect. 1, Vol. 21, Handbook of physiology* (pp. 189–228). Bethesda, MD: American Physiological Society.

Matthews, P. B. C. (1988). Proprioceptors and their contribution to somatosensory mapping: Complex messages require complex processing. *Canadian Journal of Physiology and Pharmacology*, *66*, 430–438.

McCloskey, D. I., Cross, M. J., Honner, R., & Potter, E. K. (1983). Sensory effects of pulling or vibrating exposed tendons in man. *Brain*, *106*, 21–37.

Miles, F. A., & Lisberger, S. G. (1981). Plasticity in the vestibulo-ocular reflex: A new hypothesis. *American Review of Neuroscience*, *4*, 273–299.

Pompeiano, O. (1988). Basic aspects of vestibulo-spinal function. In M. Igarashi & K. Nute (Eds)., *Proceedings of the Symposium on Vestibular Organs and Altered Force Environments* (pp. 1–23). Houston, NASA/Space Biomedical Research Institute.

Reschke, M. F., Anderson, D. J., & Homick, J. L. (1984). Vestibulo-spinal reflexes as a function of microgravity. *Science*, *225*, 212–214.

Schmitt, H. H., & Reid, D. J. (1985). Anecdotal information on space adaptation syndrome. NASA/Space Biomedical Research Institute, USRA/Division of Space Biomedicine.

Sperry, R. (1950). Neural basis of the spontaneous optokinetic response induced by visual inversion. *Journal of Comparative Physiological Psychology*, 482–489.

Stryker, M. P., & Sherk, H. (1975). Modification of cortical orientation selectivity in the cat by restricted visual experience: A re-examination. *Science*, *190*, 904–906.

Thompson, R. F. (1986). The neurobiology of learning and memory. *Science*, *233*, 941–947.

Wallach, H. (1976). *On Perception.* New York: Quadrangle.

Watt, D. G. D., Money, K., & Tomi, L. M. (1986). MIT/Canadian vestibular experiments on the Spacelab-1 mission: 3. Effects of prolonged weightlessness on a human otolith-spinal reflex. *Experimental Brain Research*, *64*, 308–315.

Welch, R. (1978). *Perceptual modification: Adapting to altered sensory environments.* New York: Academic Press.

# 19 Contribution of Skeletal and Extraocular Proprioception to Kinaesthetic Representation

J. P. Roll
J. C. Gilhodes
R. Roll
J. L. Velay
*Laboratoire de Neurobiologie Humaine*
*Université de Provence Marseille, France*

## ABSTRACT

The sensory feedback originating in muscles seems to be of major importance in the central elaboration of motor representation. The muscle spindle messages underlying movement and posture may be processed together in order to constantly elaborate static and dynamic body references on the basis of which retinal information may be specified in terms of spatial coordinates. The main arguments supporting this view are as follows:

1. Experimental manipulation of muscle spindle output by tendon vibration induces segmental or postural kinaesthetic illusory movements in the direction in which the vibrated muscles are stretched. By modifying the spatial distribution (agonists and/or antagonists), the frequency, and the duration of stimuli, it is possible to induce simple or complex kinaesthetic illusions with predictable characteristics. The microneurographic recording of single-muscle afferent activity in man indicates the exact nature of sensory receptors activated by vibration and the frequency organization of the induced messages.
2. Painless well-patterned vibrations, applied with an increasing frequency from 10 to 80 Hz to either the medial, lateral, superior, or inferior rectus of a subject's eye are found to induce directional perceptual and motor effects that are closely related to the postural context.
3. Likewise, illusory directional shifts of a visually fixed target in darkness during extraocular muscle vibration are reported by subjects. These

data suggest that extraocular proprioception contributes to the coding of eye, head, and body position in relation to postural and environmental conditions.

4. As with eye muscle stimulation, directional visual and postural effects are induced by vibration of neck and/or ankle postural muscles. These effects are found to add to each other when vibrations are applied simultaneously to the eye, neck, and ankle muscles.

The probable involvement of extraocular proprioception in the interrelating of body space with extrapersonal space in overall oriented behavior is discussed.

## INTRODUCTION

Since the beginning of the century when Sherrington (1900; p. 1006) claimed that "muscular sense is based on a specific set of sensations, obtained by specific sense organs in muscles, tendons and joints," there has been much controversy about the contribution of muscular information to the central representation of human motor activities. Without going into the details of this controversy (see Matthews, 1982; McCloskey, 1978; Roll, 1981 for reviews), we would just mention here that nowadays the role of sensory feedback in the elaboration of posture and movement conscious coding is no longer a matter of debate. What remain to be established are the respective roles played by central signals arising from the motor command, the so-called efferent copy or corollary discharges (Sperry, 1950; Von Holst & Mittelstaedt, 1950), and by the peripheral messages elicited by ongoing motor activities. The final proof that muscular feedback contributes to the awareness of movement and posture was provided by the finding that in humans it is possible to induce illusory sensations of movement by vibration, the characteristics of which (direction, speed, duration) depend on the stimulus parameters. This applies both to single-limb and whole-body illusory movements (Bonnet, Roll, & Lacour, 1973; Eklund, 1972; Goodwin, McCloskey, & Matthews, 1972; Roll & Vedel, 1982).

By manipulating muscle proprioception by means of mechanical vibration applied to one or several muscles simultaneously, it is therefore possible to induce complex kinaesthetic sensations with specific, predictable spatio-temporal characteristics (Gilhodes, Roll, & Tardy-Gervet, 1986). This suggests that each movement, or motor pattern, is closely associated with a specific sensory set that might reflect the movement execution features. The fact that artificially elicited afferent messages suffice to evoke the corresponding movement sensations supports the idea that the former at least partly mediates the awareness of the latter (Roll, 1987).

Kinaesthetic illusions of this kind can involve any of the body segments and/or the whole body, depending on the functional role of the stimulated muscle and the particular postural context. For example, we have shown that it is possible to use vibration to manipulate extraocular proprioception in order to elicit illusory head, trunk, or whole-body movements that depend on the subject's posture (subject sitting or standing, with head fixed or free) (Roll & Roll, 1987). These data suggest that the direction of the subject's gaze, via the associated extraocular proprioceptive information, may play a key role in the organization of body posture along with the other proprioceptive information arising from the various body segments.

Given that the moving human eye is carried by a set of linked body segments, namely the head, trunk, and legs, we have formed the hypothesis that a proprioceptive chain connects the retina to the ground, and mediates the awareness of the static and dynamic postural configurations that are adopted during behavior.

Along these lines, our results indicate that besides the visual and vestibular contribution, extraocular, neck, and leg proprioception may contribute together to the elaboration of a body reference frame that may also be involved in body orientation within extrapersonal space (Roll & Roll, 1988). The latter point has been confirmed by the fact that under certain conditions, visual perception can be modified by applying of vibration to eye (Roll & Roll, 1987), neck (Biguer, Donaldson, Hein, & Jeannerod, 1988) and/or leg muscles (Roll & Roll, 1988). It is now proposed to develop the above considerations in the light of recently obtained experimental results.

## RESULTS

### Microneurographic recordings of vibration-induced afferent messages

If the occurrence of conscious sensations of movement in humans subjected to vibration is to be used as evidence for the central use of muscular feedback in motor representation, it is first necessary to elucidate the exact nature and the spatio-temporal organization of the sensory messages involved. With this aim in mind, and in an attempt to extend to man results previously obtained on animals concerning the sensory effects of vibrations (Brown, Engberg, & Matthews, 1967; Clark, Matthews & Muir, 1981; Matthews & Watson, 1981), the activity of single proprioceptive fibers were recorded in man from the peroneus lateralis nerve using tungsten microelectrodes. Unitary discharges originating from muscle spindle primary and secondary endings and Golgi tendon organs were identified using classical physiological tests (see Burke, Hagbarth, Lofstedt, & Wallin, 1976a; 1976b; Vallbo, 1973). The sensitivity of these various kinds of proprioceptors to low amplitude mechanical vibration was then studied (amplitude .2 to .5 mm peak to peak). The results we obtained are given in Fig. 19.1. Muscle spindle primary endings (upper diagrams) were found to be the most sensitive to this mechanical stimulus ($n$ = 25). The discharge of some of them could be driven in a one-to-one manner up to 180 Hz. Most of them fired harmonically with the vibration up to 80 Hz and then discharged in a subharmonic manner ($\frac{1}{2}$–$\frac{1}{3}$) with increasing vibration frequency. Muscle spindle secondary endings ($n$ = 11; lower left diagram)

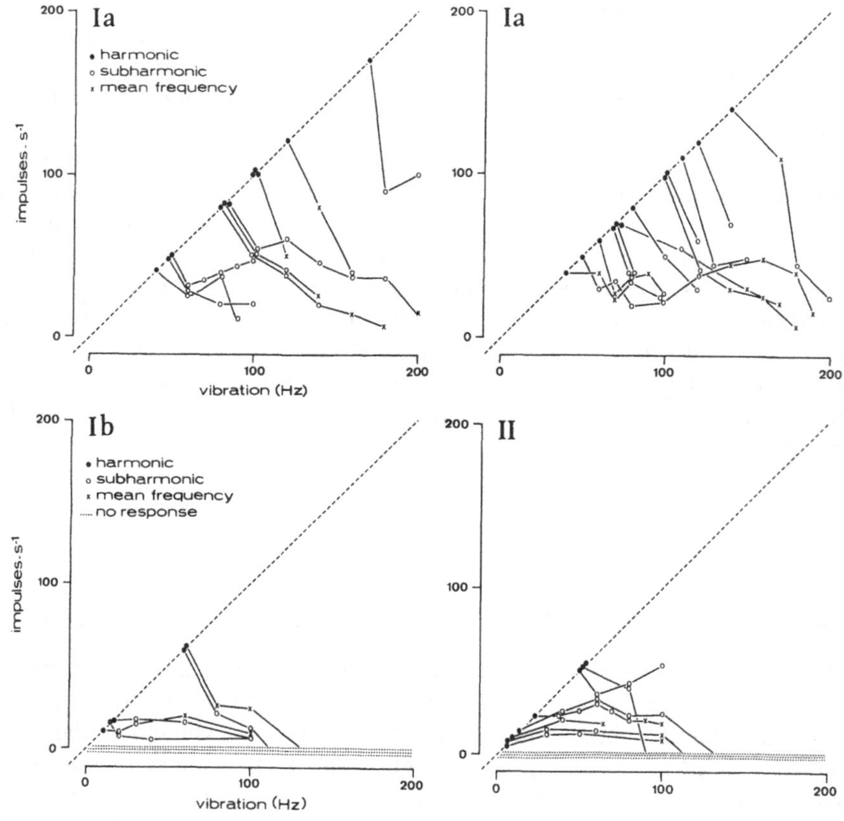


FIG. 19.1. Driving modalities in the activity of muscle spindle primary (Ia) and secondary (II) endings and tendon organs (Ib) induced by mechanical vibrations applied at various frequencies to the tendon of the receptor-bearing muscles (Tibialis anterior, Extensor digitorum longus). Black circles indicate for each unit the maximal one-to-one driven activity; open circles illustrate subharmonic responses and crosses symbolize the mean frequency of the induced sensory activity which was not phase-related to vibration frequency. Vibrations were imposed on relaxed muscles maintained slightly stretched (From Roll et al., 1989).

and Golgi tendon organs ($n = 8$; lower right diagram) were found to be either insensitive or only slightly sensitive to tendon vibration. Most of them did not follow the stimulus up to 20 to 30 Hz.

These results are consistent with data by Burke et al. (1976a; 1976b) and with the conclusion reached by Brown (1967) that in man "vibration applied to tendon, in the absence of muscle contraction, can be presumed to be a specific stimulus for primary endings." Moreover the fact that in man, most primary endings respond in phase to vibratory stimuli up to 70 to 100 Hz suggests that within this frequency range the stimulus frequency is fairly proportional to the responses of the majority of the primary endings. The existence of a one-to-one linkage within this frequency range means that by modulating the vibration frequency, it is possible to induce a proportional modulation in the primary-ending discharge frequency. This makes it possible to predict the activity level of the Ia afferent messages induced in man when muscle tendons are vibrated. This predictability makes tendon vibration a useful tool for experimentally eliciting kinaesthetic illusions based on previously known afferent messages.

## Simple and complex vibration-induced illusory movements in man

In the absence of visual information, muscle vibration induces kinaesthetic illusions the direction of which corresponds to that of an actual movement in which the vibrated muscle is stretched (Bonnet, Roll, & Lacour, 1973; Roll, Gilhodes, & Tardy-Gervet, 1980; Goodwin, McCloskey, & Matthews, 1972).

Quantification of the parameters of the vibration-induced illusory sensations of movement can be achieved by asking the subjects to track what they feel in the vibrated arm with the unvibrated one. The velocity of the illusory movements always corresponded to low-velocity displacements and was dependent on the vibration frequency (Roll & Vedel, 1982). This indicates that the total amount of Ia activity increased when the vibration frequency was increased. The results obtained using simple vibration, that is, vibration of only one muscle, with a constant frequency stimulus are summarized in Fig. 19.2. In these experiments, biceps vibration resulted in all subjects in a continuous sensation of forearm extension. Increasing the stimulus frequency in steps from 20 to 80 Hz increased the speed of the illusion, whereas an increase in vibration frequency from 80 to 120 Hz tended to decrease the velocity.

The results of these simple experiments suggest that the sensory signals from stretched antagonist muscles play a major part in the coding of movements during which the agonists have a motor function. Accompanying the increase in muscle spindle activity in the stretched muscle, however, a corresponding decrease occurred in the active muscles, even in the pres-

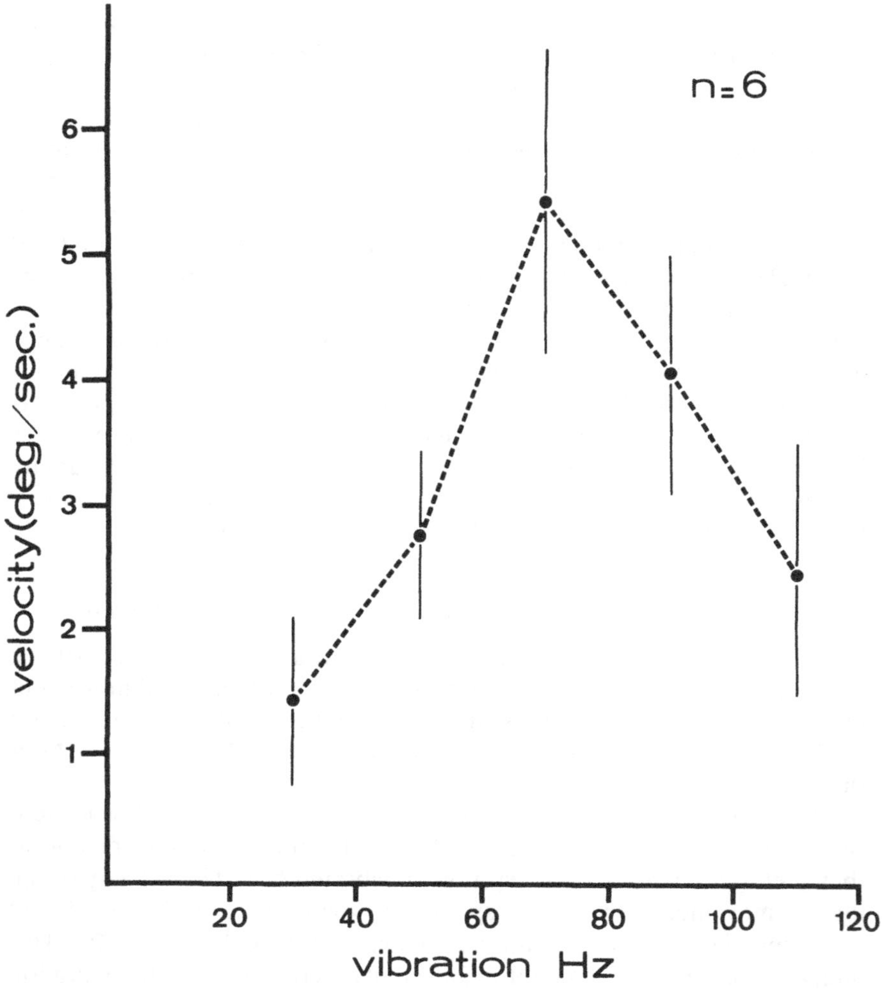


FIG. 19.2. Relationships between the vibration frequency and the velocity of the illusory movement. The curve illustrates the mean results obtained from 6 subjects. Each point represents 18 trials.

ence of an $\alpha = \gamma$ co-activation (Ribot, Roll, & Vedel, 1986). This led us to simultaneously manipulate by vibration the proprioception of two or several antagonist muscles in order to investigate the relationships between proprioceptive feedback and the corresponding sensations. By applying vibration either simultaneously or alternately to two or more muscles, one can obtain complex kinaesthetic sensations. We then established the relationships between these sensations and the underlying afferent messages. Figure 19.3 shows, for example, that by using two vibrators applied re-

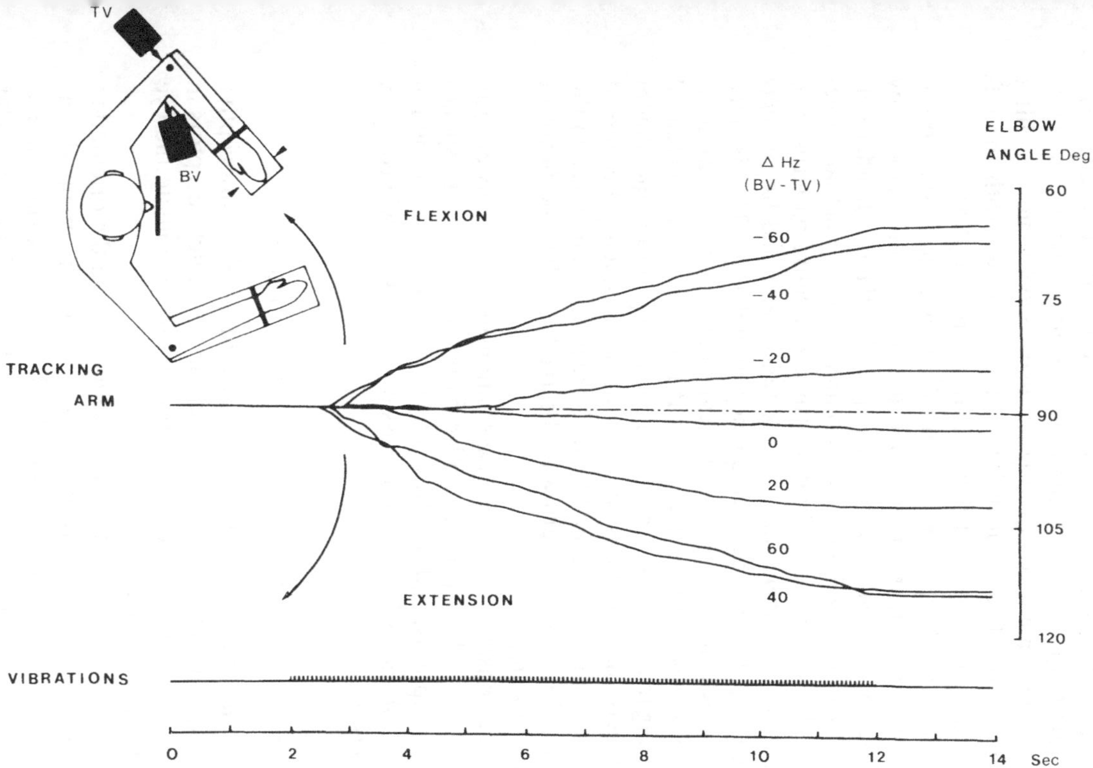


FIG. 19.3. Schematic diagram of experimental situation and recordings illustrating kinesthetic sensations. Vibrators Bv and Tv were applied to the distal tendons of the left biceps brachii and triceps brachii respectively. The left arm was resting on a fixed horizontal bracket. The contralateral arm (tracking arm) was placed on a movable bracket connected to the recording system (goniometer). Traces reproducing movement sensations in one subject have been superimposed; above each trace is marked the difference in vibration frequency between Bv and Tv. (from Gilhodes et al., 1986)

spectively to biceps and triceps muscles it is possible to elicit a sensation of either extension or of flexion. The sensation was almost always one of flexion of the forearm when the frequency of triceps vibration was higher than that of biceps vibration, and one of extension in the opposite case. When the vibration frequencies applied to the two antagonist muscles were equal, no illusory movement occurred (the mean velocity did not differ significantly from zero). The mean velocity then increased with the difference between the frequencies of biceps and triceps vibrations.

Because the vibration frequency is a crucial factor contributing to the velocity of illusory movements, it was decided to vary the velocity by varying the frequency of the stimuli applied at peripheral level. In addition, on the basis of microneurographic results obtained during actual movements, it was possible to drive the vibrators using spindlelike patterns. The results showed that the velocity of the sensations perceived by subjects depended closely on the vibration–frequency modulation. This is illustrated in Fig. 19.4, which shows an experiment where vibration was alternately applied to biceps and triceps muscles at frequencies that increased with time according to a "concave" mode: The resulting illusory movement shown in A reflects the variations in the stimulus frequency because it comprised first a slow movement and then, 2 sec later, an increase in the perceived velocity. The same pattern was observed whether the flexor or the extensor was vibrated.

These results demonstrate that a precise relationship exists between the frequency organization of the proprioceptive patterns and the corresponding kinesthetic perceptions, and that by controlling the spatial distribution and the frequency of each vibration, it is possible to evoke at will complex illusory sensations of movement.

## From eye to foot: A proprioceptive chain controlling posture

Because vibration is a means of selectively activating the proprioceptive afferent channel in man, and in view of the density of stretch receptors in extraocular muscles, we attempted to experimentally manipulate the eye muscle proprioceptors in man.

The direction of gaze is known to play a leading role in the orientation of the body and limbs during behavior but, up to now, the emphasis has mainly been placed on the retinal factors. Without overlooking the importance of outflow information and inflow from the visual and vestibular systems, we decided to focus on the role of the oculomotor proprioceptive input in postural regulation. Earlier studies indicated that in animals (Baron, 1955), as well as in man (Martins Da Cunha, 1987; Gagney, 1987), this extraretinal component might contribute to whole-body postural organization. Moreover, the close relationships that link the eye with the head

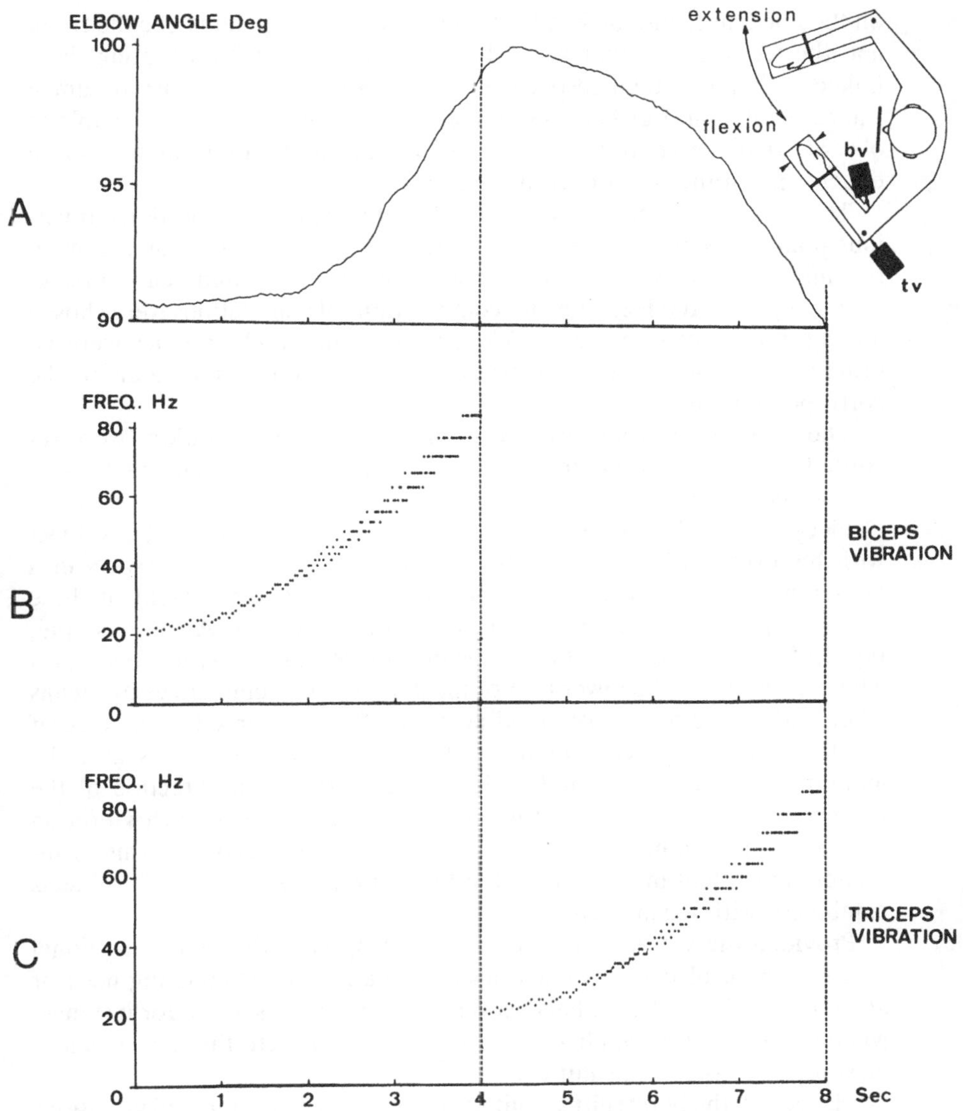


FIG. 19.4. Schematic diagram of experimental situation (upper right). The vibrated arm was fixed and the subject tracked the illusory left arm movement with his right arm. B and C show the frequencegrams of the vibratory stimulus applied to Biceps b. and Triceps b. tendons respectively (vibration duration 4 sec increasing frequency from 20 to 80 Hz). A shows the corresponding time recordings of the extension and flexion illusory movements.

on the one hand, and the head with the other body segments like the trunk and the legs on the other hand, led us to investigate, by vibrating these linked body parts, their respective proprioceptive contributions to human stance regulation. Furthermore, under specific conditions we were able to observe that the central processing of retinal information might also be dependent on the proprioceptive feedback.

The extraocular muscles were stimulated by applying vibrations to various points around the orbit. The vibrations were produced by means of two mini-vibrators set on a helmet (amplitude: .1 to .2 mm peak-to-peak; frequency : 1 to 100 Hz). Various combinations of stimulations were chosen among those shown in Fig. 19.5A. The neck and ankle muscles were vibrated by means of inertial vibrators attached with elastic bands to the corresponding tendons.

The postural effects of eye, neck or ankle muscle vibrations were recorded by means of four strain gages enclosed in a stabilometer set under each subject's foot.

Five-second vibratory stimuli applied to the eye muscles of a subject with his eyes closed gave rise to spatially oriented postural sways in a direction that depended on the stimulated muscles. The extent of these effects depended moreover on the vibration frequency. As an example, Fig. 19.5B shows that vibration of the inferior recti gave rise to a backward whole-body shift. The lower part of the diagram (C) summarizes by means of individual vectors the oriented postural effects observed in the case of the four stimulation combinations. The moduli of the vectors give the amplitude of the effects, and their angular position the direction of the postural shift. Groups of vectors occurred in the case of each vibration site. Section D summarizes the observed postural effects in terms of direction and extent in the case of the four vibration situations, all subjects combined, with mean vectors.

Provided the vibrated muscles were appropriate, the same directional whole-body displacements could also be obtained by stimulating neck or ankle postural muscles: A backward body shift was observed, for instance, when vibrations were applied either to the inferior recti, the sterno-cleido-mastoidus, or the soleus muscle.

Moreover, the body shift amplitude increased when combined vibrations were simultaneously applied to two or all of these muscle groups (Fig. 19.6).

These data suggest that proprioceptive inputs coming from the stretched muscles of various body segments can contribute jointly to a given postural configuration. This implies that a common processing of all these muscular feedbacks subserves postural regulation and that these feedbacks participate together in the elaboration of a body reference.

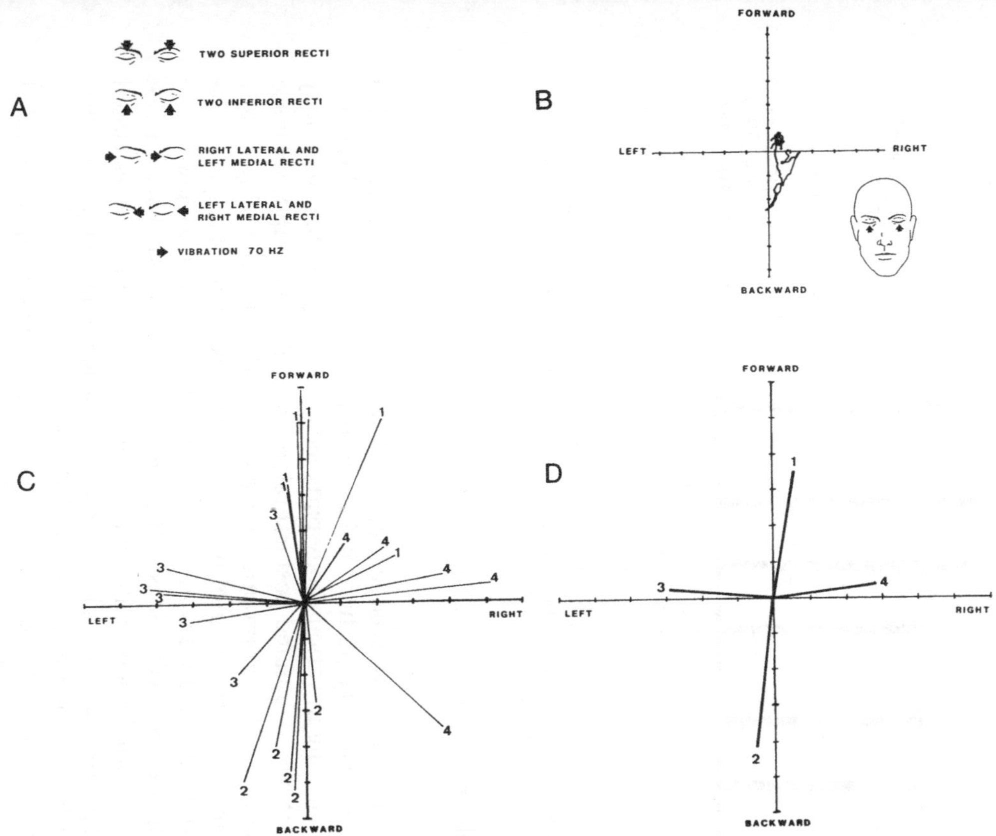


FIG. 19.5. Upper part: (A) Four different combined stimulations of both eyes. (B) Directional postural effect (backward body shift) of the vibratory stimulation of both inferior recti illustrated by one subject's stabilometric recording. Lower part: Group of individual (C) and mean (D) vectors corresponding to various combinations of extraocular muscle stimulation. 1) two superior recti, 2) two inferior recti, 3) lateral rectus of right eye together with medial rectus of left eye, 4) lateral rectus of left eye together with medial rectus of right eye (from Roll & Roll, 1987).

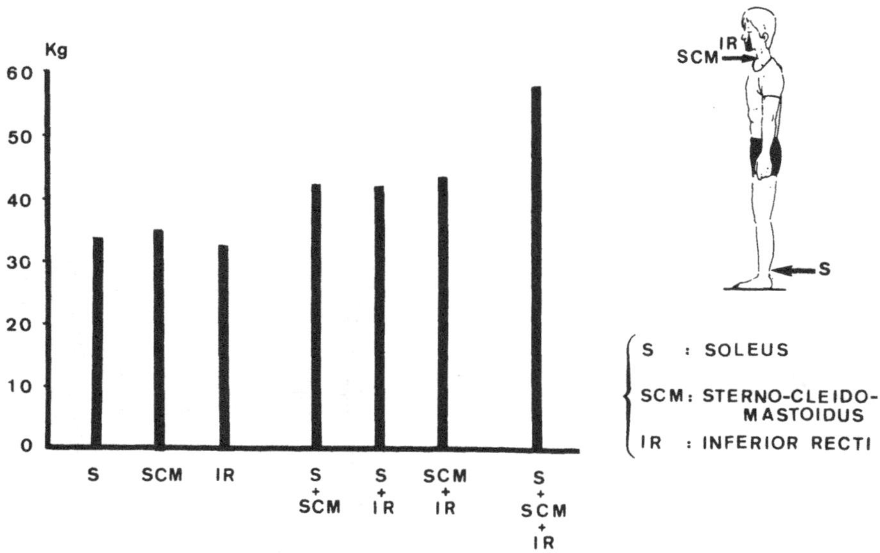


FIG. 19.6. Mean postural effects of separate and combined vibrations applied to inferior recti, sterno-cleido-mastoidus and soleus muscles (from Roll & Roll, 1988).

## Extraocular proprioception as a link between personal and extrapersonal space

As shown in the above section, the fact that extraocular proprioception can add its regulatory postural influence to that of other proprioceptive inputs originating in various body parts led us to propose that there may exist a centrally elaborated postural reference that is built up from these simultaneously collected proprioceptive cues, and that is probably mainly involved in the interrelating of body space with extrapersonal space. The last experimental data we would like to describe support this view.

The application of vibrations to the same muscles as those that gave rise to directional whole-body shifts in a standing subject with his eyes closed induced apparent directional movements of a luminous target that was monocularly fixed by the subject in darkness.

As with the postural sways, the target displacement direction depended strictly on which muscle was vibrated: Vibration of the inferior rectus induced an upward target shift, whereas a downward shift occurred under superior rectus vibration. External rectus vibration elicited an illusory target displacement in the medial direction and vibration of the medial rectus a displacement in the external direction. Likewise, a directional shift of the visual target was elicited by vibration applied to neck muscles (splenius, sterno-cleido-mastoidus, or trapezius) and, more surprisingly, to ankle pos-

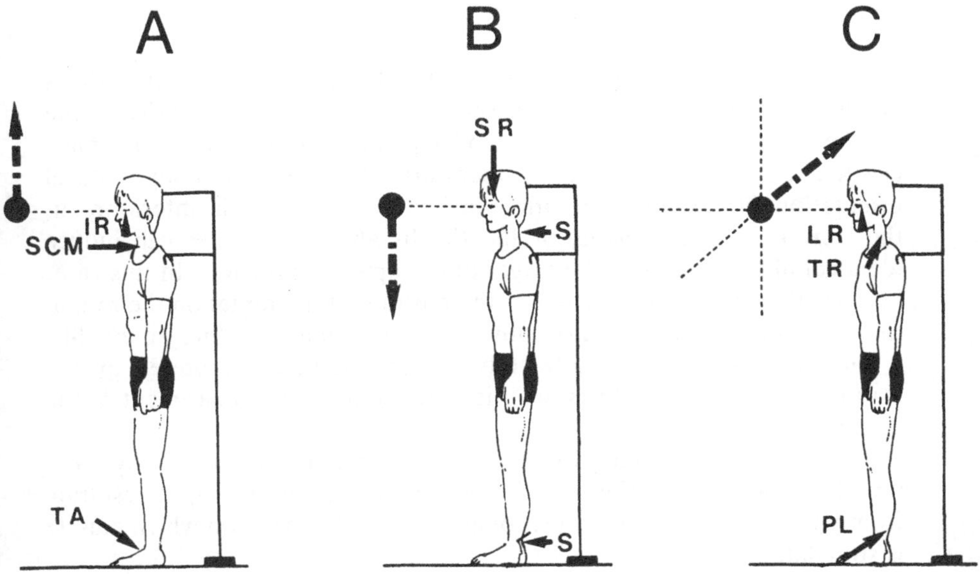


| | | | | |
|---|---|---|---|---|
| Extra-ocular Muscles | Inferior recti ↑ | Superior recti ↓ | Right Lateral Left Medial recti ← | Left Lateral Right Medial recti → |
| Neck Muscles | Sterno-cleido mastoïdeus ↑ | Splenii ↓ | Right Trapezius ← | Left Trapezius → |
| Leg Muscles | Tibialis anterior ↑ | Soleus ↓ | Right Peroneus longus Left Tibialis posterior ← | Left Peroneus longus Right Tibialis posterior → |

FIG. 19.7. Apparent movement of a visual target induced by muscle vibration (70 Hz): the three sites of application of vibrations giving rise to identical apparent movement of the target upward, downward and rightward are shown in A B and C respectively. The table summarizes the various directions of the apparent movements of the target (arrows) and the corresponding vibrated muscles (from Roll & Roll, 1988).

tural muscles (soleus, tibialis, or peroneus). The most noteworthy findings are summarized in Fig. 19.7 that shows the directional similarity of the apparent target movement obtained on applying vibrations to specific muscles that are usually stretched together during a given whole-body postural orientation. For example, vibration, in darkness, of either the inferior recti, the sterno-cleido-mastoidus, or the tibialis anterior muscles can induce identical illusory shifts of the target in the upward direction. In this case, because the only available cue was the position of the target on the retina, the vibration-induced proprioceptive pattern simulating either an eye lift, a backward head, or a whole-body movement, was interpreted by the stationary subject in darkness as if it were an upward displacement of the target.

If the subject is required to accurately locate a visual target by hand pointing, the modification of extraocular proprioceptive input resulting from vibrations can induce directional errors that depend on which muscle is stimulated.

In total darkness, the spatial processing of retinal information, particularly in the case of the manual localization of a visual target, is based on a body-centered reference. This means that, in addition to the retinal coding of the target, the extraretinal signal of position of the eye in its orbit, in relation to head, trunk, and whole-body posture may play a leading role in human spatially oriented behavior, in particular the goal-directed movements.

In assuming that this signal results from the processing of congruent proprioceptive messages arising in the various body segments that link the foot to the eye, we imply that the content of this information is mainly proprioceptive.

The accuracy of subjects' performances was tested in a visuo-manual pointing task where vibrations were applied to either the inferior, superior, or external rectus of one eye. Subjects were seated in a darkroom, facing monocularly a vertical panel on which light-emitting diodes served as visual targets. On the panel, a grid-patterned, printed circuit automatically supplied the rectangular coordinates of the point contacted by the finger. Head movements were prevented by a chin rest.

Vibrations were applied at the onset of the target; they lasted for 7 seconds, during which subjects were asked to maintain their gaze on the target. As soon as both the target and the vibrator were off, subjects had to point at the place where the target had disappeared.

As an example, Fig. 19.8 shows that vibration of the right external rectus induced a leftward shift of the hand-pointing error as compared to the control conditions, whether the vibrated eye was open or closed (A and B). The extent of the overpointing was less, however, when the vibrated eye was closed and the other eye looking at the target (B).

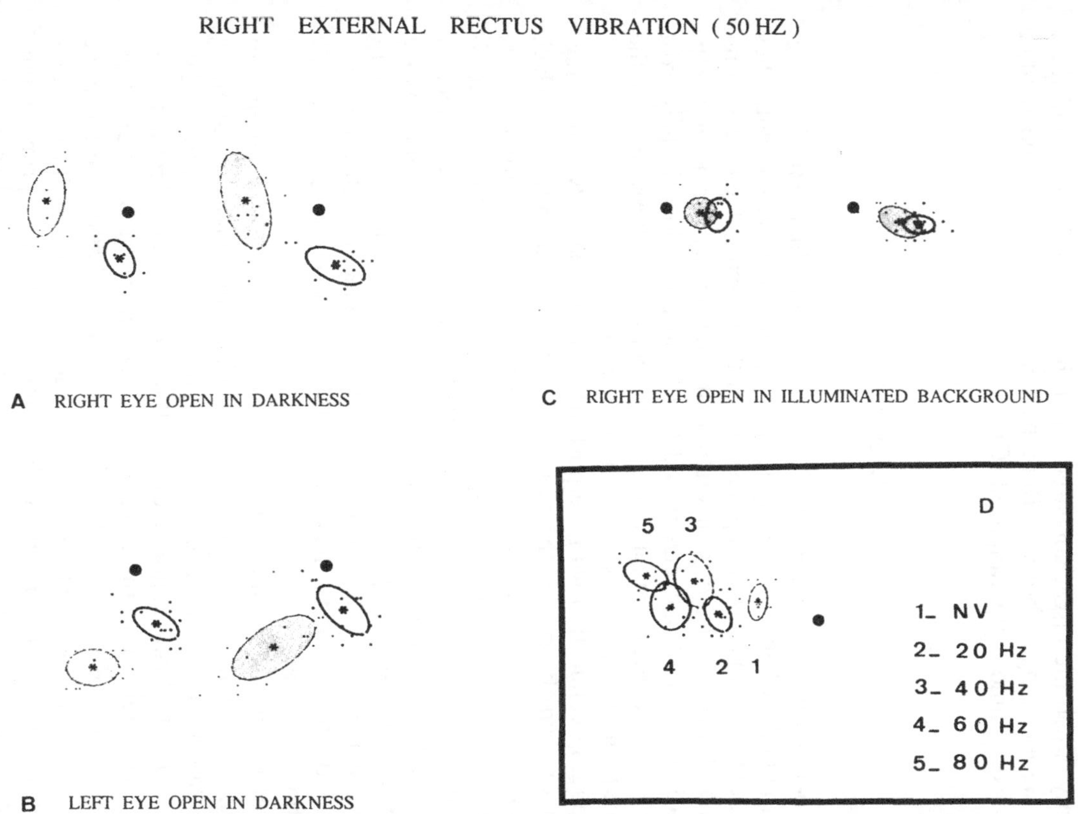


FIG. 19.8. Pointing error as shown by the confidence ellipses in normal vision (open surfaces) and when one subject's right lateral rectus was vibrated (hatched surfaces) under three experimental conditions (A, B, and C). D illustrates the variation of the pointing error as a function of the vibration frequency when the stimulation was applied to the external rectus of the right eye open in darkness.

As in the case of postural reactions, the vibration-induced effects on eye–hand accuracy were found to be frequency dependent, that is, the extent of the pointing error increased as a function of the vibration frequency between 20 Hz and 80 Hz (D).

Lastly, when the target appeared on an illuminated background these effects were no longer observed. Both the mean pointing error and pointing surface were almost superimposed (C).

## CONCLUSION

The experimental data presented here may contribute first to further specifying the body scheme concept in terms of proprioceptive content. They suggest that a common central processing of the muscular feedbacks, coming from all the muscles of a myotatic unit, may subserve conscious representation of a limb movement and that a close relationship may exist between the frequency modulation of the peripheral messages and the perceived parameters of a specific motor activity.

Second, in connection with the posture of the whole body and its regulation, one can observe, on the basis of the results of experiments in which proprioception was manipulated by applying vibrations at various levels, that extraocular, neck, and ankle proprioceptive inputs combine their effects to organize human stance. This fact is of importance because it means, on the one hand, that the proprioceptive component of gaze may play a leading role in postural regulation and, on the other hand, that a proprioceptive chain may link the various mobile body segments such as the eyes, the head, the trunk, and the legs forming the main part of what is generally called body reference.

Third, because the manipulation of proprioception in the absence of actual movement modifies the processing of visual information by inducing oriented apparent target movements, it is possible that the centrally elaborated postural scheme might be used in the spatial coding of retinal cues.

Last, the finding that the modification resulting from vibration of extraocular (Roll & Roll, 1987) or neck (Biguer et al., 1988) proprioceptive afferents alters manual pointing to visual targets suggests that the muscular proprioceptive chain linking eye to foot support may be of prime importance in the interrelating of body space with extrapersonal space. This hypothesis is in agreement with experimental data collected on kittens (Hein & Diamond, 1982) indicating that early suppression of extraocular proprioception alters the development of visually guided activity.

## ACKNOWLEDGMENTS

Financial support for this research was provided by a grant from INRS and Fondation pour la Recherche Medicale. English revision was carried out by Dr. Jessica Blanc.

## REFERENCES

Baron, J. B. (1955). Muscles moteurs oculaires, attitudes et comportements des vertébrés. *Thèse Faculté des Sciences*, Paris.

Biguer, B., Donaldson, I. M. L., Hein, A., & Jeannerod, M. (1988). Neck muscle vibration modifies the representation of visual motion and direction in man. *Brain*, *11*, 1405–1424.

Bonnet, M., Roll, J. P., & Lacour, M. (1973). High frequency activation of myotatic pathway in man and illusion of movement. *E.E.G. Clinical Neurophysiology*, *34*, 810.

Brown, M. C., Engberg, I., & Matthews, P. B. C. (1967). The relative sensitivity to vibration of muscle receptors in the cat. *Journal of Physiology*, *192*, 773–800.

Burke, D., Hagbarth, K. E., Lofstedt, L., & Wallin, B. G. (1976a). The responses of human muscle spindle endings to vibration of non-contracting muscles. *Journal of Physiology*, *261*, 673–693.

Burke, D., Hagbarth, K. E., Lofstedt, L., & Wallin, B. G. (1976b). The responses of human muscle spindle endings to vibration during isometric contraction. *Journal of Physiology*, *261*, 695–711.

Clark, F. J., Matthews, P. B. C., & Muir, R. B. (1981). Response of soleus Ia afferents to vibration in the presence of the tonic vibration reflex in the decerebrate cat. *Journal of Physiology*, *311*, 97–112.

Eklund, G. (1972). Position sense and state of contraction: the effects of vibration. *Journal of Neurology, Neurosurgery and Psychiatry*, *35*, 606–611.

Gagey, P. M. (1987). L'oculomotricité comme endo-entrée du système postural. *Agressologie*, *28*, 899–903.

Gilhodes, J. C., Roll, J. P., & Tardy-Gervet, M. F. (1986). Perceptual and motor effects of agonist-antagonist muscle vibration in man. *Experimental Brain Research*, *61*, 395–402.

Goodwin, G. M., Mc Closkey, D. I., & Matthews, P. B. C. (1972). The contribution of muscle afferents to kinaesthesia shown by vibration induced illusions of movement and by the effects of paralysing joint afferents. *Brain*, *95*, 705–748.

Hein, A., & Diamond R. (1982). Contribution of eye movement to representation of space. In A. Hein & M. Jeannerod (Eds.), *Spatially oriented behaviour* (pp. 119–133). New York: Springer-Verlag.

Martins Da Cunha, H. M. (1987). Le syndrome de déficience posturale (SDP). *Agressologie*, *28*, 941–943.

Matthews, P. B. C., & Watson, J. D. G. (1981). Action of vibration on the response of cat muscle spindle Ia afferents to low frequency sinusoidal stretching. *Journal of Physiology*, *317*, 365–381.

Matthews, P. B. C. (1982). Where does Sherrington's "muscular sense" originate? Muscles, joints, corollary discharges? *Annual Review of Neurosciences*, *5*, 189–218.

Mc Closkey, D. I. (1978). Kinaesthetic sensibility. *Physiological Review*, *58*, 763–820.

Ribot, E., Roll, J. P., & Vedel, J. P. (1986). Efferent discharges recorded from single skeletomotor fibres in man. *Journal of Physiology*, *375*, 251–268.

Roll, J. P., Gilhodes, J. C., & Tardy-Gervet, M. F. (1980). Effets perceptifs et moteurs des vibrations musculaires chez l'homme normal: Mise en evidence d'une réponse des muscles antagonistes. *Archives Italiennes de Biologie*, *118*, 51–71.

Roll, J. P. (1981). Contribution de la proprioception musculaire à la perception et au contrôle du mouvement chez l'homme. Thèse Faculté des Sciences, Marseille.

Roll, J. P., & Vedel, J. P. (1982). Kinaesthetic role of muscle afferents in man, studied by tendon vibration and microneurography. *Experimental Brain Research*, *47*, 177–190.

Roll, J. P. (1987). Les bases physiologiques des conduites: fonctions de prises d'informations et d'exploration. In *Encyclopédie de la Pleïade* (pp. 1476–1535). Paris: Ed. Gallimard.

Roll, J. P., & Roll, R. (1987). Kinaesthetic and motor effects of extraocular vibration in Man. In O. Regan, J. K. & A. Levy-Schoen (Eds.), *Eye movements: From physiology to cognition* (pp. 57–68). Amsterdam: Elsevier, North Holland.

Roll, J. P., & Roll, R. (1988). From eye to foot: a proprioceptive chain involved in postural control. In B. Amblard, A. Berthoz, & F. Clarac (Eds.), *Posture and gait: Development, adaptation and modulation* (pp. 155–164). Amsterdam: Elsevier, North Holland.

Roll, J. P., Vedel, J. P., & Ribot, E. (1989). Alteration of proprioceptive messages induced by tendon vibration in man. *Experimental Brain Research*, *76*, 213–222.

Sherrington, C. S. (1900). *Textbook of physiology*. Schäfer (Ed.), 1002–1025.

Sperry, R. W. (1950). Neural basis of the spontaneous optokinetic response produced by visual neural inversion. *Journal of Comparative and Physiological Psychology*, 482–489.

Vallbö, A. B. (1973). Muscle spindle afferent discharge from resting and contracting muscles in normal human subjects. In J. E. Desmedt (Ed.), *New developments in EMG and clinical neurophysiology* (Vol. 3, pp. 251–262). Basel: Karger.

Von Holst, E., & Mittelstaedt, H. (1950). The reafference principle. Interaction between the central nervous system and the periphery. In *Selected papers of Erich von Holst: The behavioural physiology of animals and Man* (Vol. 1, pp. 139–173). London: Methuen (From German).

# 20 Eye Movements to a Visual Stimulus Flashed Before, During, or After a Saccade

Hitoshi Honda
*Department of Psychology*
*Niigata University, Japan*

## ABSTRACT

It is well known that a target presented at about the time of a saccade is perceptually mislocalized. Some studies report, however, that, in contrast to perception, the eye is directed to the actual position of the target. We examined this discrepancy by comparing oculomotor (saccadic) and perceptual localizations of a target flashed at about the time of a prior saccade. The results showed that the perceptual mislocalization is a time function. Targets flashed before and at the beginning of the saccade were mislocalized in the saccade direction. About midway through the saccade, the targets were localized near the actual position. At the end of the saccade, the targets were mislocalized in the opposite direction from the saccade. Oculomotor localization showed a similar mislocalization to the perceptual judgment: The eye moved not to the actual target position in space, but to its perceptually judged position. The results were interpreted as showing that perceptual and oculomotor localizations were based on the same extraretinal eye position information that was not properly timed to the actual eye position. Therefore, the results were not consistent with strict "spatial models" of the saccade system, which assume that saccade target are localized with respect to head or body. The detailed analysis of the oculomotor responses supported some findings on the saccade system reported by earlier double step stimuli experiments.

## INTRODUCTION

Earlier models of the saccade system assumed that saccade targets were localized by using only retinal error, that is, the distance and direction of a target image from the fovea (Robinson, 1973; Young & Stark, 1963). Recent studies have shown, however, that saccades are made on the basis of a combination of retinal and eye position information. Mays and Sparks (1980) electrically stimulated the superior colliculus of a macaque monkey during the latency period of a saccade to the location of a briefly flashed target and found that although electrical stimulation drove the eyes away from an original fixation point, the monkey made a compensatory saccade near the actual position in space of the target. This finding suggests that saccade targets were localized by using not only retinal error but also information about the size of eye movements caused by electric stimulation; that is, spatial models of the saccade system assume that saccade targets are localized with respect to head or body and not with respect to the retina.

Another piece of evidence for spatial models was reported by Hallett and Lightstone (1976a, 1976b). They analyzed the oculomotor (saccade) responses to a visual target flashed during a prior saccade. The results showed that the saccade to the target did not occur on the basis of the position of the retinal image of the target, but was directed, rather, to the actual target position in space. In one of the experimental conditions they employed, a visual target was presented near the fovea of the moving eye during a prior saccade, and, consequently, the size of the retinal error was very small. Nevertheless, subsequent saccades were observed to reach the actual target position.

The finding of Hallett and Lightstone (1976a, 1976b) is quite important from the viewpoint of the saccade system. However, their finding is not consistent with the results of psychophysical studies on perceptual localization. It has been widely accepted that the position of a visual target briefly presented during a saccade is mislocated at different positions from its actual position (Bischof & Kramer, 1968; Mateeff, 1978; Matin, Matin, & Pearce, 1969; Matin, Matin, & Pola, 1970). Bischof and Kramer (1968), for example, reported that when a target was flashed at the beginning of a saccade, the target was mislocalized in the saccade direction. How can we explain this discrepancy?

One possible solution to this question is to admit that there is a dissociation between the perceptual and the oculomotor level of processing. One example of this dissociation is the phenomenon of blindsight (Perenin & Jeannerod, 1975; Poppel, Held, & Frost, 1973; Weiskrantz, Warrington, Sanders, & Marshall, 1974). However, before we accept this kind of explanation, it is necessary to attempt to find out other possible explanations. For this purpose, we need a complete description of the perceptual and oculomotor localizations of a visual target flashed at the time of a saccade.

The size of mislocalization of a visual target presented at the time of a saccade is known to change as a function of the time from the saccade onset (Bischof & Kramer, 1968; Mateeff, 1978). In the present experiment, therefore, visual targets were presented at various points in time before, during, or after a prior saccade. Because visual targets were presented in the dark, the subjects were not able to use any visual cue, such as a steady visual background, for judging the target position. Therefore, we can assume that the subjects judged the target position on the basis of both the position of the target image on the retina (retinal information) and the information about the position of the eye in the orbit during the saccade (extraretinal eye position information). If the eye is directed to the perceptually judged position of the target, then we can say that the saccade system uses the same information as it does for the perceptual system. On the other hand, if the eye is directed to the actual target position, this means that the saccade system has its own extraretinal eye position information.

The experimental paradigm of the present study was a modification of the double step stimuli experiments that have been employed in many behavioral studies on the saccade system (Becker & Jurgens, 1979; Komoda, Festinger, Phillips, Duckman, & Young, 1973; Westheimer, 1954; Wheeles, Boynton, & Cohen, 1966).

## METHOD

### Procedure

A subject was seated in a dark room with the head fixed by a dental bite board and a forehead rest. The subject's horizontal eye movements of the right eye were recorded by the scleral reflection technique (DC—500 Hz, −1 dB) with an accuracy of 0.5 deg. On each trial, a buzzer warning signal (1 sec in duration) was given, and then a fixation point (red LED with a diameter of 0.3 deg, 20 cd/m$^2$) was presented for 1–1.8 sec at the center of the visual field. The subject was asked to keep watching the fixation point. At the offset of the fixation point, a cue for a prior saccade was presented for 20 msec, 8 deg to the right of the fixation point. The cue consisted of two vertically arranged rectangular red LEDs (0.1 deg × 0.3 deg, 10 cd/m$^2$), the distance between the centers of the LEDs being 0.4 deg. The subject was asked to make a prior saccade (saccade $S_0$) to the cue. The cue was extinguished before the onset of saccade $S_0$. At around the time of the saccade $S_0$, a visual target (yellow LED, 0.3 deg in diameter, 40 cd/m$^2$) for perceptual and oculomotor localization was presented for 2 msec. The position of the target was selected randomly from 5 positions, that is, −4, 0, +4, +8, +12 deg, where the minus and the plus signs indicate the left and the right sides of the fixation point, respectively. The subject was asked to move the eyes (saccade $S_1$) as rapidly as possible to where the target had appeared and to maintain fixation (oculomotor localization). About 1.4 sec after the target presentation, a light spot (laser beam, 0.2 deg in diameter, 40 cd/m$^2$) was presented for 5 sec. The subject could manipulate the horizontal position of the light spot by moving a lever by hand. The subject indicated the perceptually judged position of the target by moving the light spot to that position (perceptual localization). During perceptual localization, the subject was allowed to move the eyes freely.

To present the targets during or after a prior saccade $S_0$, the output of the eye movement recording apparatus was fed into a differential circuit that triggered the onset of the targets. Targets before saccade $S_0$ were presented by presetting a shorter time interval than the latency of normal saccades between the cue for the prior saccade $S_0$ and the target.

In addition to the saccade condition described above, accuracy of perceptual localization was also examined in a condition in which the target was presented when the eye kept still. Under this control condition, the subject was asked to keep watching a visual stimulus presented at the position of the original fixation point or at the position of the cue for saccade $S_0$ (+8 deg), and just after the offset of these stimuli, a target was presented for 2 msec. The subject made a saccade to the target and reported its apparent position.

Three subjects participated in the experiment for 7 to 8 days. On each day, 5 sessions of 14 experimental (saccade) trials and 2 sessions of 10 control trials, a total of 90 trials, were given for each subject. The sessions on the first day were regarded as practice and were excluded from the data analysis.

Subject H. H. participated in an additional experiment. In the additional experiment (244 trials), the subject, after making a prior saccade $S_0$, was asked not to move the subjects' eyes to the target flashed at the time of the saccade $S_0$; that is, to keep watching the position of the cue for the prior saccade. The additional experiment was conducted to examine the effects of making the saccade $S_1$ on perceptual localization of the target.

### Data analysis

The output from the photoelectric eye movement monitor was recorded on a magnetic tape by using a data recorder (RMG-5304, Nihon Kohden), and the parameters shown in Fig. 20.1 were measured by a high-speed digital storage scope (DA-6121A, Iwatsu). Data were excluded from the data analysis if (a) The subject moved his or her eyes to the cue for a prior saccade $S_0$ by double saccades; (b) the latency of the saccade $S_0$ was shorter than 50 msec or longer than 400 msec; (c) eye movement recording was contaminated by eye blinking; (d) the subject did not detect the target.

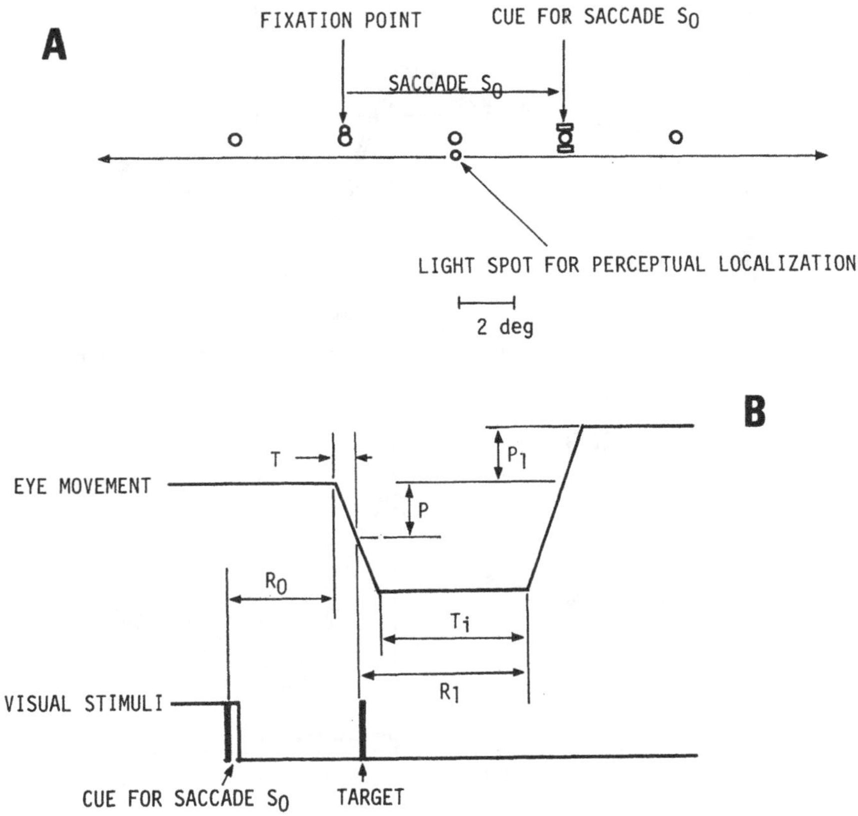


FIG. 20.1. A: Relative locations of the fixation point, the cue for a prior saccade $S_0$, 5 positions of the target, and the adjustable light spot for perceptual localization. B: Definition of eye movement response parameters. P = position of the eye when the target was presented, $P_1$ = position of the eye after saccade $S_1$, $R_0$ = latency of saccade $S_0$, $R_1$ = latency of saccade $S_1$, T = time interval between the target presentation and the onset of saccade $S_0$, $T_i$ = time interval between the end of saccade $S_0$ and the onset of saccade $S_1$.

## RESULTS AND DISCUSSION

### Perceptual mislocalization

The target sometimes appeared as a smeared, horizontally extended image. The horizontal length of the image was about 2 to 3 times that of the target diameter. In this case, the midpoint of the image was regarded as the apparent target position. Open circles in Fig. 20.2 show the size of the errors in perceptual localization as a function of the timing of the target

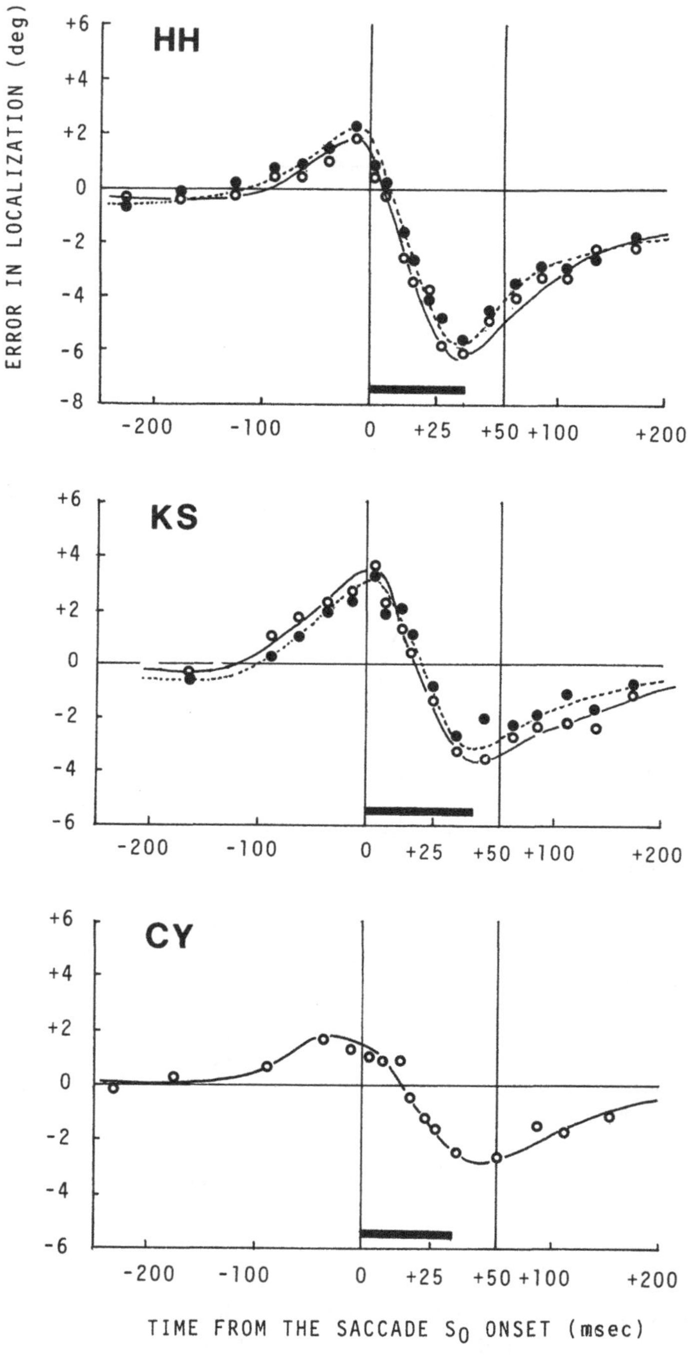


FIG. 20.2. Error in perceptual (○) and oculomotor (●) localizations as a function of time from the prior saccade $S_0$ onset (T). Plus sign in the ordinate indicates errors in the saccade $S_0$ direction. Horizontal bars indicate the duration of saccade $S_0$.

presentation (time from the saccade $S_0$ onset). When the target was presented immediately before or at the beginning of saccade $S_0$, all subjects mislocalized the targets in the direction of the saccade $S_0$, that is, to the right side of the actual target positions. The mislocalization appeared from about 100 msec before the onset of saccade $S_0$. At the end of saccade $S_0$, on the other hand, the subjects largely mislocalized the targets in the opposite direction of saccade $S_0$, that is, to the left side of the actual target positions. The mislocalization continued about 100 msec after the end of saccade $S_0$. The trends of the time course of errors were approximately the same for all actual target positions. The results were consistent with those of Bischof and Kramer (1968), although in their experiment, targets were presented on a steady visual background. Note that when targets were presented about midway through saccade $S_0$, the subjects perceptually localized the targets near the actual position.

### Oculomotor responses

When the target was presented immediately after the cue for saccade $S_0$, the eye did not move to the cue, but moved directly to the target position. In this case, therefore, saccade $S_0$ did not occur.

When saccade $S_0$ was observed, there were three possible responses: (a) only saccade $S_0$ occurred; (b) both saccades $S_0$ and $S_1$ occurred; (c) after saccade $S_0$, a pair of saccades $S_1$ and $S_2$ occurred; that is, the subjects caught the target with double saccades. In this case, the saccade $S_2$ was not considered as a so-called corrective saccade because its latency was in the range of latencies of normal saccades.

In any case, the positions finally reached by the eye were not the actual target positions, but the perceptually judged positions. Filled circles in Fig. 20.2 show mean errors in localization by the eye. The data from subject C. Y. were excluded from the following analysis of oculomotor responses because the subject showed long latencies of saccade $S_1$, usually longer than 400 msec. In calculating the means, we excluded the data when only 1 saccade was observed because in this case it was difficult to decide whether the subjects caught the target by saccade $S_0$ or the subjects failed in making saccade $S_1$. In both subjects H. H. and K. S., the trends of the two curves were quite similar. This indicates that the oculomotor responses were directed to the perceptually localized target positions.

Figure 20.3 shows the relation between final positions of the oculomotor responses to a target and its perceptually localized positions by using different symbols for three types of oculomotor responses. In subject K. S., there was a discrepancy between the positions localized by perceptual judgment and those by oculomotor responses. This was caused, however, by the following: (a) when the perceptually judged position was near the

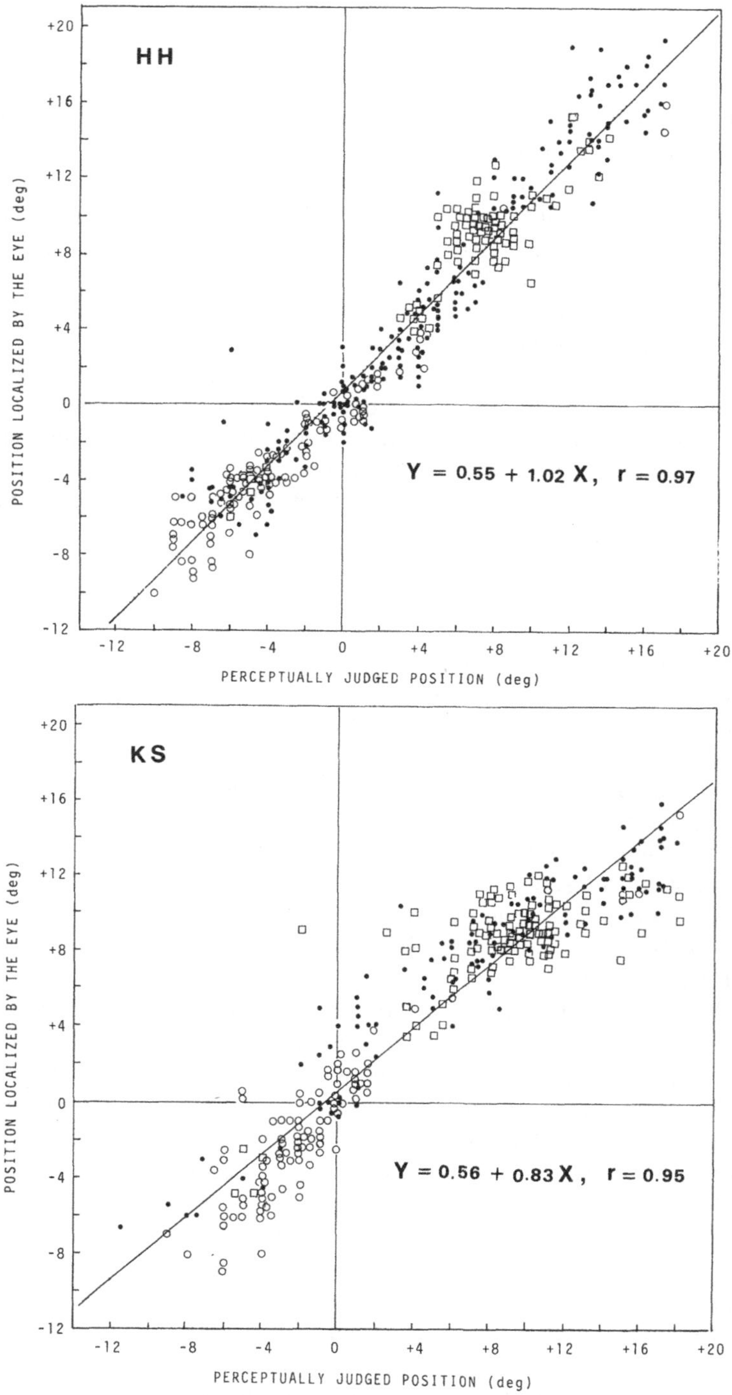


FIG. 20.3. Relation between the perceptually judged positions and the final positions of oculomotor responses. □: only one saccade (saccade $S_0$ or a saccade to the target without saccade $S_0$) was observed: ●: after saccade $S_0$, saccade $S_1$ occurred. ○: after saccade $S_0$, a pair of saccades, $S_1$ and $S_2$ occurred.

position of the cue for saccade $S_0$, only saccade $S_0$ occurred, and saccade $S_1$ was omitted; and (b) when the perceptually judged position was beyond 12 deg or so from the original fixation point, the eye did not reach that position. Therefore, it seems that the discrepancy between the perceptual and oculomotor localizations observed in subject K. S. was a result of a rough ability to localize the saccade. The oblique lines in Fig. 20.3 are regression lines. In calculating the regression line, the data when only one saccade was observed (□ in Fig. 20.3) were excluded.

Hallett and Lightstone (1976a) reported that saccade $S_1$, which occurred after a prior saccade $S_0$, was directed to the actual target position. On the other hand, the results shown in Fig. 20.2 included positions localized by saccade $S_2$. We, therefore, next analyzed only the positions localized by saccade $S_1$. The open and the filled circles in Fig. 20.4-a show the errors of localization by saccade $S_1$ and those of final positions of oculomotor responses, respectively. In subject H. H., the trends of the two sets of curves were similar. In subject K. S., however, the positions localized by saccade $S_1$ shifted upward, in the saccade $S_0$ direction, from the final positions of oculomotor responses. This was because when targets were presented at the left side of the original fixation point (−4 deg), that is, in the opposite direction of saccade $S_0$, saccade $S_1$ sometimes did not reach the perceptually judged target positions. In any case, however, the positions localized by saccade $S_1$ were not the actual target positions.

## Perceptual localization without saccade $S_1$

As described, eye movements to a visual target flashed at the time of a prior saccade $S_0$ were directed not to the actual target position, but to its perceptually judged position. However, there was a possibility that the subjects perceptually judged the target position by using eye position after saccade $S_1$. An additional experiment was conducted, therefore, in which, after finishing saccade $S_0$, the subject was not allowed to make saccade $S_1$ to the target. Figure 20.4-b shows the errors in perceptual localization obtained in the additional experiment. It is evident that the errors in perceptual localization were observed even when the subject did not move the eye to the target.

## Extraretinal eye position information

In the present experiment, visual targets were presented in the dark. We can assume, therefore, that the subjects perceptually judged the target positions (perceived position: *PP*) mainly on the basis of the information about the positions of the target image on the retina (retinal information: *RI*) and/or the information about the eye position around the time of saccade $S_0$ (extraretinal eye position information: *EEPI*).

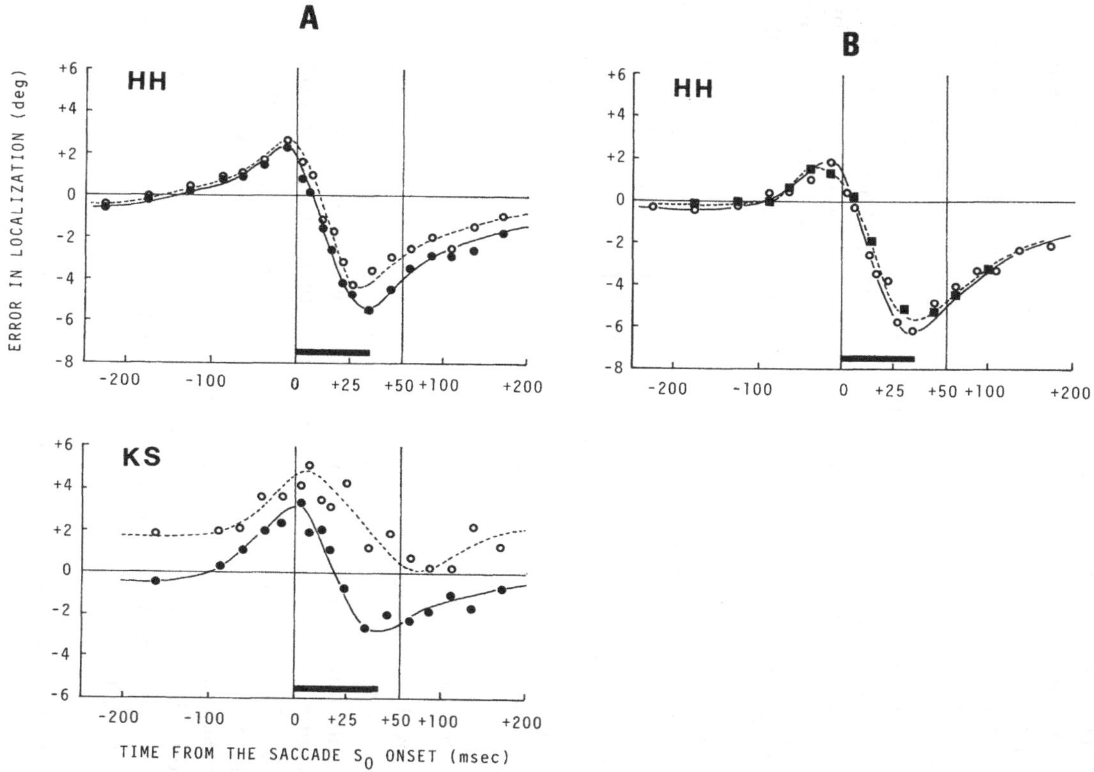


FIG. 20.4. A: a comparison of the error in localization by saccade $S_1$ (○) and that by the final position of oculomotor responses (●). Horizontal bars indicate the duration of saccade $S_0$. B: a comparison of error in perceptual localizations with (○) and without saccade $S_1$ to the target (■).

One possible assumption that *PP* was determined solely by *RI* is not consistent with the results shown in Fig. 20.2. For instance, when targets were presented at the end of saccade $S_0$, subject K. S. showed errors of about 4 deg in the opposite direction to saccade $S_0$. This means that a 12 deg right target was perceived as an 8 deg target, and, in addition, it resulted in a nearly null saccade $S_1$, in spite of the existence of *RI* of 4 deg. Similarly, an 8 deg right target flashed at the end of saccade $S_0$ fell on the fovea, that is, $RI = 0$, but was nevertheless perceived as a 4 deg target and elicited a saccade to the left. These results suggest that *PP* was determined on the basis of the combination of *EEPI* and *RI*.

On this assumption, *EEPI* is

$$EEPI = PP - RI.$$

Figure 20.5 shows the time course of *EEPI* obtained in this experiment, that is, the point-to-point subtraction of the curves *PP* and *RI*. In calculating *EEPI*, *PP*s were corrected by using the perceived positions obtained in control conditions in which targets were presented while the eye was being kept still because the results of the control conditions indicated that there was a small indigenous error, especially when the targets were presented at eccentric positions. As shown in Fig. 20.5, *EEPI* appeared about 100 msec before the saccade $S_0$ onset, but it developed so slowly that it could not catch up with the change of the eye until about 100 msec after the end of saccade $S_0$.

## Why did the eye not direct itself to the actual target position?

Hallett and Lightstone (1976a, 1976b) reported that the eye directed itself to the actual position in space of the target flashed during a prior saccade. In the present study, however, the eye directed itself to the perceptually judged positions. This discrepancy, we think, was caused by the following reasons:

First, in the Hallett and Lightstone experiment, targets were presented at a fixed point in time during a prior saccade. In one set of their experiments (Hallett & Lightstone, 1976b), for example, the target was presented for 20 msec, 10 msec after the onset of the prior saccade: The target was presented from the time that the eye had accelerated to about 330 deg/sec until it decelerated to about 295 deg/sec. This means that in their experiment the target was presented about midway through the prior saccade. It was shown by the present study, however, that the mislocalization of a target flashed at the time of saccade depended on the timing of the target presentation and that targets presented about midway through the saccade were seen near the actual target position. Therefore, we speculate that when the subjects in Hallett and Lightstone's experiment moved their

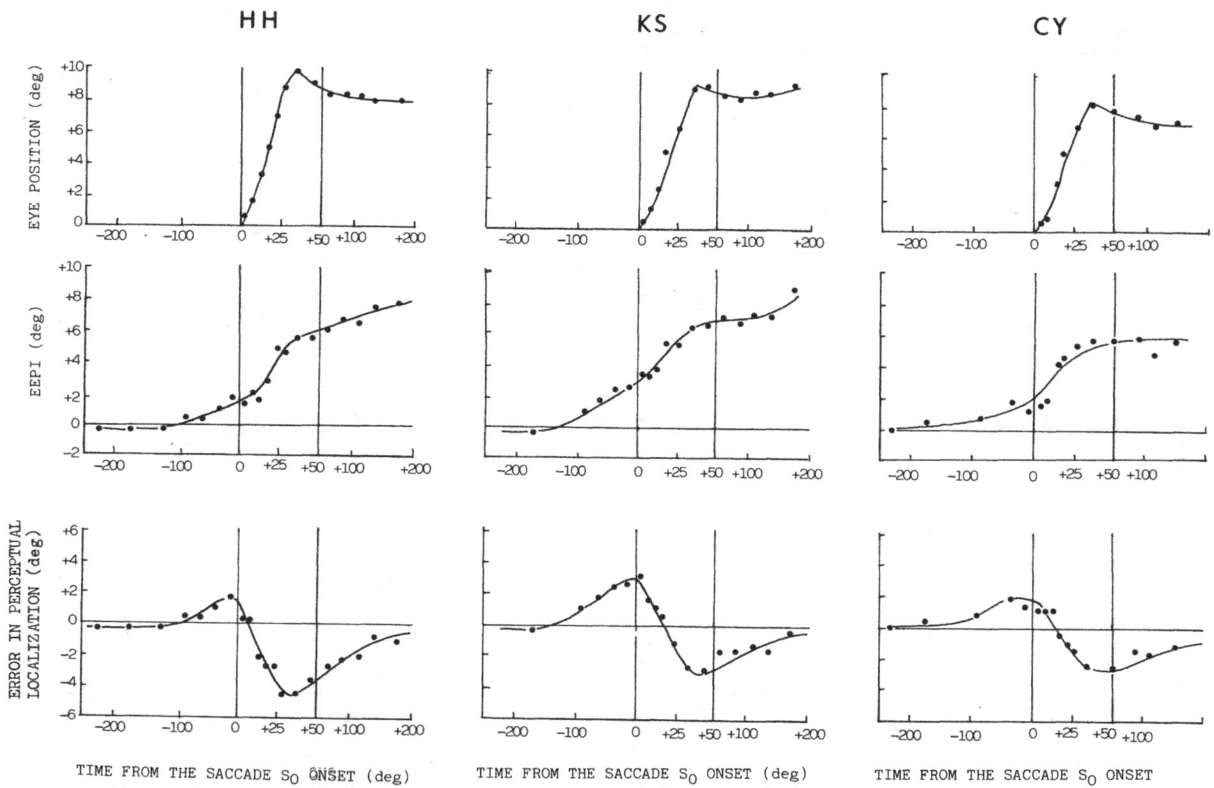


FIG. 20.5. Time course of actual eye position, extraretinal eye position information (EEPI), and corrected error in perceptual localization.

eyes to the position where the target had appeared, the position localized by the eyes happened to coincide with the actual target position. If, in their experiment, targets were presented at another point in time during a saccade, the subject's eyes might not direct themselves to the actual target positions.

Second, in Hallett and Lightstone's experiment, the subjects were given information about the actual target position on each trial after they made a saccade to the target. In one of their experiments (Hallett & Lightstone, 1976a), after a target was extinguished for 250–300 msec, it appeared again for 750 msec at the same position. Therefore, the subjects could know the accuracy of the saccade they had made to the target. Furthermore, the possible positions of the target were limited, and the subjects repeated some hundred trials. As Howard (1982) already commented, through many trials and with the knowledge of actual results, the subjects might have learned to make the appropriate saccade.

From the reasons described, it is not easy to properly interpret the results of Hallett and Lightstone's experiments. The present study showed that saccades to a target flashed around the time of a prior saccade were directed to a perceptually mislocalized position. We speculate that the perceptually localized position was caused by a mismatch between *EEPI* and the actual eye position. *EEPI* is information about eye position registered at some level in the central nervous system. Therefore, it can be said that both perceptual and oculomotor localizations were based on the same centrally registered eye position signal that was not properly timed to the actual eye position in the head.

## Other findings on the saccade system

Further analysis of the amplitudes and the latencies of the oculomotor responses supported some findings on the saccade system reported by the earlier double step stimuli experiments.

*The direction and amplitude of the first saccade.* When targets were presented immediately after the presentation of a cue for a prior saccade $S_0$, the first appearing saccade directed itself, not to the cue for saccade $S_0$ (+8 deg), but to the target position, that is, the final angle response (FAR), as termed by Becker and Jurgens (1979). We examined the relation between the positions localized by the first saccades (saccade $S_0$ and FARs) and the time of the target presentation (time from the onset of the first saccade). The timing of target presentation (T) at which FARs occurred depended on the actual positions of targets. When the actual target position was 4 deg to the left of the fixation point (−4 deg), FARs occurred only when targets were presented at more than 200 msec before the onset of the first saccade ($T < -200$ msec). On the other hand, when targets were presented at +4 deg and +12 deg, FARs occurred at $T < -100$ msec.

In addition, there was an amplitude transition period in which the amplitude of the first saccade gradually changed from the target position to the position of the cue for saccade $S_0$. The amplitude transition period, which was also reported by Becker and Jurgens (1979) and Hou and Fender (1979), suggests that if a new visual stimulus is given during the time when the saccade system is preparing for saccade $S_0$, the saccade system cancels the programming for saccade $S_0$ and begins to make new programme for the saccade to the new visual stimulus. According to Hou and Fender (1979), the amplitude transition period suggests that the direction of saccades is programmed separately from the amplitude and that there is a state in the programming process after which the direction of a saccade cannot be changed, but the amplitude can still be shortened in response to new target information.

As described above, the timing of target presentation at which FARs occurred depended on the actual position of the target. From the viewpoint of the saccade system, this means that it takes more time for the saccade system to cancel the programming for saccade $S_0$ and to make a new programming for FAR when FAR is in the opposite direction to saccade $S_0$ rather than when the two are in the same direction.

*Time interval between the end of saccade $S_0$ and the onset of saccade $S_1$ ($T_i$).* We analyzed the relation between $T_i$ and T when targets were presented before and during, not after, saccade $S_0$. Because the latency of saccade $S_1$ abruptly increased when the amplitude of the saccade was small, only $T_i$ obtained when the amplitude of saccade $S_1$ was more than 4 deg were examined. As T shifted farther in the minus direction ($T < 0$), $T_i$ gradually become shorter: At $T < -100$ msec, $T_i$ was about 100 msec, or a half of $T_i$ at $T > 0$. At $T > 0$, $T_i$ was approximately the same as $R_1$ (latency of saccade $S_1$), indicating that at $T > 0$, saccade $S_1$ was programmed independently of saccade $S_0$. Furthermore, the large decrease of $T_i$ at $T < -100$ msec suggests that when the target was presented more than 100 msec before saccade $S_0$, the programming for saccade $S_1$ began before or during saccade $S_0$, or that saccade $S_0$ and $S_1$ were integrated into a single movement. A similar finding was reported by Becker and Jurgens (1979). They found, in their double step stimuli experiment, that the time interval from the first saccade onset to the second saccade onset become shorter as the time interval from the onset of the first stimulus to the first saccade onset become longer. They interpreted this finding as indicating that the programming for the second saccade began during or even before the first saccade. Komoda et al. (1973) also reported that the time interval from the end of the first saccade to the onset of the second saccade was shorter than the latency of the first saccade. Their explanation of this finding was that there is continuous information input into the saccade system and that the computation of the second saccade began before the computation of the first saccade had been completed.

## CONCLUSION

The present study showed that when a visual target was flashed at the time of a saccade, the target was seen at different positions from its actual position, and the eye moved to the perceptually mislocalized position, not to the actual position. This result concerning oculomotor responses is not consistent with a strict *spatial model* of the saccade system. Of course, the result does not agree with *retinocentric models* either. The results of the present study suggest that the visual (perceptual) system and the oculomotor system use the same extraretinal eye position information, in other words, the same centrally registered eye position signal that was not properly timed to the actual eye position in the head.

## ACKNOWLEDGMENT

This research was supported by a 1986 Grant-in-Aid for Scientific Research (No. 61510038) from the Japanese Ministry of Education, Science and Culture. Attendance at conferences was partially supported by the Japan Foundation.

## REFERENCES

Becker, W., & Jurgens, R. (1979). An analysis of the saccade system by means of double step stimuli. *Vision Research*, *19*, 967–983.

Bischof, N., & Kramer, E. (1968) Untersuchungen und Überlegungen zur Richtungswahrnehmung bei willkürlichen sakkadischen Augenbewegungen. *Psychologische Forschung*, *32*, 185–218.

Hallett, P. E., & Lightstone, A. D. (1976a). Saccadic eye movements towards stimuli triggered by prior saccades. *Vision Research*, *16*, 99–106.

Hallett, P. E., & Lightstone, A. D. (1976b). Saccadic eye movements to flashed targets. *Vision Research*, *16*, 107–114.

Hou, R. L., & Fender, D. H. (1979). Processing of direction and magnitude by the saccadic eye movement system. *Vision Research*, *19*, 1421–1426.

Howard, I. P. (1982). *Human Visual Orientation*. Toronto: Wiley.

Komoda, M. L., Festinger, L., Phillips, L. J., Duckman, R. H., & Young, A. R. (1973). Some observations concerning saccadic eye movements. *Vision Research*, *13*, 1009–1021.

Mateeff, S. (1978). Saccadic eye movements and localization of visual stimuli. *Perception & Psychophysics*, *24*, 215–224.

Matin, L., Matin, E., & Pearce, D. G. (1969). Visual perception of direction when voluntary saccade occur: I. Relation of visual direction of fixation target extinguished before a saccade to a flash presented during the saccades. *Perception & Psychophysics*, *5*, 65–80.

Matin, L., Matin, E., & Pola, J. (1970). Visual perception of direction when voluntary saccades occur: II. Relation of visual direction of a fixation target extinguished before a saccade to a subsequent test flash presented before the saccade. *Perception & Psychophysics*, *8*, 9–14.

Mays, L. E., & Sparks, D. L. (1980). Saccades are spatially, not retinocentrically coded. *Science*, *208*, 1163–1165.

Perenin, M. T., & Jeannerod, M. (1975). Residual vision in cortically blind hemifields. *Neuropsychologia*, *13*, 1–7.

Poppel, E., Held, R., & Frost, D. (1973). Residual visual function after brain wounds involving the central visual pathways in man. *Nature*, *243*, 295–296.

Robinson, D. A. (1973). Models of the saccadic eye movement control system. *Kybernetik*, *14*, 71–83.

Weiskrantz, L., Warrington, E. K., Sanders, M. D., & Marshall, J. (1974). Visual capacity in the hemianopic field following a restricted occipital ablation. *Brain*, *97*, 709–728.

Westheimer, G. (1954). Eye movement responses to a horizontally moving visual stimulus. *Archives of Ophthalmology*, *52*, 932–941.

Wheeles, L. L., Boynton, R. M., & Cohen, G. H. (1966). Eye movement responses to step and pulse step stimuli. *Journal of the Optical Society of America*, *56*, 956–960.

Young, L. R., & Stark, L. (1963). Variable feedback experiments testing a sampled data model of eye tracking movements. *IEEE Transaction of Human Factors in Electronics*, *4*, 38–51.

# 21 Basic Perceptuo-Motor Dysfunctions in Cerebral Palsy

David N.Lee
Brigid M. Daniel
Jane Turnbull
*Department of Psychology*
*University of Edinburgh*

Michael L. Cook
*Department of Psychology*
*Australian National University*

ABSTRACT

The chapter reports analyses of perceptuo-motor performance in young children with cerebral palsy and in normal children. The studies, aimed at understanding the underlying disorders, were concerned with the following basic perceptuo-motor functions that are necessary in establishing a frame of reference for action: (a) gearing vision to the environment by coordinated head and eye movements, so that adequate information can be picked up when an object of interest and/or the person is moving; (b) gearing together the visual and nonvisual proprioceptive systems so that coordinated body movements are possible. Both these fundamental functions were found impaired in young hemiparetic cerebral-palsied children.

In stabilizing gaze, the hemiparetic children showed weak eye/target gearing, even compared with normal 11-week-old infants, and they also appeared deficient in head/target gearing. Because head–eye coordination in stabilizing gaze was found normally to develop rapidly between about 11 and 16 weeks and show quite a consistent pattern, valuable information might be obtained about basic perceptuo-motor development in young infants with known or suspected brain damage by measuring gaze stabilization performance. The information could be useful, for example, in setting up exercise programs to help overcome deficiencies in gaze stabilization—a skill on which many other skills depend.

Early assessment of proprioception is equally important. The present study revealed different forms of disorder. In two hemiparetic children the sensed position of the head/eyes was affected, which impaired visual localization of objects. In two others, active localization with the hemiparetic arm was affected. Such detailed information could be valuable for therapy.

## INTRODUCTION

Cerebral palsy is a chronic neurological condition that results in abnormal development of movement and postural control in children. It is due to nonprogressive injury, insult, or defect in the immature brain. Prevalence is likely to rise as more preterm infants survive with improved intensive care techniques (Paneth, Kiely, Stein, & Susser, 1981), because low birthweight infants show steeply increased rates of cerebral palsy (Hagberg, Hagberg, & Olow, 1984). The early diagnosis and treatment of cerebral palsy is, therefore, an urgent problem.

At present, no clinical techniques can predict with accuracy neurological outcome for high-risk neonates (Majnemer, Rosenblatt, & Riley, 1988). Although severe forms of infantile brain damage usually show obvious early signs of movement disorder, less severe damage can be difficult to diagnose with standard neurological tests in the early months, and disorders can even go undetected until the child is a year or more old.

The accurate assessment of movement and postural control is critically important, too, for devising and improving treatment programs, both at the broad institutional level and at the level of the individual patient. Physical therapy and physiotherapy are the main treatments offered in Western countries. Their aims are similar, namely to aid development of movement control and prevent musculoskeletal complications. However, the extent to which these aims are fulfilled has never been adequately measured (Palmer, Schapiro, Wachtel, Allen, Hiller, Harryman, Mosher, Meinhert, & Capute, 1988). Other forms of treatment—for example, the Hungarian method of Conductive Education (Cottam & Sutton, 1986)—have, likewise, not been adequately evaluated.

Because cerebral palsy is a developmental disorder, assessment and treatment need to be founded on principles of development of movement and postural control. With early brain damage particularly, basic functions are likely to be affected, which could later affect the development of complex skills like walking and writing. In this chapter, we consider two basic functions in perceptuo-motor development and report experiments on cerebral palsied and normal children, which were designed to measure these functions.

## TWO BASIC FUNCTIONS IN PERCEPTUO-MOTOR DEVELOPMENT

Gearing action to the environment and to objects and events is a prime need of any organism. In general, it requires coordinating limb, head, and trunk movements into an appropriate action system (Reed, 1982).

The gearing together of body movements to form an action system and the gearing of the action system to the environment both depend on information. The former requires *propriospecific* information about the position, orientation, and movement of the body parts relative to each other: The latter requires *expropriospecific* information about the position, orientation, and movement of the body parts relative to the environment.

The mechanoreceptors in the joints, muscles, and skin are most important in proprioception because they are a *constant* source of information. They are, however, subject to drift (e.g., Harris, 1965; Craske, 1967) and need to be kept in tune with vision (Lee, 1978), which can provide very accurate propriospecific information. Therefore, one basic function in perceptuo-motor development is:

1. *Linking the propriospecific information from mechanoreceptors and vision so that coordinated movements are possible.*

Vision is a major source of expropriospecific information as well as providing propriospecific information. Therefore the development of visual ability is necessary for the development of many peceptuo-motor skills. Seeing is, in fact, a complex perceptuo-motor skill in its own right. The foundation of vision is the optic flow field, the changing pattern of light incident at a moving point of observation (Lee, 1980), and the task of the visual system is to optimize the pickup of information from the optic flow field. This is far from being a passive process. It requires stabilizing gaze through coordinated eye, head, and body movements, as well as controlling accommodation of the lens and convergence of the eyes.

Gaze stabilization is necessary not only to pick up detail through the fovea but also to optimize the pickup of information throughout the whole visual field. The reason is that the optic flow field has, in general, two components: a *translatory* component resulting from translation of the eye, and a *rotary* component due to rotation of the eye relative to the environment. The translatory component affords information about the layout of the environment and about the organism's progression through it (Lee, 1980), whereas the rotary component provides information only about rotation. Hence, it is usually to an organism's advantage to minimize the rotary component so that the translatory component can be better registered. This entails rotating the eye relative to the environment by coordinated rotations of the head-on-body and/or of the eye-in-head.

Thus a second basic function in perceptuo-motor development is

2. *Gearing vision to the environment and to moving objects, including parts of the body, in order to optimize the pickup of information for controlling actions.*

These two basic functions are the subject of the experiments reported in the following sections.

## GEARING VISION TO THE ENVIRONMENT

Research on how humans stabilize gaze on the environment when the body is turning has, in the main, been confined to adults and has involved artificially locking the head to the shoulders in order to study in isolation the compensatory rotation of the eyes in the head. This is generally considered to be controlled both by the vestibular system, through the so-called vestibular ocular reflex, and by vision (Robinson, 1981). How the vestibular and visual systems cooperate in controlling the eyes is far from clear (Biguer & Prablanc, 1981; Owen & Lee, 1986). However, it does appear that the vestibular information needs to be kept in tune with the visual, just as the information from the mechanoreceptors needs to be kept in tune with vision in proprioception. First, the gain of the vestibular ocular reflex (the ratio of eye velocity to head velocity in the dark) is quite labile: It can, for instance, be increased by simply asking the subjects to do mental arithmetic or by asking them to fixate (in the dark) an imaginary point in front of them (Barr, Schultheis, & Robinson, 1976). Second, in an elegant series of experiments, Melvill Jones and his colleagues have shown how introducing discordant visual information, by wearing left–right reversing prisms, can reverse the direction of the vestibular ocular reflex so that it is in accord with the visual information (Melvill Jones, 1977; Melvill Jones, Guitton, & Berthoz, 1988). Similar plasticity in the vestibular ocular reflex has been found, for example, in the rhesus monkey (Miles & Fuller, 1974) and in the cat, where it has been shown that the direction of the vestibular ocular reflex can even be turned through 90° by introducing appropriately conflicting visual information (Baker, Wickland, & Peterson, 1987).[1]

Thus, one problem in learning to stabilize gaze is linking the visual and vestibular information into the control system. Another problem is coordinating head and eye movements. Studies with infants that have not restricted head movements have shown a variety of patterns of coordination in which the head plays an active role (Tronick & Clanton, 1971; Goodkin, 1980; Regal, Ashmead, & Salapatek, 1983; Daniel, 1987). Indeed, the study by Daniel (1987) shows that head movement plays a *precisely controlled* role in stabilizing gaze. We here summarize those developmental results and compare them with new data on cerebral palsied children that were collected to determine whether they showed disorder in gaze stabilization.

[1]Given this evidence, it is very misleading to continue to refer to the contribution of vestibular information to ocular control as a reflex and the term *vestibular ocular reflex* should arguably be dropped from the literature.

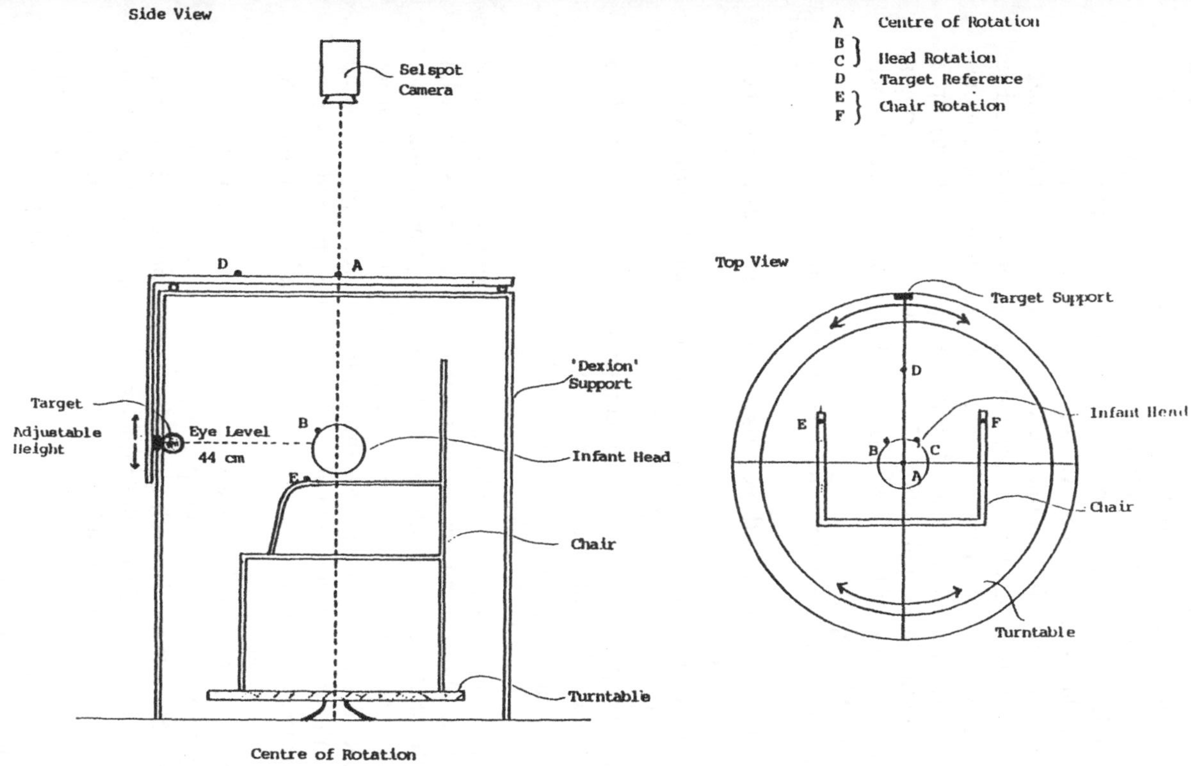


FIG. 21.1. Apparatus for measuring head–eye coordination in stabilizing gaze on a moving target or on a stationary target when the subject's chair is oscillated about a vertical axis. The Selspot camera, linked to a computer, records the movement of infrared light emitting diodes attached to a headband above the eyes (B and C), to the chair (E and F), and to a cross-wire (A and D), indicating the direction of the target. Eye movements are recorded concurrently on the computer using electro-oculography.

## Method

The apparatus is shown in Fig. 21.1. The subject sat on a chair mounted on a horizontal turntable that could be rotated to and fro. The turntable and subject's head were centred within a framework that supported a vertical adjustable strut that could be moved in an arc around the chair. On the strut, at the subject's eye level and at a distance of 44 cm, was mounted a target onto which the subject was encouraged to stabilize gaze, either with the chair moving and the target stationary or the reverse. The motions of the target and chair (which were moved by hand) were monitored together with the subject's head and eye movements (eog) on a Selspot system (detail in Daniel, 1987). The data were sampled during a period of 3 ms, at a rate of 62.5 Hz, and recorded on a computer. Each session was also videotaped.

Four groups of subjects were run in separate experiments:

1. Six normal healthy infants, three girls and three boys, tested between the ages of 11 and 29 weeks.
2. Five normal children from the Departmental Nursery School, 3 years, 2 months to 4 years, 6 months of age.
3. Five cerebral palsied children with congenital hemiplegia (2 left, 3 right), 2 years, 2 months to 7 years, 8 months of age.
4. Six normal adults, 24 years to 40 years old.

The infants sat on their mother's lap on the chair and were presented with a small toy, about 6 cm across, as target. The toy was stuck with Velcro onto the shaft of a small electric motor pointing toward the child. This meant that toys could be rapidly changed and briefly spun in order to maintain the infant's attention. For the young children, who sat alone on the chair, the target was a miniature (2 cm x 1.5 cm) video screen showing a favorite cartoon film. The adults simply had a 1 cm diameter fixation spot to look at.

Infants were tested every 3 weeks between the ages of about 11 and 29 weeks, giving a total of six sessions per baby. A session comprised five blocks of four 15-sec trials. In each block there were four conditions presented in random order: Either the target or the chair was moved approximately sinusoidally, at about 0.2 Hz, $\pm 50°$ either side of the subject's medial plane, or the movement was irregular within the same $\pm 50°$ range and with the same peak velocity of about 125°/sec.

The procedure for the adults was similar except that they were each tested in only one 20-trial session. They were simply asked to keep looking at the fixation spot. The procedure for the children was somewhat different. Only regular sinusoidal movements (0.2 Hz, $\pm 50°$) of the target and chair

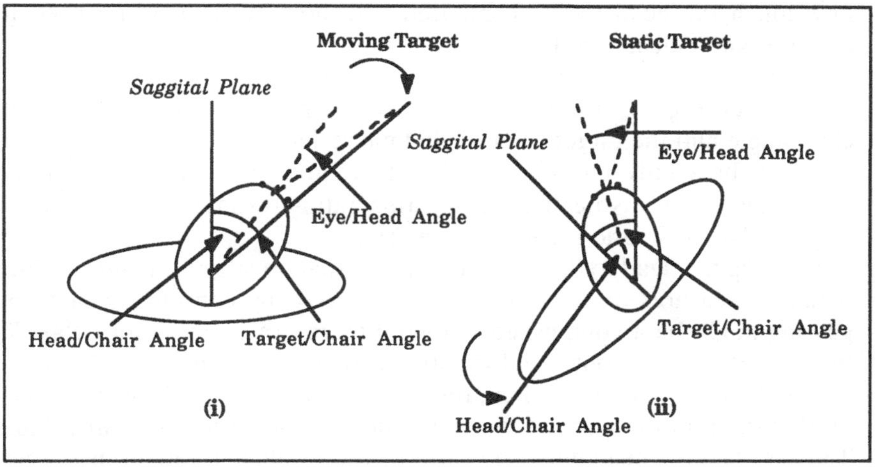


FIG. 21.2. The basic angles (all shown positive) computed from the Selspot and electro-oculographic data.

were used because the regular and irregular movements used with the infants and adults had yielded similar results. Depending on how attentive the child was, he or she was given between two and five 45-sec trials in each condition until a sufficient quantity of valid data had been collected.

## Performance measures

Analyses were based on the following two basic measures computed from the Selspot records (see Fig. 21.2).

*Target/Chair Angle:* the direction of the target with respect to the chair; the angle is zero with the target straight in front of the chair and positive when it is to the right.

*Head/Chair Angle:* the direction of facing of the head with respect to the chair; the angle is zero when facing straight ahead and positive when facing right.

To obtain valid measures of performance, sections of the records were omitted where, as judged from the video records, the attention of the infant or child subjects was not on the target. The following performance measures were then computed on each data section.

*Head/Target Movement Ratio.* The ratio of the standard deviations of the head/chair angle and the target/chair angle time series. This measures how much the head moves compared with how much it would need to move to keep facing the target.

*Head/Target Correlation.* The cross-correlation between the head/chair angle and the target/chair angle time series. This measures how precisely the head movement is geared to the relative movement between the target and chair. (Note that the gearing ratio—estimated by the head/target movement ratio—need not be 1:1.)

Gearing the head to the target requires visual prospective control. More detailed measures of that control were derived by considering how the gearing might be accomplished. Suppose there is a visuo-motor delay, T, between the visual control information arriving at the eye and the information being acted upon. Then, for accurate action, the information must predict the position and motion of the target relative to the chair at time T ahead, and the head must be geared to this predicted state of the target. There are thus two components to the prospective control, which were measured by:

*Head/Target Peak Correlation.* The maximum cross-correlation obtained by time shifting the head/chair angle and target/chair angle time series. This measures how precisely the head movement is geared to the relative movement between target and chair, independently of synchronising with it.

*Head/Target Lag.* The time shift yielding the maximum head/target correlation. This measures the error in intrinsic information on visuo-motor delay. Positive lag means the delay has been underestimated.

Finally, accuracy of eye gearing onto the target was measured by

*Gaze Velocity Error.* The root mean square difference between the time series of the calibrated eog velocity and the eye angular velocity necessary to stabilize the image of the target on the retina (eog was calibrated on the basis of sections of the experimental record that showed good eye tracking).

## Results and Discussion

Figure 21.3 shows typical records from the different subject groups and Fig. 21.4 shows the means and standard deviations of subjects' average scores on the performance measures. Individual scores were averaged across experimental conditions because there was little difference between conditions and the pattern of results in each condition matched the average shown in Fig. 21.4. For the infants, only the values at 11 weeks and 29

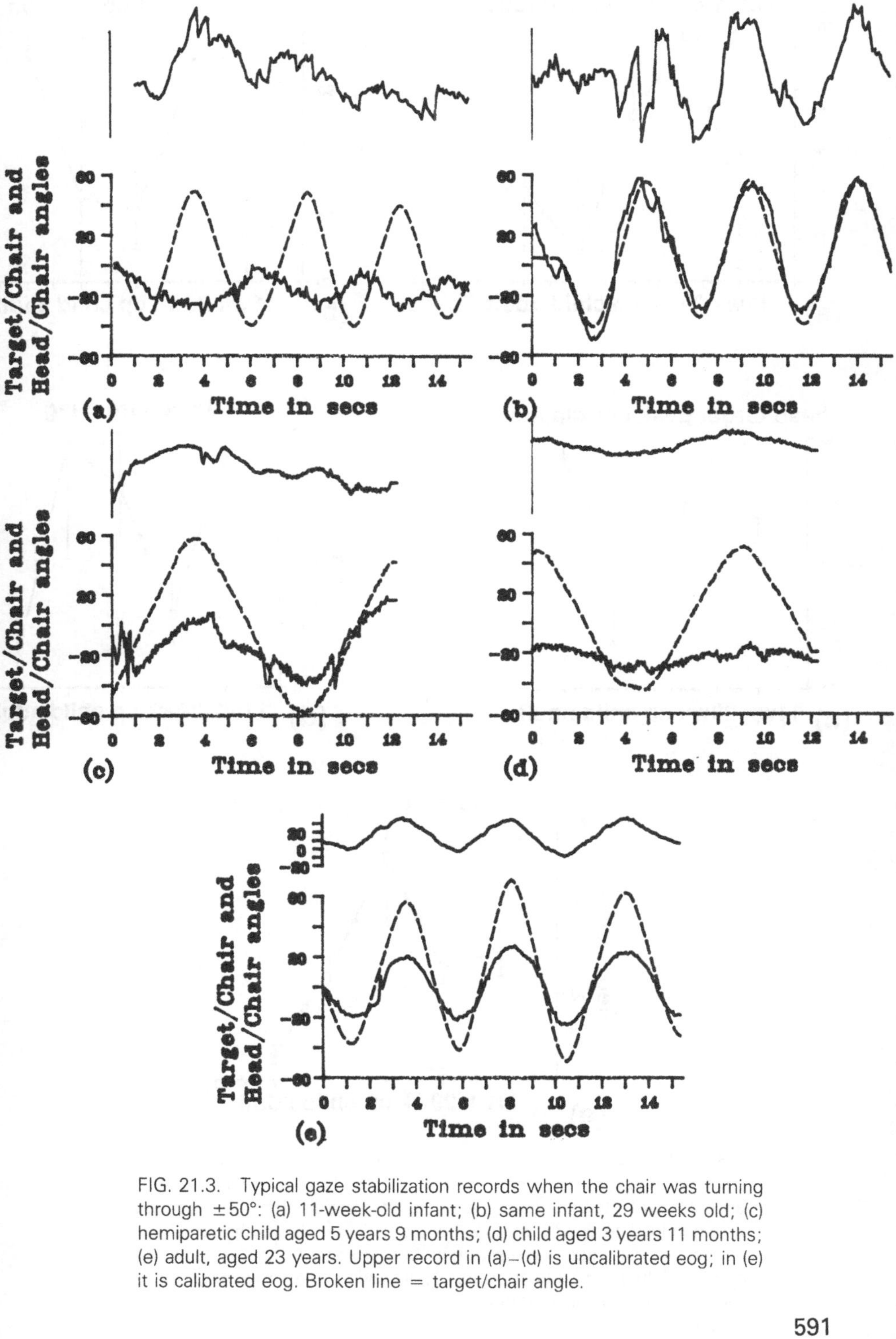


FIG. 21.3. Typical gaze stabilization records when the chair was turning through ±50°: (a) 11-week-old infant; (b) same infant, 29 weeks old; (c) hemiparetic child aged 5 years 9 months; (d) child aged 3 years 11 months; (e) adult, aged 23 years. Upper record in (a)–(d) is uncalibrated eog; in (e) it is calibrated eog. Broken line = target/chair angle.

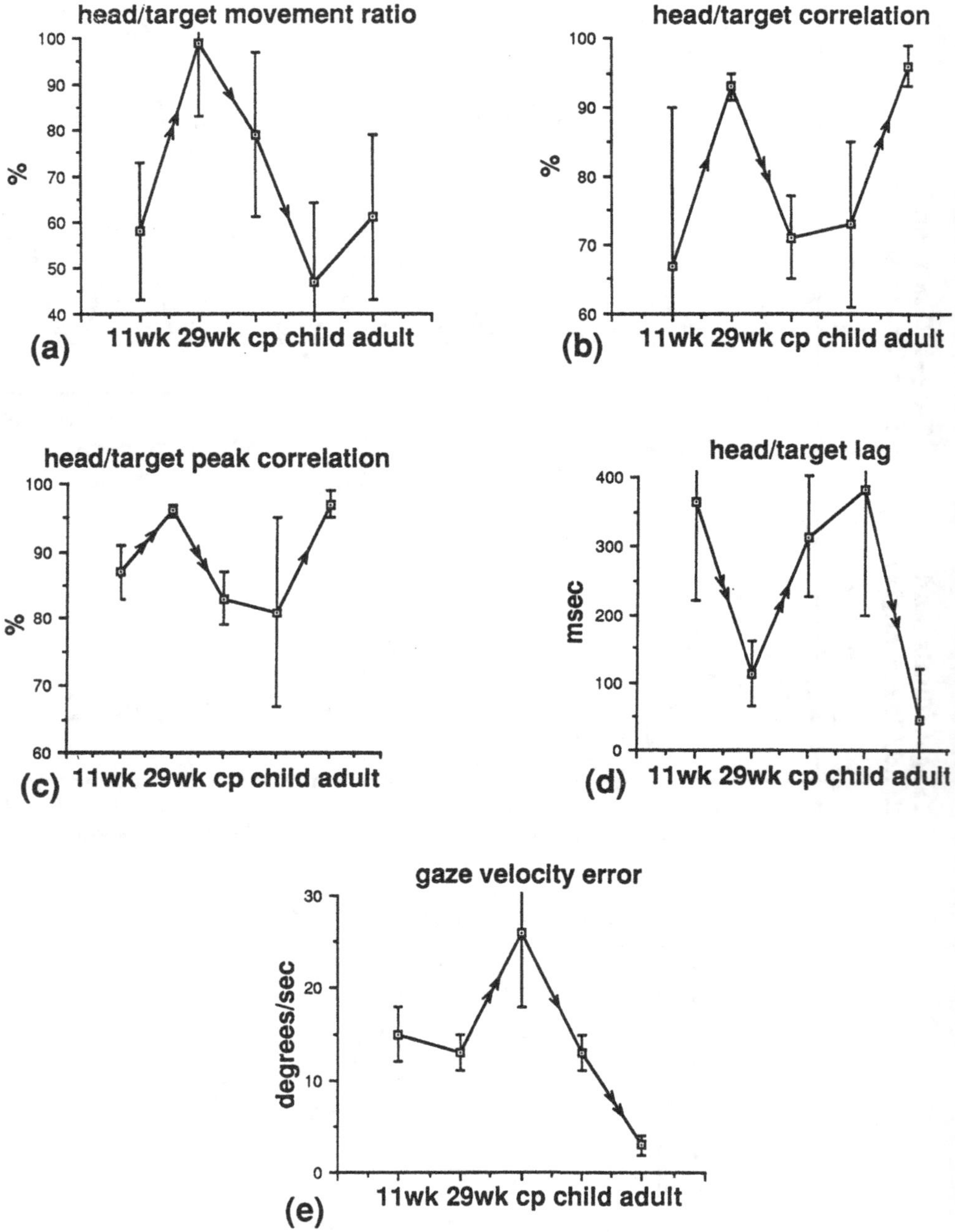


FIG. 21.4. Means and standard deviations of gaze stabilization performance measures for the different subject groups: *11 weeks and 29 weeks*—six normal infants tested at these ages; *cp*—five cerebral palsied children with hemiparesis aged 2 years 2 months to 7 years 8 months; *child*—five normal children aged 3 years 2 months to 4 years 6 months; *adult*—six normal adults aged 24 to 40 years. Single and double arrows on the graph lines indicate significant differences on *t*-test at $p < 0.05$ and $p < 0.01$ respectively.

weeks of age are plotted because intermediate ages gave intermediate values (Daniel, 1987). The results may be summarised as follows:

*Head/Target Gearing.* Between 11 and 29 weeks of age, the infants turned their heads increasingly more in stabilizing gaze (Fig. 21.4a) and, concurrently, improved to near adult level in prospectively gearing the head movement to the relative movement between the body and the target (Fig. 21.4 b, c, and d). Most of the increase in degree and precision of head movement, in fact, took place before about 16 weeks (details in Daniel, 1987). Interestingly, this is the age when visually guided reaching and catching normally starts developing (von Hofsten, 1980; 1986). The video records, in fact, frequently showed the infants at 16 weeks and older actively orienting toward the target as if to possess it.

The cerebral palsied and the normal children both showed weaker head/target gearing than the 29-week-old infants (Fig. 21.4 b, c, d; Table 21.1). For the normal children, this could have been due to an indisposition to move their heads, eye movements being largely sufficient to look at the video target. Their head/target movement ratio was, in fact, the smallest of all the subjects (Fig. 21.4a). The cerebral palsied children, on the other hand, had a relatively high head/target movement ratio, and so their weak head/target gearing is more likely to be due to inability.

*Eye/Target Gearing.* The infants' eye/target gearing did not tighten with age, and was significantly weaker than the adults' (Fig. 21.4e). The normal children's eye/target gearing was similar to the infants'. The cerebral palsied children, however, showed very weak eye/target gearing compared with all other subjects, even the 11-week-old infants (Fig. 21.4e). This weakness could have a common origin with their apparent weakness in gearing the head to the target. On the other hand, it could be that their poor eye/target gearing (which presumably was similar in infancy) handicapped the development of head/target gearing in stabilizing gaze because that requires good visual information.

## LINKING PROPRIOSPECIFIC INFORMATION

The aim of this study was to assess the ability of hemiparetic children, compared with normal children, to localize objects and their own limbs with respect to the body. In particular, we measured the precision of (a) localizing a seen target, (b) localizing the (unseen) affected hand, and (c) pointing with the (unseen) affected hand.

TABLE 21.1

Individual mean performances in gaze stabilization of five cerebral palsied children compared with mean performance of five normal children.

| *CP Subject* | *Age (yr:mth)* | *head/target mov't ratio (%)* | *head/target correl. (%)* | *head/target peak. correl. (%)* | *head/target lag (ms)* | *gaze velocity error (deg.s')* |
|---|---|---|---|---|---|---|
| SB (lt hemi) | 2:2 | 86 | 70 | 85 | 453 | 32 |
| EB (rt hemi) | 3:4 | 71 | 73 | 80 | 261 | 16 |
| NT (rt hemi) | 4:2 | 67 | 81 | 89 | 330 | 29 |
| DY (lt hemi) | 5:6 | 108 | 63 | 79 | 222 | 34 |
| NS (rt hemi) | 7:8 | 64 | 69 | 80 | 303 | 21 |
| Mean | | 79 | 71 | 83 | 314 | 26 |
| Sd | | 18 | 6 | 4 | 88 | 8 |
| Normal Children (3:2-4:6) yrs | | | | | | |
| Mean | | 47 | 73 | 81 | 382 | 13 |
| Sd | | 17 | 12 | 14 | 183 | 2 |

## Method

The apparatus is illustrated in Fig. 21.5. It consists of two parallel horizontal linear tracks, one above the other. A vertical cylindrical handle (13 cm long x 3 cm diameter) is attached rigidly to a carriage that slides freely on bearings on the lower track, the handle remaining perpendicular to the track. A horizontal plate (48 cm wide x 28 cm deep) can slide on the upper track. In the middle of the plate is a small button–switch. The child stands holding the handle, with the tracks adjusted to a comfortable height in a frontal plane, and has to move the handle until he or she perceives it to be directly beneath the button. When this point is reached, the child either presses the button with the other hand or instructs the experimenter to do so. In familiarizing children with the task, a transparent horizontal plate is used, so they can see what they are doing. In this case, to help maintain the child's interest, a small toy attached to the shaft of an electric motor

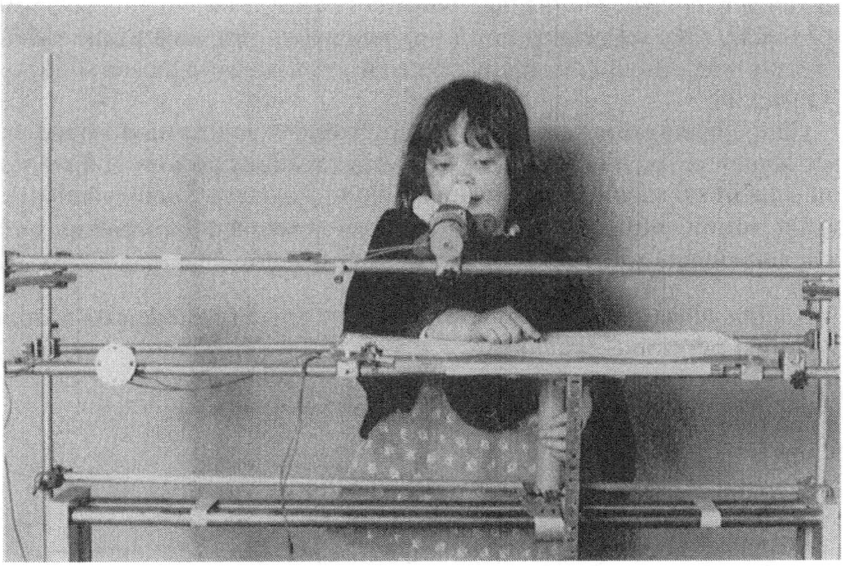

FIG. 21.5. Apparatus to measure ability to localize objects and one's own limbs. The child is wearing a cloak so that it cannot see its left hand and arm. With that hand the child is moving a slider along a track, attempting to line it up beneath the *target* button on which the right finger rests. In this case, it is a *seen and felt target*. With eyes closed, it would be a *felt target*. With eyes open and finger not on the button, it would be a *seen target*. The three types of target were used in the experiment. The target was opposite either the child's left or right shoulder, and the child used the left or right hand. Errors in positioning were read off the scale on the viewer's side of the apparatus. The toy in front of the child was rotated on 50% of trials on a random basis to keep up the child's interest.

on top of the apparatus (Fig. 21.5) is spun briefly when the child accurately positions the handle beneath the button.

In the experiment proper, an opaque sheet was placed over the horizontal plate and an opaque cloak was fastened around the child's neck and fixed to the front of the apparatus. This prevented the child from seeing any part of the lower track, carriage and handle, and any part of the arm and hand holding the handle. There were twelve ($2 \times 2 \times 3$) experimental conditions: (a) the handle was moved by either the left or right hand; (b) the button target was opposite either the child's left or right shoulder; and (c) the target was either a *seen target* (the button) or *felt target* (the forefinger of the child's other hand placed on the button, with eyes closed) or a *seen and felt target* (the forefinger on the button, with eyes open). Each child did eight trials in each of the twelve conditions. The 96 trials were run in randomized order in two 48-trial sessions. Again, to help maintain the child's interest, the spinning toy was used. However, it was no longer connected to the target switch, but instead was activated on 50% of trials by the experimenter on a random basis, unrelated to the child's success on the trial. Performance on a trial was never, in fact, made known to the child.

The (signed) error in positioning the handle was recorded by the experimenter, who sat opposite the child and read the positions of a pointer attached to the handle on a scale invisible to the child. Subsequently, the eight error recordings for each of the twelve experimental conditions were used to compute three error measures per condition:

*Root Mean Square Error (rms).* The spread of judgments around the target position—equal to

$$\sqrt{\frac{\sum x^2}{8}}$$

where x = signed error.

*Mean Error (m).* The mean judged position of the target—equal to

$$\frac{\sum x}{8}$$

*Standard Deviation Error (sd).* The spread around the mean judged position—equal to

$$\sqrt{\frac{\sum (x - m)^2}{8}}$$

Because the three error measures are related by the equation $rms^2 = sd^2 + m^2$ only two need be considered. We chose rms and sd, because they are more informative. Mean Error can be misleading as, for example, when the average of nonzero mean errors across subjects or conditions turns out to be virtually zero, thus suggesting perfect performance. Rms and sd measures are not subject to this particular problem because they are positive.

Of the five hemiparetic children who took part in the first experiment, only two (NT and DY) were able to manage the present experiment; the others had too much difficulty in using the hemiparetic hand. Therefore, two further hemiparetic children were recruited (see Table 21.2). Twelve normal children (6 girls and 6 boys) also completed the experiment. They were recruited from the Departmental Nursery School and ranged in age from 3 years 2 months to 4 years 6 months.

### Results and Discussion

The nursery school children localized the targets very accurately, with no significant difference between the hands (Table 21.2). Overall, their sd and rms errors were only 1.6 to 1.8 cm for the *seen* and *seen and felt* targets, which were not significantly different. For the *felt* target, the overall errors were slightly but significantly higher ($p < 0.005$ for sd, $p < 0.0005$ for rms; t-test). In short, the children were better at localizing what they could see than what they could feel. Von Hofsten and Rosblad (1988) obtained similar results with children from 4 to 12 years of age, as have we with adults (unpublished results).

The five cerebral palsied children performed less well than the normals (Table 21.2). We examined their handicap in more detail in terms of the following questions.

1. *How good was visual localization of targets?* This was measured by the rms and sd errors in localizing the *seen* target with the hand that was the more accurate. For the cerebral palsied children this was the unaffected hand; for the normals it was (marginally) the left hand (see Table 21.2).

2. *How good was the felt localization of the affected limb?* For the cerebral palsied children, this was measured by the rms and sd errors in localizing the affected hand with the unaffected hand in the *felt* target condition. For the purposes of comparison, the normal children's left hand was considered the affected one because they were all right hand dominant; in fact, there was virtually no difference in performance on the experiment between the hands (Table 21.2).

3. *How good was the affected limb at actively localizing targets?* This was measured by the error in localizing with the affected hand (left for the

TABLE 21.2

Means and standard deviations (in parentheses) of sd and rms errors in the proprioception experiment for 12 normal nursery school children aged 3 years 2 months to 4 years 6 months, together with the errors for four children with cerebral palsy: DY, left hemiparesis; the other three, right hemiparesis.

| *Subject* | *Target* | *Left Hand* | | *Right Hand* | | *Overall* | |
|---|---|---|---|---|---|---|---|
| *Nursery* | | *sd* | *rms* | *sd* | *rms* | *sd* | *rms* |
| | seen | 1.7 (0.9) | 1.8 (0.9) | 1.8 (1.0) | 1.8 (0.9) | 1.8 (0.9) | 1.8 (0.9) |
| | seen & felt | 1.6 (0.8) | 1.7 (0.8) | 1.6 (0.6) | 1.7 (0.6) | 1.6 (0.7) | 1.7 (0.7) |
| | felt | 2.3 (1.1) | 2.6 (1.1) | 2.3 (0.9) | 2.5 (1.0) | 2.3 (1.0) | 2.6 (1.1) |
| | | *Affected Hand* | | *Unaffected Hand* | | | |
| *Cerebral Palsy* | | *sd* | *rms* | *sd* | *rms* | *sd* | *rms* |
| NT | seen | 4.5 | 4.6 | 3.1 | 3.5 | 3.8 | 4.0 |
| 4 years 2 months | seen & felt | 2.1 | 2.1 | 2.5 | 2.7 | 2.3 | 2.4 |
| | felt | 2.3 | 2.1 | 2.7 | 2.7 | 2.5 | 2.4 |
| JH | seen | 3.3 | 4.6 | 3.8 | 3.8 | 3.5 | 4.2 |
| 3 years 9 months | seen & felt | 1.7 | 2.5 | 1.7 | 1.7 | 1.7 | 2.2 |
| | felt | 2.6 | 2.6 | 3.0 | 3.4 | 2.8 | 3.0 |
| GB | seen | 4.3 | 4.9 | 2.0 | 2.0 | 3.2 | 3.5 |
| 6 years 4 months | seen & felt | 3.4 | 3.7 | 2.1 | 2.4 | 2.8 | 3.1 |
| | felt | 3.4 | 3.7 | 3.1 | 3.2 | 3.3 | 3.5 |
| DY | seen | 3.3 | 3.7 | 1.5 | 1.8 | 2.4 | 2.7 |
| 5 years 9 months | seen & felt | 3.2 | 3.4 | 2.1 | 2.9 | 2.6 | 3.1 |
| | felt | 3.3 | 4.6 | 2.6 | 3.6 | 2.9 | 4.1 |

normals, as above) the target, *seen* or *felt*, on which the subject performed better.

Figure 21.6 shows the rms errors in visual, felt, and active localization for each hemiparetic child as a percentage of the mean rms errors of the normal children (Table 21.2). Two of the children (NT and JH) showed

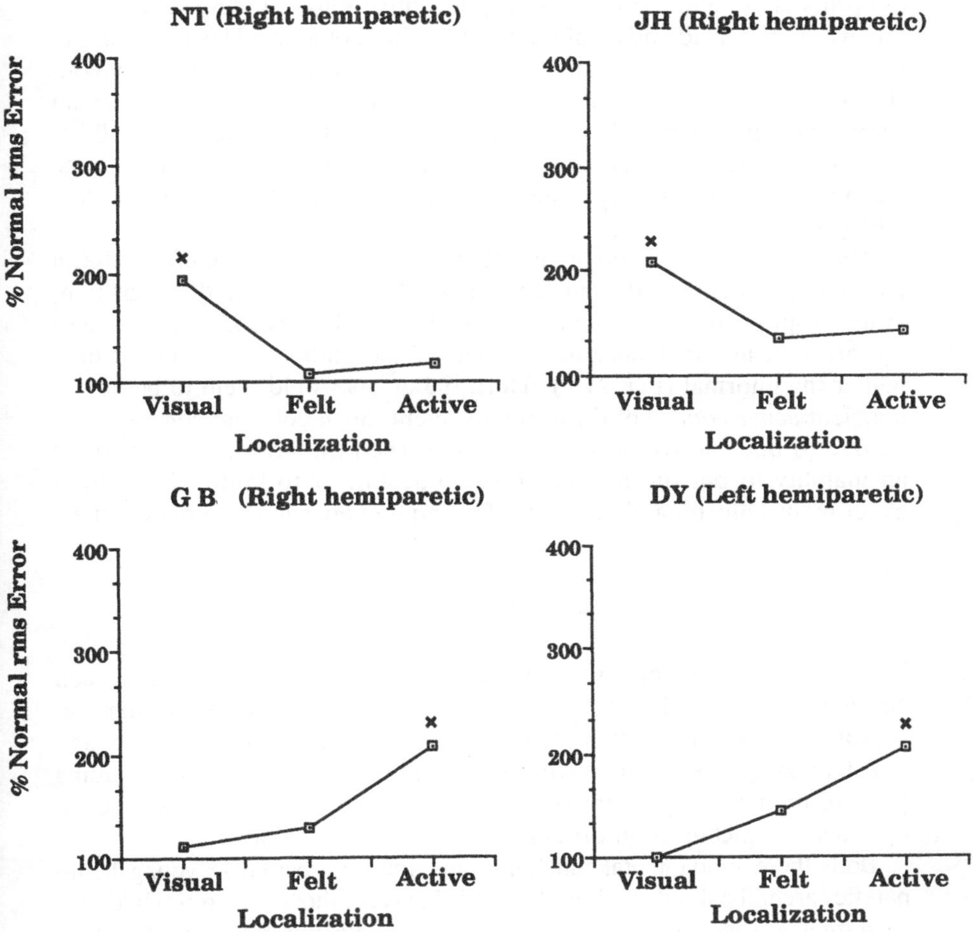


FIG. 21.6. Different error patterns of four hemiparetic children in: *visual localization* of targets (pointing with the unseen unaffected hand); *felt localization* of the affected limb (pointing to it with the unaffected hand with eyes covered); *active localization* with the affected limb (pointing with it, unseen, to a *seen* or *felt* target, whichever allowed the better performance). Rms errors are plotted as percentage of the mean rms error produced by twelve normal nursery school children. Asterisks indicate errors significantly ($p < 0.05$, normal test) higher than those of the nursery school children.

significantly high rms error in visual localization and two (GB and DY) in active localization. None showed significantly high rms error in felt localization of the affected hand. The pattern of results for sd error was similar, except that NT's visual localization error was not quite significantly higher than the normal children's error.

The high rms and sd errors in visual localization shown by NT and JH are probably due to a disturbed proprioceptive sense of the orientation of the eyes in the head and/or of the head on the shoulders. This would affect visual localization of objects with respect to the trunk. For JH, only consistency of performance was affected; there was no systematic shift (mean error was only $-0.26$). NT, however, also showed a significant ($p < 0.05$, normal test) mean shift to the left of $-1.88$ cm, compared with the average mean error for the nursery school children of 0.33 cm to the right (st.dev. 0.61 cm).

The high rms and sd errors in active localization with the affected arm and hand, shown by GB and DY, cannot be attributed to deficient nonvisual proprioception in that limb when it had reached its end point, because the error in felt *stationary* localization of the limb was not significantly higher than normal (Fig. 21.6). Thus, the errors would seem to be due to deficient *active control* of the arm movement. Such deficient control could be due to poor nonvisual proprioception when the arm is moving, or to an inability to put the information into action, or to both. The present experiment unfortunately cannot differentiate between these possibilities.

## SUMMARY AND CONCLUSIONS

Two basic requirements in gearing actions to the environment have been considered: (a) linking the propriospecific information from mechanoreceptors and vision so that coordinated movements are possible; and (b) stabilizing gaze on the environment and on moving objects (including parts of the body) by means of coupled head/eye movements, in order to optimize the pickup of information for controlling actions.

Both these fundamental functions were found impaired in young hemiparetic cerebral palsied children. In stabilizing gaze, the hemiparetic children's eye/target gearing was weak, even compared with normal 11-week-old infants', and their head/target gearing appeared to be weak too. Because it was also found that head–eye coordination in stabilizing gaze normally develops rapidly between about 11 and 16 weeks and shows quite a consistent pattern, it is likely that measures of head–eye coordination would provide valuable information about basic perceptuo-motor development in young infants with known or suspected brain damage.

The information would be useful, for example, in setting up exercise programs to help overcome deficiencies in gaze stabilization—a skill that physiotherapy tends to neglect, despite the fact that many skills depend on it. One possible way of developing the skill in young infants is to hold them upright and gently turn them to and fro while giving them something interesting to look at (e.g., one's own face). The situation, which is analogous to the moving chair condition used in the experiment, makes available more information for stabilizing gaze than moving an object, including information from whole-field optic flow and from the vestibular system. The extra information makes possible more precise gaze stabilization in normal infants and adults (Daniel, 1987) and could, therefore, help in disorders of gaze stabilization, particularly where foveal vision is impaired.

Early assessment of proprioception is equally important and the present study indicates that different forms of disorder might be expected. However, the techniques for use with infants still need to be devised. Fetal movements (Prechtl, 1986) suggest that the linking together of *non*visual parts of the proprioceptive system starts before birth, whereas hand regard and like behavior in neonates would point to vision being linked into the system soon after birth. What little evidence there is, however, suggests that visual *tuning* of the nonvisual components of the proprioceptive system does not start until around 5 months of age (Owen & Lee, 1986).

Because of visual tuning, the development of proprioception would be affected by visual development, including the development of gaze stabilization: If visual information is poor, visual tuning would be poor. Conversely, the development of gaze stabilization would be affected by the development of proprioception for controlling the head and eyes. Therefore assessment and treatment need to address these two basic functions in parallel.

## ACKNOWLEDGMENTS

We thank Deborah Kerr, Audrey van der Meer, David Young, Ruud van der Weel, and the departmental technical staff for help with the apparatus and experiments; Keith Brown and Robert Minns for helpful discussions and providing the patients; and the parents, infants, and children for their good-natured participation. The work was supported by grants to the first author from the Medical Research Council and the European Office of Aerospace Research and Development, and by a postgraduate studentship to the second author from the Science and Engineering Research Council.

## REFERENCES

Baker, J., Wickland, C., & Peterson, B. (1987). Dependence of cat vestibulo-ocular reflex direction adaptation on animal orientation during adaptation and rotation in darkness. *Brain Research*, *408*, 339–343.

Barr, C. C., Schultheis, L. W., & Robinson, D. A. (1976). Voluntary non-visual control of the human vestibulo-ocular reflex. *Acta Oto-Laryngology*, *8*, 365–375.

Biguer, B., & Prablanc, C. (1981). Modulation of the vestibulo-ocular reflex in eye–head coordination as a function of target distance in man. In *Progress in Oculomotor Research.* (eds. A. F. Fuchs & W. Becker). Amsterdam: North Holland.

Cottam, P. J., & Sutton, A. (1986). *Conductive education: A system for overcoming motor disorder.* New Hampshire: Crook Helm.

Craske, B. (1967). Adaptation to prisms: Change in internally registered eye position. *British Journal of Psychology*, *58*, 329–335.

Daniel, B. M. (1987). *Looking with the head and eyes: A developmental study.* Unpublished Doctoral Thesis, University of Edinburgh.

Goodkin, F. (1980). The development of mature patterns of head–eye coordination in the human infant. *Early Human Development*, *4*, 373–386.

Hagberg, B., Hagberg, G., & Olow, I. (1984). Prenatal and perinatal risk factors in a survey of 681 Swedish cases. In *The epidemiology of the cerebral palsies. Clinics in Developmental Medicine*, No. 87. (eds. F. Stanley & E. Alberman). London: S.I.M.P. with Blackwell Scientific; Philadelphia: Lippincott. 116–134.

Harris, C. S. (1965). Perceptual Adaptation to inverted, reversed, and displaced vision. *Psychological Review*, *72*, 419–444.

Hofsten, C. von (1980). Predictive reaching for moving objects by human infants. *Journal of Experimental Child Psychology*, *30*, 369–382.

Hofsten, C. von (1986). The emergence of manual skills. In *Motor development in children: Aspects of coordination and control.* (eds. M. G. Wade & H. T. A. Whiting). Dordrecht: Martinus Nijhoff.

Hofsten, C. von, & Rosblad, B. (1988). The integration of sensory information in the development of precise manual pointing. *Neuropsychologia*, *26*, 805–821.

Lee, D. N. (1978). The functions of vision. *In Modes of perceiving and processing information.* (eds. H. L. Pick & E. Saltzman.) Hillsdale, NJ: Lawrence Erlbaum Associates.

Lee, D. N. (1980). The optic flow-field: the foundation of vision. *Philosophical Transactions of the Royal Society of London B*, *290*, 169–179.

Majnemer, A., Rosenblatt, B., & Riley, P. (1988). Prognostic significance of the auditory brainstem evoked response in highrisk neonates. *Developmental Medicine and Child Neurology*, *30*, 43–52.

Melvill Jones, G. (1977). Plasticity in the adult vestibulo-ocular reflex arc. *Philosophical Transactions of the Royal Society of London B*, *278*, 319–334.

Melvill Jones, G., Guitton, D., and Berthoz, A. (1988). Changing patterns of eye–head coordination during 6h of optically reversed vision. *Experimental Brain Research*, *69*, 531–544.

Miles, F. A., & Fuller, J. H. (1974). Adaptive plasticity in the vestibulo-ocular responses of the rhesus monkey. *Brain Research*, *80*, 512–516.

Owen, B. M., & Lee, D. N. (1986). Establishing a frame of reference for action. In *Motor development in children: Aspects of coordination and control.* (eds. M. G. Wade & H. T. A. Whiting). Dordrecht: Martinus Nijhoff.

Palmer, F. B., Shapiro, B. K., Wachtel, R. C., Allen, M. C., Hiller, J. E., Harryman, S. E., Mosher, B. S., Meinhert, C. L., & Capute, A. J. (1988). The effects of physical therapy on cerebral palsy. *The New England Journal of Medicine 318*(13), 803–808.

Paneth, N., Kiely, J. L., Stein, Z., & Susser, M. (1981). Cerebral palsy and newborn care. *Developmental Medicine and Child Neurology*, *23*, 801–817.

Prechtl, H. F. R. (1986). Prenatal motor development. In *Motor development in children: Aspects of coordination and control.* (eds. M. G. Wade & H. T. A. Whiting). Dordrecht: Martinus Nijhoff.

Reed, E. S. (1982). An outline of a theory of action systems. *Journal of Motor Behavior*, *14*(2), 98–134.
Regal, D. M., Ashmead, D. M., & Salapatek, P. (1983). The coordination of eye and head movements during early infancy: a selective review. *Behavioral Brain Research*, *10*, 125–132.
Robinson, D. A. (1981). Control of eye movements. In *Handbook of physiology*, *Section 1*, *Volume II*, *Part 2*. Bethesda, MD: American Physiological Society.
Tronick, E., & Clanton, C. (1971). Infant looking patterns. *Vision Research*, *11*, 1479–1486.

# 22 Joint Visual Attention, Manual Pointing, and Preverbal Communication in Human Infancy

George Butterworth
Lesley Grover
*University of Stirling, Scotland and*
*Eastleigh College of Technology, England*

## ABSTRACT

In the world of the prelinguistic infant an adult's direction of gaze exerts a powerful effect in redirecting the infant's visual attention. The adult's behaviour serves to signal potentially interesting objects and events to the baby. Our experiments suggest that three cognitive mechanisms are implicated in the comprehension of the adult's head and eye movements between 6 and 18 months. These we call the "ecological," the "geometric," and the "representational" mechanisms of looking where someone else is looking. This chapter explores the relationship in development between the signalling function of joint visual attention and the infant's comprehension and production of manual pointing.

## INTRODUCTION

Gaze following and gaze aversion are important in governing social interaction and in referential communication between an infant and caretaker. Joint visual attention, simply expressed as "looking where someone else is looking," is known technically as "deictic gaze." Deictic gaze is thought to pave the way in development for deictic gestures, such as manual pointing, that draw attention to a particular object by locating it for another person. It is argued that joint attention provides the basis in shared experience for the acquisition of language (Bruner, 1983). This chapter reviews research[1] designed to obtain information about the mechanisms available to the prelinguistic infant that govern shared visual attention. It begins by reviewing briefly a program of studies concerned with the infant's comprehension of gaze. Then the relation between comprehension of gaze and comprehension of index finger pointing is considered. Finally, how both of these abilities may be related in development to the production of manual pointing by the infant are reviewed. The aim is to show that mechanisms of attention, when considered in social context, constitute a basic means of communication in infancy.

## COMPREHENSION OF GAZE DIRECTION

A pioneering study on the comprehension of gaze between adults was carried out by Gibson and Pick (1963) who suggest that information for being or not being looked at is given by the form of the light and dark regions of the eyes in relation to the form of the face. However, we know very little about the infant's comprehension of looking. Perhaps the reason why the topic has a very short history is that it has been traditional to consider the infant as totally egocentric, and therefore incapable of comprehending the possibility of alternative points of view. Most of the infant literature stresses the role of adults in monitoring the infant's attention and changing their own focus of visual attention accordingly. Furthermore, babies aged below 8 or 9 months are notoriously distractible, and this also may have seemed to preclude shared attention with an adult.

There is little doubt that adults monitor very closely the focus of the infant's attention and adjust their own gaze to maintain shared experience. Mothers vocalise at suitable moments when they can see that the baby is attending to a particular object or event, and by establishing joint visual attention they create a suitable tutorial environment. Schaffer (1984) reviewed a number of studies that show that the majority of episodes of joint attention arise as a result of the mother monitoring the infant's line of gaze.

It was therefore an important discovery when Scaife and Bruner (1975) showed that infants as young as 2 months would readjust their gaze con-

[1]The studies by Butterworth and Jarrett (1980) were carried out with the support of a grant from the Economic and Social Research Council, and the extensions of this work by Lesley Grover, (1982, 1988) were carried out under a doctoral studentship linked to the ESRC grant. Experiments on the production of pointing were also funded by the ESRC and were carried out with the assistance of Elvidina Adamson-Macedo and Fabia Franco (Butterworth & Adamson-Macedo, 1987), Franco and Butterworth (1988). Fabia Franco received financial support from the Italian National Research Council (CNR) and from the European Science Foundation.

tingent on a change in the focus of attention of an adult. It suggests that the capacity for joint attention is a reciprocal phenomenon; that the baby may be aware of a spatial objective of the mother's change of gaze and that the infant perceives the mother to be signaling the "permanent possibility of an object" available to joint attention.

The capacity to share visual attention raises important questions about the status of egocentrism at the origins of development and about the contribution of social experience to cognitive growth. These have been dealt with in detail elsewhere and are not discussed here (see Butterworth, 1987 and Butterworth and Grover (1988) on egocentrism and Bornstein and Ruddy (1984) on maternal attentional style and the development of intelligence). This chapter concentrates on the mechanisms governing joint attention and on changes with development in the mechanisms governing such processes, which lie at the heart of communication between the adult and the preverbal infant.

## GENERAL METHODOLOGY OF THE STUDIES

Butterworth and Cochran (1980) and Butterworth and Jarrett (1980) made an extensive series of studies in an attempt to establish the mechanisms serving joint visual attention. The studies were carried out under strictly controlled conditions in an undistracting environment, with identical targets placed at various positions relative to the mother and infant. These conditions allow relatively unambiguous conclusions to be drawn concerning how the baby is able to single out the referent of the mother's gaze because distractions and other possible artifacts are eliminated. In these experiments, the mother was instructed to interact naturally with the infant and then on a signal, to turn, in silence and without pointing manually, to inspect a designated member of a set of targets placed at various positions relative to the mother and baby around the room. Babies between the ages of 6 and 18 months were studied; the interaction was videotaped and subsequently scored by two independent observers who noted the direction and accuracy of the infant's response relative to the object of the mother's line of gaze. Figure 22.1 schematizes the laboratory layout.

We found evidence for three successive mechanisms of joint visual attention in the age range between 6 and 18 months. At 6 months, babies look to the correct side of the room, as if to see what the mother is looking at, but they cannot tell *on the basis of the mother's action alone* which of the two identical targets on the same side of the room the mother is attending to, even with angular separations as large as 60 degrees between the targets. Although the babies are accurate in locating the object referred to by the mother's change of gaze when the correct target is first along

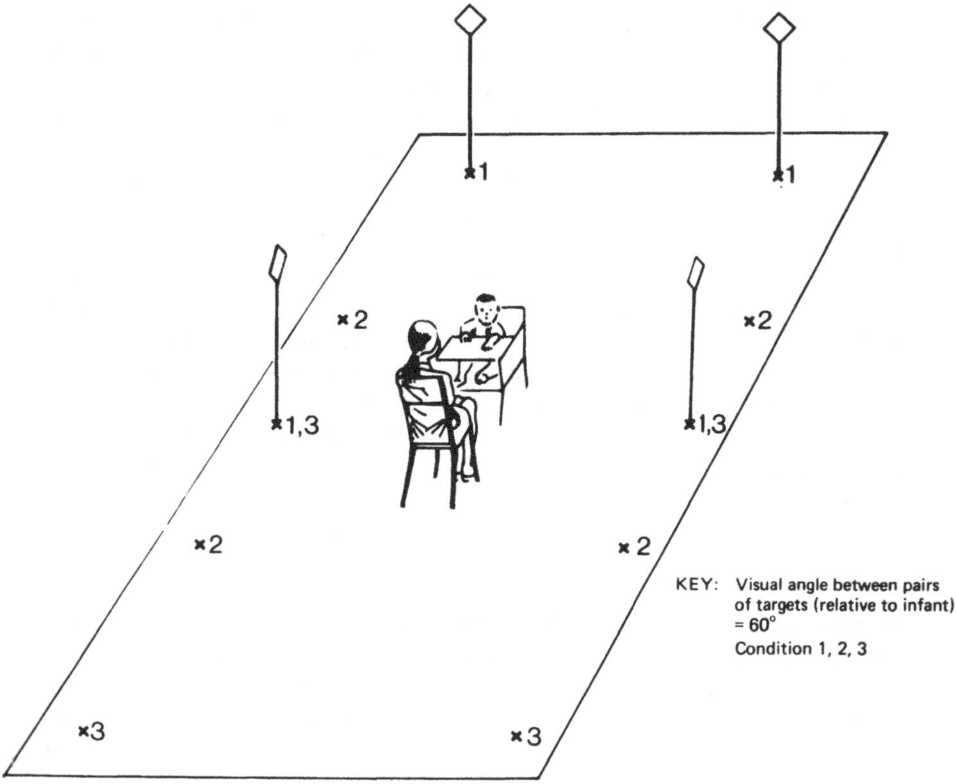


FIG. 22.1. Schematization of laboratory layout.

their path of scanning from the mother to the target, they are at chance level when the correct target is second along the scan path. Furthermore, infants only localize the targets within their own visual field and hardly ever locate targets that the mother looks at in the region behind the baby, out of view. Figure 22.2 illustrates the data for 6-month-old infants in one of our studies in which four identical targets were presented at various locations relative to the mother and baby.

If the mother looks at a target behind the baby, the infant either fixates a target in front and within the visual field or does not respond. This phenomenon is not caused by any inability of babies to turn behind them; indeed they often would turn behind them on first being seated in the laboratory or in response to some inadvertent noise. The most likely explanation is that there is a basic inability to attribute the mother's signal to the space outside the immediate visual field. This basic finding on failure to search "behind" (first reported in Butterworth & Cochran, 1980) has subsequently been replicated on several occasions in our own laboratories. It is noteworthy that a similar inability to search at locations behind the infant has recently been demonstrated in a manual search task involving rotation of the infant relative to objects that were first hidden in the field of view (Landau & Spelke, 1988).

On the other hand, so long as all the possible locations are within the infants' field of view, the infants are capable of correctly locating targets presented one at a time at visual angles that introduce separations between mother and the referent of her gaze of up to 135 degrees (Fig. 22.3). This experiment demonstrates that babies are perfectly capable of noting even very small changes in the mother's head orientation. Thus, failure to search behind cannot be attributed to failure to perceive the change of head orientation. If the targets are presented one at a time in a control condition, in the spatial arrangement schematized in Fig. 22.2a, a very similar pattern of results is obtained, again with failure to search behind. At age 6 months,

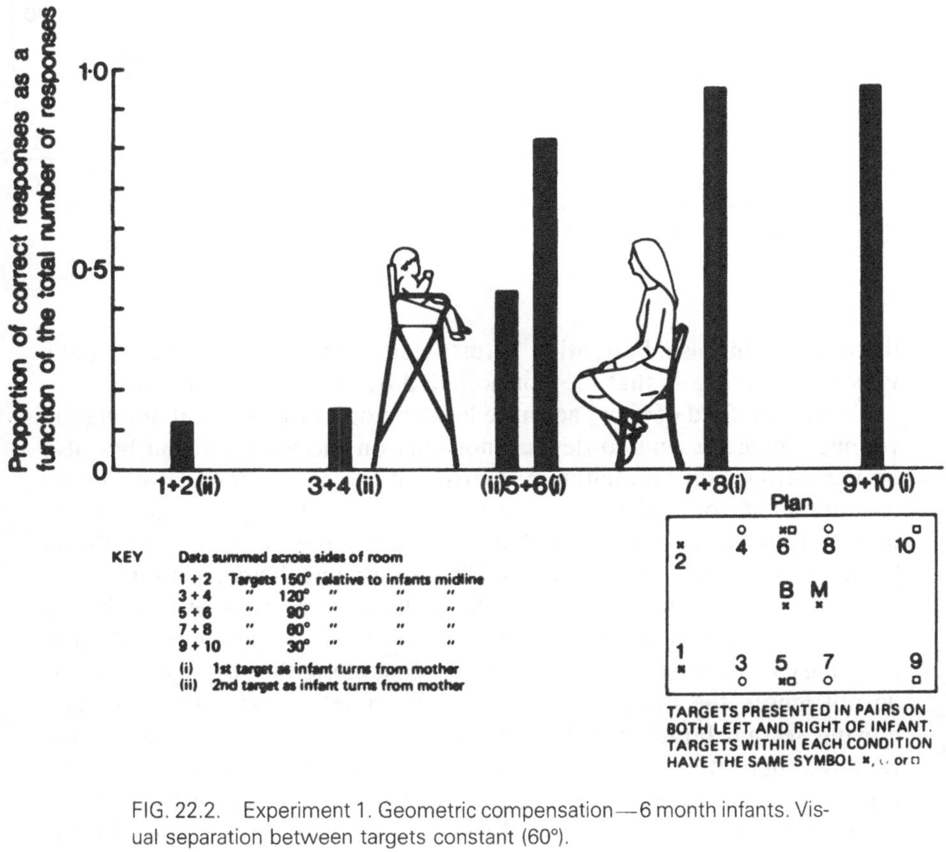


FIG. 22.2. Experiment 1. Geometric compensation—6 month infants. Visual separation between targets constant (60°).

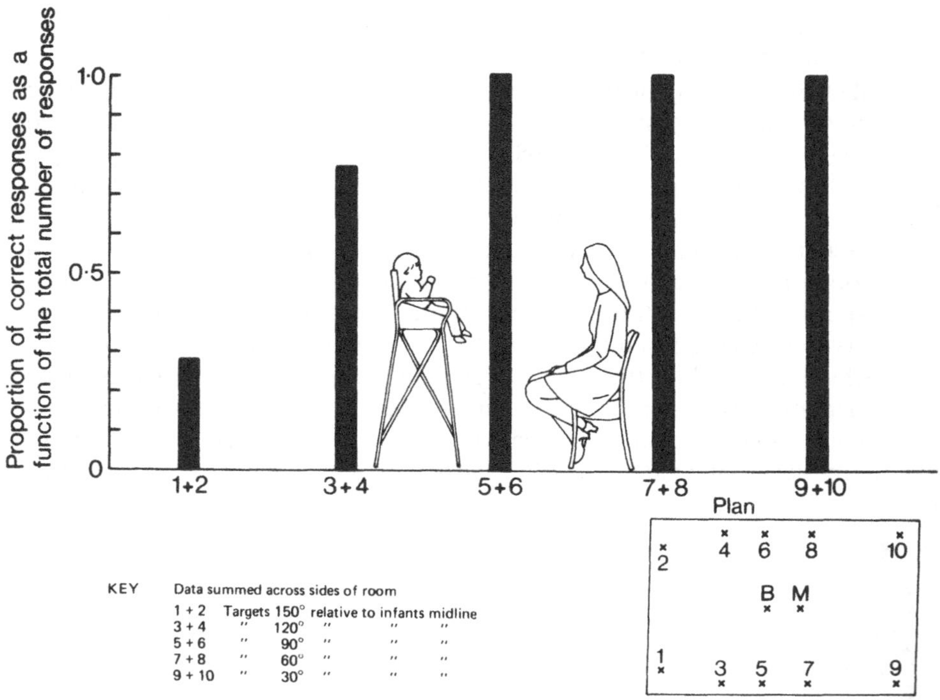


FIG. 22.2A. Experiment 3. Limits of accurate spatial localisation—6 month infants. (1 target at a time)

therefore, joint visual attention is restricted to targets within the infant's view (i.e., to targets that are not occluded by the baby's own body).

Within the field of view, accurate localization of the referent among the youngest babies seems to depend not only on the adult's signal but also on the intrinsic differentiating properties of the object being attended by the mother. Grover (1988), for example, showed that adding movement to both targets elevated the probability of a response among 9-month-old babies and also raised accuracy to 100% so long as the correct target was first along the infant's scan path. The addition of movement to only one of the two targets showed that babies correctly locate the target, whether stationary or moving, so long as it is the first along the scan path. Infants do not ignore the first, correct, stationary target to fixate the incorrect, moving second target, which shows that it is not the motion alone that captures their attention. Thus, motion in the target ensures a correct localization of the referent so long as this property coincides with the tendency to select the first salient object in the direction indicated by the mother. However, if the correct target is stationary and second along the scan path, while the incorrect target is moving and first along the scan path young infants (9 month olds) incorrectly fixate the first target. (It is not until about 15 months that infants begin to be able to overrule motion at the first target in favor of a stationary second target; Grover, 1988).

This earliest mechanism of joint visual attention we have called *ecological*, because we believe that it is the differentiated structure of the natural environment that completes for the infant the communicative function of the adult's signal. What initially attracts the mother's attention and leads her to turn is also likely, in the natural environment, to capture the attention of the infant. The ecological mechanism enables a "meeting of minds" in the self-same object.

By 12 months of age the infant is beginning to localize the targets correctly, whether first or second along the scan path, when the target is stationary in the visual field (Fig. 22.4).

The only information allowing this is the angular displacement of the mother's head and eye movement. It is interesting to note that the infant fixates intently on the mother while she is turning, then, when the mother is still, the infant makes a rapid eye and head movement in the direction of the target. The mean latency of response after the end of the mother's head movement is about 1 sec (Butterworth & Cochran, 1980, p. 268). This brief interval may be sufficient for the baby to register information about the angular orientation of the mother's head. We call this new ability the "geometric" mechanism because it seems to involve extrapolation of

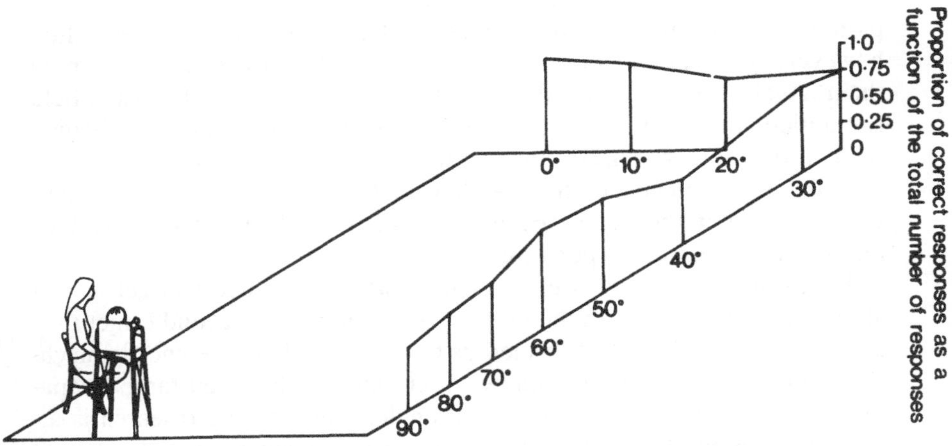


FIG. 22.3. Experiment 2. Limits of accurate, spatial localisation—6 month infants. (1 target at a time)

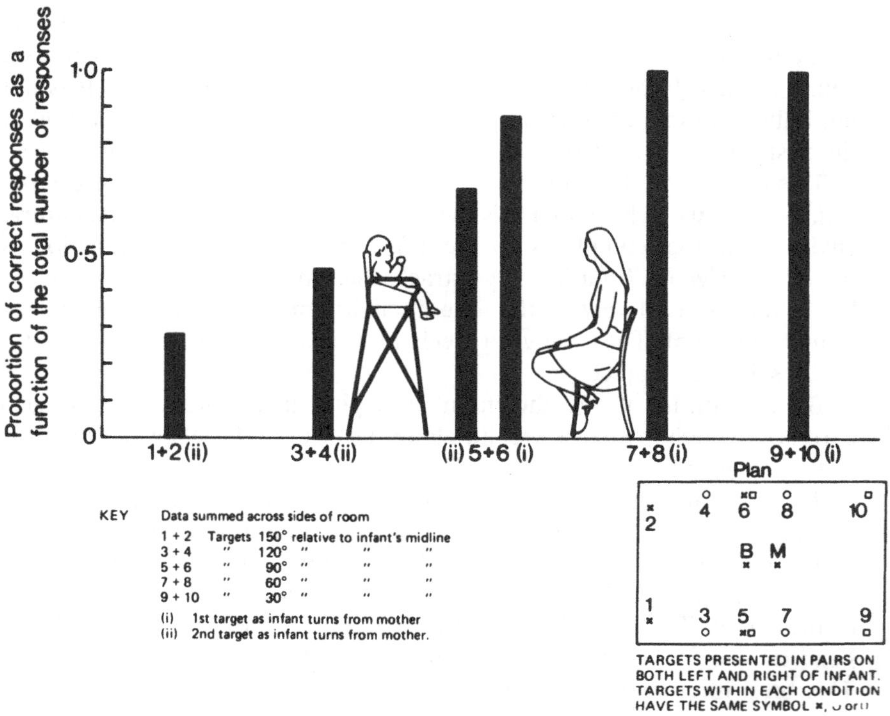


FIG. 22.4. Experiment 1. Geometric compensation—12 month infants. Visual separation between targets constant (60°).

an invisible line between the mother and the reference of her gaze, as plotted from the infants' position. Despite this newfound geometric ability, however, babies at 12 months still fail to search for targets located behind them. Again, we have carried out control studies in which the visual field is emptied completely of targets, yet babies of 1 year do not turn behind them at the mother's signal. Instead, they turn to scan to about 40 degrees of visual angle and give up the search when they fail to encounter a target. It seems that if the geometric mechanism is available, it must still be restricted to the infant's perceived space.

By 18 months, babies are as accurate when the correct target is first along their scan path from the mother, as when it is the second target they encounter (Fig. 22.5), and this led us to conclude that a geometric mechanism may be available that allows discrimination between targets separated by at least 60 degrees of visual angle. Our ongoing research is attempting to establish the minimum spatial separation between targets allowing correct discrimination between the alternate referents of the mother's signal.

Furthermore, although the babies still do not search behind them when there are targets in the field of view, they do so if the visual field is empty of targets. We found that head and eye movements to targets behind the baby would elicit turning to the correct target, so long as there was nothing in front of the 18-month-old infant, in the field of view. Thus, infants seem to be able to access the invisible portion of space at 18 months (but not at 12 months) so long as there is no competition from locations within the visual field. This leads us to postulate the development of a third "representational" spatial mechanism for controlling joint visual attention that is based on an understanding of being contained within space.

In summary, as far as the comprehension of gaze is concerned, we have evidence in the first 18 months of life that three successive mechanisms are involved in "looking where someone else is looking." The earliest, the *ecological mechanism*, depends on completion of joint attention by the intrinsic, attention—capturing properties of objects in the environment,

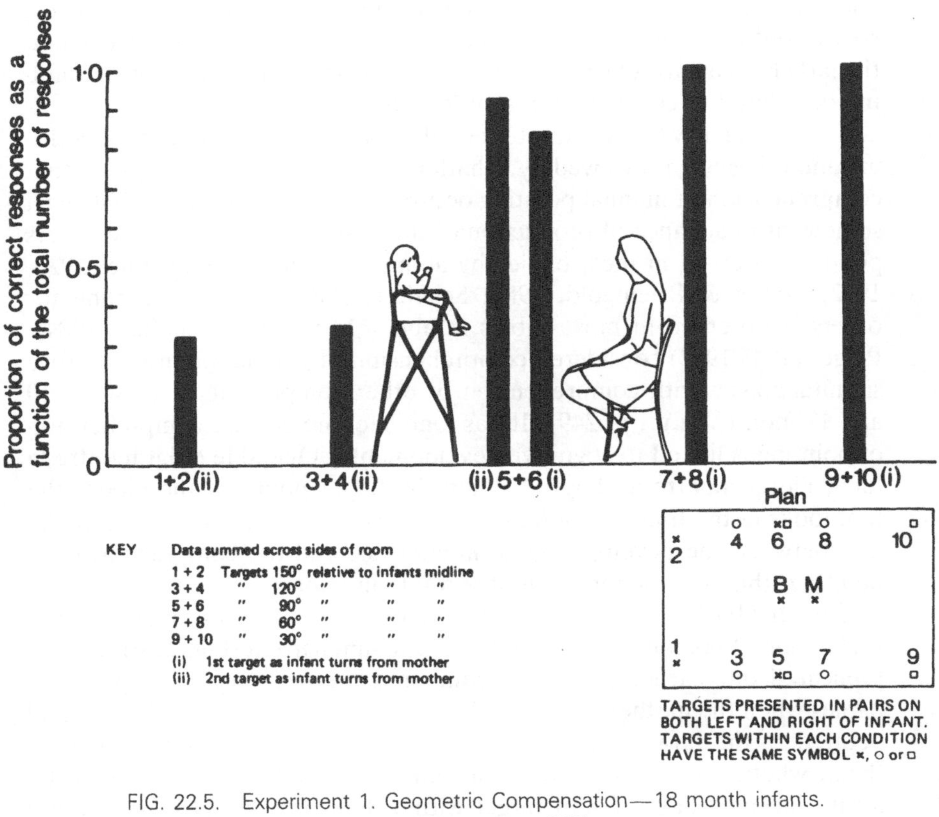


FIG. 22.5. Experiment 1. Geometric Compensation—18 month infants. Visual separation between targets constant (60°).

as well as on the change in mother's direction of gaze. At around 12 months, we have evidence for the beginning of a new mechanism, a *geometric* process, whereby the infant from its own position extrapolates from the orientation of the mother's head or gaze the intersection of a line with a relatively precise zone of visual space. Finally, at sometime between 12 and 18 months, there is an extension of joint reference to a *represented* space that contains the infant and other objects outside the immediate visual field.

## RELATION BETWEEN COMPREHENSION OF GAZE AND COMPREHENSION OF POINTING

In our more recent studies we have attempted to establish how the infant's comprehension of gaze may be related to comprehension of manual pointing. Manual pointing, defined as the use of an outstretched arm and index finger to denote an object in visual space, is species-specific to humans, and it is thought to be intimately linked to language acquisition. It is the specialized referential function that is of interest here because it is a particularly human type of social cognition. Could the comprehension of pointing be related to comprehension of looking?

There is a fairly sparse literature on the comprehension of manual pointing and it has been reviewed by Schaffer (1984). It is generally agreed that comprehension of manual pointing occurs toward the end of the first year, somewhat in advance of production of the gesture. Looking where others point is observed in most babies by about 12 months (Guillaume, 1926, 1962; Leung & Rheingold, 1981; Schaffer, 1984) whereas pointing for others is observed in most babies at about 14 months (Schaffer, 1984). Piaget (1945/1952) considered comprehension of manual pointing to arise simultaneously with comprehension of other complex signs, between 10 and 12 months (e.g., p. 249). It has long been noted that comprehension of pointing is linked to cognitive development: "Quand le doigt montre la lune, l'idiot montre le doigt" "When the finger points at the moon, the fool looks at the finger" is a famous aphorism which illustrates vividly the link between the development of intelligence and the comprehension of pointing; this is a phenomenon observable in babies.

Grover (1982, 1988) has obtained extensive data on the infant's comprehension of pointing, which can be readily summarized (Fig. 22.6). Grover replicated our earlier study and compared the accuracy of response of babies when the mother merely looked at the correct target, or looked and pointed to the target. Infants at 12 months fail to locate targets behind them, whether the mother looks or looks and points. They can correctly locate the mother's referent target within their own visual field whether

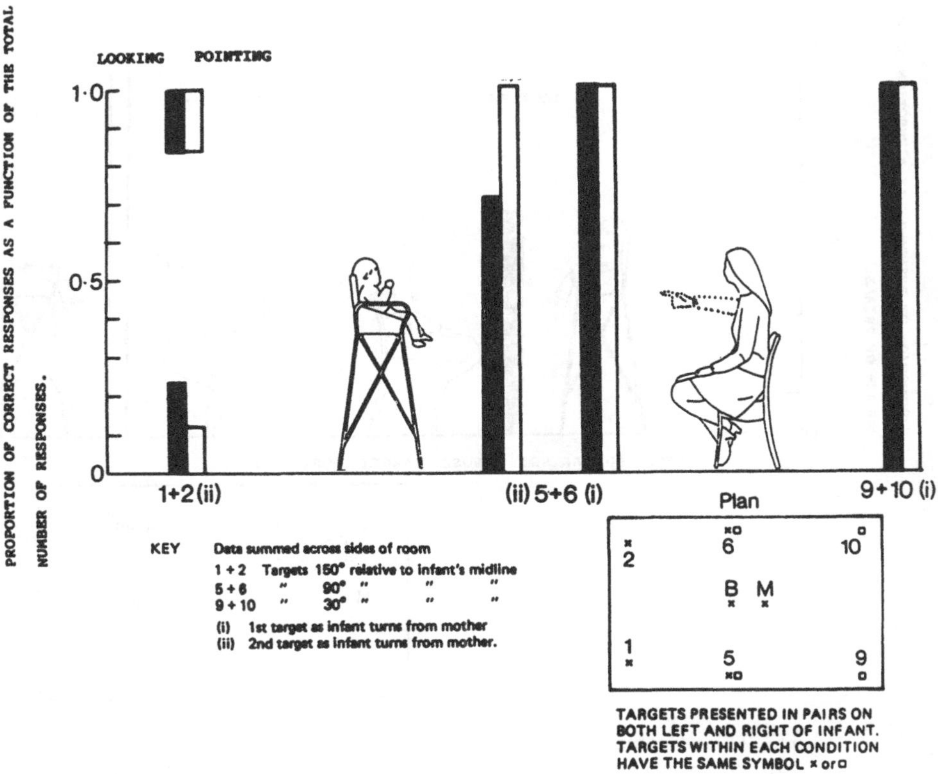


FIG. 22.6. Geometry of preverbal communication: Comprehension of looking and pointing at 12 months (Grover, 1982).

the target is first or second on their scan path. Thus, the addition of pointing by the mother does not make the space outside the field of view more accessible to the infant.

The main effect of adding the manual point is to significantly increase the probability that the infant responds. Adding manual pointing to simple change of gaze has a compelling effect on the infants' attention. The proportion of trials to which the infant responds when looking is accompanied by pointing increases to 96% (from 59% of trials when the mother only looks at the target) and perhaps this is only to be expected because the extension of the arm is an attention worthy event in itself.

Grover (1982) examined the development of comprehension of pointing at 6 months, 9 months, and 12 months. The experimental laboratory and cameras were arranged so that the direction of the infant's gaze, whether toward the mother's hand or to the referent target, could be precisely measured (Fig. 22.8). The results demonstrate that infants at 6 months and

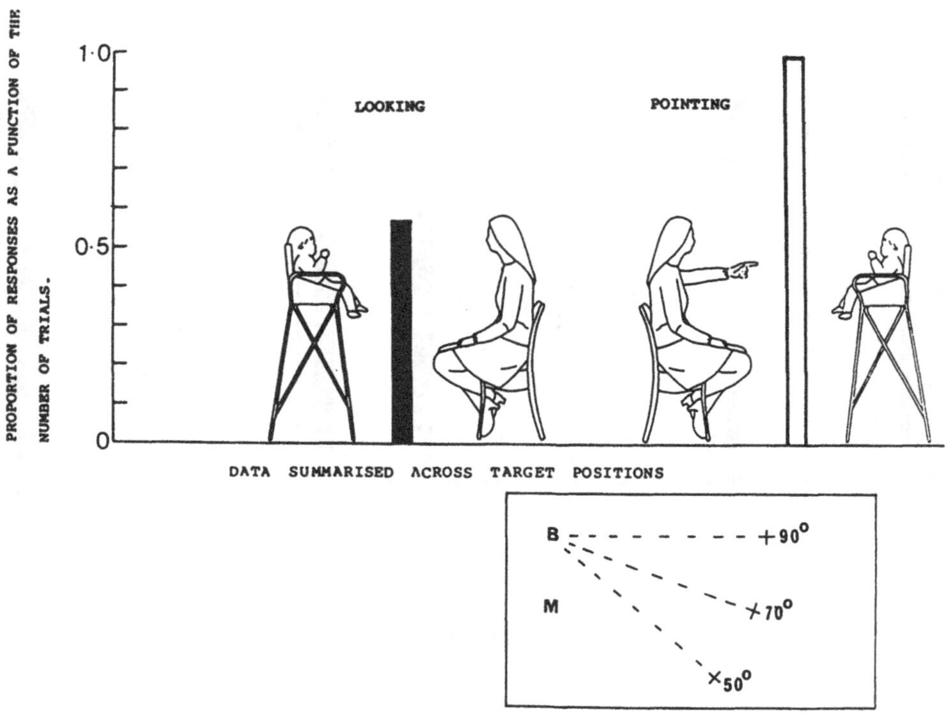


FIG. 22.7. Geometry of preverbal communication: Comparison of mother looking or pointing on response rate of infants at 12 months (Grover, 1982).

9 months do not comprehend the manual pointing gesture. They are equally likely to fixate the mother's hand as the target to which she is pointing. By 12 months, 96% of responses are to the target, and hand fixations are very few indeed. The 12-month-old infants have little difficulty in comprehending the manual point, and they swiftly and smoothly look to the target being designated. They understand that they must look on from the hand toward the location in the visual field. By contrast, 6- and 9-month-old infants, if they locate the target, do so relatively slowly, in a two-step movement in which they first gaze at the mother's hand for a second or two before moving onto the target.

These experiments are important because they demonstrate that quite extensive experience in responding to mother's direction of gaze in the first 10 months of life does not help the infant with comprehension of her manual pointing. An apprenticeship in "looking where someone else is looking" does not qualify the infant to "look where someone else is pointing." On the contrary, the evidence is much more consistent with the hypothesis that the basis for comprehension of looking acquires an additional *geometric* mechanism. This aspect of cognitive development allows comprehension of manual pointing. In our own most recent, unpublished experiments we are finding that the addition of manual pointing to simple change of gaze not only increases the probability that babies of 15 months respond but also increases the accuracy of geometric localization, at least for a limited range of spatial separations. This is what would be expected if the extended arm, like a lever, is being used by the infant as the basis for geometric extrapolation into visual space from the mother to the referent. For any given visual angle, the end of the extended index finger makes a greater lateral movement than does the head (or eyes) and, hence, the precise referent may be more accurately specified. Furthermore, in our own most recent unpublished studies, we find that addition of movement to the targets helps infants who would not ordinarily comprehend the manual point to single out the correct target. It seems possible that ecological factors may play an important part in allowing the infant to comprehend manual pointing perhaps in association with cognitive developmental changes that mark the onset of a *geometric* means of isolating the referent of the mother's signal.

In summary, the series of experiments as a whole suggest that babies begin to comprehend manual pointing at about 12 months, the age at which the new *geometric* mechanism first becomes available. The acquisition of

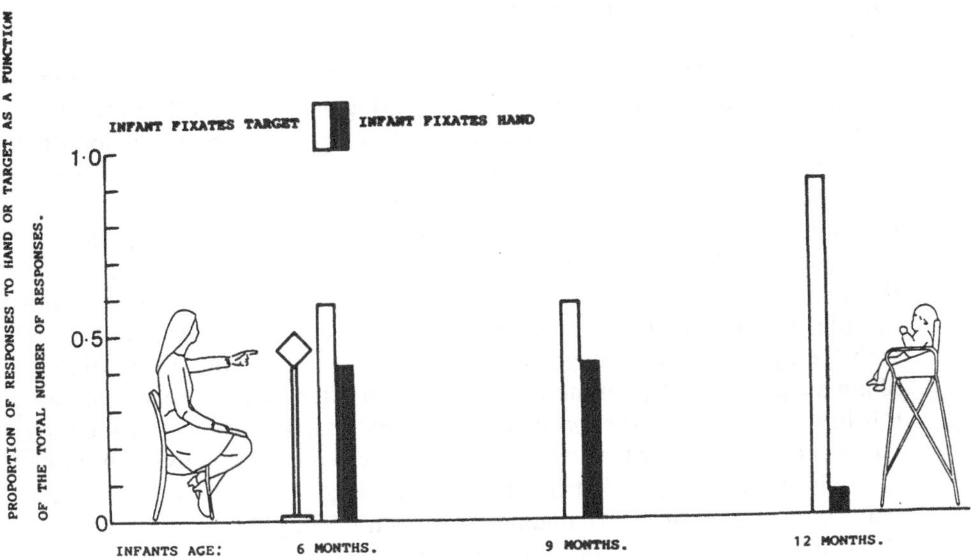


FIG. 22.8. Geometry of preverbal communication: Comprehension of pointing.

the geometric ability during the second year of life does seem to reflect a developmental discontinuity, rather than a simple progressive increase in the ability to overcome distraction, or, more accurately, to localize gaze. The geometric mechanism does not replace the ecological process in development but functions in cooperation (and sometimes in competition) with it to ensure that the infant identifies the referent of the mother's communicative signals. Longitudinal studies would be necessary to establish more precisely the nature of the transition involved.

## JOINT VISUAL ATTENTION AND PRODUCTION OF POINTING

Manual pointing has been called "the most specialised motoric means for the expression of reference" (Werner & Kaplan, 1984), and it is this species-typical referential function that is of interest here. The use of an outstretched arm and index finger to denote an object in visual space seems to reflect hominid evolutionary adaptations of the index finger and thumb (Hilton, 1986). For example, the relative ratio of index finger and thumb length, as well as the position of articulation of the thumb have changed from a chimpanzee-like hand in Australopithecus Afarensis (4 million years before present) to an arrangement akin to that of modern man by the time of Homo Erectus (1.5 million years before present). Although such data have been interpreted in relation to the evolution of the precision grip in tool manufacture and use, it is easy to speculate that the increase in precise control of the hand might also be reflected in its species, typical use in signalling by contrast with whole body orienting observed in other species.

Theorists do not agree on the ontogenetic origins of the production of pointing. Preyer (1896) considered pointing to be a movement originally expressing a wish to seize, and Vygotsky (1962/1926) similarly argued that pointing develops out of the mother's interpretation of her infant's failed attempts at prehension. Shinn (1900, quoted in Schaffer, 1984) suggested that manual pointing may develop out of conjoint visual–tactual inspection of objects. The infant of 8 or 9 months simultaneously touches an object with the tip of the index finger while engaged in close visual exploration. Shinn suggested that pointing may begin soon after by the application of this intersensory–motor coordination to an extended space. A third, contemporary, view consistent with the species typicality and universality of the pointing gesture, is that it reflects a modular, coordinative structure that serves a specialized communicative function (Fogel & Thelen, 1987). Fogel and Hannan (1985) showed that index finger extensions occur reliably in face-to-face interaction between infants as young as 2 months old and their mothers. The index finger does not single out a particular object; it

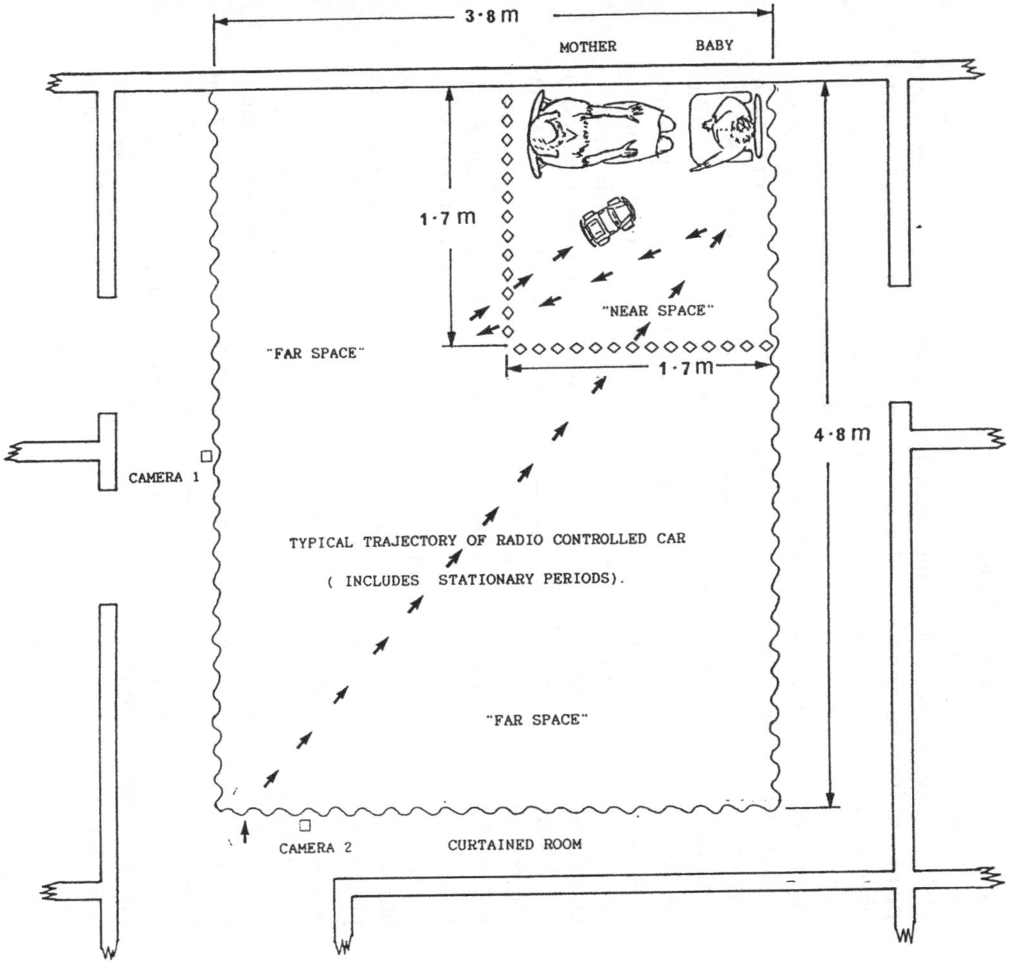


FIG. 22.9. Layout of the laboratory: Pilot study on pointing.

is not correlated with the infant's gaze direction or arm extension, but the "point" is reliably preceded or succeeded by vocalisation or mouth movement and occurs in interpersonal contexts. By 6 months, the hand may spontaneously adopt the pointing posture when an object attracts the infant's attention in a social context, but the arm is not extended. Intentional extension of the hand and index finger, in a communicative gesture, is observed at the beginning of the second year of life. Fogel and Thelen (1987) therefore argue that the specialized function of the index finger point and its relation to shared attention may be innate. Developmental changes are to be explained by the successive acquisition of arm control and fine manipulative skills, culminating in the cognitive integration of the communicative roles of adult, infant, and the objects of shared attention.

Our own research on mechanisms underlying the production of pointing has only recently begun. Butterworth and Adamson-Macedo (1987) developed a technique to elicit pointing in babies using a radio-controlled car that moves toward the infant and stops or starts at various positions (Fig. 22.9). So far, we can say that the onset of intentional manual pointing (at an average age of 13.5 months in our pilot study) clearly involves coordination of viewpoints, with the index finger holding the position of the interesting object while the baby checks to see whether the mother has located the referent. We found that the majority of pointing episodes occurred when the car was stationary and within 2 meters of the baby. Only 10% of episodes occurred when the car was more distant or moving. It seemed possible, therefore, that manual pointing may be linked to mechanisms of focal attention, specialized for identification of objects at fixed positions and somewhat suppressed by mechanisms of "ambient" attention specialized for the detection of movement (Trevarthen, 1968).

Franco and Butterworth (1988) have contrasted pointing with reaching. These studies have shown that reaching is not addressed to distal stimuli, whereas pointing is. Pointing is readily addressed to moving objects at stationary positions, and so the relations between pointing and ambient and focal attentional processes need to be carefully qualified. Even though this dichotomous approach to attention mechanisms has previously been restricted to processes controlling the fixation of gaze, it may be fruitful to explore the possibility that the pointing hand may also share in attentional processes controlling the eye.

The intrinsically social nature of the pointing gesture is suggested by the fact that the infant checks to establish whether the mother is attending to the object of interest. Checking with mother occurs on approximately 50% of pointing episodes among 12-month-old infants, whereas checking with mother accompanies reaching on only 31% of episodes in the same babies (Franco & Butterworth, 1988). Thus, reaching in the service of grasping, and pointing in the service of redirecting attention, are probably

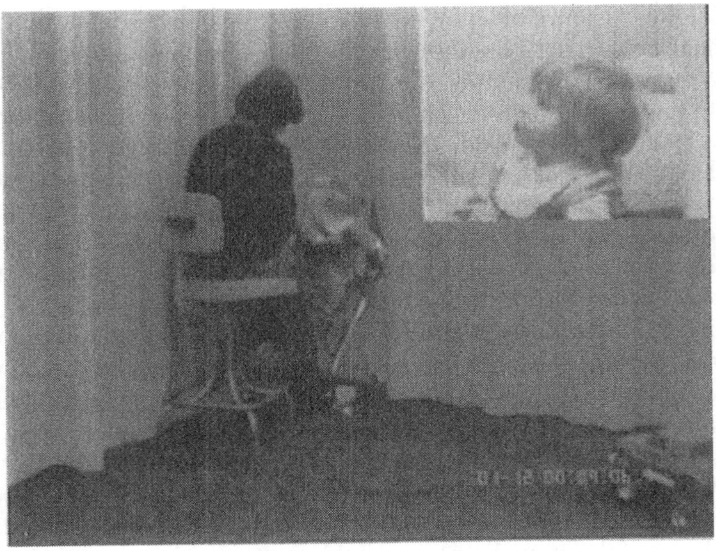

FIG. 22.10.

not ontogenetically closely related actions. We also found that vocalisation was invariably associated with pointing and checking, as if all three behaviors act as a single unit. The 12-month-old infants would first point to the interesting sight and then check to see if the mother was looking. By 15 months, infants would first check to see if the mother was looking at them and only then point to redirect her attention to the interesting object. This anticipatory checking is very clear evidence that a concern for communication is the motive for the pointing gesture (Fig. 22.10).

## CONCLUSION

Our results make it clear that even the very young infant comprehends the signal function of a change in the adult's direction of gaze; communication is not solely dependent on the greater cognitive sophistication of the adult. For the 6-month-old infant, the mother's head movement indicates the general direction (left or right) in which to look. Communication occurs because the easily distractible baby attends to the same attention-compelling features of the objects in the environment as the mother. When seen in social context, the earliest *ecological* mechanism of joint attention allows communication in relation to publicly shared objects through their common effects on intrinsic attention mechanisms of mother and baby.

A more general theoretical implication concerns the infant's comprehension of object permanence. The infant clearly takes the mother's action to signal the existence of a potentially interesting object somewhere within the visual field. That is, although the infant does not search for hidden objects before 8 or 9 months, the mother's behaviour nevertheless signals the "permanent possibility of an object" potentially *within the field of view*. Looking for an object outside the field of view, on the basis of the mother's signal does not occur until the second year of life. This more advanced ability seems to depend on acquisition of a representation of space conceived as a "container" within which the infant is located. Such a representation may depend on feedback from locomotion, vestibular activity, or auditory information; it seems not to be intrinsically visual.

During the first year, joint visual attention remains limited to locations within the infant's own visual space. The baby behaves as if its own field of vision is shared with the adult. This may be interpreted as evidence that the baby treats visual space as common to several points of view, rather than as evidence for complete egocentrism. Perception must originate at a particular viewpoint, but the infant can perceive that others also have a perspective on a space that is held in common. Although limited by the boundaries of the immediate visual field, joint attention nevertheless allows intersubjectivity in the first year of life. Cognitive development of the participants in the interaction is at very different levels, but immediate perception provides a basis for agreement on the objects of experience.

Superimposed on this basic mechanism, with cognitive development, comes precise *geometric* localization of the referent of the mother's gaze, a development that contributes to comprehension of manual pointing. The geometric mechanism lessens the ambiguity of reference because even targets that are identical in all respects except position can now be singled out by the infant. Once the geometric mechanism is available, communication does not require differential intrinsic properties of the objects being singled out; the infant chooses the correct object in relation to the angular displacement of the mother's head and arm. Ecological factors may still assist in allowing definite reference in the natural environment, but our laboratory studies suggest that they are not any longer the essential base for joint attention. This new capacity for geometric compensation coincides with the comprehension of manual pointing, itself an important social means for re-directing attention.

Finally, the production of manual pointing may be based on a modular coordinative structure that operates in the interpersonal context. It comprises a specialized posture of the index finger, vocalization, social referencing, and attentional processes involved in object identification.

In conclusion, the mechanisms of attention underpin the origins of communication and language in infancy when infant abilities are considered in the appropriate social and ecological context.

## REFERENCES


Bornstein, M. H., & Ruddy, M. G. (1984). Infant attention and maternal stimulation: prediction of cognitive and linguistic development in singletons and twins. In H. Bouma & D. G. Bouwhuis (Eds.), *Attention and Performance X: Control of Language Processes* (pp. 433–446). London: Lawrence Erlbaum Associates, Ltd.

Bruner, J. S. (1983). *Child's talk: Learning to use the language*. Oxford, England: Oxford University Press.

Butterworth, G. E. (1987). Some benefits of egocentrism. In J. S. Bruner & H. Haste (Eds.), *Making sense: The child's construction of this world* (pp. 62–80.) London and New York: Methuen.

Butterworth, G. E., & Adamson-Macedo, E. (1987). *The origins of pointing: A pilot study*. Paper presented at the Annual Conference of the Developmental Psychology Section of the British Psychological Society, York.

Butterworth, G. E., & Cochran, E. (1980). Towards a mechanism of joint visual attention in human infancy. *International Journal of Behavioural Development*, 253–272.

Butterworth, G. E., & Grover, L. (1988). The origins of referential communication in human infancy. In L. Weiskrantz (Ed.), *Thought without language* (pp. 5–21). Oxford, England: Oxford University Press.

Butterworth, G. E., & Jarrett, N. (1980). The geometry of pre-verbal communication. *Paper presented at the Annual Conference of the Developmental Psychology Section of the British Psychological Society*, Edinburgh.

Churcher, J., & Scaife, M. (1982). How infants see the point. In G. E. Butterworth & P. Light (Eds.), *Social cognition; Studies of the development of understanding*. (pp. 110–136). Brighton, England: Harvester.

Fogel, A., & Hannan, T. E. (1985). Manual actions of nine to fifteen week old human infants during face to face interactions with their mothers. *Child Development*, *56*, 1271–1279.

Fogel, A., & Thelen, E. (1987). Development of early expressive and communication action: Reinterpreting the evidence from a dynamic systems perspective. *Development Psychology*, *23*(6), 747–761.

Franco, F., & Butterworth, G. E. (1988). *The social origins of pointing in human infancy*. Paper presented at the Annual Conference of the Developmental Psychology Section, British Psychological Society, Coleg Harlech, Wales.

Gibson, J. J., & Pick, A. D. (1963). Perception of another person's looking behaviour. *American Journal of Psychology*, *76*, 386–394.

Grover, L. (1982). The comprehension and production of the pointing gesture in human infants. *Paper presented to the 40th Annual Convention of the International Council of Psychologists*, Southampton.

Grover, L. (1988). *Comprehension of the manual pointing gesture in young infants*. Unpublished doctoral thesis, University of Southampton.

Guillaume, P. (1962). *Imitation in Children*. Chicago: Chicago University Press. (First published in French 1926)

Hilton, C. E. (1986). *Hands across the old world: The changing hand morphology of the hominids*. Unpublished paper, Department of Anthropology, University of New Mexico, USA.

Landau, B., & Spelke, E. (1988). Geometric complexity and object search in infancy. *Developmental Psychology*, *24*(4), 512–521.

Leung, E. H. L., & Rheingold, H. (1981). Development of pointing as a social gesture. *Developmental Psychology*, *17*(2), 215–220.

Piaget, J. (1952). *The origins of intelligence in children.* New York: International Universities Press. (First published in French, 1945)

Preyer, W. (1896). *The senses and the will.* New York: Appleton.

Scaife, M., & Bruner, J. S. (1975). The capacity for joint visual attention in the infant. *Nature, 253*, 265.

Schaffer, H. R. (1984). *The child's entry into a social world.* New York and London: Academic Press.

Trevarthen, C. (1968). Two mechanisms of vision in primates. *Psychologisches Forschung, 31*, 299–377.

Vygotsky, L. S. (1962). *Thought and language.* Cambridge: MIT Press. (First published in Russian, 1926)

Werner, H., & Kaplan, B. (1984). *Symbol formation.* Hillsdale, NJ: Lawrence Erlbaum Associates. (First published 1963)

# VI THE ROLE OF SENSORY-BASED ADJUSTMENTS IN THE ACHIEVEMENT OF THE GOAL

# 23 Functional Contributions of Rapid and Automatic Sensory-Based Adjustments to Motor Output

James H. Abbs
Carolee J. Winstein
*University of Wisconsin–Madison*

ABSTRACT

No issue in motor control has created as much controversy as the role of afferent input in movement. From 1950 to 1970, the intensive experimental efforts to resolve this issue primarily involved acute preparations in anesthetized animals, sensory alteration procedures, or reaction time paradigms in waking humans. Although these methodologies all have serious shortcomings for addressing the natural character of sensorimotor influences, certain relatively extreme positions were nevertheless formulated. In the last two decades, single neural unit recording in waking animals, unanticipated perturbation techniques in humans, and like procedures have yielded a moderation of those extremes, with increasing development of a common ground. Specifically, advocates of central nervous system pre-specification of muscle contraction and movement in the form of motor programs have altered their constructs and definitions to accommodate the ubiquitous evidence that almost all motor output parameters are modifiable during execution. The limited and almost strawman-like conceptualization of sensory influences being implemented exclusively via closed-loop, negative feedback circuits also has been rejected. Likewise, even zealous advocates of sensory influences have recognized that the particular contributions of a given afferent input vary considerably with the motor system under study, task goals, experience, and yet other variables. In this latter respect, the challenge for the next decade is to discover principles of motorsensory control that generalize across motor systems and tasks, despite the tremendous variation in both the degree and kind of obviously critical sensory contributions.

## INTRODUCTION

Determining sensory contributions to the control of movement from the published scientific literature is similar to evaluating the condition of the U.S. economy based on election year campaign speeches. A great deal is said but there is little agreement. In fact, it is possible to consult two equally substantive papers on the motor control of movement, each published in the last decade, and get different perspectives on the apparent significance of sensory mechanisms. On the one extreme is the position that most well-learned motor behaviors, especially those involving sequences of short duration movements, are the result of individual muscle contraction and movement parameters being specified in advance, with relatively minimal changes after initiation. Conversely, it is stated, with equal conviction and comparable numbers of seemingly supporting studies, that even the smallest detail of muscle contraction and movement is influenced, on a moment-to-moment basis, by sensory input, operating throughout the neuraxis. A range of data are cited by the proponents in this controversy, varying from analytic considerations of feedback control to anecdotal observations in patients with rare sensory impairments. In some cases these strikingly different views of sensory contributions to motor output result from limited appreciation of other scientists' data or paradigms. However, in many cases the theoretical and experimental contradictions seem irreconcilable.

Unfortunately, in dealing with these respective counterarguments, the antagonists in this classic struggle often are less than responsive, and sometimes even appear to be disinterested. As a result of this situation, there is very little agreement on major theoretical issues or on which kinds of experiments are most needed in this important area of behavioral neuroscience. One of the likely reasons for the absence of a cohesive perspective on this problem is the wide range of experimental paradigms that have been utilized to evaluate sensory contributions and the consequent voluminous research and theoretical literature on the subject. Even for a scientist with a broad background it is very difficult to reconcile, for example, single afferent unit firing data obtained from an anesthetized cat, in one case, with movement patterns observed in waking human in another. A related problem is the different goals of research apparently addressing this issue. Purposes of this research range from determining the neurobiological mechanisms underlying sensorimotor contributions to work directed toward more behavioral factors regarding when and if such contributions are significant.

In this context, the purpose of this chapter is to evaluate some of the disparate observations on the *functional contributions* of *rapid* and *automatic* sensory-based motor adjustments. In the spirit of the present volume, this commentary has a behavioral orientation. As such, we wish to define *functional contributions* as those sensory influences that have been documented in a particular analysis or experiment to have significant biological and/or behavioral consequences for successful performance of the motor task under consideration. For example, demonstration that certain afferent-to-efferent connections can merely be activated does not, by itself, constitute evidence for a functional contribution of those connections. In this vein, observed responses to foreign or unusual stimuli obviously also are questionable in their functionality. Likewise, by concentrating on *rapid* and *automatic* actions we wish to focus on those influences of sensory input that are an inherent aspect of the motor behavior under study. The "deliberate" detection of a stimulus and generation of a response, as is the case in most reaction time studies, thus are not deemed either rapid or automatic.

Although it might seem necessary to generate additional general definitions at this juncture, further abstract specification of these and related terms is likely to be counterproductive. That is, terms such as *functional*, *rapid*, and *automatic* are best considered as relative rather than absolute. Although this may appear elusive, attempts at universal definitions tend to actually trivialize the complexity of the multiple processes involved. For example, it is apparent that a response with a latency of 60 msec would be rapid and functional for postural adjustments but prohibitively slow and, hence, without functional significance in the control of saccadic eye movements. As such, our particular use of these terms are specified operationally as necessary for different aspects of the discussion.

A further point of emphasis is in terms of the kinds of movements one considers. One major source of disagreement on sensory contributions is related to whether one is examining rapid versus slow movements, novel versus skilled movements, or learning versus control of highly practiced movements. The role of sensory input in the process of learning a motor skill is certainly an important topic; however, initially it is preferable to focus primarily upon fully developed motor behaviors. Likewise, movements that are primarily products of laboratory innovation raise a number of ecological issues that add undesirable complexity to an already difficult subject. The control of natural and skilled movements appears to be the optimal domain in which to evaluate the potential role of sensory information. That is, highly practiced motor actions such as those associated with speech, precision manipulation, and maintenance of postural equilibrium are perhaps the most important movements in which to consider these functional contributions. Also, common movements that reach a high degree of proficiency in a large part of the population are those in which alternatives to utilization of sensory information, such as pattern generators or motor programs, are most likely to be functional. Conversely, movements associated with many forms of athletic performance, particularly as they are manifest in the 1% to 2% of the most gifted individuals, may not be the best general material for addressing this issue. Fortunately, several lines of recent research have addressed the particular role of sensory information in control of such ecologically significant movements.

## IS THE COMMON GROUND COMMON?

If one considers the major arguments supporting or refuting the contribution of sensory processes to motor control, it seems that one source of disagreement is a lack of consensus as to what constitutes acceptable evidence for or against such contributions, coupled with a great deal of ambiguity in the theoretical positions, models, or hypotheses addressing this issue. For example, there appears to be little general agreement as to what constitutes a motor program, a central pattern generator, a sensory feedback system, and so on, despite the central importance of these constructs in considerations of possible sensory inputs. Obviously motor programs and central pattern generators are, for some scientists, the major alternatives to complete dependence on sensory information. In the discussions that immediately follow, we thus identify certain factors, variables, and concepts that we believe need to be considered and reconsidered when examining some of the old and new evidence on sensory contributions.

### Classical and Contemporary Views of Sensory Contributions

*The early perspectives.* Early concepts of utilization of sensory information in movement control tended to be relatively simple, perhaps reflecting a different view of the nervous system's complexity. For example, James (1890) proposed a reflex-chaining model of movement sequencing/coordination, which was based on a number of simplifying assumptions. In the era surrounding World War II, the early chaining notions were displaced by a particularly compelling concept, the closed-loop feedback model. The excitement with feedback-based control of movement was fueled in part by the overall concepts of closed-loop regulation that emerged in the 1940s and 1950s. Countless papers describing biological control processes operating in this manner were generated for different systems and tasks (cf. Fairbanks, 1954; Guyton, Crowell, & Moore, 1956; Krendel & McRuer, 1960; Simon & Smith, 1956; Stark & Cornsweet, 1958; Stark & Sherman, 1957). In large part, this initial view was that these were "automatic" control processes wherein a goal or reference was established and feedback would then operate automatically to minimize or eliminate error. This model of sensory influences on motor output was elaborated substantially by virtue of focusing on the contributions of one form of mechanoreceptor, the muscle spindle, that permitted active gamma motor manipulation of its sensitivity and led to hypotheses of whether movement was initiated via activation of gamma motoneurons (cf. Merton, 1953), alpha motoneurons, or both (Granit, 1955).

Many scientists who focused on potential sensory contributions to motor output maintained the view that these processes were automatic, especially for highly learned movements. However, some individuals working particularly with human subjects utilized external feedback signal manipulation to discern features of closed-loop control and learning (e.g., Adams, 1971). This latter work led some researchers to an emphasis on more deliberate subject behaviors as well as infusing this work with an information processing perspective, primarily in tracking tasks and for simple movements (cf. Bilodeau, 1966; Moray, 1981; Poulton, 1974). These studies focused on reaction time-like processes rather than the more automatic, reflexive contributions. There was some implication that automatic contributions were relatively low-level and hence of minimal interest in the acquisition and control of skilled movements. As demonstrated repeatedly in the last decade, such implied dichotomies are artificial. Almost all forms of higher and lower level sensorimotor processes appear to be subject to modification with practice, motor set, task goal, perception, and like variables. Indeed, it is now apparent that a reflex is not a definable entity and even monosynaptic stretch responses are subject to changes with training and experience (e.g., Meyer-Lohmann, Christakos, & Wolf, 1986; Prochazka, Hulliger, Zangger & Appenteng, 1985; Wetzel, 1986; Wolpaw, O'Keefe, Noonan, & Sanders, 1986, 1987).

*Response latencies, instability, and movement dynamics.* Perhaps the oldest concern regarding sensory influences is with the time lag between the instant that an error is detected and when it is possible to generate an appropriate motor response to correct that error. It is well documented that a closed-loop system becomes unstable if the loop delay is prohibitively long in relation to the duration or dynamics of the movement to be controlled. In engineering control-system parlance, if the gain of the feedback pathway exceeds zero at a movement frequency where the neural transport delay yields a phase lag of 180°, the feedback control process becomes nonfunctional. Obviously, if a correction is sufficiently delayed, it cannot usefully correct the detected error and, if sufficiently late, may create additional error or exacerbate the existing one. If this additional error elicits a correction in turn, a clonic-like instability can result, generating an oscillatory "hunting" pattern especially near the motor target, not unlike the tremor observed in certain movement disorders (or in man-made control systems; cf. Rack, 1981; Stark, 1968).

Based on these considerations, it has been argued that sensory information is of little consequence, particularly in movements of very short duration with very long feedback delays. Given the extent to which practical mathematical analyses of feedback systems are proven and tested, in principle this argument unquestionably is sound. At issue is whether actual feedback delay times and system dynamics are such that feedback is nonfunctional in control of most mammalian movements. Factors influencing the degree of functionality include: (a) timing and availability of peripheral sensory information, (b) the actual neural transport delays associated with afferent and efferent transmission, (c) the interval necessary for processing of the incoming information, and (d) the lag time between muscle contraction and movement. Using strict engineering analyses, Rack (1981) has illustrated, at least in linear systems, the interacting contributions of these multiple factors. There are some issues, however, that need to be reconsidered in more detail.

*Peripheral sensory information* presumably is available continuously prior to and during its execution. However, if one conceives of sensory input strictly in relation to its use in a negative feedback system, a correction cannot be initiated until an error is detected—after all, this is a reactionary control process. With this provision, for short duration movements in particular, the time available is especially restrictive. However, some data indicate that sensory information regarding an evolving movement is available prior to its physical initiation. If one examines single afferent discharge from the muscle spindles associated with human hand movements (cf. Vallbo, 1973, Fig. 23.5), it is apparent that spindle discharge signaling the character of the upcoming movement is available well in advance of the initial displacement. Likewise, for cutaneous mechanoreceptors (cf. Hulliger, Nordh, Thelin, & Vallbo, 1979), initial afferent discharge appears to be activated by the muscle contraction, starting anywhere from 20 to 35 msec prior to movement onset. Initiation of sensory discharge appears to coincide with contraction onset and may signal the magnitude of muscle contraction and/or dynamics of the associated skin distortion (for cutaneous mechanoreceptors). As such, it appears reasonable to suggest that some information regarding the upcoming movement, its timing, and perhaps potential errors, is available to the nervous system before movement actually starts.

Aside from the precise time of sensory information availability. neural conduction delays remain a potentially serious problem. However, the degree of this problem needs to be examined quantitatively in relation to several factors including the level of the nervous system at which the afferent-to-efferent integration is mediated, and the motor system under consideration. For example, for the orofacial system or the distal upper limbs, the afferent projections to the highest levels of the nervous system (the cerebral neocortex) are numerous and of relatively short latency, ranging from 9–15 msec in primates. Likewise, for these muscle groups there appear to be dense monosynaptic connections from pyramidal cells of the primary motor cortex to the associated motoneurons. Moreover, for the orofacial system, the more peripheral reflex latencies are even more striking—only 6 msec for the jaw-closing muscles (muscle spindle IAs to alpha motoneurons) and only 20–30 msec for longer latency responses. By contrast, for the proximal limb muscles or muscles of the trunk, the connectivity appears less direct (cf. Lenz, Tatton, & Tasker, 1983a, 1983b). Ironically, a great deal of research attempting to address the issue of sensory contributions to motor output has been conducted on the proximal upper limb; it may be inappropriate to generalize from these data to most other movements and motor systems.

One popular source for estimating feedback delays, particularly in the behavioral literature, was reaction times, especially with visual stimuli (cf. Keele & Posner, 1968). Estimates of minimal reaction times from 175–250 msec were commonly cited and led predictably to the conclusion that these latencies precluded utilization of such sensory information for motor control, other than via low-level reflex actions. It is questionable whether such reaction time estimates are generally meaningful given that many important contributions of sensory input are thought to be functioning automatically, at least in the control of skilled movements (cf. Bridgeman, Kirch, & Sperling, 1981; Hansen & Skavenski, 1977). Particular concern is that such latencies often were obtained in naive subjects performing unusual laboratory tasks with unknown stimulus–response compatibilities. Several recent studies have suggested that even motor responses obtained in reaction time paradigms may have latencies that are only a fraction of early estimates, ranging from 50–80 msec (Cole & Abbs, 1983; Crago, Houk, & Hasan, 1976; Lansing & Meyerink, 1982; Winstein, Abbs, & Petashnick, 1988). Based on these arguments, the potential role of reaction time-based sensorimotor processes in coordination of rapid sequential movements has been re-examined (cf. Abbs, Gracco, & Cole, 1984; Cole & Abbs, 1983).

Another factor thought to play a role in reducing the potential detriments of sensory feedback delay is whether there is a phase lead or advance in mechanoreceptor transduction. Phase advances up to 90° are possible (a velocity sensitive mechanoreceptor—cf. Poppele & Bowman, 1970; Rack, 1981), and hence potentially beneficial in reducing instability and extending the bandwidth over which feedback can operate. However, counteracting such phase advance, and in addition to the delays introduced by nerve conduction and sensory processing, are the influences of *muscle contraction mechanics*. Muscles act as low-pass filters, creating an additional source of "lag" in the system. These lags obviously vary with motor system, the passive biomechanical inertial properties of the body part to be moved, and the muscle dynamics involved. Movements of the eyes, lips, or perhaps the digits have relatively faster muscles and lesser inertias than the proximal upper limbs or the trunk. The engineering analyses of these multiple factors were presented by Rack (1981) in some detail.

Obviously, as described, it is necessary to consider multiple factors to evaluate the influence of feedback delays. Indeed, utilizing the simple techniques demonstrated by Rack (1981), it is possible to evaluate these claims more objectively. In this vein, several years ago we utilized such a procedure to examine the influences of neural transport delays for a particular movement system for a specific motor task. Our focus was on lip movements for speech. These analyses were undertaken not because we believed that control of the lips was accomplished by such a simplified negative feedback process, but because we wished to empirically evaluate the often-stated claims that sensory feedback was too slow for rapid, sequential movements of this kind.

Initially, lip movement bandwidth data were obtained for rapid and normal speech tasks. In no case did the bandwidth observed exceed dc-10 Hz. (Abbs, Muller, Hassul, & Netsell, 1977; DePaul, Nadler, Ishii, & Johnson, 1988). The delays of afferent feedback for perioral mechanical stimulation appear to be approximately 12–18 msec for a peripheral pathway (Abbs & Cole, 1982; Netsell & Abbs, 1975) and 25–30 msec for a long-latency response (cf. Abbs & Gracco, 1984). If one assumes that the lips are served by mechanoreceptors with sensitivity to movement velocity (cf. Caligiuri & Abbs, 1984; Larson, Folkins, McClean, & Muller, 1978), a moderate phase advance can be assumed. Additionally, because these muscles are primarily viscoelastic (cf. Muller, Milenkovic, & MacLeod, 1985), phase lags should not exceed 90°. Overall, even these minimal data thus permit one to quantitatively estimate whether a 180° phase lag as a result of these multiple system properties occurs at a frequency that prohibits contributions to lip movement control for speech.

With a bandwidth of 10 Hz, both longer and shorter pathways appear to have the capability to contribute significantly to lip movements. As can be calculated (i.e., $\phi = 360 \times \text{frequency} \times \text{time delay}$), even the longer latency sensorimotor pathways are capable of contributing to almost the full range of these very rapid movements (with a 45° phase lag due to biomechanics). This conclusion was empirically confirmed several years later with introduction of unanticipated perturbations to lip movements. As was shown in those later studies (Abbs & Gracco, 1984; Gracco & Abbs, 1985), if unanticipated perturbations are introduced on speech lip movements, functionally significant compensatory adjustments are observed in response to the first such occurrence of perturbation in naive subjects. If one undertakes comparable analyses of neural transport delays for sensory contributions to elbow joint rotation, with narrower movement bandwidths in these latter movements and feedback delays that are longer, but not proportionately so, it may be that comparable sensory contributions are possible.

Overall, the foregoing comments indicate that it is possible to obtain data to support any one of several different conclusions, depending upon movement speed, movement duration, and the sensorimotor connectivity of the motor system under consideration and its biomechanics. If one were to examine very slow movements of a fast acting system (i.e., minimal neural transport and biomechanical lags) the answer would be very different than if one were to study very rapid movements of a relatively slow system. Obviously, neither of these results would be very convincing in terms of a general conclusion for the greater proportion of important human movements. Additional and more substantial empirical analyses of the kind discussed would appear to provide at least a useful first approximation to address this issue in a substantive manner. In a teleological vein, it is intriguing to speculate whether those muscle groups most involved in rapid precise movements are also endowed with afferent-to-efferent connections of the shortest latency. The viability of such task and muscle group-specific neurophysiological designs is also apparent in subsequent discussions of these factors and the possible use of sensory information.

*Contributions outside the closed-loop model.* Often, available formalisms developed for designing man-made systems offer substantial insights into biological organisms. In other cases, these constructs, carried too far, limit the paradigms and hypotheses that would otherwise be naturally pursued. Regarding the possibility of sensory contributions to on-line motor control, the formalisms of closed-loop feedback systems have been a mixed blessing. It might be argued that in the last decade the closed-loop feedback system has become a straw man. Although analyses presented here offer some new perspectives, closed-loop feedback is not the only process by which sensory information has a moment-to-moment influence on evolving motor output. In particular, the lack of viability of closed-loop feedback processes does not automatically support an alternate interpretation that motor output is determined by preprogrammed specification.

Technically, a *feedback* system is one in which an error signal directly drives a corrective adjustment at the site where that error is introduced. Recent experimental data indicate that many (or most) important contributions of sensory information to the on line control of movement are not captured by this fundamental construct. These data include actions interpreted to be due to sensory-based feedforward, possibly *open-loop* (Abbs et al., 1984; Arbib, 1981; Houk & Rymer, 1981), precomputed responses (Rack, 1981), distributed compensatory processes (Marsden, Merton, &

Morton, 1981; 1983) as well as others. These widely varying but immediate motor adjustments to sensory input reflect a degree of complexity and flexibility that appears well beyond most current engineering models of these processes. Several features of these sensory contributions deserve special comment. First, it has been shown that the distribution and character of sensorimotor adjustments is task and motor set dependent. Specifically, it has been observed that if an unanticipated perturbation is imposed on a movement (e.g., the lower lip) and that movement is a component of a multi-articulate task (i.e., coordinated upper lip, lower lip, and jaw movements for an oral speech action) the compensatory responses are distributed, in a functional manner, among those components (Abbs & Gracco, 1984; Cole & Abbs, 1987). In this example, remote compensatory responses are seen in upper lip and jaw movements that offset or cancel the influence of the movement error induced by the perturbation on the lower lip. Further, if the tasks are varied in such a manner that one of the components is no longer critical to task achievement, the associated remote compensatory response also was absent (cf. Abbs & Gracco, 1983; Cole & Abbs, 1987; Cordo & Nashner, 1982). Of particular significance in these processes is the operation of flexible sensorimotor processes. Based on experience with certain kinds of changes in the environment, the motor goal, or in the perceived environment, compensations are functionally adjusted in response to a given induced error. These recent observations are wholly consistent with the well-studied oculomotor–head movement coadjustments observed in the control of gaze, including rapid and very short latency compensations in saccadic eye movements to head movement perturbations (see Prablanc, & Pelisson, this volume).

## Evolution of the Motor Program Alternative

An obvious alternative to on-line, moment-to-moment adjustment and readjustment of motor output by sensory input is the specification of some or all aspects of the motor output in *advance* of movement execution. In general, this prespecification of motor output parameters has been represented in the form of a motor program (e.g., Keele & Summers, 1976; Summers, 1981). Historically, from the behavioral perspective, a motor program or *neuromotor plan* has been thought to represent an abstract memory structure, implemented unconsciously (cf. Henry & Rogers, 1960). More precisely, Keele's often cited 1968 definition stated that a motor program is a set of prestructured muscle commands that executes movement in the absence of peripheral feedback. Although this definition was overinterpreted (Requin, Semjen, & Bonnet, 1984), it has, nevertheless, had substantial impact. In particular, the motor program concept was viewed, at least by many, as an alternative to a closed-loop feedback control. Although the centralist–peripheralist issue has been addressed elsewhere (e.g., Glencross, 1977; Kelso & Stelmach, 1976), it is useful to examine the apparent discrepancies between notions of prespecification of motor output with those suggesting ubiquitous sensory involvement. When examined from recent perspectives, it becomes clear that these two seemingly inconsistent positions largely have been reconciled.

What is contained or specified in the motor program has been the central issue for some time (e.g., Lashley, 1951; Marteniuk, Mackenzie, & Leavitt, 1988; Morasso, 1981; Schmidt, 1983). One version of the motor program was thought to specify the order of muscle contractions, the phasing or temporal relationships between these contractions, and the relative muscle forces involved. This framework has been termed a generalized motor program for which certain parameters, such as the muscle actions and the overall duration of the movement, are specified just prior to execution (Schmidt, 1980). There was some evidence to suggest that such generalized motor programs exist for typing, handwriting, speech, and locomotion, based on patterns of temporal regularity in the motor output, despite variations in overall movement duration (cf. Armstrong, 1970; Hollerbach, 1981; Merton, 1972; Raibert, 1977; Terzuolo & Viviani, 1979; Tuller, Kelso, & Harris, 1982). However, on closer examination, the timing characteristics of the motor output, under different total durations of these movement patterns, did not appear to be specified under a simple relative-timing rule (see Abbs & Connor, 1989, for a recent review; Gentner, 1987). More importantly, when the actual movement producing the output were examined (e.g., finger movements in typing), the timing of their initiation was highly variable, despite consistent timing of movement terminations for key presses (Gentner, Grudin, & Conway, 1980). Similarly, studies of human locomotion (cf. Winter, 1984) indicate that the torque contributed by synergistic muscles also was variable, even in the face of relatively consistent temporal patterns. These data thus suggest that the motor program does not specify the precise movements, muscle contraction parameters, or their timing, but rather the final goals of the movement process (cf. Birkenblit, Feldman, & Fukson, 1986).

Because the available evidence increasingly suggested that movements appear to be commonly influenced by peripheral feedback (e.g., gamma loop—25 msec in the arm and 6 msec in the jaw), Schmidt (1982) suggested a redefinition of a motor program as: "an abstract structure in memory that is prepared in advance of the movement; when it is executed, the result is the contraction and relaxation of muscles causing movements to occur without involvement of feedback leading to corrections for errors in selection" (p. 205–206) (of a new motor program). This change in definition allowed for feedback ("reflex") contributions during program execution in carrying out a predetermined pattern of movement (i.e., in determining details of submovements and muscle contraction patterns), but not as a means to select or initiate a "new" pattern or temporal phasing. This redefinition led to some additional complications. There is evidence from a number of sources (e.g., Abbs & Gracco, 1984; Georgopoulos, Kalaska, & Massey, 1981; Goodale, Pelisson, & Prablanc, 1986; Gordon, Bermejo, & Ghez, 1988; Johansson & Westling, 1984) showing that, in response to an external disturbance or a change in target location, the trajectory and/or the relative forces were functionally altered during movement execution. Currently, it is thus apparent that parameters of a movement are not prespecified as a "rigid set of commands" (Requin et al., 1984), but, rather, represent the specification of a given environmental objective or set of functional goals. This less explicit characterization is reminiscent of more traditional conceptions such as the "action plan" (Miller, Galanter & Pribram, 1960).

Although the control of certain skilled actions, such as typing and speech, are not well characterized by a strict motor program concept, more innate, cyclical actions, such as locomotion and chewing, have been ascribed to similar central control structures. For example, in some animals, locomotion is thought to be coordinated spirally at a relatively low level by a network of neurons or a central pattern generator (CPG). There are numerous examples of CPGs functioning quite adequately without afferent involvement. However, more recent work has stressed the importance of afferent information for various aspects of these actions (e.g., Grillner & Rossignol, 1978; Smith & Zernicke, 1987; Smith et al., 1987). Although CPGs apparently are capable of functioning without afferent involvement, this does not mean that they do (Bässler, 1986).

Overall, it appears that motor programs and CPGs "control the flow of information" (MacKay, 1985), with much of that information being sensorimotor in nature. This view of a motor program is consistent with that of Abbs et al. (1984) who similarly noted that a "motor program is more likely the representation of the dynamic processes whereby the appropriate sensorimotor contingencies are set up" (p. 214) rather than motor patterns, per se. In this same vein, Cruse, Dean, Heuer, and Schmidt, (in press) suggested that motor programs are really subsets of a "motor control structure" that subserve two functions: first, the autonomous generation of control signals to the peripheral system, and second, the processing of sensory information. Further, although these two functions have been discussed as logical alternatives, the interesting question pertains to how sensory information is integrated with autonomous control to effect functionally adaptive responses.

In keeping with our position that different motor tasks and muscle groups need be examined for differing sensorimotor processes, it is apparent that at the extremes locomotion and typing involve very different kinds of behaviors. Yet, both have been used as evidence for motor programming. It does not seem reasonable to invoke the same mechanism for both highly skilled and innate actions. The interface between peripheral input and motor output at the motoneuron level and at higher levels seems certain to be quite different. In this vein, one is reminded of Pew's (1984) statement, "we should not be searching for a unitary integrative concept of a motor program: It may have different representations at different levels in the motor system" (p. 26). As will be apparent, not only should there be a consideration of different forms of a motor program at different levels in the motor system, but also with other functional variations, as well, including task, muscle groups, and movements.

## SOME FUNCTIONAL AND ECOLOGICAL PERSPECTIVES

Although the above comments address long-standing issues and arguments in the potential role of sensory contributions to motor output, there are other related ecological and biological issues, that from our perspective, largely have been neglected. Many hypotheses apparently addressing the role of sensory information are impractically broad and/or lack sufficient detail to address the issues clearly. For example, certain workers appear to assume that all forms of sensory information (visual, auditory, proprioceptive, and so on) are utilized in the same manner for movements throughout the body. Conversely, other investigators implicitly ignore or dismiss most forms of sensory information with the exception of muscle afferents.

### Functionality Criteria in Evaluating Sensory Contributions

Some years ago, it was popular to note that in so-called motor nerves the number of axons associated with muscle afferents and gamma innervation of muscle spindles far exceeds the number of alpha motoneuron axons. This bit of morphology was intended to emphasize the importance of muscle spindle afferents for motor control. Unfortunately, such arguments by themselves do not constitute sufficient support for the contribution of sensory input any more than the presence of wings on an ostrich suggests that these birds must fly. This is not to say that teleological approaches to this problem are without value; taking an extreme position in which no evidence of natural design is acceptable would be most unfortunate. However, it is important to satisfy, or to at the least address, certain functional criteria in evaluating evidence for sensory contributions to motor output. That is, in order to argue that a given sensory input contributes to a given motor task, observed motor responses to the sensory input should:

1. Appropriately change with stimulus changes, that is, yield increased excitation with increased opposition to a contracting muscle ("load") and the opposite response with the opposite stimulus ("unload").
2. Demonstrate functional contributions the first time the experimental stimulus is experienced by the performer, hence documenting normal functionality and eliminating contamination by possible experience artifacts.
3. Be elicited by stimuli that are in the "physiological range"; that is, with movement, force, and/or other mechanical characteristics that are within the same range as those associated with the motor behavior under study.
4. Demonstrate sufficient magnitude or "gain" to contribute significantly to the movement or the behavior under consideration, namely, address the question as to whether a significant proportion of the total motoneuron discharge for a given movement or joint torque is primarily due to the afferent influences.
5. Manifest robust and consistent patterns of compensatory adjustment in natural tasks, for example, tasks with minimal or limited constraints on the number of joints to be moved in behaviors requiring functionally significant actions.

Although it is not always possible to satisfy these criteria, proponents of a significant role for afferent influences should provide at least some evidence or reasoning to support the position that the particular sensory-based responses being considered are not simply experimental curiosity or laboratory epi-phenomena. The long-standing and intensive studies on the monosynaptic influence of spindle IA afferents on motoneurons may have been more fruitfully pursued if some of the listed criteria, particularly the fourth one, had been addressed. That is, spindle IA afferent-to-efferent connections with the shortest latency and potentially with the most direct motor influence appear to lack sufficient gain to have a significant influence on motor output. Henatsch (1967) was apparently sensitive to this criticism and the need for documentation of ecological significance when noting that we have sufficient data to explain how the "elephant dances the twist," but no evidence that any of these processes are actually used.

Likewise, one must obviously be suspect of responses that adapt rapidly or are not manifest until stimuli are presented numerous times. For example, one wonders as to the significance of motor responses, sometimes reported in studies of monkey limb movements, that only occur after the animal has experienced them several times (cf. Polit & Bizzi, 1979). A comparable interpretive problem might be said to exist for perturbation of motor output in which the frequency of the induced errors is very high.

In these cases, it can be argued that after substantial experience with the perturbations, subject responses no longer represent natural error correction, but, rather, a contaminated abnormal circumstance. In this same vein, afferent-to-efferent connections that are demonstrated with intense or foreign stimuli cannot be taken too seriously as major influences in the control of the motor output. The classical example of this particular problem has been differences in the responses to electrical stimuli of nerves or skin versus natural mechanical stimuli. Rack (1981) commented specifically: "In particular it is generally unsafe to use information obtained from one sort of movement or disturbance (such as an impulsive one) to make predictions about some other pattern of movements" (page 230). As Grimm and Nashner (1978) and others have pointed out, it is possible to "create" pathways with certain kinds of stimuli. At the very least, the physiological connectivity associated with functionally normal behaviors may not be manifest if extreme degrees of excitation resulting from abnormal stimuli are used.

It is important in this context to acknowledge recent work that indicates that the nervous system channels and shapes sensory input based on task requirements and other functional criteria (cf. Chapin & Woodward, 1981; Ghez & Pisa, 1972). On the other hand, the modulation or gating of some sensory input influences does not always appear to be an active process; the literature aimed at attempting to show reflex sensitivity variation is clouded by the fact that many such adjustments are merely changes in the excitability of the motoneuron pool. So-called reflex tuning, independent of motoneuron excitability, has only rarely been demonstrated, despite its widespread popularity (see, however, Lund & Rossignol, 1981; Lund, Rossignol, & Murakami, 1981). In this context, with default sensory input gating, it clearly is possible to demonstrate erroneous connections if stimulus characteristics are extreme.

Not surprisingly, a related demonstration of this problem is that stimuli that have no normal meaning to the intended behavior appear to reveal patterns of responses and connections that actually impair function. A notable example is seen in observations by Nashner, (1976) with rotational platform perturbations (toes up). These perturbation stimuli obviously are artificial to normal experience in control of postural equilibrium, except perhaps for lumberjacks who walk on rolling logs. Not surprisingly, subjects manifested responses that were inappropriate (causing them to fall) for the destabilizing platform rotation, but appropriate for a more functional stimulus (backward–forward platform translation). Obviously, these responses to foreign stimuli are not meaningful in discerning the normal control of posture. Similar kinds of data have been reported by Lee and Aronson (1974) in the inappropriate postural responses to visual stimuli of experimental chamber wall movements.

Finally, related to this issue is the question as to whether it is appropriate to evaluate the role of sensory information with motor tasks that are highly constrained. In single joint movements the possible adjustments for induced errors are limited to a direct correction of the agonist muscle contraction via a classical feedback process. This constraint may provide some insight into the findings of often-cited elbow joint blocking experiments (e.g., Wadman, van der Gon, Geuz, & Mol, 1979) and the associated ambiguity of determining control variables generally (Stein, 1982). Indeed, as noted previously, it appears that a significant contribution of sensory input is with regard to nonautogenic compensatory actions, that is, distributed adjustments in a multi-articulate system. If movements are constrained to a single joint, perhaps the degrees of compensatory freedom normally available likewise are unnaturally constrained (cf. Abbs, 1982; Abbs et al., 1984).

### Variations with Sensory Modalities

Given the degree of complexity in the possible contributions of sensory input to motor control, it has been necessary to adopt certain simplifications, explicitly and implicitly. Aside from conceptual manageability, these simplifications also facilitate efforts at developing more unified perspectives on these contributions. The obvious dilemma that arises is whether, in brewers' terms, homogenization and pasteurization have eliminated the most important and interesting aspects of sensory contributions to motor output. Based on the rich variety of behavioral and neurobiological findings addressing this problem over the last decade, it appears that many such simplifications need to be re-examined. One of these, particularly in the behavioral literature, is an implicit assumption that it is possible to consider information from muscle spindles, golgi tendon organs, cutaneous mechanoreceptors, joint receptors, vision, and audition as contributing to motor output in the same general manner. As a result of this perspective, investigators often cite experimental evidence regarding the influence (or lack thereof) of one form of sensory input (e.g., vision) on motor control as a basis for general theories and universal hypotheses.

First, it is a neurobiological fact that these different receptor systems are distinct in (a) the kinds of information provided to the nervous system, (b) the degree to which they directly or indirectly influence motoneurons or higher levels of the final common pathway (i.e., primary motor cortex pyramidal cells), (c) provision of motor output sensing for different muscle groups (e.g., the lip muscles lack spindles), and (d) contributions for guiding different motor tasks (visual information provides little data on orofacial functions). These considerations, even in such a simplified form, compel us to assume that different sensory systems *are not equivalent* in their contributions across various muscle groups (i.e., proximal upper limb, hand, jaw, tongue, lip), for different tasks performed even by the same muscle group, for differing degrees of skill. Further, the value of having a simplified view of sensory input influences on motor output is lost if one has to take into account the nervous system transformations necessary to treat all of these inputs alike, even at some abstract level. Rather than attempt such simplification, it would seem more reasonable to carefully examine these differences. From an ultimate ecological perspective, it could be argued that each sensory modality contributes uniquely and hence must be considered separately. There is a convincing and productive history of examining differences to discern general principles in biology; this is most striking in comparative neuroanatomy, anthropology, and zoology.

A second common oversimplification has been an attempt, especially by neurobiologists conducting invasive experiments, to explain sensory contributions to motor control generally on the basis of just a limited set of receptor systems. Perhaps there are more papers in the last three decades offering explanations of motor control based on muscle spindle functions than papers addressing any other kind of sensory contribution. It wasn't until 10 years ago that there was even a serious formal attempt to incorporate inputs from golgi tendon organs with muscle spindles (cf. Houk, 1978; Houk & Rymer, 1981). A great deal of more recent work, especially in the last 5 years, documents the critical utilization of other sensory inputs, especially cutaneous mechanoreceptor influences on motor control of precision grip (cf. Abbs & Cole, 1987; Johansson & Westling, 1984; 1987; Winstein et al., 1988) and in movements of the lips for speech (Abbs & Gracco, 1984; Gracco & Abbs, 1985). Further, the subordination of muscle afferents to other inputs, depending on the motor system and movement task, (e.g., the dominating influence of vestibular inputs over neck muscle afferents in control of gaze) documents the limitation of such singular orientations.

There are several alternate perspectives to total anarchy in the manner that different sensory inputs function across the motor systems of the body. Some researchers, for example, have attempted to view different forms of sensory information as being utilized differently, but in a manner dependent on motor task and muscle group. In this vein, it has long been popular to consider certain sensory inputs as providing information during the phases of learning, particularly those that are accessible for self-monitoring, such as vision (cf. Fleishman & Rich, 1963; Mulder & Hulstijn, 1984; Rack, 1981). Likewise, other sensory inputs, such as muscle afferents, have been considered to be more involved in moment-to-moment control. Although these kinds of suspected differences have always been provocative, experimental evidence has not been optimal. Systematic consideration of these differences would appear to provide the basis for interesting experiments and more detailed hypotheses on sensory-based contributions to motor control in general.

## Variations with Muscle Groups and Motor Task

A parallel restriction to that of focusing too specifically on one or two sensory inputs is that of overreliance on data from movements of one or two motor systems (i.e., elbow rotation, forearm rotation, and so forth) or a parallel overemphasis on certain motor tasks (pointing or reaching). In certain respects, it could be argued, albeit at an empirical extreme, that there is no probability of devising a general model of motor control, that is, one that is not specific to particular muscles, movements, and tasks involved. It is clear from basic neuroanatomy (irrespective of physiology) that the connectivity of sensory inputs from the different muscle groups, their access to motor cortex pyramidal cells, and so forth varies substantially across the motor systems of the body. Similar magnitudes and degrees of variation are apparent across muscle groups and motor tasks in terms of movement velocities, required movement precision, and so on. If one considers that these muscle groups vary also in their biomechanical properties (i.e., neither the lips nor the eyes have significant inertia) and the character of the muscle contraction processes (eye and lip muscles have contraction times that are less than 20% of those for the elbow rotation muscles), the need for system-specific models and theories becomes even more apparent. Despite these substantial and functionally significant variations, many models of motor control appear to be aimed at explaining all manner of motor systems and behaviors on the one hand but fail to incorporate the muscle group-specific and task dependent sensory inputs on the other. An obvious solution to this problem is to assume that these different forms of sensory input are processed in the same manner as the more common ones. However, even at the more peripheral levels there are notable and functionally significant motor system differences in the processing of relatively similar inputs (cutaneous mechanoreceptor or muscle spindle) to motoneurons or to primary motor cortex pyramidal cells. Likewise, latencies of these various inputs differ in functionally significant ways.

To illustrate the degree of this problem further, it is useful to evaluate how a specific model might function for a range of movements and motor tasks. An almost archetypic model is the one in which the primary sources of movement information are muscle spindles, golgi tendon organs, and perhaps, joint receptors (a common and seemingly acceptable premise). The goal is then to consider how that model would predict sensory contributions to the control of say head–eye movements, lip movements, and elbow rotation, with position control as the primary task variable. Not surprisingly, given the wealth of research on the proximal upper limbs, the experimental literature suggests that elbow rotation may indeed be influenced in certain ways, particularly by muscle spindles. However, the lip muscles, which clearly appear to be influenced by some sensory input of position (cf. Abbs & Gracco, 1984) are lacking spindles and joint receptors entirely; control of lip position appears to depend rather on deformation of the facial skin by muscle contraction so as to activate cutaneous mechanoreceptors (cf. Johansson, Trulsson, Olsson, & Abbs, 1988). Similarly, although head–eye position control is influenced by inputs from neck muscle afferents, the predominant and very powerful input, particularly to eye movements, is from the vestibular system, at least for much on-line control (cf. Dichgans, Bizzi, Morasso, & Tagliasco, 1974). These are not necessarily special cases; rather, for most motor systems and tasks other than the proximal upper limb movement to targets in free space, and especially the classic elbow rotation, available models provide limited insights.

Obviously, the contribution of sensory inputs must be considered to vary with motor task, motor system, and sensory modality. This position could be viewed as advocating that the only safe position is to not make any generalizations regarding utilization of sensory input without detailed empirical verification. Although this is a caution to be considered, it is not the critical point of this particular commentary. Indeed, general hypotheses may be forthcoming quite readily. As an example, Abbs and Cole (1987) recently suggested that the differing importance of various sensory modalities is task related. In some experiments on rapid hand movements it was observed that subjects chose naturally to generate a rapid precision grip with finger and thumb contact at the distal pulpal surfaces, despite receiving no instructions to do so and independent of any consistent spatial reference. Specifically, the joint positions of the finger and thumb were variable across multiple trials, as, by necessity, were muscle lengths. On this basis, Cole and Abbs (1986) argued that distal somatic sensory surface contact on the digits was the overall basis for the organization of this behavior, especially given that other forms of sensory input and motor output were variable across trials. These authors argued further:

> The planning of motor goals in terms of sensory consequences is fundamentally pragmatic and task specific and would appear to have prerequisites for planning as well as learning of almost all voluntary movements. To this point, for speech the planned motor goals are not absolute positions of the tongue, jaw, or lips or even stereotypic shapes of the oral-pharyngeal cavity, but particular acoustic patterns that correspond to auditory–perceptual categories of human speakers and listeners. Similarly, in movements of the head and the eyes, the significant goal is reception of light on a sensitive region of the retina in the same way as movement of the head and pinna in auditory

localization are aimed at the reception of a difference in sound between the ears" (p. 1420).

From this theoretical perspective, for eye–head movements and speech movements, vision and audition, respectively, thus are argued to represent the sensory consequences of successful motor task achievement, despite their lack of contribution to moment-to-moment control. As such, this hypothesis separates the utilization of sensory input on the basis of tasks and with regard to the consequences of successful motor goal achievement, respectively. That is, for speech and head–eye movement, vision and audition are argued to play comparable roles in monitoring motor goal achievement, in guiding task learning, and as a means for motor goal representation. In certain respects, these suggestions are related to the recognition schema suggested by Schmidt (1975) and the improved estimates of accuracy observed with motor task experience.

Although such hypothesized functional variations in the utilization of various sensory inputs require additional experimental support, at the least, these hypotheses are a step beyond the unqualified generalizations that have characterized some of the literature in the past. Unfortunately, depending on the choice of motor system and sensory input, it is possible, quite inadvertently to find support (or lack thereof) for the full range of sensory contributions to motor output. Specifically, if one were to base an assessment on the role of visual input in the on-line control of eye movements, the general conclusion would be very different than if vision were assessed based on its similar contributions to postural equilibrium. These kinds of conclusions seem best drawn from experimental evaluation in specific cases, with attempts at generalization aimed at principles that are documented to transcend muscle groups and motor tasks. One could argue that these kinds of developments are in the true spirit of an ecological approach to motor control, one that recognizes differences and identifies associated neurobiological and behavior adaptations.

## A CELEBRATION OF DIFFERENCES

In the spirit of a more rigorous yet natural approach to sensorimotor control, it is apparent that the past tendency for broad generalization needs to be balanced by a more systematic consideration of detailed differences and similarities among muscle groups and motor systems. Many general principles that need to be addressed are apparent. For example, it appears increasingly that in those motor systems or tasks requiring interfaces between the organism and the environment, cutaneous mechanoreceptors play a prominent role in movement and force control (e.g., for actions of the digits, lips, tongue). These functional differences are paralleled by physiological and neuroanatomical correlates, especially in regard to cerebral representations of somatic sensory projections. Similarly, based on observations in humans and nonhuman primates, it is apparent that long latency reflex actions are more prominent in muscle groups receiving direct (monosynaptic) projections from the primary motor cortex; these latter muscle groups, we think not coincidentally, are the ones involved in some of the most precise motor control behaviors. To verify and extend these and like generalizations requires that investigators attempt to conduct studies that involve comparisons among muscle groups and for different tasks; such studies are a necessary prerequisite to general models wherein common attributes among motor systems, or as a function of certain classes of motor task variation, can be addressed with attendant experimental support.

The default alternative to such comparative approaches has been to analyze a verified subcomponent of a movement in substantial detail and on that basis develop a grand model. These attempts have primarily served, from the standpoint of the present authors, to highlight the fact that the whole of an everyday multi-joint movement, with an organism-significant goal, is much more than simply the sum of its single-joint parts.

## ACKNOWLEDGMENTS

The authors express thanks to Geraldine Torti for her careful assistance in manuscript preparation. This work was supported in part by National Institutes of Health Grants NS-16373, NS-13274, and HD-03352.

## REFERENCES

Abbs, J. H. (1982). A speech motor system perspective on nervous system control variables. *Behavioral and Brain Sciences*, *5*, 541–542.

Abbs, J. H., & Cole, K. J. (1982). Consideration of bulbar and suprabulbar afferent influences upon speech motor coordination and programming. In S. Grillner, B. Lindblom, J. Lubker, & A. Persson (Eds.), *Speech motor control* (pp. 159–186). Oxford: Pergamon Press.

Abbs, J. H., & Cole, K. J. (1987). Neural mechanisms of motor equivalence and goal achievement. In S. Wise & E. Evarts (Eds.), *Higher brain functions: Recent explorations of the brain's emergent properties* (pp. 15–43). New York: Wiley.

Abbs, J. H., & Connor, N. P. (1989). Motor coordination for functional human behaviors: Perspectives from a speech motor data base. In S. Wallace (Ed.), *Perspectives on the coordination of movement* (pp. 157–183). Amsterdam: North-Holland.

Abbs, J. H., & Gracco, V. L. (1983). Sensorimotor actions in the control of multi-movement speech gestures. *Trends in Neuroscience*, *6*, 391–395.

Abbs, J. H., & Gracco, V. L. (1984). Control of complex motor gestures: Orofacial muscle responses to load perturbations of lip during speech. *Journal of Neurophysiology*, *51*(4), 705–723.

Abbs, J. H., Gracco, V. L., & Cole, K. J. (1984). Control of multi-joint movement coordination: Sensorimotor mechanisms in speech motor programming. *Journal of Motor Behavior*, *16*, 195–231.

Abbs, J. H., Muller, E. M., Hassul, M., & Netsell, R. (1977, November). *A systems analysis of possible afferent contributions to lip movement control.* Paper presented at the meeting of the American Speech and Hearing Association, Chicago.

Adams, J. A. (1971). A closed-loop theory of motor learning. *Journal of Motor Behavior*, *3*, 111–149.

Arbib, M. A. (1981). Perceptual structures and distributed motor control. In V. B. Brooks (Ed.), *Handbook of physiology, Vol. II, Motor control, Part 2* (pp. 1449–1480). Bethesda, MD: American Physiological Society.

Armstrong, T. R. (1970). *Training for the production of memorized movement patterns.* (Tech. Rep. No. 26). Ann Arbor: University of Michigan, Human Performance Center.

Bässler, U. (1986). On the definition of central pattern generator and its sensory control. *Biological Cybernetics*, *54*, 65–69.

Bernstein, N. (1967). *The co-ordination and regulation of movements.* Oxford, England: Pergamon Press.

Bilodeau, I. M. (1966). Information feedback. In E. A. Bilodeau (Ed.), *Acquisition of skill* (pp. 255–296). New York: Academic Press.

Birkenblit, M. B., Feldman, A. G., & Fukson, O. I. (1986). Adaptability of innate motor patterns and motor control mechanisms. *Behavioral and Brain Sciences*, *9*, 585–638.

Bridgeman, B., Kirch, M., & Sperling, A. (1981). Segregation of cognitive and motor aspects of visual information using induced motion. *Perception and Psychophysics*, *29*, 336–342.

Caligiuri, M. P., & Abbs, J. H. (1984). The functional significance of the perioral reflex in voluntary motor control of the lips. *Society for Neuroscience Abstracts*, *10*, 331.

Chapin, J. K., & Woodward, D. J. (1981). Modulation of sensory responsiveness of single somatosensory cortical cells during movement and arousal behaviors. *Experimental Neurology*, *72*, 164–178.

Cole, K. J., & Abbs, J. H. (1983). Intentional responses to kinesthetic stimuli in orofacial muscles: Implications for the coordination of speech movements. *Journal of Neuroscience*, *3*, 2660–2669.

Cole, K. J., & Abbs, J. H. (1986). Coordination of three-joint digit movements for rapid finger–thumb grasp. *Journal of Neurophysiology*, *55*(6), 1407–1423.

Cole, K. J., & Abbs, J. H. (1987). Kinematic and electromyographic responses to perturbation of a rapid grasp. *Journal of Neurophysiology*, *57*(5), 1498–1510.

Cordo, P. J., & Nashner, L. M. (1982). Properties of postural adjustments associated with rapid arm movements. *Journal of Neurophysiology*, *47*, 287–302.

Crago, P. E., Houk, J. C., & Hasan, Z. (1976). Regulatory actions of the human stretch reflex. *Journal of Neurophysiology*, *39*, 925–935.

Cruse, H., Dean, J., Heuer, H., & Schmidt, R. A. (in press). Information for motor control. In O. Neumann & W. Prinz (Eds.), *Relations between perception and action: Current approaches.* Berlin: Springer.

DePaul, R., Nadler, R., Ishii, T., & Johnson, C. (1988, Spring). Wisconsin x-ray microbeam system: Subject variables and experimental procedures. *Journal of the Acoustical Society of America*, *83*, (Suppl. 1), S112.

Dichgans, J., Bizzi, E., Morasso, P., & Tagliasco, V. (1974). The role of vestibular and neck afferents during eye–head coordination in the monkey. *Brain Research*, *71*, 225–232.

Fairbanks, G. (1954). Systematic research in experimental phonetics: 1. A theory of the speech mechanism as a servosystem. *Journal of Speech and Hearing Disorders*, *19*, 133–139.

Fleishman, E. A., & Rich, S. (1963). Role of kinesthetic and spatial-visual abilities in perceptual motor skill learning. *Journal of Experimental Psychology*, 66, 6–11.

Gentner, D. R. (1987). Timing of skilled motor performance: Tests of the proportional duration model. *Psychological Review*, *94*(2), 255–276.

Gentner, D. R., Grudin, J., & Conway, E. (1980). *Finger movements in transcription typing*. LaJolla, CA: Center for Human Information Processing (Report No. 8001).

Georgopoulos, A. P., Kalaska, J. F., & Massey, J. T. (1981). Spatial trajectories and reaction times of aimed movement: Effect of practice, uncertainty, and change in target location. *Journal of Neurophysiology*, *46*, 725–743.

Ghez, C. & Pisa, M. (1972). Inhibition of afferent transmission in cuneate nucleus during voluntary movement in the cat. *Brain Research*, *40*, 145–151.

Glencross, D. J. (1977). Control of skilled movements. *Psychological Bulletin*, *84*(1), 14–29.

Goodale, M. A., Pelisson, D., & Prablanc, C. (1986, April). Large adjustments in visually guided reaching do not depend on vision of the hand or perception of target displacement. *Nature*, *320*, 748–750.

Gordon, J., Bermejo, R., & Ghez, C. (1988). Updating of isometric force trajectories: Corrections for unexpected changes in target amplitude. *Society for Neuroscience Abstracts*, *14*, 467.

Gracco, V. L., & Abbs, J. H. (1985). Dynamic control of the perioral system during speech: Kinematic analyses of autogenic and nonautogenic sensorimotor processes. *Journal of Neurophysiology*, *54*, 418–432.

Gracco, V. L. & Abbs, J. H. (1986). Variant and invariant characteristics of speech movements. *Experimental Brain Research*, *65*, 156–166.

Granit, R. (1955). *Receptors and sensory perception*. New Haven: Yale University Press.

Grillner, S., & Rossignol, S. (1978). On the initiation of the swing phase of locomotion in chronic spinal cats. *Brain Research*, *146*, 269–277.

Grimm, R. J., & Nashner, L. M. (1978). Long loop dyscontrol. In J. E. Desmedt (Ed.), *Progress in neurophysiology Vol. 4: Cerebral motor control in man: Long loop mechanisms* (pp. 70–84). Basel: S. Karger.

Guyton, A. C., Crowell, J. W., & Moore, J. W. (1956). Basic oscillating mechanism of Cheyne-Stokes breathing. *American Journal of Physiology*, *187*, 395–398.

Henatsch, H. O. (1967). Instability of the proprioceptive length servo: Its possible role in tremor phenomena. In M. D. Yahr & D. P. Purpura (Eds.), *Neurophysiological basis of normal and abnormal motor activities* (pp. 75–90). Hewlett, NY: Raven Press.

Hansen, R. M., & Skavenski, A. A. (1977). Accuracy of eye position information for motor control. *Vision Research*, *17*, 919–926.

Henry, F. M., & Rogers, D. E. (1960). Increased response latency for complicated movements and a "memory drum" theory of neuromotor reaction. *Research Quarterly*, *31*, 448–458.

Hollerbach, J. M. (1981). As oscillation theory of handwriting. *Biological Cybernetics*, *39*, 139–156.

Houk, J. C. (1978). Participation of reflex mechanisms and reaction-time processes in the compensatory adjustments to mechanical disturbances. In J. E. Desmedt (Ed.), *Cerebral motor control in man: Long loop mechanisms. Progress in Clinical Neurophysiology*, *4* (pp. 193–215). Basel: Karger.

Houk, J. C., & Rymer, W. Z. (1981). Neural control of muscle length and tension. In V. B. Brooks (Ed.), *Handbook of physiology: Sec. 1: Vol. 2. motor control* (pp. 257–323). Bethesda, MD: American Physiological Society.

Hulliger, M., Nordh, E., Thelin, A. E., & Vallbo, A. B. (1979). The responses of afferent fibers from the flabrous skin of the hand during voluntary finger movements in man. *Journal of Physiology*, *291*, 233–249.

James, E. (1890). *The Principles of Psychology*, *Vol. 1*. New York: Holt.

Johansson, R. S., Trulsson, M., Olsson, K. Å., & Abbs, J. H. (1988). Mechanoreceptive afferent activity in the infraorbital nerve in man during speech and chewing movements. *Experimental Brain Research, 72*, 209–214.

Johansson, R. S., & Westling, G. (1984). Roles of glabrous skin receptors and sensorimotor memory in automatic control of precision grip when lifting rougher or more slippery objects. *Experimental Brain Research, 56*, 550–564.

Johansson, R. S., & Westling, G. (1987). Signals in tactile afferents from the fingers eliciting adaptive motor responses during precision grip. *Experimental Brain Research, 66*(1), 141–154.

Keele, S. W. (1968). Movement control in skilled motor performance. *Psychological Bulletin, 70*, 387–403.

Keele, S. W. (1973). Movement control. In *Attention and human performance* (pp. 111–134). Pacific Palisades, CA: Goodyear Publishing.

Keele, S. W., & Posner, M. I. (1968). Processing of visual feedback in rapid movements. *Journal of Experimental Psychology, 77*, 155–158.

Keele, S. W., & Summers, J. J. (1976). The structure of motor programs. In G. E. Stelmach (Ed.), *Motor control: Issues and trends* (pp. 109–142). New York: Academic Press.

Kelso, J. A. S., & Stelmach, G. E. (1976). Central and peripheral mechanisms in motor control. In G. E. Stelmach (Ed.), *Motor control: Issues and trends* (pp. 1–40). New York: Academic Press.

Krendel, E. S., & McRuer, D. T. (1960). A servomechanisms approach to skill development. *Journal of the Franklin Institute, 269*, 24.

Lansing, R. W., & Meyerink, L. (1982). Load compensating responses of human abdominal muscles. *Journal of Physiology, 320*, 253–268.

Larson, C. R., Folkins, M., McClean, M., & Muller, E. (1978). Sensitivity of the human perioral reflex to parameters of mechanical stretch. *Brain Research, 146*, 159–164.

Lashley, K. S. (1951). The problem of serial order in behavior. In L. A. Jeffress (Ed.), *Cerebral mechanisms in behavior (the Hixon Symposium)* (pp. 112–137). New York: Wiley.

Lee, D. N., & Aronson, E. (1974). Visual proprioceptive control of standing in human infants. *Perception and Psychophysics, 15*, 527–532.

Lenz, F. A., Tatton, W. G., & Tasker, R. R. (1983a). The effect of cortical lesions on the electromyographic response to joint displacement in the squirrel monkey forelimb. *Journal of Neuroscience, 3*(4), 795–805.

Lenz, F. A., Tatton, W. G., & Tasker, R. R. (1983b). Electromyographic response to displacement of different forelimb joints in the squirrel monkey. *Journal of Neuroscience, 3*(4), 783–794.

Lund, J. P., & Rossignol, S. (1981). Modulation of the amplitude of the digastric jaw opening reflex during the masticatory reflex. *Neuroscience, 6*, 95–98.

Lund, J. P., Rossignol, S., & Murakami, T. (1981). Interactions between the jaw-opening reflex and mastication. *Canadian Journal of Physiology & Pharmacology, 59*, 683–690.

MacKay, W. A. (1985). The motor program: Back to the computer. In E. V. Evarts, S. P. Wise, & D. Bousfield (Eds.), *The motor system in neurobiology* (pp. 101–105). Amsterdam: Elsevier.

Marsden, C. D., Merton, P. A., & Morton, H. B. (1981). Human postural responses. *Brain, 104*, 513–534.

Marsden, C. D., Merton, P. A., & Morton, H. B. (1983). Long-latency automatic responses to muscle stretch in man: Origin and function. In J. E. Desmedt (Ed.), *Motor control mechanisms in health and disease* (pp. 509–539). New York: Raven Press.

Marteniuk, R. G., Mackenzie, C. L., & Leavitt, J. L. (1988). Representational and physical accounts of motor control and learning: Can they account for the data? In A. M. Colley & J. R. Beech (Eds.), *Cognition and action in skilled behavior* (pp. 173–190). Amsterdam: North-Holland.

Merton, P. A. (1953). Speculations on the servo-control of movement. In G. W. Wolstenholmme (Ed.), *The spinal cord* (pp. 183–198). Boston: Little, Brown.

Merton, P. A. (1972). How we control the contractions of our muscles. *Scientific American, 226*, 30–37.

Meyer-Lohmann, J., Christakos, C. N., & Wolf, H. (1986). Dominance of the short-latency component in perturbation induced electromyographic responses of long-trained monkeys. *Experimental Brain Research, 64*, 393–399.

Miller, G. A., Galanter, E., & Pribram, K. H. (1960). *Plans and the structure of behavior.* New York: Holt, Rinehart & Winston.

Morasso, P. (1981). Spatial control of arm movements. *Experimental Brain Research, 42*, 223–227.

Moray, N. (1981). Feedback and the control of skilled behavior. In D. Holding (Ed.), *Human skills* (pp. 15–39). New York: Wiley.

Mulder, T., & Hulstijn, W. (1985). Sensory feedback in the learning of a novel motor task. *Journal of Motor Behavior, 17*, 13–24.

Muller, E. M., Milenkovic, P. H., & MacLeod, G. E. (1985). Perioral tissue mechanics during speech production. In J. Eisenfeld & C. DeLisi (Eds.), *Mathematics and computers in biomedical applications* (pp. 363–372). Amsterdam: North-Holland.

Nashner, L. M. (1976). Adapting reflexes controlling the human posture. *Experimental Brain Research, 26*, 59–72.

Netsell, R., & Abbs, J. H. (1975, November). *Modulation of perioral reflex sensitivity during speech movements.* Paper presented at the meeting of the Acoustical Society of America, San Francisco.

Pew, R. W. (1984). A distributed processing view of human motor control. In W. Prinz & A. F. Sanders (Eds.), *Cognition and motor processes* (pp. 19–27). Berlin: Springer-Verlag.

Polit, A., & Bizzi, E. (1979). Characteristics of motor programs underlying arm movements in monkeys. *Journal of Neurophysiology, 42*(1), 183–194.

Poppele, R., & Bowman, R. J. (1970). Quantitative description of linear behavior of mammalian muscle spindles. *Journal of neurophysiology, 33*, 59–72.

Poulton, E. C. (1974). *Tracking skill and manual control.* New York: Academic Press.

Prochazka, A., Hulliger, M., Zangger, P., & Appenteng, K. (1985). "Fusimotor set": New evidence for α-independent control of motoneurons during movement in the awake cat. *Brain Research, 339*, 136–140.

Rack, P. M. (1981). Limitations of somatosensory feedback in control of posture and movement. In V. B. Brooks (Ed.), *Handbook of physiology section 1: The nervous system, motor control, Vol. 1* (pp. 229–256). Bethesda, MD: American Psychological Society.

Raibert, M. H. (1977). *Motor control and learning by the state-space model* (Tech. Rep.). MIT: Artificial Intelligence Laboratory.

Requin, J., Semjen, A., & Bonnet, M. (1984). Bernstein's purposeful brain. In H. T. A. Whiting (Ed.), *Human motor actions: Bernstein reassessed* (pp. 467–504). Amsterdam: North-Holland.

Schmidt, R. A. (1975). A schema theory of discrete motor skill learning. *Psychological Review, 82*, 225–260.

Schmidt, R. A. (1980). On the theoretical status of time in motor program representations. In G. E. Stelmach & J. Requin (Eds.), *Tutorials in motor behavior* (pp. 145–166). Amsterdam: North-Holland.

Schmidt, R. A. (1982). More on motor programs. In J. A. S. Kelso (Ed.), *Human motor behavior: An introduction* (pp. 189–217). Hillsdale, NJ: Lawrence Erlbaum Associates.

Schmidt, R. A. (1983). On the underlying structure of well-learned motor responses: A discussion of Namikas and Schneider and Fisk. In R. A. Magill (Ed.), *Memory and control of action* (pp. 145–165). Amsterdam: North-Holland.

Simon, J. R., & Smith, K. U. (1956). Theory and analysis of component errors in aided pursuit tracking in relation to target speed and aided-tracking time constant. *Journal of Applied Psychology, 40*, 367–370.

Smith, J. L., Bradley, N. S., Carter, M. C., Giuliani, C. A., Hoy, M. G., Koshland, G. F., & Zernicke, R. F. (1987). Rhythmical movements of the hindlimbs in spinal cat: Considerations for a controlling network. In M. Golberger, A. Gorio, & M. Murray (Eds.), *Development and plasticity of the mammalian spinal cord* (pp. 347–361). Padova: Liviana Press.

Smith, J. L., & Zernicke, R. F. (1987). Predictions for neural control based on limb dynamics. *Trends in Neuroscience, 10*(3), 123–128.

Stark, L. (Ed.). (1968). *Neurological control systems: Studies in bioengineering.* New York: Plenum.

Stark, L., & Cornsweet, T. N. (1958). Testing a servoanalytic hypothesis for pupil oscillations. *Science, 127*, 588.

Stark, L., & Sherman, P. M. (1957). A servoanalytic study of consensual pupil reflex to light. *Journal of Neurophysiology, 20*, 17–26.

Stein, R. B. (1982). What muscle variable(s) does the nervous system control in limb movements? *Behavioral and Brain Sciences, 5*, 535–577.

Summers, J. J. (1981). Motor programs. In D. Holding (Ed.), *Human skills* (pp. 41–64). New York: Wiley.

Terzuolo, C. A., & Viviani, P. (1979). The control representation of learned motor patterns. In R. E. Talbott & D. R. Humphrey (Eds.), *Posture and movement* (pp. 133–121). New York: Raven Press.

Tuller, B., Kelso, J. A. S., & Harris, K. S. (1982). Interarticulator phasing as an index of temporal regularity in speech. *Journal of Experimental Psychology: Human Perception and Performance, 8*, 460–472.

Vallbo, A. B. (1973). Muscle spindle afferent discharge from resting and contracting muscles in normal human subjects. In J. E. Desmedt (Ed.), *New developments in electromyography and clinical neurophysiology, Vol. 3: Human reflexes, pathophysiology of motor systems, methodology of human reflexes* (pp. 251–262). Basel: S. Karger.

Wadman, W. J., van der Gon, D., Geuze, J. J., & Mol, C. R. (1979). Control of fast goal-directed arm movements. *Journal of Human Movement Studies, 5*, 3–17.

Wetzel, M. C. (1986). Operant conditioning in motor and neural integration. *Neuroscience & Behavioral Reviews, 10*, 387–429.

Winstein, C. J., Abbs, J. H., & Petashnick, D. E. (1988). Effect of instructions on compensatory adjustments during precision grip. *Society for Neuroscience Abstracts, 14*, 1234.

Winter, D. A. (1984). Kinematic and kinetic patterns in human gait: Variability and compensating effects. *Human Movement Science, 3*, 51–76.

Wolpaw, J. R., O'Keefe, J. A., Noonan, P. A., & Sanders, M. G. (1986). Adaptive plasticity in primate spinal stretch reflex: Persistence. *Journal of Neurophysiology, 55*(2), 272–279.

Wolpaw, J. R. (1987). Operant conditioning of primate spinal reflexes: The H-reflex. *Journal of Neurophysiology, 57*(2), 443–459.

# 24 Gaze Saccade Orienting and Hand Pointing are Locked to Their Goal by Quick Internal Loops

Claude Prablanc
Denis Pelisson
*Laboratoire de Neuropsychologie Expérimentale*
*Bron, France*

ABSTRACT

It has been commonly accepted that gaze saccade orienting toward a visual target and hand pointing toward the same target are two fundamentally different processes based on specific organization rules. Orientation of gaze is considered to be an open-loop system in which the vestibular signal, through the vestibulo-ocular reflex, mathematically adds to the saccadic command when the head is free to move. By contrast, hand movements have been thought to operate under two sequential modes of control, with a first pre-programmed command distributing the synergies among muscles, followed by a closed-loop process under visual control using the simultaneous vision of hand and target as an error signal.

Two main experiments have been performed, which reveal basic properties common to both systems. It is first shown through passive motion of the head that a gaze saccade is not an open loop process, and that it is very likely controlled by a central feedback loop. In hand pointing, random artificial errors are generated at the onset of the response, by changing the target location unknown to the subject; these errors are automatically corrected by smooth and early modifications, without adding any additional reaction time. These results suggest that continuous internal feedback loops correct unconsciously the dynamic errors throughout the movement.

To understand how we guide quick orientating actions in space and how the different motor systems of eye, head, and limbs are coordinated with the sensory systems became a challenge well before the beginning of this century. A number of ideas concerning sensorimotor organization have developed from Sherrington's ideas on the organization of complex actions seen as a succession of reflexes and Helmoltz intuition of sense of effort, but few attempts have been made to reach a synthesis between peripheralist and centralist theories. On one hand, the inflow theories state that an organized movement needs feedback from the peripheral senses (tendinous, fusorial, cutaneous, skeletamotor, retinal, and so on) in order to be performed correctly (Adams, 1977). On the other hand, outflow theories have put less emphasis on sensory signals and argued that preprogramming of preset patterns of muscle innervation can account for the accuracy of single and multijoint movements (Feldman, 1966; Polit & Bizzi, 1979; Taub, Goldberg, & Taub, 1975). These outflow theories were strengthened when Von Holst (1950) introduced the notion of efference copy, that is, a central and accurate copy of the command sent to the effectors, which was supposed to feed in the sensory brain structures receiving their direct input from receptors mechanically linked to the above effectors. Associating with the efference copy, perception became the cross product of a raw sensory input and of an intended command. Above all, the efference copy allowed the introduction of the notion of feedback at the central level.

The oculomotor system (responsible for orienting the eyes toward a visual target) and the hand-pointing system have been viewed as being too fundamentally different mechanisms. Because the eye has nearly linear biomechanical characteristics with no external load to modify the response to a neural control signal, and because its quick orienting behavior appears highly stereotyped, the saccadic system has been considered as a preprogrammed open-loop system. By contrast, the hand-pointing system with its nonlinear multijoint structure and its varying load has been thought of as hybrid system combining a preprogrammed initial component and a final closed-loop component; the latter has been considered to be under visual control and to be fed by the simultaneous vision of the hand and target for error correction.

The aim of the present chapter is to examine the possibility that despite large anatomical and biomechanical differences, both systems share common properties based on internal feedback control. Although the processing of external feedback such as vision of the hand is relatively slow, the specificity of internal feedback control seems to be autonomous and quick. It appears that the automatic modulations of the ongoing responses are organized in such a way as to instantaneously match the goal. These corrections are carried out without any conscious perception of the manipulations and without any sensation of their occurence. The results suggest that quick internal servo loops control these corrections at low levels.

## EYE–HEAD ORIENTATION

The limited number of degrees of freedom of the eye has made the oculomotor system a priviledged system in which new concepts have been tested. Within 20 years since the first rudimentary models were built, cooperation among neuroanatomists, neurophysiologists, and engineers has made impressive breakthroughs in which control system models fit the neurophysiological data. The notion of the saccadic system as an open-loop control system based on preprogramming was introduced by Young & Stark (1963). In the sampled data model, the retinal error between the directions of eye and target was supposed to drive an open-loop command to the oculomotor system uninfluenced by retinal input during the whole saccadic response. Later, Robinson showed that this apparent open-loop process was in fact centrally controlled by an internal "local feedback loop" that compares the eye position command (the so-called efference copy of Holst) and the representation of the spatial goal. This study was corroborated by both experimental and clinical observations (Becker & Jürgens, 1979; Hallett & Lighstone, 1976a, 1976b; Mays & Sparks, 1980; Robinson, 1975).

The simultaneous, quick eye and head orientation toward a target has long been considered a simple and elegant mechanism of addition at the neuronal level of a saccade command and of a vestibular reflex input counteracting the physical head displacement (Morasso, Bizzi, & Dichgans, 1973). According to this *additivity hypothesis*, the kinematics of an eye saccade with head fixed should be the same as the kinematics of a gaze saccade accompanied by a simultaneous head orientation. However, some observations in man (Robinson & Zee, 1981) and cat (Guitton, Douglas, & Volle, 1984; Haddad & Robinson, 1977) have cast some doubt on the *additivity hypothesis*. Furthermore, in the last few years, decisive experiments have been conducted that have shown that an inhibition of the vestibulo-ocular reflex (VOR) does occur in primate and man during a saccade (Guitton & Volle, 1987; Haddad & Robinson, 1977; Laurutis & Robinson, 1986; Pélisson & Prablanc, 1986; Tomlinson, & Bahra, 1986). We describe here a quantitative investigation of this mechanism in man. The study raises at least two new questions: What is the trade-off between eye and head kinematics during a simultaneous, quick eye–head orientation toward a visual target? How is the accuracy of the response preserved despite a partial inhibition of the VOR?

If preservation of gaze saccade accuracy under a normal VOR (i.e., with a unity gain) is a trivial problem, keeping the accuracy of a gaze response under a partial inhibition of the VOR is difficult to imagine either with an open-loop oculomotor system, or with a local feedback loop that does not take into account the position of the head but only the position of the eye within the orbit.

The reason for undertaking the present study was the following. If the *additivity hypothesis* is true, then a gaze saccade results from the pure addition of a preprogrammed command with a unity gain VOR. Thus, if a passive motion of the head is produced during the orienting saccade to a target, the gaze kinematics should be the same as the eye kinematics of a similar orienting saccade performed with the head fixed. This invariance of gaze kinematics should be maintained whatever the direction of the head motion, either ipsilateral or contralateral to the direction of the saccade.

Alternatively, if there is a partial VOR inhibition during the saccade and the gaze saccade still results from the pure addition of a preprogrammed saccadic command and from the VOR with its reduced gain, then for a 30° saccade (which duration is about 80 ms) occurring during a constant motion of the head at 200 deg/s, with a VOR turned down to 0.5, a systematic gaze saccade error of about $200 \times 0.5 \times 80 \times 10^{-3} = 8°$ and a change in gaze velocity of $200 \times 0.5 = 100$ deg/sec can be predicted.

If, contrary to these two open-loop hypotheses, the kinematics of the gaze saccade depend on the direction of head motion and the accuracy is always preserved, a central feedback loop carrying gaze direction in space is necessary to explain the control of the eye movements.

## APPARATUS AND METHODS

Subjects were sitting in a darkened room in front of a horizontal semicircular screen whose radius of curvature corresponded to the distance between the screen and the axis of rotation of the head (1.2 m). Peripheral visual targets (red light-emitting diodes) were located at 30° and 20° to the right and to the left of a central target, along a horizontal line of the screen. In one experimental session, a torque-motor was used to drive the subjects' head in a horizontal plane during their saccadic ocular response. A helmet and a bite-board rigidly linked the head to the servo-driven torque-motor. The measured dead zone between head position command and actual head position was less than 1°. The entire experiment was controlled by computer: rotation of the torque-motor controlling head position, random selection of target position, and sampling of eye and head position signals at 1,000 Hz. Head position was derived from a potentiometer and eye position signal from an electro-oculographic (EOG) technique (with an analog filter of 80 Hz cut-off frequency and an automatic counter-drift device). After an off-line linearisation of the EOG signal, absolute gaze position was reconstructed by adding the two signals sampled by the computer: eye position in the orbit and head position.

In the two following experimental sessions every trial consisted of a target step from the initial point to a randomly selected peripheral position, followed by another target step back to the center. Four targets were presented 8 times at 20° and 30° from the initial point on each side, in a random order. Five subjects participated in the experiment. They were instructed to track the target when it jumped from the initial point to a peripheral one, irrespective of the imposed head movement, when it occurred.

In the first experimental session (the control condition), the head was immobilized in a straight ahead position, whereas in the second session, a passive rotation of the head was applied during subjects' saccadic response to the peripheral target. In half the trials of the second session, the head was rotated toward the peripheral target (*target ipsi* condition); in the remaining trials (*target contra* condition) the head was rotated in the direction opposite to the location of the peripheral target. In order to quantify interactions between VOR and saccades, a constant angular head velocity (200 deg/s) had to be maintained.

Head rotation started from a 40° eccentric position (the eye fixating the center and thus starting from a 40° eccentric position in the orbit) and ended either at the symmetrical opposite position for the *target ipsi* condition or at a 10° opposite position for the *target contra* condition. The latter asymmetry allowed subjects to fixate the more peripheral target until the end of the trial by keeping the eye within its physiological range (see Fig. 24.1). To avoid any kind of adaptation, during this second session, the size (20° and 30°) and the direction of target steps (ipsilateral or contralateral) with respect to the direction of head motion were randomly mixed. Care was also taken to avoid effects of fatigue by restricting the total number of trials to 32 in both sessions.

In the *target ipsi* condition, gaze-orienting behavior predicted from the hypothesis of VOR inhibition and gaze feedback loop corresponds to a reduced saccadic duration and an increased gaze velocity with respect to the head fixed behavior. The same hypothesis predicts exactly the opposite effects in the *target contra* condition. Gaze saccade accuracy is expected to be little if at all influenced by conditions.

## RESULTS

The three single responses shown in Fig. 24.1 are representative of the general behavior of the five subjects. From top to bottom are represented the oculomotor responses to a 30° target step in *target ipsi*, *target contra*, and *head-fixed* conditions. In both head-driven trials (*ipsi* and *contra*) slow eye movement compensates for nearly all head movement: It keeps gaze on the central position before the saccadic response and on the peripheral target position after it. The comparison of gaze trajectories between *target ipsi* and *head-fixed* conditions reveals an obvious difference in the shape of gaze shifts. This detail is even more striking on gaze velocity profiles: When head and target move in the same direction, maximum gaze velocity

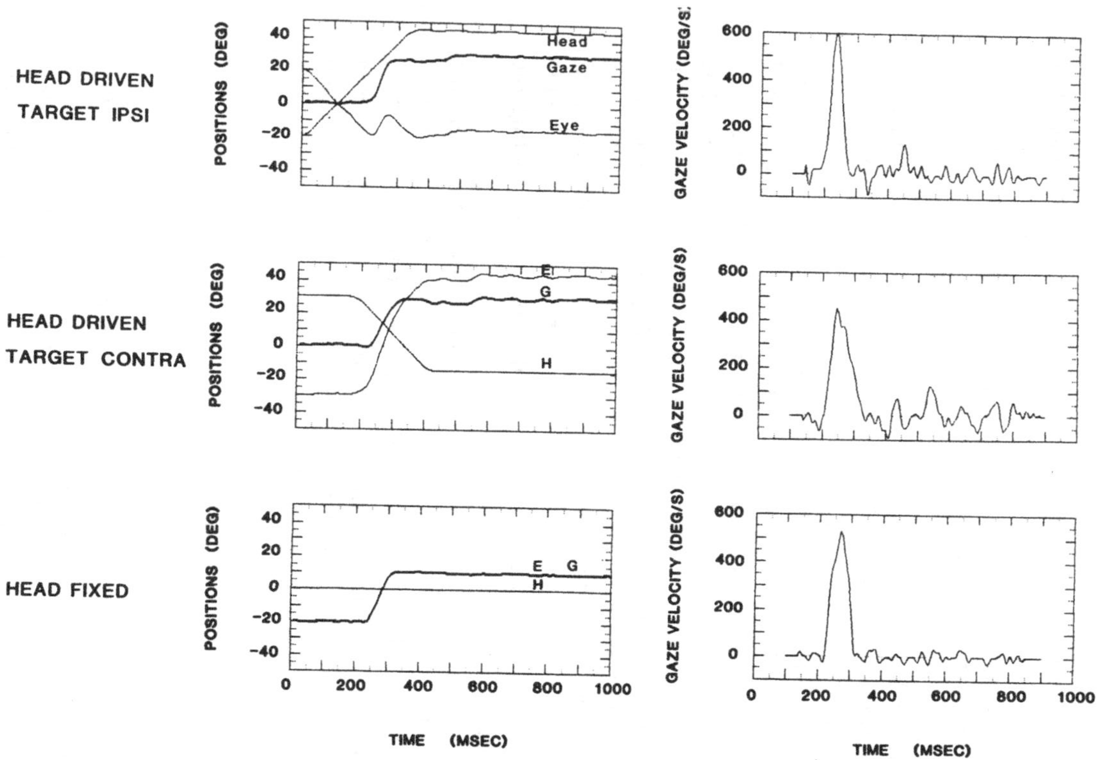


FIG. 24.1. Individual responses from one subject to a 30° target step on the right side, in the three experimental conditions: *target ipsi*, *target contra* and *head fixed* (from top to bottom). Gaze position was computed merely by adding signals of eye position (linearised EOG signal) and of head position, and was differentiated (gaze velocity shown on the right). Eye and gaze signals were numerically filtered with a 30 Hz cut-off frequency. Note the clear increase in dynamics of gaze shift between *target ipsi* and *head fixed* conditions, and their decrease between *target contra* and *head fixed* conditions. For the *head-fixed* condition the initial position of the eye is 20° left within the orbit, which corresponds to the same

is greater and saccade duration is shorter than in *head-fixed* condition. The same held true when comparing the dynamics of gaze shifts (either peak gaze velocity or saccade duration) between *head-fixed* and *target contra* conditions. Because it was difficult to rely on accurate measurement of saccade duration (as seen in *target ipsi* response of Fig. 24.1), maximum gaze velocity seemed to be more reliable than mean gaze velocity and was used to quantify the interaction between VOR and saccadic ocular command. The averaged results of this computation are summarized in Fig. 24.2. Three main points are worth noting:

1. A constant increase of maximum gaze velocity across all subjects and target step sizes ($p < 0.01$) can be observed when the head is rotated toward the target position (*target ipsi* condition) with respect to *head fixed* condition.
2. The same phenomenon, showing a decrease in maximum velocity under *target contra* condition can be observed.
3. The mean amplitudes of gaze saccades under the three conditions do not differ significantly, though under the two head-moving conditions they are more scattered.

The quantitative observations as well as the previous qualitative ones do not support the *additivity hypothesis*. Instead, the increase in the maximum gaze velocity when head and eye move in the same direction, and its decrease when head and saccade move in opposite directions, favors the hypothesis of a VOR inhibition during the saccade. A last observation can be made from Fig. 24.2. Although all subjects seem to exhibit roughly the same behavior with respect to experimental conditions and target eccentricities, there is a large intersubject variation in maximum gaze velocity that suggests that the percentage of VOR inhibition during the saccade varies with subjects.

Despite a VOR inhibition ranging from 15% to 70% (mean individual computed values) final gaze accuracy is not affected by head movement. This is shown in Fig. 24.3 where gaze shift errors are displayed in the three experimental conditions. A large variability of percentage error in both *head driven* conditions is observed, due to both intrasubject and intersubject variabilities. However, the main result is the absence of significant differences between the three means. In particular, no trend in overshooting can be noticed in the *target ipsi* condition, and the preservation of gaze accuracy is efficiently achieved by a slight reduction of saccadic duration (see, for example, Fig. 24.1). Thus, when head and eyes are both directed to the target, the same gaze shift amplitude is executed as when the head is fixed, although gaze velocity is faster in the first case. When eye and head move in opposite directions, the mean amplitude of gaze saccade is

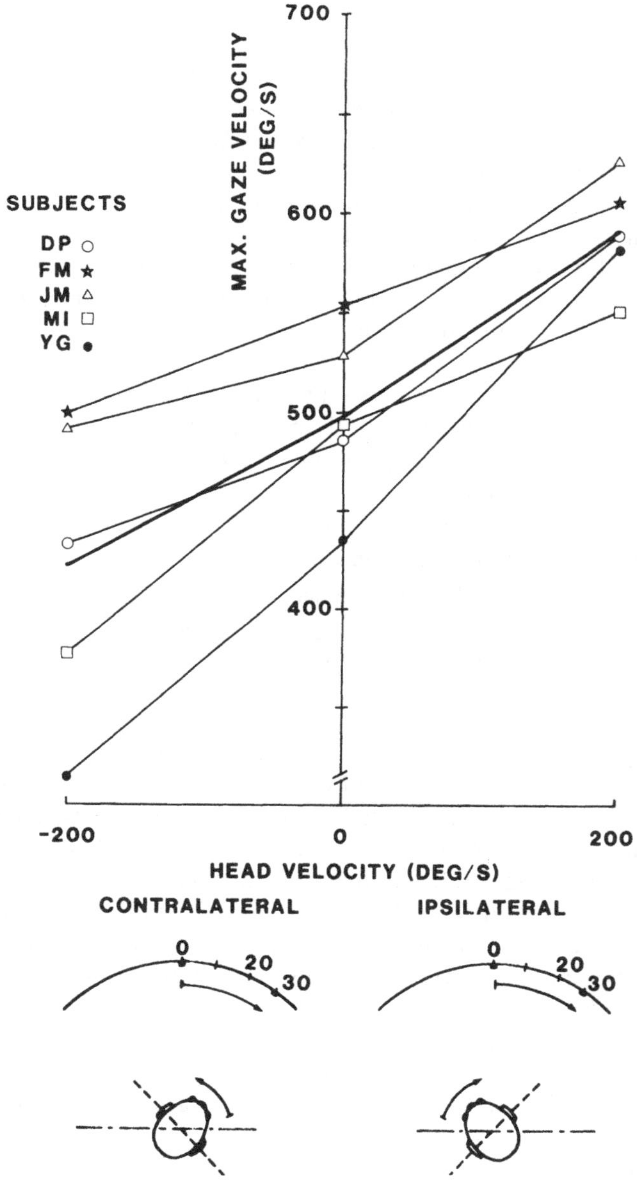


FIG. 24.2. Effect of passive head velocity on maximum gaze velocity. Pooled responses (16 repetitions) to 20° and 30° target steps for each subject and the overall means (thick line) are plotted against head velocity. Negative head velocity (−200 deg/s) indicates that the head is rotated away from the target (*target contra* condition). For each subject, maximum gaze velocity is statistically significantly different ($p < 0.05$ in all cases) when the head is passively rotated as opposed to when it is fixed. Note the linearity of the overall relationship.

unchanged while it is performed with a longer duration and a slower peak velocity.

## DISCUSSION

The observed results showing a modulation of peak gaze velocity, the sign of which corresponds to the sign of the vestibulo-ocular input, allow us to conclude that there is partial VOR inhibition. These findings confirm the conclusions of other studies in man and monkey (Guitton & Volle, 1987; Laurutis & Robinson, 1986; Tomlinson & Bahra, 1986). Our last main conclusion is that a central feedback loop exists that controls the gaze kinematics regardless of the vestibular input. The results preclude the

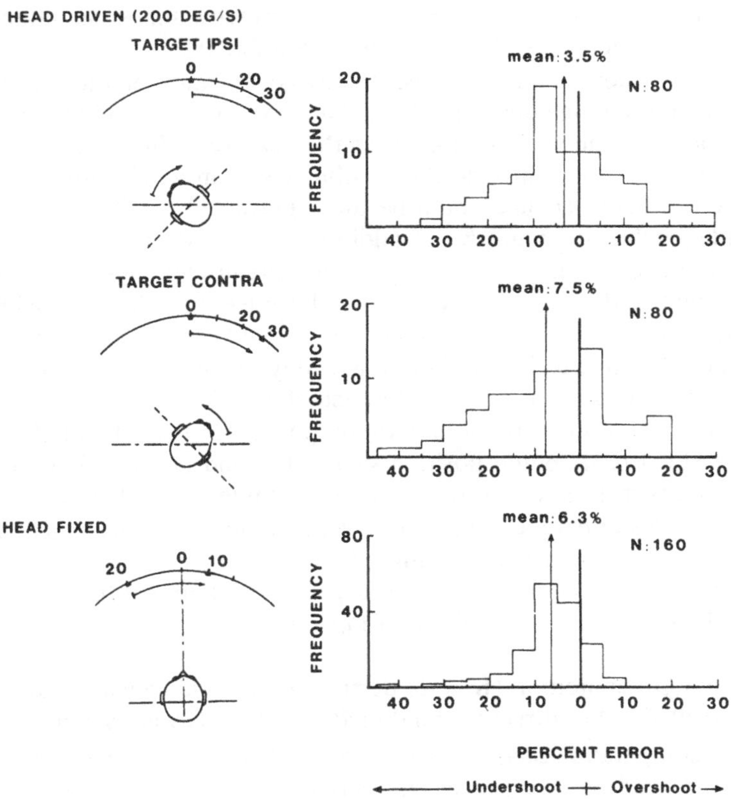


FIG. 24.3. Accuracy of gaze saccadic responses. Percentage error after the gaze main saccade [defined as (gaze saccade amplitude − target step)/ target step], for the pooled responses over the two target steps and the five subjects is depicted for each condition.

preprogramming of eye and head command, because the head was passively moved and the direction of saccade was unpredictable. We present here a model that could account for most discrepant results. Basically, it uses an inhibition that is a function of the initial retinal error prior to the saccadic response, and a feedback loop carrying gaze command.

This model, which is similar to that of Laurutis and Robinson (1986), modifies the concepts developed by Bizzi's group in the *additivity hypothesis* by adding two additional pathways (dashed lines) to explain the present results.

First, we consider the mechanisms suggested by the *additivity hypothesis* illustrated by the solid lines (so that $\hat{G} = E_c =$ estimate of eye position) and further assume that VOR gain $K = 1$. When a step of target position occurs, the peripheral target creates a retinal error ($\varepsilon$) that is sent to both saccadic and head motor systems. The output signals or neural commands ($E_c$ and $H_c$) of the two systems reach the muscles almost simultaneously (Biguer, Jeannerod, & Prablanc, 1982; Warabi, 1977).

When the head is free to move, the two commands lead to the classical eye–head orienting behavior: The global gaze shift is thought to be similar to saccadic eye movement alone with the head restrained, because a negative feedback loop through the vestibular system would provide a head position estimate $\hat{H}$ that would be fully subtracted (as $K = 1$) from the saccadic ocular command $E_c$. In addition, as the external visual feedback loop is opened during the saccade, a widespread hypothesis suggests that an internal feedback loop (the so-called local feedback loop) providing an estimate of eye position (assume $\hat{G} = E_c$) would control saccadic accuracy (Jürgens, Becker, & Kornhuber, 1981; Mays & Sparks, 1980; Robinson, 1975). Although we know that the actual signals might combine at the level of velocities before an integration phase, for the sake of clarity, we have instead represented position signals ($E_c$ and $\hat{H}$). In addition, other mechanisms such as cervico-ocular reflex, optokinetic, and pursuit systems, which to a lesser extent contribute to gaze stabilization during head movements, have voluntarily been omitted.

The results of the present study suggest two additional functional pathways shown by the dashed lines on Fig. 24.4:

1. a VOR inhibition ($K < 1$) during the saccade which could be controlled either directly or indirectly by the saccadic system;
2. a computation of an estimate of absolute gaze position ($\hat{G}$) by adding the head position signal ($\hat{H}$) derived from the vestibular system to the saccadic eye position command ($Ec$).

The suggestion that the phasic inhibition is controlled in some way by the saccadic system is based on the observation of perfect stabilization of

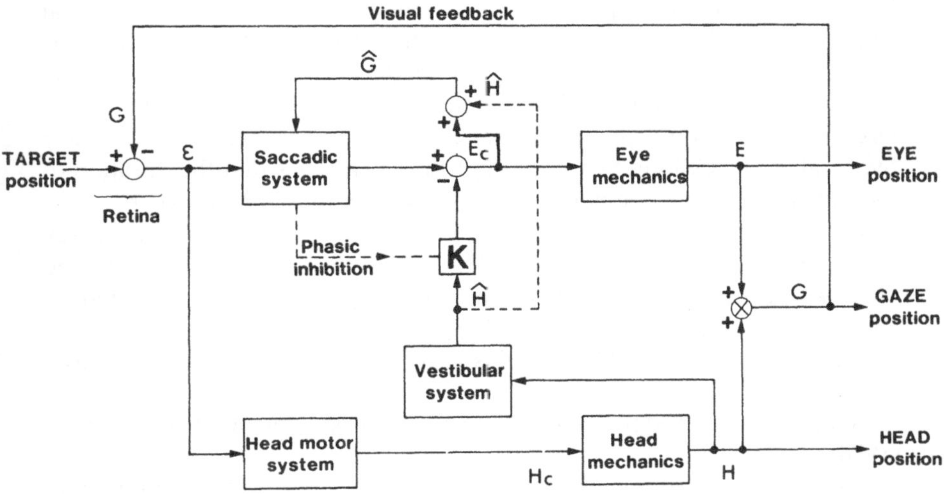


K: VOR gain
ε: retinal error
E,H,G: eye, head and gaze positions
$E_c$, $H_c$: eye and head position commands
$\hat{H}$, $\hat{G}$: estimates of head and gaze position

FIG. 24.4. Simplified model of partial additivity and gaze internal loop hypothesis. For the sake of clarity, the boxes labeled *saccadic system* and *vestibular system* are supposed to contain their own neural integrator and thus to provide position output commands: *Ec* and *H* (see text). The diagram shows clearly the adding junction between saccadic ocular command (*Ec*) and head position estimated ($\hat{H}$) through the vestibular system assuming a unity VOR gain ($K = 1$); the solid lines representing the full *additivity hypothesis*. Two additional pathways (dashed lines) are added to account for the present results. The first is an inhibition of VOR ($K < 1$) that is phasically triggered by the saccadic system, and the second involves a computation of a gaze position estimate ($\hat{G}$) by adding actual head position ($\hat{H}$) of vestibular origin and saccadic command.

gaze in space achieved by a unity gain VOR before and after the saccade, despite a VOR inhibition during the saccade. This implies good timing between VOR inhibition and saccadic eye movement. The hypothesis that an estimate of absolute gaze position is computed suggests how an additional loop in the model can explain the accuracy of gaze shifts in the *head-driven* conditions. Consider the situation of a head movement in the same direction as the saccadic eye movement: If the eye movement were purely preprogrammed and the duration of phasic activity were the same when head is fixed and during passive rotation, the gaze would overshoot its goal when VOR inhibition is large. A similar degree of overshoot should occur if saccadic duration were controlled by the local feedback loop conveying only eye position command. The absence of a trend in overshooting in the *target ipsi* condition supports neither the preprogramming hypothesis nor the local feedback loop hypothesis but rather suggests that the duration of saccadic activity must be determined by a continuous comparison between target position and actual gaze position command. Thus a local feedback loop carrying gaze position estimate ($\hat{G}$), rather than eye position command ($E_c$), would be more compatible with the data, as already suggested by some authors (Fuller, Maldonado, & Schlag, 1983; Guitton & Volle, 1987; Laurutis & Robinson, 1986; Robinson & Zee, 1981; Roucoux, Guitton, & Crommelinck, 1980).

A more detailed description of this model is given in Pélisson et al. (1988).

The finding that gaze endpoints were much more scattered around target position in the *head driven* condition than in the *head-fixed* condition could be accounted for by the passive rotation of the head. Indeed under natural conditions, a better coding of head position might be accomplished by using additional signals derived from the head rotation command (such as an efferent copy) not available with passive head movements. Further experiments in head-free conditions are required to investigate this issue.

Additional observations can be made in the light of the present results. The properties of the vestibular pause neurons (Baker & Berthoz, 1974; King, Lisberger, & Fuchs, 1976; Pola & Robinson, 1978) and those of gaze burst neurons described by Lestienne, Whittington, and Bizzi (1981), suggest that these neurons might mediate the features of the model represented by dashed lines. They could represent a neural substrate of the variable gain VOR because they carry a vestibular signal that pauses during saccadic eye movements; they could drive the output of the summator between head and eye position ($\hat{H}$ and $Ec$) in the local feedback loop, because their firing rate is related to gaze position. Another indirect argument for a central computation of a gaze position signal comes from the recent observation of cat tectal neurons firing in relation to the error between gaze and target positions (Munoz & Guitton, 1985).

Previous conceptions of the operation of the saccadic system were based on a servoloop whose role was to bring the eye to a given position in the orbit. The above results and those obtained recently by Laurutis and Robinson (1986) and Guitton and Volle (1987) suggest, instead, the existence of an internal representation of the absolute coordinates of the target and of a copy of gaze position in space that are the two ultimate branches of a fast servoloop whose goal is it to saccadically drive the gaze onto the target, whatever the head or body movements.

## EYE–HAND POINTING

The problems of coordinated motor responses considered in the previous section for eye–head orienting are far more complex for the hand motor control, even for a simple act such as reaching a point in space with the finger. First, there is redundancy in the degrees of freedom and nonlinearities, such as variable moments of inertia, and changes in stiffness. Movements are less stereotyped than eye movements and can be carried out, for the same trajectory, with very different velocities.

The first analytical studies on arm control were conducted on a single-joint system; they reached interesting conclusions such as the notion of a control specifying only the final equilibrium point (Feldman, 1966; Bizzi, 1975). Using deafferentation of the whole limb, Polit and Bizzi (1979) showed that a monkey could perform a single-joint response without afferent feedback. However, the generalization of this concept to multijoint movements has failed.

Critical time processing in motor control has been investigated through a series of experiments using isometric control force by Ghez and colleagues in order to measure reaction time to changing targets. It was shown that rapid isometric force responses could be continuously modified in response to changing sensory information and that faulty initial specification of trajectory was corrected by internal feedback (Gordon & Ghez, 1987; Henning, Favilla, & Ghez, 1988; Vicario & Ghez, 1984). However, because the above experiments used isometric forces, they eliminated the biomechanical consequences of force production and were limited to single-joint control. Generalization of the notions of fast internal loops to real multijoints movements, which imply a processing of complex biomechanical interactions, were part of the goal of the following eye–hand study.

The capabilities of reprogramming real movement responses following changes in the visual goal have been extensively investigated in man and primate through the classical double step paradigm at the level of the oculomotor system (Becker & Jurgens, 1979; Taümer, 1975) or at the level of the hand movement with two joint movements—mainly shoulder and elbow (Georgopoulos, Kalaska, & Massey, 1981; Soechting & Lacquaniti, 1983). The main conclusion derived from these studies is that the existence of a refractory period (or intermittent serial control) is not a general rule. It appears that a sudden change in the goal for the eye or for the hand occurring prior to the execution of the response can be taken into account before the completion of the initial response; particularly, the reaction time to the second stimulus can be reduced to as low a value as the normal reaction time to the first initial stimulus.

For the guidance of the hand in space toward a single target, most studies have regarded the movement as composed of two major phases: a preprogrammed ballistic phase and a final phase under visual guidance of the seen hand toward the target; among the many studies carried out, the controversy has centered more on the time at which visual feedback from the hand becomes effective than on the nature of the corrective feedback.

If the existence of a finely tuned motor control under simultaneous vision of the hand and target during the homing phase is not cast into doubt, the notion of a preprogrammed set of muscle activity for a given movement is not so clear; in a control experiment Prablanc, Pélisson, and Goodale (1986) have shown that the persistence of the foveal target increases the accuracy of a movement performed without vision of the hand, by comparison with a situation where the foveal target is turned off at the onset of the movement. Thus the external feedback loop is likely not to be the only one that can amend erroneous response; another internal loop may also control the ongoing movement all along the hand trajectory.

We now present some evidence in favor of such a smooth control of neural commands adjusted to the goal (accuracy of reaching or gaze capture) and its optimization (with respect to duration). Such an operation needs the intervention of some neural on line computation. Preprogramming can hardly account for those large adjustments that occur in the very early part of the movement and do not need a conscious action. The idea developed here, without excluding a minimum number of preset patterns, is that the motor patterns are modifiable throughout the response; this would require a hierarchy in motor control organization in which once a goal has been defined (a path, its velocity, and the intended accuracy of the reach), an automatic process is started that compares the initial intent, the goal and its execution, and an error reducing process like a servo.

The aim of the following experiment is to demonstrate that, although the hand is made experimentally nonvisible during a movement of pointing toward a visual target, the vision of that target is nonetheless unconsciously used to quickly amend the trajectory if it had been initiated incorrectly.

In carrying out this work, we took advantage of the fact that during a reaching movement the actual eye movement is much earlier than the hand movement. It was possible, therefore, to change the position of the target during the point of highest velocity of the first saccade, just as the hand had begun to move. This procedure was applied to create an artificial error around the beginning of the pointing movement without the subject perceiving a change in the location of the target. Because subjects typically undershot targets with their first saccade, they had an opportunity to make a larger or smaller correction saccade and to modify the trajectory of their limb movement to compensate for the change in the position of the target.

## Method and Procedure

The schema of the set up has been presented elsewhere (Pélisson, Prablanc, Goodale, & Jeannerod, 1986). The experiment was carried out in the

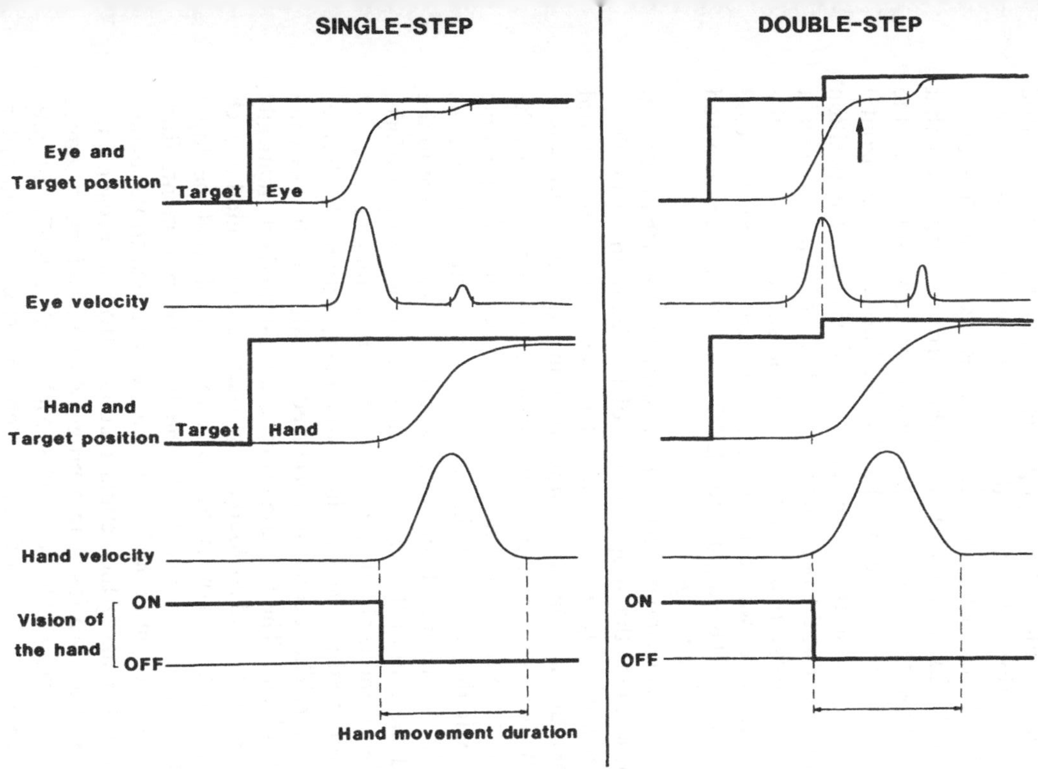


FIG. 24.5. Schematic diagram illustrating the spatiotemporal organization of the movements of eyes and hand in response to single- (left) and double-step (right) displacements of the target. As the diagram indicates, vision of the hand was prevented at the onset of hand movement in all trials. The second target displacement on double-step trials (restricted to 10% of the initial displacement) occurred just around the point of peak velocity of the first saccade, that is, after the onset of hand movement. Subjects were thus not aware of the existence of double-step trials, and both kinds of trials were randomly mixed during the same experimental session (single-step targets to 30, 40, and 50 cm; double-step targets to 30–32 cm, 40–44 cm, and 50–54 cm). The arrow indicates the point at which visual information about new target position can be processed.

following way. Four naive subjects were seated in turn in a darkened room with the index finger of the right hand resting on a small visual target located in a central position on a platform directly in front of them. Each was instructed to move the index finger as quickly and as accurately as possible to the new location of the target as it stepped from its central position to a more peripheral one. At the beginning of a trial, the subject viewed the hand and arm through a large half-reflecting mirror. The target consisted of a small LED positioned above the mirror in such a way that its virtual image appeared below the mirror on the same surface as the subject's hand. The subject's head was fixed with a bite bar, and horizontal eye movements were monitored by DC electro-oculography. The position of subject's index finger at the beginning and end of a pointing movement was derived from resistive paper that covered the surface on which the virtual image of the target was located. The trajectory was reconstructed from a photograph of a flashing infrared LED (2 ms at 100 Hz) attached to the end of the subject's finger.

As soon as the subject's index finger left the surface on which the central target had been located, the subject's view of the hand and arm disappeared. It reappeared only after the target moved back to the central position and when the hand began to move back in preparation for a new trial. This meant that as soon as the subject moved the limb toward the peripheral target, the subject had no visual feedback about his pointing movement. On half the trials (single-step), the target stepped from its central position to a randomly determined position 30, 40 or 50 cm to the right and stayed there until the subject's finger recontacted the surface. On the remainder of the trials (double-step), the target jumped to one of these positions, followed by a second jump to a new position 10% farther away from the center. The second step occurred just after the first saccade reached its peak velocity. The two conditions are illustrated schematically in Fig. 24.5. Thirty single- and 30 double-step trials were presented to each of the four subjects in random order. Subjects were not told that the target would sometimes change position during the first eye movement. Randomization of the single- and double-step trials was critical to the experimental procedure. Changing the retinal feedback from a saccade (in a systematic way) is known to result in a progressive and adaptive change in the gain of the saccades (Henson, 1978; McLaughlin, 1967). By presenting the subjects with single- or double-step targets in random order, we were able to keep the gain of the oculomotor response constant and could discard the adaptive process that occurs with modified feedback. Moreover, should the subjects have been able somehow to detect the second jump, they would have been unable to predict when such a trial would occur. The entire experiment was controlled and monitored by an on line computer.

## Results

At the end of the experiment, each subject had to report sensations and whether any change in target position during the pointing movement was experienced: Subject never noticed the occurrence of double steps. The quantitative results of this experiment were clear and unambiguous. As Fig. 24.6a illustrates, although the subjects consistently undershot the position of the target, the distributions of their final positions in the single- and double-step conditions were separated by a distance nearly equal to the size of the second displacement of the target (an analysis of variance yielded no significant differences between undershoot errors at any of the 3 pairs of amplitudes). In other words, the subjects were correcting the trajectory of their limb movement to fully accommodate this displacement. Moreover, the duration of a limb movement made to a double-step target corresponded to the duration of movement that would have been made to a single-step target located at the same final position (Fig. 24.6b). Thus, no additional processing time was required during the double-step trials. Following these clear results, planar finger trajectory was recorded from other subjects during the same pointing task. To compare kinematics of pointings at a given target location between single-step and double-step trials, trajectories were recorded for a 44 cm single-step target and for a 40 to 44 cm double-step target. The horizontal velocity and acceleration profiles (Fig. 24.7) were very similar in both conditions; the differences in finger deceleration between single- and double-step responses being not statistically significant. In particular, velocity profiles of finger trajectory to a double step target showed no "inflection point" to indicate a reprogramming of the manual response.

## Discussion

These results differ from those of previous studies concerning double-step stimuli that have shown that either oculomotor responses (Becker & Jürgens, 1979; Taümer, 1975) or limb responses (Georgopoulos et al., 1981; Megaw, 1974; Soechting & Lacquaniti, 1983) to a double-step stimulus could overlap. Even when midflight modification of saccade trajectories can be induced by double-step stimuli under particular conditions (Van Gisbergen, Van Opstal, & Roebroek, 1987) such oculomotor responses do not look like saccades at single-step targets. Indeed, except for the delayed responses reaching directly the second location of the target, their trajectory always showed a clear "inflexion point" whose latency with respect to the second step increased significantly the total duration of the response. In such instances, the visuomotor system in the best cases could work with a constant delay. Thus, classical double-step experiments deal more with the reprogramming capabilities of the sensorimotor system,

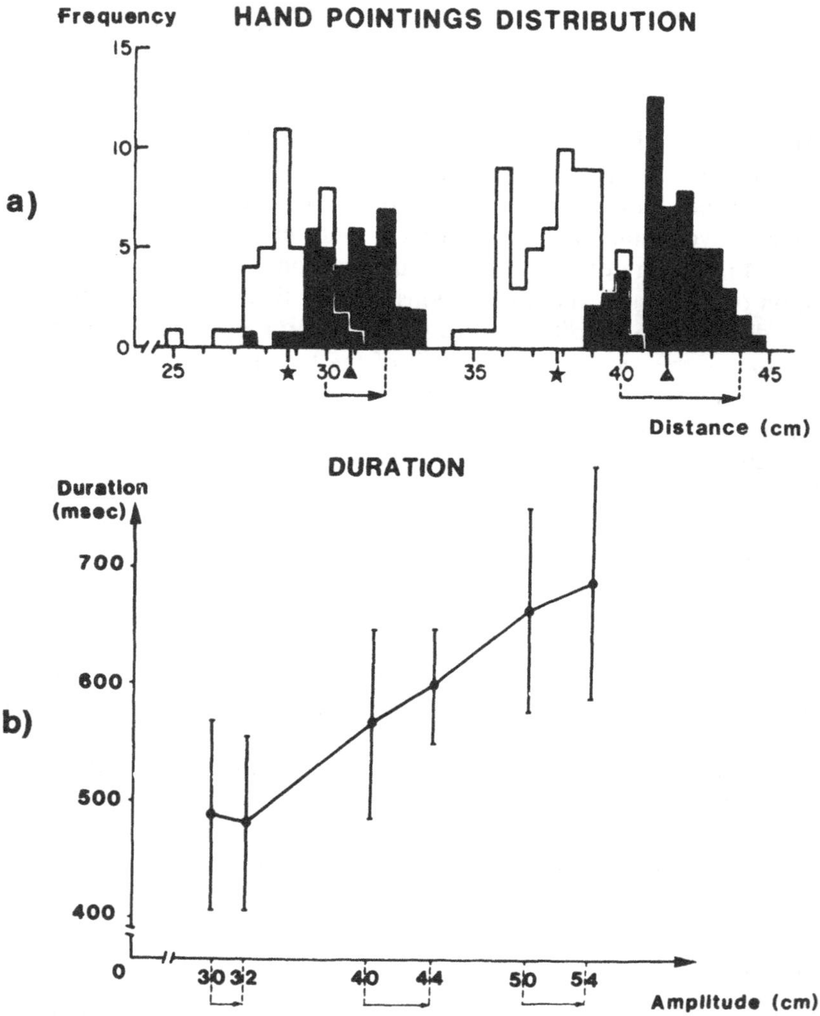


FIG. 24.6. (a) Frequency distributions of the final positions of the index finger on single- (white histograms) and double-step (black histograms) trials for the 30 cm and 32 cm and the 40 cm and 44 cm targets. The size of the second target displacement is indicated by the small arrows on the abscissa. For a single-step to 30 cm, the distribution of pointings represented by the white histogram is undershooting as shown by the mean response (star). In trials on which this target was further displaced during the saccade from 30 cm to 32 cm, mean response (triangle) indicates also an undershooting; but the distance between the two histograms is equal to the target shift. The same phenomenon applies for target shifts from 40 cm to 44 cm and from 50 cm to 54 cm (not shown for an insufficient number of data). (b) Mean duration of the hand movements plotted as a function of amplitude of the target final position. Vertical lines indicate standard deviations. Note that pointings on double-step trials belong to the same duration–amplitude relationship as pointings on single-step trials. Thus, no additional reaction time appears within the total duration of responses to double-step targets and no increase in variability of movement duration can be observed.

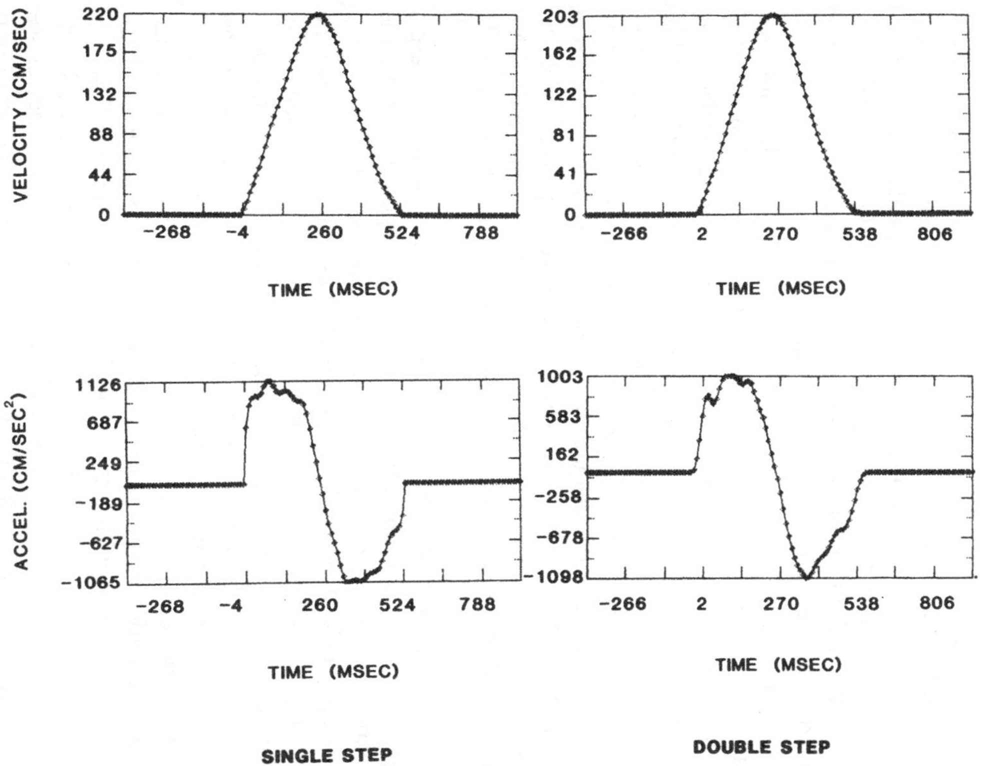


FIG. 24.7. Time course of horizontal hand trajectory for a 44 cm single-step target (left) and for a 40 cm to 44 cm double-step target (right). Velocity and acceleration profiles were computed using a polynomial approximation method with a 12 Hz cut-off frequency; they were then averaged for each subject and each condition (for that subject, 9 and 8 repetitions were averaged for single- and double-step conditions, respectively).

whereas ours was more directed toward understanding how retinal feedback from the saccade allows an uptake of stimulus location and thus can influence the ongoing limb motor output.

The present results strongly suggest that when rapid eye and hand movements are directed at a visual target, an initial set of signals is sent to the muscles controlling both the eye and the hand based on information about the position of the target that is available on the peripheral retina. After the first saccade, however, the combined input from retinal and extraretinal signals updates the information about target position (Goodale, Pélisson, & Prablanc, 1986; Pélisson et al., 1986; Prablanc et al., 1986), which is then immediately used for "fine-tuning" the hand trajectory. If this hypothesis is correct, then it becomes much clearer why the duration of a movement is the same for a given position of the target, independent of whether that position was achieved by a single or a double displacement of the target. In either case, the same postsaccadic information would be used to update the motor signals controlling the hand trajectory (at least if the amplitude of the second displacement in the double-step condition is less than 10% of the initial displacement). In other words, the apparent correction in the trajectory that occurred on double-step trials could be nothing more than the normal updating of the motor programming that occurs at the end of the first saccade on an ordinary trial (Pélisson et al., 1986).

At no time in this experiment did any of the subjects notice the target had stepped to a new location during their pointing response. Nor did the subjects experience any abnormal sensation when pointing to double-step as compared to single-step conditions. In short, they failed to perceive the change in target position even though the reaches were modified to accommodate that change. Indeed, had the subjects perceived the change in target position that occurred on double-step trials and, on the basis of this perception, quite consciously altered the trajectory of their reach, then the duration of the movements on these trials would have fallen well outside the amplitude–duration curve obtained on single-step trials and would have given the same results as with classical double-step experiments. The fact that this did not occur (see Fig. 24.6b) is additional evidence that subjects were treating single- and double-step trials in the same way, and, as we have already argued, suggests that the type of "correction" on double-step trials was not different from the normal fine-tuning of the trajectory that occurs at the end of the first saccade on single-step trials.

All these modifications in the trajectory of the pointing movement were made without the aid of any visual feedback about the limb position. Thus, the new information about the location of the target in visual space that was obtained at the end of the first saccade must have been combined with information about the current trajectory of the hand that was derived from nonvisual sources. However, our experiment says nothing about the nature of this nonvisual information that was used. Two possible candidates are proprioceptive feedback from the hand and limb and/or efference-copy of the signals sent to hand and limb. Either or both of these sources of information about the position of the limb could have been compared to the central spatial representation of the target provided by retinal and extraretinal signals at the end of the first saccade. Whatever the mechanisms might be, our experiment suggests that a quick and efficient loop can modulate the limb motor output in the absence of visual hand reafference.

Concerning the absence of reaction time, the nonconsciousness of the target jump could have been considered a prerequisite for an error processing without a call for reprogramming. Such was not the case, because in a recent experiment (unpublished data) the same kind of readjustments were observed when the jump occurred after the orienting saccade, the subject being thus fully aware of the target jump. However, its knowledge induced a change in the general strategy; even for the stationary targets, the velocity curve had a longer deceleration tail than acceleration tail.

In order to know at which point in time a trajectory begins to depart from its original path, additional data are needed because even for stationary targets the velocity curves exhibit large variability. Increasing the size of the jump is not a good method to dissociate the normal and "perturbed" response. Indeed, beyond 15% or 20% of the initial amplitude, the nature of the correction changes, and (as described in most double step experiments) a classical reaction time added to the normal duration is observed (unpublished data).

## GENERAL CONCLUSION

Through the two types of experiments, eye–head orientation and eye–hand pointing, we found that gaze saccadic control toward a target and hand-pointing control toward a target share some common properties. Once the goal has been defined, any type of unnatural stimulation or artificial error is automatically compensated for or nullified through quick and unconscious corrections.

The hypothesis we formulate is that an internal representation of absolute position in space of target and gaze is available at a lower level than the one where decisions are made and that an internal feedback loop controls the selected final effector (gaze or finger) toward its goal. The way these spatial modifications are translated into individual joint motion modifications is still an unsolved problem.

Further experiments are necessary to explore the time required to process errors of an ongoing response and the way these corrections are implemented; the technique of on line modifications of sensory cues is a useful tool to move, like a "cursor," the time of occurrence of sensory perturbations during the response. In order not to modify higher processes such as reprogramming, these perturbations must stay within the limits of flexibility of the initial response.

## REFERENCES


Adams, J. A. (1977). Feedback theory of how joint receptors regulate the timing and positioning of a limb. *Psychological Reviews*, *84*, 504–523.

Baker, R., & Berthoz, A. (1974). Organization of vestibular nystagmus in oblique oculomotor system. *Journal of Neurophysiology*, *37*, 195–217.

Becker, W., & Jürgens, B. (1979). An analysis of the saccadic system by means of double-step stimuli. *Vision Research*, *19*, 967–983.

Biguer, B., Jeannerod, M., & Prablanc, C. (1982). The coordination of eye, head and arm movements during reaching at a single visual target. *Experimental Brain Research*, *46*, 301–304.

Bizzi, E. (1975). Motor coordination: Central and peripheral control during eye-head movement. In M. S. Gazzaniga and C. Blakemore (Eds.), *Handbook of Psychobiology* (pp. 427–437). New York: Academic Press.

Feldman, A. G. (1966). Functional tuning of the nervous system during control of movement of a steady posture. II. Controllable parameters of the muscle. *Biophysics*, *11*, 566–578.

Fuller, J. H., Maldonado, H., & Schlag, J. (1983). Vestibular-oculomotor interaction in cat eye-head movements. *Brain Research*, *271*, 241–250.

Georgopoulos, A. P., Kalaska, J. F., & Massey, J. T. (1981). Spatial trajectories and reaction times of aimed movements: Effects of practice, uncertainty and change in target location. *Journal of Neurophysiology*, *46*, 725–743.

Gisbergen, J. A. M. van, Opstal, A. J. van, & Roebroek, J. G. H. (1987). Stimulus-induced midflight modification of saccade trajectories. In J. K. O'Reagan & A. Levy-Schoen (Eds.), *Eye movements: From physiology to cognition.* Amsterdam: Elsevier North-Holland.

Goodale, M. A., Pelisson, D., & Prablanc, C. (1986). Large adjustments in visually guided reaching do not depend on vision of the hand or perception of target displacement. *Nature*, *320*, 748–750.

Gordon, J., & Ghez, C. (1987). Trajectory control in targeted force impulses. III. Compensatory adjustments for initial errors. *Experimental Brain Research*, *67*, 253–269.

Guitton, D., Douglas, R. M., & Volle, M. (1984). Eye–head coordination in the cat. *Journal of Neurophysiology*, *52*, 1030–1050.

Guitton, D., & Volle, M. (1987). Gaze control in humans: Eye–head coordination during orienting movements to targets within and beyond the oculomotor range. *Journal of Neurophysiology*, *58*, 427–459.

Haddad, G. M., & Robinson, D. A. (1977). Cancellation of the vestibulo-ocular reflex during active and passive head movements in the normal cat. *Society for Neuroscience Abstract*, *3*, 155.

Hallet, P. E., & Lightstone, A. D. (1976a). Saccadic eye movements towards stimuli triggered by prior saccades. *Vision Research*, *16*, 99–106.

Hallet, P. E., and Lightstone, A. D. (1976b). Saccadic eye movements to flashed targets. *Vision Research*, *16*, 107–114.

Hening, W., Favilla, M., & Ghez, C. (1988). Trajectory control in targeted force impulses.

V. Gradual specification of response amplitude. *Experimental Brain Research*, *71*, 116–128.

Henson, D. B. (1978). Corrective saccades: Effects of altering visual feedback. *Vision Research*, *18*, 63–67.

Holst, E. von, and Mittelstaedt, H. (1950). Das Reafferenzprinzip. Wechselwirkungen zwischen zentralnervensystem und Peripherie. *Naturwissenchaften*, *37*, 464–476.

Jürgens, R., Becker, W., & Kornhuber, H. H. (1981). Natural drug-induced variations of velocity and duration of human saccadic eye movements: Evidence for a control of the neural pulse generator by local feedback. *Biological Cybernetics*, *39*, 87–96.

King, W. M., Lisberger, S. G., & Fuchs, A. F. (1976). Response of fibers in medial longitudinal fasciculus (MLF) of alert monkey during horizontal and vertical conjugate eye movements evoked by vestibular or visual stimuli. *Journal of Neurophysiology*, *39*, 1135–1149.

Laurutis, V. P., & Robinson, D. A. (1986). The vestibulo-ocular reflex during human saccadic eye movements. *Journal of Physiology* (London), *373*, 209–233.

Lestienne, F., Whittington, D. A., & Bizzi, E. (1981). Single cell recording from the pontine reticular formation in monkey: Behavior of preoculomotor neurons during eye-head coordination. In A. F. Fuchs and W. Becker (Eds.), *Progress in oculomotor research* (pp. 325–334). Amsterdam: Elsevier North-Holland.

Mays, L. E., & Sparks, D. L. (1980). Saccades are spatially, not retinocentrically, coded. *Science*, *208*, 1163–1165.

Mc Laughlin, S. C. (1967). Parametric adjustment in saccadic eye movements. *Perception & Psychophysics*, *2*, 359–362.

Megaw, E. D. (1974). Possible modification to a rapid on-going programmed manual response. *Brain Research*, *71*, 425–441.

Morasso, P., Bizzi, E., & Dichgans, J. (1973). Adjustment of saccade characteristics during head movements. *Experimental Brain Research*, *16*, 492–500.

Munoz, D. P., & Guitton, D. (1985). Tectospinal neurons in the cat have discharges coding gaze position error. *Brain Research*, *341*, 184–188.

Pelisson, D., & Prablanc, C. (1986). Vestibulo-ocular reflex (VOR) induced by passive head rotation and goal-directed saccadic eye movements do not simply add in man. *Brain Research*, *380*, 397–400.

Pelisson, D., Prablanc, C., Goodale, M. A., & Jeannerod, M. (1986). Visual control of reaching movements without vision of the limb: II. Evidence of fast unconscious processes correcting the trajectory of the hand to the final position of a target step. *Experimental Brain Research*, *62*, 303–311.

Pelisson, D., Prablanc, C., & Urquizar, C. (1988). Vestibulo-ocular reflex inhibition and gaze saccade control characteristics during eye–head orientations in humans. *Journal of Neurophysiology*, *59*, 997–1013.

Pola, J., & Robinson, D. A. (1978). Oculomotor signals in medial longitudinal fasciculus of the monkey. *Journal of Neurophysiology*, *41*, 245–259.

Polit, A., & Bizzi, E. (1979). Characteristics of motor programs underlying arm movements in monkeys. *Journal of Neurophysiology*, *42*, 183–194.

Prablanc, C., Pelisson, D., & Goodale, M. A. (1986). Visual control of reaching movements without vision of the limb. I. Role of retinal feedback of target position in guiding the hand. *Experimental Brain Research*, *62*, 293–302.

Robinson, D. A. Oculomotor control signals. In G. Lennerstrand and P. Bach-y-Rita (Eds.), *Basic mechanisms of ocular motility and clinical implications* (pp. 337–378). Oxford: Pergamon Press.

Robinson, D. A., & Zee, D. S. (1981). Theoretical considerations of the function and circuitry of various rapid eye movements. In A. F. Fuchs and W. Becker (Eds.), *Progress in oculomotor research* (pp. 3–9). Amsterdam: Elsevier North Holland.

Roucoux, A., Guitton, D., & Crommelinck, M. (1980). Stimulation of the superior colliculus in the alert cat. II. Eye and head movements evoked when the head is unrestrained. *Experimental Brain Research*, *39*, 75–85.

Soechting, J. F., & Lacquaniti, E. (1983). Modification of trajectory of a pointing movement in response to a change in target location. *Journal of Neurophysiology*, *49*, 548–564.

Taub, E., Goldberg, I. A., & Taub, P. (1975). Deafferentation in monkeys: Pointing at a target without visual feedback. *Experimental Neurology*, *46*, 178–186.

Täumer, R. (1975). Three reaction mechanisms of the saccadic system in response to a double-jump. In G. Lennerstrand and P. Bach-y-Rita (Eds.), *Basic mechanisms of ocular motility and clinical implications*. Oxford: Pergamon Press.

Tomlinson, R. D., & Bahra, P. S. (1986). Combined eye–head gaze shifts in the primate. II. Interactions between saccades and the vestibulo-ocular reflex. *Journal of Neurophysiology*, *56*, 1558–1570.

Vicario, D. S., & Ghez, C. (1984). The control of rapid limb movement in the cat. IV. Updating of ongoing isometric responses. *Experimental Brain Research*, *55*, 134–144.

Warabi, T. (1977). The reaction time of eye–head coordination in man. *Neuroscience Letters*, *6*, 47–51.

Young, L. R., & Stark, L. (1963). A discrete model for eye tracking movements. *IEEE Transactions on biomedical engineering*, *MIL7*, 113–115.

# 25 Tactile Afferent Signals in the Control of Precision Grip

Roland S. Johansson
Göran Westling
*Department of Physiology*
*University of Umeå, Sweden*

## ABSTRACT

For successful manipulation using the precision grip, the motor commands have to be precisely tuned to the relevant physical properties of the manipulated objects. This appears to primarily take place on the basis of previous manipulative experiences with the current object or similar objects, that is, the relevant information related to various objects we have experienced are internally represented in sensorimotor memories. If an erroneous internal model results in inappropriate motor programming, somatosensory signals may automatically trigger compensatory actions and update relevant parameters of the internal model. Tactile afferent innervation, which conveys information about mechanical events particularly at the interface between the hand and the manipulated object, plays a crucial role in this context. This chapter specifically addresses (a) how tactile afferent units innervating the skin of the fingers actually respond during manipulative movements, and (b) the significance of those responses for the control of the manipulation. These questions are considered in relation to a fairly simple but fundamental manipulative task: to lift, hold in the air, and replace a small test object using the precision grip formed between the tips of the thumb and an opposing finger. It is shown how tactile afferent signals may be utilized in the adaptation of the motor commands to the weight of the object and its frictional characteristics in relation to the skin.

## INTRODUCTION

During the last decades, neurophysiological studies in various mammals, including primates and man, have supplied a wealth of experimental data and theories that constitute the basis of our present understanding of the neural basis and function of tactile afferent mechanisms (see Burgess & Perl, 1973; Iggo, 1974, 1977; Johnson, 1983; Vallbo, Haybarth, Torebjork, & Wallin, 1979; Vallbo & Johansson, 1984). In most studies, however, tactile afferent mechanisms have essentially been viewed only in relation to the perceptual experiences of cutaneous stimulation in passive subjects. Consequently, a number of important problems remain to be investigated to fully appreciate the function of tactile afferent mechanisms. For instance, the analysis of cutaneous input during joint movements and its possible role in kinesthesia and position sense is an area that has barely been opened (e.g. Burke, Gandevia, and Macefield, 1988; Clark, Burgess, & Chapin, 1986; Nordh, Thelm, & Vallbo, 1979; Moberg, 1983). Another newly opened field, which is the topic of the present contribution, is the analysis of the information provided by the tactile sense during manipulative finger movements and its utilization in the motor control. A related field is the afferent input during precise facial movements such as speech gestures (Johansson, Trulsson, Olsson, & Westberg, 1988; Johansson, Trulsson, Olsson, & Abbs, 1988b).

The importance of cutaneous afferent input in the control of hand movements is well recognized (Denny-Brown, 1966; Moberg, 1962; Mott & Sherrington, 1895; Twithchell, 1954). Recent laboratory research suggests that one role of cutaneous input (and possible joint input) from fingers is to provide a general facilitation of motor commands activating finger muscles during various voluntary actions—particularly those involving flexion movements of the receptor-bearing fingers (e.g., Day & Marsden, 1982; Gandevia & McCloskey, 1977a, 1977b; Garnett & Stephens, 1981; Kanda & Desmedt, 1983; Marsden, Merton, & Morton, 1977; Torebjörk, Hagbarth, & Eklund, 1978; see also, Rood, 1860). Cutaneous input may also contribute to "long-latency" reflex responses in intrinsic hand muscles elicited by mechanical perturbations or weak electrical stimulation of the fingers (Darton, Lippold, Shahani, & Shahani, 1985; Caccia, McComas, Upton, & Blogg, 1973; Garnett & Stephens, 1980; Marsden, Merton, & Morton, 1985; Marsden et al., 1977).

Despite this variety of studies of tactile influences on the activity in hand muscles, it is far from clear (a) how tactile afferent units respond during manipulative movements, and (b) how these responses may be utilized in the control of manipulative actions. In the present contribution, these questions are explored in relation to a series of studies involving a fairly simple but fundamental manipulative task: to lift, hold in the air and replace a small test object using the precision grip between the tips of the thumb and a finger. Obviously, to accomplish the task, the executing motor commands must have been compiled on the basis of (a) voluntary signals reflecting the intention and the manipulative plan, and (b) on the basis of information related to the relevant physical properties of the object to be manipulated. It is shown that tactile afferent signals may have a decisive role in providing this information.

## Tactile Afferent Innervation

The tactile afferent units in the glabrous skin (nonhairy skin) on the inside of the hand signal information related to the mechanical status of the skin to the central nervous system (CNS) through fast myelinated fibres conducting at 35–80 m/s (Johansson & Vallbo, 1983). Their exquisite sensitivity to skin deformation changes and various mechanical transients, combined with their location close to the points of contact with the manipulated object should make them the most efficient sensors to provide information about various mechanical events at the object/hand interface during manipulation. As illustrated in Fig. 25.1, four different types of tactile units can be distinguished on the basis of the adaptation to sustained skin deformation and the structure of the cutaneous receptive field (see Johansson & Vallbo, 1983; Vallbo & Johansson, 1984). All four types have strong similarities with well-defined classes described in other species (e.g., Burgess & Perl, 1973; Iggo, 1974; Pubols & Pubols, 1983).

Two types (FA) show fast adaptation to maintained skin deformation, the FA I and FA II units, that is, they only respond to skin deformation changes. The other two types adapt slowly (SA), the SA I and SA II units. In addition to being dynamically sensitive (in particular, the SA Is) the SA units exhibit a response related to the strength of maintained skin deformation.

The FA I and SA I units have small and well-defined cutaneous receptive fields (typically 10 $mm^2$). Their field properties and their high densities in the skin of the fingers, especially in the tips, indicate that they can encode detailed spatial information. In contrast, the FA II and SA II units, whose densities are lower and more uniform, have receptive fields that are larger and less well defined, that is, they are responsive to more remote stimuli.

The FA IIs are responsive to transient mechanical stimulation (ca. 50 Hz–500 Hz vibrations) often originating at remote sites, whereas the SA IIs show a considerable sensitivity to lateral stretching of the skin, often with an appreciable directionality: Their response may increase for stretch in one direction and decrease for stretch applied at right angles or opposite to this direction (Knibestöl, 1975; Knibestöl & Vallbo, 1970). Thus, the population of SA II units constitutes a set of sense organs that provides information about direction and magnitude of lateral tension in the skin, as well as between the skin and deeper structures (Johansson, 1978). Information of this nature may be significant to monitor not only shearing forces between the skin surface and hand-held objects but also to sense

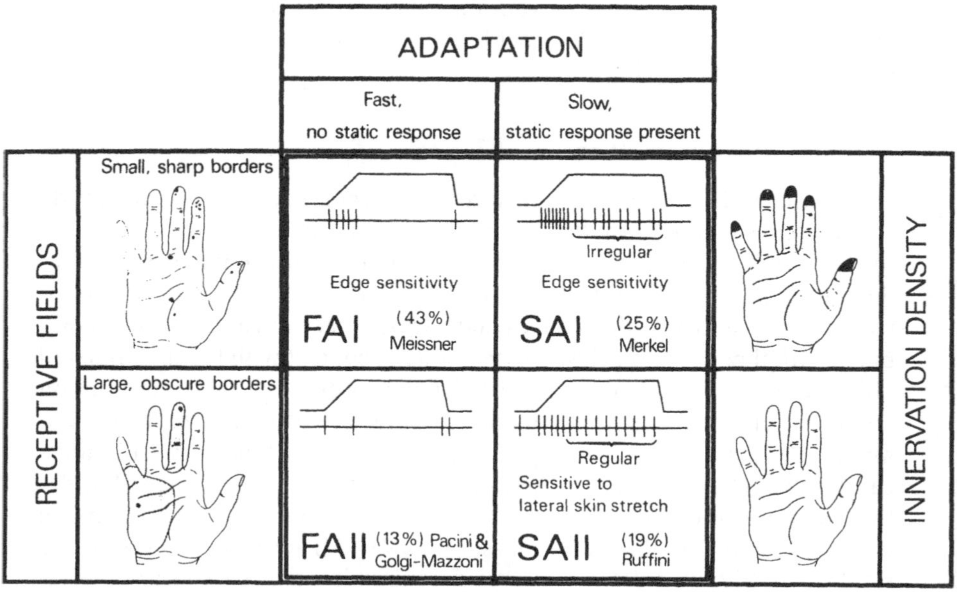


FIG. 25.1. Types of mechanoreceptive afferent units in the glabrous skin of the human hand and some of their distinguishing properties. The graphs in the middle schematically show the impulse discharge (lower traces) to perpendicular ramp indentations of the skin (upper traces) for each unit type. The black patches and the dashed areas of the left drawings of the hand show the extent of typical receptive fields for type I and type II units, respectively. The right drawings of the hand show the average density of type I and type II units, respectively. Each dot in the drawings represents a single sensory unit. In addition, the relative frequency of occurrence and the probable morphological correlate are indicated for each unit type respectively. It should be noted that the terminology of the tactile units in the present account follows the one proposed by Johansson and Vallbo (1983). Thus, the earlier terms, RA (rapidly adapting) or QA (quickly adapting), are replaced by FA I. This is done because the previous terms are being used more and more loosely, for example, to characterize neurons at higher levels in the somatosensory system. Further, the physiological term FA II has replaced the morphological term PC (Pacini corpuscle) as a notation for the primary afferent unit. For further details see text.

joint position. Indeed, both the FA II and the SA II units may respond readily to finger movements without involving direct touch (Hulliger et al., 1979).

## METHODOLOGICAL CONSIDERATIONS

### Task of the Subject

Healthy human subjects were asked to grasp a test object that rested on a table top; lift it about 2 cm above the table top; hold it still in the air for about 10 s; and then replace and release it. The object was grasped between the tips of the index or long finger and thumb. For the purpose of timing the 10 s hold phase, the subject watched an illuminated clock with a second hand.

### Test Object

The test object (Fig. 25.2) was touched at two surfaces placed on each side in two parallel planes. The *grip force*, the *load force* (vertical lifting force), and the *vertical movement* were measured. An *accelerometer* recorded *vibrations* in the object, for example, caused by slips.

The object's *surface structure* (and thus the friction in relation to the skin) and the *weight* of the object (100–1000 g) could be changed between consecutive lifts. The surface structures used were sandpaper, suede, or silk, which provided coefficients of static friction in relation to the skin of 1.21, 0.68 and 0.35, respectively (mean values, Johansson & Westling, 1984a). These materials were selected as representative for the frictional range of common materials. A gray color combined with reduced illumination *precluded visual discrimination* of the structures. Likewise, to ensure that the subject could not visually discriminate the object's weight, it was altered between lifts by attaching weights to an extension of the object that remained below the tabletop at all times (Fig. 25.2).

### Neurophysiological Methods

In separate experiments, using the method of *microneurography* (Vallbo & Hagbarth, 1968), impulses were recorded in single tactile afferent units innervating the glabrous skin of the fingers. A tungsten microelectrode was inserted percutaneously into the median nerve 5–10 cm proximal to the elbow. Activity in hand/arm muscles were monitored electrically (*electromyography*). The amplified surface EMG signals were root mean square

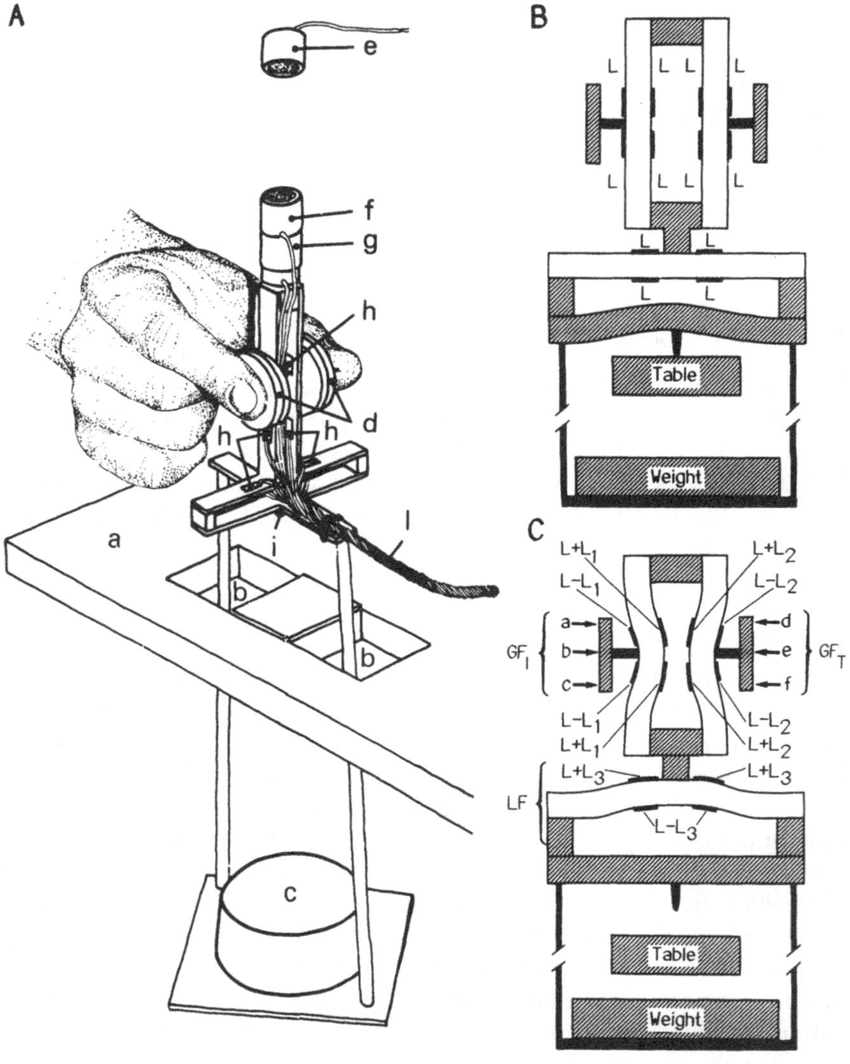


FIG. 25.2. **A** Schematic drawing of the apparatus. **a** - table; **b** - holes in table; **c** - exchangeable weight shielded from the subject's view by the table; **d** - exchangeable discs which constituted the touch surfaces; **e** and **f** - vertical position transducer (dc—560 Hz) with an ultrasonic receiver (**e**) and an ultrasonic transmitter (**f**); **g** - accelerometer (10–600 Hz); **h** - strain-gauge force transducers for measurement of grip force and vertical lifting force also denoted load force (dc—120 Hz); **i** - peg with a hemispherical tip on which the object rests while standing on the table. To reliably define the point in time when the object took off from the table and when it made contact again, the contact with the table was electrically detected. (From

processed with rise and decay time constants of 1 ms and 3 ms, respectively. In additional experiments, the skin of the fingertips was anesthetized during lifts.

## BEHAVIORAL OBSERVATIONS

### General Features of the Lifting Trials

The present manipulative action, to lift an object from a table, hold it in the air for some time and then replace it, may be divided into a series of distinct phases, or *sequentially coordinated events* (Johansson & Westling, 1984b, 1988a). Figure 25.3A shows the sequence of events. At the *preload phase* (a) the subject just grips the object for about 0.1 s. At the following *loading phase* (b) the load force (i.e., the vertical lifting force) and the grip force increase in parallel during isometric conditions. Soon the load force overcomes the force of gravity, and the lifting movement (c) (elbow flexion) takes place (*transitional phase*). The grip force shows a peak while the object accelerates, followed by a decay. When the lifting movement is completed, the *static phase* (d) is reached. Following the *replacement* (e), when the object contacts the table, there is a short *delay* (f), after which the *unloading phase* commenced. During this phase the two forces decline in parallel until the object is *released.*

Thus, except during the preload phase, the grip and the load force vary in *parallel*, that is, the two forces are generated on the basis of coordinated muscle commands. This *parallel coordination* is further emphasized in Fig. 25.3C showing the two forces displayed against each other. Furthermore,

---

Johansson and Westling, 1984b.) **B** and **C**. Schematic illustration of the arrangement of the strain gauges and levers constituting the force transducers. Three groups of four strain gauges (with length = L) were glued to both sides of three levers (nonfilled in fig.). Each group constituted a full electric bridge measuring the load force and the two grip forces, respectively. **B** The object rests on the table with unloaded transducers. **C** Loaded transducers while the object is lifted. The initial length of the strain gauges are labeled L, and the lengths after loading L + Ln; and L − Ln ($n$ = 1, 2 and 3 for the three transducers, respectively). Two of the strain gauges in each group were lengthened whereas two were shortened. Because the load force bridge was balanced when the object freely rested on the table, any load force at the touched surfaces was reliably measured. (From Westling, 1986.)

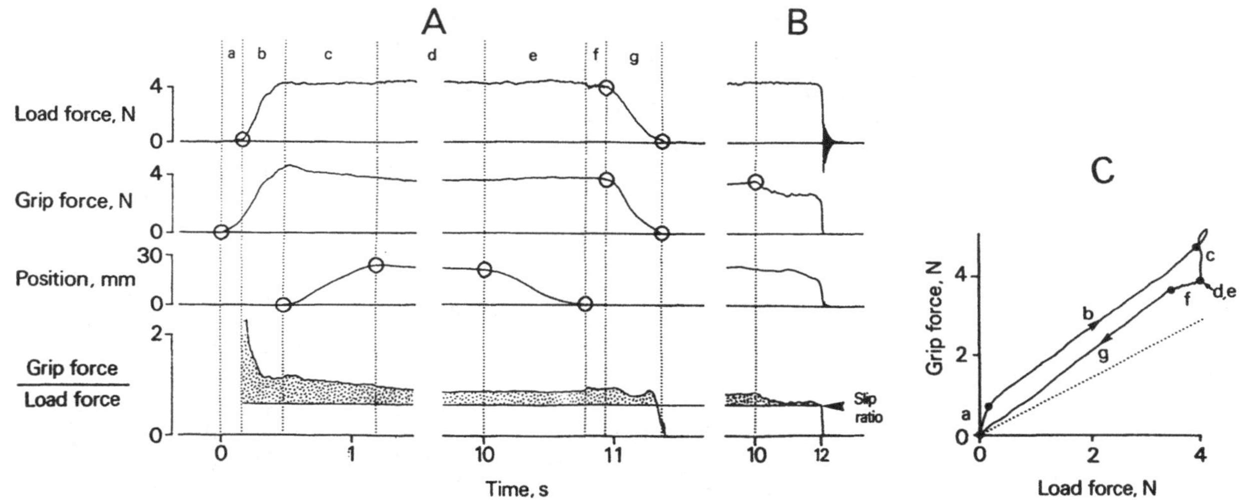


FIG. 25.3. The principal phases of a lifting trial. **A** Load force, grip force, vertical position and ration between grip and load force as a function of time for a sample trial (weight 400 g, surface structure, sandpaper). The periods indicated are: **a** - preload phase (increase of grip force from initial contact until object is securely grasped); **b** - loading phase (parallel isometric increase of the forces); **c** - transitional phase (while the object is elevated); **d** - static phase (the object is held stationary in space); **e** - replacement phase; **f** - delay; **g** - unloading phase (parallel isometric decrease of the forces). Note the interrupted time scale. **B** Slip ratio measurements at the end of a separate trial subsequent to the trial in A by asking the subject to hold the object in air and decrease the grip force until slips occurred. (the same surface structure and weight). The vertical dashed line in B indicates the start of a slow voluntary grip force decrease. Horizontal dashed lines in A and B indicate the slip ratio obtained, and the shaded areas indicate the safety margin to avoid slip. **C** Grip force as a function of load force for a similar lift as in A. The phases indicated as in A. The dashed line indicates the slip ratio. (**A** and **B** adapted from Johansson & Westling, 1984b).

the *ratio grip force/load force* (Fig. 25.3A) is fairly constant throughout the lift except for the steep decline early during the loading phase, which is due to the "grip force offset" gained during the preload phase.

To avoid slips, the ratio grip force/load force must exceed a minimal value, the *slip ratio* (horizontal line in Figs. 25.3A and B), determined by the coefficient of friction between the skin and the object. The more slippery the structure, the higher the slip ratio. The *safety margin* to prevent slips is the difference between the slip ratio and the force ratio automatically coordinated by the subject (for details, see Johansson & Westling, 1984a).

## Frictional and Weight Adaptation

To overcome forces counteracting the manipulative goal (e.g., gravity and kinetic forces), the subject had to apply a vertical lifting force (load force) to the object. However, for smooth and successful manipulation, the development of the load force must match the weight of the object. Likewise, the grip force/load force ratio must be adjusted to the frictional condition between the skin and the object; a grip that is too weak may allow the object to slip, whereas a grip that is too strong may damage the object or the hand, and/or result in unnecessary muscular fatigue hampering further manipulatory movements superimposed on the basic grip.

*Weight adaptation.* The weight of the object principally modified the duration of the loading and unloading phases and the force rates during these phases. As can be seen in Fig. 25.4A, the heavier the object the more extended these phases of parallel force increase and the higher the force rates. However, the ratio between the two forces was not influenced.

*Frictional adaptation.* By analyzing the force coordination with the various surface structures having different frictional conditions in relation to the skin, we found that the *parallel* coordination was influenced. The ratio between the two forces was changed. Figure 25.4B, shows that the more slippery the material, the higher the rate of grip force increase, and the higher the final grip force. In contrast, the course of the lifting movement and the rate of development of the load force are essentially unaffected by the frictional condition. A comparison between the force ratios employed for the three materials, and the corresponding slip ratios (bottom graph of Fig. 25.4B), indicates that the force coordination was adjusted to the frictional conditions providing a fairly small safety margin to prevent slips. (For interindividual differences in safety margin, see Westling & Johansson, 1984.)

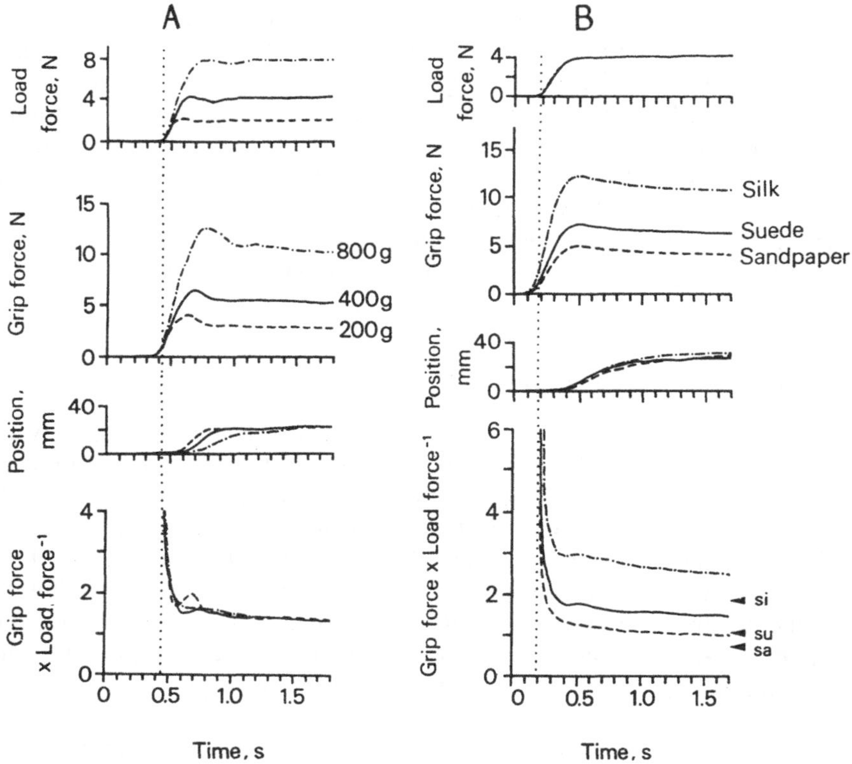


FIG. 25.4. Force coordination during the initial part of lifting trials. Load force, grip force, vertical position and ratio between grip and load force as a function of time. **A** Influence of different weights: 800 g, 400 g, and 200 g. Surface structure constant (suede). Average data from 24 trials (single subject). **B** Influence of different surface structures: finely textured silk—most slippery, suede—less slippery, and sandpaper—least slippery. Weight constant at 400 g. Average data from 120 trials (9 subjects). **A** and **B** Arrowheads indicate mean slip ratios for the three structures. Illustrations represent standardized conditions in which the previous trials were carried out with the same weight and friction, respectively. (Modified from Johansson & Westling, 1984b.)

The weight of the object and the frictional condition during the *previous lifts* influenced the programming of the force coordination. The illustrations in Figs. 25.4A and B represent lifts in which the previous trials were carried out with the same friction and weight, respectively.

*"Everyday" tasks.* Various experiments were performed to examine force coordination in more general contexts, that is, in "everyday situa-

tions." For instance, it was shown that the force coordination adapts to the considerable *frictional changes* occurring after washing the hands with ordinary soap and water (Johansson & Westling, 1984a). With washing, sweat is removed from the skin, and the friction in relation to any given object temporarily decreases. As can be seen in Fig. 25.5A, the static grip force follows these frictional changes. This finding further indicates that the adaptation is made to the friction per se, rather than on the basis of different texture properties of the touched materials.

In various experiments involving *manipulation superimposed on the basic precision grip* it was again shown that the coordination of the motor commands accounting for the grip and load forces is critically regulated to the weight and the current frictional conditions. One of these experiments—a bimanual task—is schematically shown in Fig. 25.5B. An empty glass was placed at the weight carrier of our test object. While holding this "transducer-equipped glass" in air, the subject was asked to fill the glass using the free hand. This was performed with each of the three surface materials, silk, suede, and sandpaper, respectively. As can be seen, the grip force increases in parallel with the increasing load force while the glass is filled and, again, the force ratio is maintained at a remarkably constant level adequately adjusted to the current frictional demands.

In another study, the subject held a test object with a precision grip. The load was now increased rapidly by dropping a small ball onto the test object. If the ball was dropped by the subject during a bimanual task, sophisticated preparatory actions occurred soon before the impact. These were adequately scaled to the various kinetic and frictional factors such that an adequate safety margin was maintained at the critical moment of impact (Johansson & Westling, 1988b).

## TACTILE AFFERENT SIGNALS AND THEIR FUNCTIONAL RELEVANCE

Three independent sets of evidence indicate that input from tactile afferents with endings in the area in contact with the manipulated object are decisive for the proper adaptation of the grip force/load force ratio to the friction between the object and the skin: (a) the adaptation failed if the afferent signals from the skin areas in contact with the object were abolished by local anesthesia (Johansson & Westling, 1984b; Westling & Johansson, 1984); (b) force ratio changes could be triggered by electric stimulation of tactile afferents; and (c) microneurography recordings revealed that there were signals in tactile afferent units to account for this adaptation, that is,

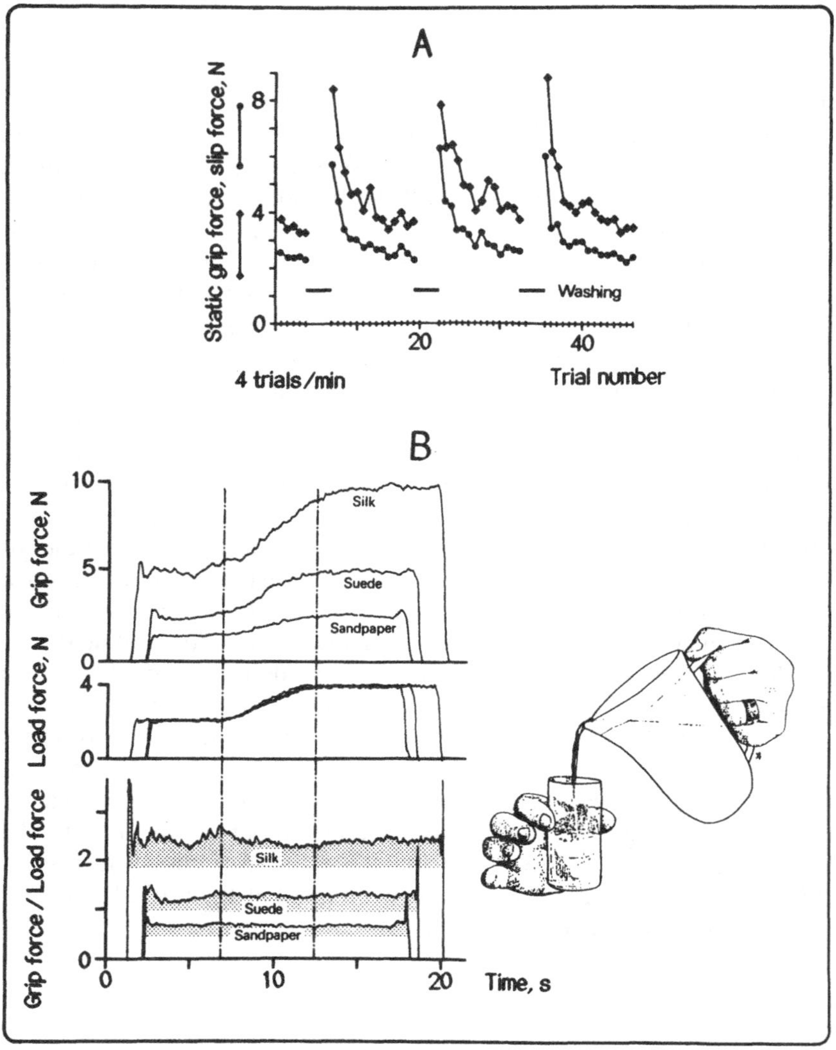


FIG. 25.5. Adjustment of the grip force to the frictional changes in everyday situations. **A** Static grip force (diamonds) and the corresponding slip force (dots) before and after the subject's hand was washed (soap and water) and dried (towel); 48 consecutive lifts with suede as surface structure (single subject, weight constant at 400 g). Each trial was terminated by slip force measurement (From Johansson & Westling 1984a). **B** Coordination between grip and load forces during a bimanual task. An empty glass was attached to the test object that was lifted and held stationary in space (weight: 200 g). Load force, grip force, and the ratio between grip and load force as a function of time (time scale with arbitrary origin). The changes in the load force during the period between the two vertical lines were caused by the subject pouring water into the glass (ca 0.2*l*) using a pitcher held by the contralateral hand. Three superimposed trials with three different structures in contact with the skin (three friction conditions) are shown. Dashed areas illustrate safety margins. (From Johansson & Westling, 1987b.)

distinctive afferent responses were observed prior to adaptive changes in the force balance (Johansson & Westling, 1987a; Westling & Johansson, 1987). Likewise, there is evidence that tactile afferent signals may provide information necessary for the sequential coordination during the present multiphasic lifting task, that is, information necessary to control when to switch from one phase to the next (Johansson & Westling, 1988a).

In the following, the various tactile afferent responses observed during the lifts are described and their functional roles examined. Figure 25.6 schematically summarizes the various responses and some of the motor effects resulting from these responses. Note that the responses indicated were *not* all observed in every trial.

## Initial Afferent and Motor Responses

### *Initial Afferent Responses*

After being virtually silent between the lifts, all FA I units showed a distinct burst discharge initially during the gripping (ca. 30—200 imp/s). This *initial response*, which virtually appeared at the very moment of contact, showed its peak discharge rate at grip forces below ca. 1 N, that is, during the preload phase or early during the loading phase (Fig. 25.6 and Fig. 25.7). Then the firing rate declined during the following loading phase and sometimes the units became silent. Similar initial responses were also observed with many of the SA I units. Many FA II units also responded during the same period. The strength of the FA II responses in general was weaker and varied from trial to trial mainly depending upon the abruptness of the touch. In contrast, the SA II units showed no initial responses.

The marked initial responses in the FA I and SA I units, together with their high innervation densities in the finger tips, suggest that the tactile input during the lift was most intensive during the preload phase and early during the loading phase. Indeed, it was estimated that ca. 450 type I units would have been stimulated at each finger contacting the object. About $\frac{2}{3}$ of these would be FA I units. For further details concerning the initial responses, see Westling & Johansson (1987).

### *Functional Significance of Initial Afferent Responses*

*Triggering of the motor commands accounting for the loading phase* Signals in cutaneous afferent arising during the preload phase seem important for the release of the motor commands accounting for the loading phase: During topical anesthesia of the skin areas contacting the object, the start of the loading phase takes place at abnormally high grip forces (Johansson & Westling, 1984b). It seems likely that the decisive afferent

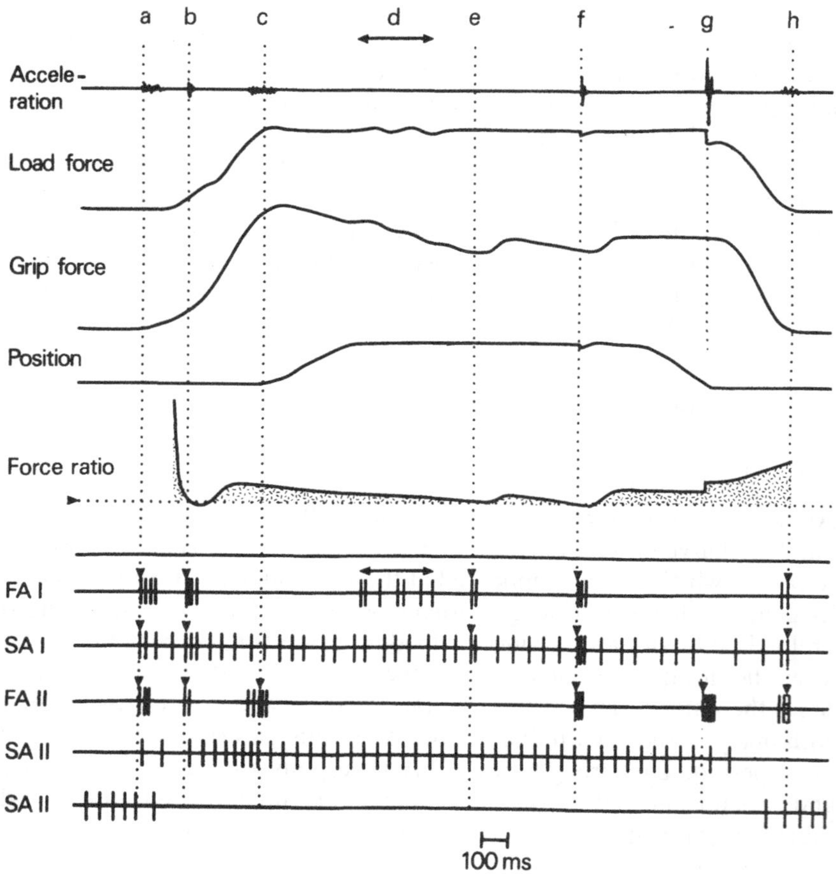


FIG. 25.6. Schematic illustration of the various kinds of afferent responses that may occur in the four types of tactile units (FA I, SA I, FA II and SA II) during lifts. Indicated events are: **a** - initial responses in FA I, SA I and FA II units; **b** - afferent slip responses during the loading phase in FA I, SA I and FA II units. Subsequent to the responses there was an upgrading of the force ratio; **c** - FA II discharges related to the start of the vertical movement; **d** - FA I responses during periods of pronounced physiological muscle tremor; **e** - localized slip responses in FA I and SA I units (slips not reflected in the accelerometer trace) followed by an upgrading of the force ratio; **f** - afferent slip responses in FA I, SA I and FA II units eliciting a subsequent upgrading of the force ratio during the static phase; **g** - FA II burst at the moment of table contact; **h** - release responses in the FA I, SA I and FA II units. Note that the responses indicated at **a**, **c**, **g**, and **h** were obligatory, whereas the other dynamic responses only occurred intermittently (see text). (From Westling, 1986.)

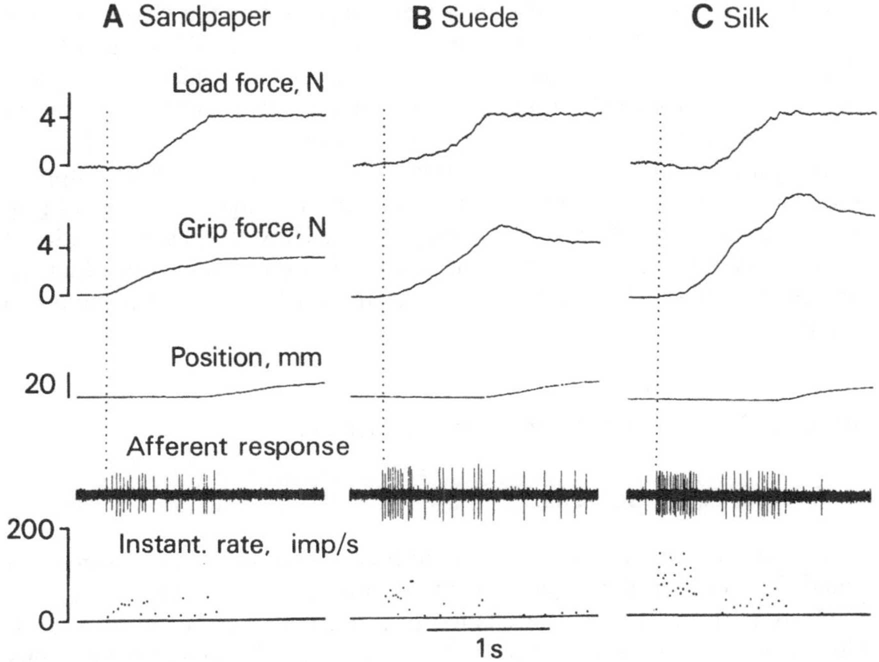


FIG. 25.7. Initial responses in an FA I unit, and the influence of the surface structure. **A** - **C**, single trials with the three different surface structures. Note the different response intensities. Vertical lines indicate the initial touch. (Adapted from Johansson & Westling, 1987a.)

information is encoded in the initial responses from the FA I and/or SA I units because many FA II units, with their endings outside of the anesthetized areas, probably still would have shown initial responses due to their high sensitivity to remote transient mechanical stimuli.

Speaking in general terms, it seems obvious that the CNS would need information indicating a reliable contact before releasing the muscle commands leading to further manipulation, that is, the loading phase in the present task.

*Initial adjustment to frictional changes.* In lifts following an unexpected change in the frictional condition, the motor output is influenced by the new friction at different phases of the lift (Johansson & Westling, 1984b). Initially, the force development is adequate for the friction during the previous lift ("frictional memory"), but an early adjustment to the new friction can already appear about 0.1 s after the gripping (Johansson & Westling, 1984b). Hence, somatosensory information related to the frictional condition already enters during the preload phase. There is evidence that the appropriate afferent signals were conveyed by the vigorous initial responses observed in the FA I units (cf. Johansson & Westling, 1987a). The initial responses in some of the FA I units were influenced by the surface material strongly enough to predominate over other factors causing response variation, for instance, rate of grip force increase. An example of responses in such a unit is represented in Fig. 25.7. The more slippery the material, the higher the discharge rate and the higher the rate of grip force increase later during the loading phase. However, as is considered ahead, this initial adjustment to the new friction sometimes is insufficient, and further "*secondary*" adjustments occur later in the lift in response to small slips.

## Afferent Slip Responses and Motor Responses

### *Afferent Slip Responses.*

Despite the initial adjustment to the frictional condition, occasionally a small slip between the object and the skin occurred later in the lift, particularly if there had been a change to a more slippery condition. Such slips, caused by an insufficient safety margin, readily excited tactile units (see Fig. 25.8). This was true although most slips were so small that they were rarely noticed by the subject and could not be visually detected. Such *afferent slip responses* were observed in FA I, FA II, and SA I units, and each of the three unit types appeared capable of reliably signalling the occurrence of slips with each of the three materials. The intensity of the slip responses ranged from a single impulse to brief bursts with peak frequencies up to ca. 300 imp/s. In contrast, slip responses were virtually absent in the SA II units. For more details, see Johansson and Westling (1987a).

The afferent slip responses considered so far were associated with slips that were clearly detectable in the accelerometer signal. Sometimes, however, during periods of declining force ratio or at force ratios close to the slip ratio, more or less marked afferent burst responses occurred in the *absence* of detectable acceleration events. Because these responses apparently originated from small slips localized to only a part of the skin area in contact with the object, they were denoted as *localized (afferent) slip responses* (Johansson & Westling, 1987a). Localized slip responses were only observed in the FA I and SA I units. Moreover, for ca. 10% of the afferent slip responses in FA I and SA I units, localized slip responses were present at the decrease of the force ratio *prior* to the overt slips causing accelerometer responses.

### *Functional Significance of Afferent Slip Responses*

Following the *afferent slip response*, there was an upgrading in the grip force/load force ratio to a higher and stable value, that is, the safety margin was restored preventing further slips (Johansson & Westling, 1984b). Thus, a *secondary adjustment* took place that compensated for an insufficient initial adjustment to a new frictional condition. Secondary adjustments occurred most frequently during the loading phases while the object was partly supported by the table (Fig. 25.8A), but they were also observed while the object was held in air (Fig. 25.8B and C). Similar upgradings of the force ratio could, during certain conditions, be elicited by electrical stimulation of tactile afferents innervating the skin areas in contact with the object. This indicated that tactile input by itself could trigger the adjustments of the force balance (Johansson & Westling, 1987a).

Different *patterns of muscle commands* caused the slip-triggered upgrading of the force ratio during the static and loading phases, respectively. During the static phase, the grip force increased to a higher, stable value, whereas during the loading phase, changes in the load force rate were involved, sometimes accompanied by an increased grip force rate. Similar dependence on the *phase of movement* or on the postural situation have been described previously with other multiarticulate actions triggered by cutaneous input in man (Crenna & Frigo, 1984; Hagbarth & Finer, 1963) and in animals (Forssberg, Grillner, & Rossignol, 1975, 1977). Moreover, this is in agreement with the vast observations indicating that a variety of animal and human reflexes are purposefully modified to the particular motor task, the postural situation, and the specific phase of movement (Bonnet & Requin, 1982; Evarts & Tanji, 1974; Grillner, 1981; Hammond, 1956; Lund & Olsson, 1983; Tatton & Bruce, 1981).

That the resulting new force balance was maintained supports the idea that the force coordination was updated via a *sensorimotor memory* rather than being somehow regulated on the basis of a continuous closed feedback loop involving tactile receptors. Accordingly, the central nervous system appears to completely rely on memory information in the absence of the cutaneous input (Johansson & Westling, 1984b).

The *latencies* between the onset of the slips and the changes in the force ratio were 74 +/− 9 ms (mean +/− SD; ca. 10 ms shorter from analysis of the EMG signals) which is about twice the latency of the most rapid spinal reflex in intrinsic hand muscles (Darton et al., 1985; Matthews, 1984b), but less than half the voluntary reaction time for a grip force response triggered by cutaneous input. Reflex responses with similar latencies in intrinsic hand muscles elicited by stimulation of cutaneous afferents has been previously reported (Darton et al., 1985; Garnett & Ste-

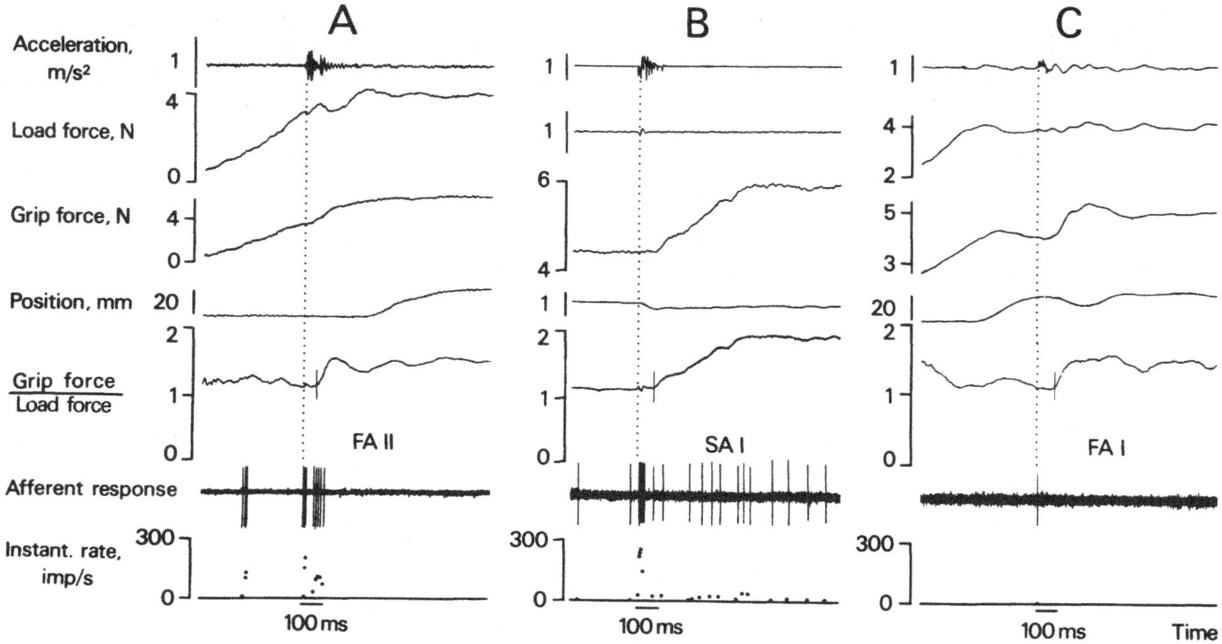


FIG. 25.8. Examples of afferent slip responses and upgradings of the grip force/load force ratio elicited by small but distinct frictional slips. Acceleration, load force, grip force, vertical position and force ratio are shown as functions of time, as well as the impulse discharges in single afferents and the corresponding instantaneous impulse frequencies. Vertical dashed lines indicate the onset of the slips as revealed by the acceleration signal, and the short solid lines indicate the onset of the upgrading of the force ratio. **A**, slip during the loading phase, FA II unit. **B**, slip during the static phase, SA I unit. **C**, gentle slip early during the static phase, FA I unit. **A** - **C**, surface structure silk. (From Johansson & Westling, 1987a.)

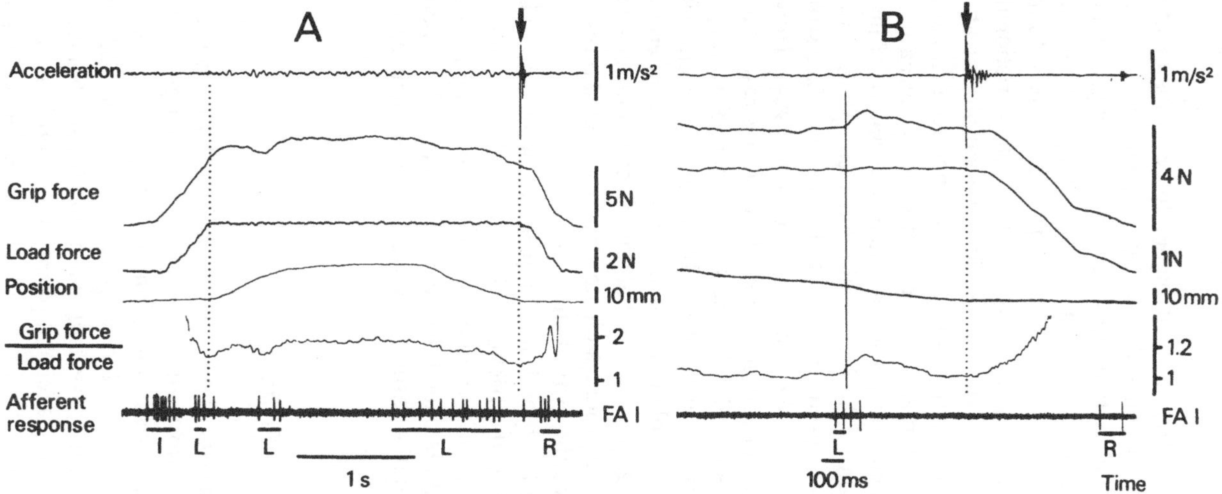


FIG. 25.9. Examples of localized slip responses in two separate FA I units during ordinary lifts. Responses indicated by L were regarded as localized slip responses. I and R indicate the initial response and the release response, respectively (see Westling & Johansson, 1987). **A** Entire lift. Vertical lines indicate the start and the stop of the vertical movement. **B** Modest upgrading of the grip force/load force ratio during the replacement phase starting (solid line) ca. 50 ms after the onset of a weak afferent burst. Dashed line indicates the moment of table contact. **A** and **B**, note the different time scales. Arrows indicate the strong acceleration event when the object contacts the table. Note the absence of afferent responses to this event in the FA I unit. Surface structure suede, weight 400 g. For further details see text.

phens, 1980; Marsden et al., 1985). Moreover, there is evidence that such reflexes may be supraspinally mediated (Jenner & Stephens, 1982; but, see Darton et al., 1985).

The *localized afferent slip responses* at higher force ratios or at small dips in the force ratio were often followed by an upgrading of the grip force/load force ratio similar to that triggered by overt slips, albeit less marked (Fig. 25.9A and B). Thus, as with the upgrading of the force ratio triggered by afferent slip responses at overt slips, the *localized* afferent slip responses would aid the adaptation to a *more slippery* frictional condition. These mechanisms, however, would not compensate for an inappropriate initial adjustment if there had been a change to a *less slippery* condition, that is, if the safety margin to prevent slips was already excessively high. Indeed, this high safety margin remains during brief lifts (ca. 10–15 s) and the adaptation of the force ratio is not complete until after the initial adjustment in the subsequent lift with the same surface structure (Johansson & Westling, 1984b; Westling & Johansson, 1984). However, slow adjustments toward lower grip force/load force ratios can take place in the static phase of prolonged lifts lasting for minutes. This adaptation probably depends on localized slip responses because it can occur without slip indications. A plausible explanation is that there is an inherent tendency to slowly weaken the grip force and that this effect is counterbalanced by small intermittent force upgradings elicited by localized slip responses at force ratios clearly above the critical slip ratios.

In conclusion, signals in the type I units appears to be indispensable to the adjustment of the grip to the friction: These were the only units that showed both initial responses, slip responses, and localized slip responses. It might be that the FA I units play a decisive role because it was only in units of this type that the initial responses were clearly influenced by the object's surface material.

## Afferent Responses at Phase Shifts

The present lifting task consists of a series of distinct phases as shown in Fig. 25.3. There were obligatory and quite distinct discharges in tactile afferent units at the onset and/or termination of these phases as schematically indicated in Fig. 25.6: (a) *initial responses*, as considered above, responses at the (b) *start of the movement* and at the (c) sudden *cessation of movement* terminating the replacement phase, and (d) *release responses* at the final release of the object. There is experimental evidence that these signals may be utilized as triggers for the shifts from one phase to the next, that is, causing the sequential release of "new" motor commands.

The *initial responses*, which also may be denoted as *contact responses*, most likely play a role in triggering of the release the motor commands responsible the loading phase as previously described.

*Responses at Start and Termination of the Movement.*

The most striking feature of the FA II units—that is, Pacinian corpuscle units—was distinct burst responses at the *start of the movement* and at the sudden *cessation of the movement* ending the replacement phase (Fig. 25.10). Remotely located FA II units also responded readily to these transient mechanical events. Even FA II units with endings located at the transition between the palm and the wrist responded. This is in agreement with the abundance of evidence that Pacinian corpuscle units are easily excited by high-frequency vibrations spreading through the tissues (e.g. Johansson, 1978; Johansson, Landström, & Lundström, 1982). The other three types of tactile units in the glabrous skin were virtually indifferent to the mechanical transients present at these moments.

The FA II responses at the start of the movement were probably caused

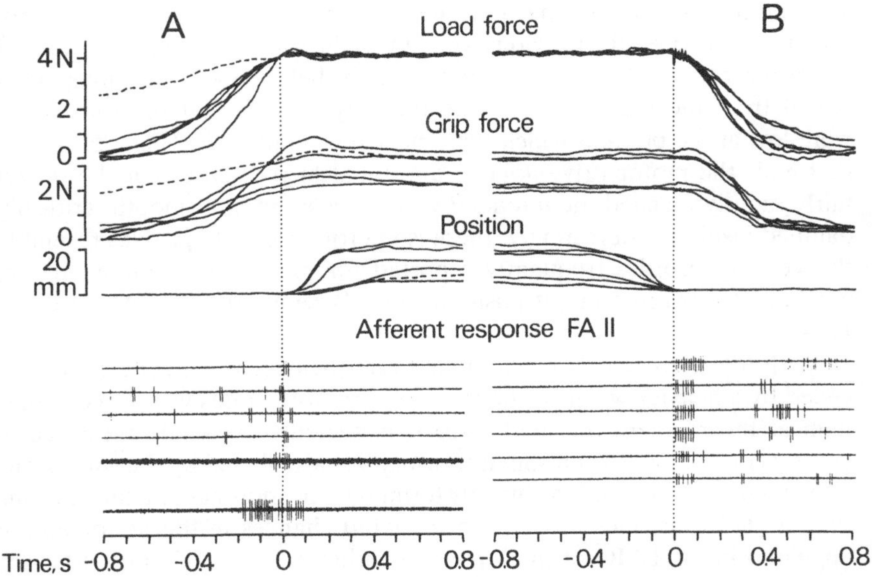


FIG. 25.10. Afferent responses in a single FA II unit during five lifting trials. Load force, grip force and vertical position are superimposed. Vertical lines indicate the start of the vertical movement and moment of terminal table contact in **A** and **B**, respectively. Dashed curves and lowermost nerve trace in **A** indicate lift carried out during extremely slow force changes. The receptor for this unit was located in the palm close to the base of the thumb and the test object was gripped using the tips of the index finger and thumb.

by vibrations due to small lateral sliding movements between the object and the table occurring during the period the lifting force almost counterbalanced a force of gravity. The FA II units responded vigorously, also, in trials with very slow force rates, and the responses during these conditions were prolonged as would be expected due to the prolongation of the period of nearly counterbalancing forces (see Fig. 25.10).

### *Functional Significance of the Responses at the Start and Finish of the Movement*

*Triggering of the termination of the loading phase.* The motor drive accounting for the increase in the grip and load forces in the loading phase was terminated following the first of the two points at which the FA II units responded most distinctly and reliably, that is, after the start of the object's movement.

In lifting series with unexpected weight changes between lifts, it was established that the pattern of force development during the loading phase was programmed to handle the same weight as in the previous lift, that is, a sensorimotor memory specified the force rates. During adequately programmed lifts (i.e., the previous weight was the same as the current weight) the force rate profiles were approximately bell shaped and adequately scaled from the force onset to the final force, such that the force rates were low at the point at which the object started to move (Fig. 25.11). To that end, the motor drive appeared optimal in the sense that the object fairly rapidly reached the intended vertical position in a smooth, critically damped fashion. Interestingly, these isometric force rate profiles resemble the velocity profiles frequently reported for programmed intended arm movements toward a target position (e.g., Bizzi & Abend, 1983; Brooks, 1984).

With lifts erroneously programmed for a heavier weight (i.e., lifts preceded by a heavier weight), the load and grip force rates were exceedingly high at the moment the load force unexpectedly overcame the force of gravity (Fig. 25.12). Then somatosensory signals elicited by the start of the movement were utilized to abruptly terminate the motor commands causing the parallel force increase. These triggered changes in the motor output appeared in the EMG of the intrinsic hand muscles 100–110 ms after the lift-off and at ca. 20-ms shorter latencies for the wrist and elbow muscles. However, the lag between the actual moment of lift-off and the cessation of the force increase caused a pronounced positional overshoot and a high grip force peak, respectively.

With lifts erroneously programmed for lighter weight (i.e., lifts preceded by a lighter weight) the rates of the grip and load forces were prematurely decreased to low values at the point the lift-off was expected (Fig. 25.13).

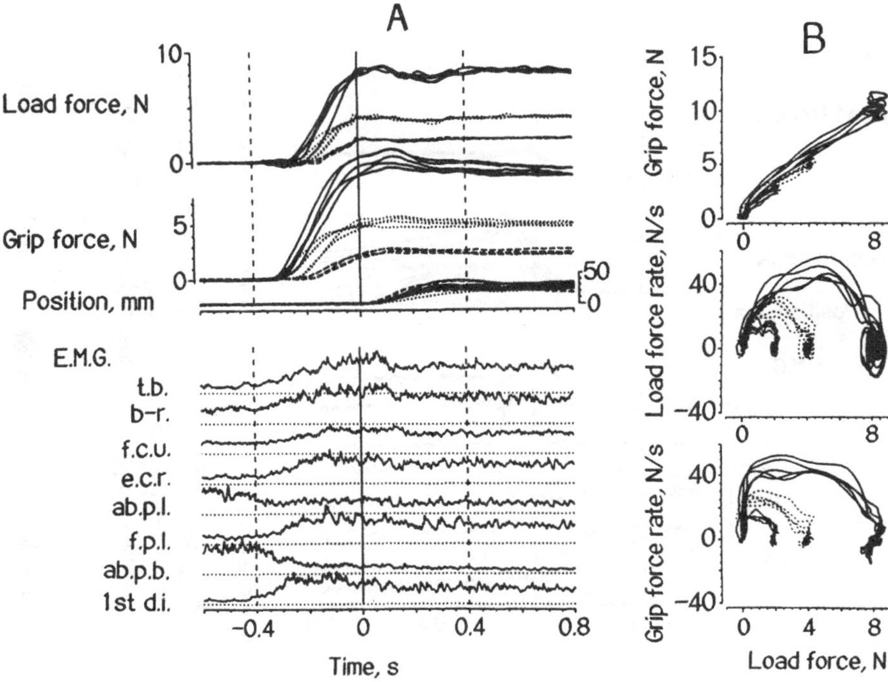


FIG. 25.11. Initial parts of adequately-programmed lifts with 200 g (-----), 400 g (......) and 800 g (______). A Load force, grip force, vertical position as a function of time for 15 lifts (superimposed) by a single subject. EMG signals refer to the 800-g lifts (averaged data). Abbreviations: t.b.—triceps, brachii, b-r.—brachioradialis, f.c.u.—flexor carpi ulnaris, e.c.r.—extensor carpi radiales, ab.p.l.—abductor pollicis longus, f.p.l.—flexor pollicis longus, ab.p.b.—abductor pollicis brevis, 1st d.i.—1st dorsal interosseous. Note that all four pair of antagonist muscles in the hand and arm co-contracted to differing degrees providing the appropriate stiffness of the hand/arm system. Trials synchronized in time at the moment the object started to move (time = 0). **B** Grip force, load force rate, and grip force rate displayed in relation to the load force for the same trials as in A. Note that the grip and load force rate trajectories were mainly single peaked, bell shaped, and roughly proportional to the final force. (Modified from Johansson & Westling, 1988a.)

However, the fact that the object did not move was obviously recorded by somatosensory systems because the parallel force increase then persisted, although more slowly and in a discontinuous fashion. Once the force of gravity was overcome, sensory information was available signaling that the movement started, and this was utilized to terminate the slow force increase. Interestingly, such force rate profiles resemble the multiphased, "discontinuous," velocity profiles observed during "minimally pro-

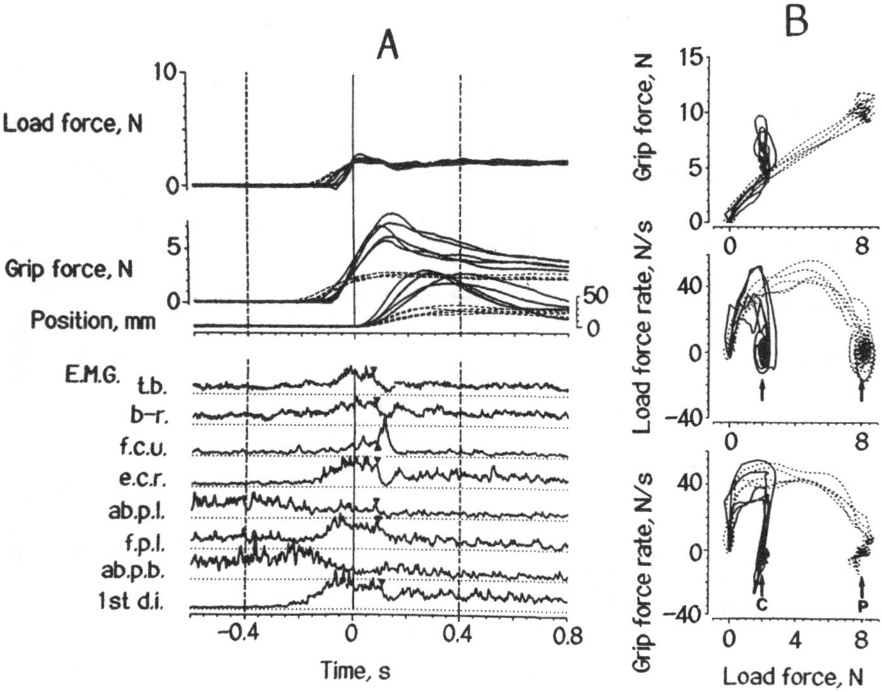


FIG. 25.12. Initial parts of lifts erroneously programmed for a heavier weight. **A** Load force, grip force, vertical position as a function of time for 5 lifts with 200 g programmed for 800 g (______) and for 5 adequately programmed 200-g trials (-----). Note the pronounced overshoots in the grip force and position signals for the erroneously programmed trials. EMG signals refer to the erroneously programmed 200-g lifts (average data). Arrowheads indicate points with fairly abrupt changes in the EMG signals. Trials synchronized in time at the moment the object started to move (time = 0). **B** Grip force, load force rate and grip force rate displayed in relation to the load force for the erroneously programmed trials in A (____) and for 5 adequately programmed 800-g trials (-----). Arrows labelled C and P indicate the load force at which the object started to move in the current and previous trial, respectively. Single subject. For further details see legend to Fig. 25.11 and text. (Modified from Johansson & Westling, 1988a).

grammed" movements during target-reaching tasks (e.g., Abend, Bizzi, & Morasso, 1982; Brooks, 1979, 1984).

Thus, with erroneously programmed lifts, the termination of the loading phase was triggered by sensory signals related to the moment of lift-off. The muscle activation patterns of the loading phase were disrupted, and new sets of muscle commands accounting for the vertical positioning of the object must have been released.

*Triggering of the unloading phase.* A similar triggering by somatosensory information applied to the release of programmed motor commands accounting for the unloading phase (Johansson & Westling, 1988a). The activity changes in intrinsic hand muscles appeared 60–70 ms after the moment of table contact, and the latencies for activity changes for muscles operating over the wrist and elbow were ca. 20 ms shorter. The release commands were always adequately programmed for the weight because unambiguous information about the current weight had been gained at the lift-off.

*Somatosensory signals.* For the somatosensory input that provided the triggers at the lift-off and the terminal table contact, the CNS probably relied on the afferent system that most reliably, quickly, and accurately signaled these mechanical transients. According to a conservative estimate,

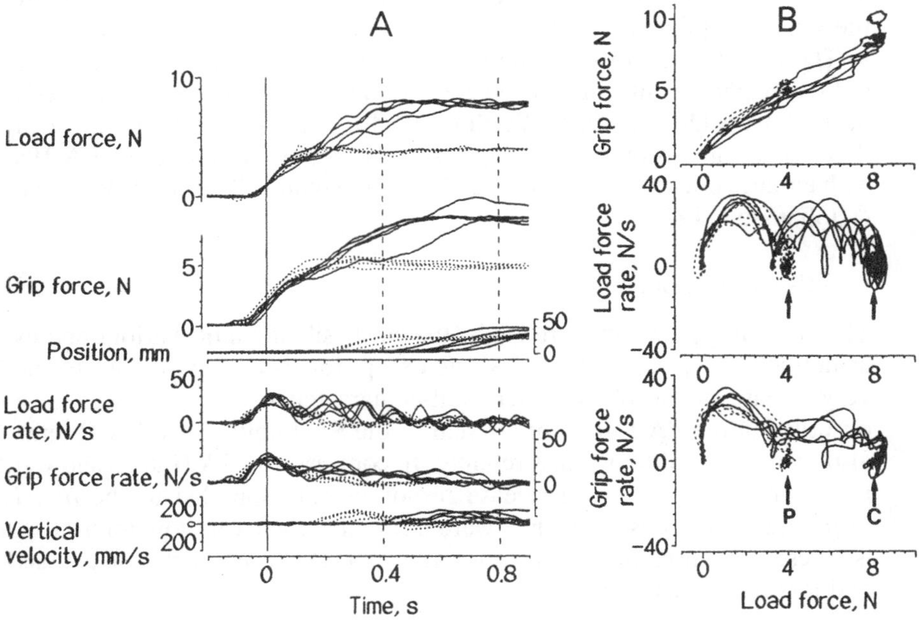


FIG. 25.13. Initial parts of lifts with 800 g erroneously programmed for a lighter weight (400 g). **A** Load force, grip force, vertical position and their time derivatives as a function of time for 5 sample trials with 800 g (______) preceded by adequately programmed 400-g lifts (-----). Trials synchronized in time at the moment the load force reached 1 N (time = 0). **B** Grip force, load force rate and grip force rate displayed in relation to the load force for the same data as in A. Arrows labelled C and P indicate the load force at which the object started to move in the current and previous trial, respectively. Single subject. For further details see legend to Fig. 25.11 and text. (Modified from Johansson & Westling, 1988a.)

about 500–1,000 FA IIs would inject a nearly synchronous impulse volley into the CNS during these events (Westling & Johansson, 1987). Hence, it appears that the FA II afferents provided eminent trigger signals. Available data on the sensitivity of musculotendinous afferents in human finger muscles to small transient external disturbances indicate that these afferents would hardly match the sensitivity of the FA II units (Vallbo, personal communication, 1988). Thus, the high innervation density of FA IIs (compared with musculotendinous receptors; cf. Johansson & Vallbo, 1979), their exquisite sensitivity to mechanical transients, and their location close to the points to contact with the manipulated object should make them the most efficient to provide feedback signals about various mechanical transient events at the object–hand interface. Not surprisingly, the muscle spindles (as well as the Golgi tendon organs) seem to be much more concerned with proprioceptive functions during actively generated movements than with exteroceptive functions (which are primarily mediated by the sensory equipment in the skin) (see, also, Evarts, 1981, 1985).

The fact that the termination of the loading phase and the start of the unloading phase contacting the object may appear normal during anesthesia of the fingers (Johansson & Westling, 1984b) does not contradict the idea that FA II units provide the trigger signals because a number of FA II's with endings outside the anesthetized regions would still respond (Westling & Johansson, 1987).

### *Release Responses*

The unloading phase appeared to be the most "silent" of the various phases as defined in Fig. 25.3. This was true except for its final part close to the very release of the object when well-demarcated *release responses* were observed in the FA I units and in half of the SA I units. Most FA II units also responded, although less reliably. In contrast, the SA II units showed no release responses. The release responses corresponded to the *initial responses* in the sense that they were both elicited at very low force, that is, these responses provided information about the moments of contact and loss of contact with an object.

### *Functional Significance of the Release Responses*

As already pointed out, the CNS seems to require sensory information that an adequate contact is established between the fingers and the object before releasing the muscle commands leading to further manipulation, for example, the loading phase in our lifting task. However, signals related to lost contact might be equally important. For instance, during manipulation they might be used in determining which parts of the fingers are disengaged and free for further tasks or as forewarning that the contact may be lost, which, in turn, might be used to elicit compensatory motor responses.

The need for detailed spatial information about the contact and release patterns may be particularly pronounced while manipulating small objects. The extremely high densities of the tactile afferent units showing distinct contact and release responses in the finger tips (FA I and SA I units) provide the necessary high-spatial acuity (Johansson & Vallbo, 1980). During finger numbness, when the brain has scant information in this respect, manipulative forces often cause small items to elude the grip because the forces are applied at inappropriate directions in relation to the contact areas. During stereognostic discrimination (i.e., gross form discrimination, see Johnson, 1983) the importance of contact and release responses is obvious, especially in units providing high spatial resolution, that is, the FA I and SA I units.

## Additional Responses

### *Responses During Holding*

As schematically indicated in Fig. 25.6, during the static and replacement phases, there was a tonic input from tactile units that seemed to be dominated by a fairly regular impulse activity in the SA II units. During the same periods, many SA I units also continued firing but at lower and irregular rates. Likewise, a minority of the FA I units discharged in response to the physiological muscle tremor. In some trials the impulse frequency increased if the grip force was decreased and *vice versa* which suggests that the responses were actually caused by localized slips triggered by the tremor.

In agreement with the exquisite directional sensitivity of the SA II units to skin stretch, they were strongly influenced by lateral (shear) forces with directional preferences (Westling & Johansson, 1987). Hence, typically, the gripping of the object evoked an ongoing discharge and the discharge rate was either increased or decreased (depending on the properties of the individual unit). As illustrated in Fig. 25.14, loading in the opposite direction (e.g., negative loading) showed the opposite pattern.

At present, it is not at all clear whether the sensitivity of the SA II units to shearing forces across the skin is important in perception. However, as with the FA II units, there are reasons to believe that the SA II units are involved in motor control processes. In addition to exteroceptive attributes, the SA II units clearly show proprioceptive features (Hulliger, et al., 1979; Johansson, 1978; Knibestöl, 1975), and units with discharge properties and end-organs similar to the SA II units in the skin are widely distributed in

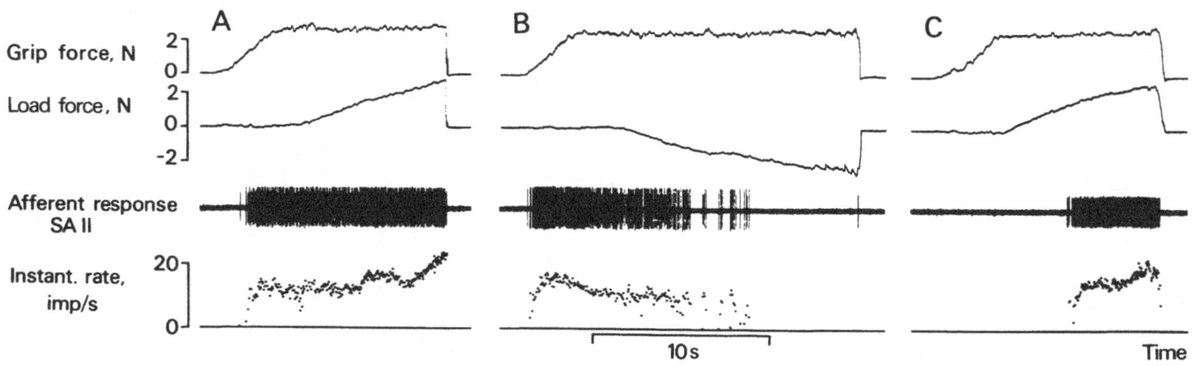


FIG. 25.14. Influence of the load force on the responses in a single SA II unit. The subject was instructed to grip the object and hold a constant grip force (visual force feedback). The experimenter, who supported the object, manually produced slow load force ramps starting after the object had been gripped for a few s. **A** and **B**, the experimenter produced a load directed downward or upward causing a positive or a negative load force, respectively. **C**, the object was re-gripped such that the unit was not excited by the grip force, and a downward load was applied. Note the response to the load causing lateral skin strain. **A** - **C** Surface structure suede. (From Westling & Johansson, 1987.)

the deeper fibrous tissues, for example, joint capsules, tendon sheets, and various fascias (see Burke et al., 1988; Johansson, 1978).

Tonic input from receptors in the fingers are considered to furnish general facilitation of motor commands accounting for finger movements, particularly for flexion movements (see Introduction). However, the pronounced three dimensional force sensitivity of the SA IIs suggests more specific functions. For instance, during manipulation, the population of SA II units may play a role in registering the magnitudes and directions of load forces and the balance between the grip forces and other manipulative forces. Preliminary findings indicate that such information is utilized to regulate the grip forces and opposing forces while holding and further manipulating objects subjected to unpredictable external loads (Riso, Häger, Johansson, & Westling, 1988).

#### *Responses During Isotonic Finger Movement*

During the isotonic finger movements between the lifts (corresponding to ca. 1-cm movements of the finger tips), virtually no responses were observed in the SA I and FA I units innervating the glabrous skin. In contrast, about half of the SA II units responded. These responses were generally modulated by the finger movements. Likewise, the majority of FA II units responded with some impulses. The receptive field characteristics and related properties of the type II units probably account for their high responsiveness to joint movements as well as to different joint positions (Hulliger et al., 1979). Psycho-physical findings indicate that responses in cutaneous afferents might be relevant for kinesthesia and position sense in the human hand (cf. Burgess, Wei, Clark, & Simon, 1982; Clark et al., 1986; McCloskey, 1978). Preliminary findings suggest, however, that signals from receptors in the skin of dorsal aspect of the fingers and palm provide the most detailed information in this respect (Edin, personal communication, 1989).

## GENERAL DISCUSSION AND CONCLUDING REMARKS

### Sensorimotor Set

The triggered changes in the muscle commands following the initial touch, the slips, and the moments of liftoff and terminal table contact clearly involved automatic *task-specific actions*. The various latencies between the triggers and the motor responses were all within the ranges of *automatically initiated* responses. For instance, the latencies of the grip force responses to slips were 74 +/− 9 ms (mean +/− SD), whereas the corresponding voluntary reaction times when subjects were asked to promptly tighten the grip in response to cutaneous sensations from the gripped surfaces were more than twice as long (Johansson & Westling, 1987a).

The automatically triggered responses showed latencies similar to those generally reported for automatic digital motor responses elicited by exteroceptive input (Darton et al., 1985; Garnett & Stephens, 1980; Marsden et al., 1985). These are also similar to response latencies for various "long-latency" mono-articulate stretch reflexes (or M2 responses) in hand and arm muscles (Hammond, 1956; Lee, Murphy, & Tatton, 1983; Lee & Tatton, 1975; Marsden, Merton, & Morton, 1976; Marsden, Rothwell, & Day, 1983; Matthews, 1984a) and more complex (and more natural) automatic compensatory actions to external disturbances requiring coordinated control among multiple hand/arm muscles and joints (Cole & Abbs, 1987; Cole, Gracco, & Abbs, 1984; Cordo & Nashner, 1982; Lacquaniti & Soechting, 1984; Marsden, Merton, & Morton, 1981; Traub, Rothwell, & Marsden, 1980; see also Johansson & Westling, 1988b). It seems plausible that "long-latency" reactions in general represent the expression of the triggered release of prestructured sets of motor commands (cf. Crago, Houk, & Hasan, 1976), and that the function of the neural mechanisms accounting for these reactions in the hand/arm system would be to organize task-dependent muscle synergies constituting various automatic components of voluntary movements. (The autogenic response components merely represent particular ingredients of these commands.) Furthermore, such task-related synergies would represent a high level of adaptability based on integration among multimodal sensory information and previous experiences (for further arguments along these lines, see Johansson & Westling, 1988b). Indeed, there is plenty of evidence that the current "*sensorimotor set*" (which depends on the goal, the context, the phase of the specific movement, and previous experience) determine both how sensory input is to be processed in sensorimotor pathways (e.g., Abbruzzese, Ratto, Favali, & Abbruzzese, 1981; Nelson, 1985; Phillips & Porter, 1977), and the nature of the elicited motor responses (e.g., Evarts & Tanji, 1974; Ghez, Vicario, Martin, & Yumiga, 1983; Gracco & Abbs, 1985; Horak & Nashner, 1986; Houk & Rymers, 1981). Hence, the "sensorimotor set" established by the task involves task-specific relationships between sensory inputs and motor responses.

These relationships, however, may be quite readily modified to "filter out" nonessential or irrelevant stimuli initially eliciting strong but inappropriate motor responses. A striking example occurred when weak cutaneous electric stimulation was delivered through the object while it was held in the air. Initially, motor responses similar to those following slips could be elicited. Such responses, though, gradually declined with repeated shocks and virtually disappeared after only 3 to 4 shocks. The decrease and the final cessation of the response occurred after the nervous system had recognized the task-related irrelevance of the stimuli. The capacity of the CNS to identify and specifically disregard the artificial stimuli was further emphasized by the fact that spontaneously occurring slips still elicited normal motor responses. (For further details, see Johansson & Westling, 1987a.)

The automatic regulatory processes of the current, "sensorimotor set" may strongly interfere with voluntary actions superimposed on the grip. Specifically, the mechanisms that automatically regulate the grip force/load force ratio (to maintain a safety margin to prevent slips) appeared to interfere with the drive to voluntarily decrease the grip force while the test object was held in the air (Johansson & Westling, 1987a). During the period of decreasing force, the subjects felt as if their fingers adhered to the object. The effort required to overcome this *magnet phenomenon* was especially pronounced with sandpaper as the surface structure, and became stronger the closer the grip force was to the minimum grip force required to prevent slips. Indeed, there were slip-triggered grip force responses similar to those observed during spontaneous slips of ordinary lifts. In striking contrast to the ordinary slip responses, however, they did not result in a maintained change of the force balance. Thus, the voluntary command to separate the fingers appeared to selectively suppress one component of the motor response, that is, the maintained, memory-based upgrading of the grip force. That an early brief force response still remained suggests that two different CNS mechanisms mediate the motor response to slips.

## General View of the Control of the Lifting Task

According to traditional concepts, the production and release of coordinated muscle commands during voluntary actions may be regarded as controlled by stored motor programs (Keele, 1968; Lashley, 1917), most of which are acquired during fumbling experiences in early childhood. However, such movement templates, capable of specifying the sequential and parallel aspects of the required muscle coordination, would necessarily contain certain variant features to provide flexibility to adapt the motor responses to the specific environmental demands (see Brooks, 1979; Keele & Summers, 1976; Pew, 1974; Schmidt, 1976). In the present lifting task, one rather invariant feature of the programs utilized was the parallel grip and load force changes. In contrast, the ratio of grip force/load force and the force rate profiles during the phases of parallel force change were flexible parameters that were adapted to the frictional condition and to the object's weight, respectively. These parameters were defined by information in sensorimotor memories related to the frictional and weight conditions of the current object. It is quite likely that other physical parameters might be represented as well that would be relevant for the successful control of manipulation, for instance, the object's moment of inertia and its center of gravity.

Because of this *internal representation*, based on previous experiences with the current object or similar objects, the motor system primarily works in an anticipatory control mode. For instance, reliance on programmed coordination, adjusted to the current friction, would allow the CNS to simultaneously change the grip and load forces while automatically maintaining an appropriate balance between the two. The intermittent resetting of an open-loop operation would avoid the time lag that would be involved if the grip force was regulated exclusively on the basis of a continuous closed-feedback loop using cutaneous receptors, for example, a continuous tactile feedback about incipient slip (cf. Rack, 1981). Moreover, a programmed force coordination that automatically maintains an appropriate safety margin to prevent slips probably disengages a larger fraction of the sensorimotor apparatus for the higher level exploratory and/or manipulatory tasks usually superimposed on the basic precision grip. Likewise, during complex manipulation involving bimanual interactions with objects, there are sophisticated anticipatory actions involved that are all dynamically scaled on the basis of information about the relevant physical parameters of the system manipulated (Johansson & Westling, 1988b).

With an erroneous internal representation resulting in erroneous programming, somatosensory signals may cause compensatory actions and update various parameters of the internal model. The tactile sense, which conveys information about mechanical events particularly in the interface between the hand and the manipulated object, plays a crucial role in this context. It is well known that the dominating symptoms in patients with deficits in the sensibility of the fingers are problems during precise manipulative tasks. The sensory signals automatically evoke various compensatory actions: (a) immediate corrections, that is, "safety mechanisms" programmed as a part of the motor set are triggered, (b) the proposed internal representation of the object is automatically updated/corrected, that is, adaptive capacity is provided, and (c) sensory information must be used during sensorimotor learning when new models of unknown objects are created.

## ACKNOWLEDGMENTS

This study was supported by the Swedish Medical Research Council (project 14X-08667), Gunvor and Josef Aners Stiftelse, and the University of Umeå, and is gratefully acknowledged.

## REFERENCES


Abbruzzese, G., Ratto, S., Favale, E., & Abbruzzese, M. (1981). Proprioceptive modulation of somatosensory evoked potentials during active or passive finger movements in man. *Journal of Neurology, Neurosurgery and Psychiatry, 44*, 942–949.

Abend, W., Bizzi, E., & Morasso, P. (Eds.). (1982). Human arm trajectory formation. *Brain, 105*, 331–348.

Bonnet, M., & Requin, J. (1982). Long loop and spinal reflexes in man during preparation for intended directional hand movements. *Journal of Neuroscience, 2*, 90–96.

Brooks, V. B. (1979). Motor programs revisited. In R. E. Talbott & D. R. Humphrey (Eds.), *Posture and movement* (pp. 13–49). New York: Raven.

Brooks, V. B. (1984). How are "move" and "hold" programs matched. In Bloedel et al. (Ed.), *Cerebellar functions* (pp. 1–23). Berlin: Springer-Verlag.

Burgess, P. R., & Perl, E. R. (1973). Cutaneous mechanoreceptors and nociceptors. In A. Iggo (Ed.), *Somatosensory system. Handbook of sensory physiology* (Vol. II, pp. 29–78). Berlin: Springer-Verlag.

Burgess, P. R., Wei, J. K., Clark, F. J., & Simon, J. (1982). Signaling of kinesthetic information by peripheral sensory receptors. *Annual Review of Neuroscience, 5*, 171–187.

Burke, D., Gandevia, S. C., & Macefield, G. (1988). Responses to passive movements of receptors in joint, skin and muscle of the human hand. *Journal of Physiology (London), 402*, 347–361.

Caccia, M. R., McComas, A. J., Upton, A. M., & Blogg, T. (1973). Cutaneous reflexes in small muscles of the hand. *Journal of Neurology, Neurosurgery and Psychiatry, 36*, 960–977.

Clark, F. J., Burgess, R. C., & Chapin, J. W. (1986). Proprioception with the proximal interphalangeal joint of the index finger. *Brain, 109*, 1195–1208.

Cole, K. J., & Abbs, J. H. (1987). Kinematic and electromyographic responses to perturbation of a rapid grasp. *Journal of Neurophysiology, 57*, 1498–1510.

Cole, K. J., Gracco, V. L., & Abbs, J. H. (1984). Autogenic and nonautogenic sensorimotor actions in the control of multiarticulate hand movements. *Experimental Brain Research, 56*, 582–585.

Cordo, P. J., & Nashner, L. M. (1982). Properties of postural adjustments associated with rapid arm movements. *Journal of Neurophysiology, 47*, 287–302.

Crago, P. E., Houk, J. C., & Hasan, Z. (1976). Regulatory actions of human stretch reflex. *Journal of Neurophysiology, 39*, 925–935.

Crenna, P., & Frigo, C. (1984). Evidence of phase-dependent nociceptive reflexes during locomotion in man. *Experimental Neurology, 85*, 336–345.

Darton, K., Lippold, O. C. J., Shahani, M., & Shahani, U. (1985). Long-latency spinal reflexes in humans. *Journal of Neurophysiology, 53*, 1604–1618.

Day, B. L., & Marsden, C. D. (1982). Accurate repositioning of the human thumb against unpredictable loads is dependent upon peripheral feed-back. *Journal of Physiology (London), 327*, 393–407.

Denny-Brown, D. (1966). *The cerebral control of movement.* Liverpool: The Liverpool University Press.

Evarts, E. V., & Tanji, J. (1974). Gating of motor cortex reflexes by prior instruction. *Brain Research, 71*, 479–494.

Evarts, E. V. (1981). Sherrington's concepts of proprioception. *Trends in Neuroscience, 4*, 44–46.

Evarts, E. V. (1985). Transcortical reflexes: Their properties and functional significance. Hand Function and the neocortex. *Experimental Brain Research, Suppl., 10*, 130–154.

Forssberg, H., Grillner, S., & Rossignol, S. (1975). Phasic dependent reflex reversal during walking in chronic spinal cats. *Brain Research*, *85*, 103–107.

Forssberg, H., Grillner, S., & Rossignol, S. (1977). Phasic gain control of reflexes from the dorsum of the paw during locomotion. *Brain Research*, *132*, 121–139.

Gandevia, S. C., & McCloskey, D. I. (1977a). Effects of related sensory inputs on motor performance in man studied through changes in perceived heaviness. *Journal of Physiology (London)*, *272*, 653–672.

Gandevia, S. C., & McCloskey, D. I. (1977b). Changes in motor commands, as shown by changes in perceived heaviness, during partial curarization and peripheral anaesthesia in man. *Journal of Physiology (London)*, *272*, 673–689.

Garnett, R., & Stephens, J. A. (1980). The reflex responses of single motor units in human first dorsal interosseous muscle following cutaneous afferent stimulation. *Journal of Physiology (London)*, *303*, 351–364.

Garnett, R., & Stephens, J. A. (1981). Changes in the recruitment threshold of motor units in the first human dorsal interosseous muscle following cutaneous afferent stimulation. *Journal of Physiology (London), 303,* 351–364.

Ghez, C., Vicario, D., Martin, J. H., & Yumiya, H. (1983). Sensory motor processing of target movements in motor cortex. In J. E. Desmedt (Ed.), *Motor control mechanisms in health and disease* (pp. 61–92). New York: Raven.

Gracco, V. L., & Abbs, J. H. (1985). Dynamic control of the perioral system during speech: Kinematic analyses of autogenic and nonautogenic sensorimotor processes. *Journal of Neurophysiology*, *54*, 418–432.

Grillner, S. (1981). Control of locomotion in bipeds, tetrapods, and fish. In V. B. Brooks (Ed.), *Handbook of physiology: The nervous system* (Vol. 2, pp. 1179–1236). Bethesda, MD: American Physiology Society.

Hagbarth, K. E., & Finer, B. L. (1963). The plasticity of human withdrawal reflexes to noxious skin stimuli in lower limbs. In G. Moruzzi, A. Fessard, & H. H. Jasper (Eds.), Brain Mechanisms: *Progress in brain research*, (Vol. 19, pp. 65–81). Amsterdam: Elsevier.

Hammond, P. H. (1956). The influences of prior instructions to the subject on an apparently involuntary neuro-muscular response. *Journal of Physiology (London)*, *132*, 17P–18P.

Horak, F. B., & Nashner, L. M. (1986). Central programming of postural movements: Adaptation to altered support surface configurations. *Journal of Neurophysiology*, *55*, 1369–1381.

Houk, J. C., & Rymers, W. Z. (1981). Neural control of muscle length and tension. In V. B. Brooks (Ed.), *Handbook of physiology: The nervous system*, (Vol. 2, pp. 257–323). Bethesda, MD: American Physiology Society.

Hulliger, M., Nordh, E., Thelin, A. E., & Vallbo, Å. B. (1979). The responses of afferent fibres from the glabrous skin of the hand during voluntary finger movements in man. *Journal of Physiology (London)*, *291*, 233–249.

Iggo, A. (1974). Cutaneous receptors. In J. I. Hubbard (Ed.), *Peripheral nervous system* (pp. 347–404). New York: Plenum Press.

Jenner, J. R., & Stephens, J. A. (1982). Cutaneous reflex responses and their central nervous pathways studied in man. *Journal of Physiology (London)*, *333*, 405–419.

Johansson, R. S. (1978). Tactile sensibility in the human hand: Receptive field characteristics of mechanoreceptive units in the glabrous skin area. *Journal of Physiology (London)*, *281*, 101–123.

Johansson, R. S., Landström, U., & Lundström, R. (1982). Responses of mechanoreceptive afferent units in the glabrous skin of the human hand to sinusoidal skin displacements. *Brain Research*, *244*, 17–25.

Johansson, R. S., & Vallbo, Å. B. (1979). Tactile sensibility in the human hand: Relative and absolute densities of four types of mechanoreceptive units in the glabrous skin. *Journal of Physiology (London)*, *286*, 283–300.

Johansson, R. S., & Vallbo, Å. B. (1980). Spatial properties of the population of mechanoreceptive units in the glabrous skin of the human hand. *Brain Research*, *184*, 353–366.

Johansson, R. S., & Vallbo, Å. B. (1983). Tactile sensory coding in the glabrous skin of the human hand. *Trends in Neuroscience*, *6*, 27–31.

Johansson, R. S., & Westling, G. (1984a). Influences of cutaneous sensory input on the motor coordination during precision manipulation. In C. von Euler, O. Franzen, U. Lindblom, & D. Ottoson (Eds.), *Somatosensory Mechanisms* (pp. 249–260). London: Macmillan Press.

Johansson, R. S., & Westling, G. (1984b). Roles of glabrous skin receptors and sensorimotor memory in automatic control of precision grip when lifting rougher or more slippery objects. *Experimental Brain Research*, *56*, 550–564.

Johansson, R. S., & Westling, G. (1987a). Signals in tactile afferents from the fingers eliciting adaptive motor responses during precision grip. *Experimental Brain Research*, *66*, 141–154.

Johansson, R. S., & Westling, G. (1987b). Tactile afferent input influencing motor coordination during precision grip. In A. Struppler & A. Weindl (Eds.), *Clinical aspects of sensory motor integration.* (pp. 3–13). Berlin: Springer-Verlag.

Johansson, R. S., & Westling, G. (1988a). Coordinated isometric muscle commands adequately and erroneously programmed for the weight during lifting tasks with precision grip. *Experimental Brain Research*, *71*, 59–71.

Johansson, R. S., & Westling, G. (1988b). Programmed and reflex actions to rapid load changes during precision grip. *Experimental Brain Research*, *71*, 72–86.

Johansson, R. S., Trulsson, M., Olsson, K. Å., & Westberg, K. G. (1988). Mechanoreceptor activity from the human face and oral mucosa in man. *Experimental Brain Research*, *72*, 204–208.

Johansson, R. S., Trulsson, M., Olsson, K. Å., & Abbs, J. H. (1988). Mechanoreceptive afferent activity in the infraorbital nerve in man during speech and chewing movements. *Experimental Brain Research*, *72*, 209–214.

Johnson, K. O. (1983). Neural mechanisms of tactual form and texture discrimination. *Federation Proceedings*, *42*, 2542–2527.

Kanda, K., & Desmedt, J. E. (1983). Cutaneous facilitation of large motor units and motor control of human fingers in precision grip. In J. E. Desmedt (Ed.), *Motor control mechanisms in health and disease.* (pp. 253–261). New York: Raven.

Keele, S. W. (1968). Movement control in skilled motor performance. *Psychological Bulletin*, *70*, 387–404.

Keele, S. W., & Summers, J. J. (1976). The structure of motor programs. In G. E. Stelmach (Ed.), *Motor control: Trends and issues.* (pp. 109–142). New York: Academic Press.

Knibestöl, M. (1975). Stimulus-response functions of slowly adapting mechanoreceptors in the human glabrous skin area. *Journal of Physiology (London)*, *245*, 63–80.

Knibestöl, M., & Vallbo, Å. B. (1970). Single unit analysis of mechanoreceptor activity from the human glabrous skin. *Acta Physiologica Scandinavica*, *80*, 178–195.

Lacquaniti, F., & Soechting, J. F. (1984). Behavior of the stretch reflex in a multi-jointed limb. *Brain Research*, *311*, 161–166.

Lashley, K. S. (1917). The accuracy of movement in the absence of excitation from the moving organ. *Journal of Physiology (London)*, *43*, 169–194.

Lee, R. G., Murphy, J. T., & Tatton, W. G. (1983). Long-latency myotatic reflexes in man: Mechanisms, functional significance, and changes in patients with Parkinson's disease or hemiplegia. In J. E. Desmedt (Ed.), *Motor control mechanisms in health and disease* (pp. 489–508). New York: Raven Press.

Lee, R. G., & Tatton, W. G. (1975). Motor responses to sudden limb displacements in primates with specific CNS disorders and in human patients with motor system disorders. *Canadian Journal of Neurological Sciences*, *2*, 285–293.

Lund, J. P., & Olsson, K. Å. (1983). The importance of reflexes and their control during jaw movement. *Trends in Neuroscience, 6,* 458–463.

Marsden, C. D., Merton, P. A. & Morton, H. B. (1981). Human postural responses. *Brain, 104*, 513–534.

Marsden, C. D., Merton, P. A., & Morton, H. B. (1976). Servo action in the human thumb. *Journal of Physiology (London)*, *257*, 1–44.

Marsden, C. D., Merton, P. A., & Morton, H. B. (1977). The sensory mechanisms of servoaction in human muscle. *Journal of Physiology (London)*, *265*, 521–535.

Marsden, C. D., Merton, P. A., & Morton, H. B. (1985). New observations on the human stretch reflex. *Journal of Physiology (London)*, *360*, 51P.

Marsden, C. D., Rothwell, J. C., & Day, B. L. (1983). Long-latency automatic responses to muscle stretch in man: Origin and function. In J. E. Desmedt (Ed.), *Motor control mechanisms in health and disease* (pp. 509–539). New York: Raven Press.

Matthews, P. B. C. (1984a). Evidence from the use of vibration that the human long-latency stretch reflex depends upon spindle secondary afferents. *Journal of Physiology (London)*, *348*, 383–415.

Matthews, P. B. C. (1984b). The contrasting stretch reflex responses of the long and short flexor muscles in the human thumb. *Journal of Physiology (London)*, *348*, 545–558.

McCloskey, D. I. (1978). Kinesthetic sensibility. *Physiological Reviews*, *58*, 763–820.

Moberg, E. (1962). Criticism and study of methods for examining sensibility in the hand. *Neurology*, *12*, 8–19.

Moberg, E. (1983). The role of cutaneous afferents in position sense, kineasthesia, and motor functions of the hand. *Brain*, *106*, 1–19.

Mott, F. W., & Sherrington, C. S. (1895). Experiments upon the influence of sensory nerves upon movement and nutrition of the limbs. *Proceedings of the Royal Society of London, B.*, *57*, 481–488.

Nelson, R. J. (1985). Sensorimotor cortex responses to vibrotactile stimuli during initiation and execution of hand movement. In A. W. Goodwin & I. Darian-Smith (Eds.), *Hand function and the neocortex* (pp. 59–76). Berlin: Springer-Verlag.

Phillips, C. G., & Porter, R. (1977). *Corticospinal Neurones.* London: Academic Press.

Pew, R. W. (1974). Human perceptual motor performance. In B. H. Kantowitz (Ed.), *Human information processing: Tutorials in performance and cognition* (pp. 1–39). Hillsdale, NJ: Lawrence Erlbaum Associates.

Pubols, B., & Pubols, L. (1983). Tactile receptor discharge and mechanical properties of the glabrous skin. *Federation Proceedings*, *42*, 2528–2535.

Rack, P. M. H. (1981). Limitations of somatosensory feedback in control of posture and movements. In V. B. Brooks (Ed.), *Handbook of physiology: The nervous system* (Vol. 2, pp. 229–256) Bethesda, MD: American Physiology Society.

Riso, R., Häger, C., Johansson, R. S., & Westling, G. (1988). Grasp force response in man matches the characteristics of pulling loads. *Soc Neuroscience Abstracts.*

Rood, O. N. (1860). On contraction of the muscles induced by contact with bodies in vibration. *American Journal of Sci Arts*, *24*, 449.

Schmidt, R. A. (1976). The schema as a solution to some persistent problems in motor learning theory. In G. E. Stelmach (Ed.), *Motor control: Trends and issues* (pp. 41–65) New York: Academic Press.

Tatton, W. G., & Bruce, I. C. (1981). Comment: A schema for the interactions between motor programs and sensory input. *Canadian Journal of Physiology*, *59*, 691–699.

Torebjörk, H. E., Hagbarth, K. E., & Eklund, G. (1978). Tonic finger flexion reflex induced by vibratory activation of digital mechanoreceptors. In G. Gordon (Ed.), *Active touch* (pp. 197–203). Oxford: Pergamon.

Traub, J. C., Rothwell, J. C., & Marsden, C. D. (1980). A grab reflex in the human hand. *Brain*, *103*, 869–884.

Twitchell, T. E. (1954). Sensory factors in purposive movement. *Journal of Neurophysiology*, *17*, 239–252.
Vallbo, Å. B., & Hagbarth, K. E. (1968). Activity from skin mechanoreceptors recorded percutaneously in awake human subjects. *Experimental Neurology*, *21*, 270–289.
Vallbo, Å. B., Hagbarth, K-E., Torebjörk, H. E., & Wallin, B. G. (1979). Somatosensory, proprioceptive and sympathetic activity in human peripheral nerves. *Physiological Reviews*, *59*, 919–957.
Vallbo, Å. B., & Johansson, R. S. (1984). Properties of cutaneous mechanoreceptors in the human hand related to touch sensation. *Human Neurobiology*, *3*, 3–14.
Westling, G. K. (1986). Sensorimotor mechanisms during precision grip in man. *Umeå University Medical Dissertations*, (New Series No. 171, pp. 1–20).
Westling, G., & Johansson, R. S. (1984). Factors influencing the force control during precision grip. *Experimental Brain Research*, *53*, 277–284.
Westling, G., & Johansson, R. S. (1987). Responses in glabrous skin mechanoreceptors during precision grip in humans. *Experimental Brain Research*, *66*, 128–140.

# 26 Motor Representations in Deafferented Humans: A Mechanism for Disordered Movement Performance

Jerome N. Sanes
*Human Motor Control Section, Medical Neurology Branch*
*National Institute of Neurological Disorders and Stroke*
*National Institutes of Health*

## Abstract

Humans with a large-fiber sensory neuropathy exhibit several abnormalities in movement control, encompassing deficits in postural control, locomotion, and purposive simple or complex limb movements. Accompanying the disordered movements are abnormal patterns of muscle activity often manifested as extremes in co-contraction levels during successful maintenance of steady postures. This indicates that constant motor output is poorly regulated in deafferented patients. It was hypothesized that the fluctuations in motor output exhibited by deafferented patients, as seen in the electromyogram, were related to a poor perceptual appreciation of central nervous system output signals. This was tested by evaluating the sense of muscular effort. Deafferented patients showed an impaired effort sense on several different types of tests and a corresponding lower concordance between psychophysical decisions and the observed muscle activity. This result suggests that appreciation of motor output is mediated in part by peripheral return from somatic sensory afferent systems.

## INTRODUCTION

There is a wealth of clinical and experimental data that supports the idea that somesthetic information has a significant role in regulating all phases of limb movement control and hand manipulation (Cole & Abbs, 1988;

---

The current address of the author is: Jerome N. Sanes, Center for Neural Sciences, Brown University, Box 1953, Providence, RI 02912.

Johansson & Westling, 1987; Mott & Sherrington, 1895; Rothwell, Traub, Day, Obeso, Thomas, & Marsden, 1982; Sanes, Mauritz, Dalakas, & Evarts, 1985). Perhaps the best experimental model by which to demonstrate global effects of somesthetic afferent control of movement has been the deafferented preparation. Although both deafferented monkeys (surgically) and humans (pathologically) are capable of initiating and terminating voluntary movements, there is not a consensus as to the characteristics of these movements. Despite a long history of experimental and clinical investigations of deafferented nonhuman primates and deafferented humans (for reviews see Nathan & Sears, 1960; Sanes et al., 1985), the conclusions from such studies must be taken with caution because the methodologies employed were often primitive and incomplete. More recently, Bizzi and his colleagues (Accornero, Chapple, & Hogan, 1984; Bizzi, Dev, Morasso, & Polit, 1978; Bizzi, Polit, & Morasso, 1976; Polit & Bizzi, 1979) observed that the learned end point of voluntary movements performed by surgically deafferented monkeys was well achieved, but with a relatively large variability. Additionally, new movements requiring a different set of kinesthetic cues and learned movements performed under mechanical constraints different from those encountered during the task learning were performed inaccurately. These results were, at first, explained with regard to the equilibrium point model of voluntary movement control that was derived from the theoretical and experimental results of Fel'dman (Asatryan & Fel'dman, 1965; Fel'dman 1966a; 1966b) and then according to a virtual trajectory model (Hogan, 1984). In these models, there is a central neural specification of either the muscle activities required to attain the posture at the end point of the movement (Lestienne, Polit, & Bizzi, 1981) or the point-to-point location of a projected movement path (Bizzi et al., 1984). The evidence for such models of movement control has been reasonably good from experiments that have used nonhuman primates. However, data derived from human subjects have generally not supported the equilibrium point model of voluntary movement control. In particular, there has been a failure to observe the appropriate reciprocal levels of muscle activity at "final" positions (Matheson, Hallett, Berardelli, Weinhaus, & Inzucchi, 1985), inaccuracies have been observed in final positioning following application of dynamic loads that trigger short latency muscle responses (Day & Marsden, 1982; Sanes, 1986), and poor control of movement trajectories and end point accuracy occurs in patients with a pathological loss of proprioception (Forget & Lamarre, 1987; Rothwell et al., 1982; Sanes et al., 1985). The latter results are of particular interest because evaluation of humans with a selective neuropathy of sensory fibers provides an opportunity to study the contribution of cutaneous and proprioceptive afferents to voluntary movement control. Indeed, movement control in such patients has been described as generally crude and uncoordinated and often interrupted by pseudo-athetoid involuntary movements (Rothwell et al., 1982; Sanes et al., 1985). However, the neural mechanisms contributing to the disordered movement control in deafferented humans remains unclear. The subject of the present report is to present psychophysical and physiological data that suggest that the movement disorders in deafferented subjects are related to deficient monitoring of the signals used to control muscle activity.

## DISORDERS OF VOLUNTARY MOVEMENTS IN DEAFFERENTED HUMANS

Patients with a large-fiber peripheral neuropathy that selectively attacks somesthetic afferents have a variety of disorders in voluntary movement (Forget & Lamarre, 1987; Rothwell et al., 1982; Sanes et al., 1985). Figure 26.1A illustrates what is probably the fundamental defect in movement control of deafferented humans. In contrast to normal subjects, humans with proprioceptive loss typically fail to maintain a static hand posture in the absence of visual guidance. The response by the deafferented subjects is not necessarily what would be predicted on the basis of mechanical stimuli applied to the hand. The observed positional drift can be in the direction that an external load is pushing the hand (e.g., toward extension when a flexor load is present) or can be against the direction the load is pushing the hand. Furthermore, the absolute magnitude of the positional drift is not simply related to the magnitude of the applied load; the largest errors in postural control do not necessarily occur when the lightest constant loads are applied (Sanes et al., 1985). Thus, deafferented patients do not necessarily maintain a static position and in not doing so, exhibit positional responses that would not necessarily be predicted on the basis on simple mechanical constraints.

Prior results illustrating patterns of muscle activity accompanying postural maintenance (Fig. 26.1B, upper section, Sanes et al., 1985) might indicate the mechanism by which postural control is impaired in the deafferented patients. The lower section of Fig. 26.1B illustrates records from a normal subject showing the typical electromyogram (EMG) exhibited during postural maintenance. Through central nervous system computations about the initial conditions of limb position and external loading, the subject sets levels of activity in antagonist pairs of muscles that tend to maintain the hand in the correct position. This EMG pattern is typically maintained, with minor fluctuations, despite repeated, and sometimes long-term, withdrawal of visual guidance. The mechanism by which normal subjects maintain the steady posture despite challenges of absent visual guidance and sustained muscle contractions resulting in changes in muscle properties could be related to a slowly decaying output signal or to periodic updates of the output signal by comparing the sensory inputs, visual and somesthetic, with the expected output. In contrast to the steady state EMG of normal subjects during postural maintenance, the deafferented subjects, when they were able to maintain a relative amount of postural stability, exhibited a varying EMG pattern. The EMG showed widely different levels of co-contraction in antagonist muscle pairs, but with all yielding a similar end point. The fluctuations in the EMG were independent of the external loading or of the availability of visual guidance. It is possible that the widely divergent EMG patterns yielding similar kinematic outputs in the deafferented patients reflect a mismatch between the output command from the central nervous system to the muscles and the expected inputs from somesthetic receptors. Of course, the deafferented patients have impoverished or absent somesthetic inputs. Thus, the output system for muscle control is working essentially open loop. Other models of feedback, such as vision, not readily associated with muscle activity can assist in position control but apparently do not become easily associated with muscular output.

## SENSE OF MUSCULAR EFFORT AND LARGE-FIBER SENSORY NEUROPATHY

A prevalent argument concerning the neural substrate of the sense of muscular effort is that it is mediated primarily by internal monitoring of CNS signals directed toward muscles. These results were based primarily on the effects of reversible or pathological limb paralysis on the perception of muscular effort (Gandevia, 1982; Gandevia & McCloskey, 1977b). In addition, observations from a single deafferented human indicated an intact sense of effort in the absence of proprioception, though these results may have been contaminated by the patients' ability to detect the speed of movement from external cueing (Rothwell et al., 1982). However, there are other indications that somesthetic afferents, if not completely mediating, do contribute to effort sense (Cafarelli & Kostka, 1981; Gandevia & McCloskey, 1977a; Marsden, Rothwell, & Traub, 1979; Roland & Ladegaard-Pedersen, 1977). The observations that deafferented patients were unable to control muscular output on a moment-to-moment basis during a postural task and cannot appropriately scale the amplitude or timing of muscle activity (Sanes et al., 1985; Forget & Lamarre, 1987) suggested that in these patients the neural commands to muscles were inappropriately calibrated or decomposed rapidly. This in turn might suggest that deafferented patients have a poor ability to recognize the magnitude of the motor output command which has been suggested to be related to the sense

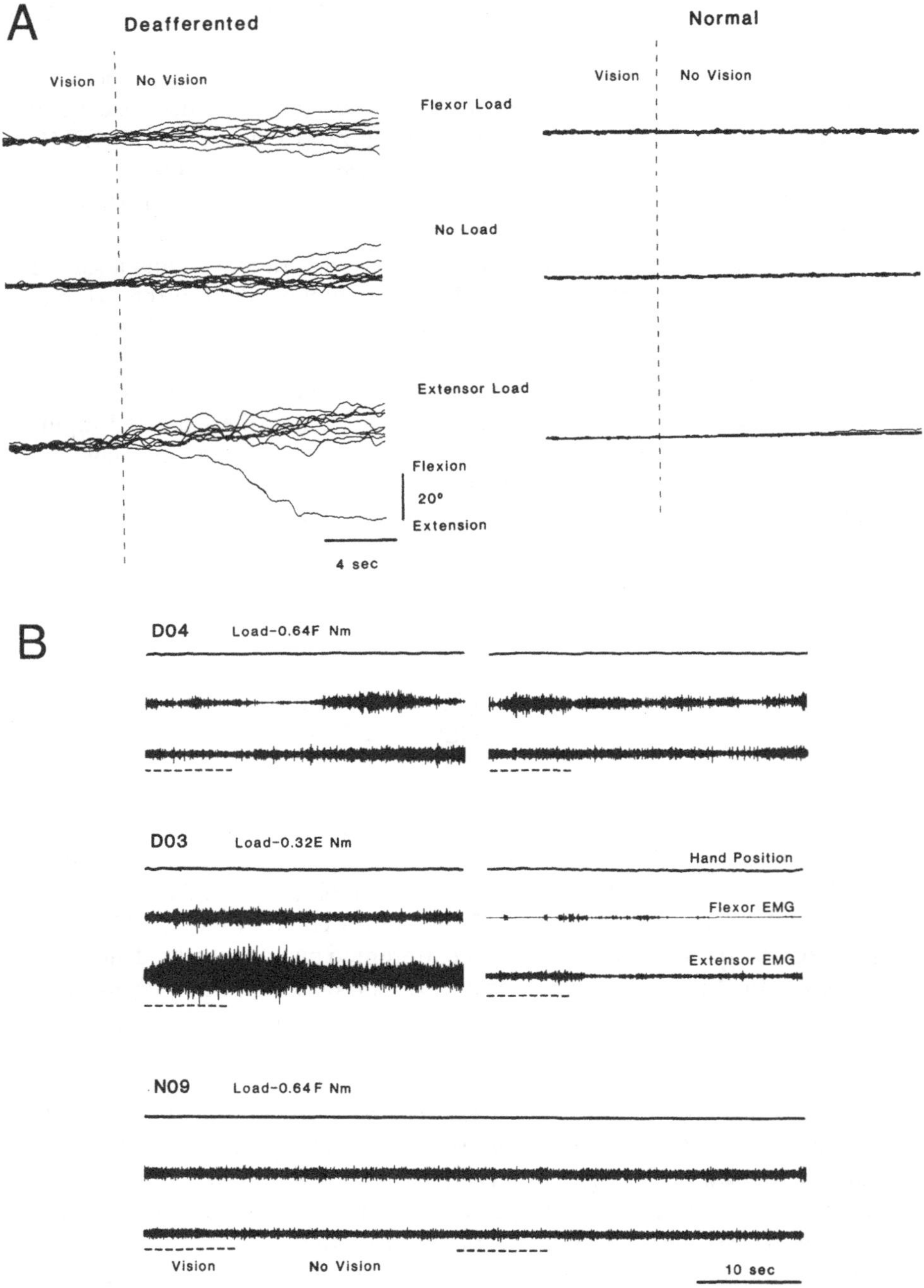

A
Deafferented
Normal
Vision
No Vision
Vision
No Vision
Flexor Load
No Load
Extensor Load
Flexion
20°
Extension
4 sec
B
D04
Load-0.64F Nm
D03
Load-0.32E Nm
Hand Position
Flexor EMG
Extensor EMG
N09
Load-0.64F Nm
Vision
No Vision
10 sec

of effort. With this in mind, the issue of perceptual appreciation of muscular effort was reexamined in deafferented patients with psychophysical and physiological methods.

## METHODS

### Patients

The patients ($n = 6$) evaluated in this study presented originally with a progressive and predominant large-fiber sensory neuropathy. Duration of the disease process ranged from 5 to about 30 years. A standard neurological examination indicated reduced sensation to light touch, temperature, pin prick, and limb position and vibration sense. The severity and extent of the sensory deficit varied from patient to patient; in some, it extended only up the elbows and knees, whereas in others, the sensory deficits had progressed up to and including the shoulders and groin area. Muscle strength was normal or near normal and deep tendon reflexes were absent in all patients. Graphesthesia and stereognosis were typically impaired. The clinical neurophysiology examination showed normal motor nerve conduction velocities in all but one patient and absent sensory potentials in all patients. A nerve biopsy showed a predominant loss of large myelinated fibers and in two patients evidence of mild denervation. More details about the neurological and diagnostic examinations can be found elsewhere (Sanes et al., 1985). Normal control subjects ($n = 7$) were studied for comparison to the deafferented patients.

### Apparatus

All studies were done with a torque motor system that has been previously described (Sanes, 1986). Briefly, subjects were seated with the lower arm(s) and hand(s) restrained between lightly padded supports with the elbow

---

FIG. 26.1 (on facing page). Postural control and muscle responses in a normal and a deafferented human. **A.** Superimposed hand position records of several attempts by a deafferented patient (left) and a normal control (right) to maintain postural stability against a load opposing flexion (top) or extension (bottom) or no load. To the left of the vertical dashed line the subjects had continuous visual guidance of hand position, whereas to the right of the line the video screen was blanked. **B.** Hand position and wrist flexor and extensor muscle activity for two deafferented patients (D03 and D04) and one normal subject (N09) during successful maintenance of a static posture. The dashed lines underneath the extensor EMG indicate when subjects had visual guidance of the hand position.

angle at approximately 90°. The hand was positioned such that flexion–extension movements about the wrist could be executed in the horizontal plane. The wrist was colinear with an axle of a torque motor that artificially produced external constant torques of varying magnitudes that passively pushed the hand toward wrist flexion or extension. The torques could be stepped instantaneously or gradually to a new level. A video monitor was placed approximately 1 m in front of the seated subjects. A cursor that represented the instantaneous orientation of the hand and a target cursor were presented on the video screen. The subjects were required to align the two cursors with the aide of the video display but were prevented from seeing their hand directly. On some tests, a second torque motor system, identical to the first, was used. With this motor system, the subjects were allowed to see the hand, and alignment of the hand at 0° was accomplished by means of visible markings on the motor housing.

### Procedures

Three separate procedures were used to assess the perception of muscular effort. Two involved assessments using one hand while the magnitude of the externally applied torques varied, and the third involved matching external torques between a static magnitude torque applied to one hand and a variable magnitude torque applied to the other hand. The right hand was used for the unimanual tasks and for the hand for which the torque varied in the bimanual task. The experimental procedures for the two unimanual tasks are illustrated in Fig. 26.2. Subjects performed 240 trials for both of these tasks, and on the first trial no load opposed postural maintenance or active movement. In the first unimanual task, subjects were required to maintain a steady wrist position of 0° ± 2–4° for 1.5 to 2.5 sec by aligning the hand position cursor with the target cursor that appeared on the video screen. During this period, one of six static loads, ranging from 0.0 to 0.8 Nm, in 0.16 Nm steps and applied in a random fashion, opposed wrist flexion continuously. At the termination of the alignment period, the subject was required to verbally state whether the current load was the "same," "more," or "less" than the previous load. After the response, the video screen was blanked and the load was changed to a new magnitude via a ramp in torque lasting from 1 to 5 sec. The load remained unchanged on 20% of the trials; these trials were chosen randomly. The video screen was restored when the new load was attained. For the second unimanual task, the hand was pushed against a mechanical stop at 45° of wrist extension by the external load. After 1.5 to 2.5 sec, subjects were required to move to 0° of wrist rotation and then to make an immediate comparison, as described, between the perception of muscular effort required to move from 45° of extension to 0° on the current

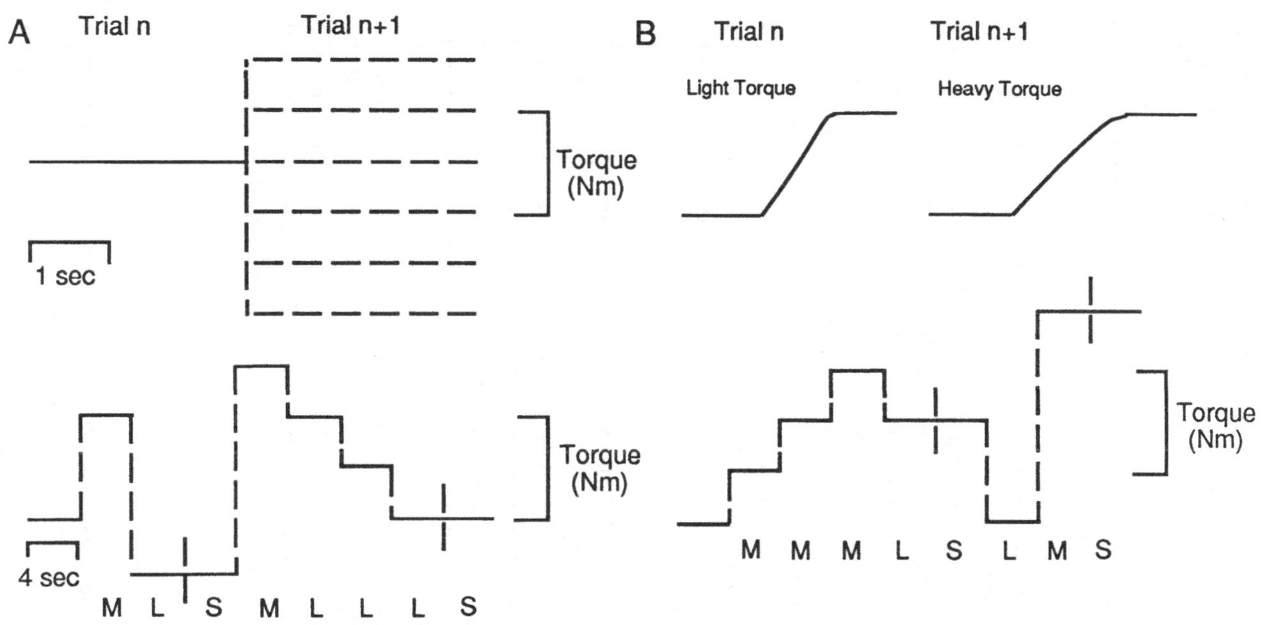


FIG. 26.2. Methods employed for assessment of unimanual torque perception. **A.** Unimanual posture method. The top section illustrates that the loads were changed on successive trials. The bottom section depicts a possible series of trials as the torque was stepped among different magnitudes. The correct decisions are abbreviated below; M = more, L = less, S = Same. **B.** Unimanual active movement method. The top section illustrates that when the loads were changed on successive trials, the movement characteristics (hypothetical position records are displayed) reflected the alteration in the mechanical conditions. The bottom section depicts a possible series of trials, and the correct psychophysical decisions, as the torque was stepped to different magnitudes.

trial with the muscular effort expended on the previous trial. After the judgment was made, the motor was switched into a position-servo mode, and the hand was passively returned to 45° of extension. The new load was applied coincidental with the release of the position-servo. The method of random application of load magnitude and the proportion of trials in which the load did not change was similar to that described first. For both unimanual experiments, the proportion of trials in which a correct decision was made was derived.

Bimanual sense of muscular effort was assessed using a tracking procedure that employed both the ascending and descending Method of Limits (Fig. 26.3A). One of 5 loads opposing flexion (0.16 to 0.8 Nm in 0.16 Nm steps) was applied continuously to the left hand. A variable load was applied to the right hand, and the subject reported whether the applied load was greater on the matching (right) or the reference (left) hand or whether the loads were the same on the two hands. The load on the right hand, first at 0.0 Nm, was incremented in 0.08 Nm steps until the subject indicated on three successive trials that the load applied on the right hand was greater than the load applied to the left hand. When subjects reached this criterion, or if subjects failed to reach this criterion and the load applied to the right hand reached 1.28 Nm, the load was then decreased in steps of 0.08 Nm until either the subject responded on three successive trials that the load applied to the left hand was greater than that applied to the right hand or that the load magnitude applied to the right hand reached 0.0 Nm. This up–down procedure was repeated for 10 reversals of torque. The load magnitudes at which the subjects reported that the loads were the same were averaged across the 10 reversals independently for the ascending and descending limbs of the load staircase. If subjects failed to report that the loads were similar on any limb of the load staircase, then the thresholds were taken as the average of the load magnitudes at which the subject switched reporting hands for which the load was greater. If the highest load was reached without attainment of the criterion, then the ascending threshold was taken as 1.28 Nm, and if the floor load was reached without criterion attainment, then the descending threshold was taken as 0.0 Nm.

## Data Recording and Analysis

For all experiments, the muscle activities and position records were recorded on analog tape and digitized subsequently at 200 Hz for off line analysis. Each trial was inspected visually on a computer graphics screen, and suitable data epochs were marked with electronic cursors. Within these epochs, muscle activities were integrated and normalized to the duration of the integration period. Psychophysical decisions and muscle activity of the flexor and the ratio of the flexor and extensor muscle activities were

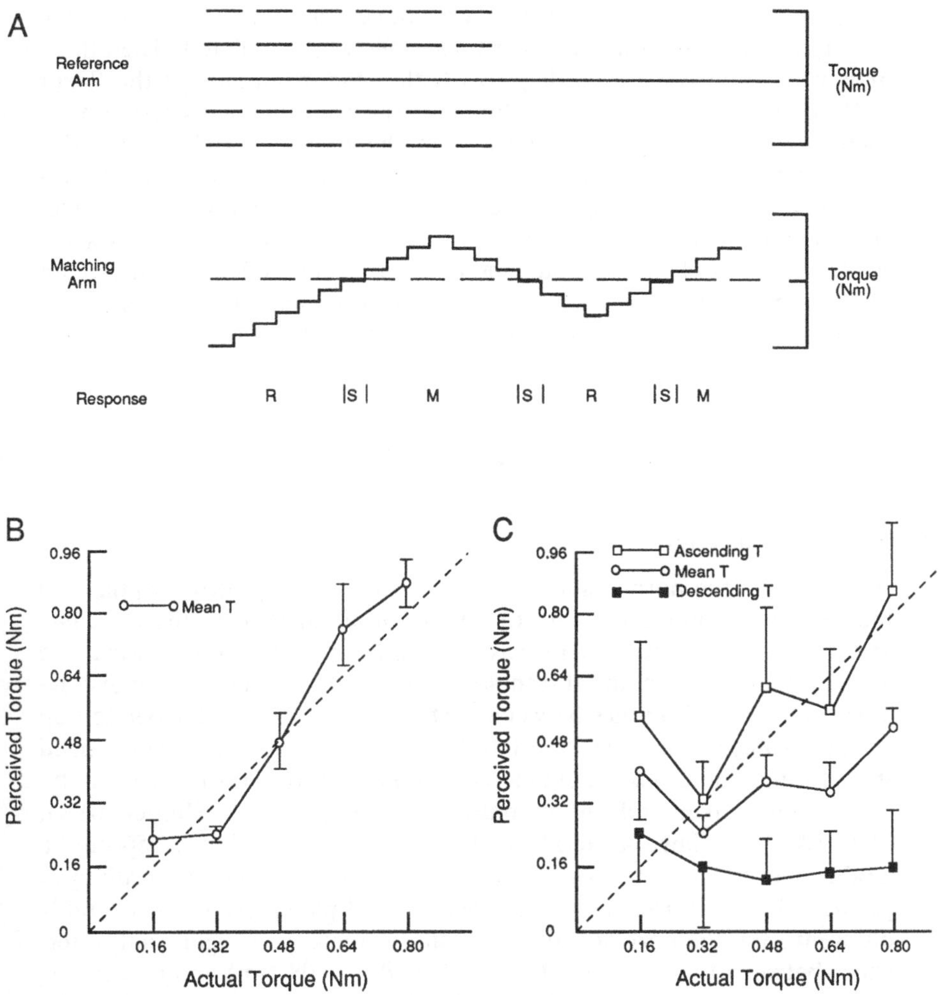


FIG. 26.3. Bimanual muscular effort sense. **A.** Methods employed. The top section depicts the possible loads that could be applied to the reference (left) arm. The solid line indicates that the middle load, 0.48 Nm, was chosen. The bottom section illustrates an up–down staircase of loads applied to the matching (right) arm with correct psychophysical decisions noted below—R = reference arm has more applied torque; S = same torque applied to both arms; M = matching arm has more applied torque. **B.** Perceived torque for the normal subjects. The ascending and descending thresholds were averaged. Error bars are SEMs in this and subsequent graphs. **C.** Perceived torque for the deafferented subjects. The ascending, descending and mean thresholds are illustrated.

analyzed for concordance. A four-way table was devised in which the occurrences of correct and incorrect decisions were tabulated. Then these trials were subdivided according to whether the magnitude of the flexor EMG or the flexor–extensor ratio from the trial under observation, in comparison to the muscle activity from the previous trial, predicted a correct psychophysical decision. The predictive capability of the EMG was determined by first deriving half the average difference in EMG magnitude, or EMG ratios, between the various constant flexor loads and then assaying whether the muscle activity on consecutive trials exceeded this threshold. The number of trials in each category were summed and expressed as a proportion of the total number of trials in which the decision was correct or incorrect.

## RESULTS

### Bimanual Matching

As expected, the normal subjects matched the varying torque applied to the reference arm closely with the static torque applied to the matching arm (Fig. 26.3B, left). There was no significant difference between the ascending and descending thresholds of perceived torque; the largest average individual difference between the two measures was 0.176 Nm (group absolute value mean = 0.075 ± 0.011 Nm). At the higher levels of applied static torque, there was a slight increase in the perceived torque from unity. This is most likely related to muscle fatigue (Jones & Hunter, 1983; McCloskey, Ebeling, & Goodwin, 1974) because the subjects reported a modicum of fatigue after holding the higher magnitude loads for prolonged periods. The performance of the deafferented patients was considerably different from the normal subjects insofar as there were significant differences between the ascending (mean = 0.58 ± 0.08 Nm) and descending (mean = 0.17 ± 0.05 Nm) thresholds, $t(19) = 3.77$, $p \leq .001$. With the exception of the lightest load, the mean ascending threshold of perceived torque for the deafferented patients closely approximated the torque applied to the reference arm, and therefore was not different than the ascending threshold of the normal subjects. However, the ascending thresholds of all patients did not always show a clear linear relationship with changes in the reference torque and the perceived torque. The descending threshold of the perceived torque for the deafferented patients deviated substantially from the reference torque and was significantly different from the descending threshold of the normal controls, $t(38) = 4.41$, $p \leq .0005$. In most individual cases, the average perceived torque was close to the lowest level of torque (0.0 Nm) used on the descending limb of the load staircase, though one patient had a consistent descending threshold of 0.45–0.61 Nm regardless of the reference load. The mean (absolute value) discrepancy between the ascending and descending thresholds was 0.58 ± 0.06 Nm (range = −0.51 to 1.09 Nm, signed values). The large discrepancies between the ascending and the descending thresholds in the deafferented patients were significantly different from the threshold discrepancies in the normal subjects, $t(38) = 7.96$, $p \leq .0005$.

### Unimanual Matching: Psychophysical Decisions

The deafferented patients made more errors than the normal subjects (Fig. 26.4) when the correct decision would have been that the applied torque had changed (postural task: $t(11) = 3.04$, $p \leq .025$; movement task: $t(10) = 4.75$, $p \leq .001$). The error rate for the trials in which the load remained unchanged (catch trials) was not different between the two groups. Further analyses of the data from the postural experiment revealed that the deafferented patients made more decision errors than the normal subjects only when the torque shift was less than 0.48 Nm, $t(11) \geq 2.26$, $p \leq 0.05$ (Fig. 26.5A). For the active movement experiment, the deafferented patients made more decision errors for all gradations of the torque shifts, $t(10) \geq 2.74$, $p \leq .025$ (Fig. 26.5B).

### Comparison Between Muscle Activity and Psychophysical Decisions

The relationship between muscle activity and psychophysical decision processes was examined only for the unimanual postural maintenance task. For both the deafferented and normal subjects, there was a roughly linear relationship between mean flexor muscle activity recorded during postural maintenance and the magnitude of the constant loads opposing flexion (repeated measure of load, $F(5, 40) = 121.13$, $p \leq .0001$, Fig. 26.6A). However, the deafferented patients showed a slightly higher relative flexor EMG at the loads less than 0.48 Nm than the normal subjects (patient by load interaction, $F(5, 40) = 3.17$, $p \leq .025$). The magnitude of the extensor EMG was also related to the magnitude of the applied constant load, $F(5, 40) = 9.71$, $p \leq .0001$ (Fig. 26.6B), but there was no difference in the extensor EMG in relation to the applied load between the deafferented and the normal subjects. The ratio of the flexor to extensor EMG (not illustrated) to the applied load paralleled the EMG-to-load relationship observed for the flexor muscle activity in both the normal and deafferented subjects.

Samples of the muscle activity in the flexor and the extensor muscles, position records, and corresponding psychophysical decisions on a series of consecutive trials taken from one deafferented patient are shown in Fig.

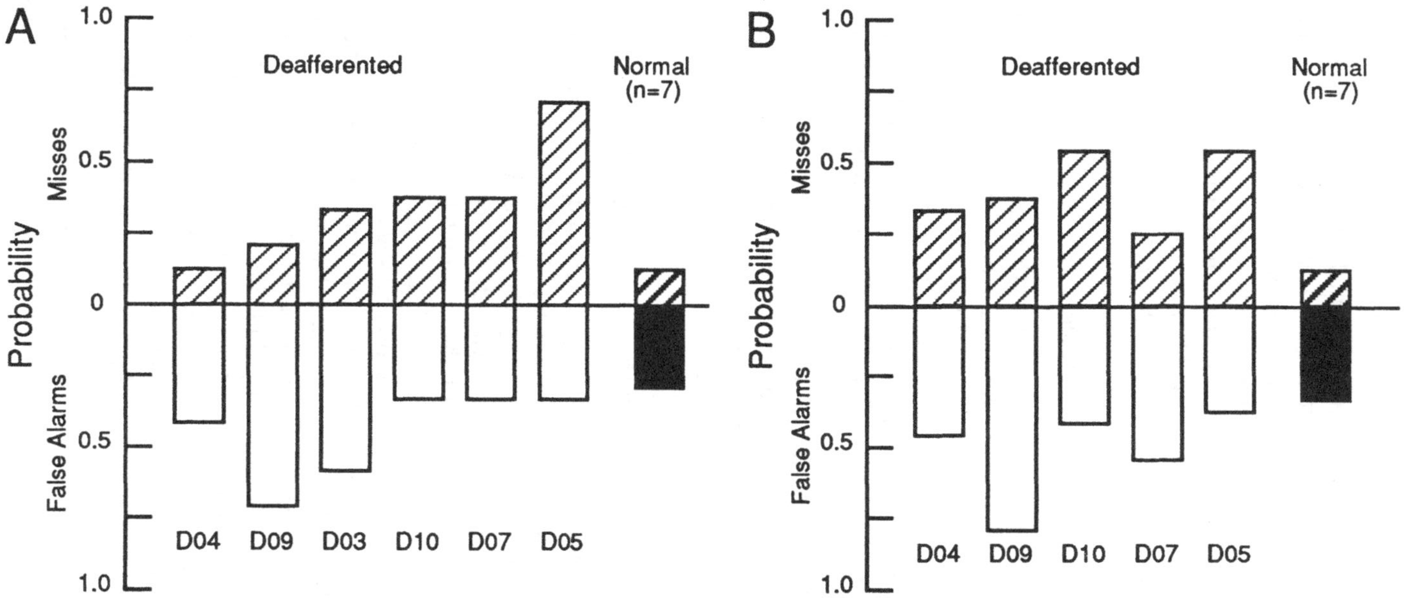


FIG. 26.4. Psychophysical decisions for normal and deafferented subjects during the unimanual posture (**A**) and the active movement (**B**) tasks. The probability of making incorrect decisions when a torque shift did (misses) or did not (false alarms) occur are plotted for each patient and for the group of normal subjects.

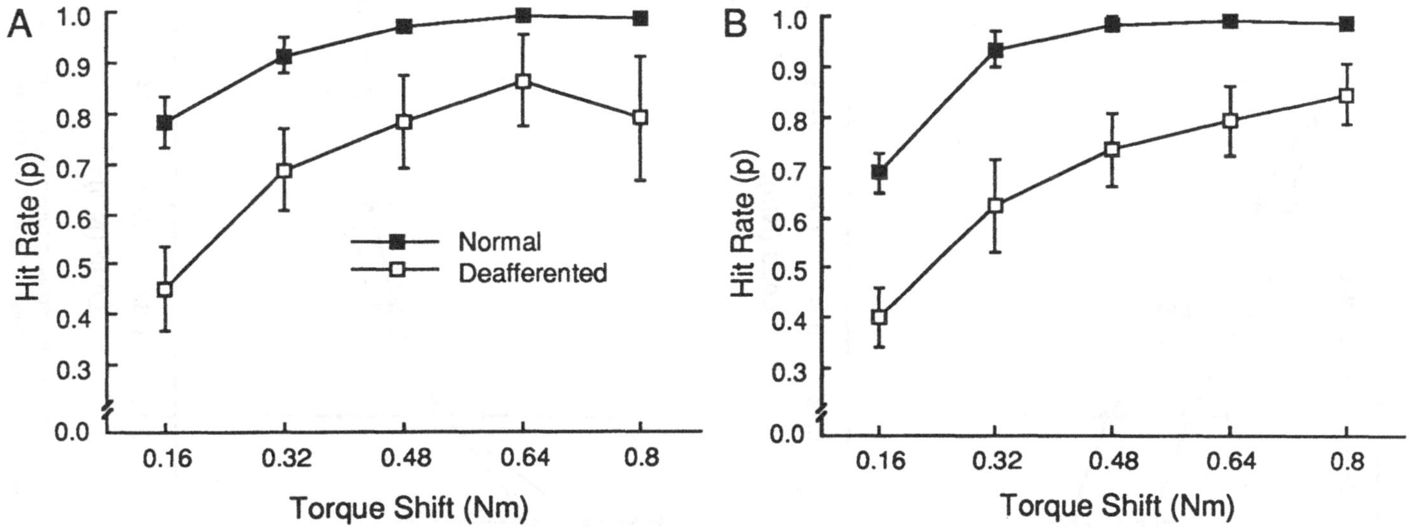


FIG. 26.5. Frequency of correct decisions versus amount of the torque shift for the unimanual posture (**A**) and for the active movement (**B**) tasks. The probability of making correct decisions is plotted for the normal and deafferented subjects.

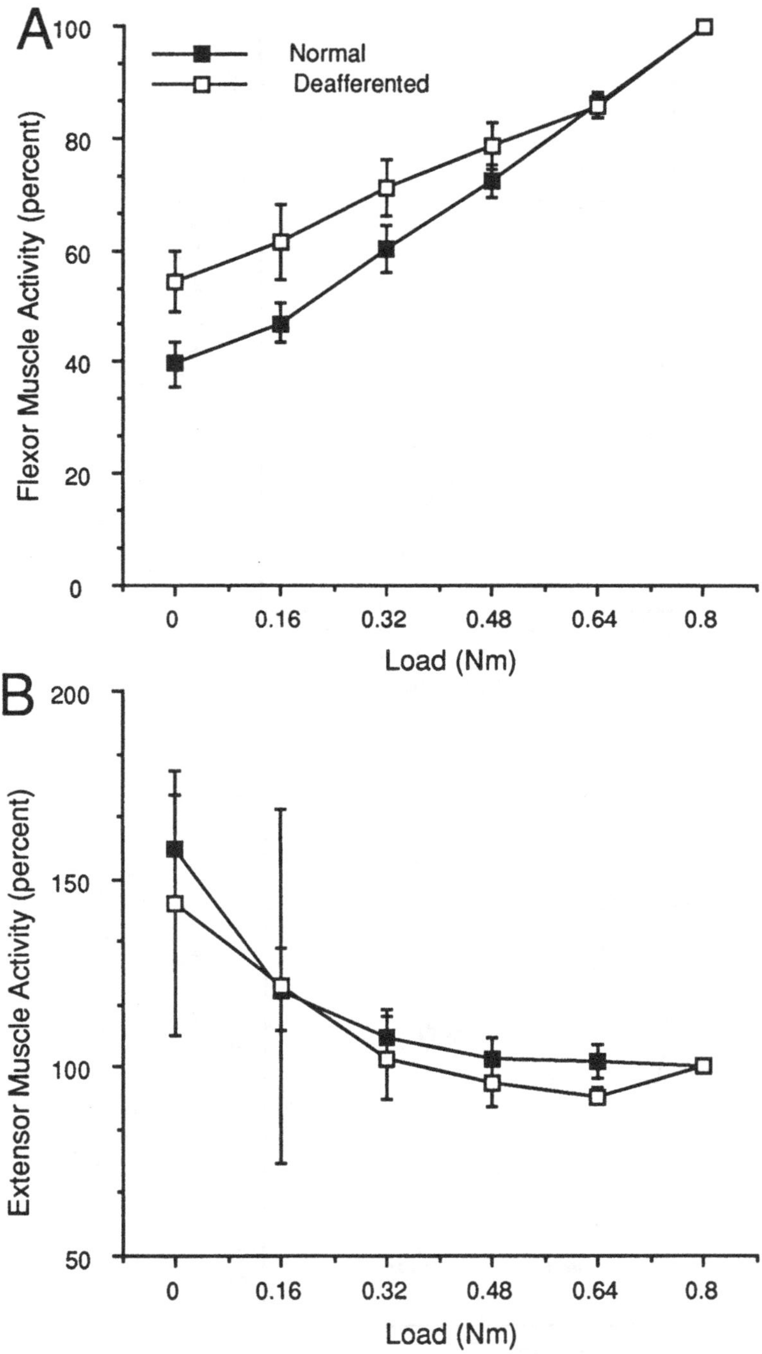


FIG. 26.6. Muscle activity versus opposing torque in wrist flexor (**A**) and wrist extensor (**B**) muscles in deafferented and normal subjects.

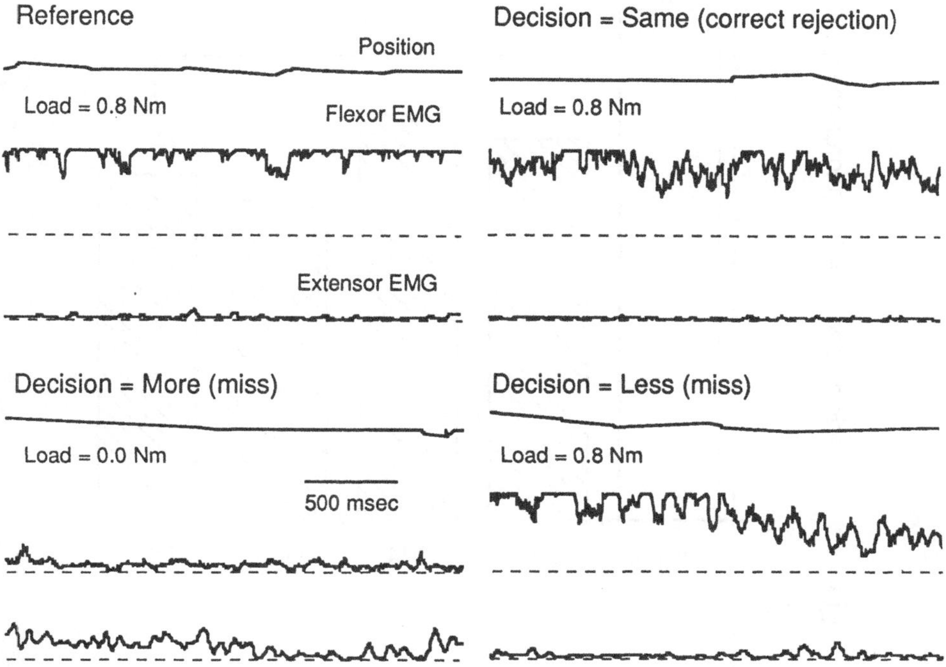


FIG. 26.7. Sample successive trials during the unimanual posture task. In each panel the position (top) and flexor (middle) and extensor (bottom) muscle activity records are displayed. The hatched lines indicate the baseline level of muscle activity. The opposing load and the psychophysical decisions are also shown. The first trial of the series was considered a reference trial.

26.7. On these successive trials, there was first a good correspondence between the patient's decision about the absence of a change in the applied load and the muscle activity (compare top left panel with top right panel). The patient made an incorrect decision on the third trial, even though the flexor muscle activity decreased in accord with the opposing load (compare top right with bottom left). On the last trial, there was again a noncongruence between the psychophysical decision process and the muscle activity.

Normal subjects had a high concordance between the correct identification that the load had or had not changed and the accompanying flexor EMG or flexor–extensor EMG ratio. There was a tendency for the EMG ratio to be more related to the psychophysical decisions than the flexor EMG, $t(5) = 2.22$, $p = .077$; five of the six normal subjects had a higher concordance between the EMG ratio and the perceptual decisions, the sixth subject had an equal concordance between the EMG ratio or the

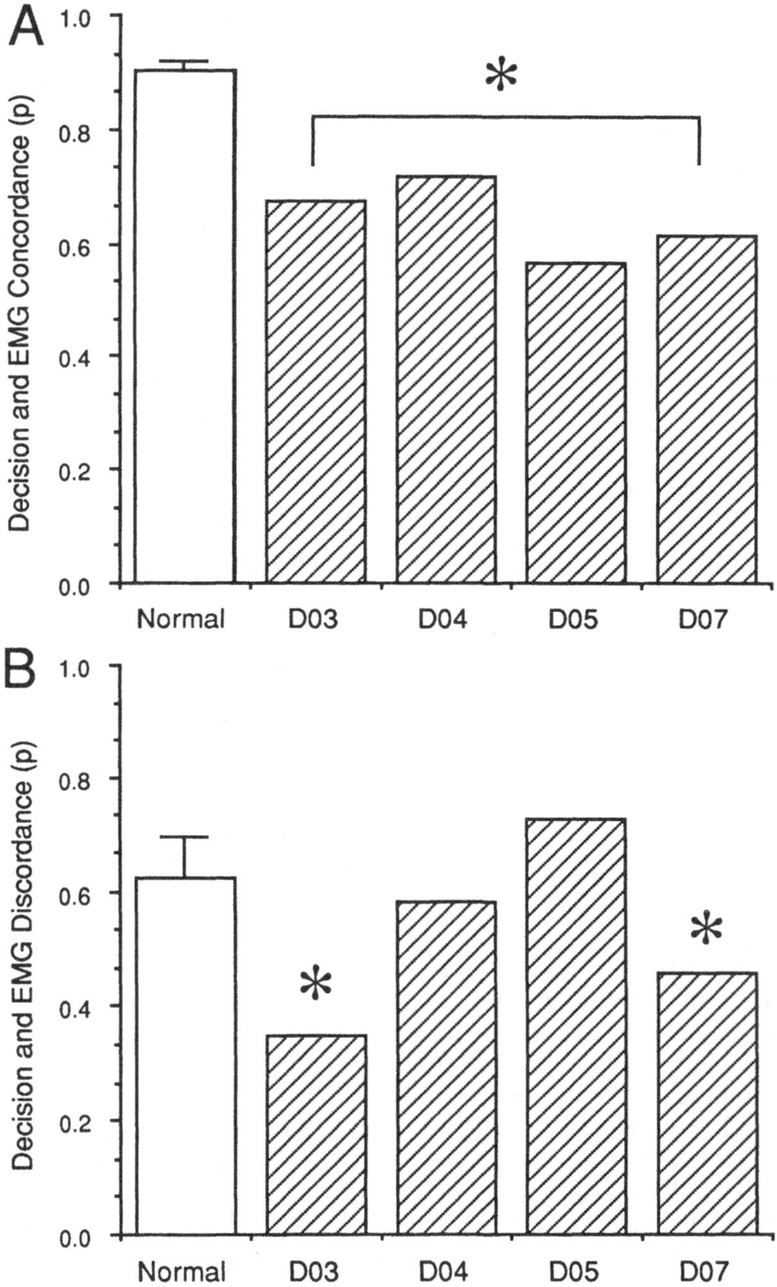


FIG. 26.8. Correlation between psychophysical decisions and flexor muscle activity. (**A**) The concordance between the decisions and the muscle activity when the subjects made the correct perceptual decision. (**B**) The discordance between the decisions and the muscle activity when the subjects made the incorrect perceptual decision.

flexor EMG and the psychophysical decisions. As a group, the deafferented patients had a poorer concordance than normal subjects between correct psychophysical decisions and the corresponding predictive flexor EMG, $t(8) = 7.78, p \leq .0001$, or flexor–extensor ratio, $t(8) = 5.91, p \leq .0005$, (Fig. 26.8A). When normal subjects made the incorrect psychophysical decision (Fig. 26.8B), there was a poorer correspondence between the EMG responses and the perceptual processes than for what was observed for concordance between correct psychophysical decisions and the observed EMG, $t(5) = 4.38, p \leq .005$. Two of the four patients had similar performance to the normal subjects, whereas two others had a poorer correspondence than the normal subjects on this measure (Fig. 26.8B).

## DISCUSSION

The results of this experiment indicated that patients with a large-fiber sensory neuropathy had an impaired sense of muscular effort. In addition, there was a lower correspondence between the muscular output and the psychophysical decision process in deafferented patients in comparison to normal subjects.

The results of the present experiment are seemingly at variance with those of Rothwell et al. (1982) and of the general conclusions of Gandevia and McCloskey (1977a; 1977b; 1977c; Gandevia, 1982). In the studies of Rothwell et al. (1982), the sense of effort of a single deafferented human was evaluated by matching torques applied to the top joint of both thumbs while the subject moved a required distance. The deafferented patient in that study appeared to be able to adjust a variable torque applied to one thumb to match a static torque applied to the opposite thumb. Because thumb movement was a task requirement in those studies, it was noted by Rothwell et al. (1982) that a potential confounding variable was that the subject assessed the elapsed time of the movement excursion instead of judging the applied loads to make the appropriate matches in torque between the two thumbs. The postural tasks used in the present study eliminated velocity detection as a potential source of feedback for matching of different loads and, as noted, demonstrated marked abnormalities by the deafferented patients in detecting changes in applied loads.

Gandevia and McCloskey (1977a; 1977b; 1977c; Gandevia, 1982; Gandevia, McCloskey, & Potter, 1980) have argued that the sense of muscular effort is related primarily to a corollary discharge signal originating within the central nervous system. The primary experimental results used to support this argument have been that subjects with the experimental or pathological induced weakness or paralysis overestimate the muscular effort needed to lift a weight. Although experiments that have anesthetized the skin surfaces and joints that are used to judge applied weight have shown a diminished sense of effort (Marsden et al., 1979), Gandevia and McCloskey (1977c) have argued that dimunition of the tonic facilitation of output EMG by stretch reflexes following peripheral anesthesia should yield a stronger motor command that would then be interpreted as more muscular effort. This mechanism cannot operate because reflex properties were not changed significantly, or were not concurrent with changes in effort sense, when the distal skin surfaces were anesthetized (Marsden et al., 1979). However, even when reflex properties are altered—for example, when muscle spindle information is disrupted by tendon vibration—the perceived force is reduced (Cafarelli & Kostka, 1981). This occurred despite the absence of a tonic vibration reflex that would tend to enhance the EMG. An additional general problem with the anesthesia and motor conduction block experiments of Gandevia and McCloskey is that the role of the Golgi tendon organs (GTO) is not fully considered. GTO output would be expected to be reduced in experimentally or pathologically weakened subjects because of the lowered maximum contractile force (Binder & Osborn, 1985; Houk & Henneman, 1967). Therefore, GTO discharge characteristics by themselves would not be predictive of the enhanced sensation of muscular effort in such subjects. However, as noted by Marsden et al. (1979), inputs from different submodalities of somesthetic afferents could sum and yield important information about the perception of muscular effort.

It is generally assumed that the corollary discharge signal thought to mediate effort sense emanates from cerebral cortical structures, particularly the primary motor cortex, and is sent to structures, also possibly located in the cerebral cortex, concerned with sensation of muscular output. One critical assay of the cortical location of the sense of muscular effort would be that direct stimulation of the corticospinal pathway would yield conscious sensations of muscular effort. The noninvasive transcranial electric or magnetic activation of the primary motor cortex that evokes muscle responses has been employed recently in human subjects (Barker, Freeston, Jalinous, Merton, & Morton, 1985; Merton & Morton, 1980). Although there has not been a formal examination of the sensations evoked from such stimulation, mass activation of the output elements in the primary motor cortex of humans does not appear to result in any clear, conscious sensation related to muscular effort (Day, Rothwell, Thompson, Maertens de Noordhout, Nakashima, Shannon, & Marsden, 1989). It is unclear which elements of the pyramidal tract are being activated by transcranial stimulation of the primary motor cortex, though is is likely that the largest pyramidal cells located in layer V are being activated. Thus, it is possible that the lack of conscious sensation from such stimulation is because the appropriate motor cortical cells have not been activated. Alternatively, and because other cerebral cortical areas such as the premotor cortex (Weinrich & Wise, 1982), supplementary motor cortex (Kurata & Tanji, 1985), and parietal cortex (Chapman, Spidalieri, & Lamarre, 1984) play major roles in the initiation and execution of movement, the putative internal mechanism for the sense of effort could be located in a brain structure other than the primary motor cortex.

Even in the most severely impaired deafferented patients there remained some modest ability to detect changes in applied loads, especially when the shift in load exceeded 0.5 Nm (see Fig. 26.5). Indeed, for the postural maintenance task, the deafferented patients, as a group, did not differ from the normal subjects when the load shift was greater than 0.48 Nm. The explanation for this seemingly bimodal performance of deafferented patients could be that a mechanism of corollary discharge operates at high levels of motor output but that somesthetic information is critical for detection of changes in motor commands for muscle activity at low levels of motor output. Indirect evidence in support of such a mechanism would be that small movements tend to be more dependent than large movements on, and are more disrupted by, somesthetic afferent inputs (Sanes, 1986; Sanes & Evarts, 1983). Additionally, neurophysiological studies indicate that pyramidal tract neurons are more sensitive to somesthetic inputs when monkeys perform small in comparison to large movements (Fromm & Evarts, 1977; 1978).

In normal volunteers, there was a reasonably good correspondence between the psychophysical decision process and the observed changes in muscle activity on successive trials. These results would tend to support the notion that monitoring of efferent output is sufficient for correct perception of muscular effort (Aniss, Gandevia, & Milne, 1988). However, even in normal subjects there were trials in which there were clear discrepancies between the EMG and correct or incorrect psychophysical decisions. For these trials, an alternate mechanism for perceptual decisions must be invoked. The discrepancies between the EMG and decisions about the applied load were exacerbated in the deafferented patients, especially for trials in which there were correct assessments of the change in applied load. Together these results lend further support to the notion that corollary discharge cannot be the only mechanism by which humans appreciate changes in required muscular effort.

## REFERENCES

Aniss, A. M., Gandevia, S. C., & Milne, R. J. (1988). Changes in perceived heaviness and motor commands produced by cutaneous reflexes in man. *Journal of Physiology, 397*, 113–126.

Asatryan, D. G., & Fel'dman, A. G. (1965). Functional tuning of nervous system with control

of movement or maintenance of a steady posture–I. Mechano-graphical analysis of the work of the point on execution of a postural task. *Biophysics, 10*, 925–935.
Barker, A. T., Freeston, I. L., Jalinous, R., Merton, P. A., & Morton, H. B. (1985). Magnetic stimulation of the human brain. *Journal of Physiology, 369*, 3P.
Binder, M. D., & Osborn, C. E. (1985). Interactions between motor units and Golgi tendon organs in the tibialis posterior muscle of the cat. *Journal of Physiology, 364*: 199–215.
Bizzi, E., Accornero, N., Chapple, W., & Hogan, N. (1984). Posture control and trajectory formation during arm movement. *Journal of Neuroscience, 4*, 2738–2744.
Bizzi, E., Dev, P., Morasso, P., & Polit, A. (1978). The effect of load disturbances during centrally initiated movement. *Journal of Neurophysiology, 41*, 542–556.
Bizzi, E., Polit, A., & Morasso, P. (1976). Mechanisms underlying achievement of final head position. *Journal of Neurophysiology, 39*, 435–444.
Cafarelli, E., & Kostka, C. E. (1981). Effect of vibration on static force sensation in man. *Experimental Neurology, 74*, 331–340.
Chapman, C. E., Spidalieri, G., & Lamarre, Y. (1984). Discharge properties of area 5 neurones during arm movements triggered by sensory stimuli in the monkey. *Brain Research, 309*, 63–77.
Cole, K. J., & Abbs, J. H. (1988). Grip force adjustments evoked by load force perturbations of a grasped object. *Journal of Neurophysiology, 60*, 1513–1522.
Day, B. L., & Marsden, C. D. (1982). Accurate repositioning of the human thumb against unpredictable dynamic loads is dependent upon peripheral feedback. *Journal of Physiology, 327*, 393–407.
Day, B. L., Rothwell, J. C., Thompson, P. D., Maertens De Noordhout, A., Nakashima, K., Shannon, K., & Marsden, C. D. (1989). Delay in the execution of voluntary movement by electrical or magnetic brain stimulation in intact man: Evidence for the storage of motor programmes in the brain. *Brain, 112*, 649–663.
Fel'dman, A. G. (1966a). Functional tuning of nervous system with control of movement of maintenance of a steady posture–II. Controllable parameters of the muscles. *Biophysics, 11*, 565–578.
Fel'dman, A. G. (1966b). Functional tuning of nervous system with control of movement or maintenance of a steady posture–III. Mechanographic analysis of the execution by man of the simplest motor task. *Biophysics, 11*, 766–775.
Forget, R., & Lamarre, Y. (1987). Rapid elbow flexion in the absence of proprioceptive and cutaneous feedback. *Human Neurobiology, 7*, 27–37.
Fromm, C., & Evarts, E. V. (1977). Relation of motor cortex neurons to precisely controlled and ballistic movements. *Neuroscience Letters, 5*, 259–265.
Fromm, C., & Evarts, E. V. (1978). Motor cortex responses to kinesthetic inputs during postural stability, precise fine movement and ballistic movement in the conscious monkey. In G. Gordon (Ed.) *Active touch: The mechanism of recognition of objects by manipulation* (pp. 105–117). Oxford: Pergamon Press.
Gandevia, S. C. (1982). The perception of motor commands or effort during muscular paralysis. *Brain, 105*, 151–159.
Gandevia, S. C., & McCloskey, D. I. (1977a). Sensations of heaviness. *Brain, 100*, 345–354.
Gandevia, S. C., & McCloskey, D. I. (1977b). Effects of related sensory inputs on motor performances in man studied through changes in perceived heaviness. *Journal of Physiology, 272*, 653–672.
Gandevia, S. C., & McCloskey, D. I. (1977c). Changes in motor commands, as shown by changes in perceived heaviness, during partial curarization and peripheral anaesthesia in man. *Journal of Physiology, 272*, 673–689.
Gandevia, S. C., McCloskey, D. I., & Potter, E. K. (1980). Alterations in perceived heaviness during digital anaesthesia. *Journal of Physiology, 306*, 365–375.

Hogan, N. (1984). An organizing principle for a class of voluntary movements. *Journal of Neuroscience, 4*, 2745–2754.

Houk, J., & Henneman, E. (1967). Responses of Golgi tendon organs to active contractions of the soleus muscle of the cat. *Journal of Neurophysiology, 30*, 466–481.

Johansson, R. S., & Westling, G. (1987). Signals in tactile afferents from the fingers eliciting adaptive motor-responses during precision grip. *Experimental Brain Research, 66*, 141–154.

Jones, L. A., & Hunter, I. W. (1983). Effect of fatigue on force sensation. *Experimental Neurology, 81*, 640–650.

Kurata, K., & Tanji, J. (1985). Contrasting neuronal activity in supplementary and precentral motor cortex of monkeys. II. Responses to movement triggering vs. nontriggering sensory signals. *Journal of Neurophysiology, 53*, 142–152.

Lestienne, F., Polit, A., & Bizzi, E. (1981). Functional organization of the motor process underlying the transition from movement to posture. *Brain Research, 230*, 121–131.

Marsden, C. D., Rothwell, J. C., & Traub, M. M. (1979). Effect of thumb anaesthesia on weight perception, muscle activity and the stretch reflex in man. *Journal of Physiology, 294*, 303–315.

Matheson, J., Hallett, M., Berardelli, A., Weinhaus, R., & Inzucchi, S. (1985). Failure to confirm a correlation between electromyogram and final position. *Human Neurobiology, 4*, 257–260.

McCloskey, D. I., Ebeling, P., & Goodwin, G. M. (1974). Estimation of weights and tensions and apparent involvement of a "sense of effort". *Experimental Neurology, 42*, 220–232.

Merton, P. A., & Morton, H. B. (1980). Stimulation of the cerebral cortex in the intact human subject. *Nature, 285*, 200.

Mott, F. W., & Sherrington, C. S. (1895). Experiments upon the influence of sensory nerves upon movement and nutrition of the limbs. *Proceedings of the Royal Society, 57*, 481–488.

Nathan, P. W., & Sears, T. A. (1960). Effects of posterior root section on the activity of some muscles in man. *Journal of Neurology, Neurosurgery, and Psychiatry, 23*, 10–22.

Polit, A., & Bizzi, E. (1979). Characteristics of motor programs underlying arm movements in monkeys. *Journal of Neurophysiology, 42*, 183–194.

Roland, P. E., & Ladegaard-Pedersen, H. (1977). A quantitative analysis of sensations of tension and of kinaesthesia in man. Evidence for a peripherally originating muscular sense and for a sense of effort. *Brain, 100*, 671–692.

Rothwell, J. C., Traub, M. M., Day, B. L., Obeso, J. A., Thomas, P. K., & Marsden, C. D. (1982). Manual motor performance in a deafferented man. *Brain, 105*, 515–542.

Sanes, J. N., & Evarts, E. V. (1983). Effects of perturbations on accuracy of arm movements. *Journal of Neuroscience, 3*, 977–986.

Sanes, J. N. (1986). Kinematics and end-point control of arm movements are modified by unexpected changes in viscous loading. *Journal of Neuroscience, 6*, 3120–3127.

Sanes, J. N., Mauritz, K.-H., Dalakas, M. C., & Evarts, E. V. (1985). Motor control in humans with large-fiber sensory neuropathy. *Human Neurobiology, 4*, 101–114.

Weinrich, M., & Wise, S. P. (1982). The premotor cortex of the monkey. *Journal of Neuroscience, 2*, 1329–1345.

# VII CONSTRAINTS ON MOTOR LEARNING AND DEVELOPMENT

# 27 A Perception–Action Perspective on the Development of Manual Movements

Claes von Hofsten
*Department of Psychology,*
*Umeå University*

ABSTRACT

The perception–action approach stresses two aspects of motor behavior that often have been overlooked in the past. First, it stresses the fact that motor behavior has to be controlled by perception and that any motor act is as much an accomplishment of the perceptual system as it is an accomplishment of the motor system. It also stresses the point that actions are functionally specified, inherently purposeful and that prospective control plays a role in any motor act. In this chapter the perception–action approach is applied to the problem of manual development. The implications of the approach are discussed.

The manual capabilities of the human adult are amazingly sophisticated and flexible. There seems to be almost no task too difficult for the human hand. The same hands perform surgery under a microscope, catch fast baseballs, and chop firewood with both force and precision. What are the underlying principles for the construction of this remarkable sensorimotor system? To be able to understand manual behavior we, first of all, need to consider how biology has solved the manual problem during the course of evolution. Evolution works by solving practical problems, and complex systems emerge from numerous such concrete adaptations. In this process, old solutions are not thrown away when new ones are needed. The old structures are typically used and altered and maybe supplemented with new structures superimposed on the old ones (Gould, 1982).

The successful solution of action problems in evolution is always a question of both developing the right tool for the action and getting access to the right information for controlling the tool in achieving the action (Arbib, 1981). In other words, perception has been evolved to serve action (Gibson, 1979) and is as essential a part of any adaptive system as the implementation structure itself. The wings of a bird are no more essential for flying than the perceptual system that guides the movements of those wings. The perception part and the motor part are tailored to each other for optimal functioning. In the case of manual action, the upper limbs seem to have been designed to work optimally within the visual field. On the perceptual side, the emergence of binocular stereopsis in man may have been the result of a specific manual adaptation to provide more exact information about near space. Binocular vision gives superior spatial resolution within manual space, but as the binocular resolution diminishes with the square of the viewing distance, it becomes rapidly less useful at larger distances.

The view that motor skills are made up of perception–action systems has important implications for the study of sensorimotor development. The problem is not how the motor system makes contact with perception because perception is already, from the start, a basic nondetachable part of the action system. The developmental changes of an action system are as likely to occur in the perceptual domain as in the motor domain. They are complementary in the sense that when any part of the system undergoes differentiation and refinement, it necessitates a new synthesis (Fentress, 1986). Another important implication is that action systems are functionally, and not anatomically or mechanically, specific (Hofsten, 1985; Reed, 1982). Reaching for a target, for instance, could be carried out by either right or left hand, the starting position of the approach could change from one instance to the next, and no two reaching trajectories look exactly the same. These movements are still classified as the same action, though, because they share the same function. Actions, finally, are always purposeful and goal directed. Information is used to plan and predict the upcoming action and to control it. Actions are never just triggered or elicited. Although triggered movements have commonly been the subject of intensive study in attempts to understand motor control, such movements are probably extremely rare in everyday life (Reed, 1982).

## THE SENSORY BASIS OF MANUAL BEHAVIOR

Manual tasks require three kinds of information for their solutions. First, they require information about the object to be manipulated, its position and orientation in space, its size and form, and its substance and texture. Second, they require information about the positions and movements of the arms and hands relative to the subject's own body. Finally, they require information about the positions and movements of the arms and hands relative to the object to be manipulated. Vision as well as proprioception and haptics contribute to the efficiency of manual action. They all have their weaknesses and strengths, but through evolution they have come to collaborate with and supplement each other in an optimal way to provide an integrated informational support for manual action. One and the same action system may exploit several varieties of information as well as several muscles groups.

Vision gives detailed information about the position and orientation in space of a manual target before it is encountered and about its form and size. It gives equally good information about the positions of the hands in space. Vision is the superior sense in close guidance of the hands relative to a target in manipulative and explorative action. The shortening of vision is that information is only available within the visual field, more detailed in the center and less detailed in the periphery. This means that vision, if it is to monitor the positions and movements of a body part in any precise way, has to be directed toward that part of the body. However, in reaching, the subject needs to look at the target toward which the action is directed. If the target and the body part used for the action are much separated, vision encounters problems in monitoring both. For instance, in throwing a ball toward a distant target, the throwing arm is typically extended in the direction opposite to the target to attain optimal throwing force. The subject typically focuses on the target and controls the movement of the arm proprioceptively.

Proprioception provides means for defining the position and movements of the hands and arms relative to the body. Thus, proprioception makes it possible to detach the acting hand from the visual field while directing it toward a visual target, as in the preceding example. It also makes manual action possible in darkness and around corners where vision does not apply.

Haptics, finally, provides unique information about object properties and about the precise relation between the hand and the manipulated object and is therefore of great importance for the guidance of manual action. For instance, it provides information about object substance and how an object is deformed during grasping. It regulates grip force by monitoring microslips between the hand and the object not detectable through vision (Johansson & Westling, 1987).

## NEWBORN REACHING

From a perception–action perspective it is expected that the apperence of extended arm movements in young infants is accompanied by an ability to control these movements. It has long been recognized that neonates make extensive arm movements, but their ability to control these movement has been questioned. Neonatal arm movements have traditionally been characterized as random thrashing (White, Castle, & Held, 1964). Piaget (1953) believed that vision would take over the control of grasping only after a process of mutual assimilation between the visual–motor schema of looking at the hand and looking at the object and the tactual motor schema of feeling with the hand. According to Piaget (1953), this development would occur between the 4th and the 6th month of life.

Contrary to these traditional beliefs (Piaget, 1953), Hofsten (1982) found that forward extended arm movements performed while the newborn fixated a target were aimed much closer to it than movements performed while the infant looked elsewhere or closed its eyes. Each forward extension performed during a 7-minute period was recorded with two video cameras placed 90° relative to each other, which allowed a 3-D analysis of the movements. To be able to calculate aiming, the movements were subdivided into functional units, each of which consisted of one acceleration and one deceleration phase (see Brooks, Cooke, & Thomas, 1973). The unit that carried the hand closest to the trajectory of the object was examined further. This part of the movement should be aimed at the object if there is visual control, but not otherwise.

The result showed that when the infants fixated the target, the movement of the hand was off the target by, on the average, 32°. In comparison, the hand was, on the average, off the target by 52° when the infant did not fixate it and by 54° when the infant had its eyes closed. Table 27.1 shows the results from the five most frequent reachers in the study. The effect of fixation is substantial in each of the five subjects.

There is evidence that newborn arm movements may be controlled by proprioception only, that is, the position of the arm as well as the target (the infant's mouth) being specified proprioceptively. Butterworth (1986) and Rochat, Blass, and Hoffmeyer (1987) found evidence of directed arm

TABLE 27.1
Mean approach angles for different individuals and different kinds of looking behavior.

| *Subject* | *Nonfixation and Closed Eyes* | *Fixation* |
|---|---|---|
| G.H. | 51.4 (13) | 21.9 (12) |
| G.K. | 52.4 (16) | 32.8 (14) |
| F.L. | 62.0 (5) | 28.4 (10) |
| F.S. | 50.0 (13) | 21.9 (8) |
| F.N. | 53.5 (12) | 28.8 (7) |

Note. The number of reaches on which each mean is based is shown in parenthesis. (from Hofsten, 1982).

movements toward the mouth in the neonate. Butterworth reported that the mouth was significantly more likely to be open before and throughout the arm movement when the hand ended up in the mouth than when the hand did not end up in the mouth. Butterworth also found that although the hand did not always go directly to the mouth but sometimes contacted other parts of the face first, it still had a distinct goal directed character. He found no evidence of rooting after contact; the head was held still and the hand moved "immediately in the direction of the mouth" (p. 28).

## REACHING AND GRASPING

Reaching and grasping seem to constitute separate perception–action systems. They clearly serve different functions, require different kinds of movements, and use different kinds of information for their guidance. When initiating a reach, the position of the object is defined visually and, most commonly, the position and movement of the reaching arm proprioceptively. Because vision is not used in controlling the movement of the arm, this phase of the approach has often been described as open loop. However, recent evidence indicates that proprioceptive feedback may be used to control such movements (Prablanc & Pelissom, this volume). Whereas the movement of the arm is defined in body-related coordinates during the early phase of the approach, it is defined in object-related coordinates during its later phase. As the hand enters the visual field toward the end of the approach, its position and movement are defined visually relative to the object. This enables the subject to prepare the grasping of the object by performing delicate and precise adjustments of the hand.

There are different motor pathways serving reaching and grasping. Knowledge of these structures have mainly been gained through elaborate studies on the rhesus monkey by Kuypers and associates (Kuypers, 1973; Lawrence & Kuypers, 1968b). The manual system of the rhesus monkey is in several respects similar to the human manual system. It has an opposing thumb and an index finger that can move independently of the other fingers.

Lawrece and Kuypers (1968b) found it appropriate to distinguish between two subcortical pathways: one ventromedial and one lateral, grouped according to their termination in the spinal grey matter. They studied how interruption of these two systems affected manual behavior in monkeys with prior bilateral interruption of the pyramidal tracts. The animals with lesion of the lateral brainstem could not control hand movements independent of arm movements. However, movements involving the whole limb and the body, like walking and climbing, were only minimally affected. Interruption of the ventromedial pathways produced severe impairments of axial and proximal limb movements and the maintenance of body posture. However, despite these impairments, they could pick up pieces of food with their hands if the limbs were appropriately supported and brought to the food. In summary, the ventromedial pathway seems to be especially concerned with reaching movements and the maintenance of balance during reaching. The lateral pathway superimposes on that control the capacity for grasping objects.

On a more functional level, there is further evidence for separate organization of reaching and grasping. Jeannerod (1981) has shown that reaching and grasping really constitute two different modes of control deeply engraved into the sensorimotor system. The deceleration of the reaching movement toward the end of the approach is marked by a discontinuity where tangental velocity tends to become constant or even increase. The time at which this discontinuity occurred invariably corresponded to 70%-80% of the total movement time. The possibility that the discontinuity was simply an effect of visual correction for undershooting was found to be less likely because re-accelerations also occurred in the absence of visual feedback from the reaching hand (Jeannerod & Biguer, 1982). At the same time as the re-acceleration occurred, the hand was found to have its peak aperture (Jeannerod, 1981). In other words, preparation for grasping the object is an integrated aspect of the visually guided secondary part of the approach.

## EARLY DEVELOPMENT OF REACHING AND GRASPING

Newborn infants do not grasp objects reached for. The movements of their arms and hands are synergistically coupled. As the arm is extended forward, the hand has a tendency to open, and as the arm is flexed toward the body, the hand has a tendency to close. These synergies can be observed in some of the classical newborn "reflexes." In the vestibularly elicited Moro response, for instance (Prechtl, 1965), both arm and hand extend together at first, after which they flex together. In the traction response (Twitchell, 1965), the neonate's hand is pulled by the wrist stretching shoulder adductors and flexors. The response is a synergistic flexing of all joints, including the joints of the hand. Hofsten (1984) has shown that the same synergistic properties are also typical for spontaneous arm–hand movements. Out of 225 observed reaches performed by newborn infants, the hand was coded as open when it was farthest away from the body in more than 60% of the time, and as fisted in less than 10% of the time. The hand was actually observed as opening up during the approach in 35% of the cases. Performance was similar whether the infant looked at the target or not. The hand was observed to open up in 34% of the reaches performed while the infant closed his or her eyes or looked away from the target.

The corresponding figure for reaches performed while the infant looked at the target was 38%.

The performance is very similar to the reaching behavior of newborn rhesus monkeys as described by Lawrence and Hopkins (1976). They found that the direct cortico-motoneuronal connections were not yet established in the newborn rhesus monkey and that the two subcortical systems seemed relatively undifferentiated. Earliest reaching for food occurred around 4 weeks of age. The aim was rather inaccurate and the arm movements unsteady, so the hand frequently missed the food. If the food was grasped at all, it was done by closure of the fingers altogether and the subsequent releasing of it in the mouth when the arm was flexed was reported to be difficult.

Around 2 months of age, the described synergy between arm and hand movements starts to break up. Hofsten (1984) found that, instead of opening the hand during the extension of the arm, 2-month-old infants typically fisted the hand, often vigorously, in the extended phase of the arm movement. This kind of behavior increased from less than 10% in the 1-week-olds to more than 70% in the 7-week-olds. This is shown in Fig. 27.1.

At around 3 months of age, the infants studied by Hofsten (1984) started to open the hand again when extending the arm, but this time only when fixating the target. At this age, the opening of the hand can no longer just be described as a part of an extension synergy but is, rather, a preparation for grasping the object. At the same age, the number of reaching attempts increased greatly.

Parallel to this development, better means of obtaining precise visual information about spatial relations in reaching space emerge. The sensitivity to binocular disparity develops very rapidly between 3 and 5 months of age. Fox, Aslin, Shea, and Dumais (1980) found that $3\frac{1}{2}$-month-olds, but not $2\frac{1}{2}$-month-olds would track a moving virtual object specified by binocular disparity in a dynamic random-dot stereogram. Held and colleagues (Birch, Gwiazda, & Held 1982; Held, Birch, & Gwiazda, 1980) showed the same developmental trend using a modified preferential-looking technique. They showed rapid rise in detection of fine disparities from 3 months of age to an adultlike level between 5 and 7 months of age.

As the hand is gaining independent status and the binocular system is maturing, the child is starting to master the visually guided grasping phase. At this age, infants spend considerable time watching their hands as they move them in the visual field. This kind of behavior has earlier been described by Piaget (1953) and White, Castle and Held (1964). White et al. (1964) wrote: "Occasionally, one hand will be raised, looked at, and brought slowly to the stimulus while the glance shifts from hand to object repeatedly" (p. 356). Piaget thought that this behavior reflected the mutual assimilation between the visual motor schema of eye–hand and eye–object

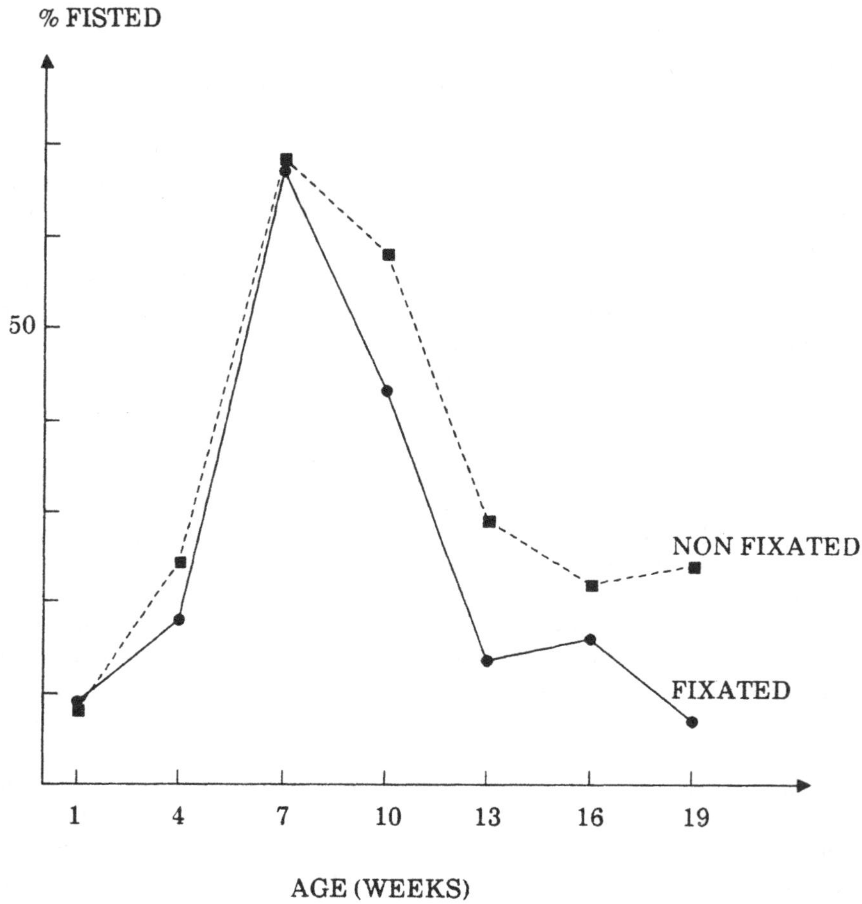


FIG. 27.1. Percentage of fixated movements (F) and nonfixated movements (NF) at which the hand is fisted for different prereaching age levels. (from Hofsten, 1984).

and the tactual motor schema of the hand. In other words, Piaget thought that the looking back and forth between the hand and the target somehow connected the eye to the hand. This is most probably wrong, but, on the other hand, Piaget was correct in concluding that some experience was necessary before successful reaching and grasping becomes possible.

By 4 months of age an infant corrects a reach for a target seen through horizontally displacing prisms (McDonnell, 1975), showing that by this age infants can visually guide the final path of their reaching. During the months to follow, vision comes to play an increasingly important role in manual action. Relevant information about this phenomenon comes from Lasky

(1977). Lasky had infants reach for objects seen through a horizontally placed mirror. At the proper place underneath the mirror was an object identical to the reflected one. In the control condition, a panel of clear plastic replaced the mirror. Lasky found that being able to see the reaching hand had very little effect on the performance of the 4-month-old infants. They retrieved the object in 23% of the attempts performed in the mirror condition and in 28% of the attemps performed in the control condition. The 5- and 6-month-olds showed similar performance in the mirror condition, but in the clear plastic condition they were substantially better. In this condition, the 5-month-olds retrieved the object in 48% of the cases and the 6-month-olds in 68% of the cases.

## THE REFINEMENT OF THE APPROACH

When the infant first starts to reach successfully for objects, the approach looks rather different from that of the adult. The movement often consists of several units or phases, and the approach is awkward and crooked (Hofsten, 1979). In only about half the reaches, the first unit of the approach is also the largest unit. However, this changes very quickly with age. The first step grows in importance. More and more of the approach was found to be concentrated in this step. Subsequent steps became more and more subordinate to the first one, having less to do with the approach and more with increasing the precision of the reach. At around 6 months of age, the adultlike reaching pattern started to dominate. From Fig. 27.2 it can be seen that from that age on, reaches consist of, at most, two movement elements; presumably one reaching unit and one grasping unit.

## THE GRASPING ACT

One of the most important aspects of grasping skill is to what degree the grasp is prepared beforehand. As described, the second phase of the approach seems to be a part of the grasping act, although the object is not yet encountered. Jeannerod and Biguer (1982) found that at the same time the re-acceleration occurred, the hand started to close. All preparatory adjustments like this one are visually controlled. They may be of two kinds. First, there are spatial adjustments of the reaching hand to the orientation, form, and size of the object. Second, the securing of the target should be timed in such a way that the hand starts to close around the target in anticipation of and not as a reaction to the encounter with the object.

Young infants adjust the orientation of the hand to the orientation of the object before it is encountered (Hofsten & Fazel-Zancy, 1984; Lock-

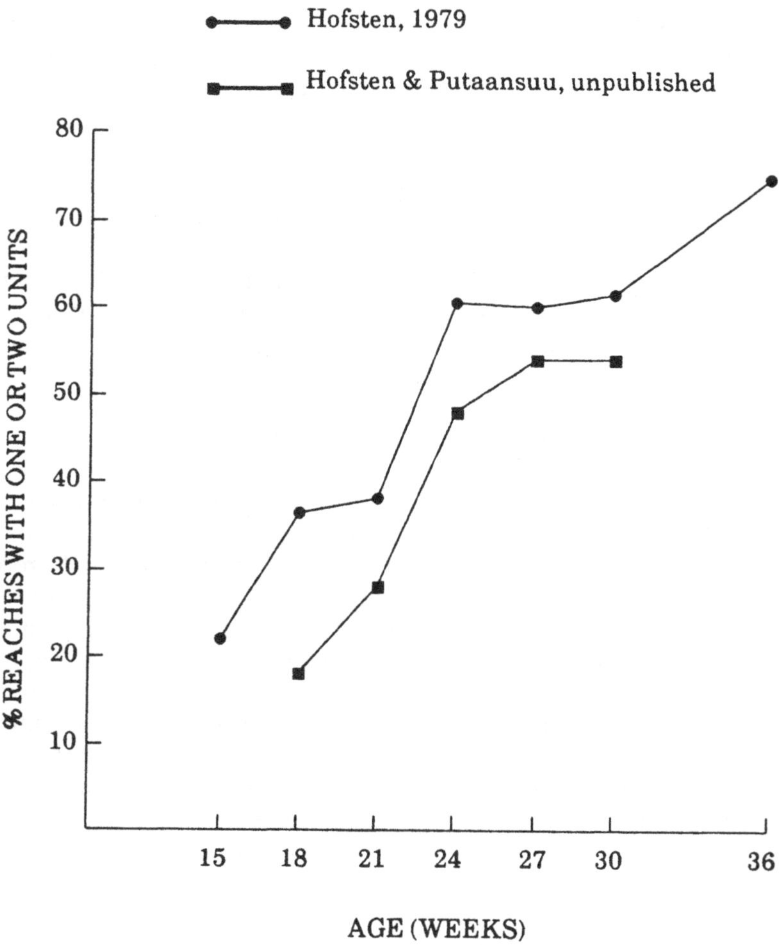


FIG. 27.2. Proportion of reaches consisting of one or two movement units as a function of age of the subjects, CIRCLES = data from Hofsten (1979), SQUARES = data from three subjects recorded with Selspot by Hofsten and Putaansuu, unpublished.

man, Ashmead & Bushnell, 1984; Morrongiello & Rocca, 1986). These adjustments are of great advantage to the child. They need to be performed if the hand is going to close around the object in an adequate way. Hofsten and Fazel-Zandy (1984) and Morrongiello and Rocca (1986) found that infants as young as 5 months made such preparatory adjustments of hand orientation. At this age the preparation was found to be rather crude and mainly done in the early phase of the reach or before the start of the approach. After 6 months of age, these adjustments improved greatly and were often performed continuously during the approach. However, strategies varied between individuals, and some 5-month-olds showed the same tendency to adjust the orientation of the hand continuously during the approach (Hofsten & Fazel-Zandy, 1984).

Adults also alter the opening of the hand as a function of the size of the object to be reached for (Jeannerod, 1981). However, preparatory adjustments to object size are less crucial than preparatory adjustments to object orientation. Instead of adjusting the opening of the hand to the size of the object, it would also be possible to open the hand fully during the approach. The disadvantage of that strategy is the additional time it takes to close a fully opened hand relative to a semi-opened hand. The advantage of opening the hand more fully is that it implies decreased demands for endpoint accuracy. Therefore, adults use this strategy when having to make a fast reach (Wing, Turton, & Fraser, 1986).

Hofsten and Rönnqvist (1988) monitored the distance between thumb and index finger with an optoelectronic technique (Selspot) in reaches performed by 5- to 6-month-olds, 9-month-olds, and 13-month-olds. Spherical targets of three different sizes were used. They were 15 mm, 25 mm, and 35 mm in diameter. It was found that the infants in the two older age groups did adjust the opening of the hand to the size of the target reached for. However, this was not true for the youngest age group. This can be seen from Fig. 27.3 that shows the maximum opening of the hand during the approach and the opening at touch as a function of target size for the three different age groups.

Why is the opening of the hand not systematicaly related to target size at 5 to 6 months? One possibility is that at 5 to 6 months of age, infants are not predominantly using the thumb and the index finger in grasping objects (see, e.g., Halverson, 1931). Instead, they use the medial part of the hand and the palm. There is therefore a risk that monitoring the distance between the thumb and the index finger is not reflecting the most central properties of the spatial adjustments of the grasp at this age.

Although the older infants would adjust the opening of the hand to the size of the target, they would do this in a less differentiated way than adults. The hand was more fully opened during the approach for all the different target sizes used. There may be at least two possible reasons for this. It may be a reflection of an incomplete differentiation between the motor systems of the arm and the hand. As the arm extends, extension of the digits may be facilitated, too. From a neurological perspective, Twitchell (1970) observed and discussed this abduction or dorsiflection of the fingers during the approach in terms of "approach and avoidance" responses constraining the movements of the upper limbs in the infant. Another interpretation of this behavior relies on the fact that it is adaptive. A fully opened hand optimizes the possibility of capturing the object if the

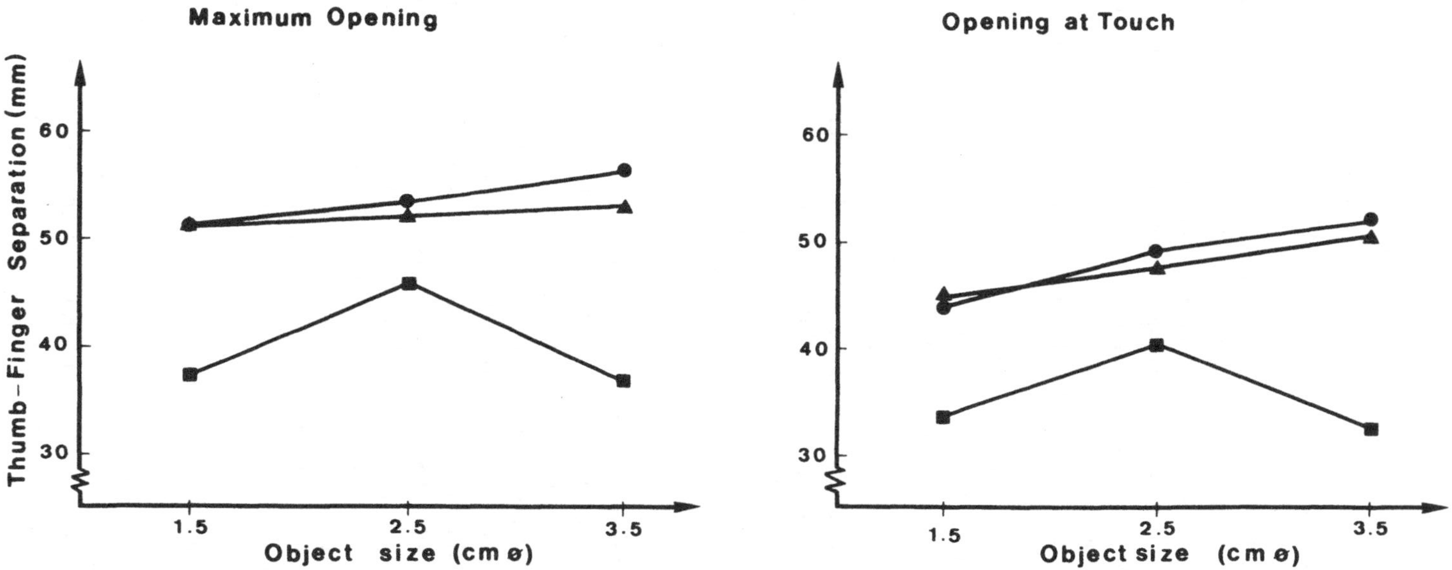


FIG. 27.3. Mean maximum opening of the hand during the approach and opening at touch as a function of object size and age, SQUARES = 5- to 6-month-old infants, TRIANGLES = 9-month-old infants, CIRCLES = 13-month-old infants.

movement is spatially less precise. Both interpretations could be valid, of course.

A smooth reaching action requires that the grasp be adequately timed relative to the encounter with the object. This is especially true if the object to be grasped is moving. If the hand closes too late, the object just bounces on the palm and is lost. If the hand closes too early, the object hits the knuckles. Alderson, Sully, and Sully (1974) found that the time precision in grasping a lightly thrown ball had to be around 14 msec. Because most adults master this task with ease it can be concluded that precise timing is a prominent feature of adult grasping.

Precise timing requires that the grasp be planned for and initiated in anticipation of the encounter with the target. Such planning can only occur under visual control. Tactual control of the grasping act implies that the grasp is initiated after the target has been touched, which would be too late for securing the target if, for instance, it is moving. Because tactually controlled grasping is initiated after contact, it by necessity also induces an interruption in the reach-and-grasp act. Thus, the emergence of visually controlled, well-timed grasping is crucial for the development of manual skill. It is well known that grasping also can be tactually controlled in the prereaching infant (Twitchell, 1970), but little is known about when visual control of grasping starts to be effective. When during development does the child start to use visual information for timing the grasping act?

Hofsten and Rönnqvist (1988) studied the question of timing of the grasp in infant reaching. They monitored the distance between thumb and index finger as the infant approached the target and determined precisely when this distance started to diminish. They also monitored the position of the target. When the hand encountered the target it was immediately displaced, and the time of this event could be precisely determined, too. An example of this is shown in Fig. 27.4. At all three age levels, the closing of the hand was well timed in relation to the encounter with the object. This is shown in Fig. 27.5. Figure 5 depicts the distribution of time differences for individual reaches between starting to close the hand and touch. In the two younger age groups the distribution of timing centers around touch. For the 5- to 6-month-olds, 75% of the grasps started within ± 0.1 sec of the encounter with the target. For the 13-month-olds, the distribution of timing did not center around touch anymore. At this age, the grasping action typically started well before touch. When studying adult grasping, Hofsten and Rönnqvist (1988) found that subjects would only partly adjust the maximum opening of the hand to the size of the object. Instead, they would open the hand much more than needed and adjust the timing of the grasp to the size of the target. In other words, they would start closing the hand earlier for the smaller than for the larger target. There was no age

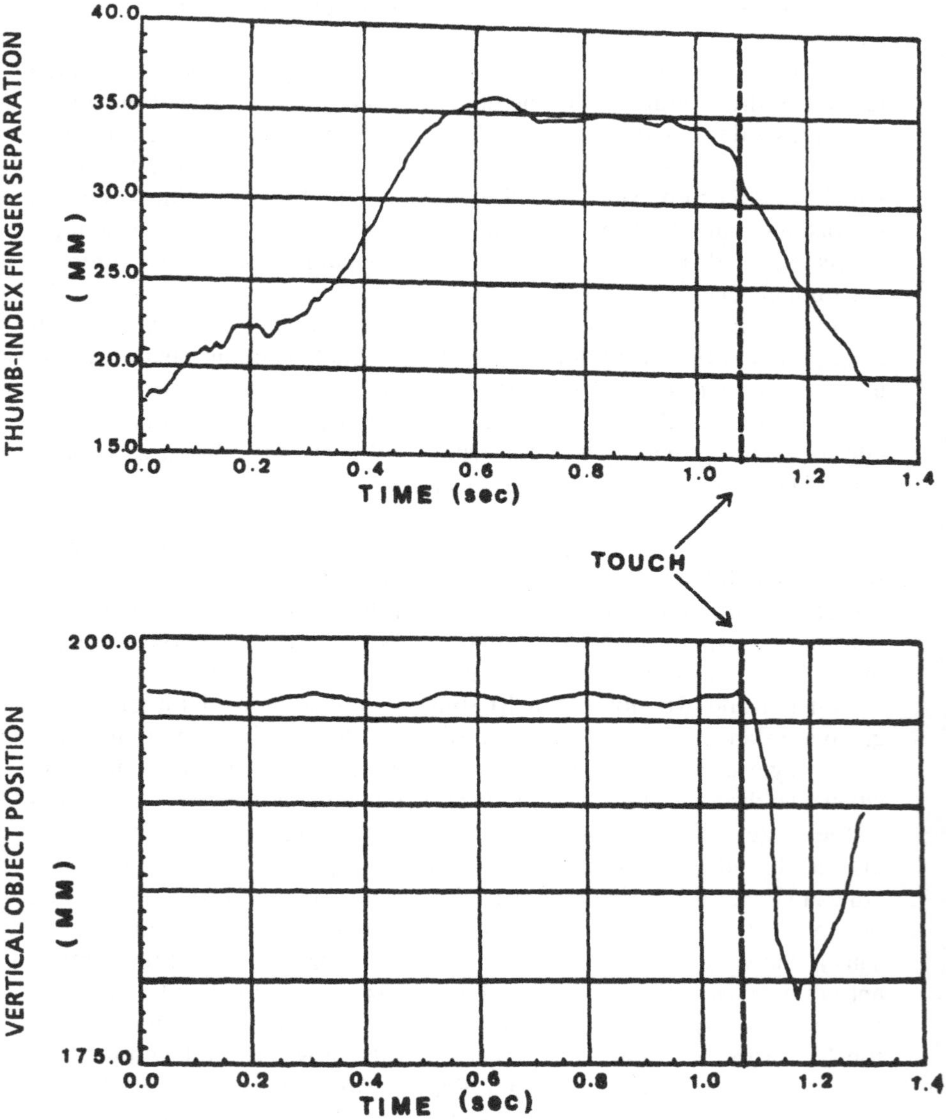


FIG. 27.4. An example of a reach performed by a 5-month-old infant. The target size was 15 mm in diameter. The upper graph shows the distance between the thumb and indexfinger as a function of time. The lower graph shows the position of the target as a function of time and how it is displaced when contacted. Note that the hand starts to close before contact.

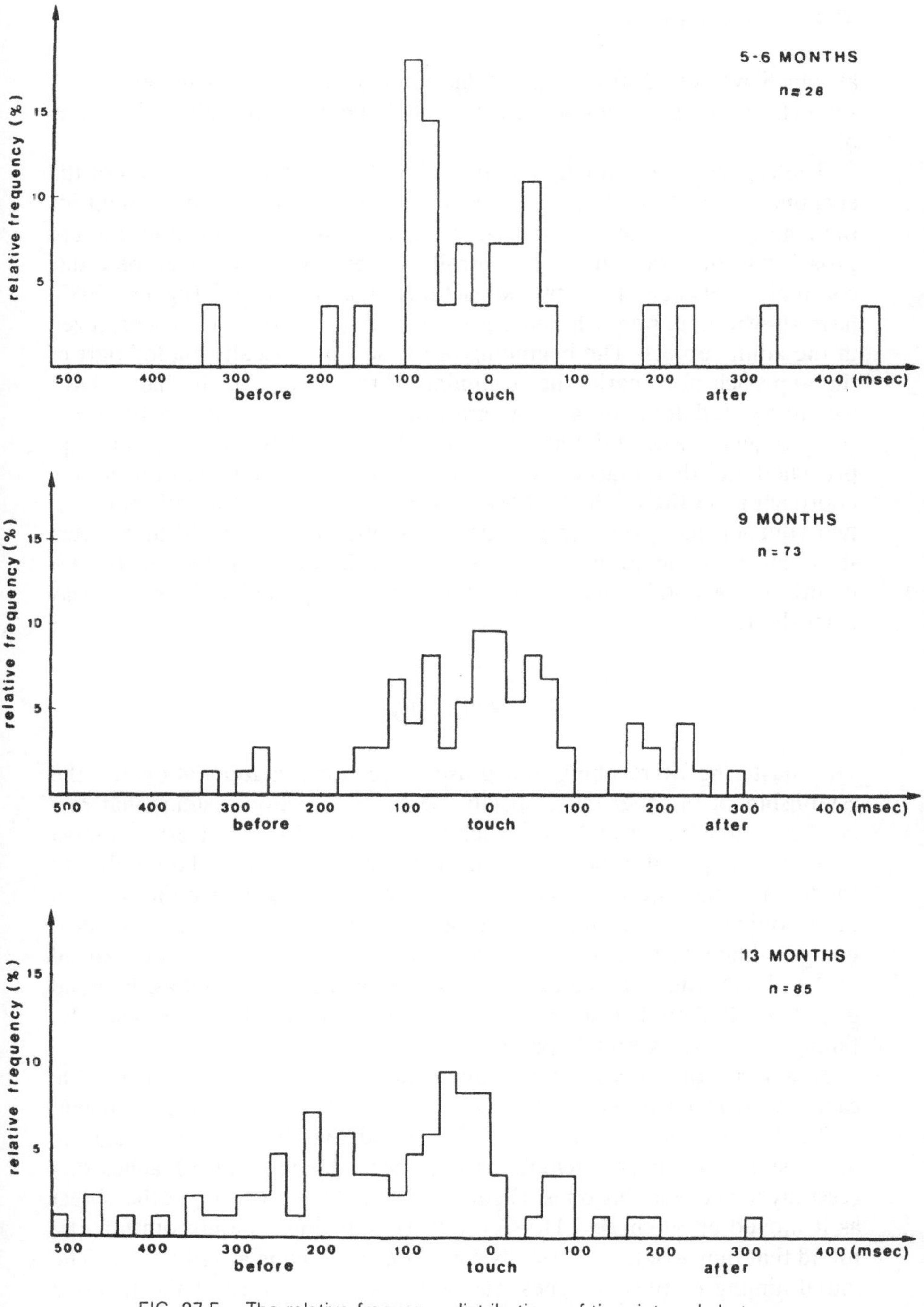


FIG. 27.5. The relative frequency distributions of time intervals between starting to close the hand and touch for individual reaches performed by the 5- to 6-month-old infants, the 9-month-old infants, and the 13-month-old infants in Experiment 2.

at which we found any signs of this kind of adjustments in our infant subjects. This refinement of grasping is yet to be developed after 13 months of age.

The capacity for planning and initiating the grasp in anticipation of the encounter with the object is a necessary but not sufficient requirement for the emergence of integrated reach-and-grasp action. Even if both the approach and the grasp are well controlled, there might very well be a discontinuity between these two subactions. Jeannerod and Biguer (1982) have shown that approach and grasp are well integrated and synchronized in the adult subject. The beginning of the second, visually guided part of the approach also marks the beginning of the closing of the hand. How well integrated are approach and grasp when infants start to reach for and grasp objects successfully at 4 to 5 months of age? Is the target first approached and then grasped, or is the grasping action initiated during the approach as in the adult. Hofsten and Rönnqvist (1988) found that their two younger age groups (5 to 6 and 9 months of age) would most often start closing the hand in the close vicinity of the target. However, the 13-month-olds would be more inclined to start closing the hand farther away from the target.

## CATCHING

The mastering of reaching and grasping provides a foundation for the establishment of other more specific perception–action systems that rely on these capacities. Catching is such a case. It is also the nicest and most striking example of anticipation in infants' manual action. To be able to catch a moving object, the subject not only needs to perceive the position of the object at an instant but also needs to perceive where the object is going and how fast. A successful catch has to aimed for some point ahead of the object where the hand and the object would meet, and as the hand gets there, it should close around the object at the right time. Obviously, timing has to be extremely precise.

In a series of studies, I have found that infants possess a remarkable capacity to catch objects (Hofsten, 1980, 1983; Hofsten & Lindhagen, 1979). Hofsten and Lindhagen (1979) found that from the very age an infant starts to master reaching for stationary objects, it also reaches successfully for fast moving ones. Eighteen-week-old infants caught the object as it moved at 30 cm/sec. Hofsten (1983), studying 8-month-old infants, found that they would successfully catch objects moving at 120 cm/sec. The initial aiming of these reaches was within a few degrees of the meeting point with the target, and the variable timing error was between 50 and 60 msec.

## FINE MANUAL SKILLS

Skilled hand movements serve many diverse functions and there are probably several perception–action specializations contributing to these abilities. They serve both manipulative and explorative functions. Some explorative actions, such as rubbing, squeezing, and banging, are systematically used by infants shortly after they have mastered reaching and grasping (Palmer, 1985; Ruff, 1984). Feeding is another class of manual movements that seems specialized. Rizzolatti, Gentilucci, Fogassi, Luppino, Matelli and Ponzoni-Maggi (1987) found evidence for rostral parts of inferior area 6 in the macaque monkey to be specialized for serving various kind of manual actions, for example, hand–mouth behavior.

When the child first starts to reach successfully for objects, the medial part of the hand plays a prominent role, and the whole hand is typically used in grasping any object (Halverson, 1931). During the second half of the first year of life, the focus of the grasp moves toward the radial part of the hand. The child starts to be able to pick up small objects using only the thumb and the index finger and to control relatively independent finger movements in manipulating objects (Gesell, 1928; Halverson, 1933). These highly sophisticated skills presuppose delicate visual guidance.

The motor pathways responsible for the control of relatively independent finger movements seem to be the pyramidal tracts. Lawrence and Kuypers (1968a) showed that interruption of both pyramidal tracts caused permanent loss of independent finger movements in the rhesus monkey. The pyramidal pathway seems to be the most recently evolved motor pathway. Cortico-motoneuronal connections occur in increasing numbers in the rhesus monkey, the chimpanzee, and man (Kuypers, 1973). In development, the pyramidal pathway matures after the maturation of the two subcortical pathways (Lawrence & Hopkins, 1976).

The dependence of manual movements on vision not only incerases during this period, but it also decreases. Bushnell (1982) has reported that subjects are less dependent on seeing the reaching hand. The subject might even look away while reaching and does not seem to be as bothered as the 6-month-olds when sight of the reaching hand is disrupted (Bushnell, 1982). This indicates an increasing ability of the infant to construct and support manual actions independent of vision that is needed in skilled manipulation. By the end of the first year, infants start to be able to make detours reaching around a barrier for objects they see through the barrier, thus being able to separate seeing an object at a specific location and the reaching strategy for it (Diamond, 1988; Lockman, 1984).

At the end of the first year of life, many of the essential basic requirements for manual actions are established. This does in no way mean that development is over. Manual skills develop all through childhood.

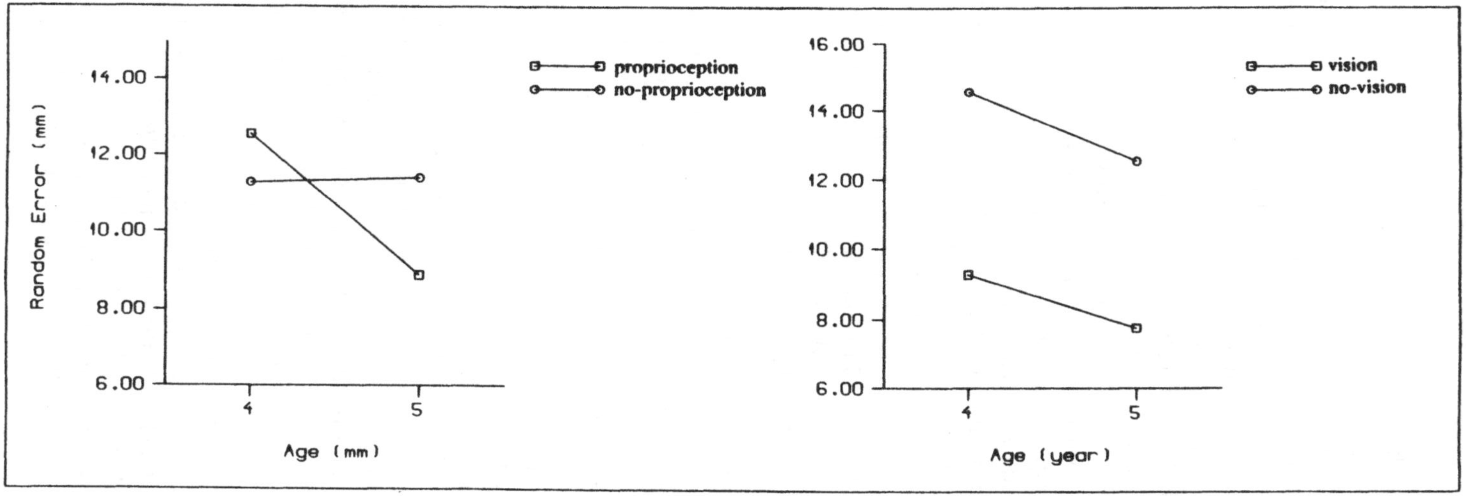


FIG. 27.6. Mean random errors of the 4- and 5-year-olds on the two closest dots. The left figure depicts random error as a function of age and availability of visual information (the two visual conditions are collapsed and the two nonvisual conditions are collapsed), and the right figure depicts the random error as a function of age and availability of proprioceptive informatiion (the two proprioceptive conditions are collapsed and the two non-proprioceptive conditions are collapsed).

Hofsten and Rösblad (1988) studied visual and proprioceptive control of precise pointing movements in children 4 to 12 years of age. The task of the subject was to place drawing pins underneath a table top at the locations of four dots placed on the upper side of the table top. Information about target position was varied. In the "visual" condition, the subject could see the positions of the dots at which he or she pointed with the unseen hand. In the proprioceptive condition, the subject placed the index finger of the other hand on the dot to be pointed to, closed the eyes, and placed the drawing pin underneath the dot. Two additional conditions were included. In one of them, the dots to be pointed at were both seen and felt with the index finger of the other hand. In the other, the subject had to look at the target, close the eyes, and point to the remembered position of the target. Thirty subjects, 15 boys and 15 girls, were tested at each age level. The results were analyzed both in terms of random and systematic errors.

The most dramatic effect of the whole study is the superiority of visual control in younger children. When the target to be pointed at was visually defined, the systematic as well as the random error was reduced, but especially so in the younger children.

In the proprioceptive and memory conditions between 5 and 8 years of age, the random error is reduced by 36% and 44%, respectively. This agrees well with Laszlo and Bairstow's data (1980). These data suggest that the nonvisual means of controlling fine movements of the hand are substantially developed during the late preschool age. The analysis of random errors for the two closest dots at 4 and 5 years of age is illustrative on this point. Figure 27.6 shows that all the improvement in performance between these two ages can be attributed to an improved proprioceptive system. A recent study by Cook and Griffith (1988) suggests that the improvement of proprioceptive control of fine hand movements may continue into adulthood.

## DISCUSSION

The perception–action approach stresses two aspects of motor behavior that often have been overlooked in the past. First, it stresses the fact that motor behavior has to be controlled by perception and that any motor act is as much an accomplishment of the perceptual system as it is an accomplishment of the muscles implementing the movement. Second, it stresses the point that movements are inherently purposeful and that prospective control plays a role in any motor act. I have tried to illustrate these points by applying them to the development of manual action in children.

The perception–action approach is focused on functional aspects of movements rather than on structural ones. It is therefore very different from the informational processing approach (cf. Sanders, 1980, Sternberg, 1969) that has the opposite bias. In the present approach, movements cannot be considered in isolation of the context in which they are performed, because it is the context that carries the function of the act and supplies the information that drives the act. If the motor system has been designed to accomplish certain functions, then the performance on apparently similar, but contextually different tasks might not be comparable. An experiment by McLeod, McLaughlin and Nimmo-Smith (1985) is illustrative on this point. McLeod et al. had subjects either trying to hit a falling ball with a bat or pressing a button coincident with the ball passing a defined point in space. Although they are both coincidence-anticipation tasks, the context is quite different. McLeod et. al. (in press) found that the temporal standard deviation was 6 times higher on the button-pressing task than on the ball-hitting task. From an informational processing point of view, this an unexpected finding.

The present approach deals with the problem of what motor development is about. It has less to say about the developmental process itself. Sensorimotor development constitutes a complex phenomenon and is determined by several parallel and converging processes that are of several different origins. Thelen (1986), in discussing development of locomotion in the child, mentions the following determinants for that development: tonus control, articulator differentiation, muscle strength, postural control, visual flow sensitivity, body constraints, and motivation. They are all essential in one way or another, not only for the development Thelen focuses on but also for manual development.

In early infancy, it seems to be mainly the maturation of the brain that sets the limits of the manual system. In the sensory domain, this was exemplified by the emergence of binocular stereopsis and in the motor domain by the maturation of the three pathways regulating the movements of the upper limbs. As the brain matures, more and more possibilities for manual action emerge, and the child seems to be ready for them. When the infants begin to reach for and grasp objects, they do it intelligently from the start. The hand gets oriented to the placement of the target, and the grasp is prepared beforehand and timed relative to the encounter. Even such a complex and highly specific skill as catching (Hofsten, 1980, 1983) is mastered, in principle, as soon as the sensory and motor system are ready for it.

The fact that sensory and neuromotor development set the limits of the manual system in development does not mean that experience does not play an important role. On the contrary, experience is probably a crucial factor in development at all stages. However, the environment does not set the limits, and experience in itself does not constrain development. The opportunities are always there when the child is ready for them. Therefore, it is hard to trace the impact of experience on an age curve. Only in cases when the availability of experience is manipulated is this possible. Bauer and Held's (1975) experiment is an example of that. In infant rhesus monkeys, they manipulated the availability of visual information about the movements of the upper limbs and showed that access to such information was necessary for the establishing eye–hand coordination.

Finally, the influence of body constraints should be mentioned. It is obvious that the physical growth of the child affects the movement production problem in both quantitative and qualitative ways (Kugler, Kelso, & Turvey, 1982). As the body is scaled up, there has to be continuous readjustment of the perception–action systems, to keep them calibrated to each other and to the external world. In the manual domain, the necessity for such calibration was demonstrated by Bauer and Held (1975). There is also the possibility that at least some sensorimotor systems are based on body scale information, insensitive to growth. In the manual domain, vergence-controlled arm movements may be such a case. The growth of the arm is almost completely compensated for by the increase over age in the interocular distance and the decrease over age in the angular displacement of the foveas relative to the visual axes (Hofsten, 1977). Convergence at arm's length is approximately the same throughout the first year of life. Another growth problem concerns the size relation between the hand and external objects. An object that for a small child has to grasped with both hands may for an older child constitute a unilateral task.

## ACKNOWLEDGMENTS

This paper was prepared while I was a fellow at the Center for Advanced Study in the Behavioral Sciences, Stanford. I am grateful for the financial support provided by National Science Foundation #BNS87-00864.

## REFERENCES

Alderson, G. J. K., Sully, D. J., & Sully, H. G. (1974). An operational analysis of a one-handed catching task using high speed photography. *Journal of Motor Behavior, 6*, 217–226.

Arbib, M. A. (1981). Perceptual structures and disturbed motor control. In V. B. Brooks (Ed.) *Handbook of Physiology, The Nervous System II. Motor Control* (pp. 1449–1480). Bethesda, MD; American Physiological Society.

Bauer, J., & Held, R. (1975) Comparison of visually guided reaching in normal and deprived infant monkeys. *Journal of Experimental Psychology: Animal Behavior Processes, 1*, 298–308.

Birch, E. E., Gwiazda, J., & Held, R. (1982). Stereoacuity development for crossed and uncrossed disparities in human infants. *Vision Research, 22*, 507–513.

Brooks, V. B., Cooke, J. C., & Thomas, J. S. (1973). The continuity of movements. In R. B. Stein, K. G. Pearson, R. S. Smith, & J. B. Redford (Eds.), *Control of posture and locomotion* (pp. 257–272). New York: Plenum Press.

Bushnell, E. (1982). The decline of visually guided reaching during infancy. *Infant Behavior and Development, 8*, 139–156.

Butterworth, G. (1986). Some problems in explaining the origins of movement control. In M. G. Wade & H. T. A. Whiting (Eds.). *Motor Development in Children: Problems of Coordination and Control* (pp. 23–32). Dordrecht: Martinus Nijhoff.

Cook, M., & Griffith, K. (1988). *The accuracy of performance in a finger alignment task*. Manuscript.

Diamond, A. (1988). Differences between adult an infant cognition: Is the crucial variable presence or absence of language? In L. Weiskrantz (Ed.), *Thought without language* (pp. 337–370). Oxford: Oxford University Press.

Fentress, J. C. (1986). Development of coordinated movement. Dynamic, relational and multileveled perspectives. In M. G. Wade & H. T. A. Whiting (Eds.). *Motor Development in Children: Problems of Coordination and Control* (pp. 77–106). Dordrecht: Martinus Nijhoff.

Fox, R., Aslin, R. N., Shea, S. L., & Dumais, S. T. (1980). Stereopsis in human infants. *Science, 207*, 323–324.

Gesell, A. (1928). *Infancy and human growth*. New York: MacMillan.

Gibson, J. J. (1979). *The Ecological Approach to Visual Perception*. Boston: Houghton Mifflin.

Gould, S. J. (1982). *The Panda's Thumb*. New York: Norton.

Halverson, H. M. (1931). Study of prehension in infants. *Genetic Psychology Monographs, 10*, 107–285.

Halverson, H. M. (1933). The acquisition of skill in infancy. *Journal of Genetic Psychology, 43*, 3–48.

Held, R., Birch, E. E., & Gwiazda, J. (1980). Stereoacuity of human infants. *Proceedings of the National Academy of Science U.S.A., 77*, 5572–5574.

Hofsten, C. von. (1977). Binocular convergence as a determinant of reaching behavior in infancy. *Perception, 6*, 139–144.

Hofsten, C. von. (1979). Development of visually guided reaching: the approach phase. *Journal of Human Movement Studies, 5*, 160–178.

Hofsten, C. von. (1980). Predictive reaching for moving objects by human infants. *Journal of Experimental Child Psychology, 30*, 369–382.

Hofsten, C. von. (1982). Eye–hand coordination in newborns. *Developmental Psychology, 18*, 450–461.

Hofsten, C. von. (1983). Catching skills in infancy. *Journal of Experimental Psychology: Human Perception and Performance, 9*, 75–85.

Hofsten, C. von. (1984). Developmental changes in the organization of prereaching movements. *Developmental Psychology, 20*, 378–388.

Hofsten, C. von. (1985). Perception and action. In M. Frese & J. Sabini (Eds.), *Goal Directed Behavior: The concept of action in psychology* (pp. 80–96). Hillsdale, NJ: Lawrence Erlbaum Associates.

Hofsten, C. von., & Fazel-Zandy, S. (1984). Development of visually guided hand orientation in reaching. *Journal of Experimental Child Psychology, 38*, 208–219.

Hofsten, C. von., & Lindhagen, K. (1979). Observations on the development of reaching for moving objects. *Journal of Experimental Child Psychology, 28*, 158–173.

Hofsten, C. von, & Rönnqvist, L. (1988). Preparation for grasping an object: A developmental study. *Journal of Experimental Psychology: Human Perception and Performance, 14*, 610–621.

Hofsten, C. von, & Rösblad, B. (1988). The integration of sensory information in the development of precise manual pointing. *Neuropsychologica, 26*, 805–821.

Hofsten, C. von., and Spelke, E. S. (1985). Object perception and object directed reaching in infancy. *Journal of Experimental Psychology: General, 114*, 198–212.

Jeannerod, M. (1981). Intersegmental coordination during reaching at natural visual objects. In J. Long & A. Baddeley (Eds.), *Attention and Performance IX* (pp. 153–168). Hillsdale, NJ: Lawrence Erlbaum Associates.

Jeannerod, M., & Biguer, B. (1982). Visumotor mechanisms in reaching within extrapersonal space. In D. J. Ingle, M. A. Goodale and R. J. W. Mansfield (Eds.), *Analysis of Visual Behavior* (pp. 387–409). Cambridge: MIT Press.

Johansson, R., & Westling, G. (1987). Signals in tactile afferents from the fingers eliciting adaptive motor responses during precision grip. *Experimental Brain Research*, *66*, 141–154.

Kugler, P. N., Kelso, J. A. S. & Turvey, M. T. (1982). On the control and coordination of naturally developing system. In J. A. S. Kelso and J. E. Clark (Eds.), *The Development of Movement control and Coordination* (pp. 5–78). New York: Wiley.

Kuypers, H. G. J. M. (1973). The anatomical organization of the descending pathways and their contribution to motor control especially in primates. In J. E. Desmedt (Ed.), *New developments in Electromyography and Clinical Neurophysiology: Vol. 3*, (pp. 38–68). New York: S. Krager.

Laskey, R. E. (1977). The effect of visual feedback of the hand on reaching and retrieval behavior of young infants. *Child Development, 48*, 112–117.

Laszlo, J. I., & Bairstow, P. J. (1980). The measurement of kinestethic sensitivity in children and adults. *Developmental Medicine and Child Neurology, 22*, 454–464.

Lawrence, D. G., & Hopkins, D. A. (1976). The development of the motor control in the rhesus monkey: Evidence concerning the role of corticomotorneuronal connections. *Brain, 99*, 235–254.

Lawrence, D. G., & Kuypers, H. G. J. M. (1968a). The functional organization of the motor system in the monkey: I. The effects of bilateral pyramidal lesions. *Brain, 91*, 1–14.

Lawrence, D. G., & Kuypers, H. G. J. M. (1968b). The functional organization of the motor system in the monkey: II. The effects of the descending brainstem pathways. *Brain, 91*, 15–36.

Lockman, J. J. (1984). Development of detour abilities in infants. *Child Development, 55*, 482–491.

Lockman, J. J., Ashmead, D. H., & Bushnell, E. W. (1984) The development of anticipatory hand orientation during infancy. *Journal of Experimental Child Psychology, 37*, 176–186.

McDonnell, P. (1975). The development of visually guided reaching. *Perception & Psychophysics, 18*, 181–185.

McLeod, P., McLaughlin, & Nimmo-Smith, I. (1985). Informational encapsulation and automaticity: Evidence from the visual control of finely timed actions. *Attention and Performance XI* (pp. 391–400). Hillsdale, NJ: Lawrence Erlbaum Associates.

Morrongiello, B., & Rocca, P. (1986, April). *Age-related changes in reaching behavior*. Paper presented at ICIS, Los Angeles.

Palmer, C. F. (1985). *Infants' exploration of objects: Relations between perceiving and action*. Unpublished doctoral dissertation, University of Minnesota, Minneapolis.

Piaget, J. (1953). *The Origins of Intelligence in the Child*. New York: Routledge.

Piaget, J. (1970). Piaget's theory. In P. H. Mussen (Ed.), *Carmichel's Manual of Child Psychology. Vol. 1*. (pp. 703–732). New York: Wiley.

Prablanc, C., & Pelison, D. (in press). Gaze saccade orientation and hand pointing are locked to their goal by quick internal loops. In M. Jeannerod (Ed.), *Motor Representations and Control: Attention and Performance XIII*. Hillsdale, NJ: Lawrence Erlbaum Associates, Inc.

Prechtl, H. F. R. (1965). Problems of behavioral studies in the newborn infant. In *Advances in the Study of Behavior, Volume 1*. New York: Academic Press.

Reed, E. S. (1982). An outline of a theory of action systems. *Journal of Motor Behavior, 14*, 98–134.

Rizzolatti, G., Gentilucci, M. Fogassi, L., Luppino, G., Matelli, M., & Ponzoni-Maggi, S. (1987). Neurons related to goal-directed motor acts in inferior area 6 of the Macaque monkey. *Experimental Brain Research, 67*, 220–224.

Rochat, P., Blass, E. M., & Hoffmeyer, L. B. (1987). *Oropharyngeal control of hand–mouth coordination in newborn infants*. Manuscript.

Ruff, H. (1984). Infants' manipulative exploration of objects: Effects of age and object characteristics. *Developmental Psychology, 20*, 9–20.

Sanders, (1980). Stage analysis of reaction processes. In Stelmach and J. Requin (Eds.) *Tutorials in Motor Behavior*. Amsterdam: North-Holland.

Sternberg, (1969). The discovery of processing stages: Extension of Donder's method. *Acta Psychologica, 30*, 276–315.

Thelen, E. (1985). Developmental origins of motor coordination: Leg movements in human infants. *Developmental Psychobiology, 18*, 1–18.

Twitchell (1965). The automatic grasping responses in infants. *Neuropsychologia, 3*, 247–259.

Twitchell, T. E. (1970). Reflex mechanisms and the development of prehension. In K. Connolly (Ed.) *Mechanisms of Motor Skill Development* (pp. 25–37). London: Academic Press.

White, B. L., Castle, P., & Held, R. (1964). Observations on the development of visually directed reaching. *Child Development, 35*, 349–364.

Wing, A. M., Turton, A., & Fraser, C. (1986). Grasp size and accuracy of approach in reaching. *Journal of Motor Behavior, 18*, 245–261.

# 28 Units of Motor Behavior: Modifications With Practice and Feedback

Douglas E. Young
*Motor Behavior Laboratory*
*Department of Physical Education*
*California State University, Long Beach*

Richard A. Schmidt
*Motor Control Laboratory*
*Department of Psychology*
*University of California at Los Angeles*

ABSTRACT

Alterations in movement control across practice and the influence of augmented information feedback—knowledge of results (KR)—represent two fundamental concerns in skill acquisition research. In this chapter, these areas were joined together in a paradigm that involves a coincident-timing task in which a moving target is intercepted by a level controlled by a flexion–extension movement at the elbow joint. With analyses that allow the description of programmed units of behavior, we show that practice with KR resulted in a large and systematic reorganization in movement control. In early practice, subjects produced a single, preprogrammed bi-directional action. However, in later practice, subjects segmented this action into two independently controlled parts. This reorganization was characterized by a pause in movement after the backswing (flexion) a variable period in which subjects presumably processed visual feedback from the display and then a second programmed forward swing (extension) movement to strike the target. These segmented patterns acquired via practice and KR suggested changes in motor control with practice that were counter to predictions from several current viewpoints about skill acquisition.

At a casual or superficial level of observation, we are all strongly aware of major alterations in movement control that occur when performers engage in long-term practice or training of complex skills. In addition to the obvious changes in goal attainment indexed by the outcome score for the task, numerous subtle changes seem to occur, such as alterations in the smoothness and "ease" with which the actions are generated, modifications in the movement trajectories, and, in some cases, rather clear alterations in the patterns of coordination. It is tempting to suggest that these alterations in global movement activity have their counterparts in terms of the underlying control processes for the action, particularly considering the kinematic variables that are controlled and the nature of movement organization. However, despite the interest in these variables and their obvious importance for understanding learning, it is surprising that these problems have received relatively little study. There have been several attempts to ask about practice-related changes in movement kinematics, and these experiments have provided a good start toward an understanding of how practice influences motor control (e.g., Hobart, Kelly, & Bradley, 1975; Hobart & Vorro, 1974; Landa, 1979; McGrain, 1980, 1986; Moore & Marteniuk, 1987; Vorro & Hobart, 1974). But the usefulness of these studies has been somewhat limited, in our view, by the fact that the tasks involved only a small number of kinematic degrees of freedom (but, see Higgins & Spaeth, 1972, for an exception), so that only minor modifications in movement control were possible across practice. This general lack of emphasis on kinematically "complex" actions is understandable, particularly given the technical difficulties in recording kinematic variables from several moving limb segments; only recently have high-speed LED-based recording systems been developed that minimize these problems.

Almost in parallel to the efforts to document kinematic changes are the efforts to understand how information feedback operates in skill acquisition. Feedback about the outcome of a movement in terms of the environmental goal—usually termed knowledge of results (KR)—is widely recognized as critical to learning (e.g., Adams, 1971; Bilodeau, 1966; Schmidt, 1988). Yet, despite KR's importance, it has also been typically studied in tasks with a small number of degrees of freedom to be controlled, whose patterns of improvement occur very quickly across trials. With very few exceptions (e.g., Adams, 1985), the result is that this tradition of research has not rendered very much insight into the alterations in movement control with feedback and practice.

One of our ongoing research projects over the past few years has addressed these two fundamental research questions simultaneously. Our approach has been different than approaches seen in the earlier work in these areas, mainly because of our impatience with the relatively simple movement behaviors that have been previously examined. First, we have begun to ask how practice alters the kinematic patterns that describe relatively complicated movement control, searching for generalizations that could be stated in terms of the alterations in motor-control processes that covary with improvements in skill. Second, we have begun to address how augmented information feedback regarding the outcome (i.e., KR), as one of the most important variables for motor learning, plays a role in these alterations. In the course of these investigations, we have developed a paradigm that seems sufficiently "rich" to provide a way to examine kinematic changes that occur with practice and feedback.

Through our initial work several years ago, we discovered a number of important changes in the structure of movement control as practiced continued. Of particular interest was evidence that seemed to reveal fundamental, sequentially organized, *units* of movement behavior. Discovering such units has been of interest for decades. Recent candidates for such units of behavior have been the stress group as a programmed element in speech (Sternberg, Monsell, Knoll, & Wright, 1978), whole words as units in typing (Terzuolo & Viviani, 1979, 1980), and portions of drawing movements during which certain relational invariants have been identified (e.g., Lacquaniti, Terzuolo, & Viviani, 1983; Viviani, 1986). The notion that such a unit-structure might exist in the kind of skilled actions described here provides new insight into how these underlying structures of the movement are controlled. Subsequent work with out paradigm provided information about how the units of these actions was altered across practice, and has yielded an initial understanding about how feedback interacts with practice in the formation of these units. This early work, focusing on the evidence for units and how they are altered with practice and feedback, provides the basis of this chapter.

We have several goals for this chapter. First, we describe a paradigm, and the rationale for it, that provides a way to uncover alterations in movement control in relatively "complex" tasks as a function of practice and feedback. Next, we describe analyses that allow a principled, objective description of fundamental units of action in this task. These units are altered markedly across practice—particularly when feedback is provided to the learners—and we provide evidence describing the nature of these changes in the underlying structure of the action. Finally, several commonly held viewpoints about the role of practice in altering movement control are challenged by the findings reported here, and these issues are discussed in the final section.

## PARADIGM DEVELOPMENT

Our original goal in designing this paradigm was to provide a way to evaluate the role of kinematic feedback about the pattern of action that leads to goals achievement (i.e., knowledge of performance, KP). As such, there were three features of these actions that we considered essential for our laboratory paradigm. First, the action needed to be reasonably complex, in that more than a single degree of freedom should be under the subjects' control, unlike the relatively "simple" actions that have been used in learning research in the past (e.g., the ballistic-timing task, and others); thus the task should be somewhat more representative of everyday motor skills seen in sport or industry. Second, the actions of the limb(s) relevant to performance should be computer monitored, so that kinematic features could be measured rapidly and fed back to the subject as KP without long delays. Third, and probably most important, we demanded that the goal of the movement be essentially separate from the patterns of action that produce the goal. In the ballistic-timing task, for example, the goal is to move a slide along a trackway to produce a movement time of a given duration. Providing KP about some kinematic feature of this simple action (e.g., peak velocity, time of maximum acceleration, and so on) involves information that is almost completely redundant with the overall goal score (e.g., Newell & Carlton, 1987; Newell & McGinnis, 1985; Newell & Walter, 1981). That is, a particular kinematic pattern in such tasks *is* (or is correlated highly with) the performer's goal, so providing information about these kinematics necessarily provides information about the goal. In most real-world activities, on the other hand, because very similar outcomes can be produced with a variety of kinematic patterns, we find that the kinematic features of the action are easily separated from the outcome the pattern achieved in the environment. For example, hundreds of separate descriptors of a golf swing's kinematic features (e.g., time of downswing initiation, height of backswing) are possible, and these need not be associated in any simple way to the outcome that the swing achieves in the environment—that is, the distance the ball traveled. However, it seems that these features must be related to the outcome in some way that is not well understood at present.

## Apparatus and Task

The task can be thought of as a laboratory analog of hitting a moving object (such as a ball) with a hand-held object (such as a bat), a common action in several sport activities (e.g., American baseball). The subject was seated at a desk and viewed a display made up of 32 red LEDs aligned in a row, 4.5 cm apart. They were oriented to the right of the subject's midline, with the nearest end of the row being on the desk surface, and the far end being elevated by 45°. When these lamps were illuminated sequentially, the subject received the impression that an object was moving toward the midline of the body, with a velocity of 134 cm/s (3 mph). The duration of this apparent movement was 1.176 s (range = 1.172 to 1.180 s). The LED located third from the end of the row nearest the subject was designated at the target LED and was marked clearly with red tape.

A vertical axle was attached via ball-bearing mounts to the desk, and an aluminum level (39 cm long) was attached to the axle so that the lever was free to move in the horizontal plane 3 cm above the desktop. The axle was positioned in line with the extension of the row of LEDs, so that the lever was in the 90° position when directly over the LEDs, as shown in Fig. 28.1. A vertical handle protruded upward from the other end of the

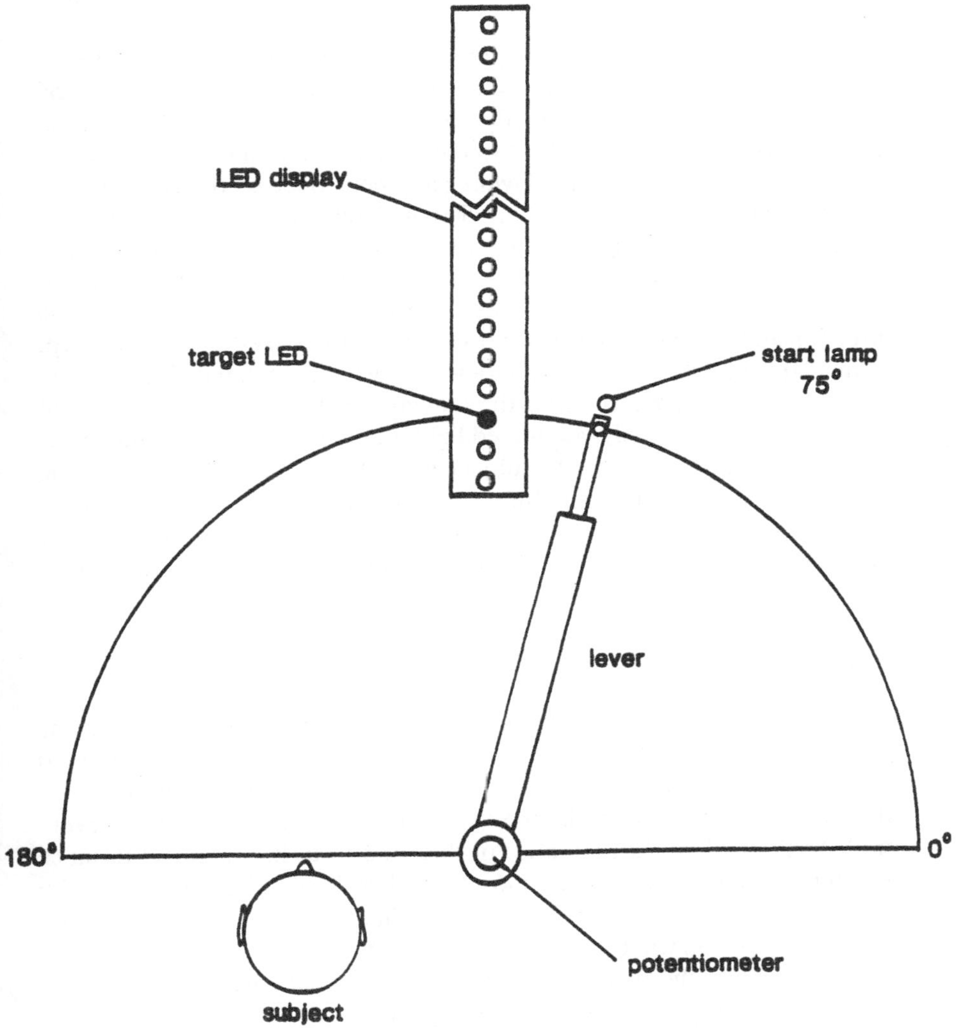


FIG. 28.1. A schematic overhead view of the coincident-timing apparatus. The subject begins with the lever at the starting position, backswings (counterclockwise) past the row of LEDs, and then makes a forward swing to "hit" the moving LEDs as they pass the coincidence-point.

lever, and a small red lamp was located on an extension of the lever, visible to the subject. The lever and row of LEDs were oriented so that, when the lever was moved past the row of LEDs, the red lamp was in the same (i.e., horizontal) plane as the target LED. At the lower end of the axle was a potentiometer for recording lever position. The potentiometer output, and a pulse indicating the time of illumination of the target light were delivered to a DEC 11/23 laboratory computer located in an adjacent room for later analysis.

The subject began seated slightly to the left of the row of LEDs (Fig. 28.1), engaging the lever so that the right hand grasped the vertical handle, and with the elbow directly over the vertical axle. The lever began in a starting position oriented 75° to the subject's frontal plane (Fig. 28.1), with the lever pointed at a small white lamp positioned on the desk surface. A warning tone was provided, and 500 ms later the first LED in the row was illuminated, starting the apparent motion. The subject's task was to move the lever counterclockwise, as in a backswing, past the row of LEDs to some position to the subject's left (but not beyond 180°), then reverse direction and move with a forward stroke to the right to attempt to intercept ("hit") the target LED at the moment it was illuminated. The subject understood that "contact" with the moving LEDs could only be made at the target LED, and only via the red light mounted at the end of the lever. Aside from the above rules, the subject was not constrained. The backswing movement could begin at any time, the backswing distance could be from 90° to 180° at the subject's choice, the forward swing could begin at any time, and the pattern of action in making the forward swing was free to vary as well.

## Scoring Procedures

The subject's goal in this task was to maximize an outcome *score*. In the analogy of hitting a moving object such as a ball, this score represented the distance that the ball was propelled in the horizontal dimension, and the subject's job was to make the "ball" travel as far as possible on every trial. In the real-world examples, of this kind of task, this travel distance is determined by essentially two interdependent factors. First, lever velocity is a factor, in that the faster the bat is traveling when (if) it strikes the ball, the farther the ball will travel. But second, if there is an error in the location of the bat when the ball is at the coincidence point, there could be only indirect contact with the ball, resulting in a shorter travel distance. Or, if the spatial errors are sufficiently large, there would be no contact at all, resulting in a ball-travel distance of zero. Furthermore, these two factors are related to each other, in that when subjects attempt to produce movements with increased velocity, there is a systematic tendency for the spatial errors to increase at the same time—a common application of the findings from the literature on speed–accuracy trade-offs in movement control (e.g., Newell, Hoshizaki, Carlton, & Halbert, 1979; Schmidt, Zelaznik, & Frank, 1978; Schmidt, Zelaznik, Hawkins, Frank & Quinn,1979)[1].

We incorporated all of these aspects into the scoring system. First, the score on each trial was determined by the product of two factors–the velocity at the coincidence-point, and a weighting factor that was itself a function of the absolute spatial error on that trial, as follows:

$$\text{Score} = \text{Velocity} \times \text{Weight}. \quad (1)$$

Velocity was determined on line from the computer and was the instantaneous angular velocity computed at the moment the lever passed the coincidence point. The Weight was computed by an algorithm in which larger absolute spatial errors—that is, the absolute distance in degrees of the lever from the coincidence point when the moving LEDs arrived there—produced systematically decreased weights.[2] This relationship is plotted in Fig. 28.2. Here, the algorithm maximized the Weight at 1.0 when absolute spatial error was equal to 0°. The Weight dropped sharply to 0.5 as the absolute spatial errors increased from 0° to 20°, decreased more gradually toward 0.0 as the spatial errors approached 60°, and was 0.0 for all absolute spatial errors greater than 60°. This scheme corresponded to a physical analog of this task in which absolute spatial errors of less than 20° were scored as relatively "direct hits," with a very strong benefit as the spatial errors approached 0°. Errors between 20° and 60° were considered "glancing contacts," with a maximum Weight of 0.5, and a systematically reduced value as the errors increased. Spatial errors greater than 60° were treated as "misses," where the Weight = 0.0. Various weighting algorithms have been used in the development of this task (see, e.g., Schmidt, Young, Swinnen, & Shapiro, 1989; Young, 1988), each allowing a slightly different emphasis on speed versus accuracy. The present one has the characteristic that it minimizes the number of zero scores produced (errors are seldom greater than 60°), while still providing a strong benefit as errors are reduced toward zero. Subjects were clearly aware of this scoring procedure through

---

[1]Actually, two opposite effects of increased velocity on spatial errors have been identified. Particularly when the velocities are very low (outside the range used in the present task), increased velocity leads to decreased spatial errors. However, when velocities are larger, then the reverse effect dominates, where spatial errors increase with increased velocity (see Newell et al., 1979; Schmidt, 1988).

[2]In this method, positive errors, where the lever is beyond (i.e., to the subjects right of) the coincidence-point when the LEDs arrived, and negative spatial errors are weighted identically. In other applications, however, there may be a desire to weight positive and negative spatial errors differentially, and this could be done easily with changes in the algorithm.

instructions that involved a presentation of a graph like that in Fig. 28.2 as well as a discussion of its implications for subjects' performance.

Generally, the task required the subjects to search for an optimum combination of velocity and spatial error that would maximize the task score, just as one must do when performing a real-world batting task of this type. Furthermore, unlike many motor responses used in the learning literature, the subject was free to choose nearly any pattern of action that he or she desired in order to achieve this goal—for example, making the backswing distance larger or smaller, or the forward swing earlier or later, and so on. Thus, the subject's only goal was to maximize the score as determined from the relationships shown in Equation 1. Also, neither the apparatus for this task, nor the instructions to the subjects about how to perform, necessarily required a particular pattern of action—the same score being possible with wide-ranging combinations of spatial error and velocity, and with many different kinematic patterns. Hence, KR about the score did not inform the subject very directly about which aspects of the movement *pattern* were deficient (see Fowler & Turvey, 1978), and thus, feedback did not seem to provide the kind of trial-to-trial guidance that is seen in the more simple learning tasks. In terms of the real-world analogy again, KR about the score gives the subject the same kind of information as seeing how far a batted ball was propelled. Presumably, for maximizing learning, a teacher or coach might attempt to give information about those aspects of the pattern that require modification (i.e., KP).

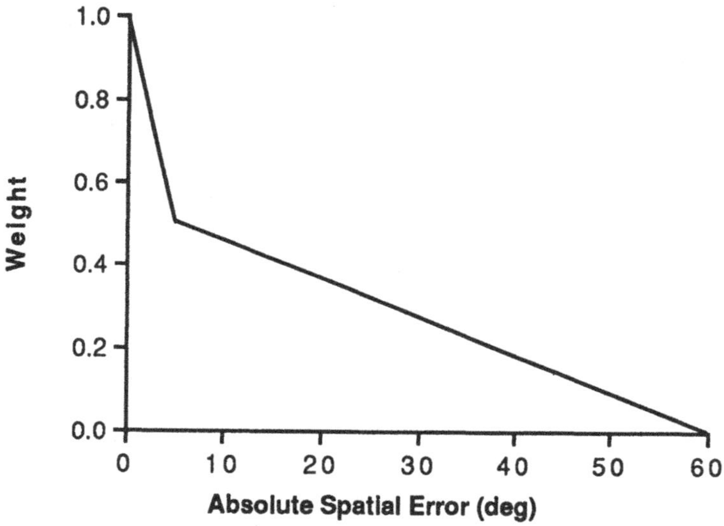


FIG. 28.2. The relationship between spatial errors and the weighting coefficient used to compute the score.

As such, this paradigm carries strong potential as an effective way to study how kinematic feedback (i.e., KP) from the pattern of action is involved in the learning process, and this emphasis continues to be our major motivation in the development of this paradigm (see Schmidt & Young, 1989; Young, 1988). Over the several years required for this KP-paradigm development, we have gained experience with a number of experimental manipulations, become familiar with the patterns of action to which subjects converge for maximizing performance, and noted several important changes that occur over practice with feedback. These observations suggest some important principles of movement organization and learning. Perhaps most importantly, however, these data provide strong suggestions about fundamental *units* of behavior that underlie subjects' responding in tasks like this one. The rationale for these ideas, and the data that support them, are provided in the next sections.

## DISCOVERING OF UNITS OF ACTION IN COINCIDENT TIMING

In one of the earliest exploratory experiments using this paradigm, we were concerned simply with the role that KR about the movement outcome (i.e., the score only) plays in learning. We suspected that subjects may be able to obtain information about their proficiency on their own, as spatial errors and velocity—the two components of the score—can presumably both be detected by subjects during (or after) a movement, and the score "computed" subjectively. We therefore asked whether—as would be the case in a real-word analog of this task—subjects could learn this response without any KR at all, relying exclusively on their own intrinsic sensory feedback. This study, then, simply contrasted a group of 10 university undergraduates who received KR (about the score only) after each trial with another group of 10 undergraduates who received no KR information at all (but could use intrinsic feedback mechanisms as a way to obtain information about the outcome). No augmented information about the components of the score (velocity or spatial error), nor about any of the kinematic measures, was ever provided as feedback. Subjects practiced for 105 trails during an initial acquisition phase, then after a delay of 10 min received a 20-trial no-KR retention test, and, then, after 24 hours received another 20-trial no-KR retention test.

### Variations in Performance with Practice

The mean performance scores for the two groups of subjects are shown in Fig. 28.3, where the mean score (Velocity × Weight) is plotted for successive 15-trial blocks in acquisition, and for 20-trial blocks in the retention

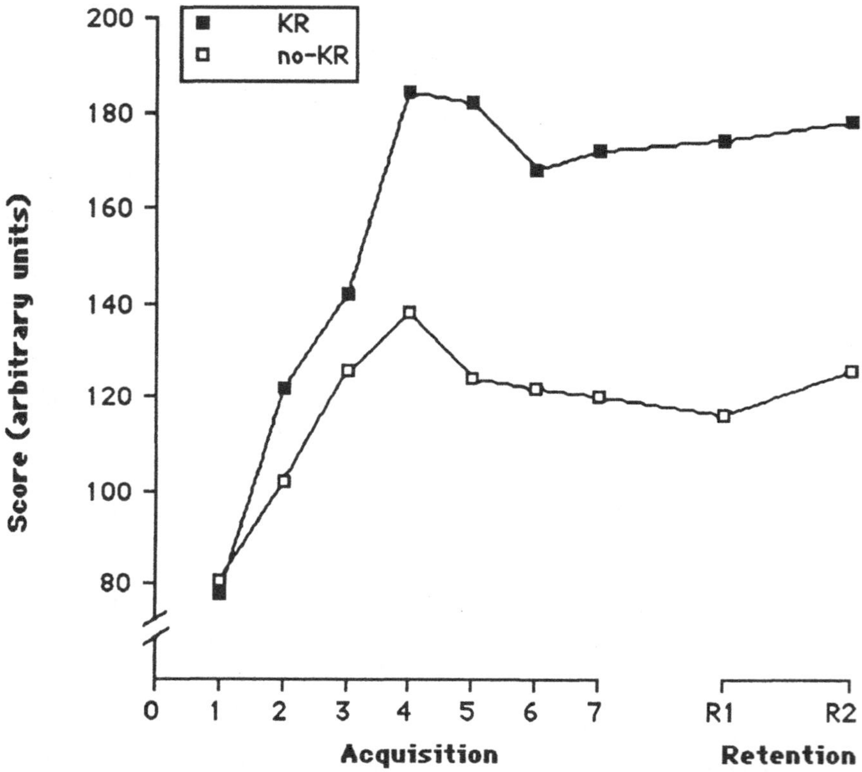


FIG. 28.3. The average scores (Velocity × Weight) for the KR and no-KR groups shown for 15-trial blocks in acquisition, and 20-trial retention tests after 10 min and 24 hours.

tests. For the KR subjects, scores increased from around 80 units in the first block to around 180 units in the final block and were essentially stable across the retention tests. Scores increased also for the no-KR subjects, but the final level of proficiency was about 60 units lower than for the KR subjects, both at the end of initial acquisition and during the retention tests. In acquisition, the groups effect, $F(1,18) = 9.8$, and the blocks and the blocks × groups interaction effects $Fs(6,108) = 11.4$ and 2.6, respectively, were significant, all $ps < .05$. It was clear that subjects could learn this task without KR, but that extrinsic information about the score aided considerably. These trends in the overall score were paralleled almost exactly by the velocity measures, but with relatively little change in spatial errors either as a function of groups or practice. Thus, proficiency increased in this task mainly through subjects moving more rapidly, but doing so in such a way that there were no corresponding increases in error. In other words, subjects learned a way to hold errors under control while making faster movements, and the KR subjects were more successful in this respect than the no-KR subjects.[3]

### Two Fundamentally Different Kinematic Patterns Emerge with Practice

In conjunction with the analysis of these KR effects, we also examined some of the kinematic patterns produced by subjects in these two groups. Essentially, two distinct patterns of actions emerged, as defined next. Figure 28.4 contains position-, velocity-, and acceleration-time traces taken from two representative subjects. The position-time data were digitized at 500 Hz, and smoothed with a Butterworth filter (12-Hz cutoff frequency) before being differentiated to obtain velocity- and acceleration-time traces. The traces were then averaged across all 20 trials in the delayed retention test. The upper traces in the figure were from a subject in the no-KR group, whereas the lower ones were from a subject in the KR group. A time marker is shown between the two sets of curves. The position-time traces show that the subjects waited at the starting position until the LEDs had traveled about 400–500 ms before initiating the backswing, the arm reached maximum backswing amplitude at about 900–1000 ms, and then they produced a forward swing toward the target (1200 ms).

*Criteria for classification.* Whereas there are only minimal differences between the two patterns of position-time traces, these two patterns differ substantially when the velocity-time, and particularly the acceleration-time, traces are considered. In the velocity-time traces, notice that the subject in the no-KR group (top) showed a biphasic velocity pattern, with no apparent break between the positive and negative peaks. The KR subject (lower), however, showed a positive velocity, then a flattening at about 800 ms, followed by the negative-going trace beginning at about 1100 ms. This difference in pattern is even more clear when the acceleration-time trace is considered. For the no-KR subject at the top, the acceleration pattern has two phases—a positive acceleration in the direction of the backswing followed by a negative acceleration in the direction of the target, again with no apparent break between them. For the KR subject (lower), however, we see a positive acceleration toward the backswing, followed by a small reversal in direction (from about 700 ms through 900 ms), followed by a sharp negative-going acceleration pattern to move the lever toward the target area. This latter deviation in acceleration-time traces—

[3]This does not, as it might at first appear, provide evidence against the notion that there are speed–accuracy trade-off effects in this task. In other studies with this paradigm, in which (a) we asked subjects to move rapidly versus slowly in hitting the target, and (b) we measured the spatial accuracy of naturally generated (i.e., without instructions) fast versus slow movements, we were able to demonstrate clear increases in spatial errors as the velocities increase.

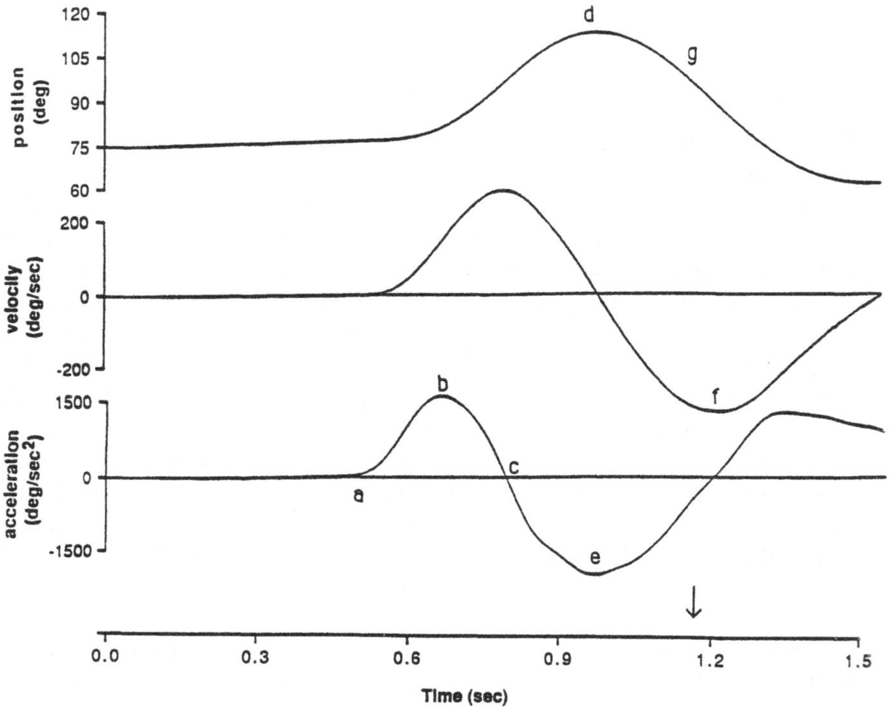

120
105
90
75
60
position (deg)
d
g
200
0
-200
velocity (deg/sec)
1500
0
-1500
acceleration (deg/sec$^2$)
a
b
c
e
f
0.0
0.3
0.6
0.9
1.2
1.5
Time (sec)


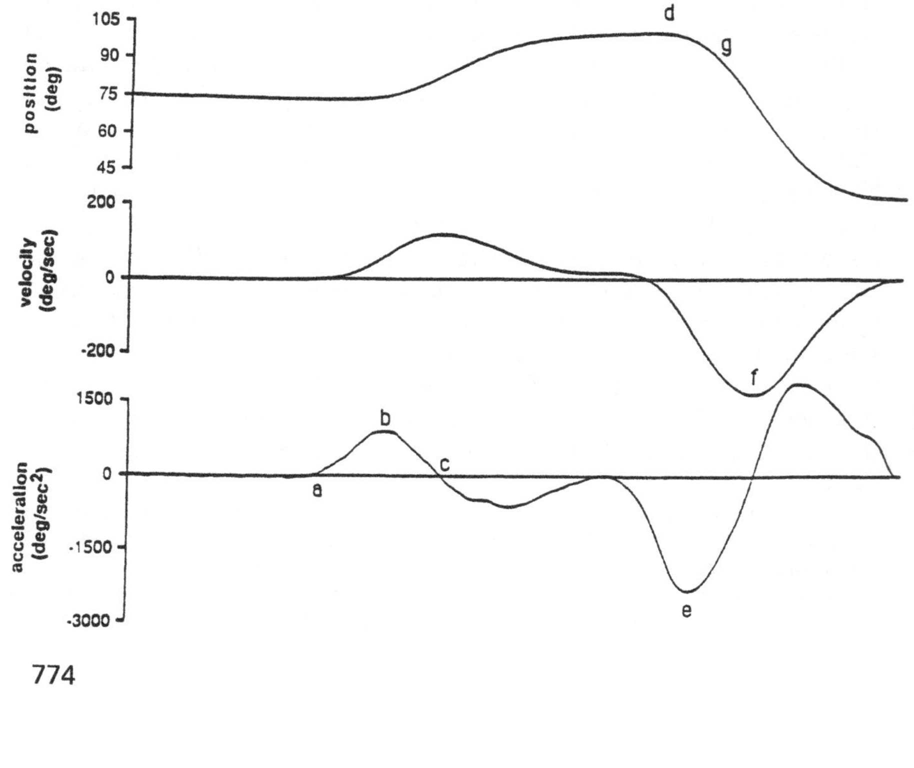

105
90
75
60
45
position (deg)
d
g
200
0
-200
velocity (deg/sec)
f
1500
0
-1500
-3000
acceleration (deg/sec$^2$)
a
b
c
e

TABLE 28.1
Proportion of Type II Patterns During Acquisition and Retention Trials for KR and no-KR Groups

| | *Acquisition Blocks* | | | | | | | *Retention Tests* | |
|---|---|---|---|---|---|---|---|---|---|
| | 1 | 2 | 3 | 4 | 5 | 6 | 7 | 10 min | 1 day |
| Groups | | | | | | | | | |
| KR | .44 | .44 | .46 | .54 | .52 | .54 | .56 | .51 | .68 |
| No-KR | .42 | .24 | .20 | .20 | .18 | .14 | .18 | .22 | .15 |

Note: Type II patterns were classified as those with two additional reversals in acceleration, as defined in Fig. 28.4

Type I with a continuous biphasic pattern, versus Type II with a pattern demonstrating a reversal in acceleration between the two phases at either end—is of particular interest here, and forms the objective basis for classifying these patterns into two distinct types. This classification was easily done "by eye" with interactive graphics, and this rule could be unambiguously applied to all trials for all subjects in either group.

*Changes in patterning with practice.* One initial analysis we performed involved simply counting the number of patterns of each type as they occurred across trials, separately for the KR and no-KR subject. The results are shown in Table 28.1, where the proportion of Type II trials (those with additional acceleration fluctuations as described above) is shown for the seven acquisition blocks and the two retention tests. On the first block of trials, both groups showed about the same proportion of Type II trials (proportions of .44 and .42, respectively). As practice continued, the proportion of Type II trials for the KR subjects increased to about .56 in acquisition, finally maximizing at about .68 on the delayed retention test. Essentially, the opposite pattern was shown for the no-KR subjects, where the proportion of Type II trials dropped from .42 to about .15 in the delayed retention test. Thus, although the two groups of subjects began the experiment with essentially the same proportion of Type II trials, they diverged systematically to the point that the KR subjects were producing

FIG. 28.4 (on facing page). Position-, velocity-, and acceleration-time functions for single trials from one subject in the no-KR group (top panel) and another subject in the KR group (lower panel), showing the temporal locations of Landmarks *a* through *g* (see text for their exact definitions); Type I movement (top) are defined as those with biphasic acceleration-time functions, whereas Type II movements (lower) have an extra acceleration reversal after Point c and before Point d.

Type II responses more than four times as often as the no-KR subjects in the final retention test, $\chi^2(2) = 12.5$, $p <$ 05.

*Pattern classification, group membership, and performance.* Next, we asked about the distributions of these two response types across subjects, and how these related to the group to which the subject had been assigned (KR versus no-KR). We counted the number of responses (out of 20 trials on the second retention test) that were classified as Type II, and Fig. 28.5 contains a frequency histogram of these counts for all 20 subjects in both groups. This distribution is strongly bimodal, with six subjects producing no Type II responses in 20 trials, whereas at the right, five subjects produced 20 Type II responses in 20 trials; about half the subjects produced mixtures of these kinds of responses. In addition, though, the six subjects shown in the right half of the figure (18 or more Type II responses in 20 trials) were all members of the KR group, whereas the seven subjects who produced two or less Type II responses in 20 trials were all members of the no-KR group. Furthermore, as perhaps would be expected from the strong relationship between group membership and performance on the retention test (see the right side of Fig. 28.3, for example), there was a tendency for those subjects who produced the largest number of Type II responses to be the best performers in the task (as defined by performance on the delayed retention test). This correlation, computed across all 20 subjects in both groups, between the number of Type II responses and the score on the retention test was .69. These observations make it clear that the presence or absence of KR in acquisition clearly affected the way in which subject performed this task, with the KR condition producing far more Type II responses than the no-KR condition. Furthermore, these were *functional* differences in patterning, in that Type II responses tended to be associated with far larger scores on the task than were Type I responses.

## Kinematic Patterns Provide Evidence about Units of Motor Control

We next began to ask about what these different kinematic patterns might mean for understanding the processes of movement control in these actions, particularly because these patterns emerged with practice and KR and were highly related to task proficiency. One suspicion was that the Type I and Type II patterns represented movements that were fundamentally different in their underlying representations, or structures. Specifically, we thought that the Type I pattern might represent a movement governed by a single unit was initiated when the backswing began and ran its course essentially unmodified by further sensory information. After all, the pattern of the lights' movement in predictable and unvarying across trials, and there

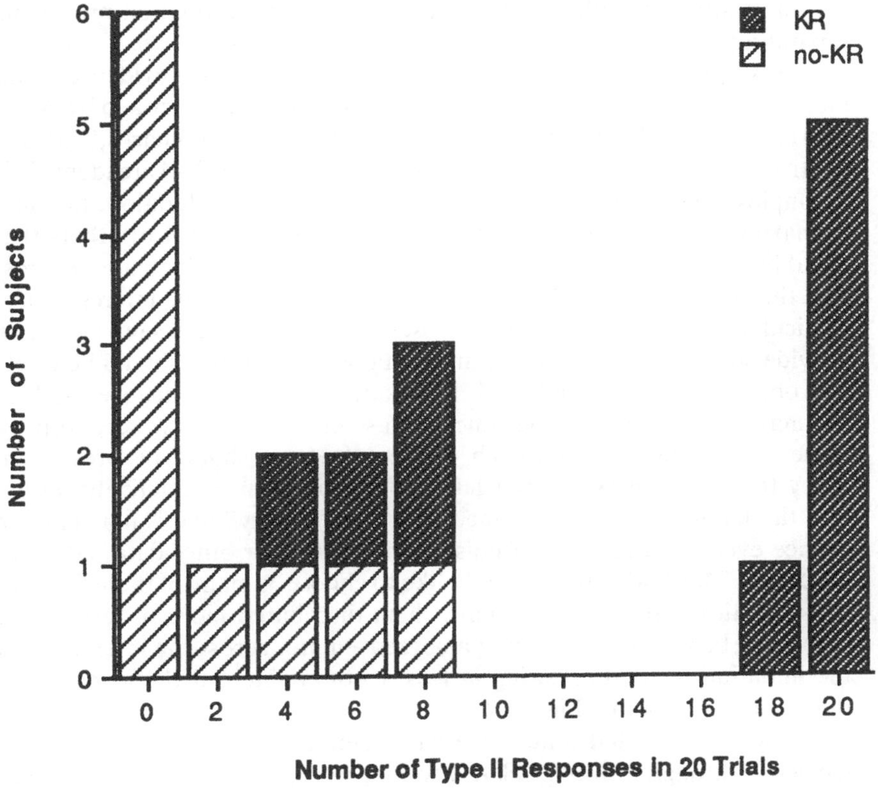


FIG. 28.5. The number of subjects producing different number of Type II responses during the 20-trial delayed no-KR retention test; all six subjects with 18 or 20 Type II movements were in the KR group, and all seven subjects with two or fewer Type II movements were in the no-KR group. (On the Abscissa, 0 = 0 or 1 response, 2 = 2 or 3 responses, . . . , 18 = 18 or 19 responses, and 20 = 20 responses.)

would perhaps be little information the subject could detect within a trial that would be helpful. On the other hand, we suspected that Type II movements were governed by two units—one for the production of the backswing and another for the forward swing—and that these two units were "connected" by a period of information processing during which the movements of the LEDs were sensed and final decisions about when to initiate the forward swing were made. Under this view, the backswing portion of the action is somewhat "irrelevant," serving mainly to position the limb properly for the forward swing, and then the important decisions about when to initiate the forward swing are made quite late in the movement sequence. The essential idea, then, was that Type I movements are produced by a single unit, whereas Type II movements are produced by two units strung together, with a period of visual information processing separating them.

Two kinds of analyses of our kinematic data can provide evidence about these views, following suggestions made almost a half century ago by Weiss (1941). In both these methods, various kinematic "landmarks," or distinctive features, of the kinematic traces as shown in Fig. 28.4 are identified. Examples might be the time of initiation of positive acceleration, the time of maximum velocity in the backswing direction, or a host of others that could be identified. These features are detected via the kinematic records, and the *time* of their occurrence is taken as the data of interest.[4] The particular features to be chosen are essentially arbitrary, but several criteria provide some guidance in their choice. First, the features should be easily and objectively recognized, and, thus, one would prefer features such as the maximum value of some function, its zero crossing, the second peak value, and so on. Second, each of these features should be present on every trial—or almost every trial—so that one can be relatively certain that the feature chosen has some behavioral "reality" and is not simply a chance event or a product of noise from filtering or other data reduction processes. Third, ideally, we would like to choose as many of these features as is practically possible, and have them distributed more or less evenly across the behavior that is being analyzed—here, across the interval from the initiation of the backswing at least to the point that the lever passes the coincidence point (Fig. 28.4).

In the analyses that follow, we have identified seven such landmarks, labeled *a* through *g* in Fig. 28.4. The first feature, Point a, is the time of positive acceleration onset, and represents the earliest indication of the action. Point b is the time of peak positive acceleration, and Point c is the time that the acceleration returns to the zero baseline; thus Point c corresponds to the peak velocity in the backswing direction. Point d is the time of maximum backswing amplitude, and Point e is the time of peak negative (i.e., in the direction of the target) acceleration. Point f is the time of peak negative (toward the target) velocity, and Point g is the time that the lever crosses the coincidence point.

Using these basis data—that is, the temporal occurrences of these various features—one of the methods involves an analysis of the within-subject correlations among the times for these distinctive features. Another method, which is undoubtedly related to the first, concerns the patterns of within-subject standard deviations (SDs) of the temporal intervals sep-

[4]Other dimensions of these features—chiefly representing the ordinate values of the time series shown in Fig. 28.4—could have been examined as well, such as the amplitude of peak velocity, or the position of the maximum backswing. But the analyses here are restricted to the time these particular events occurred.

arating these various landmarks. These two styles of analysis, together with their rationale, are discussed in the following sections.

*Rationale for correlational analyses.* Assume for a moment that the Type II patterns come from movements with two distinct sequential units, and these values are not exactly constant from trial to trail so there is some nonzero within-subject variance. If so, then Weiss (1941) argued that the correlations among any two landmarks that were within a given one of these units would be relatively high, whereas the correlations between any two landmarks lying in different units should be relatively low. This is so, in part, because these units are supposed to be separated by a variable interval of time—here that time used for information processing between units. Now, if these correlations are computed for each of the pairs of landmarks measured, then the patterns of correlations that emerge should reflect the unit-structure of the action. That is, there should be a group of correlations with relatively uniform, high values associated with the landmarks in a certain portion of the movement, and another, essentially separate set of correlations among units associated with some other portion of the movement. Further, should none of the landmarks in one portion of the record correlate well with any other landmark in a different portion of the record, then this would provide evidence that two distinct units were produced. If, on the other hand, all of the correlations were uniformly high (or at least moderately high), then this would be evidence that only a single unit was involved.

Essentially the same prediction, but for somewhat different reasons, can be made from the theory of generalized motor programs (e.g., Schmidt, 1985a). Here, we need only add that the units—if they exist—are really just examples of short generalized motor programs that are strung together to form a complete action. In each of these motor programs, it is assumed that the temporal structure is invariant; this means that the *relative timing* is constant—that is, for a particular subunit, the proportion of its duration occupied by the interval between two particular kinematic segments contained in it would be a constant value. Nevertheless, the overall (i.e., absolute) durations of these subunits are assumed to be variable from trial to trial (see Schmidt, 1985b or 1988 [Chap. 8], for rationale for this idea, and see Gentner, 1987, for a criticism). If the relative timing is invariant, but the overall duration of the unit is variable, then it is easily derived that the correlations between the times of *any* two kinematic landmarks contained in it should be essentially 1.0 (attenuated somewhat by errors in measurement and other sources of noise, of course). If there are two such units separated by a variable interval where no program is "running," then the temporal location of a feature in one unit would correlate essentially zero with any feature contained in the other unit. If there is only one unit governed by a single motor program, then the correlations should be uniformly high across the entire action. And, if there is no program structure at all, then the correlations should be uniformly low across the action. Thus, it is possible, through examining the patterns of correlations found in such situations, to estimate the unit structure of the actions making up a sequence of behavior such as used in the coincidence-timing task described earlier here.

*Correlational analysis.* Using the seven temporal measures described earlier, we examined the patterns of correlations among them in subgroups of subjects who showed the strongest tendencies for either Type I or Type II responses. Thus, we used five subjects from the KR group and five subjects from the no-KR group; they produced either 20 out of 20 Type II responses or 0 out of 20 Type II responses on the delayed retention test, respectively. First, consider the values shown in the upper portion of Fig. 28.6. Here are plotted the average correlations (via $Z'$ transform) between the time of the start of the backswing (Point a) and each successive point in time—that is, for the various intervals a–b, a–c, a–d . . . a–g. These data are shown separately for the no-KR subjects on the left, and for the KR subjects on the right. For the no-KR subjects, these values were uniformly around .70, but they did decrease very slightly (to around .65) as the landmarks became more separated in time from each other. From the rationale presented above, this suggests a movement governed by a single unit across its total duration. However, for the KR subjects, we see a very different pattern. Here, the correlations between the time of the first part of the movement (Point a) and successively later points are very high (from .93 to .83) for the landmarks up to Point d, but the correlations with Point a drop markedly to about .25 for the landmarks located in the last part of the action (Points e, f, and g). A groups by intervals ANOVA on the $Z'$-transformed correlations showed a significant effect of groups, $F(1,18) = 9.3$, of intervals, $F(5,40) = 7.0$, and the interaction between groups and intervals, $F(5,40) = 4.0$, all $ps < .05$. The simple main effects analysis showed that the intervals for the no-KR subjects produced correlations that were not different from each other, but the a–b, a–c, and a–d intervals had significantly larger correlations than the a–e, a–f, and a–g intervals for the KR group. Our interpretation is that the Points a through d were members of a given unit, and were not members of the part of the action represented by Points e, f, and g. We take this as evidence for at least one unit isolated in the initial part of the action, tentatively from Points a through d.

Next, we performed this analysis in a similar way, but with the common landmark being the point at which the lever crossed the coincidence point, or Point g. Now the correlations are taken with successively earlier and

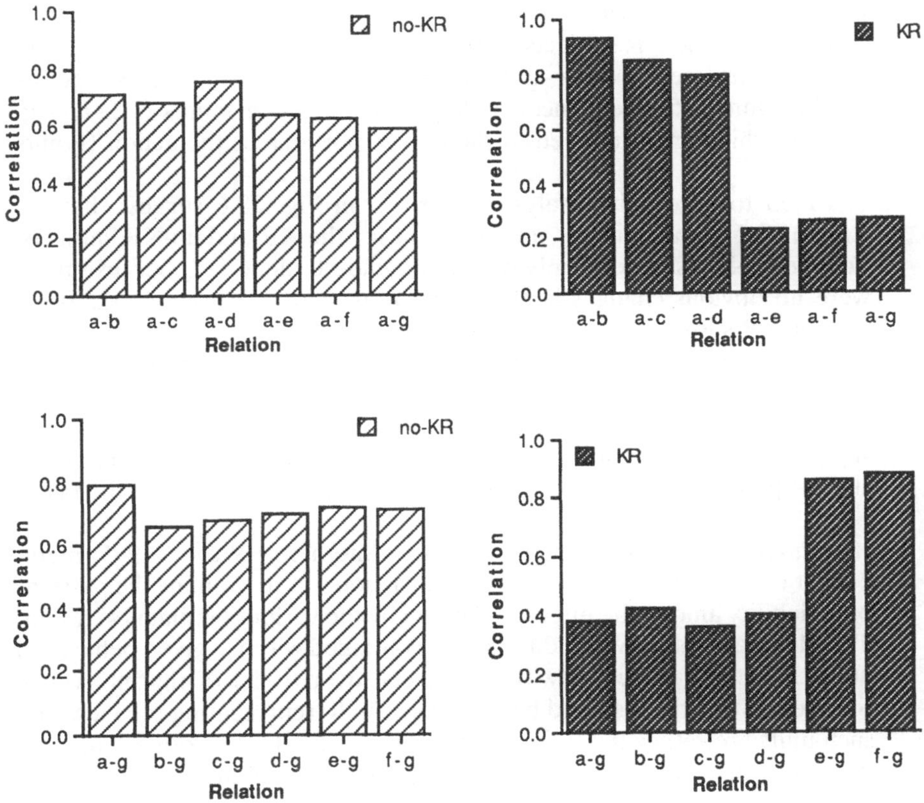


FIG. 28.6. In the upper panel, averaged within-subject correlations between the time of the first landmark (a) and the times of the times of the later Landmarks (b through g), separately for the no-KR group (left) and the KR group. In the lower panel, averaged within-subject correlations between the time of the last landmark (Point g) and the times of the earlier landmarks (a through f), separately for the no-KR group (left) and the KR group (right).

earlier landmarks, and are shown in the lower portion of Fig. 28.6. For the no-KR subjects on the left, again the correlations among the temporal landmarks were uniformly around .70 to .75, with almost no tendency for changes across landmarks, again suggesting that the entire movement was governed by a single unit for the no-KR subjects. For the KR subjects, on the other hand, the correlations between Point g and landmarks late in the action (e and f) were around .95 or higher, but correlations abruptly dropped to about .40 as landmarks in the earlier portion of the movement were paired with Point g. The ANOVA showed significant effects of group, $F(1,18) = 3.9$, intervals, $F(5,40) = 29.4$, and the group by intervals interaction, $F(5,40) = 3.5$, all $ps < .05$. Analysis of simple main effects again showed that the no-KR subjects' correlations did not change across intervals, whereas the KR subjects' correlations were different for the e–g and f–g landmarks versus the a–g, b–g, c–g, and d–g intervals. This suggests another unit in the sequence, perhaps encompassing Points e, f, and g, none of which are contained in the earlier portions of the action (Points a–d).

Taken together, the analyses described in Fig. 28.6 suggest that the structure of this action was very different for the KR and no-KR subjects. For the no-KR subjects, only one unit could be identified, because there were no obvious changes in the correlations across the landmarks used. For the KR subjects, on the other hand, there were apparently two units—one made up of parts of the action contain Landmarks a through d, and another made up of parts of the action containing Landmarks e through g. Further, this unit structure emerged with practice and KR—Table 28.1 suggests that the probability of these two kinds of action was similar at the start of practice, and diverged systematically with experience on the tasks.

*Rationale for the variability analyses.* Another of Weiss' (1941) methods for understanding the structure of an action was to examine the patterns of variability among the intervals defined by the various landmarks chosen. The idea was that if two landmarks are contained within a given behavioral unit, then the variability (within subjects, over trials) of the interval of time separating them should be relatively small, representing the view that such units are rather tightly organized temporally. On the other hand, for two landmarks in different units, the temporal interval separating them should have considerably more variability. This variability between landmarks in different units would be expected to be even higher if the two units are separated by a period where no unit is acting, such as in the present case where visual information processing is thought to occur between the two units of muscular activity. This logic leads naturally to an analysis of the SDs of the various intervals among the landmarks that were defined earlier in Fig. 28.4, and provides potentially converging evidence about the differential structures of the two kinds of actions discussed here.

*Analysis of SDs.* Figure 28.7 (upper portion) provides the mean within-subject SDs of various intervals among the landmarks described earlier, again graphed separately for the no-KR (left) and KR (right) subjects. These intervals are chosen to be "parallel" to those in the analysis of the correlations (Fig. 28.6), in that the intervals have a common starting point (Point a), but successively later and later endpoints, so that the intervals are a–b, a–c, a–d . . . a–g. For the no-KR subjects on the left, these SDs increased slightly and gradually as the separation from Point a increased, with the SDs ranging from 60 ms for Interval a–b to about 80 ms

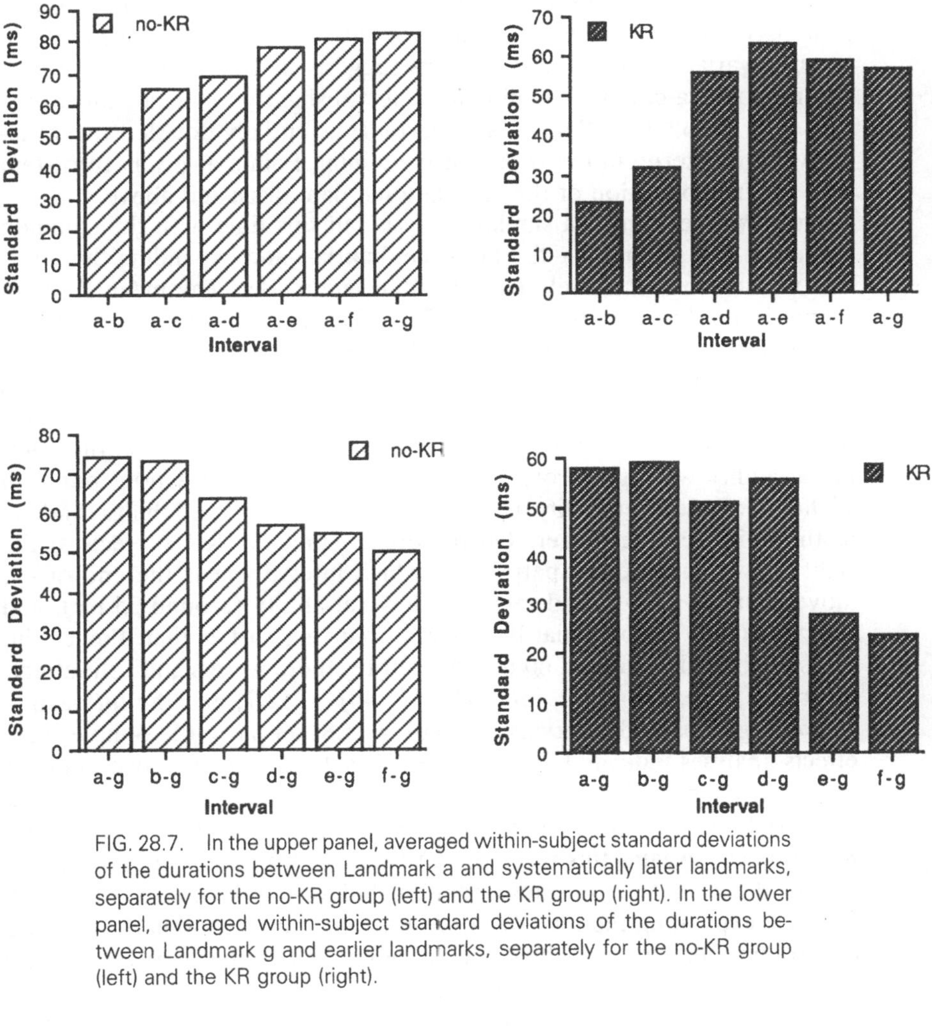


FIG. 28.7. In the upper panel, averaged within-subject standard deviations of the durations between Landmark a and systematically later landmarks, separately for the no-KR group (left) and the KR group (right). In the lower panel, averaged within-subject standard deviations of the durations between Landmark g and earlier landmarks, separately for the no-KR group (left) and the KR group (right).

for Interval a–g. This would be expected from a single-unit structure where the cumulative effects of errors in muscular contraction force and timing (and/or small errors in measurement) systematically raise the variabilities slightly as longer and longer intervals are measured (Schmidt et al., 1979). On the other hand, the analogous analysis for the KR subjects shows relatively small SDs for the Intervals a–b and a–c (around 25–30 ms), and then an abrupt increase to about 60 ms for the intervals containing later endpoints (i.e., a–d through a–g). An ANOVA of these data showed a significant effect of groups, $F(1,18) = 3.8$, intervals, $F(5,40) = 11.4$, and the groups by intervals interaction, $F(5,40) = 5.8$, all $ps < .05$. Simple main effects analysis indicated significant differences only for the KR subjects between the first two intervals and the later ones. This abrupt transition in variability in the KR subjects suggests that there was a unit containing Points a-c, but expanding the interval slightly to include any of the next later points (i.e, Point d and beyond) essentially doubled the variability. This interpretation of a unit containing Points a–c was consistent with the interpretation of the correlational data discussed above.

Next, we performed a similar analysis but this time with the common landmark being Point g—the latest landmark—and expanding the intervals to include successively earlier initial landmarks. The mean SDs are shown in the lower portion of Fig. 28.7. Thus, the shortest intervals are shown at the right in the figure, and they increase in length as one moves to the left of the abscissa. For the no-KR subjects at the left, there was a pattern of slightly increasing variability as the intervals became longer, ranging from about 60 ms to 70 ms. The failure to detect an abrupt shift in variability was again consistent with a single unit of action, but with slightly increased variability as more and more "error" variability is added as the units became longer. On the other hand, the KR subjects at the right showed a different pattern, in which intervals near the end of the movement (e–g, f–g) had relatively low variability (around 25 ms), and intervals made up of initial landmarks occurring earlier than Point e increased abruptly to about 60 ms. As before, there were significant effects of groups, $F(1,18) = 2.9$, intervals, $F(5,40) = 12.7$, and the interaction between groups and intervals, $F(5,40) = 5.8$, all $ps < 05$. Simple main effects analyses indicated that, for the KR subjects, the last two intervals (e–g, f–g) had significantly lower variability than the remaining ones, but the variabilities of these intervals were not reliably different for the no-KR subjects. Again, this result is consistent with the notion that there is a unit at the end of the movement (including Landmarks e, f, and g). Because expanding the interval to include any earlier landmarks (Point d or earlier) more than doubled the variability, the initial part of the final unit occurs approximately at Point e.

Taken together, the data from Fig. 28.7 provide evidence of a single unit of behavior for the no-KR subjects, as there were no abrupt transitions between the sizes of the SDs as the intervals were expanded to include landmarks distributed across the action. On the other hand, the KR subjects showed strong evidence of producing two units of behavior, one located approximately at the beginning of the action (including Landmarks a through c), and another located at the end of the action (including Landmarks e through g). In this analysis, Landmark d seemed to be included in neither unit, as any interval that included Landmark d had relatively high variabilities compared to intervals wholly at the beginning or at the end of the action.

*Relative SDs of intervals between adjacent landmarks.* Finally, Weiss (1941) suggested that, if there were several segments in an action, then the SDs of the durations of intervals between landmarks contained within one segment should be small, and the SDs of the durations of intervals between landmarks in different segments should be large. As an extension of this notion, the variabilities of each successive interval defined by the sequence of *adjacent* landmarks should reveal the end of one segment, and the beginning of another. Here, progressing through the movement "forward" in time, we should expect that the variabilities should be relatively low until the last landmark in the first segment is encountered, and then the variabilities for the next interval should be abruptly higher, because it contains landmarks in the first *and* second units. Any large, abrupt increase in the variability should therefore provide evidence about the segment boundaries. However, one problem with the analysis of interval SDs as defined here is that the SD of an interval (even within one segment) generally increases in proportion to the mean duration of the interval (Schmidt et al., 1979). Thus, any abrupt increases in the SDs of the adjacent intervals might be confounded with the mean other durations of the different intervals. An approximate solution to this problem is to use the *relative variabilities* (i.e., the SD/mean), which tends to correct for differences in the durations of the different intervals in the movement. Here, though, care must be taken to ensure that all interval durations are positive; that is, variability across trials can sometimes result in a given interval being positive on some trials and negative on others, which could result in artifactually small mean interval durations. Our solution was to use the mean *absolute* interval duration as the denominator in producing relative variability measures. This final analysis examined the relative variabilities of the adjacent intervals throughout the action.

The relative variabilities of the adjacent intervals are shown in Fig. 28.8, the data from the no-KR subjects on the left and the KR subjects on the right. For the no-KR subjects, the relative variabilities increased slightly from about .25 to .30 as successively later intervals were considered. However, for the KR subjects, these relative variabilities increased slightly from .12 to .18 for the first three adjacent intervals (a–b, b–c, and c–d), but increased sharply to about .46 for the d–e interval; this variability was some three to four times as large as those for the previous intervals. The relative variabilities for the intervals followed the d–e interval decreased considerably again, to approximately .25. In the ANOVA, the group effect was not significant, $F(1,8) < 1$, but the interval effect and the intervals by group interaction were both significant, $Fs(5,40) = 3.3$ and 2.9, $ps < .05$. *Post-hoc* analyses showed that, for the KR group, the relative variabilities for the d–e interval were larger than all earlier and later intervals, with none of the other intervals being reliably different from each other. For

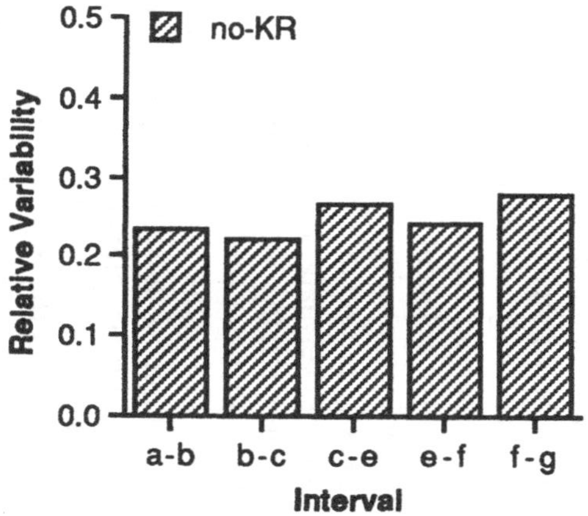


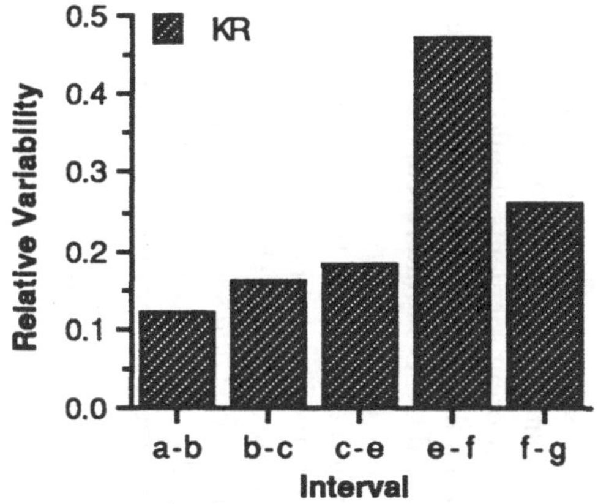


FIG. 28.8. Average within-subject relative variabilities (SDs/mean) for intervals defined by successive pairs of adjacent landmarks in the movement, separately for the no-KR group (left) and the KR group (right).

the no-KR group, the relative variabilities for the various intervals were not reliably different from each other.

Taken together, these findings converge with the earlier analyses to support the hypothesis that the no-KR subjects produced these actions with a single unit (with a positive and negative acceleration-time impulse in its structure), based on the finding of no obviously elevated relative variability for any of the adjacent intervals. The KR subjects, on the other hand, showed a clearly elevated relative variability for the d–e interval, indicating that they segmented the movement into two parts, with the d–e interval containing the boundary between the two behavioral units. These adjacent-interval analyses of relative variabilities appear to provide a powerful basis for locating the unit boundaries in such actions.

## DISCUSSION

The primary focus here has been the interaction of practice and feedback in the development of skilled performance, particularly in terms of how these critical variables influence the patterns of kinematics demonstrated. At a superficial level of analysis in terms of the gradual increases in the overall score for the task, through watching the performers, and to some extent even with the aid of certain kinds of kinematic data (e.g., position-time information), it is difficult to see dramatic or revolutionary changes associated with learning in this action. All subjects appear to make a relatively "continuous" backswing followed by a forward swing, with a rapid follow-through beyond. However, when other kinematic variables are considered—particularly velocity-time, and especially acceleration-time, records—it is possible to identify underlying units of action that give insight into the nature of movement control in these activities. With these methods, it has also been possible to document rather marked changes in the organization of these actions with practice.

Early in practice, learners in this task produce a pattern involving a single, biphasic action that consists of a single movement unit. Later in practice, learners with KR acquired a completely different action to produce the same environmental goal, this involving a sequence of two distinguishable movement units. We suspect that the two units are separated by a highly variable interval during which various kinds of information processing occurs, such as the detection and correction of errors produced in the first unit as well as visual information processing related to the time to contact remaining before the pattern of LEDs arrives at the coincidence point. Practice per se was not sufficient for this two-unit sequence of action to develop, and this pattern apparently required extrinsic *feedback* about subjects' success in meeting the environmental goal. Furthermore, no KR information was even given about the patterns of action used to achieve these environmental goals, yet subjects modified their kinematic patterns markedly with practice. These findings are related to numerous movement control and learning issues, some of which are addressed in more detail in the following sections.

## Issues in the Identification of Units of Action

Several approaches to the identification of the basic units of action have been attempted recently, such as reaction time methods for understanding the structure of speech (Sternberg et al., 1980), and the use of various relational invariants in drawing movements (e.g., Viviani, 1986). We choose to approach the unit structure in limb actions based on several ideas from Weiss (1941), together with recent thinking and considerable empirical support for the theory of generalized motor programs (e.g., Schmidt, 1985a, 1985b, 1988). Under this view, a unit is simply considered to be a motor program that governs behavior for several hundred milliseconds, having all of the properties generally attributed to them (e.g., Schmidt, 1988, Chap. 8). If so, than these programs also should obey the principles of *generalized* motor programs; most importantly, over the time they are operating, they should control actions so that the temporal structure displays essentially invariant relative timing, even though the actual duration of the unit may vary. Invariant relative timing implies the following property: For any interval defined by two kinematics landmarks within the unit, the ratio of this interval divided by the total duration of the unit should be invariant across trial-to-trial changes in the overall duration of the unit. This property of *invariant relative time* predicts that the linear correlation between the durations of any two intervals within the unit, or between any interval within the unit and the total unit duration, should be 1.0. (Of course, in practice, errors in measurement or other "noisy" processes in the muscular contractions would tend to make these correlations be slightly less than 1.0). Finally, we assume that the two different generalized motor programs would demonstrate different patterns of relative timing, with each showing its own invariance across changes in overall unit duration.

If these assumptions can be accepted, then we can use this principle of invariant relative timing as a criterion to determine whether a given action is governed by a generalized motor program. Conversely, if we observe that the relative timing of a whole action does not demonstrate invariant relative timing across its total duration, but does demonstrate a set of local invariances in relative timing for various smaller segments, then each of these segments can be argued to be controlled by separate units, each based on a different motor program. Boundaries between units are those parts of the action where the invariance in relative timing breaks down, either to form a new relative timing structure in the next unit, or to form no consistent relative timing structure indicative of no motor program being executed during that time. In either case, a boundary is indicated by abrupt departures from invariance in relative timing that are detectable in patterns of correlations and SDs. Although the overall concept underlying the objective identification of units of behavior has been suggested previously (e.g., Schmidt, 1985a; Weiss, 1941), it has not, to our knowledge, been used to identify the unit structure, or boundaries between units, in relatively long-duration actions as studied here. When applied to the data in the present experiment, the method generates evidence for two different kinds of unit structures—one for the subjects who had received KR during practice (Type II patterns), and a second for subjects who had not received KR during practice (Type I patterns).

### Advantages of Movements Controlled by a Two-Unit Structure

For the no-KR subjects, the patterns indicated a two-phase movement, consisting of the backswing and the forward swing that were both controlled as part of a single motor program. This unit was initiated when the backswing began, approximately 700 ms before the LEDs' arrival at the coincidence point; the unit was run off more or less unmodified until the coincidence point was reached.[5] On the other hand, the KR subjects, who were far more successful in producing high scores in this task than the no-KR subjects, produced the action as two sequential units of behavior. The first unit was a backswing needed to move the limb to the left of the coincidence point. Then the subject appeared to terminate this unit, probably while information about the travel of the light pattern—perhaps together with kinesthetic information about the location of the limb at or about the desired backswing position—was processed. Finally, the second unit was initiated, resulting in the limb being driven through the coincidence point. Thus, here the initiation of the unit that mainly determined the arrival time at the coincidence point was at approximately the point of maximum backswing, or approximately 200 ms before the LEDs' arrival at the coincidence point.

Because the performance score in this task was heavily penalized for making spatial errors, an analysis of the spatial–variability components of the task in terms of the concepts of impulse variability (Schmidt et al., 1979) reveals several possible reasons why the two-unit structure was far

[5]It is also possible, although unlikely, that no-KR subjects produced behaviors that were controlled by 2-unit structures, with essentially no temporal break between them. However, there is no evidence to support this view because the higher order derivatives show no break between units. Also, there is a necessity for parameters to be the same in the same unit for correlations to be high across the whole action, which is relatively unlikely.

more effective for performance. These advantages were probably clearly revealed through subjects' trial and error behavior in early practice. First, a major advantage for the two-unit pattern is the fact that the *duration* of the movement governed by the final unit is shorter. It is well known that, for quick actions, the within-subject variability in movement time is nearly proportional to the average movement time; therefore, decreasing the movement duration by shortening the final, programmed segment (here, from about 700 to 200 ms) would have strong effects on temporal variability. Of course, variability in when the lever arrives at the coincidence point is strongly related to variability in spatial position of the lever at the moment the target arrives. Because the latter is a component in the overall score, shorter movement times increase the score by decreasing absolute spatial errors (see Schmidt et al., 1988).

Second, absolute spatial errors are decreased for the two-unit relative to the one-unit pattern because the distance to be moved across the duration of the unit is smaller. For the one-unit case, the unit must control the limb for the backswing and the forward swing, whereas the final unit in the two-unit case only controls the forward swing. Impulse–variability analyses of this kind of task show that reducing the movement distance decreases the spatial errors at the endpoint almost proportionally (Sherwood, Schmidt, & Walter, 1988). Therefore, spatial variability at the coincidence point is presumably reduced in the two-unit structure because the distance to be traveled after the final unit is triggered is approximately half that in the one-unit pattern.

Third, the visual information that serves as the "trigger" for the final program occurs closer in time to the coincidence point in the two-unit as opposed to the one-unit case. Therefore, the estimation of the target arrival time is shorter in the two-unit case, making the perceptual anticipation (Poulton, 1957) more accurate (e.g., Schmidt, 1969, 1988). In addition, the light pattern is spatially closer to the subject when this final response in triggered, probably allowing more stable determinations of the time to trigger the action. These last effects might be relatively small ones, however, based on recent evaluations of coincidence-timing behavior based on optical flow variables, such as time to contact (e.g., Lee & Young, 1985). Even so, to the extent that there is some advantage in having the time-to-contact information generated when the moving object is closer to the subject, there might be additional benefits from the two-unit control strategy.

## Some Additional Theoretical Issues

*Involvement of muscle dynamics across practice.* A commonly held generalization about skill acquisition, based on ideas from Bernstein (1967) and others, is that the subject learns to exploit the limb dynamics—passive, inertial, and mechanical properties of the limb to be moved—to simplify and smooth the control, to increase movement velocity, and to decrease energy costs. One important class of muscle dynamics is the so-called "stretch–shorten cycle," in which the performer uses the spring like properties of muscle to increase the accelerations associated with reversing direction, as in the transition from the backswing to the forward swing in the present task. Placing the muscle on stretch, initiating the contraction as the muscle is lengthening, has long been thought of as underlying the usefulness of "wind-ups" or backswings in forceful actions.

However, in the present task, the successful performers apparently learned a pattern of action that systematically *eliminated* the stretch–shorten cycle. That is, the two-unit pattern involved a prolonged slowing of the lever as the maximum backswing was approached—almost a stop at the maximum backswing position—before the forward swing was initiated (see Fig. 28.4). This behavior appears to reduce the involvement of muscle dynamics and the stretch–shorten cycle, rather than increase it as might be expected. This generalization should be taken cautiously, however. Simply because the *lever* was not moving at the point of maximum backswing position does not necessarily mean that the relevant muscles were not being placed on a stretch; and, there could be other joints "upstream" from the action (e.g., the shoulder) that are utilizing the stretch–shorten cycle, and these processes are not seen in our measures of the elbow joint. Analyses with EMG, and/or with additional analyses of intersegmental dynamics (e.g., Schneider, Zernicke, Schmidt, & Hart, in press), could be used to resolve these issues in future work. However, if the change in motor control seen here does decrease the involvement of the stretch–shorten cycle, it seems to place a limitation on the generalization that practice typically increases the role of passive, dynamical muscle properties.

The reasons these dynamical properties might have been systematically eliminated (if they were) are perhaps related to the heavy penalties imposed on spatial errors here. The use of the stretch–shorten cycle to increase forward-swing velocity probably occurs at considerable cost in terms of spatial errors; forceful muscle contractions are relatively "noisy" processes that result in considerable variability in the resulting trajectory (Schmidt et al., 1979). This noise is acceptable in many tasks in which high velocity is a major goal (e.g., as in Schneider et al., in press), but it does not appear to be consistent with the concomitant production of high movement precision as demanded here. This view suggests at least one boundary condition on the use of such dynamical properties in movement control—the demand for high precision—and motivates a search for a systematic evaluation of various other boundary conditions that might present themselves in human skills.

*Learning and "longer" motor programs—the gearshift analogy.* An interesting notion is that learning involves systematically longer and longer open-loop control of behavior. There have been several statements of this position. One is Pew's (1966) idea of increased lower level control with practice, with decreases in the involvement of environmental information for moment-to-moment control. Keele (1976) introduced the concept of the "gearshift analogy," where several small programs are integrated into a larger one with practice, essentially as we might expect in the task of learning to control the feet and hands in shifting gears in a car (Schmidt, 1988). However, our data tend to show the opposite effect, where the learners changed from a long (one-unit) pattern in early practice to a behavior employing two distinct movement units. One interpretation is that employing two movement units allowed the involvement of sensory information processing *during* the action, with the second unit being triggered visually at about the maximum backswing position in order to make use of last moment data about the LEDs position, and to compensate for movement errors produced in the first unit. Again, this modification seems consistent with the extreme accuracy requirement in the task, and on this basis might not be particularly surprising. However, the notion that learning involves a systematic increase in the use of environmental information *during* such actions has not had very much empirical support, and scientists have instead tended to argue for increased open-loop processes, or increased automaticity, as a basis for increased skill with practice (e.g., Schmidt, 1988). These data suggest that either of these kinds of processes could occur, depending on the particular task constraints imposed.

*Knowledge of results and altered kinematic patterns.* A final point concerns notions about how such new patterns of action are learned, and the informational basis that is necessary to generate them. Fowler and Turvey (1978), for example, have argued for the inadequacy of KR—as information only about movement outcome—to allow the learner to change the movement pattern kinematics that lead to the outcome (see also Newell & McGinnis, 1985). Their point is that except in the simplest of tasks, KR as information about the singular product of a complex kinematic pattern does not instruct the learner about which aspect of this pattern should be modified on a subsequent trial in order to improve that outcome next time. They go on to claim that aspects of the pattern would be altered only to the extent that (kinematic) information about that aspect of the pattern is provided.

However, our data tend to contradict Fowler and Turvey's view. Note that subjects in the KR group received information only about the movement outcome, and never received information about the movement patterning that led to that outcome. Nevertheless, these subjects altered the movement patterning radically, shifting from a one-unit structure to a two-unit structure, probably with the inclusion of visual information processing between the two units. That KR was *necessary* to produce such changes is evident from the behavior of the no-KR subjects, who did not develop this two-unit pattern to any important extent. This does not deny, of course, that information about certain aspects of the movement pattern would be useful in altering it, as such informational supports in the form of knowledge of performance (KP) or so-called kinetic or kinematic feedback have been shown recently to have a strong role in movement learning in this task (Young, 1988; Young & Schmidt, 1989). In the present case, though, outcome information was certainly sufficient for such modifications, and probably occurred via trial-and-error alterations in movement production associated with the outcomes that this "experimentation" produced (reported via KR).

## ACKNOWLEDGMENTS

Supported by Contract No. MDA 903-85-K-0225 to R. A. Schmidt and D. C. Shapiro. We thank David E. Meyer and Karl M. Newell for comments on an earlier draft, and Claudia A. Lange for help with the preparation of the manuscript.

## REFERENCES

Adams, J. A. (1971). A closed-loop theory of motor learning. *Journal of Motor Behavior, 3*, 111–150.

Adams, J. A. (1985). The use of a model of movement sequences for the study of knowledge of results and the training of experts. *Journal of Human Movement Studies, 11*, 223–236.

Bernstein, N. (1967). *The co-ordination and regulation of movements*. Oxford: Pergamon Press.

Bilodeau, I. M. (1966). Information feedback. In E. A. Bilodeau (Ed.), *Acquisition of skill* (pp. 255–296). New York: Academic Press.

Fowler, C. A., & Turvey, M. T. (1978). Skill acquisition: An event approach with special reference to searching for the optimum of a function of several variables. In G. Stelmach (Ed.), *Information processing in motor control and learning* (pp. 1–40). New York: Academic Press.

Gentner, D. R. (1987). Timing of skilled motor performance: Tests of the proportional duration model. *Psychological Review, 94*, 255–276.

Higgins, J. R. & Spaeth, R. K. (1972). Relationship between consistency of movement and environmental condition. *Quest, 17*, 61–69.

Hobart, D. J., Kelly, D. L., & Bradley, L. S. (1975). Modifications occurring during acquisition of a novel throwing task. *American Journal of Physical Medicine, 54*, 1–24.

Hobart, D. J., & Vorro, J. R. (1974). Electromyographic analysis of the intermittent modifications occurring during the acquisition of a novel throwing skill. In R. C. Nelson & C. Morehouse (Eds.), *Biomechanics IV* (pp. 559–566). Baltimore: University Park Press.

Keele, S. W. (1976). Unpublished observations, University of Oregon.

Lacquaniti, F., Terzuolo, C., & Viviani, P. (1983). The law relating kinematic and figural aspects of drawing movements. *Acta Psychologica, 54*, 115–130.

Landa, J. (1979). Analysis of skill acquisition on a novel throwing task in terms of biomechanical factors. *Journal of Human Movement Studies, 5*, 53–60.

Lee, D. N., & Young, D. S. (1985). Visual timing of interceptive action. In D. Ingle, M. Jeannerod, & D. N. Lee (Eds.), *Brain mechanisms and spatial vision* (pp. 1–30). Dordrecht: Martinus Nijhoff.

McGrain, P. (1980). Trends in select kinematic and myoelectric variables associated with learning a novel motor task. *Research Quarterly for Exercise and Sport, 51*, 509–520.

McGrain, P. (1986). Myotemporal changes associated with learning a simple motor task using two different control strategies. *Journal of Human Movement Studies, 12*, 59–70.

Moore, S. P., & Marteniuk, R. (1987). Kinematic and electromyographic changes that occur as a function of learning a time-constrained aiming task. *Journal of Motor Behavior, 18*, 397–426.

Newell, K. M., & Carlton, M. J. (1987). Augmented information and the acquisition of isometric tasks. *Journal of Motor Behavior, 19*, 4–12.

Newell, K. M., Hoshizaki, L. E. F., Carlton, M. J., & Halbert, J. A. (1979). Movement time and velocity as determinants of movement timing accuracy. *Journal of Motor Behavior, 11*, 49–58.

Newell, K. M., & McGinnis, P. M. (1985). Kinematic information feedback for skilled performance. *Human Learning, 4*, 39–56.

Newell, K. M., & Walter, C. B. (1981). Kinematic and kinetic parameters as information feedback in motor skill acquisition *Journal of Human Movement Science, 2*, 255–270.

Pew, R. W. (1966). Acquisition of hierarchical control over the temporal organization of a skill. *Journal of Experimental Psychology, 71*, 764–771.

Poulton, E. C. (1957). On prediction in skilled movements. *Psychological Bulletin, 54*, 467–478.

Salmoni, A. W., Schmidt, R. A., & Walter, C. B. (1984). Knowledge of results and motor learning: A review and critical reappraisal. *Psychological Bulletin, 95*, 355–386.

Schmidt, R. A. (1969). Movement time as a determiner of timing accuracy. *Journal of Experimental Psychology, 79*, 43–47.

Schmidt, R. A. (1985a). Identifying units of motor behavior. *Behavioral and Brain Sciences, 8*, 163–164.

Schmidt, R. A. (1985b). The search for invariance in skilled movement behavior. *Research Quarterly for Exercise and Sport, 56*, 188–200.

Schmidt, R. A. (1988). *Motor control and learning: A behavioral emphasis* (2nd ed). Champaign, IL: Human Kinetics.

Schmidt, R. A., Young, D. E., Swinnen, S., & Shapiro, D. C. (1989). Summary knowledge of results: Evidence for the guidance hypothesis. *Journal of Experimental Psychology: Learning, Memory and Cognition, 15*(2), 352–359.

Schmidt, R. A., Zelaznik, H. N., & Frank, J. S. (1978). Sources of inaccuracy in rapid movement. In G. E. Stelmach (Ed.), *Information processing in motor control and learning* (pp. 183–203). New York: Academic Press.

Schmidt, R. A., Zelaznik, H. N., Hawkins, B., Frank, J. S., & Quinn, J. T. (1979). Motor-output variability: A theory for the accuracy of rapid motor acts. *Psychological Review, 86*, 415–451.

Schneider, K., Zernicke, R. F., Schmidt, R. A., & Hart, T. J. (in press). Learning unrestrained rapid arm movements: Coordination of intersegmental dynamics. *Journal of Biomechanics*.

Sherwood, D. E., Schmidt, R. A., & Walter, C. B. (1988). Rapid movements with reversals in direction: II. Control of movement amplitude and inertial load. *Experimental Brain Research, 86*, 355–367.

Sternberg, S., Monsell, S., Knoll, R. L., & Wright, C. E. (1978). The latency and duration of rapid movement sequences: Comparisons of speech and typewriting. In G. E. Stelmach (Ed.), *Information processing in motor control and learning* (pp. 117–152). New York: Academic Press.

Terzuolo, C. A. & Viviani, P. (1979). The central representation of learning motor programs. In R. E. Talbot & D. R. Humphrey (Eds.), *Posture and movement* (pp. 113–121). New York: Raven Press.

Terzuolo, C. A., & Viviani, P. (1980). Determinants and characteristics of motor patterns used for typing. *Neuroscience, 5*, 1085–1103.

Viviani, P. (1986). Do units of motor action really exist? *Experimental Brain Research Series, 15*, 201–216.

Vorro, J. R., & Hobart, D. J. (1974). Cinematic analysis of the intermittent modifications occurring during the acquisition of a novel throwing skill. In R. C. Nelson & C. A. Morehouse (Eds.), *Biomechanics IV* (pp. 553–558). Baltimore: University Park Press.

Weiss, P. (1941). Self-differentiation of the basic patterns of coordination. *Comparative Psychology Monographs, 17*, Whole No. 4.

Young, D. E. (1988). *Knowledge of performance and motor learning*. Unpublished doctoral dissertation, UCLA.

Young, D. E., & Schmidt, R. A. [Manuscript in preparation, UCLA.]

# 29 Motor Learning and the Degrees of Freedom Problem

Michael I. Jordan
*Massachusetts Institute of Technology*

ABSTRACT

A computational approach to motor skill learning is presented, with an emphasis on problems involving excess degrees of freedom. The approach is based on the notions of (a) an internal forward model of the environment, and (b) intrinsic constraints on motor learning. A forward model allows constraints in distal, task-based coordinate systems to be converted into constraints in motoric coordinate systems, even when the mapping between these coordinate systems is many-to-one. Task constraints are augmented by a set of internal intrinsic constraints on learning. The intrinsic constraints are a relatively small class of functional forms that are assumed to apply broadly across tasks. I discuss how this approach can be implemented using network learning algorithms and demonstrate applications of the approach to inverse kinematic learning, trajectory formation, composite feedback-feedforward control, and the learning of input parameters.

Theories of action that take seriously the need for internal control structure have had difficulty dealing with the flexibility and adaptability of human action (Abbs & Gracco, 1983; Lindblom, Lubker, & Gay, 1979; Kelso, Tuller, Vatikiotis-Bateson, & Fowler, 1984; Öhman, 1966). Although the concept of a motor program has long since emerged from its origins in the physiologist's fixed action pattern, motor program structures have yet to be demonstrated that can cope adequately with the degrees-of-freedom problem and with the context-sensitive nature of action patterns.

There have been two major kinds of response to this problem—the generalized motor program perspective and the dynamical systems perspective. According to the generalized motor program approach (Schmidt,

1975), motor programs are composed of schemata, which are learned mappings relating movement outcomes, control signals, and boundary conditions. Schemata are parameterized so as to be applicable to a variety of related activities. Although recent evidence has belied the earlier hope that the parameters would be inferrable from linear scaling laws (Gentner, 1987), the notions of parameter and schema continue to provide useful constructs for characterizing flexible motor programs (Meyer, Abrahms, Kornblum, Wright, & Smith, 1988; Rosenbaum, Gordon, Stillings, & Feinstein, 1987; Shapiro & Schmidt, 1982; Stelmach, Mullins, & Teulings, 1984).

The dynamical systems perspective (Kugler, Kelso, & Turvey, 1980; Saltzman & Kelso, 1987; Schöner & Kelso; 1988) focuses on the dynamical structure (vector field) that a successful composite system must have. The idea is that action is a matter of setting up the interaction between biomechanical variables and environmental variables so that the resulting nonlinear dynamical system is characterized by multiple stable manifolds of low dimensionality. This approach has had success with trajectory-formation and degrees-of-freedom problems (Saltzman, 1986; Schöner & Kelso, 1988) and has begun to provide useful characterizations of internal control structure (Saltzman & Kelso, 1987).

This chapter describes ongoing work on a theory of action that combines elements of both the generalized motor program perspective and the dynamical systems perspective. The structure studied in this theory are parameterized nonlinear dynamical systems that have elsewhere been demonstrated to be characterized by multi-stability (Jordan, in press). In the current paper, I discuss the problem of motor learning within this framework. The approach that is developed here has similar goals to those of generalized motor program theory but is more suited to problems involving excess degrees of freedom and goal-directed learning (Jordan & Rosenbaum, 1989). The approach can be characterized as a dynamical systems approach to describing the flow through parameter space. After presenting the general framework, its implementation using network learning algorithms, and an example, I present simulation experiments that illustrate applications of the approach to inverse kinematic learning, a composite feedback-feedforward system, trajectory formation for a dynamical arm, and the learning of input parameters.

## TERMINOLOGY

The following terminology is used throughout the chapter. *Task space* is the vector space in which environmental specifications of tasks are given to the learner (see Saltzman & Kelso, 1987, for a related use of the term).

Task space vectors are movement outcomes such as perceived location, perceived reasonace, or perceived duration, and may be expressed using data from a variety of modalities, including vision, audition, and kinesthesia. *Articulatory space* is the vector space of possible control signals to the effectors.

## MOTOR LEARNING AS CONSTRAINED OPTIMIZATION

The approach I am pursuing treats the learning of motor programs as a constrained optimization problem. Learning is described as the optimization of a quantity—a cost functional—defined in task space, subject to various additional constraints. The major theoretical problem in this framework is that of ensuring that all of the terms needed for optimization are in fact available to the learner. There are several cases in which this assumption seems problematic. In particular, when the controller must interact with an unknown external system or when the biophysical plant undergoes change due to maturation or injury than the solutions derived by optimization change accordingly. Thus if motor learning is to be characterized as a process of optimization, then the underlying learning algorithm must itself be reconfigurable and sensitive to a changing environment. In the current framework, this problem is addressed via the notion of a *forward model* of the environment (Jordan & Rumelhart, 1989). In terms of its input/output behavior, a forward model is closely related to notions of "internal simulation," "efference copy," or "predictive model" that have been considered by many authors (e.g., Eccles, 1977; Gurfinkel & Levik, 1979; Lindblom, Lubker, & Gay, 1979): A forward model is a learned internal mapping that allows the system to predict the results it expects to obtain, given the state of the environment and the output of the motor program. Somewhat surprisingly, a forward model can also be used to allow the system to go backwards, that is, to find a motor program output that achieves a desired result. Consider, for example, a simple positioning problem in which the system must achieve a task-space target $\mathbf{z}^*$. We define a cost functional that is the weighted sum of squares of the different between the desired result $\mathbf{z}^*$ and the actual result $\mathbf{z}$:

$$J = \frac{1}{2}(\mathbf{z}^* - \mathbf{z})^T\mathbf{H}(\mathbf{z}^* - \mathbf{z}), \tag{1}$$

where $\mathbf{z}^*$ and $\mathbf{z}$ are vectors in task space and $\mathbf{H}$ is a positive definite matrix.[1]

[1]Readers not familiar with the vector calculus may wish to follow through the equations in the one-variable case, where $J = \frac{1}{2}h(z^* - z)^2$. Also, a worked-out example is given in a following section.

The vector $\mathbf{z}$ is a function of the environmental state $\mathbf{s}$, and the motor program output $\mathbf{x}$:

$$\mathbf{z} = \mathbf{f}(\mathbf{s}, \mathbf{x}). \tag{2}$$

The motor program output $\mathbf{x}$ is in turn a function of the internal state $\mathbf{q}$, an input vector $\mathbf{p}$ (the "plan"), and a set of structural parameters $\mathbf{w}$ that describe the current configuration of the motor program.[2]

$$\mathbf{x} = \mathbf{g}(\mathbf{q},\mathbf{p},\mathbf{w}). \tag{3}$$

The problem is to find a set of parameters $\mathbf{w}$ so that $J$ is minimized, that is, so that the desired result $\mathbf{z}^*$ and the actual result $\mathbf{z}$ are equal. To solve this optimization problem, we first find the gradient of $J$ with respect to the parameters $\mathbf{w}$:

$$\begin{aligned} \nabla_{\mathbf{w}} J &= \frac{\partial}{\partial \mathbf{w}} \frac{1}{2}(\mathbf{z}^* - \mathbf{z})^T \mathbf{H}(\mathbf{z}^* - \mathbf{z}), \\ &= -\frac{\partial \mathbf{z}^T}{\partial \mathbf{w}} \mathbf{H}(\mathbf{z}^* - \mathbf{z}) \\ &= -\frac{\partial \mathbf{x}^T}{\partial \mathbf{w}} \frac{\partial \mathbf{z}^T}{\partial \mathbf{x}} \mathbf{H}(\mathbf{z}^* - \mathbf{z}). \end{aligned} \tag{4}$$

For this gradient to be useful in a learning rule, each of its terms must be available to the learner, either separately or in combination. The vector $\mathbf{z}^* - \mathbf{z}$ is the task space error and can be assumed to be provided by the environment. The matrix $\partial\mathbf{x}/\partial\mathbf{w}$ describes a credit assignment process internal to the motor program. The Jacobian matrix $\partial\mathbf{z}/\partial\mathbf{x}$ is the only term whose origin appears to be problematic. It is the matrix that relates small changes in the motor output to small changes in the task space results, and cannot be assumed to be available a priori, or provided by the environment. However, all of the derivatives in the matrix are *forward* derivatives. They are easily obtained by differentiation if a forward model is available. To accommodate the possibility of an unknown or changing environment, the forward model must itself be learned, but this can be achieved directly by system identification (e.g., by "babbling"). Once the model is accurate over a particular domain, its derivatives provide a learning operator that allows the system to convert errors in task space into errors in articulatory space, and thereby change the motor program.

The basic transformations that characterize the forward modeling approach are summarized in Fig. 29.1. The top diagram shows the relationship between the environment and the forward model. The model is adjusted

[2]*Structural* parameters are to be distinguished from *input* parameters (which are components of the plan vector).

based on the discrepancy between the actual result $\mathbf{z}$ and the predicted result $\hat{\mathbf{z}}$. The bottom diagram depicts the learning of the motor program. The discrepancy between the desired result $\mathbf{z}^*$ and the actual result $\mathbf{z}$ is transformed by the learning operator $\widehat{\partial \mathbf{z}/\partial \mathbf{x}}$ obtained by differentiating the forward model. The transformed error is then used to adjust the motor program. The link between the two diagrams is provided by the forward model; as the modeling process in the top diagram progresses, the optimization process in the lower diagram becomes increasingly accurate.[3]

The focus in the current paper is on the problem of excess degrees of freedom, that is, the case in which the transformation from articulatory space to task space is many-to-one. A virtue of the forward modeling approach is that is it applicable whether or not there are excess degrees of freedom. This is true because forward derivatives exist irrespective of whether a transformation has a unique inverse. Moreover, a cost functional in task space is a straightforward way of capturing the notion of "motor equivalence" (Hebb, 1949). The set of motor program outputs that achieve a particular task space goal is simply that set for which the cost functional is minimal. This set is an equivalence class in articulatory space.

## TASK CONSTRAINTS

It is important to distinguish between action-based descriptions of skilled performance and the manner in which tasks are actually specified to the learner by the environment (Fowler & Turvey, 1978; Newell, 1986). Although little is known in general about task specification, it seems reasonable to assume that the environment provides an incomplete description of skilled performance to the learner. There are at least two reasons for this conjecture. First, tasks are generally specified in perceptual coordinate systems that have fewer degrees of freedom than needed to describe the details of movement. Second, many aspects of performance are due to properties of the performer rather than the task. Clearly, biomechanical properties of the performer are relevant to determining what can be left unspecified. There also appear to be "soft" constraints on skill acquisition. For example, humans do not generally need to be told to attempt to make

---

[3]It is important to stress that the forward model serves only as an intermediate construct in the derivation of a learning operator; in particular, its input/output behavior is not needed in learning the motor program. Of course, the output of a forward model may be useful for other purposes such as the detection of errors, prediction, and the reduction of delay in closed-loop control (Eccles, 1977; Gurfinkel & Levik, 1979; Lindblom, Lubker, & Gay, 1979). This functionality provides additional motivation for the use of a forward model as a component in a control system but is orthogonal to its use in deriving a learning operator.

reasonably rapid, smooth movements that take advantage of excess degrees of freedom. That such properties of skill are not necessary consequences of the biomechanics or the controller is demonstrated by the performance of novices. The implication is that these properties arise from internal constraints on the learning process. Evidence supporting this assertion was provided in a tracking task by Newell and Carlton (1987). They showed that the pointwise specification of a criterion trajectory in addition to a global criterion did not improve the learning rate for a "natural trajectory" (a Gaussian curve), but did improve learning for an unfamiliar trajectory (see also Brooks & Watts, 1988; Pew, 1974).

The distinction between properties of the performer and properties of the task specification is particularly important in the ubiquitous case of learning by imitation. If performers are characterized by shared internal constraints on the learning process, then fewer of the details available on a learning trial need to be transferred explicitly to the learner. Thus, to take an example from speech production, not all of the details of coarticulation would need to be sorted out from the acoustic waveform in order for the child to acquire the appropriate articulatory patterns. In particular, the presence of coarticulatory movements having little perceptual effect would not need to be inferred.

In the remainder of the chapter, I present examples of cost functional forms of internal constraints on learning. These constraints are assumed to reflect generally desirable properties of performance, such as smoothness, speed, robustness, and communicability. In addition, by regularizing the control signal, such intrinsic constraints may help maintain system stability as the parameters are changed during learning.

When the intrinsic constraints on learning are factored away, we are left with a minimal task specification that must be provided by the environment. Such a task specification can take many forms, depending on the coordinate systems that are used and the level of detail needed for the particular task. The task may specify conditions on sequences, trajectories, final configurations, time, or other quantities to be optimized. In the current framework, such minimal specifications are referred to as *task constraints* and are assumed to be expressed by a term in the overall cost functional. For example, a target sequence $\mathbf{z}_1, \mathbf{z}_2, \ldots, \mathbf{z}_N$ can be written in terms of a quadratic cost functional $J_{\mathbf{z}}$:

$$J_{\mathbf{z}} = \frac{1}{2} \sum_{i=1}^{N} (\mathbf{z}_i^* - \mathbf{z}_i)^T \mathbf{H}_i (\mathbf{z}_i^* - \mathbf{z}_i).$$

This form of task-level constraint both defines desired performance (as the equivalence class that minimizes the cost functional) and provides infor-

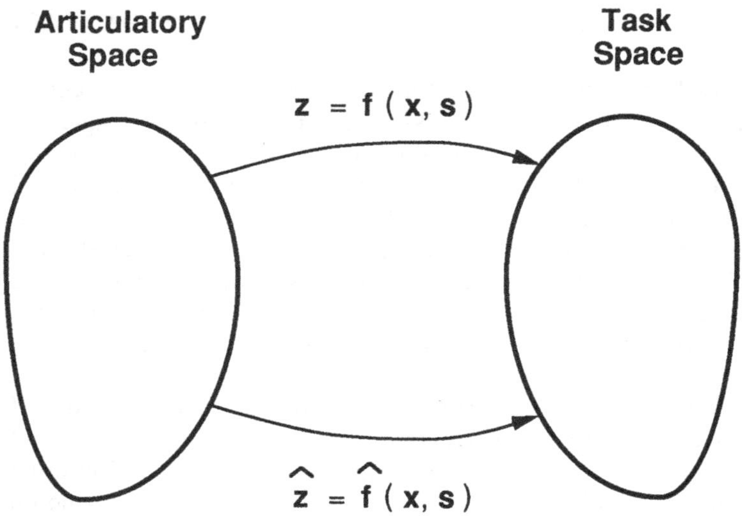


(a) **Modeling**

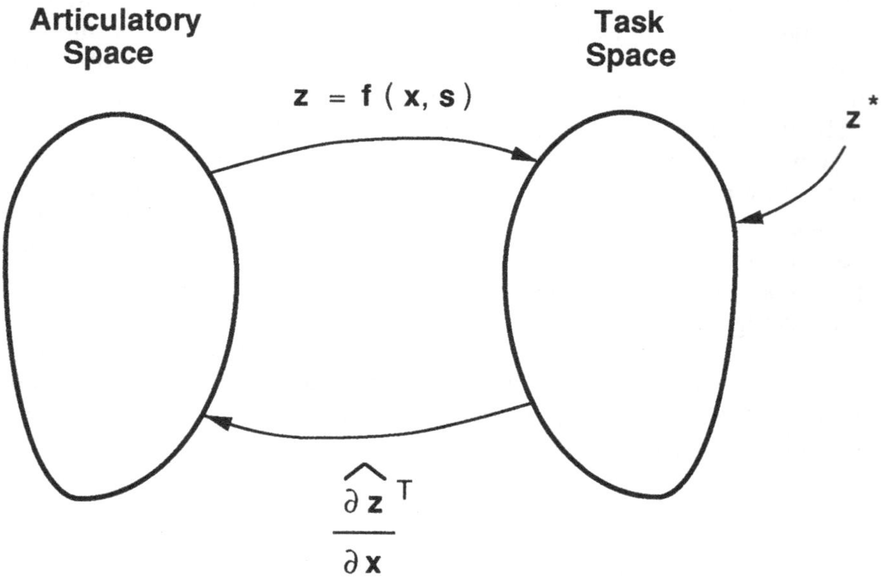


(b) **Optimization**

mation about how to achieve the desired performance (the gradient of the cost functional).

## INTRINSIC CONSTRAINTS ON MOTOR LEARNING

Many authors have proposed that action can be characterized in terms of the optimization of quantities such as energy, time, smoothness, distinctiveness, and information (e.g., Hogan, 1984; Lindblom, 1983; Lindblom, 1986; Mussa-Ivaldi, McIntyre, & Bizzi, 1988; Nelson, 1983; Rosenbaum, Marchak, Barnes, Vaughan, Slotta, & Jorgensen, this volume). What distinguishes the current approach is that such principles are viewed not only as descriptions of the endstate of skill, but also as active constraints on the process of learning. In this section, I discuss three examples of intrinsic constraints on motor learning. As is demonstrated in the sequel, it is of particular interest to study the effects of these constraints when they apply simultaneously in different coordinate systems.

### Smoothness Constraints

Least-effort principles abound in the physical sciences, and it seems sensible that a system that evolved to manipulate the physical world should itself be characterized by some form of least-effort principle. Recent work on arm movements has provided evidence that the control signal ("virtual trajectory") is maximally smooth in end-effector coordinates (Flash, 1987), and it has been suggested that smoothness allows for a reduction in the amount of information needed to specify a trajectory (Hogan, 1984).

Smoothness constraints are clearly applicable to accounting for the form of the transitions between boundary conditions. In addition, smoothness constraints can account for the context-sensitive form that actions take in systems with excess degrees of freedom (cf. Gentner, Grudin, & Conway, 1981; Öhman, 1966). Consider a sequence of targets in task space, where each target can be achieved with many different articulatory configurations. A smoothness constraint in *articulatory* space implies that targets nearby in time are achieved with a similar articulatory configurations. Thus the choice of articulatory configuration for a given target depends on the tem-

---

FIG. 29.1. The transformations proposed by the forward modeling approach. (a) Modeling. The top arrow represents the transformation performed by the environment; the bottom arrow represents the internal forward model. (b) Optimization. The top arrow represents the transformation performed by the environment; the bottom arrow represents the internal learning operator.

poral context.[4] A variety of forms of context sensitivity can result from this general principle, particularly in multi-effector systems such as the hands or the speech articulators. For example, context sensitivity can take the form of "feature spreading" or "parallelism" when actions in a sequence make use of disjoint sets of articulators, and "compensation" when actions share articulators. Such forms of context sensitivity allow the system to move more rapidly through a sequence of task space goals than would otherwise be possible.

A general form of smoothness constraint can be written as a condition on the weighted mean squared velocity of the articulatory vector $\mathbf{x}$:

$$J_s = \frac{1}{2}\sum_{i=2}^{N} (\mathbf{x}_i - \mathbf{x}_{i-1})^T\mathbf{K}(\mathbf{x}_i - \mathbf{x}_{i-1}), \tag{5}$$

where the subscript $i$ denotes the time step. Note that if $\mathbf{x}$ includes components that are themselves velocities or accelerations of underlying variable, then $J_s$ may be interpreted in terms of mean square acceleration, or jerk.

Assuming that $\mathbf{x}$ depends on a set of parameters $\mathbf{w}$, as in Equation 3, we find the gradient of $J_s$:

$$\nabla_{\mathbf{w}} J_s = -\sum_{i=2}^{N}\{\frac{\partial \mathbf{x}_{i-1}}{\partial \mathbf{w}}^T \mathbf{K}(\mathbf{x}_i - \mathbf{x}_{i-1}) + \frac{\partial \mathbf{x}_i^T}{\partial \mathbf{w}} \mathbf{K}(\mathbf{x}_{i-1} - \mathbf{x}_i)\}, \tag{6}$$

The first term in this equation depends on partial derivatives at the $i - 1^{st}$ time step and causes $\mathbf{x}_{i-1}$ to approach $\mathbf{x}_i$. Thus, this term leads to anticipatory behavior. Similarly, the second term causes $\mathbf{x}_i$ to approach $\mathbf{x}_{i-1}$, and therefore yields perseveratory behavior. Note that these two terms can be weighted differently (although the resulting form is no longer a gradient).

A smoothness constraint can also be invoked in distal coordinate systems. In this case, as in the case for task constraints, a forward model is needed to provide the transpose Jacobian matrix that arises in computing the gradient.

### Distinctiveness Constraints

When actions have a communicatory purpose, it is sensible that they be distinct, so that they may be easily interpreted. Distinctiveness can be viewed as the opposite of smoothness, which suggests writing the constraint as the negative of a smoothness term. Such a term leads to instability, however, because the gradient grows larger as nearby vectors become more

[4]This argument applies in either continuous-time or discrete-time systems. The latter case is of particular interest because it allows context sensitivity to be analyzed separately from transitional behavior.

distinct. A more suitable distinctiveness constraint is obtained by enclosing a negative smoothness term in an exponential:

$$J_d = \frac{1}{2} \sum_{i=2}^{N} e^{-(\mathbf{y}_i - \mathbf{y}_{i-1})^T \mathbf{L}(\mathbf{y}_i - \mathbf{y}_{i-1})}. \tag{7}$$

Differentiating this term with respect to the parameters $\mathbf{w}$ yields the following expression:

$$\nabla_{\mathbf{w}} J_d = \sum_{i=2}^{N} z_i \{ \frac{\partial \mathbf{y}_{i-1}{}^T}{\partial \mathbf{w}} \mathbf{L}(\mathbf{y}_i - \mathbf{y}_{i-1}) + \frac{\partial \mathbf{y}_i^T}{\partial \mathbf{w}} \mathbf{L}(\mathbf{y}_{i-1} - \mathbf{y}_i) \}, \tag{8}$$

where

$$z_i = e^{-(\mathbf{y}_i - \mathbf{y}_{i-1})^T \mathbf{L}(\mathbf{y}_i - \mathbf{y}_{i-1})}.$$

The distinctiveness gradient is the negative of the gradient of the smoothness constraint, multiplied by a scalar that goes to zero as vectors become more distinct.

Distinctiveness and smoothness are antagonistic but need not be completely incompatible with one another if they are applied in different coordinate systems. A system may attempt to produce outputs that have distinct perceptual effects, but are nonetheless smooth in articulatory space (Kiparsky, 1975; Lieberman, 1975; Lindblom, 1986; Martinet, 1968). Excess degrees of freedom should generally be helpful in allowing these constraints to coexist.

## Rest Configuration Constraints

Whereas smoothness constraints and distinctiveness constraints impose costs on the transitions between successive configurations, it is also to be expected that there are absolute costs associated with particular configurations (Lindblom, 1986; Mussa-Ivaldi, McIntyre, & Bizzi, 1988; Saltzman & Munhall, 1989). In particular the system may penalize deviations of the articulatory configuration from a rest configuration $\mathbf{x}^{(0)}$, where the rest configuration may itself be learned or specified by the task. Such a constraint has an interpretation in terms of energy expenditure or loss of predictability for large deviations. A simple form of rest configuration constraint is given as the weighted sum of squares of these deviations:

$$J_r = \frac{1}{2} \sum_{i=1}^{N} (\mathbf{x}_i - \mathbf{x}^{(0)})^T \mathbf{M}(\mathbf{x}_i - \mathbf{x}^{(0)}). \tag{9}$$

Differentiation of $J_r$ yields:

$$\nabla \mathbf{w} J_r = - \sum_{i=1}^{N} \frac{\partial \mathbf{x}_i^T}{\partial \mathbf{w}} \mathbf{M}(\mathbf{x}^{(0)} - \mathbf{x}_i), \tag{10}$$

which is analogous to an elastic force.

## COMPOSITE COST FUNCTIONALS

The full specification of the active constraints on the learning of a particular task is assumed to be given as a composite cost functional. This cost functional is a sum of task constraints and intrinsic constraints. Consider, for example, the following functional:

$$J = \frac{1}{2}\{\sum_{i=1}^{N} (\mathbf{z}_i^* - \mathbf{z}_i)^T \mathbf{H}_i(\mathbf{z}_i^* - \mathbf{z}_i) + \gamma_1 \sum_{i=2}^{N} e^{-(\mathbf{y}_i - \mathbf{y}_{i-1})^T\mathbf{K}(\mathbf{y} - \mathbf{y}_{i-1})}$$
$$+ \gamma_2 \sum_{i=2}^{N} (\mathbf{x}_i - \mathbf{x}_{i-1})^T \mathbf{L}(\mathbf{x}_i - \mathbf{x}_{i-1}) + \gamma_3 \sum_{i=1}^{N} (\mathbf{x}_i - \mathbf{x}^{(0)})^T \mathbf{M}(\mathbf{x}_i - \mathbf{x}^{(0)})\}, \quad (11)$$

where $\mathbf{z}$ is a task space vector and $\mathbf{y}$ is a vector in a perceptual coordinate system intermediate between articulatory space and task space. Each intrinsic constraint is weighted by a parameter $\gamma_i$ that determines the salience of the constraint. The $\gamma_i$ are assumed to decrease over learning trials so that the intrinsic constraints do not prevent the task constraint terms $J_{\mathbf{z}}$ (the first term in $J$) from approaching zero. One simple way to achieve this is to define $\gamma_i$ to be linearly related to the task constraint term:

$$\gamma_i = \epsilon_i + \rho_i J_z.$$

A constraint is selected by the task specification if either $\epsilon_i$ or $\rho_i$ is nonzero. Another possibility is to treat the $\gamma_i$ as Lagrange multipliers (in which case it is necessary to compute the gradient $\nabla_{\gamma} J$ as well as $\nabla_{\mathbf{w}} J$).

## CONNECTIONIST ALGORITHMS

Jordan (1988) and Jordan and Rumelhart (1989) have shown how the forward modeling approach can be implemented in a connectionist network. The network implementation makes it possible to study phenomena of generalization and interference when many related tasks are learned in the same structure. Also, issues relating to the dynamics of an active motor program can be studied. In this section, I review the implementation of the forward modeling approach and show how intrinsic constraints can be implemented in a network learning rule.

## THE NETWORK ARCHITECTURE

The general network architecture is shown in Fig. 29.2. The network incorporates a set of connections that implement motor programs, as well as a set of connections that serve as a forward model. Learning algorithms change the weights on these sets of connections so that the motor programs eventually come to produce outputs that yield desired results. As is generally the case with this kind of network, the memory implemented by the weights integrates particular experiences rather that storing them separately, much like the motor schemata proposed by Schmidt (1975).[5]

The network in Figure 29.2 is a concatenation of two subnetworks, namely, the motor program subnetwork and the forward model subnetwork. The motor program subnetwork connects the *plan* units and the *state* units to the *articulatory* units. The plan units are the input units of the network; they receive activation from higher levels and designate the particular motor program to be activated. As is discussed below, particular components of the plan vector may be analogs of generalized motor program parameters (Schmidt, 1975). The state units receive recurrent connections from within the network and have a dual role: They provide internal state for the system, thereby allowing the network to autonomously generate sequences of actions (Jordan, in press), and they also estimate the state of the environment, thereby giving the network the capability to control the environmental dynamics. The motor program subnetwork also includes intermediate "hidden" units, so that a nonlinear function may be learned (Rumelhart, Hinton, & Williams, 1986). The articulatory units are the output units of the network; they are connected to the effectors.

The forward model subnetwork takes its input from the articulatory units and from those state units that encode the environmental state (the connections implementing the latter dependency are not shown in the figure). The output of the forward model is a pattern of activation on the *task* units. This activation pattern is the network's estimate of the consequences in task space of a particular articulatory output.

## LEARNING THE FORWARD MODEL

Learning of the forward model is a straightforward application of supervised learning techniques. Given a particular environmental state $\mathbf{s}$ and a particular action $\mathbf{x}$, the perceptual result (that is, the task space vector $\mathbf{z}$) is used as a target for the activity vector $\hat{\mathbf{z}}$ on the task units. The weights in the forward model subnetwork are changed in order to reduce the difference between these vectors. In the simulations reported here, the backpropagation algorithm (LeCun, 1985; Parker, 1985; Rumelhart, Hinton, & Williams, 1986; Werbos, 1974) is used to compute the appropriate incremental changes to the weights. The algorithm is a steepest descent

[5]More precisely, the network is a dynamical system and changes in connection strength affect the local vector field rather than storing particular trajectories.

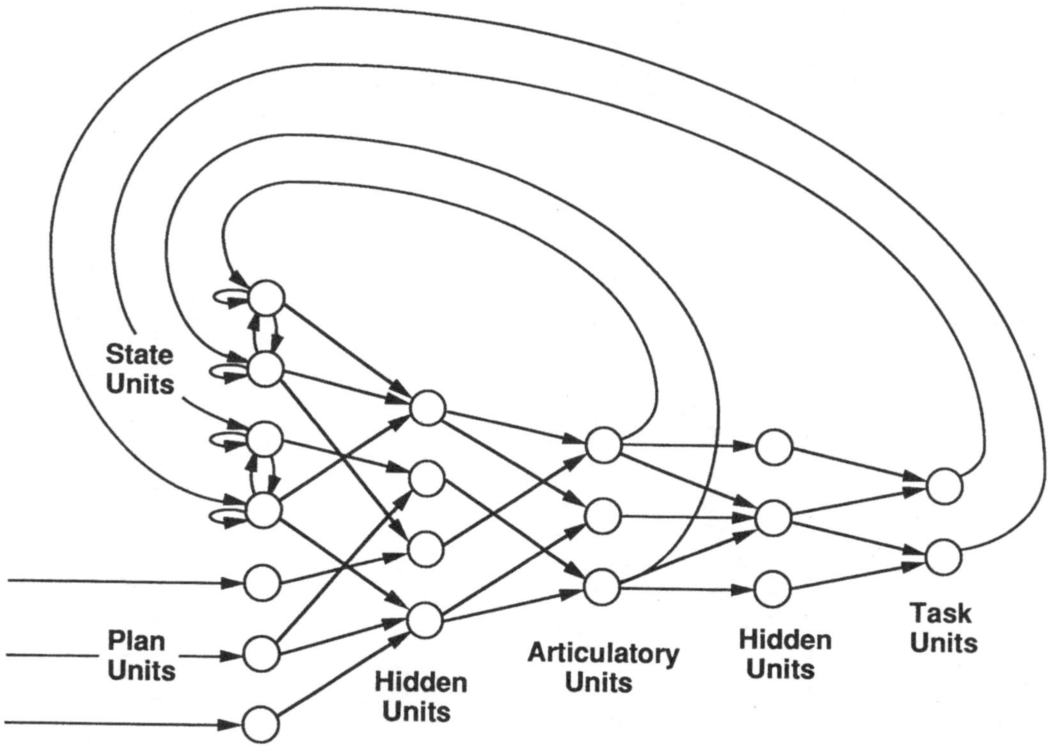


FIG. 29.2. The general network architecture. The articulatory units are the output units of the network.

algorithm in the following error measure:

$$E = \frac{1}{2}(\mathbf{z} - \hat{\mathbf{z}})^T(\mathbf{z} - \hat{\mathbf{z}}),$$

which goes to zero as the network becomes a more accurate predictor of task space results.

## LEARNING MOTOR PROGRAMS

### Task Constraints

The backpropagation algorithm can also be used to compute the gradients associated with the various constraints on the motor program output (Equation 11). Consider first the case of constraints defined in task space. It should be recalled that the problem is to transform a task space error vector into an articulatory space error vector that can be used to make the appropriate changes to the motor program.[6] The transformation involves a multiplication by the transpose Jacobian matrix of the forward model (Equation 4). In the network, this transformation is achieved by propagating the task space error vector backwards from the task units to the articulatory units; this process differentiates the forward model. The propagated vector need not depend on the model output $\hat{\mathbf{z}}$, although activation must pass forward in the forward model subnetwork so as to instantiate the correct transpose Jacobian matrix (which varies as a function of its domain) for the backward pass. From the articulatory units, the backward propagation process continues into the motor program subnetwork, where the weights are changed. (If a target trajectory $\mathbf{z}^*(t)$, $0 < t < T$, is provided in task space, or if the environmental is memoryless, then the backward propagation process can stop at the state units. Otherwise, the process must in effect continue backwards through the recurrent connections and repeatedly back through both subnetworks in order to compute the necessary gradients. An example of such a case is presented in a following section).

It is important to note that the gradients computed by differentiating a forward model are estimates of a gradient, and are exact only if the forward model is exact (over a particular domain). Exact gradients, however, are not necessary for the system to find an exact solution to the learning problem (Jordan, 1988). This is a consequence of the use of the actual task

---

[6]It is not actually necessary that the task space vector be an *error* vector. For example, a quantity such as time, energy, or distance may be optimized directly without comparison to a target.

space vector $\mathbf{z}$ rather than the estimated task space vector $\hat{\mathbf{z}}$ in the error terms (see, for example, Equation 4). A corollary of this fact is that the forward model and the motor program can be learned concurrently, with a concomitant speedup in learning time if some mechanism (e.g., a feedback controller) is available to guide the system into the appropriate domain (see Atkeson & Reinkensmeyer, 1988, and Kawato, Furukawa, & Suzuki, 1987, for related use of a feedback controller). If no such mechanism is available, then the forward model must be at least partially learned in a phase of random exploration before learning of the motor program can begin.

### Intrinsic Constraints

Additional constraints on the motor program output simply involve additional error terms introduced at various locations in the network. The weights are changed according to the sum of all of the backpropagated error terms. In the case of smoothness constraints and distinctiveness constraints, it is necessary for units to save their activation at the preceding time step so that the appropriate error terms can be formed (Equation 6 and Equation 8). Units that participate in a rest configuration constraint compute an error term based on a comparison with the rest configuration vector (Equation 10); they must therefore have access to the appropriate component of that vector. For further details, see Jordan (1988).

## AN EXAMPLE

A simple linear kinematic example may help clarify the approach under consideration here. Suppose that we have an organism with three output degrees of freedom—a tongue, a jaw, and a lower lip, whose positions are represented as $x_1$, $x_2$, and $x_3$, respectively. Each of the variables can have a value between zero and one. Because the tongue and lip ride on the jaw, sums of these values are more relevant to the acoustic output than are the absolute values themselves. For example, the sum of $x_1$ and $x_2$ specifies how high the tongue is with respect to the fixed palate. Also, the sum of $x_2$ and $x_3$ specifies the openness of the lower lip with respect to the fixed upper lip (cf. the "tract variables" of Saltzman & Munhall, 1989). For simplicity, we assume that these linear relationships have direct correlates in the acoustics, so that task space is defined by the two variables $z_1 = x_1 + x_2$ and $z_2 = x_2 + x_3$. In vector notation, we have $\mathbf{z} = \mathbf{Cx}$, where

$$\mathbf{C} = \begin{bmatrix} 1 & 1 & 0 \\ 0 & 1 & 1 \end{bmatrix}$$

Now suppose that the organism is required to learn a sequence of two actions specified in task coordinates as follows:

$$\mathbf{z}_1^* = \begin{bmatrix} 1.6 \\ * \end{bmatrix}, \quad \mathbf{z}_2^* = \begin{bmatrix} * \\ 0.4 \end{bmatrix},$$

where an asterisk denotes a "don't-care condition" (don't–care conditions correspond to zeroes on the diagonal of the matrices $\mathbf{H}_i$ in Equation 11). This target sequence requires the tongue to be high against the palate at time one, with no requirement made of the lower lip, and at time two, requires the lip aperture to be relatively open, with no requirement made of the tongue position. There are an infinite number of solutions to this problem, involving various combinations of values for the degrees of freedom. However, some solutions are better than others, in terms of the smoothness of the articulatory transition between the actions. Consider, for example, the solution

$$\begin{bmatrix} 1.0 \\ 0.6 \\ 0.2 \end{bmatrix}, \begin{bmatrix} 0.8 \\ 0.4 \\ 0.0 \end{bmatrix}, \tag{12}$$

where the jaw undergoes a small change from 0.6 to 0.4. At the first time step, the tongue has a value of 1.0, so that the requirement that the sum be 1.6 is met. Likewise, at the second step, the lip has a value of 0.0, so that the requirement that the sum be 0.4 is met. Notice further that the don't-care conditions can be used to advantage: The lip must be 0.0 at the second time step, so this value can be anticipated at the first time step with a value of 0.2. Similarly, the value of 1.0 for the tongue at the first time step can be perseverated, given the don't–care condition at the second time step. In general, this solution is very smooth and parallel; no degree of freedom need change by more than 0.2 in value. By way of contrast, consider the solution

$$\begin{bmatrix} 0.6 \\ 1.0 \\ 1.0 \end{bmatrix}, \begin{bmatrix} 0.0 \\ 0.0 \\ 0.4 \end{bmatrix}, \tag{13}$$

in which every degree of freedom must change by at least 0.6 between the time steps. This solution also meets the task constraints that are imposed; however, it is awkward and would be slow to perform.

## TASK CONSTRAINTS

Fig. 29.3 shows the forward model subnetwork of a network that can find a preferred solution such as that given in Equation 12. The network has three articulatory units, corresponding to $x_1$, $x_2$, and $x_3$ and two (linear)

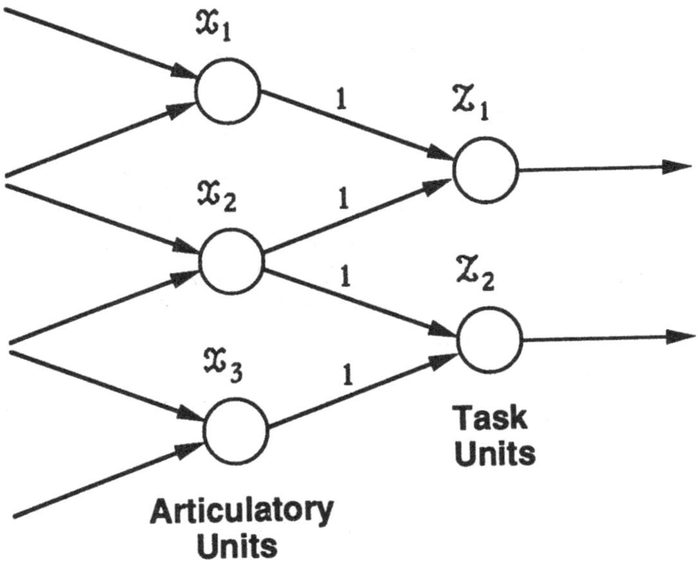


FIG. 29.3. A forward model subnetwork that computes a linear function of the articulatory vector.

tasks units, corresponding to $z_1$ and $z_2$. Hidden units are not needed because the function from articulatory space to task space is linear. The weights of unity and zero between the articulatory units and the task units constitute the forward model. They are the entries in the **C** matrix and are assumed to have been learned in a preceding exploratory phase.

At each time step, an articulatory pattern **x** is generated by the motor program subnetwork, and errors are formed at the task units between components of **z*** and **z.**[7] These errors are propagated back to the articulatory units, a computation which corresponds to multiplying the task error vector by the transpose of the **C** matrix.

Suppose, for example, that at time step one, the values of $x_1$ and $x_2$ are 0.5. The sum $z_1 = x_1 + x_2$ is compared to 1.6, and the resulting error value 0.6 is propagated back equally to both articulatory units. From the articulatory units, the errors are propagated back to the weights of the motor program subnetwork, producing changes that tend to make the activations of these articulatory units larger on future trials. No errors are formed at the third articulatory unit, due to the don't-care condition on $z_2$; thus, the activation of this unit is unconstrained at this time step. At the second time step, a comparison is made between the sum of $x_2$ and $x_3$

[7]Zero errors are always generated for don't-care conditions. For the mathematical treatment of don't-care conditions and inequalities, see Jordan (1988).

and the target value 0.4. An error of $-0.6$ drives the system to decrease the activations $x_2$ and $x_3$ on future trials. Note that in each case, the changes made to the weights of the underlying motor program subnetwork are relative to the time-varying context provided by the state vector. This preserves the serial order of the patterns being learned (Jordan, in press).

## INTRINSIC CONSTRAINTS

In addition to the task constraints, the network is assumed to have both a smoothness constraint and a rest configuration constraint on the articulatory units (the rest value for each unit is taken to be 0.5). Additional errors are generated at the articulatory units due to these constraints. In particular, temporal error vectors are formed at the second time step by comparing the current and saved activation values at each articulatory unit. The temporal errors act to pull these activation values closer together. In the case of the second articulatory unit, this pull competes with the tendency of the task constraints to drive the activation of this unit in different directions at the two time steps. Given that $x_1$ and $x_3$ are underconstrained and can move in directions that take up the slack in the task constraints involving $x_2$, the network is able to find a solution in which $x_2$ changes little in time.

The temporal errors also allow the don't-care conditions to be filled in appropriately. For example, the value of $x_3$ at time one is unconstrained, and the temporal error causes this value to track the constrained value taken by $x_3$ at time two, leading to anticipatory behavior. As the value moves further from 0.5, the error due to the rest configuration constraint starts to compete with the temporal error, and the system eventually equilibrates at a comprise value (0.8 in Equation 12). If the sequence of actions being learned were longer, this equilibration point would gradually drop off toward 0.5 for actions earlier in time.

## DISCUSSION

The major difference between this example and more realistic nonlinear problems is the transpose Jacobian matrix that converts task space errors into articulatory space errors. In nonlinear problems, this matrix is no longer constant, but varies according to the articulatory configuration. Although this would seem to require varying the weights in the forward model subnetwork, in fact it suffices to have nonlinear hidden units in the path between the articulatory units and the task units. (The derivative of the activation function of these units is variable and is used in backpropagation).

Two kinds of contextual phenomena are in evidence in the example. The first is a process of "feature spreading" or "parallelism," by which the don't-care conditions are filled in. The second is a process of contextually driven "compensation" between variables that participate in a task constraint. In the current framework, these two effects of context are due to a single underlying constraint. The difference between these phenomena is not one of process, but rather of representation, as can be seen by considering the optimization problem geometrically. In geometrical terms, optimization consists in requiring that the system move between implicitly specified regions in articulatory space in a smooth manner. When the regions are orthogonal to the axes, as is the case with don't-care conditions, smoothness yields the phenomenon of feature spreading. When the regions are oblique with respect to the coordinate axes, as is the case with the linear constraints in the example, compensatory relations are obtained. A simple change of coordinate system changes one phenomenon into the other, which suggests that they are, in fact, the same. It is also worth pointing out that neither of these phenomena should be expected to be observable in pure form in the general case, in which the regions in articulatory space are nonlinear manifolds.

## SIMULATIONS

This section presents the results of simulation experiments that demonstrate some of the properties of the general approach. Examples are given of problems in kinematics, dynamics, the use of feedback, and the learning of input parameters. In all examples, the emphasis is on the problem of indeterminacy.

## SMOOTHNESS IN A MULTI-EFFECTOR SYSTEM

The first example is an inverse kinematic learning problem. The manipulator shown in Fig. 29.4 is a two-fingered "hand" with two translational degrees of freedom and four revolute joints, yielding a 6-dimensional articulatory space. The task the system must learn is a sequential positioning task in which either one of the endpoints is required to touch a target at each time step. Target positions are specified in Cartesian coordinates, therefore task space is 2-dimensional. There is also an intermediate 4-dimensional coordinate system that encodes the positions of the endpoints of the fingers. The mapping from endpoint space to task space utilizes an

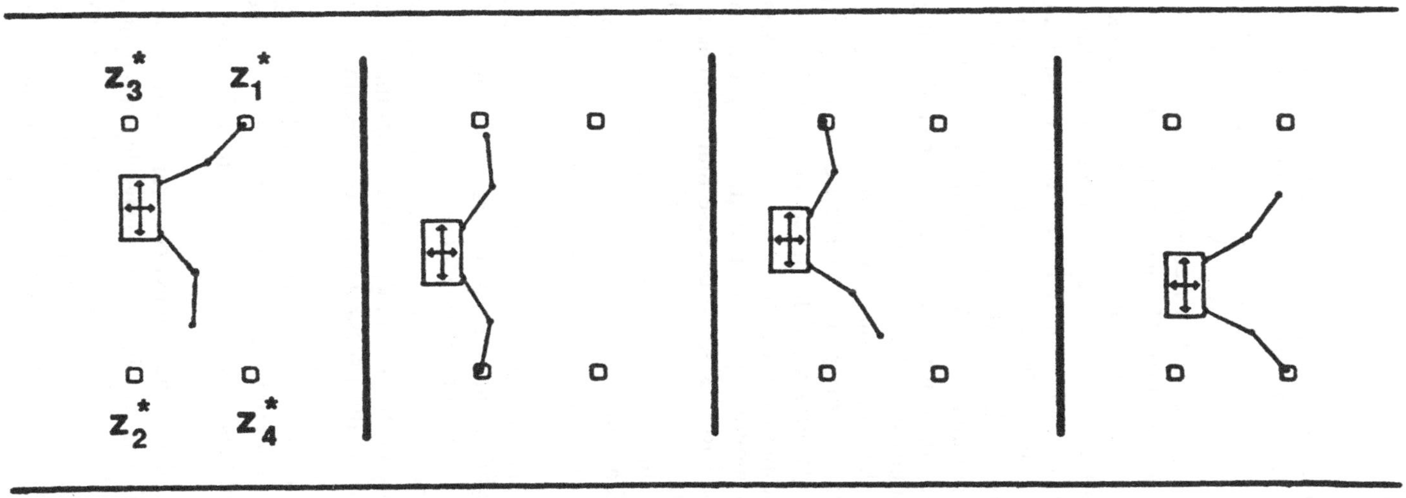


FIG. 29.4. A six-degree-of-freedom manipulator learning a cyclic positioning task. The system returns to the initial frame from the final frame.

additional binary variable to specify the "active" finger (the finger that is assigned to the current target). The assignment of fingers to targets is made dynamically during learning, based on a distance comparison.

The network used to control this manipulator is of the general form shown in Fig. 29.2. The network has six articulatory units for the manipulator's six articulatory degrees of freedom and an additional articulatory unit to encode the active finger. The activations of the first two articulatory units are interpreted as translational displacements and the next four articulatory units represent the sines of the angles of the four joints. These units are connected to a set of four endpoint units through an intermediate layer of six hidden units. The endpoint units and the active-finger unit are connected to two task units through a layer of four hidden units. The network has six plan units, and four hidden units in the motor program subnetwork. There are recurrent connections from the endpoint units to a set of four state units so that serially ordered behavior can be produced.

In a simulation, learning of the forward model subnetwork was carried out in a separate exploratory phase. This was done by repeatedly choosing random activation patterns $\mathbf{x}$ for the articulatory units (chose from a uniform distribution), and learning the forward kinematic mapping from articulatory space to endpoint space. It is not necessary to learn the mapping from endpoint space to task space because that mapping is independent of the environment. Furthermore, appropriate values for the weights in the path from the endpoint units and active-finger unit to the task units can be determined a priori. After approximately 2,000 exploration trials, the weights in the forward model subnetwork were held fixed, while a set of cyclic sequences was learned.

During the sequence-learning phase, targets were presented sequentially to the task units, while a smoothness constraint was imposed on the articulatory units. The results for a particular sequence are shown in Fig. 29.4. As can be seen, the network finds a solution to the problem that meets the task constraints. Furthermore, due to the smoothness constraint, the network makes use of the excess degrees of freedom to anticipate and perseverate other actions in the sequence. Each configuration involves a nonlinear compensatory relation between the translational degrees of freedom and the joint angles of the active finger; furthermore, the joint angles of the inactive finger show parallel anticipation of the following configuration.

A consequence of the smoothness constraint is that the same target position is generally achieved in different ways depending on the temporal context. This is shown in Fig. 29.5, where the first frame from Fig. 29.4 is overlaid with the first frame from a sequence in which the following target position differs. The figure shows that performance is variable in articulatory coordinates while invariant in task coordinates. Context-con-

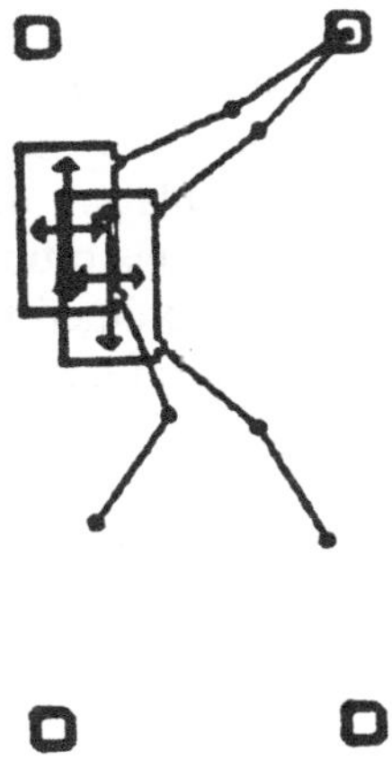

FIG. 29.5. The first frame from two different sequences having the same initial target.

ditioned variability would be observed in task coordinates if a tolerance range were allowed around the task space targets.

The results of this simulation are suggestive of phenomena of coarticulation in speech production (Kent & Minifie, 1977; Lindblom, 1983; Öhman, 1966). In speech production, the articulatory configuration associated with the production of a given phoneme changes according to its temporal context. Not all aspects of the configuration are all equally malleable, however. For example, in the production of an [*s*], there are strong constraints on the position of the tongue tip in the vicinity of the alveolar ridge, so that fricative output in a certain frequency band is achieved. There are weaker constraints on other articulators, such as the tongue dorsum and the lips. In the production of the velar fricative [*x*], there are strong constraints on the position of the tongue dorsum and weaker constraints on the tongue tip and the lips. In the current framework, all such constraints are assumed to be implicit in the specification of a task constraint in acoustic terms, in conjunction with a learned forward model of the nonlinear mapping from articulation to acoustics. The smoothness constraint leads to varying amounts of anticipation by each of the articulators, within the implicit bounds established by the inverse images of the acoustic constraints.

Behavior similar to that shown in Fig. 29.4 can be obtained without an explicit smoothness constraint if the activation of the state units changes smoothly in time (Jordan, in press). The two forms of smoothness are closely related, although they make different predictions about the details of coarticulation and so can be distinguished empirically.

Another point worth noting concerns the stability of the learned cyclic sequence. Simulations show that solution found by the network is stable; in particular, it is a periodic attractor both in articulatory coordinates and in endpoint coordinates. When the system is perturbed by adding transient noise to the activations of the articulatory units or the endpoint units, it returns to the learned sequence, demonstrating the stability of the solution. Furthermore, the same results are obtained for the other three sequences learned by the network. Thus, several stable trajectories can be learned together in the network.

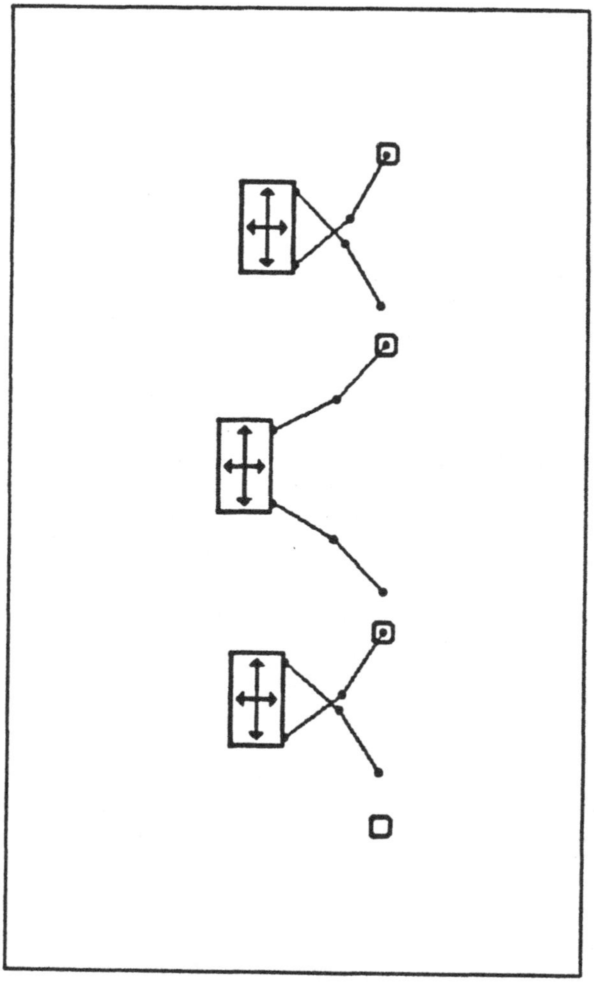

FIG. 29.6. Learning a sequential positioning task with a smoothness constraint on the endpoint units. Three successive time frames are overlaid in this figure.

## SMOOTHNESS IN THE ENDPOINT SPACE OF A MULTI-EFFECTOR SYSTEM

Consider the case of a musician playing a lengthy ascending scale on the piano. In learning such scales, students are told to cross the thumb under the index finger and the middle finger, so that every third note is played by the thumb.

In an analog of the piano-playing task, the network used in the previous experiment learned a linear sequence in which the finger assignments were alternated (the assignments were imposed from outside the network). Simulations revealed that the network readily learned the task, but was unable to find solutions in which one of the fingers crossed over the other. Rather, the solutions exhibited a small tendency toward crossing over and a larger tendency to use the translational degrees of freedom. Changing the gains in favor of the joint angles simply abolished whatever anticipatory behavior was present. The reason for these results is clear—in articulatory space, a posture with the fingers crossed is very different from an uncrossed posture. Thus, a smoothness constraint in articulatory space is inconsistent with a transition between these types of posture. Why then do piano players adopt the solution of crossing the thumb under?

Reconsideration of the task reveals that it is not smoothness in articulatory space that is relevant to the scale-playing task, but smoothness in *endpoint* space. The closer the endpoint of an unused finger is to the next target in sequence, the more quickly the transition between targets can be made. That is, given the nature of the task, the potential speed of transitions in endpoint space is more important than their cost in articulatory terms.

For the current example, smoothness in endpoint space can potentially yield contextual effects because there are four degrees of freedom in endpoint space versus the two degrees of freedom needed to specify a target. Fig. 29.6 shows the results of a simulation in which a smoothness constraint was applied to the endpoint units of the network, rather than to the articulatory units. As can be seen, the network finds a solution in which the endpoint positions are anticipated even though articulatory positions are not anticipated.

This simulation demonstrates that smoothness in different coordinate systems may lead to qualitatively different kinds of behavior. Part of the learning process may be the discovery of the appropriate coordinate systems in which to apply such constraints.

## DISTINCTIVENESS IN TASK SPACE; SMOOTHNESS IN ARTICULATORY SPACE

Rapid communication would seem to require articulatory representations to be smooth in time, whereas the resulting perceptual representations should be distinct. In a system with excess degrees of freedom, these two tendencies need not be incompatible with one another. This possibility was investigated in a simulation using a distinctiveness constraint in task space and a smoothness constraint in articulatory space.

The network used for this experiment is the same as in the previous two experiments. Targets for the task are rectangular regions in task space (this is achieved by using inequality constraints on the task units). The results for a particular sequence are shown in Fig. 29.7. As can be seen, due to the distinctiveness constraint, the endpoints tend to move toward the outer corners of the rectangles defined by the task constraints. Nevertheless, the effects of the smoothness constraint on the excess degrees of freedom are still in evidence; in particular, the position of the unconstrained endpoint tends to anticipate its position later in the sequence.

## FEEDBACK AND FEEDFORWARD

In control system terminology, the network used in the previous simulations is a feedforward controller with state (i.e., internal dynamics). I have proposed that the feedforward controller is learned by processes that estimate the parameters of a forward model of the controlled plant, and then use those parameters in forming gradients of a cost functional defined in terms of the output of the plant.

Traditionally, a feedforward controller is augmented with an error-correcting feedback controller (Arbib, 1981). The feedback controller corrects for unanticipated disturbances and provides control in those regions of state space for which the feedforward signal has yet to be learned. An example of such a composite control system for a dynamic arm is given in the following section. The current section provides an example of a composite feedforward-feedback system for a kinematic plant. The focus here is on the way in which excess degrees of freedom are utilized over the course of learning. To this end, I consider a network that can perform a sequence correctly, although slowly and awkwardly, on the first trial.

A feedback controller forms an error vector based on the difference between a reference vector and the output of the plant. This error vector is then transformed into a control vector. Note that this process is very similar to the forward modeling learning rule developed earlier. In fact, a feedback controller for a kinematic plant can be obtained by taking the gradient of the cost functional in Equation 1 with respect to the control vector **x** rather than with respect to the weights **w**. The resulting control law is the locally linear law given by:

$$\Delta \mathbf{x} = -\mu \frac{\partial \mathbf{z}^T}{\partial \mathbf{x}} \mathbf{H}(\mathbf{z}^* - \mathbf{z}), \qquad (14)$$

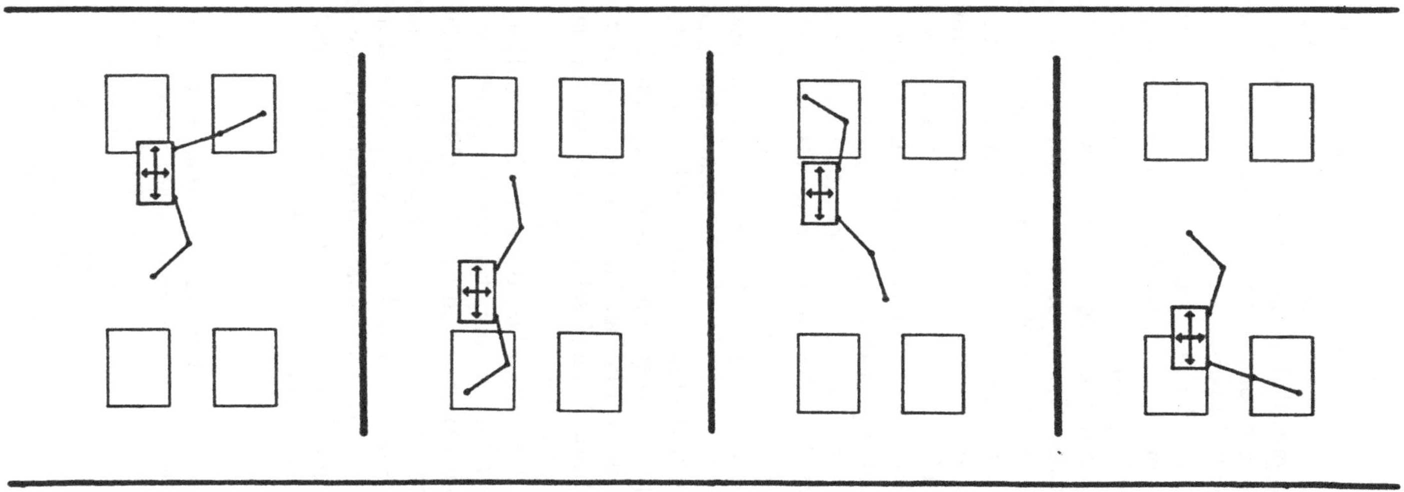

FIG. 29.7. The combined effects of a distinctiveness constraint and a smoothness constraint.

where $\mu$ is a gain. As before, the transformation due to the transpose Jacobian matrix can be computed using a learned forward model. The initial value of **x** is provided by the feedforward controller. The equation describes an error-correcting scheme that iteratively updates the control vector **x** so as to null the performance error $\mathbf{z}^* - \mathbf{z}$.[8]

The feedback controller makes it possible for the system to perform a sequence correctly on the first trial. Consider an overall system structure containing an outer loop and inner loop. The outer loop initialize the target $\mathbf{z}^*$ to the current vector in the sequence being learned and passes it to the feedforward controller. The output of the feedforward controller is a vector **x** that provides the initial control signal for the inner loop. The inner loop simply applies Equation 14 (the feedback controller) repeatedly until the magnitude of the error vector $\mathbf{z}^* - \mathbf{z}$ is less than some criterion value. At this point, the outer loop resets the target to the next vector in the sequence.

As the feedforward controller is changed through learning, the initial control vectors that it provides are more nearly correct, so that fewer iterations through the feedback controller are needed to reach the targets. This is demonstrated by the simulation results shown in Fig. 29.8 and Fig. 29.9, which display the output of the system on the last iteration of the inner loop for each target. Fig. 29.8 shows the results before any learning of the feedforward controller has taken place, whereas Fig. 29.9 displays the results after twenty learning trials. A total of 42 iterations of feedback were required to perform the sequence in Fig. 29.8, whereas only 22 iterations were needed in Fig. 29.9. Of course, once the feedforward control has been learned completely, as in the simulation shown in Fig. 29.4, no feedback iterations are required.

Learning of the feedforward controller leads not only to improvements in the speed of performance, but also increasingly effective use of excess degrees of freedom. As shown in Fig. 29.8, 29.9, and 29.4, the system shows an increasing ability to use excess degrees of freedom in anticipating future actions. This occurs even though the system is at a stage in which performance on the task is essentially correct with respect to the task constraints. Such optimization has often been noted in qualitative descriptions of learning (Fitts, 1964; Greene, 1982).

[8]An essentially identical control law has been proposed by Berkenblit, Fel'dman, and Fucson (1986) in modeling frog limb movements, although they do not address the developmental issue of an unknown or changing kinematic transformation. It is also of interest to consider a closely related control law in which the left-hand side is in units of generalized force, rather than displacement. Such a law can be used for a dynamic plant, due to the static relationship holding between the transpose Jacobian matrix $\partial \mathbf{z}^T/\partial \mathbf{x}$ and the generalized forces in the domain and range of the corresponding function (Hogan, 1985, Koditscek, 1984, Salisbury, 1980).

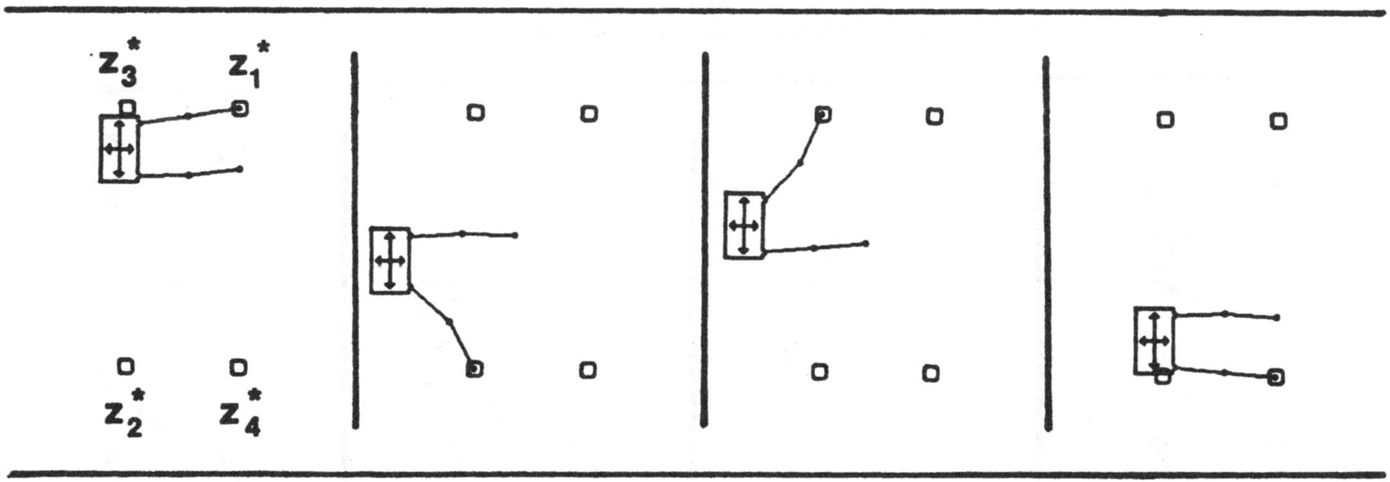


FIG. 29.8. Performance of a feedback controller. The figure shows the configurations found by the feedback controller for each target in the sequence.

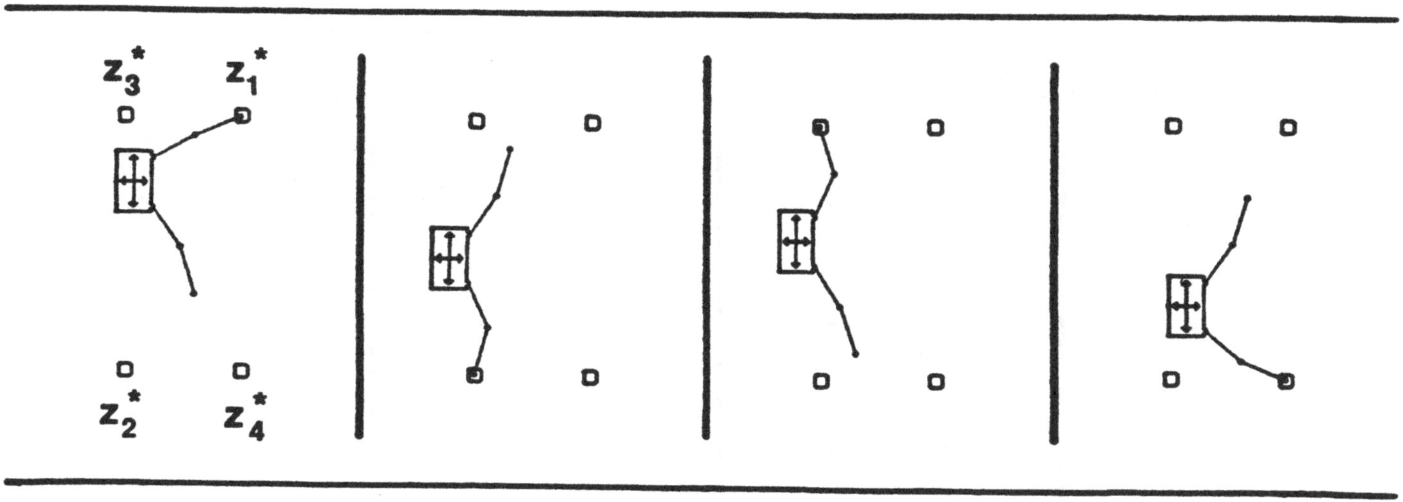


FIG. 29.9 Performance of the composite system, after partial learning of the feedforward controller.

## TRAJECTORY FORMATION FOR A DYNAMICAL ARM

Problems of indeterminacy are not restricted to issues of kinematic redundancy but also arise when trajectories must be generated. For example, a physical arm has mass and therefore moves along a spatially continuous trajectory. If a task specifies only a final position for the endpoint of the arm, then an infinite number of solutions—that is, trajectories that arrive at the target—are possible. In this section, I discuss how this indeterminacy problem can be resolved within the same computational framework as that demonstrated for the kinematic case in previous sections.[9]

Due to the elastic properties of muscles, the control of arm movement can be characterized in terms of a (virtual) equilibrium position to which the arm would move if efferent activity were held constant (Hogan, 1984). Evidence from arm movements in monkeys suggest that this virtual equilibrium position shifts gradually over the course of a movement toward the final position (Bizzi, Accornero, Chapple, & Hogan, 1984); thus, control can be treated in terms of a "virtual trajectory" (Hogan, 1984). If the muscles are modeled as second-order linear systems, then the torque $\tau$ is generated as a function of the virtual trajectory $u(t)$ according to:

$$\tau = -\beta\dot{\theta} - k(\theta - u), \tag{15}$$

where $\theta$ is the actual position of the arm, $\beta$ is a viscosity parameter, and $k$ is a stiffness parameter. For the simulations presented in this section, the control law in this equation is assumed, and the task is to learn a virtual trajectory $u(t)$, given a specification of a desired final position.

In Jordan and Rumelhart (1989), the problem of learning a controller for a two-joint arm is used to illustrate how the forward modeling approach applies in the case of environments with state. In such a case, the forward model is a learned mapping from the current state and control signal (positions, velocities, and torques) to the next state (accelerations, integrated twice). Backpropagation though the forward dynamic model allows the motor program subnetwork to find torques that cause the arm to follow desired trajectories.

The current problem differs from that of inverse dynamics in that we are not assuming that a desired trajectory is available. When the only signal available for learning is a final-position error, it is necessary to propagate the error backwards "in time" so that the entire control trajectory can be changed. Algorithms that do this either save past values of activation (Rumelhart, Hinton, & Williams, 1986; Widrow & Stearns, 1985) or compute ongoing traces of activations (Pearlmutter, 1988; Williams & Zipser, 1988). To apply such algorithms, it is necessary that the learning system be an

[9] A closely related approach to trajectory formation has been proposed by Kawato (1989).

integrated recurrent network. This can be accomplished in the current framework by identifying the state units of the forward model subnetwork with the state units of the motor program subnetwork. An example of such a network is shown in Fig. 29.10 for the case of a one-joint arm. In the diagram, the state units have been drawn to the right, to more easily depict their interconnection structure. The network is nonetheless a special case of the general network shown in Fig. 29.2. In this network, the forward model is implemented by weighted links arriving at a single task unit representing the joint acceleration. In the figure, I have assumed that the weights on these links have already been learned. As can be seen, the output of the forward model is therefore given by:

$$\ddot{\theta} = -\frac{\beta}{I}\dot{\theta}-\frac{k}{I}(\theta-u),$$

which is the forward dynamics of the one-joint arm.

Backpropagation of the final-position error in this network can be augmented with a smoothness constraint to determine a unique virtual trajectory. Smoothness in task space corresponds to the minimum jerk principles suggested by Hogan (1984). In this case, the functional minimized by the network is the follows:

$$J = \frac{1}{2}[\theta^* - \theta(\mathrm{N})]^2 + \frac{\gamma}{2}\sum_{i=1}^{N}[\ddot{\theta}(i)-\ddot{\theta}(i-1)]^2, \qquad (16)$$

where $\theta^*$ is the desired final position.

Simulations reveal that the network is able to find virtual trajectories that minimize jerk, based on only the final-position error. The time needed to learn such trajectories can be lengthy, however. A network learning four trajectories required approximately 19,000 iterations. One approach to decreasing this learning time is to use the desired final position as a control signal, as was originally proposed in the mass-spring model of arm movement (Fel'dmann, 1980; Kelso & Holt, 1980; Polit & Bizzi, 1979). The virtual trajectory is a sum of this constant signal and the time-varying being learned:

$$u(t) = \theta^* + u'(t),$$

where $u'(t)$ is the output of the articulatory unit.[10] Simulations show that this approach leads to a learning time nearly an order of magnitude faster than before (1,950 iterations).

It is straightforward to generalize the network in Fig. 29.10 to the case of a multijoint arm. Multiple units are required for vector variables and a

[10]Note that final-position control is mathematically equivalent to a proportional-derivative feedback controller; thus, this approach is conceptually a composite feedforward-feedback system.

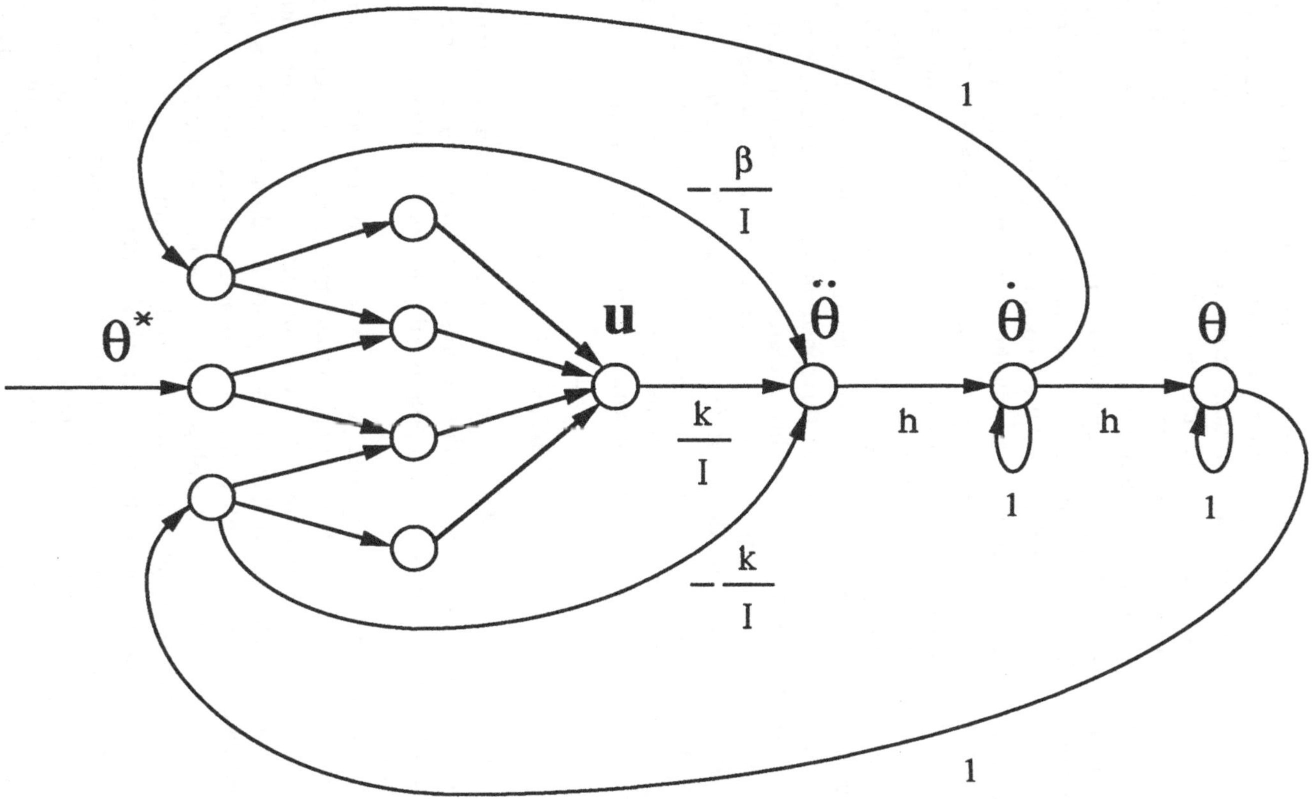


FIG. 29.10. A network that learns virtual trajectories for a one-joint arm.

hidden layer is required in the forward model to represent the nonlinear interaction terms. Learning based on the final-position error proceeds as before. However, new considerations arise concerning the choice of a smoothness constraint. Let us assume that a constraint is to be preferred if it requires no extra machinery beyond that needed to learn the unconstrained movement. In the multijoint case, this desideratum rules out the minimum jerk principle proposed by Flash and Hogan (1985), because jerk is defined in Cartesian coordinates—a kinematic transformation of accelerations using the time derivative of the Jacobian matrix would be required. It is much easier to implement a smoothness criterion on the Cartesian coordinates of the virtual trajectory; this simply requires a forward kinematic model, which is needed independently so that task constraints can be specified in Cartesian coordinates. In fact, this criterion also appears to give a better account of the data from arm movements, as has been argued by Flash (1987). The particular constraint suggested by Flash is a minimum jerk constraint; however, this constraint requires the acceleration of the virtual trajectory. Because this acceleration is not needed for learning the unconstrained movement, a constraint on a lower-order derivative of the virtual trajectory would be preferred. Tentative support for this argument is provided by data from monkey arm movements (see Fig. 2 in Hogan, 1984); the derivative of the inferred virtual trajectory is non-zero at the origin, which is inconsistent with a minimum jerk criterion.

## LEARNING PARAMETERIZED PLANS

An important component of a theory of motor learning is a description of the retrieval process. In this section, I present an approach to retrieval that is related to Schmidt's (1975) proposal that motor schemata are parameterized.

Learned motor programs must present a set of control parameters to higher level processes so that the appropriate motor program can be retrieved in particular situations. These control parameters must constitute a reasonably small set, so that the degrees-of-freedom problem is not recapitulated on the input side of the motor program. The parameters must also have a predictable relationship to the motor program output, given that environmental situations are never exactly the same.

In the current approach, the plan vector is the interface between motor programs and higher level processes. The plan provides global information that designates the motor program output to be generated; in dynamical terms, it instantiates a particular vector field. The relationship of "designation" is abstract—it is not necessary for the play to encode the details of the motor program output because these details are already embedded in the weights of the motor program subnetwork.

During learning, the error signals that are propagated through the motor program subnetwork lead to internal adjustments that produce a more correct output in the context of the current state and plan vectors. Error signals also arrive at the plan units. These signals are the partial derivatives of the cost functional $J$ with respect to the activations of the plan units; they therefore provide information about how the plan vector itself could be changed to minimize the cost functional. This information is potentially useful to encoding relationships between sequences, because sequences with similar subsequences tend to produce similar partial derivatives at the plan units. Thus, the possibility exists that appropriate constraints on plan representations are inherent in the transformed error signals.

To investigate this possibility a network with modifiable plan representations learned to control a six-degree-of-freedom planar manipulator with a single endpoint. To insure that the external input provided no information about how to form the plan representations, the technique of using standard basis input vectors was borrowed from Hinton (1986). The network consisted of a set of four *local* plan units, one and only one of which was turned on for each of the four sequences being learned.[11] These units were connected to a set of four *distributed* plan units, which were then connected to the remainder of the network. The weights between the local and distributed plan units were initially set to small random values, and were learned based on the error signals that were propagated back through the distributed plan units.

The results after 650 learning trials are shown in Fig. 29.11. The figure shows the final performance of the network for each sequence and also displays the activations of the distributed plan units. These activations have reasonably clear interpretations, which are supported by a sensitivity analysis. The first distributed plan unit clearly encodes the vertical dimension of the workspace. This is useful given the translational articulatory degree of freedom in the direction. The second unit appears to take on three values that encode the average direction and amount of curvature of the arm. For the top-left and bottom-left sequences this curvature is large, in opposite directions, whereas for the rightmost pair of sequences, little curvature is required. For the third unit, investigation of the weights on the connections to the hidden units reveals that the unit serves to gate particular hidden unit clusters that vary in opposing directions in time. Thus, this unit encodes the direction of movement—the top-left and bottom-right sequences are counterclockwise, whereas the other two sequences are clockwise. Finally, the fourth distributed plan unit appear to redundantly encode the vertical dimension of the workplace. In general,

[11]The local plan units should be thought of as placeholders for the output units of a higher level network.

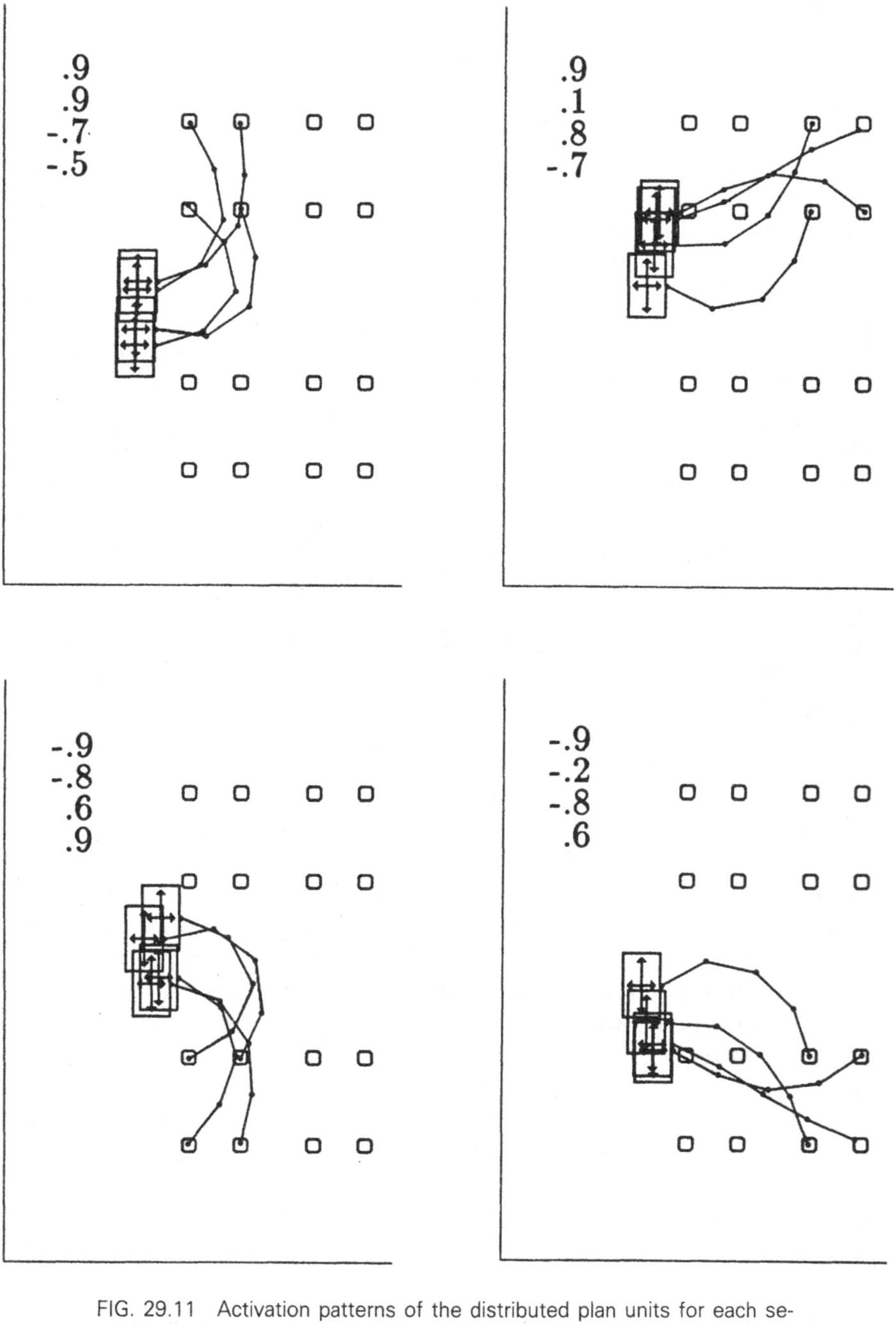


FIG. 29.11 Activation patterns of the distributed plan units for each sequence and corresponding performance of the network.

these units developed representations that appear to be useful parameterizations of the sequences to be performed.

An interesting issue arises on further consideration of these results: Is the network that produces the behavior shown in Fig. 29.11 best described in terms of a single motor program with three parameters, a pair of motor programs each with a single parameter, or perhaps four different motor programs? This question highlights the often fuzzy distinction made in generalized motor program theory between different generalized motor programs and different parameterizations of the same generalized motor program. The current results suggest that with respect to learning, this distinction is perhaps better treated as quantitative rather than qualitative. Performances may simply cluster differently in the subspaces corresponding to parameters that activate a motor program and parameters that apply to an active motor program. In general, the form of the clustering must depend on constraints inherent in the biomechanics and the controller; this is achieved in the network by propagating errors backward through the forward model subnetwork and the motor program subnetwork.

A final point that is clearly demonstrated in the simulation is that the parameters do not need to encode the details of the movement; rather, they need only make the relevant distinctions among the set of behaviors that the lower level network is capable of producing. The information needed to make these distinctions arises from errors present in the learning process that cannot be accounted for by contextual (state) variables. This systemic view of the notion of a motor program parameter is of course consistent with a venerable tradition in phonology—that of distinctive feature theory (Jakobson, Fant, & Halle, 1951).

## CONCLUSIONS

This chapter has presented a computational framework within which a number of phenomena relevant to motor learning and performance may be studied. The central idea of the approach is that motor learning can be described as the optimization of a cost functional embodying two kinds of constraints: constraints related to explicit task specifications, and intrinsic constraints that apply broadly to a range of tasks. When constraints are expressed in distal coordinate systems, an internal forward model is required so that they can be transformed into motoric coordinate systems. By combining forward models and intrinsic constraints, the approach is able to deal with indeterminant motor learning problems in a principled manner.

The problem of indeterminacy has played a large role in the development of theories of action over the past two decades. There are two strands in this development that it is useful to distinguish. The first strand, represented by the mass-spring model and its successors (see Saltzman & Kelso,

1987; Schöner & Kelso, 1988), is concerned with the problem of trajectory formation: What principles underlie the production of continuous trajectories that are constrained in only a finite number of ways? Should these principles refer to biomechanical principles, or should they refer only to movement kinematics? The second strand of theory is concerned with the issue of coordinate systems: Do task coordinate systems have fewer degrees of freedom than articulatory coordinate systems? If so, what principles determine the inverse mapping between these systems? This problem has mainly been addressed by appeal to the notion of a synergy or coordinate structure (Bernstein, 1967; Eason, 1972; Green, 1972; Kugler, Kelso, & Turvey, 1980; Saltzman, 1979).

Constrained optimization is a general strategy for dealing with indeterminate problems that has begun to see application in the study of action. For problems involving excess degrees of freedom, approaches based on optimization have generally invoked some form of pseudoinverse (Mussa-Ivaldi, McIntyre, & Bizzi, 1988; Pellionisz & Llinas, 1985; Saltzman & Kelso, 1987). Trajectory formation problems have been addressed with smoothness constraints (Hogan, 1984; Nelson, 1983; Uno, Kawato, & Suzuki, 1987). The current paper demonstrates that such optimization principles can form part of a theory of motor learning when combined with the notion of a forward model. The paper also stresses the application of smoothness constraints to problems involving both trajectory formation and excess degrees of freedom. Smoothness in the appropriate coordinate system can determine not only the transitions between action, but also the articulatory form in which those actions are expressed. Other influences on articulatory form, such as perceptual distinctiveness constraints, can be incorporated through the intermediary of the forward model.

Indeterminacies arise at a variety of levels and scales in motor control, suggesting the need for a variety of adaptive mechanisms. Those mechanisms that operate on the slowest time scales—learning mechanisms—can offer only a partial account of adaptability, given the evident ability of the motor system to adapt in complex, task-specific ways to unexpected perturbations (Abbs & Gracco, 1983; Kelso, Tuller, Vatikiotis-Bateson, & Fowler, 1984). With the disadvantage of a slow time constant, however, come certain computational advantages. Learning mechanisms can make use of iterative techniques that would be too slow for on-line processes. Furthermore, relatively simple operators can be employed at each iterative step and least commitment strategies can be used. Also, relatively complex temporal dependencies can be utilized because the entire time series is available to the adaptive process. Thus learning would appear to be able make relatively sophisticated use of excess degrees of freedom using relatively unsophisticated computation. We currently know very little, however, about limitations on the ability of on-line processes to resolve indeterminacy in task-specific ways, and we also know very little about the interactions between fast and slow adaptive processes. These are clearly topics that need further clarification.

## ACKNOWLEDGMENTS

This project was supported in part by BRSG 2 S07 RR07047-23 awarded by the Biomedical Research Support Grant Program, Division of Research Resources, National Institutes of Health. I want to thank Emilio Bizzi, Carol Fowler, Robert Jacobs, Lina Massone, Steven Pinker, Molly Potter, and Elliot Saltzman for helpful comments on the manuscript.

## REFERENCES

Abbs, J. H., & Gracco, V. L. (1983). Sensorimotor actions in the control of multimovement speech gestures. *Trends in Neuroscience, 6*, 393–395.

Arbib, M. A. (1981). Perceptual structures and distributed motor control. In V. B. Brooks (Ed.), *Handbook of physiology—The nervous system II* (pp. 1449–1480). Bethesda, MD: American Physiological Society.

Atkeson, C. G., & Reinkensmeyer, D. J. (1988). Using associative content-addressable memories to control robots. *IEEE Conference on Decision and Control*. San Francisco, CA.

Bernstein, N. (1967). *The coordination and regulation of movements*. London: Pergamon.

Berkenblit, M. B., Fel'dman, A. G., & Fucson, O. L. (1986). Adaptability of innate motor patterns and motor control. *Behavioral and Brain Sciences, 9*, 585–638.

Bizzi, E., Accornero, N., Chapple, W., & Hogan, N. (1984). Posture control and trajectory formation during arm movement. *Journal of Neuroscience, 4*, 2738–2745.

Brooks, V. B., & Watts, S. L. (1988). Adaptive programming of arm movements. *Journal of Motor Behavior, 20*, 117–132.

Easton, T. A. (1972). On the normal use of reflexes. *American Scientist, 60*, 591–599.

Eccles, J. C. (1977). *The Understanding of the Brain*. New York: McGraw-Hill.

Fel'dman, A. G. (1980). Superposition of motor programs. I. Rhythmic forearm movements in man. *Neuroscience, 5*, 544–548.

Fitts, P. M. (1964). Perceptual-motor skill learning. In A. W. Melton (Ed.), *Categories of Human Learning* (pp. 243–285). New York: Academic Press.

Flash, T. (1987). The control of hand equilibrium trajectories in multi-joint arm movements. *Biological Cybernetics, 57*, 257–274.

Flash, T., & Hogan, N. (1985). The coordination of arm movements: An experimentally confirmed mathematical model. *Journal of Neuroscience, 5*, 1688–1703.

Fowler, C. A., & Turvey, M. T. (1978). Skill acquisition: An event approach with special reference to searching for the optimum of a function of several variables. In G. Stelmach (Ed.), *Information processing in motor control and learning* (pp. 1–40). New York: Academic Press.

Gentner, D. R. (1987). Timing of skilled motor performance: Tests of the proportional duration model. *Psychological Review, 94*, 255–276.

Gentner, D. R., Grudin, J., & Conway, E. (1981). *Finger movements in transcription typing* (Tech. Rep. 8001). San Diego: University of California, Center for Human Information Processing.

Greene, P. H. (1972). Problems of organization of motor systems. *Progress in Theoretical Biology, 2*, 303–338.

Greene, P. H. (1982). Why is it easy to control your arms? *Journal of Motor Behavior, 9*, 2–42.

Gurfinkel, V. S., & Levik, Y. S. (1979). Sensory complexes and sensorimotor integration. *Fiziologiya Cheloveka, 5*, 399–414.

Hebb, D. O. (1949). *The Organization of Behavior*. New York: Wiley.

Hinton, G. E. (1986). A network which learns distributed representations of concepts. *Proceedings of the Eighth Annual Conference of the Cognitive Science Society*. Hillsdale, NJ: Lawrence Erlbaum Associates.

Hogan, N. (1984). An organising principle for a class of voluntary movements. *Journal of Neuroscience, 4*, 2745–2754.

Hogan, N. (1985). Impedance control: An approach to manipulation: Part I-Theory. *ASME Journal of Dynamic Systems, Measurement, and Control, 107*, 1–24.

Jakobson, R., Fant, G., & Halle, M. (1951). *Preliminaries to speech analysis*. Cambridge, MA: MIT Press.

Jordan, M. I. (1988). *Supervised learning and systems with excess degress of freedom*. (COINS Tech. Rep. 88-27). Amherst, MA: University of Massachusetts, Computer and Information Sciences.

Jordan, M. I. (in press). Serial order: A parallel, distributed processing approach. In J. L. Elman & D. E. Rumelhart, (Eds.), *Advances in Connectionist Theory: Speech*. Hillsdale, NJ: Lawrence Erlbaum Associates.

Jordan, M. I., & Rosenbaum, D. A. (1989). Action. In M. I. Posner (Ed.), *Foundations of Cognitive Science*. Cambridge, MA: MIT Press.

Jordan, M. I., & Rumelhart, D. E. (1989). *Supervised learning with a distal teacher*. Manuscript submitted for publication.

Kawato, M. (1989). Computational schemes and neural network models for formation and control of multijoint arm trajectory. *Proceedings of the NSF Workshop on Application of Neural Networks to Robotics and Control*. Cambridge, MA: MIT Press.

Kawato, M., Furukawa, K., & Suzuki, R. (1987). A hierarchical neural-network model for control and learning of voluntary movement. *Biological Cybernetics, 57*, 169–185.

Kelso, J. A. S., & Holt, K. G. (1980). Exploring a vibratory systems analysis of human movement production. *Journal of Neurophysiology, 43*, 1183–1196.

Kelso, J. A. S., Tuller, B., Vatikiotis-Bateson, E., & Fowler, C. A. (1984). Functionally specific articulatory cooperation following jaw perturbations during speech: Evidence for coordinative structures. *Journal of Experimental Psychology: Human Perception and Performance, 10*, 812–832.

Kent, R. D., & Minifie, F. D. (1977). Coarticulation in recent speech production models. *Journal of Phonetics, 5*, 115–117.

Kiparsky, P. (1975). Comments on the role of phonology in language. In J. F. Kavanagh & J. E. Cutting, (Eds.), *The Role of Speech in Language*. Cambridge, MA: MIT Press.

Koditschek, D. E. (1984). Natural control of robot arms. *IEEE Proceedings of the 23rd Conference on Decision and Control* (pp. 733–735).

Kugler, P. N., Kelso, J. A. S., & Turvey, M. T. (1980). On the concept of coordinative structures as dissipative structures: I. Theoretical lines of convergence. In G. E. Stelmach & J. Requin (Eds.), *Tutorials in motor behavior* (pp. 49–70). Amsterdam: North-Holland.

LeCun, Y. (1985). A learning scheme for asymmetric threshold networks. *Proceedings of Cognitiva 85*. Paris, France.

Lieberman, P. (1975). *On the origins of language: An introduction to the evolution of speech*. New York: Macmillan.

Lindblom, B. (1983). Economy of speech gestures. In P. F. MacNeilage (Ed.), *The Production of Speech* (pp. 217–245). New York: Springer-Verlag.

Lindblom, B. (1986). On the origin and purpose of discreteness and invariance in sound patterns. In J. Perkell & D. H. Klatt (Eds.), *Invariance and Variability in Speech Processes*. Hillsdale, NJ: Lawrence Erlbaum Associates.

Lindblom, B., Lubker, J., & Gay, T. (1979). Format frequencies of some fixed-mandible vowels and a model of speech motor programming by predictive simulation. *Journal of Phonetics, 7*, 147–161.

Martinet, A. (1968). Phonetics and linguistic evolution. In B. Malmberg (Ed.), *Manual of Phoentics*. Amsterdam: North-Holland.

Meyer, D. E., Abrahms, R. A., Kornblum, S., Wright, C. E., & Smith, J. E. K. (1988). Optimality in human motor performance: Ideal control of rapid aimed movements. *Psychological Review, 95*, 340–370.

Mussa-Ivalid, F. A., McIntyre, J., & Bizzi, E. (1988). Theoretical and experimental perspectives on arm trajectory formation: A distributed model for motor redundancy. In. E. Clementi & S. Chin (Eds.), *Biological and Artificial Intelligence Systems* (pp. 563–577). Escom.

Nelson, W. (1983). Physical principles for economies of skilled movements. *Biological Cybernetics, 46*, 135–147.

Newell, K. M. (1986). Constraints on the development of coordination. In M. G. Wade & H. T. A. Whiting (Eds.), *Motor skill acquisition in children: Aspects of coordination and control* (pp. 341–360). Amsterdam: Martinus Nijhoff.

Newell, K. M., & Carlton, M. J. (1987). Augmented information and the acquisition of isometric tasks. *Journal of Motor Behavior, 19*, 4–12.

Öhman, S. E. G. (1966). Coarticulation in VCV utterances: Spectrographic measurements. *Journal of the Acoustical Society of America, 39*, 151–168.

Parker, D. (1985). *Learning-logic* (Tech. Rep. TR-47). Cambridge, MA: MIT Sloan School of Management.

Pearlmutter, B. A. (1988). Learning state space trajectories in recurrent neural networks. In D. S. Touretzky, G. E. Hinton, & T. J. Sejnowski, (Eds.), *Proceedings of the 1988 Connectionist Models Summer School*. San Mateo, CA: Morgan Kaufmann.

Pellionisz, A., & Llinas, R. (1985). Tensor network theory of the metaorganization of functional geometries in the central nervous system. *Neuroscience, 16*, 245–274.

Pew, R. W. (1974). Human perceptual-motor performance. In B. H. Kantowitz (Ed.), *Human Information Processing: Tutorials in Performance and Cognition* (pp. 1–39). Hillsdale, NJ: Lawrence Erlbaum Associates.

Polit, A., & Bizzi, E. (1979). Processes controlling arm movements in monkeys. *Science, 201*, 1235–1237.

Rosenbaum, D. A., Gordon, A. M., Stillings, N. A., & Feinstein, M. H. (1987). Stimulus-response compatibility in the programming of speech. *Memory and Cognition, 15*, 217–224.

Rumelhart, D. E., Hinton, G. E., Williams, R. J. (1986). Learning internal representations by error propagation. In D. E. Rumelhart & J. L. McClelland (Eds.), *Parallel distributed processing* (Volume 1, pp. 318–363). Cambridge, MA: MIT Press.

Salisbury, J. K. (1980). Active stiffness control of a manipulator in Cartesian coordinates. *Proceedings of the 19th IEEE Conference on Decision and Control* (pp. 95–100).

Saltzman, E. L. (1979). Levels of sensorimotor representation. *Journal of Mathematical Psychology, 20*, 91–163.

Saltzman, E. L. (1986). Task-dynamic coordination of the speech articulators: A preliminary model. In H. Heuer & C. Fromm (Eds.), *Generation and modulation of action patterns* (Experimental Brain Research Series 15, pp. 129–144). New York: Springer-Verlag.

Saltzman, E. L., & Kelso, J. A. S. (1987). Skilled actions: A task-dynamic approach. *Psychological Review, 94*, 84–106.

Saltzman, E. L., & Munhall, K. G. (1989). *A dynamical approach to gestural patterning in*

*speech production*. Manuscript submitted for publication.
Schmidt, R. A. (1975). A schema theory of discrete motor skill learning. *Psychological Review, 82*, 225–260.
Schöner, G., & Kelso, J. A. S. (1988). Dynamic pattern generation in behavioral and neural systems. *Science, 239*, 1513–1520.
Shapiro, D. C., & Schmidt, R. A. (1982). The schema theory: Recent evidence and developmental implications. In J. A. S. Kelso & J. E. Clark (Eds.), *The development of movement control and coordination* (pp. 113–150). New York: Wiley.
Stelmach, G. E., Mullins, P. A., & Teulings, H. -L. (1984). Motor programming and temporal patterns in handwriting. In J. Gibbon & L. Allan (Eds.), *Timing and Time Perception*. New York: The New York Academy of Sciences.
Uno, Y., Kawato, M., & Suzuki, R. (1987). *Formation of optimum trajectory in control of arm movement-minimum torque-change model* (Japan IEICE Tech. Rep. MBE 86-79). Osaka, Japan.
Werbos, P. (1974). *Beyond regression: New tools for prediction and analysis in the behavioral sciences*. Unpublished doctoral dissertation, Harvard University.
Widrow, B., & Stearns, S. D. (1985). *Adaptive signal processing*. Englewood Cliffs, NJ: Prentice-Hall.
Williams, R. J., & Zipser, D. (1988). *A learning algorithm for continually running fully recurrent neural networks*. (Tech. Rep. ICS 8805). University of California, San Diego, Institute for Cognitive Science.

# 30 Gesture Learning and Apraxia

Pietro Faglioni
*Clinica Neurologica Università di Modena*

Anna Basso
*Clinica Neurologica Università di Milano*

Claudio Botti
*Centro Interdipartimentale di Calcolo Automatico e Informatica Applicata Università di Modena*

Salvatore Aglioti
*Istituto Fisiologia Università di Verona*

Cristina Saetti
*Clinica Neurologica Università di Modena*

ABSTRACT

Apraxia has repeatedly been considered a pathology of the learned gesture, but there are not many studies that document a defective capacity of motor learning in apraxic patients. The functions of encoding, retrieval, and storage have seldom been separately studied, and the evidence for their failure collected up to now is scarce and only indirect.

Our purpose was to identify whether and which of these functions are impaired in apraxic patients. We used a two-stage learning model. The parameters of the learning function were estimated by the method of maximum likelihood in a Markov chain where the learning process is viewed as a series of changes of state governed by transition probabilities, from a state of complete ignorance to that of complete memorization.

Three experimental groups were studied: control subject without damage to the central nervous system, right-hemisphere-damaged patients, and left-hemisphere-damaged patients, subdivided for presence/absence of apraxia. Subjects had to learn an 8-item list of gestures (4 symbolic and 4 nonsymbolic), and a list of 4 paired-associate gestures (2 symbolic and 2 nonsymbolic). The selective reminding technique of Buschke-Fuld was used. Results indicate that: (a) Right-hemisphere-damaged patients do not perform differently from normal controls; (b) symbolic and nonsymbolic gestures are not differentially learned; (c) failure of apraxic patients in learning gestures is specifically related to failure in encoding and storage of the engram of the gesture.

## INTRODUCTION

Apraxia is the inability to make the gesture that the subject intends to accomplish despite normal motor, sensory, and intellectual functions. For example, instead of correctly showing how to use a saw when requested, the apraxic patient either makes some other movement appropriate to a different tool or performs a series of hesitant and amorphous movements.

Reflex and automatic gestures are always spared in apraxia that manifests itself in the learning of gestures or when the patient performs previously acquired gestures in a new or unusual context. In the first case, the difficulty may source from encoding or storage, and in the second case, from retrieving. Few studies document a defective capacity of motor learning in left-brain-damaged patients, or at least not in apraxic patients. Some studies have considered the learning of simple movements (Heilman, Schwartz, & Geschwind, 1975; Wyke, 1971), others the learning of movements in a fixed sequence of a few elements (De Renzi, Faglioni, Lodesani, & Vecchi, 1983; Jason, 1983a, 1983b, 1985; Kimura, 1977; Roy, 1981), and others the learning of series of gestures (Rothi & Heilman, 1984).

Learning of simple movements and of movements in a fixed sequence has been studied by asking the subject to repeat the same movement or the same short sequence under the same conditions until it might be presumed that the gesture had become automatic. Learning of series of gestures independent of their order and whether or not they were previously practiced, appears to engage at maximum the 2 fundamental functions with which we are concerned: the encoding or creation of a memory trace or engram, and its storage in long term memory (Melton, 1963). The classic interpretation of apraxia is ambiguous about relating the disorder to the failure of one function rather than to the other. Apraxia as a pathology of the learned gesture (Geschwind, 1975) has been considered to be either an encoding (Liepmann, 1920), or a storage defect (Heilman, Schwartz, & Geschwind, 1975).

Our aim was to investigate which of the mnesic functions—encoding, storage, and recall—are specifically impaired in apraxic patients when consciously learning gestures. To do that, learning of gestures was studied by a model-based approach, the advantages of which are stressed by Brainerd (1985). The two-stage learning model we adopted for the learning of lists of gestures is based on a Markov chain (Taylor & Karlin, 1984) with four states, and initial and transition probabilities defined as functions of the parameters that refer to the basic psychological abilities of learning, that is, encoding, storage, and recall before storage has taken place (Kraemer, Peaboby, Tinklenberg, & Yesavage, 1983).

We studied learning of paired associated gestures and learning of a list of single gestures. Paired-associated learning was the first paradigm used in the study of learning, but it has also been suggested that list learning reveals the learning process in a simpler and more straightforward way (Greeno, 1968).

## MATERIAL AND METHODS

### Subjects

Seventy-four patients and 21 control subjects participated in this study. All patients had vascular lesions: in the right hemisphere in 12 of the patients and in the left hemisphere in the remaining 62. Site of lesion was ascertained by neurological examination and CT scan. Patients with transient ischemic attacks or epilepsy and patients unable to collaborate in the neuropsychological tasks wère excluded from the study. Control subjects were inpatients without lesions of the central nervous system. Only patients able to correctly reproduce one-by-one the gestures of the learning lists were included in the study. Education ranged from 3 to 17 years of schooling and ages from 20 to 75 years.

### Tests

Subjects were submitted to the following preliminary tests:

1. Handedness (for all subjects) was ascertained by a slightly modified version of the Edinburgh Inventory (Oldfield, 1971). A Laterality Quotient was calculated for each subject according to the formula $(R - L/R + L) \times 100$, where R is the number of acts performed with the right hand and L with the left hand. A strong right hander has a Laterality Quotient of $+100$, a strong left hander of $-100$.

2. A standard neurological examination was given to brain-damaged patients. Severity of motor, somato-sensory, and visual field impairment was evaluated. For each modality, the scores ranged from 0 to 6 (0 to 3 for upper limb or upper quadrant, and 0 to 3 for the lower limb or lower quadrant), in which 0 corresponds to no impairment and 6 to maximum impairment. Maximum neurological severity score = 18.

3. Length of illness (for brain-damaged patients) was computed in days.

4. Token Test (for all subjects): the shortened, 36-item version of the Token Test of De Renzi, and Faglioni (1978).

5. Ideomotor Apraxia (for all subjects), evaluated by a 24-item test devised by De Renzi, Motti, and Nichelli (1980). The Examiner (E) performed the movement slowly in front of the Subject (S), who was asked to reproduce it immediately with the arm ipsilateral to the lesion. Half the control Ss performed the test with the right hand and half with the left hand. If an item was not correctly reproduced after the first presentation, a second trial was given and, when necessary, a third. Correct performance on the first trial was scored 3, on the second trial 2, on the third trial 1, and not correct on three trials 0. Maximum score: 72.

## Learning Tasks

Two gesture-learning tasks were used: a single-gesture list and a paired-associate gesture list.

*Single gestures list.* The list was comprised of 8 gestures, different from those of the Ideomotor Apraxia test. Four gestures were symbolic and 4 were not. The gestures were the following: (a) money; (b) place finger tips on chin; (c) tap palm on the table; (d) bye-bye; (e) tap fist on opposite shoulder; (f) crazy; (g) pretend to use a saw; (h) pinch thumb and index finger together, other fingers flexed.

Training: The E first performed each gesture slowly in front of the S using the hand opposite to that to be used by S who was asked to imitate the gesture immediately after presentation, with the arm ipsilateral to the lesion. The E demonstrated each item until the S imitated it correctly twice consecutively without approximation and hesitation. Criterion of 2 consecutive correct performances makes it extremely probable that the S can easily correctly imitate the single gesture. If this is the case, the learning task only requires intentional evocation of the gestures in order to link them in a unique sequence. If the S could not correctly imitate the items after 10 demonstrations that S was excluded from the study. Three left-hemisphere-damaged patients were excluded in this way. For each gesture the mean number of demonstrations ranged from 1.3 for the control group to 2.5 for the apraxic group. The examination was then conducted according to the selective reminding technique (Buschke & Fuld, 1974). Criteria for correct performance of each gesture were strict. The gesture had to be integrally reproduced without hesitation and in the same spatial orientation demonstrated by the E.

Trial 1: E demonstrated the eight gestures in a fixed order, at a rate of one gesture every 2 sec. S had to reproduce all the gestures remembered, without regard to the order. Trials 2–10: E demonstrated only the gestures that S had not reproduced in the preceding trial. S had to reproduce all the gestures remembered, again without regard to the order. The test was interrupted after 3 consecutive correct reproductions of the list or after a maximum of 10 trials.

E noted the gestures correctly and incorrectly reproduced for every trial.

*Paired-associate gestures list.* The list comprised 4 paired-associated gestures—2 symbolic and 2 nonsymbolic—different from those used in the previous test.
The gestures were the following:

1. A: "stop" gesture: B: military salute.
2. A: stretched thumb, flexed fingers with knuckles on the table; B: fingers extended with ulnar side resting on the table.
3. A: drinking gesture; B: silence gesture.
4. A: thumb between index and middle finger, fist on the table, B: index and middle fingers extended and wide apart, touching the face, other fingers flexed.

Training: The E, sitting in front of S, first performed each paired-associated gesture slowly with the hand opposite to that to be used by S who was asked to repeat it immediately after presentation with the hand ipsilateral to the lesion. E demonstrated each paired-associate gesture until S correctly reproduced it twice consecutively without approximation and hesitation. If the S could not correctly imitate the paired associate after 10 demonstrations that S was excluded from the study. Ten (1 right-hemisphere-damaged and 9 left-hemisphere-damaged) patients were excluded. For each paired-associate gesture the mean number of demonstrations ranged from 1.6 for the control group to 6.1 for the apraxic group. The examination was then conducted according to the selective reminding technique (Buschke & Fuld, 1974). Criteria for correct performance were the same as for the single-gesture list.

Trials 1–10: E demonstrated the first gesture of the first paired associate and S was asked to reproduce it and execute the second gesture. If S failed, E presented both gestures of the paired associate and S was again asked to reproduce it. This was done until correct execution by S. Errors and correct responses were noted. Subsequently, E presented the first gesture of the second paired associate and S had to execute both gestures of the paired associate. This procedure was repeated for all 4 paired associates. The test was interrupted after 3 consecutive correct executions of the paired-associate list or after a maximum of 10 trials.

For the learning tasks the score for each of the 8 single gestures and the 4 paired associates was based on the sequence of failures and successes in the 10 trials. Half of the Ss were given the learning tasks in the reported order and half in the reverse order. The learning tasks were always given first, followed by the preliminary tests in the same order in which they have been described.

## ANALYSES AND RESULTS

### Manual Preference

Laterality Quotients were +20, +40, +60, and +90 for 4 controls, +90 for 3 left-brain-damaged apraxic patients and +80 for 1 left-brain-damaged borderline patient. All others were +100.

### Ideomotor Apraxia Test

The Apraxia Test was used to identify those left-brain-damaged patients who were "reliably nonapraxic" and "reliably apraxic." Performance was considered to be apraxic or nonapraxic in relation to the results for 150 control Ss comparable to our experimental groups for age and education (Faglioni & Scarpa, in press). The multiple regression of the Apraxia score on age and education was calculated. Both had significant effects on apraxia performance ($F=8.0642$ with 1 and 147 *df*, $p < .005$ 1-tailed; $F = 3.5055$ with 1 and 147 *df*, $p < .03$ 1-tailed). Raw apraxia scores were adjusted for age and education according to the following equation: corrected score = raw score − 1.948 + 0.058 years of age − 0.180 years of schooling.

Adjusted scores were used to compute one-sided, nonparametric tolerance intervals with 98% content and 95% confidence (Ackermann, 1985). The outer and inner tolerance limits were 53.0 and 63.3. This means that, with 95% confidence, Ss who score below 53 have less than 2% probability of belonging to the normal population, whereas those who score 63.3 or better have less than 2% probability of not belonging to the normal population. Performance in the Apraxia test was considered apraxic when lower than 53.0, nonapraxic when equal to or higher than 63.3 and borderline when in-between. Among the left-brain-damaged patients, 26 were nonapraxic, 20 borderline, and 16 apraxic. Control subjects and right-brain-damaged patients always scored 63 or more.

For all subsequent analyses, Ss were divided into 5 subgroups: G1 = normal controls ($n=21$); G2 = right-brain-damaged patients ($n = 12$); G3 = left-brain-damaged nonapraxic patients ($n=26$); G4 = left-brain-damaged borderline patients ($n=20$) and G5 = left-brain-damaged apraxic patients ($n=16$).

## Age, Education, Length of Illness and Neurological Severity

Table 30.1 lists the means for age, education, length of illness, and neurological severity for the 5 groups (4 for length of illness, and neurological severity). Analysis of variance showed no significant differences among the groups in age ($F<1$), education ($F=1.078$ with 4 and 90 *df*, $p=ns$) or length of illness ($F=1.772$, with 3 and 70 *df*, $p=ns$). Neurological severity was significantly different ($F=3.766$ with 3 and 70 *df*, $p<.05$). Multiple comparisons by Scheffé's (1953) method showed that apraxic patients were neurologically more impaired than left-brain-damaged nonapraxic ($F=2.8523$, with 3 and 70 *df*, $p<.05$), but did not differ from the other 2 brain-damaged groups ($F<1$).

## Token Test

The means for the 5 groups are shown in Table 30.1. Analysis of variance demonstrated significant differences between groups ($F=19.914$, with 4 and 90 *df*, $p<.0001$). Multiple comparisons with Scheffè's (1953) method showed that controls and right-brain-damaged patients have significantly higher scores than left-brain-damaged nonapraxic and borderline patients ($F=8.3040$ with 4 and 90 *df*, $p<.0001$), and that these last two groups have significantly higher scores than left-brain-damaged apraxic patients ($F=4.5460$, with 4 and 90 df, $p<.005$).

## Learning Tasks

The whole learning process, with its 4 states and the probabilities of transition between them is represented in Fig. 30.1. Until the S has created the engram, that is, before encoding, S gives only incorrect responses. The engram of the gesture is defined as being in the *out-of-memory* state (OM). In the following stage, after the engram has been encoded, the S sometimes produces incorrect and sometimes correct responses. The engram is defined as being in the *inconsistent memory* state (IM), or more precisely in the *inconsistent memory with erroneous responses* state (IME) in the first case or in the *inconsistent memory with correct responses* state (IMC) in the second case. At the terminal stage, the engram becomes permanently available and the S always gives correct responses. The engram is then defined as being in the *consistent memory* state (CM), and the learning process has been completed. The learning process can be viewed as a series of jumps from one state to another, governed by transition probabilities, which are functions of the ability to encode, store, and recall the engram before storage. The probability of leaving OM measures the encoding ability ($e$). The probability of entering CM measures the ability to store ($s$). The

TABLE 30.1
Means of Age, Education and Preliminary Tests and Comparison Between the 5 Experimental Groups

| Group | G1 (21) | G2 (12)* | Means G3 (26) | G4 (20) | G5 (16) | Analysis of variance df | F | P |
|---|---|---|---|---|---|---|---|---|
| age (years) | 59 | 59 | 57 | 56 | 61 | 4/90 | 1 | ns |
| education (years) | 6 | 7 | 8 | 9 | 7 | 4/90 | 1.078 | ns |
| neurol.severity | / | 7 | 4 | 5 | 7 | 3/70 | 3.766 | <0.05 |
| length of illness (days) | / | 153 | 483 | 178 | 390 | 3/70 | 1.772 | ns |
| Token Test score | 31.7 | 32.9 | 24.5 | 20.4 | 13.4 | 4/90 | 19.914 | <0.0001 |

G1 = normal controls; G2 = Right-damaged patients; G3 = Left-damaged nonapraxic patients; G4 = Left-damaged borderline patients; G5 = Left-damaged apraxic patients.

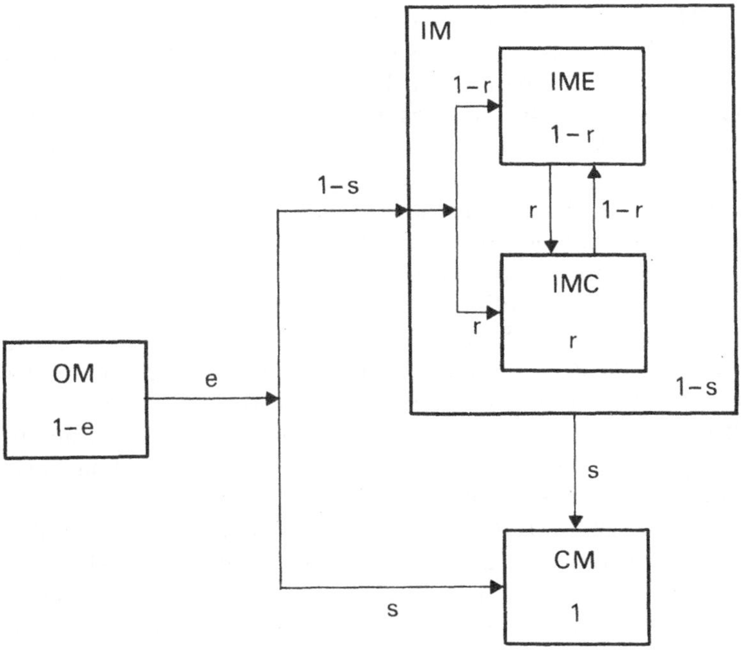


FIG. 30.1. Diagram of the 4 states Markov chain representing the learning process.
OM = out of memory state
IM = inconsistent memory state
IME = inconsistent memory with erroneous responses state
IMC = inconsistent memory with correct responses state
CM = consistent memory state
e = encoding
s = storage
r = recall

probability of recalling the engram in IM measures the ability to recall before storage has taken place ($r$).
The model contains some assumptions that need to be justified.

1. Before the first presentation of the list of gestures, the engram is in the OM state, that is, the process starts from OM.
2. Once encoding has taken place, the engram is considered to be definitively encoded for the duration of the experiment. Back transitions from IM or CM to OM are ruled out and have zero probability.
3. Once storage has taken place, it cannot be lost. For this reason the transition from CM to IM is impossible and has zero probability whereas the permanence in CM has probability 1.

4. The function of storage is unique, whether the process starts from OM, or from IME and IMC. Insofar as storage is concerned, presentation of the model on the part of the E or its execution on the part of the S without presentation by the E, as happens in the selective reminding technique of Buschke-Fuld (1974), are considered to be equivalent. This assumption is, however, not always accepted. For the learning of word lists and also for other learning paradigms, it is possible that storage only takes place after presentation of the model and not after execution by the S, that is, only starting from OM or IME but not from IMC (Greeno & Scandura, 1966; Halff, 1977; Polson, 1972; Polson, Restle, & Polson, 1965; Restle, 1962, 1964, 1965). But it is also possible that storage occurs principally after practice—preferably or only after correct recall—that is, preferably or only starting from IMC but not IME or OM (Adams & Bray, 1970; Izawa, 1971). The model that postulates a unique storage function has, however, strong experimental support (Lachman & Laughery, 1968; Tulving, 1967) and is provisionally adopted here because of its simplicity and its use in clinical research with verbal stimuli (Kraemer et al., 1983).

5. The function of recall-before-storage is also considered to be unique. That is, recall has the same probability of success when leaving OM or IME as from IMC. This assumption has generally been shown to be valid, although exceptions were found in some studies of paired-associate learning (Desrochers, & Howe, 1981; Brainerd, Howe & Desrochers, 1980).

6. The functions of encoding, storage and recall-before-storage are measured on the same rational scale that goes from 0 to 1, and are immediately comparable.

For each single and each paired-associate gesture we recorded the sequence of failures and successes in the 10 trials. The probability of occurrence of a sequence was expressed as a function of the parameters *e*, *s*, and *r* and calculated by means of the algorithm described by Millward (1969) for sequential statistics. The probability of the co-occurrence of multiple sequences (of the 8 single gestures for each subject or for the whole group of Ss, for example) was calculated as the product of the single sequential probabilities. This was the likelihood function.

The value of the parameters was estimated by maximizing the likelihood function by Chandler's (1965) direct search optimization routine. In all the cases considered, the likelihood function always had a single global maximum, making it possible to identify all the parameters (Polson & Huizinga, 1974). Table 30.2 gives the estimates of the parameters separately for each group and for single and paired gestures; no distinction is made between symbolic and nonsymbolic gestures. The statistical hypotheses were tested by the likelihood ratio method. We tested the hypotheses hierarchically from the more general to the more specific, according to Brainerd, Howe

TABLE 30.2
Estimates of the 3 Parameters in the 5 Experimental Groups

| | | G1 | G2 | G3 | G4 | G5 |
|---|---|---|---|---|---|---|
| | e | .75 | .78 | .95 | .39 | .26 |
| single gest. | s | .28 | .22 | 11 | .06 | .02 |
| | r | .63 | .66 | .48 | .51 | .42 |
| | e | 1.00 | 1.00 | .41 | .26 | .13 |
| paired gest. | s | .34 | .32 | .26 | .19 | .08 |
| | r | .42 | .32 | .46 | .62 | .44 |

e = encoding; s = storage; r = recall
G1-G5 = see legend Table 30.1

and Desrochers' (1982) suggestions. Whenever the null hypothesis was rejected, we proceeded to the next stage.

### Stage 1: Goodness-of-fit

Goodness-of-fit tests were first performed to ascertain whether the model with 4 states and 3 parameters we adopted adequately and efficiently accounted for our data (Theios, Leonard, & Brelsford, 1977).

Following Greeno (1968) we took into consideration 2 questions: (a) are 2 stages and 3 parameters *necessary*, or would a 1-stage and 2-parameter model be acceptable? (b) are 2 stages and 3 parameters *sufficient*, or are more stages and parameters required? Sufficiency of a 2-stage and 3-parameter model implies that the learning process shares the Markovian property that escape from states OM and IM is all-or-none and not gradual.

1. As far as the *necessity* question is concerned, we tested the 2-stage model against 2 single-stage models, 1 without the OM state and 1 without the CM state. These single-stage models can be viewed as 2-stage models where $e=1$ or $s=0$ respectively, and these are the hypotheses we tested. The tests were performed separately for the 2 learning tasks and simultaneously for the 5 experimental groups.

The OM necessity test yielded significant chi square values for both tasks (single gestures: chi square = 54.25, $df$ = 5, $p<.0005$; paired gestures: chi square = 115.23, $df$ = 5, $p<.0005$), thus confirming that the process starts from OM, which means that the engrams were not always encoded before the learning task started. The CM necessity test also yielded significant chi square values for both learning tasks (single gestures: chi square = 542.46, $df$ = 5, $p<.0005$; paired gestures; chi square = 381.57,

$df = 5$, $p < .0005$), thus confirming that the process can end in CM, that is, that the engrams are sometimes consistently stored.

2. As far as the *sufficiency* question is concerned, we considered 4 statistics: (a) the trial number of the first success; (b) the number of errors between the first success and the last error; (c) the trial number of the last error; and (d) the total number of errors.

The statistic (a) focuses on OM state and *e* parameter. The statistic (b) focuses on IM state and *r* parameter. The statistics (c) and (d) provide information about the model's capacity to account for the learning process as a whole.

The observed and the expected distribution functions—this last one being based on the model and the parameter estimates—have been compared by means of the Kolmogorov-Smirnov test for goodness-of-fit (Massey, 1951). Only 4 of the 40 tests performed reached the .05 significance level. They refer to the statistics (a) and (b) but never to (c) and (d). This result allows us to confidently accept the 2-stage model with 3 parameters.

### Stage 2: Comparison Between Tasks within Groups

Performance on single and paired symbolic and nonsymbolic gestures was compared within the 5 experimental groups.

1. The four parameter triplets ($e$, $s$, $r$) significantly ($p < .05$) differ in all groups (the chi square values with 9 *df* were 16.93, 21.80, 49.71, 51.38, 26.55) and further analyses were done to identify the source of the difference.
2. The parameters relative to symbolic and nonsymbolic single gestures never significantly differ in the 5 groups, even when checked for type 1 single hypothesis error rate (Miller, 1966) and the same was true for symbolic and nonsymbolic paired gestures. Therefore, symbolic and nonsymbolic gestures have been combined hereafter.
3. The parameters relative to single and paired gestures were finally compared within each group. Five hypotheses with 3 *df* each were tested. A family-wise .05, type 1 error probability corresponds to a $.05/5 = .01$ hypothesis-wise significance level (Bonferroni in Miller, 1966). Except in the control group, our chi square values (5.11, 12.28, 33.89, 46.33, 28.91) are beyond the tabulated chi square (.01/3) value $= 11.34$.
4. Within-groups comparisons for single parameters between single and paired gestures revealed significant differences for *e* and *s* in all brain-damaged groups, and for *r* in right-brain-damaged patients only.

Single and paired gestures were, therefore, hereafter separately analyzed.

### Stage 3: Comparisons between Groups within Tasks

The analyses of Stage 3 were aimed at testing separately for single and paired gestures if and how the 3 parameters, *e*, *s*, and *r*, differ among the 5 experimental groups. Because all parameters would be expected to worsen from Group 1 to Group 5, hereafter we adopted a one-tailed .05 significance level (i.e., .10 two-tailed).

1. The 3 parameters considered together were compared between groups, separately for single and paired gestures. Two hypotheses with 12 *df* each were tested. A family-wise .10, type-1 error probability corresponds to a .10/2 = .05 hypothesis-wise significance level (Bonferroni in Miller, 1966). Both our chi square values (539.75, 260.58) are far beyond the tabulated value of 21.03. We can therefore accept that the parameter triplets differ between the 5 Groups for both single and paired gestures.

2. Tests to identify which of the 3 parameters differ among the groups were carried out separately for single and paired gestures. Six hypotheses with 4 *df* each were tested. A family-wise, .10 type-1 error probability corresponds to a .10/6 =.016667 hypothesis-wise significance level (Bonferroni in Miller, 1966).
All the parameters, except *r* for paired gestures, significantly differ among the groups. Their chi square values (Table 30.3) are far beyond the tabulated value of 12.09.

3. Finally we tested whether each pair of Groups differ for *e* or *s* or *r*, separately for single and paired gestures. Sixty hypotheses with 1 *df* each were tested. A family-wise, .10 type-1 error probability corresponds to a .10/60 = .016667 hypothesis-wise significance level (Bonferroni, in Miller, 1966). The chi square values are reported in Table 30.4. Significant comparisons are marked by an asterisk. As already found at Stage 3b, param-

TABLE 30.3
Between Groups Comparisions of the 3 Parameters for Single and Paired Gestures

| *Parameter* | *Single Gesture* | | Paired Gestures | |
|---|---|---|---|---|
| | *df* | $\chi^2$ | *df* | $\chi^2$ |
| encoding | 4 | 41.469 | 4 | 122.132 |
| storage | 4 | 166.849 | 4 | 51.752 |
| recall | 4 | 31.189 | 4 | 11.192 |

$\chi^2(0.016667/4) = 12.094$

TABLE 30.4
Between Groups Pair-wise Comparisions of the 3 Parameters for Single and Paired Gestures

| *Single Gestures* | | *G2* | *G3* | *G4* | *G5* |
|---|---|---|---|---|---|
| G1 | e | 0.039 | 1.598 | 17.818* | 19.838* |
| | s | 2.589 | 42.198* | 72.228* | 125.718* |
| | r | 0.089 | 0 | 2.708 | 12.360* |
| G2 | e | | 2.271 | 18.361* | 20.001* |
| | s | | 17.031* | 38.161* | 62.191* |
| | r | | 1.571 | 8.201 | 22.181* |
| G3 | e | | | 20.790* | 20.570* |
| | s | | | 11.490* | 56.440* |
| | r | | | 1.420 | 4.350 |
| G4 | e | | | | 9.250 |
| | s | | | | 42.680* |
| | r | | | | 9.470 |
| Paired gestures | | G2 | G3 | G4 | G5 |
| G1 | e | 0 | 22.956* | 61.589* | 54.175* |
| | s | 0.110 | 3.616 | 11.619* | 42.770* |
| | r | 0.817 | 0.176 | 2.739 | 0.055 |
| G2 | e | | 22.821* | 25.657* | 53.909* |
| | s | | 1.638 | 7.062 | 30.402* |
| | r | | 1.319 | 4.136 | 1.102 |
| G3 | e | | | 10.855* | 41.379* |
| | s | | | 2.985 | 17.416* |
| | r | | | 5.055 | 0.076 |
| G4 | e | | | | 12.779* |
| | s | | | | 10.189* |
| | r | | | | 7.659 |

$\chi^2(0.001667/1) = 9.885$
e = encoding; s = storage; r = recall; G1–G5 = see legend Table 30.1

eter *r* for paired gestures does not differ between the groups. The results of the analyses at Stage 3c are shown in Fig. 30.2, in which the lines connect the Groups that do not differ significantly.

As can be seen, normal controls and right-brain-damaged patients do not differ in any of the parameters. Left-brain-damaged apraxic and borderline patients are always impaired in encoding and storage: The apraxic patients generally more so than borderline patients, except for single gesture encoding where the difference between the two groups is not significant. Left-brain-damaged nonapraxic patients are significantly impaired in encoding of paired (and not of single) gestures and in storage of single (and not paired) gestures. In left-brain-damaged apraxic and borderline patients, recall-before-storage of the engram for the gesture is less impaired than encoding and storage. Apraxic patients are, however, impaired in recall of single gestures as compared with right-brain-damaged patients and control subjects.

## DISCUSSION

The necessity and sufficiency analyses prove that the gesture-learning process does in fact start from OM, that the engrams can be consistently stored in CM, and, finally, that gesture learning can be viewed as an all-or-none process and need not be gradual. The main result of our research is failure of left-brain-damaged apraxic and, to a lesser degree, of borderline patients in encoding and storage: Failure in recall is mild and not always present (see Fig. 30.2). For many reasons, failure of apraxic patient in encoding and storage cannot be accounted for by a specific impairment due to age, lack of education, or brain damage. First, the 5 experimental groups were of similar age and education. Second, right-brain-damaged patients, who did not differ from apraxic patients with respect to neurological severity or length of illness, did not perform worse than control Ss. Third, apraxic patients' failure was not general: They failed in encoding and storage, but not, or inconsistently, in recall. Encoding and storage of gestures, as well as apraxia, depend crucially on the left hemisphere, as already thought by Liepmann (1920), and in accord with Jason's (1983a) finding that memory for gestures is specifically impaired in left-brain-damaged patients.

Aphasia in apraxic patients raises a more serious problem. It is possible that failure in the gesture-learning tasks is not specifically related to apraxia, but to a more general impairment consequent to aphasia. However, to be taken as evidence against considering failure of apraxic patients to be due to aphasia is the lack of difference between nonsymbolic and symbolic gestures. If failure in the gesture-learning task were to be explained by impairment in verbalization, symbolic gestures that are easily verbalized would be expected to be more impaired than nonsymbolic gestures that cannot be precisely verbalized or can only be verbalized with difficulty. Our result is in agreement with the repeated observation that nonsymbolic and symbolic gestures are not differently susceptible to apraxia (De Renzi et al., 1980; Goodglass & Kaplan, 1963; Kimura & Archibald, 1974; Lehmkuhl, Poeck, & Willmes, 1983; Pieczuro & Vignolo, 1967), and supports interpretation of apraxia as a primary "motor" disorder (Geschwind, 1975; Goodglass & Kaplan, 1963; Kimura, 1977, 1982; Kimura & Archibald, 1974; Lehmkuhl et al, 1983; Liepmann, 1920).

Our data partly confirm and expand the conclusions of earlier studies that distinguished encoding and storage in gesture learning. Heilman, Schwartz, and Geschwind (1975) studied motor learning in 9 aphasic aprax-

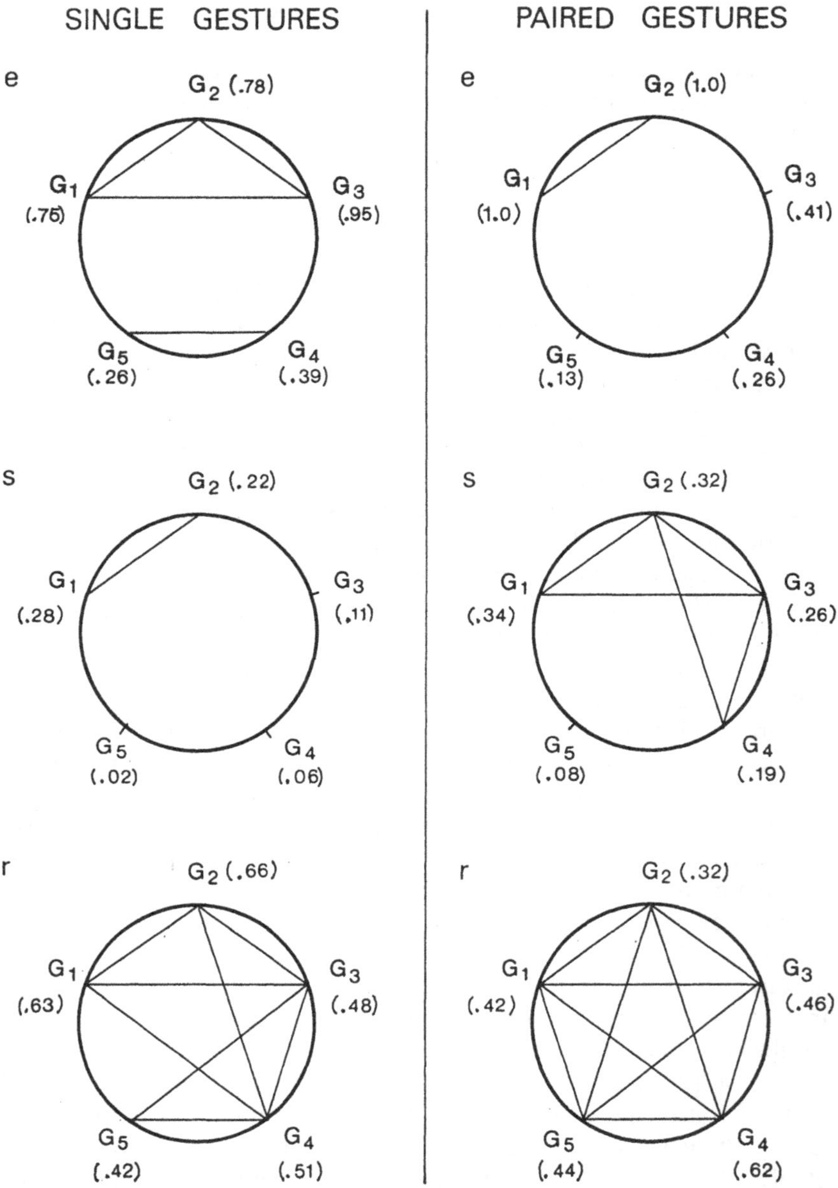


FIG. 30.2. Between groups pair-wise comparisons of the 3 parameters for single and paired gestures. The lines connect the Groups that do not differ significantly. Estimates of the 3 parameters are reported in parentheses.
G = Group
e = encoding
s = storage
r = recall

ics, with 8 aphasic nonapraxics as controls. Patients were given 6 trials on a rotary pursuit apparatus. Apraxic and nonapraxic patients showed constant increments in performance from trial 1 to trial 5. Fifteen minutes later a 6th trial was given. Although nonapraxic patients still improved their performance, that of the apraxic patients deteriorated and did not differ from their performance on trial 1, as if they had lost the newly acquired motor ability. Rothi and Heilman (1984) used four 12-single-gesture lists to study acquisition and retention of gestures by 7 apraxic patients, 6 left-brain-damaged nonapraxic patients and 6 control Ss. Apraxic patients gave a poorer performance than the other two groups in a "consolidation index" (*D*) and in the number of recalls produced without reminding (*F*). Both these indexes can be considered to be related to storage ability. Two encoding indexes (*B* and *E*) were lower for the apraxic group than for normal controls, but not for left-brain-damaged nonapraxic patients. Both studies suggest storage impairment, although the latter also indicates a mild encoding impairment in apraxic patients. Jason (1985) also found that left-brain-damaged patients were impaired in learning manual sequences compared to right-brain-damaged patients. However, once the groups had been equated on level of learning, group differences in delayed recall were not observed. For that reason, Jason suggested that left-brain-damaged patients' difficulty lay in acquisition rather than storage.

Encoding and storage differ between single and paired-associated gestures in all brain-damaged patients. In the apraxic and borderline groups, encoding ability is more impaired in the learning of paired rather than single gestures, whereas the opposite appears to be the case for storage ability. Abilities in learning seem to be to some extent selectively engaged by the 2 tests.

Our values for encoding parameters were always higher than those for storage, but this is probably due to the way we tested gesture learning, forcing storage to be underestimated. For a finer estimate of storage ability, learning should have been continued until all the engrams were in the CM state, that is, when a sufficiently large number of consecutive correct responses had been given, when some "criterion" had been achieved. For a sufficiently high criterion, in fact, the probability of a failure in successive trails is near 0 (Greeno & Steiner, 1964). However, if normal subjects can easily reach a criterion of 2 or 3 in 10 trials, brain-damaged patients would need enough trials to exceed their attention spans, and responses would no longer be reliable. Therefore, we fixed the number of trials (10) but not the learning criterion, and the estimation of parameters related to the final phases of learning (as storage) does not have the same prominence as the estimation of parameters related to the initial phases of learning, such as encoding.

Assuming the accuracy of our premise that Ss can imitate the gesture of the learning lists and inasmuch as the parameters we studied refer to the psychological abilities to encode, store, and recall before storage has taken place (Kraemer et al., 1983; Rothi & Heilman 1985), we can conclude by saying that in right-handed people, encoding and storage of gesture engrams are functions of the left hemisphere independent of aphasia, and that apraxia probably arises when both are severely impaired.

## REFERENCES

Ackermann, H. (1985). *Mehrdimensionale nicht-parametrische Normhereiche*. Berlin: Springer-Verlag.

Adams, J. A., & Bray, N. W. (1970). A closed loop theory of paired-associated verbal learning. *Psychological Review, 77*, 385–405.

Brainerd, C. J. (1985). Model-based approaches to storage and retrieval development. In C. J. Brainerd & M. Pressley (Eds.), *Basic processes in memory development* (pp. 143–207). New York: Springer-Verlag.

Brainerd, C. J., Desrochers, A., & Howe, M. L. (1981). Stages of learning analysis of picture–word effects in associative memory. *Journal of Experimental Psychology, Human Learning and Memory, 7*, 1–14.

Brainerd, C. J., Howe, M. L., & Desrochers, A. (1980). Interpreting associative-learning stages. *Journal of Experimental Psychology, Human Learning and Memory, 6*, 754–765.

Brainerd, C. J., Howe, M. L., & Desrochers, A. (1982). The general theory of two-stage learning: A mathematical review with illustrations from memory development. *Psychological Bulletin, 91*, 634–669.

Buschke, H., & Fuld, P. A. (1974). Evaluating storage, retention and retrieval in disordered memory and learning. *Neurology, 24*, 1019–1025.

Chandler, J. P. (1965). *Subroutine STEPIT* (QCPE Program No. 66). Bloomington: Indiana University, Quantum Chemistry Program Exchange.

De Renzi, E., & Faglioni, P. (1978). Normative data and screening power of a shortened version of the Token Test. *Cortex, 14*, 41–49.

De Renzi, E., Faglioni, P., Lodesani, M., & Vecchi, A. (1983). Performance of left-brain-damaged patients on imitation of single movements and sequences. Frontal and parietal-injured patients compared. *Cortex, 19*, 333–343.

De Renzi, E., Motti, F., & Nichelli, P. (1980). Imitating gestures. A quantitative approach to ideomotor apraxia. *Archives of Neurology, 37*, 6–10.

Faglioni, P., & Scarpa, M. (in press). Skull asymmetries bear no relation to the occurrence of apraxia. A clinical and CI-scan study in patients with unilateral brain damage. *Cortex*.

Geschwind, N. (1975). The apraxias: Neural mechanisms of disorders of learned movement. *American Scientist, 63*, 188–195.

Goodglass, H., & Kaplan, E. (1963). Disturbance of gesture and pantomime in aphasia. *Brain, 86*, 703–720.

Greeno, J. G. (1968). Identifiability and statistical properties of two-stage learning with no successes in the initial stage. *Psychometrika, 33*, 173–216.

Greeno, J. G., & Scandura, J. M. (1966). All-or-none transfer based on verbally, mediated concepts. *Journal of Mathematical Psychology, 3*, 388–411.

Greeno, J. G., & Steiner, T. E. (1964). Markovian processes with identifiable states: general considerations and applications to all-or-none learning. *Psychometrika, 29*, 309–333.

Halff, H. M. (1977). The role of opportunities for recall in learning to retrieve. *American Journal of Psychology, 8*, 200–224.
Heilman, K. M., Schwartz, H. D., & Geschwind, N. (1975). Defective motor learning in ideomotor apraxia. *Neurology, 25*, 1018–1020.
Izawa, C. (1971). The test trial potentiating model. *Journal of Mathematical Psychology, 8*, 200–224.
Jason, G. W. (1983a). Hemispheric asymmetries in motor function: 1. Left hemisphere specialization for memory but not performance. *Neuropsychologia, 21*, 35–45.
Jason, G. W. (1983b). Hemispheric asymmetries in motor function: 2. Ordering does not contribute to left-hemisphere specialization. *Neuropsychologia, 21*, 47–58.
Jason, G. W. (1985). Manual sequences learning after focal cortical lesion. *Neuropsychologia, 23*, 483–496.
Kimura, D. (1977). Acquisition of a motor skill after left-hemisphere damage. *Brain, 100*, 527–542.
Kimura, D. (1982). Left-hemisphere control of oral and brachial movements and their relation to communication. *Philosophical Transactions of the Royal Society of London, B, 298*, 135–149.
Kimura, D., & Archibald, Y. (1974). Motor functions of the left hemisphere. *Brain, 97*, 337–350.
Kramer, M. C., Peabody, C. A., Tinklenberg, J. R., & Yesavage, J. A. (1983). Mathematical and empirical development of a test of memory for clinical and research use. *Psychological Bulletin, 94*, 367–380.
Lachman, R., & Laughery, K. R. (1968). Is a test trial a training trial in free recall learning? *Journal of Experimental Psychology, 76*, 40–50.
Lehmkuhl, G., Poeck, K., & Williams, K. (1983). Ideomotor apraxia and aphasia: An examination of types and manifestations of apraxic symptoms. *Neuropsychologia, 21*, 199–212.
Liepmann, H. (1920). Apraxie. *Ergebnisse der gesamten Medizin, 1*, 416–453.
Massey, F. J. (1951). Kolmogorov-Smirnov test for goodness of fit. *Journal of the American Statistical Association, 46*, 68–78.
Melton, A. (1963). Implications of short-term memory for a general theory of memory. *Journal of Verbal Learning and Verbal Behavior, 2*, 1–21.
Miller, R. G. (1966) *Simultaneous statistical inference*. New York: McGraw-Hill.
Millward, R. B. (1969). Derivations of learning statistics from absorbing Markov chains. *Psychometrica, 34*, 215–232.
Oldfield, R. C. (1971). The assessment and analysis of handedness: the Edinburgh Inventory. *Neuropsychologia, 9*, 97–113.
Pieczuro, A., & Vignolo, L. A. (1967). Studio sperimentale sull'aprassia ideomotoria. *Sistema Nervoso, 19*, 131–143.
Polson, P. G. (1972). Presolution performance functions of Markov models. *Psychometrica, 37*, 453–459.
Polson, P. G., & Huizinga, (1974). Statistical methods for absorbing Markov chain models for learning: estimation and identification. *Psychometrika, 39*, 3–22.
Polson, M.C., Restle, F., & Polson, P. G. (1965). Association and discrimination in paired associates learning. *Journal of Experimental Psychology, 69*, 47–55.
Restle, F. (1962). The selection of strategies in cue learning. *Psychological Review, 69*, 329–343.
Restle, F. (1964). Learning paired associates. In R. C. Atkinson (Ed) *Studies in mathematical psychology* (pp. 116–172). Stanford: Stanford University Press.
Restle, F. (1965). Significance of all-or-none learning. *Psychological Bulletin, 64*, 313–325.
Rothi, L. J., & Heilman, K. (1984). Acquisition and retention of gestures by apractic patients. *Brain and Cognition, 3*, 426–437.

Roy, E. A. (1981). Action sequences and lateralized cerebral damage. Evidence for asymmetries in control. In J. Long & A. Baddeley (Eds.), *Attention and Performance, Vol. 9* (pp. 487–499) Hillsdale, NJ: Lawrence Erlbaum Associates.

Scheffé, H. (1953). A method for judging all contrasts in the analysis of variance. *Biometrika, 40*, 87–104.

Taylor, H. M., & Karlin, S. (1984). *An introduction to stochastic modeling*. New York: Academic Press.

Theios, J., Leonard, D. W., & Brelsdorf, J. W. Jr. (1977). Hierarchies of learning models that permit likelihood ratio comparisons. *Journal of Experimental Psychology, General, 106*, 213–225.

Tulving, E. (1967). The effects of presentation and recall of material in free-recall learning. *Journal of Verbal Learning and Verbal Behavior, 6*, 175–184.

Wyke, M. (1971). The effects of brain lesions on the learning performance of a bimanual co-ordination task. *Cortex, 7*, 59–72.

# AUTHOR INDEX

**L**

**N**

**T**

# Subject Index

## A



## B



## C

For Product Safety Concerns and Information please contact our EU representative GPSR@taylorandfrancis.com Taylor & Francis Verlag GmbH, Kaufingerstraße 24, 80331 München, Germany